Methods in Enzymology

Volume 107
POSTTRANSLATIONAL MODIFICATIONS
Part B

METHODS IN ENZYMOLOGY

EDITORS-IN-CHIEF

Sidney P. Colowick   Nathan O. Kaplan

Methods in Enzymology

Volume 107

Posttranslational Modifications

Part B

EDITED BY

Finn Wold
DEPARTMENT OF BIOCHEMISTRY
AND MOLECULAR BIOLOGY
UNIVERSITY OF TEXAS MEDICAL SCHOOL
AT HOUSTON
HOUSTON, TEXAS

Kivie Moldave
DEPARTMENT OF BIOLOGICAL CHEMISTRY
CALIFORNIA COLLEGE OF MEDICINE
UNIVERSITY OF CALIFORNIA
IRVINE, CALIFORNIA

1984

ACADEMIC PRESS, INC.

(Harcourt Brace Jovanovich, Publishers)

Orlando   San Diego   San Francisco   New York   London
Toronto   Montreal   Sydney   Tokyo   São Paulo

ACADEMIC PRESS, INC.
Orlando, Florida 32887

United Kingdom Edition published by
ACADEMIC PRESS, INC. (LONDON) LTD.
24/28 Oval Road, London NW1 7DX

LIBRARY OF CONGRESS CATALOG CARD NUMBER: 54-9110
ISBN 0-12-182007-6

PRINTED IN THE UNITED STATES OF AMERICA

84 85 86 87   9 8 7 6 5 4 3 2 1

Table of Contents

Section 1. Protein Acylations/Deacylations

v

Section II. Oxidations, Hydroxylations, and Halogenations

Contributors to Volume 107

Article numbers are in parentheses following the names of contributors.
Affiliations listed are current.

ROBERT AESCHBACH (24), *Nestlé Products Technical Assistance Company, Ltd., CH-1814 La Tour-de-Peilz, Switzerland*

VINCENT G. ALLFREY (12), *Laboratory of Cell Biology, The Rockefeller University, New York, New York 10021*

RENATO AMADÒ (24), *Department of Food Science, Swiss Federal Institute of Technology, ETH-Zentrum, CH-8092 Zurich, Switzerland*

VALERIE J. ATMAR (7), *Department of Chemistry, New Mexico State University, Las Cruces, New Mexico 88003*

ROBERT M. BARKLEY (38), *Department of Chemistry, University of Colorado, Boulder, Colorado 80309*

CHRISTINE V. BENEDICT (26), *Division of Orthopuedics, University of Connecticut, Farmington, Connecticut 06032*

DENISE BOHEMIER (38), *Department of Chemistry, University of Colorado, Boulder, Colorado 80309*

MANFRED BRAUER (3), *MRC Group on Protein Structure and Function, and Department of Biochemistry, University of Alberta, Edmonton, Alberta T6G 2H7, Canada*

TOM BRODY (37), *Department of Biochemistry, University of Wisconsin-Madison, Madison, Wisconsin 53706*

NATHAN BROT (21), *Roche Institute of Molecular Biology, Roche Research Center, Nutley, New Jersey 07110*

MAARTEN J. CHRISPEELS (22), *Department of Biology, University of California, San Diego, La Jolla, California 92093*

M. ROBERT CHRISTY (38), *Dow Chemical Company, Midland, Michigan 48640*

TIMOTHY COLEMAN (21), *Roche Institute of Molecular Biology, Roche Research Center, Nutley, New Jersey 07110*

THOMAS E. CREIGHTON (19), *Laboratory of Molecular Biology, Medical Research Council Centre, Cambridge CB2 2QH, England*

EUGENE A. DI PAOLA (12), *Field of Cell Biology and Genetics, Cornell University Graduate School of Medical Sciences, New York, New York 10021*

PIA EKMAN (6), *Institute of Medical and Physiological Chemistry, Biomedical Center, University of Uppsala, S-751 23 Uppsala, Sweden*

LORENTZ ENGSTRÖM (6), *Institute of Medical and Physiological Chemistry, Biomedical Center, University of Uppsala, S-751 23 Uppsala, Sweden*

HENRY FLISS (21), *Roche Institute of Molecular Biology, Roche Research Center, Nutley, New Jersey 07110*

S. J. FOSTER (40), *Department of Nutritional Sciences, University of Wisconsin-Madison, Madison, Wisconsin 53706*

ROBERT B. FREEDMAN (16), *Biological Laboratory, University of Kent, Canterbury, Kent CT2 7NJ, England*

STEPHEN C. FRY (25), *Department of Botany, University of Edinburgh, Edinburgh EH9 3JH, Scotland*

JAMES M. FUJITAKI (2), *Department of Chemistry and Biochemistry, and Molecular Biology Institute, University of California, Los Angeles, Los Angeles, California 90024*

P. M. GALLOP (34), *Children's Hospital and Harvard Schools of Medicine and Dental Medicine, Boston, Massachusetts 02115*

JOHN E. GANDER (9), *Department of Microbiology and Cell Science, University of Florida, Gainesville, Florida 32611*

H. E. GANTHER (40, 41), *Department of Nutritional Sciences, University of Wisconsin-Madison, Madison, Wisconsin 53706*

HIRAM F. GILBERT (20), *Marrs McLean Department of Biochemistry, Baylor College of Medicine, Houston, Texas 77030*

NEIL H. GOSS (15), *INGENE, 1701 Colorado Avenue, Santa Monica, California 90404*

AJIT GOSWAMI (31), *Department of Medi-*

cine, *Boston University School of Medicine, Boston, Massachusetts 02118*

C. M. GUNDBERG (34), *Children's Hospital and Harvard School of Dental Medicine, Boston, Massachusetts 02115*

GARY L. GUSTAFSON (9), *Department of Microbiology, University of Montana, Missoula, Montana 59812*

L. P. HAGER (28), *Department of Biochemistry, University of Illinois, Urbana, Illinois 61801*

P. V. HAUSCHKA (34), *Children's Hospital and Harvard School of Dental Medicine, Boston, Massachusetts 02115*

WAYNE C. HAWKES (42), *Department of Food Science and Technology, University of California, Davis, Davis, California 95616*

DAVID A. HILLSON (16), *Department of Biophysics, Portsmouth Polytechnical School, Portsmouth, Hampshire PO1 2DT, England*

ARNE HOLMGREN (17), *Department of Chemistry, Karolinska Institutet, S-104 01 Stockholm, Sweden*

ELISABET HUMBLE (6), *Institute of Medical and Physiological Chemistry, Biomedical Center, University of Uppsala, S-751 23 Uppsala, Sweden*

S. HUNT (27), *Department of Biological Sciences, University of Lancaster, Lancaster LA1 4YQ, England*

WIELAND B. HUTTNER (11), *Department of Neurochemistry, Max-Planck-Institute for Psychiatry, D-8033 Martinsried, Munich, Federal Republic of Germany*

T. S. INGEBRITSEN (5), *Department of Pharmacology, University of Colorado School of Medicine, Denver, Colorado 80262*

WOLFF M. KIRSCH (38), *Division of Neurosurgery, University of New Mexico School of Medicine, Albuquerque, New Mexico 87131*

H. KLOSTERMEYER (14), *Dairy Science Institute, Munich Institute of Technology, D-8050 Freising-Weihenstephan, Munich, Federal Republic of Germany*

TAD H. KOCH (38), *Department of Chemistry, University of Colorado, Boulder, Colorado 80309*

STUART KORNFELD (8), *Division of Hematology–Oncology, Washington University School of Medicine, St. Louis, Missouri 63110*

R. J. KRAUS (40, 41), *Department of Nutritional Sciences, University of Wisconsin-Madison, Madison, Wisconsin 53706*

GLENN D. KUEHN (7), *Department of Chemistry, New Mexico State University, Las Cruces, New Mexico 88003*

NIGEL LAMBERT (16), *Inorganic Chemistry Laboratory, University of Oxford, Oxford OX1 3QU, England*

LES LANG (8), *Division of Hematology–Oncology, Washington University School of Medicine, St. Louis, Missouri 63110*

RODNEY L. LEVINE (23), *Laboratory of Biochemistry, National Heart, Lung, and Blood Institute, National Institutes of Health, Bethesda, Maryland 20205*

J. B. LIAN (34), *Children's Hospital and Harvard Medical School, Boston, Massachusetts 02115*

ARIEL G. LOEWY (13), *Department of Biology, Haverford College, Haverford, Pennsylvania 19041*

FARAHE MALOOF (29), *Thyroid Unit, Massachusetts General Hospital and Harvard Medical School, Boston, Massachusetts 02114*

J. A. MANTHEY (28), *Department of Biochemistry, University of Illinois, Urbana, Illinois 61801*

TODD M. MARTENSEN (1), *Laboratory of Biochemistry, National Heart, Lung, and Blood Institute, National Institutes of Health, Bethesda, Maryland 20205*

K. D. McELVANY (28), *Mallinckrodt Institute of Radiobiology, Washington University School of Medicine, St. Louis, Missouri 63110*

MARVIN A. MOTSENBOCKER (42), *Immunodiagnostics Research and Development, Ames Division, Miles Elkhart, Indiana 46514*

JOSEPH T. NEARY (29), *Laboratory of Biophysics, NINCDS, IRP, Marine Biological Laboratory, Woods Hole, Massachusetts 02543*

GARY L. NELSESTUEN (32, 33), *Department*

of Biochemistry, University of Minnesota, St. Paul, Minnesota 55108

HANS NEUKOM (24), *Department of Food Science, Swiss Federal Institute of Technology, ETH-Zentrum, CH-8092 Zurich, Switzerland*

J. NUNEZ (30), *INSERM CNRS U 35, Hôpital Henri Mondor, 94010 Creteil, France*

PAUL A. PRICE (35, 36), *Department of Biology, University of California, San Diego, La Jolla, California 92093*

ULF RAGNARSSON (6), *Institute of Biochemistry, Biomedical Center, University of Uppsala, S-751 23 Uppsala, Sweden*

MARC L. REITMAN (8), *Division of Hematology–Oncology, Washington University School of Medicine, St. Louis, Missouri 63110*

SUE GOO RHEE (10), *Laboratory of Biochemistry, National Heart, Lung, and Blood Institute, National Institutes of Health, Bethesda, Maryland 20205*

PETER J. ROACH (4), *Department of Biochemistry, Indiana University School of Medicine, Indianapolis, Indiana 46223*

G. E. ROGERS (44), *Department of Biochemistry, The University of Adelaide, Adelaide, South Australia 5001*

ISADORE N. ROSENBERG (31), *Department of Medicine, Framingham Union Hospital, Framingham, Massachusetts 01701, and Boston University School of Medicine, Boston, Massachusetts 02118*

J. A. ROTHNAGEL (44), *Department of Bio chemistry, The University of Adelaide, Adelaide, South Australia 5001*

VAITHILINGAM SEKAR (7), *Department of Molecular and Population Genetics, University of Georgia, Athens, Georgia 30602*

S. SHENOLIKAR (5), *Department of Pharmacology, University of South Alabama College of Medicine, Mobile, Alabama 36688*

MARK X. SLIWKOWSKI (43), *Laboratory of Biochemistry, National Heart, Lung, and Blood Institute, National Institutes of Health, Bethesda, Maryland 20205*

RICHARD SLUSKI (38), *Department of Chemistry, University of Colorado, Boulder, Colorado 80309*

ROBERTS A. SMITH (2), *Department of*

Chemistry and Biochemistry, and Molecular Biology Institute, University of California, Los Angeles, Los Angeles, California 90024

MORRIS SOODAK[1] (29), *Department of Biochemistry, Brandeis University, Waltham, Massachusetts 02154*

THRESSA C. STADTMAN (39), *National Heart, Lung, and Blood Institute, National Institutes of Health, Bethesda, Maryland 20205*

RICHARD STERNER (12), *Laboratory of Cell Biology, The Rockefeller University, New York, New York 10021*

J. W. SUTTIE (37), *Department of Biochemistry, University of Wisconsin-Madison, Madison, Wisconsin 53706*

BRIAN D. SYKES (3), *MRC Group on Protein Structure and Function, and Department of Biochemistry, University of Alberta, Edmonton, Alberta T6G 2H7, Canada*

AL L. TAPPEL (42), *Department of Food Science and Technology, University of California, Davis, Davis, California 95616*

JOHN J. VAN BUSKIRK (38), *Division of Neurosurgery, University of New Mexico School of Medicine, Albuquerque, New Mexico 87131*

J. HERBERT WAITE (26), *Division of Orthopaedics, University of Connecticut, Farmington, Connecticut 06032*

HERBERT WEISSBACH (21), *Roche Institute of Molecular Biology, Roche Research Center, Nutley, New Jersey 07110*

DONALD B. WETLAUFER (18), *Department of Chemistry, University of Delaware, Newark, Delaware 19711*

ERIC C. WILHELMSEN (42), *Castle and Cook Technical Center, 2102 Commerce Drive, San Jose, California 95131*

HARLAND G. WOOD (15), *Department of Biochemistry, Case Western Reserve University School of Medicine, Cleveland, Ohio 44106*

ÖRJAN ZETTERQVIST (6), *Institute of Medical and Physiological Chemistry, Biomedical Center, University of Uppsala, S-751 23 Uppsala, Sweden*

[1] Deceased.

Introduction: A Short Stroll through the Posttranslational Zoo

The area of posttranslational modifications of proteins is often, more or less affectionately, referred to as the "amino acid zoo," and in assembling these volumes we have encountered what we imagine might be many of the zookeepers' problems, first in classifying and then in properly housing the animals in proper relation to the other existing vivariums. The present menagerie of posttranslational modifications underwent quite a few changes from the planning stage to the present exposition. Most of the changes were derived from concessions to the fact that many of the posttranslational reactions are very much a part of other subdivisions of biochemistry, indeed, that some classes of posttranslational reactions (e.g., glycosylations, phosphorylations) have become subdivisions in their own right.

With the starting premise that duplication of other volumes in this series should be kept to a minimum, the desired goal of having Volumes 106 and 107 as complete as possible represented a dilemma. We tried to resolve this dilemma by the extensive use of cross-references to other volumes in the *Methods in Enzymology* series. Since these cross-references were to replace planned chapters in these volumes, the next question was where the cross-references should appear, and that problem was finally resolved by the inclusion of this Introduction.

The outline that follows represents our current classification as it has evolved from the original outline of the various posttranslational reactions. When appropriate, a brief statement explains how a particular section has been handled in these volumes and where the omitted components can be found in other volumes.

The meaning of "posttranslational modification" needs to be briefly considered to apprise the readers of the definition that has been used in these volumes. Several authors, when first approached to contribute their chapters, pointed out that their particular derivative or reaction was not in fact "posttranslational" but "cotranslational." Our uniform response was that for the purpose of these volumes we did not feel it desirable to distinguish between the two. In other words, our usage of the term posttranslational goes back to the original usage of "translation" as a broad description of the entire process by which the polymer of three-letter codons in the mRNA is "translated" into a polypeptide chain of 20 "primary" amino acids. This process includes (1) the synthesis of the individual aminoacyl-tRNA derivatives as the commitment step in the translation, (2) the polymerization of the amino acids on the polysome using these aminoacyl-tRNA derivatives as the monomer building blocks, and (3) the release of the completed polyamino acid product. The processing of this polymer, either by shortening or lengthening the original encoded sequence or by covalently modifying the original 20 amino acids into unique noncoded amino acid derivatives, should consequently be considered posttranslational whether the processing occurs before (at the level of aminoacyl-tRNA), during, or after polymerization. According to current usage, these three stages in protein synthesis should be designated pretranslational (presumably?), cotranslational, and posttranslational, respectively. This now well-established nomenclature has unquestionably been useful in communication among the informed, and in fact one of the arguments advanced for making the distinction between co- and posttranslational processes is that it sharpens the definition of the "sloppy" term "translation." However, implicit in this refinement to distinguish the temporal stages of protein processing is the brand-new definition of translation as being synonymous with polymerization, and the benefits of the refinement should be balanced against the requirement to replace the perfectly good chemical term polymerization with the rather vague and equivocal term translation. The choice of these terms must obviously be left to the individual readers. The important point to emphasize is that in these volumes posttranslational modification will include all processing steps by which the direct translation product is altered from the structure specified by the gene.

As the list of topics grew, it became obvious that the space limitation of a single volume would be exceeded. Consequently, the list had to be cut roughly in half for distribution into two volumes. There was no obvious place to make the division, so the arbitrary guide was simply to make Volumes 106 and 107 approximately the same size.

Posttranslational Covalent Modification of Proteins: A List of Reaction Types and General Topics

VOLUME 106

Reactions Involving the Polypeptide Backbone (Limited Proteolysis)

Such reactions have been covered extensively in several recent volumes of *Methods in Enzymology,* and consequently were not included. Major general treatments of proteases, including those involved in zymogen activation, can be found in Volumes 19, 45, and 80 on proteolytic enzymes; the more specialized ones on proteolytic processing in the transport of proteins into and through membranes in Volume 96; on blood clotting, complement, the plasmin system in Volume 80; and on peptide hormones in Volume 37.

General Aspects (Naming, Releasing, and Analyzing Posttranslationally Derivatized Amino Acids)

This section will provide an overview of most of the known amino acid derivatives found to date in proteins and general starting points for proper nomenclature, analytical procedures, and the enzymatic hydrolysis of proteins. The major and most obvious cross-references for this section can be found in Volumes 25, 26, 27, 47, 48, 49, 61, and 91 on enzyme structure.

Nonenzymatic Modifications

Many individuals question the inclusion of nonenzymatic reactions as true posttranslational reactions. We feel that any reaction taking place under physiological conditions to alter the gene-specified polymer should be included. In fact, some reactions (e.g., disulfide interchange, P-450-catalyzed oxygenations) proceed either with or without specific enzyme involvement.

Modifications at the Level of Aminoacyl-tRNA
Covalent Modification of the α-Amino and γ-Carboxyl Groups of Proteins
Reactions Involving Covalent Modification of the Side Chains of Amino Acids in Proteins
   Protein Alkylations/Dealkylations (Arylations)
   Protein Glycosylations; ADP-Ribosylation

Glycosylation reactions have also been treated extensively in previous volumes of *Methods in Enzymology*. The principal reference sources are Volumes 8, 28, 50, and 83 on complex carbohydrates and Volumes 96 and 98 on membrane biogenesis. Glycosylation of one posttranslational hydroxylated amino acid, hydroxylysine, has been presented in Volume 82 on collagen and elastin, but others (hydroxyproline, β-hydroxytyrosine) are included here. The main emphasis of this section is the glycosylation reactions involving ADP-ribosylation.

VOLUME 107

Reactions Involving Covalent Modification of the Side Chains of Amino Acids (continued)
   Protein Acylations/Deacylations (other than phosphorylations)

A major component of the acylation reactions, protein phosphorylation, has been treated in many previous volumes (e.g., 68, 90, and 102) and a recent one, Volume 99, is devoted entirely to protein kinases. To avoid major overlaps with the previous volumes, this section consequently does not treat specific systems, but rather focuses on a broader overview of phosphorylations/dephosphorylation systems and analytical procedures. Some derivatives involving phosphate diesters through the incorporation into proteins or glycoproteins of sugar phosphoryl or nucleotidyl groups are included as special cases of phosphorylations. Other acylation reactions are included.

Oxidations/(Reductions), Hydroxylations, and Halogenations

This is a rather heterogeneous section, justified as a unit only by the common feature that an oxidation/reduction is part of the process by which these derivatives are made. Many of the reactions and products covered in this section have been treated in other volumes, which are cross-referenced in the individual contributions. Many of the posttranslational hydroxylated amino acids which should represent a major component of this section, such as HyL and HyP, have been considered as components of mammalian structural proteins in Volume 82. Hypusin is covered in Volume 106 (see Section V).

Miscellaneous Derivatives

The content of these two volumes, even with the cross-references to other volumes, obviously does not represent a complete display of all known posttranslational reactions and reaction products. [The chapter on nomenclature (Chapter 1, Volume 106) contains a tabulation of most of the posttranslational derivatives known by mid-1983.] The main purpose of compiling this heterogeneous and complex set of reactions into one unit is to emphasize the well-established fact that there is more to protein synthesis than the polymerization of twenty primary amino acids on the mRNA template transcript of the structural gene. The subsequent covalent manipulations of the genetically specified polypeptide chain represent an exciting set of current problems in biochemistry. The problems involve the common challenge of trying to understand all the different ways in which the information of the primary amino acid sequence can be translated into specificity signals for the modification reactions as well as the separate challenges of trying to understand each individual reaction in terms of its chemical mechanisms and biological functions.

FINN WOLD
KIVIE MOLDAVE

METHODS IN ENZYMOLOGY

EDITED BY

Sidney P. Colowick and Nathan O. Kaplan

VANDERBILT UNIVERSITY
SCHOOL OF MEDICINE
NASHVILLE, TENNESSEE

DEPARTMENT OF CHEMISTRY
UNIVERSITY OF CALIFORNIA
AT SAN DIEGO
LA JOLLA, CALIFORNIA

METHODS IN ENZYMOLOGY

EDITORS-IN-CHIEF

Sidney P. Colowick   Nathan O. Kaplan

Methods in Enzymology

Volume 107
POSTTRANSLATIONAL MODIFICATIONS
Part B

Section I

Protein Acylations/Deacylations

[1] Chemical Properties, Isolation, and Analysis of *O*-Phosphates in Proteins

By Todd M. Martensen

The association of phosphate with proteins has been known for over a century. Characterization of proteins of milk (caseins) and egg (phosvitin) showed that phosphoryl groups were covalently attached to protein. The stability of the *O*-phosphate bonds in proteins to partial acid digestion was an essential feature that enabled the detection of phosphorylated amino acids. Phosphoserine was found in the egg yolk protein vitellinic acid in 1932 by Lipmann and Levene.[1] Phosphothreonine was isolated from casein by deVerdier[2] in 1953. Phosphotyrosine was found in the middle T antigen of polyoma virus by Eckhart, Hutchinson, and Hunter in 1979.[3] Phosphotyrosine was found as the free amino acid in *Drosophila* larvae by Mitchell and Lunan in 1964.[4] In 1955 the nature of the linkage of phosphate to proteins was reviewed by Perlmann.[5] At that time a major concern was whether chemical treatments (acid/base) could cause migration of phosphoryl groups, and whether phosphodiesters constituted a portion of phosphoprotein phosphate.

At present we know that cells contain a great variety of phosphoproteins. The researcher now is often confronted with characterizing phosphorylated residues of ^{32}P-labeled proteins found in immunoprecipitates, one- or two-dimensional polyacrylamide gels, and eluates from HPLC columns where only radioactivity can be utilized for classification. Phosphorylated amino acid residues in proteins are commonly classified into *O*-phosphates, *N*-phosphates, and acyl phosphates. The *O*-phosphates are the major class in terms of quantity and extent of distribution in proteins. They are represented by phosphoserine (Ser-P), phosphothreonine (Thr-P), and phosphotyrosine (Tyr-P). The properties of the phosphoramidate class are discussed in this volume [2]. In addition, phosphoryl groups can act as bridging moieties in phosphodiester linkages of nucleotide to protein (this volume [10]). Thus, the researcher who finds ^{32}P associated with a protein should consider a variety of possible linkages before devising chemical or enzymatic characterization strategies. Dem-

[1] F. Lipmann and P. A. Levene, *J. Biol. Chem.* **98**, 109 (1932).
[2] C.-H. deVerdier, *Acta Chem. Scand.* **7**, 196 (1953).
[3] W. Eckhart, M. A. Hutchinson, and T. Hunter, *Cell* **18**, 925 (1979).
[4] H. K. Mitchell and K. D. Lunan, *Arch. Biochem. Biophys.* **106**, 219 (1964).
[5] G. E. Perlmann, *Adv. Protein Chem.* **10**, 1 (1955).

METHODS IN ENZYMOLOGY, VOL. 107

onstration that a protein contains O-phosphate residues requires analytical or radioactive evidence for a covalent attachment. The O-phosphate linkages of serine and threonine residues are defined by their stability to acid and lability to base. Tyr-P linkages are acid and base stable. The phosphoprotein fraction of cells and tissues was traditionally defined as the residue remaining after the perchloric acid precipitate was washed first with ethanol–ether, second with chloroform–methanol, and third with boiling trichloroacetic acid.[6] These treatments removed most of the phospholipid and nucleic acid contaminants. Nucleic acids and phospholipid are often contaminants of proteins isolated from cell homogenates by trichloroacetic precipitation or by immunoprecipitation. Proof that precipitated protein contains O-phosphorylated amino acids requires radioactive or analytical evidence such as comigration of ^{32}P radioactivity with protein on sodium dodecyl sulfate (SDS)–polyacrylamide gel electrophoresis and/or detection of O-phosphates after partial acid hydrolysis of the phosphoprotein.

Procedures for detection and quantification of O-phosphorylated residues in proteins have been described (see footnote 6) and in this series.[7–10] This chapter presents selected methods that can be applied to characterize O-phosphorylated amino acids encountered in laboratory research.

Properties of O-Phosphates, Preliminary Characterization

Chemical Stability of O-Phosphorylated Proteins to Acid

Perchloric and trichloroacetic acids are routinely used to precipitate O-phosphorylated proteins with no loss of phosphoryl groups. Incubation of ^{32}P-labeled ribosomal protein in 0.1 N HCl at 55° for 2 hr released less than 10% of the associated radioactivity after precipitation of protein by 20% trichloroacetic acid.[8] Treatment with 5% trichloroacetic acid at 90° for 15 min followed by incubation on ice (30 min) led to 80% recovery of the ^{32}P-labeled protein in the precipitate. Using these conditions, characterization of a [^{32}P]phosphoprotein of unknown linkage should be carried out with sufficient carrier protein to give 0.5 mg/ml in the trichloroacetic acid precipitation step. Certain O-phosphoproteins possess modification sites at N- or C-terminal regions that may be cleaved during chemical

[6] W. C. Schneider, *J. Biol. Chem.* **164**, 241 (1946). Adapted from M. Weller, in "Protein Phosphorylation" (J. R. Lagnado, ed.). Pion, London, 1979.
[7] N. K. Schaffer, this series, Vol. 11, p. 702.
[8] L. Bitte and D. Kabat, this series, Vol. 30, p. 563.
[9] J. A. Cooper, B. M. Sefton, and T. Hunter, this series, Vol. 99, p. 387.
[10] T. M. Martensen and R. L. Levine, this series, Vol. 99, p. 402.

characterization yielding acid-soluble peptides. It is useful to employ the $^{32}P_i$ extraction technique described below to confirm that the release of ^{32}P is due to production of P_i.

Lability of O-Phosphates to Base

It was early recognized that the phosphoryl content of "phosphoprotein" was quite labile to treatment with 0.25 N NaOH at 37°.[11] Later it was found that the phosphate produced was the result not of hydrolysis, but of alkali-induced β-elimination yielding dehydroalanine residues where Ser-P existed.[12] The phosphoryl groups of Ser-P and Thr-P in peptide linkage are much more labile to base than the phosphoryl groups of free amino acids.[13-15] The classification of Ser-P and Thr-P residues of proteins as "alkali labile" does not apply in all cases, since variation exists in the phosphoryl group base lability of both Ser-P and Thr-P in peptides and proteins. The lability of the phosphoryl group can be strongly influenced by neighboring proline residues.[16,17] The Ser-P of phosphorylase a, Ile-Ser(P)-Val, is very labile to base ($t_{1/2}$ <1 hr; 0.1 N NaOH, 37°) while the Thr-P in phosphoprotein phosphatase inhibitor 1, pro-Thr(P)-pro, is much more stable ($t_{1/2}$ 7 hr; 3 N NaOH, 50°).[17] The O^4-phosphotyrosine phosphoryl bond in proteins appears to be sufficiently stable to survive the conditions necessary to remove resistant Thr-P phosphoryl bonds. Treatment of four proteins containing Tyr-P with 5 N KOH at 155° for 30 min resulted in 80% recovery of free Tyr-P.[18] The base stability of Tyr-P in proteins appears to be the same as that of free phosphotyrosine; this suggests that incorporation of Tyr-P into peptide bonds does not alter the lability of the phosphoryl group to base as in the case with Ser-P and Thr-P. The stability of Tyr-P to base is similar to that of N-phosphates; however, Tyr-P is stable to acidic conditions that destroy N-phosphates.

Measurement of the lability of a ^{32}P associated with a protein to base is easily done. In certain cases the alkaline lability can provide some information regarding the type of residue modified and whether the sample contains more than a single modification site. The standard method is to incubate the protein in base and measure the trichloroacetic acid-soluble radioactivity appearance with time. It is useful to distinguish $^{32}P_i$ from

[11] R. H. Plimmer and W. M. Bayliss, *J. Physiol. (London)* 33, 455 (1906).
[12] D. K. Mecham and H. S. Olcott, *J. Am. Chem. Soc.* 71, 3670 (1949).
[13] R. H. A. Plimmer, *Biochem. J.* 35, 461 (1941).
[14] L. Anderson and G. G. Kelley, *J. Am. Chem. Soc.* 81, 2275 (1959).
[15] E. B. Kalan and M. Telka, *Arch. Biochem. Biophys.* 85, 273 (1959).
[16] P. J. Parker, F. B. Caudwell, N. Embi, and P. Cohen, *Eur. J. Biochem.* 124, 47 (1982).
[17] B. E. Kemp, *FEBS Lett.* 110, 308 (1980).
[18] T. M. Martensen, *J. Biol. Chem.* 257, 9648 (1982).

acid-soluble ^{32}P-labeled peptides or ^{32}P-labeled nucleotides released during the incubation. This can be done by utilization of the solubility of P_i as the phosphomolybdate complex in organic solvent.[19]

Procedure for Determination of the Base Lability of O-Phosphoproteins

1. Add [^{32}P]phosphoprotein to a 1.5-ml Eppendorf tube. Add carrier protein (albumin) to give 0.2–0.5 mg.
2. Precipitate protein by adding trichloroacetic acid to 20% in a combined volume of 0.5–1.0 ml. Leave samples on ice for 10–15 min prior to centrifugation at 10,000 g for ~2 min. Remove supernatant to a scintillation vial, and measure radioactivity (see below).
3. Add 0.65 ml of 1.0 N NaOH on ice to the sample and leave on ice for a few minutes. Dissolve the sample by vortexing. Remove an aliquot (0.05 ml) of the sample to a scintillation vial (add an equal volume of 1 N HCl) to determine total radioactivity in the aliquot. Another aliquot (0.10 ml) is removed to an Eppendorf tube for 0 time $^{32}P_i$ analysis, diluted with an equal volume of 1 N HCl, and placed on ice.
4. Place the base-solubilized sample in a water bath at 37°. Remove aliquots at 10, 20, 30, and 60 min and treat as the 0 time sample.
5. Precipitate protein in the aliquots by adding 0.5 mg of carrier protein (0.05 ml, 10 mg/ml) and 0.85 ml of 20% trichloroacetic acid as in step 2; centrifuge and remove an aliquot for measurement of trichloroacetic acid-soluble radioactivity and/or $^{32}P_i$ by the organic extraction method.

Procedure for the Measurement of $^{32}P_i$ Released from Chemical or
  Enzymatic Treatment of Phosphoprotein or Phosphopeptides

Reagents. These reagents, stored at room temperature, are renewed every 6 months.

  Silicotungstic acid, 5%, in 2 N H_2SO_4–silicotungstic acid $n \cdot$ hydrate
    is dissolved in 2 N H_2SO_4 (1/18 dilution of concentrated H_2SO_4) to
    give a 5% solution
  5% ammonium molybdate $\cdot$ $4H_2O$ in 2 N H_2SO_4
  Isobutanol/benzene (1 : 1, v/v)

Procedure

1. Add 0.5 ml of sample to a 12- to 15-ml conical centrifuge tube.
2. Add 0.25 ml of silicotungstic acid–H_2SO_4–50 μM P_i.
3. Add 1 ml of butanol–benzene to each tube, then 0.25 ml of ammonium molybdate.

[19] J. B. Martin and D. M. Doty, *Anal. Chem.* **21**, 965 (1949).

4. Vortex each tube for 5 sec, and centrifuge in a clinical centrifuge (~1000 *g*) for 1 min to separate the phases.

5. Remove 500–700 µl of the upper phase, and determine the radioactivity by scintillation counting. The volume of the organic phase is considered to be 1 ml for calculation of total radioactivity extracted.

The preparation of [^{32}P]phosphoprotein for alkaline lability studies requires washing. If the radioactivity of the acid supernatant used for washing exceeds 10% of the counts contained in the sample, the sample should be taken up in 0.1 *N* NaOH on ice and immediately precipitated with trichloroacetic acid as before. The supernatant is again counted; appearance of the same percentage of radioactivity in the supernatant may indicate incomplete precipitation of the protein during the washing steps. Carrier protein is used to give visible precipitates after trichloroacetic acid precipitation. This reduces the possibility of contamination of the supernatant with protein. The possibility that acid-soluble ^{32}P-labeled peptides are formed during base incubation is real (see Fig. 1). The organic extraction method should be utilized to demonstrate that the acid-soluble radioactivity is ^{32}P$_i$. The volume of the acid supernatant enables 0.5 ml to be used for both direct scintillation counting and organic extraction methods. Failure to release >50% of the radioactivity after 1 hr in 1 *N* NaOH at 37° suggests the phosphoryl bond is resistant to base. Repeat the experiment using a higher temperature, 65°, and remove samples over a 10- to 15-hr period to measure release from resistant Thr-P bonds. No release under these latter conditions suggests a Tyr-P bond or a phosphodiester bond to tyrosine.[20] Figure 1 (bottom panel) shows a semilogarithmic plot of the phosphoryl bond lability of [^{32}P]phosphoserine residues of three proteins in 1 *N* NaOH at 37°. Both methods for measuring alkali lability give similar results. The upper panel shows the phosphoryl bond resistance of the phosphotyrosine residue to 1 *N* NaOH at 65°. The long half-lives of the phosphoryl bonds of phosphothreonine residues adjacent to proline in two proteins incubated under the same conditions is also shown. Considerable peptide bond hydrolysis under these conditions is apparent.

Characterization of *O*-Phosphomonoester Linkage in Proteins by Enzymatic Methods

Demonstration that ^{32}P associated with protein is alkali labile and that it comigrates with Ser-P or Thr-P after partial acid digestion is considered to be proof for an *O*-phosphate linkage. If the ^{32}P is base stable and

---

[20] B. M. Shapiro and E. R. Stadtman, *J. Biol. Chem.* **243**, 3769 (1968).

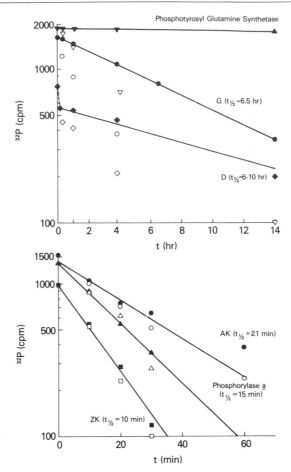

FIG. 1. Stability of O-phosphorylated residues in protein to 1 N base. *Top:* Semilogarithmic plot of the phosphoryl bond hydrolysis of phosphotyrosyl glutamine synthetase (▼) and two phosphothreonine-containing proteins to 1 N NaOH at 65°. The primary sequence of the Thr-P residue is G, Asp-Thr(P)-Pro; D, Pro-Thr(P)-Pro. Samples G (●) and D (◆) were provided by Dr. H. G. Hemmings, Rockefeller University. *Bottom:* Semilogarithmic plot of the hydrolysis of three phosphoserine-containing proteins in 1 N NaOH at 37°. Proteins AK (●) and ZK (■) were provided by Dr. H. G. Hemmings. Each sample of [^{32}P]Ser(P), -Thr(P), and -Tyr(P) phosphoprotein (15–100 pmol) was added to 0.2 ml of 10 mM HCl/albumin (10 mg/ml). A 0.1-ml aliquot was removed from each sample and added to 0.5 ml of 20% trichloroacetic acid on ice. After centrifugation the supernatant was removed and counted; in each case the acid-soluble radioactivity was <5% of the total. Each pellet was solubilized in 0.65 ml of 1 N NaOH on ice, and 0.05 ml was removed to a vial for total ^{32}P measurement. Samples of [^{32}P]Ser(P) phosphoproteins were incubated at 37°; the others at 65°. Aliquots of 0.1 ml were removed at the indicated times and added to 0.15 ml of 1 N HCl/5 mg/ml albumin on ice. At the end of the time course, 0.85 ml of 20% trichloroacetic acid was added to each aliquot. After centrifugation 0.5 ml of the supernatant was added to a

comigrates with Tyr-P after partial acid hydrolysis, this linkage is also assumed. However, ^{32}P-labeled phosphoprotein isolated from cells incubated with [32]P$_i$ may contain phosphodiester linkages to serine or tyrosine residues. These modifications have been found in prokaryotes,[20,21] in viral genome-linked proteins,[22,23] and as reaction intermediates of topoisomerases.[24,25] Partial acid digests of these phosphodiesters produce O-phosphates.[22,23] Treatment of ^{32}P-labeled protein with phosphatases can be used to demonstrate an O-phosphomonoester linkage. Specific protein phosphatases are not readily available for these studies. Nonspecific acid and alkaline phosphatases can be used for characterization studies. Failure to obtain P$_i$ release after incubation with phosphatase may be due to several features other than a phosphodiester linkage. Proteins are poor substrates for these enzymes. Characterization of O-phosphomonoester linkages is more easily accomplished by using denatured phosphoprotein or phosphopeptides produced by proteolytic digestion. It is important to demonstrate that P$_i$ is the product released during phosphatase digestion of proteins, since these preparations may contain proteases. This can be done by using the organic extraction technique described above.

Phosphoprotein concentrations ~10 μM in reaction mixtures should be used. When this is not possible, use minimal volumes and phosphatase concentrations of 0.1–1 unit/ml. To confirm that a phosphotyrosine monoester linkage to protein existed, a chymotryptic digest of ^{32}P-labeled P120 (produced by Abelson murine leukemia virus-infected cells) was boiled prior to digestion with alkaline phosphatase.[26] Electrophoresis of the reaction mixture showed that [32]P$_i$ was the reaction product. Failure to remove [32]P$_i$ by phosphatase digestion of ^{32}P-labeled denatured protein or peptides suggests a diester linkage. Snake venom nuclease will remove ^{32}P-labeled nucleotide ligated through a 5′-phosphodiester to tyrosine.[20] Micrococcal nuclease will remove nucleoside attached through this linkage, leaving a tyrosine phosphate monoester.[27]

[21] S. P. Adler, D. Purich, and E. R. Stadtman, *J. Biol. Chem.* **250**, 6264 (1975).
[22] P. G. Rothberg, T. J. R. Harris, A. Nomoto, and E. Wimmer, *Proc. Natl. Acad. Sci. U.S.A.* **75**, 4868 (1978).
[23] V. Ambros and D. Baltimore, *J. Biol. Chem.* **253**, 5263 (1978).
[24] Y.-C. Tse, K. Kirkegaard, and J. C. Wang, *J. Biol. Chem.* **255**, 5560 (1980).
[25] J. J. Champoux, *J. Biol. Chem.* **256**, 4805 (1981).
[26] O. N. Witte, A. Dasgupta, and D. Baltimore, *Nature (London)* **283**, 826 (1980).
[27] T. M. Martensen and E. R. Stadtman, *Proc. Natl. Acad. Sci. U.S.A.* **79**, 6458 (1982).

scintillation vial and counted (open symbols); another 0.5 ml of supernatant was added to a conical test tube containing 0.25 ml of 2 N H$_2$SO$_4$: 0.1 mM P$_i$ for measurement of [32]P$_i$ by the organic extraction technique (filled symbols).

Detection, Isolation, and Quantification of O-Phosphates in Proteins after
Chemical or Proteolytic Treatment

Release of O-Phosphates by Partial Acid Hydrolysis of Phosphoprotein

A variety of conditions have been utilized to produce phosphorylated amino acids through partial acid digestion of phosphorylated proteins.[7–9] The end results are similar; about 25% of the phosphorylated amino acid residues are recovered as the free phosphoamino acids, 50% are hydrolyzed yielding P_i, and the remaining residues exist in small phosphopeptides. The stability of the O-phosphates to acid is Thr-P > Ser-P > Tyr-P. The half-life of the O-phosphates in 1 N HCl at 100° was Tyr-P ~5 hr and Thr-P and Ser-P ~18 hr.[13] The stability of Ser-P and Thr-P in 6 N HCl at 105° was reported in an earlier volume.[8] Thr-P ($t_{1/2}$ ~ 25 hr) was found to be more stable than Ser-P ($t_{1/2}$ ~ 8 hr). Using cellular [^{32}P]phosphoprotein, the proportion of [^{32}P]Tyr-P to total free O-phosphates was found to be highest after 1 hr of digestion in 5.7 N HCl at 110°; after 4 hr, the proportion of Tyr-P falls to one-third the 1-hr value, and the proportion of Thr-P rises 3-fold.[9]

Procedure for Partial Acid Hydrolysis of Phosphoproteins

1. Add [^{32}P]phosphoprotein to a 1-ml Reacti-Vial (Pierce) for samples of <1 mg. Samples of >1 mg should be worked up in a 1-dram vial (Wheaton). Remove buffer salts, etc. (see below), by the addition of trichloroacetic acid to 20%. Place a Teflon liner (Pierce) atop the vials (Teflon side down) and secure the cap. Leave the sample on ice for 15 min prior to centrifugation at 10,000 g for 5 min. Carefully remove the supernatant with a fine-tipped Pasteur pipette. Resuspend the pellet in 0.5 ml of ethanol–ether (1 : 1) by vortexing; centrifuge as before. Remove the supernatant and add 0.5 ml of ether; then vortex, centrifuge, and remove the supernatant. Let the samples air dry for a few minutes before proceeding to the hydrolysis step.

2. Add 0.5 ml of 6 N HCl to the vial containing the dry protein; firmly secure the liner–cap assembly, and vortex.

3. Place the vials in a *covered* heating block (type A-1, Pierce) seated in a bench-top heater (or oven) set at 110°. Replace cover. Remove samples at 0.25, 1, 2, and 3 hr. For a single time-point, incubate for 1–2 hr.

4. Remove vials from heating block with forceps and place in an unheated block to cool. Remove the HCl from the vials by rotary evaporation, by evaporation in a vacuum desiccator over KOH, or by evaporation with an air source. After the samples are dried, add 0.05 ml of water to the

vial, rinse the sides, and take the sample to dryness. The sample is now ready to use with a detection method.

It is desirable to remove buffer salts to avoid their interference in detection methodologies for O-phosphates. Metal ions are believed to accelerate phosphoryl bond hydrolysis. Trichloroacetic acid precipitation has been the method of choice for "washing" proteins prior to acid hydrolysis. It is removed by the two organic washes. However, in dealing with immunoprecipitates or proteins eluted from polyacrylamide gels, the quantity of protein may be insufficient ($<10\ \mu g$) for trichloroacetic acid precipitation. Protein is dialyzed with $10–20\ \mu g$ of carrier protein against a volatile buffer (NH_4HCO_3) containing SDS; the sample is lyophilized twice to remove buffer, and the residue is extracted with acetone to remove sodium dodecyl sulfate prior to acid hydrolysis. Partial acid hydrolyses of dried gel pieces and electroblots can be used for rapid screening purposes.

It is useful to carry out the entire process of washing and hydrolysis in the same container. The Reacti-Vial can withstand centrifugation (in a standard 15-ml Corex tube adapter containing two type 1 rubber stoppers) and high temperature. Protein adheres to the glass surface more avidly than to polypropylene snap-cap tubes during the washing steps, thus minimizing losses. The tapered end allows the hydrolyzed sample to be taken up in a minimum volume ($2–5\ \mu l$). The washing steps still can be carried out in snap-cap tubes if desired. If hydrolyses are carried out in these containers, the tops should be secured.

Aerobic hydrolyses and the use of $6\ N$ instead of azeotropic HCl has not been shown to affect the yield of O-phosphorylated amino acids. The use of elevated temperature ($110°$) is preferred, thus reducing working time for an analysis. Removal of acid by rotary evaporation using a Speed-Vac (Savant) is useful when a large number of samples are worked up. An air stream through a Pasteur pipette will rapidly remove HCl from a sample on a hot plate at $60°$; this procedure is routinely used in this laboratory.

Semipurification of O-Phosphates on Cation Ion Exchanges

Chromatography of partial acid hydrolysates on Dowex 50 (H^+) at pH 2 ($0.1\ M$ HCOOH) results in the adsorption of the normal amino acids. Ser-P and Thr-P are essentially unadsorbed, and the elution of Tyr-P is somewhat retarded. This purification eliminates the need for regeneration of amino acid analyzer columns after each application of sample and facilitates electrophoretic analysis of hydrolysates from large amounts of protein containing small amounts of radioactivity.

Procedure

1. Tamp down a *small* amount of glass wool in a 1-ml disposable pipette tip. Pipette 0.2 ml of a slurry of 1 g of Dowex 50-X8 (H⁺), 100–200 mesh, in 1 ml of water into the tip placed atop a 10×75 mm test tube. Wash the resin with 1.0 ml of 0.1 N HCOOH. Place the tip above a 3-ml conical tube (Bellco).

2. Add 0.2 ml of the acid to the dried partial acid hydrolysate and redry. Standards of each phosphoamino acids (2–10 nmol) can be added here if desired. Redissolve the residue in 0.2 ml of the acid and carefully apply to the resin.

3. Wash the resin with 0.2, 0.2, and 0.6 ml of the formic acid.

4. Take the eluate ($\sim$1.2 ml) to dryness.

The table shows that with quantities of protein up to 10 mg, 94% of the ¹⁴C-labeled amino acids and peptides formed during 1.5 hr hydrolysis (6 N HCl, 110°) of [¹⁴C]glutamine synthetase (isolated from bacteria grown on ¹⁴C-labeled amino acids) were adsorbed to 0.1 g of resin. The yield of unadsorbed radioactivity from partial acid digests of ³²P-labeled phosphorylase *a*, HMW microtubule-associated protein (MAP), and phosphotyrosyl glutamine synthetase was 96, 102, and 84%, respectively. Figure 2 shows that prepurification of partial acid hydrolysates of milligram quantities of protein will prevent streaking during electrophoresis.

SEMIPURIFICATION OF *O*-PHOSPHATES AFTER PARTIAL ACID HYDROLYSIS[a]

Sample	Total protein hydrolyzed (mg)	% Total cpm in first 0.6 ml of eluate	% Total cpm in 1 ml of eluate
[¹⁴C]Glutamine synthetase (GS)	0.1	—	0.8
GS + bovine albumin, 1 mg	1.1	—	1.5
GS + bovine albumin, 2 mg	2.6	—	1.7
GS + bovine albumin, 5 mg	5.1	—	4.7
GS + bovine albumin, 10 mg	10.1	—	6.1
[³²P]Phosphorylase *a*	1.1	92	96
[³²P]MAP	1.8	98	102
[³²P]Phosphotyrosylglutamine synthetase	0.75	75	84

[a] Each radiolabeled protein was treated with trichloroacetic acid, ethanol–ether, and ether prior to hydrolysis in 6 N HCl for 1.5 hr at 110°. After hydrolysis each sample was taken to dryness, then dissolved in 0.1 N HCOOH. An aliquot of the [¹⁴C]glutamine synthetase (20×10^3 cpm) was added to a quantity of albumin hydrolysate prior to chromatography on Dowex 50 (0.1 g). ¹⁴C radioactivity was measured by scintillation counting; ³²P was measured by Cerenkov counting to allow recovery of samples for electrophoresis.

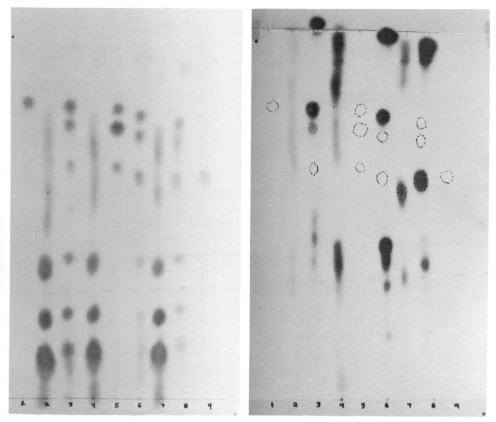

FIG. 2. Electrophoresis of partial acid hydrolysates at pH 3.5. Partial acid hydrolysates of [³²P]MAP, [³²P]phosphorylase *a*, and [³²P]phosphotyrosylglutamine synthetase were prepared in 6 *N* HCl at 110° for 1 hr. One-fourth of each hydrolysate was removed and taken to dryness. The remainder (3/4) was semipurified (see the table), then taken to dryness. All of each sample was applied to 3 MM paper with standards (10 nmol) and electrophoresed at 1500 V for 3 hr in a high-voltage electrophorator (Gilson). (+) denotes purified material; (−) denotes unpurified material. 1, Ser-P; 2, MAP(−); 3, MAP(+); 4, phosphorylase *a* (−); 5, Ser-P, Thr-P, Tyr-P; 6, phosphorylase *a* (+); 7, phosphotyrosylglutamine synthetase (−); 8, phosphotyrosylglutamine synthetase (+); 9, Tyr-P. Left: ninhydrin stain. Right: autoradiogram with phosphoamino positions outlined. Overlap position obtained by marker radioactivity on two corners.

Detection of O-Phosphates by High-Pressure Liquid Chromatography (HPLC), Electrophoresis, and Amino Acid Analysis after Chemical or Proteolytic Digestion

The identification of O-phosphates after chemical or proteolytic treatment of [³²P]phosphoprotein is achieved by comigration of ³²P-labeled

O-phosphates with authentic standards using high-voltage electrophoresis,[8,9] ion-exchange resins,[18,28] and HPLC.[29,30] When partial acid hydrolysates of semipurified [^{32}P]phosphoproteins are subjected to electrophoresis at pH 3.5 (pyridine/glacial acetic acid/H_2O; 10 : 100 : 1890), the individual O-phosphates can be identified. [^{32}P]Phosphopeptides and ^{32}P$_i$ will also migrate toward the anode. The rate of migration is P$_i$ > Ser-P > Thr-P > Tyr-P > phosphopeptides (see Fig. 2). Both thin-layer cellulose plates and 3 MM paper are used for supports. Paper allows longer runs and greater sample load compared to the thin-layer cellulose plates, which yield more discrete spots. Electrophoretic separation of Ser-P from Thr-P is better at pH 1.9 (88% formic acid–glacial acetic acid–H_2O; 5 : 15.6 : 179). However, Tyr-P will usually comigrate with Thr-P under this condition. If the [^{32}P]phosphoprotein is contaminated with ^{32}P-labeled cellular constituents (nucleotides, phospholipids), it is necessary to use a two-dimensional technique for separating the O-phosphates from other ^{32}P-labeled constituents. An unidentified cellular compound and 3'-UMP were found to migrate closely with Tyr-P under a variety of conditions.[9,22] A detailed methodology for the two-dimensional separation of O-phosphates in phosphoproteins obtained from ^{32}P-labeled cells appears in an earlier volume.[9] That chapter should be consulted for measurement of the relative levels of O-phosphates in ^{32}P-labeled cells. It should be emphasized that careful attention should be paid to identification methods, and the use of two different techniques is always desirable.

An HPLC anion-exchange column was used to detect O-phosphates in histones treated with protein-serine and protein-tyrosine kinases[29] after acid or proteolytic hydrolysis. Isocratic elution with 15 mM potassium phosphate, pH 3.8, separated the O-phosphates, whose order of elution was Thr-P, Ser-P, and Tyr-P. Better resolution was obtained using 10 mM P$_i$, pH 3.0, with 12.5% methanol during a 50-min run.[30] Neither system was used to resolve O-phosphates obtained from partial acid hydrolysates of crude phosphoprotein from labeled cell extracts.

O-Phosphates elute in the void with cysteic acid on cation-exchange resins under standard conditions used for amino acid analyses. Modification of elution buffer by reducing the pH and salt concentration enables the resolution of Ser-P, Thr-P, and Tyr-P. Trifluoroacetic acid (10 mM) was used[28] to separate O-phosphates: Ser-P, 19 min; Thr-P, 28 min; and Tyr-P, 64 min. Amino acid analyzers offer the advantage of simultaneously quantifying O-phosphates at levels of ~50 pmol. However, the

[28] J. P. Capony and J. G. Demaille, *Anal. Biochem.* **128,** 206 (1983).
[29] G. Swarup, S. Cohen, and D. L. Garbers, *J. Biol. Chem.* **256,** 8197 (1981).
[30] J. C. Yang, J. M. Fujitaki, and R. A. Smith, *Anal. Biochem.* **122,** 360 (1982).

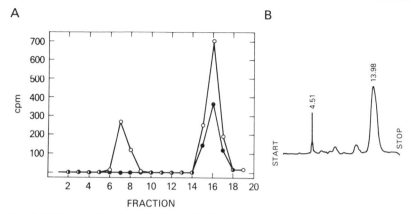

FIG. 3. Elution profile of a hydrolysate of phosphotyrosylglutamine synthetase applied to an amino acid analyzer. A sample of [^{32}P]phosphotyrosylglutamine synthetase (0.35 mg) was digested in 0.15 ml of 5.0 N KOH containing 0.12 μM [^{14}C]Tyr-P for 35 min at 155°, neutralized with HClO$_4$, and applied to the amino acid analyzer. Upon injection, a fraction collector was simultaneously activated to collect the eluate from the analyzer. (A) ^{32}P (○) and ^{14}C (●) radioactivity found in eluate from analyzer (~0.45 ml) collected at 1.0-min intervals. (B) Recorder tracing from analyzer. From Martensen.[18]

yield of O-phosphates after partial acid hydrolysis (~25%) may vary. The stability of phosphotyrosine to base hydrolysis enables 80% recovery[10,18] of the phosphorylated amino acid under conditions that result in the complete hydrolysis of protein (see below). Tyr-P was detected with an amino acid analyzer equilibrated at pH 2.0 using fluorometric and radioactive detection (Fig. 3). Conditions for the complete digestion of proteins in base (5 N KOH, 155°, 30 min) destroy Ser-P and Thr-P residues in proteins.

Quantification of O-Phosphates in Proteins by Analysis for P$_i$ Produced by Alkali or Total Digestion

Quantification of Ser-P and Thr-P in proteins is done by measuring P$_i$ after incubation in base under conditions that lead to complete removal of phosphoryl groups. Normally these residues are quantitatively dephosphorylated in 1 N NaOH at 37° for 18–20 hr.[31,32] Resistant Thr-P residues require higher temperatures (65°) for 18–20 hr.[16] If the base lability of the phosphoryl group has been determined (see introductory section of this

[31] H. G. Nimmo, C. G. Proud, and P. Cohen, *Eur. J. Biochem.* **68,** 21 (1976).
[32] B. Ames, this series, Vol. 8, p. 115.

chapter), use incubation times for 6–10 half-lives. Total phosphate analyses should agree with alkali-labile phosphate analyses if phosphoryl groups are located on serine and threonine residues.

Analytical determination of phosphate requires 1–10 nmol of phosphate per assay. For a phosphoprotein of $M_r \sim 50,000$ containing a single phosphorylated site, this is a 50–500 μg. Samples should be assayed in duplicate or triplicate. Prior to analysis, phosphoprotein is "washed" with trichloroacetic acid to remove contaminating phosphoryl-containing compounds. Special conditions may be necessary[8] for ribosomal proteins (polynucleotide contamination). Protein concentrations during trichloroacetic acid precipitation and washings should be at least 50–100 μg/ml to minimize losses. Sample preparation and analyses can easily be carried out by using Eppendorf 1.5-ml snap-cap polypropylene tubes and a benchtop microfuge for centrifugation. Glass is not recommended for alkalilabile phosphate analyses, since phosphate may be leached by base.

Procedure for the Analysis of Alkali-Labile Phosphate in Proteins

1. Add protein solution ($\leq$1 ml) containing 2–10 nmol of phosphoryl groups to an Eppendorf tube. Add 100% trichloroacetic acid (w/v) to $\sim$20%. Centrifuge after 10 min on ice (10,000 g for 2 min) and remove supernatant.

2. Add 0.5 ml of 0.1 N NaOH on ice to pellet, vortex to solubilize it, and immediately precipitate as in step 1.

3. Wash the pellet twice by suspension in 0.5 ml of 20% trichloroacetic acid and centrifugation.

4. Solubilize the pellet in 0.3 ml of 1 N NaOH, cap the tubes, and incubate for the desired time. After incubation, add 0.075 ml of 100% trichloroacetic acid and leave on ice for 10 min prior to centrifugation.

5. Remove 0.3 ml of the supernatant for P_i analysis using standards and blanks (0.3 ml) prepared with 20% trichloroacetic acid. The phosphate standard is KH_2PO_4, 0.5 mM in 5 mM HCl.

Measurement of Total Phosphate Content of Proteins

Protein is washed with trichloroacetic in borosilicate tubes (10 $\times$ 75 mm) as described above. Triplicate samples of protein containing 0.5–10 nmol of phosphate should be assayed. After the final acid wash, the supernatants are carefully removed and 0.03 ml of 10% magnesium nitrate in ethanol is added to sample, blanks, and standards.[32] The tubes are placed in a bench-top heater or an oven at 110° to dry. This procedure prevents spattering that occurs when flame-drying and leaves a residue that can easily be ashed over an intense flame in 15–30 sec. Ashing is complete after the evolution of NO_2 ceases, and a *white* residue remains. Add 0.05 ml of 1 N HCl to samples and place them in a bench-top heater at 110° in a hood until dry. Remove samples and add 0.30 ml of 1 N HCl. The samples can be analyzed by the malachite green method or the ascorbate procedure of Ames.

Phosphate Analysis

The procedure of Ames[37] is straightforward and easily reproduced. Ten nanomoles of P_i give an absorbance of 0.24 in a 1-ml volume, and the assay is linear to $A_{820} > 1$. The assay below is modified for use with 0.3 ml of alkali labile or total phosphate samples ($P_i \geq 2$ nmol), to give 0.4 ml total volume.

1. Prepare a solution of 2% ascorbic acid in 0.5% ammonium molybdate/1.2 N H_2SO_4 and place on ice. The solution should be prepared fresh daily. The molybdate/H_2SO_4 solution is stable.

2. Mix 0.10 ml reagent with 0.30 sample, blanks, and standards; seal the tube tops with parafilm.

3. Measure absorbance in microcuvettes at 820 nm after color development (20 min at 45°, 60 min at 37°).

Malachite Green Phosphate Analysis

Stull and Buss[33] adapted an assay for P_i in biological samples using malachite green[34] with the ashing procedure described by Ames to develop a more sensitive assay for total phosphate of phosphoprotein. Details of this procedure have been described (this series, Vol. 99, [7]). The absorbancy of 1 nmol in 0.4 ml is ~0.15. Though more sensitive and rapid care must be taken to avoid problems with the precipitation of the molybdate dye complex. The dye reagent used here is 0.025% malachite green instead of 0.15%. This modification lowers the blank to ~0.04; loss of linearity of absorbancy to phosphate is observed above 4 nmol. Dilution of samples with reagent/HCl allows P_i values greater than 4 nmol to be assayed.

Reagents. These reagents, stored at room temperature, are renewed every 6 months.

   Malachite green, 0.2% in 1.0 N HCl
   10% Ammonium molybdate · $4H_2O$ in 3 N HCl (10 g of molybdate is
      dissolved in 75 ml of water prior to addition of 25 ml of concen-

---

[33] J. T. Stull and J. E. Buss, *J. Biol. Chem.* **252,** 851 (1977). *Methods Enzymol.* **99,** 7–14 (1983).

[34] K. Itaya and M. Ui, *Clin. Chim. Acta* **14,** 361 (1966).

trated HCl and mixing. A fresh solution may sometimes correct
blank and turbidity problems

HCl, 1.0 N

Prepare malachite green molybdate reagent by adding 1 volume of dye
to 5 volumes of HCl, followed by 2 volumes of molybdate. The reagent
should be prepared about 30 min before use; just prior to use it should be
centrifuged at ~5000 g at room temperature for a few minutes to remove a
fine precipitate. Add 0.1 ml of reagent to 0.3 ml of sample in 1.0 N HCl
and measure the absorbance after 5 min at 660 nm. The color appears to
be stable for 30 min. The reagent can be used throughout the day providing it is centrifuged prior to use. Plasticware is used to store and transfer
the reagent. HCl (1 N) is used to rinse the microcuvettes. After use the
cuvettes are stored or washed in 50% 1 N HCl/50% ethanol. This method
cannot be used to estimate P_i in supernatants containing trichloroacetic
acid without the following modifications: (1) The trichloroacetic acid concentration of samples and standards must be $\leq 10\%$, and (2) 25 μl of
concentrated HCl should be added to each assay (prior to reagent addition).

Quantification of Phosphotyrosine in Proteins and Protein Tyrosine Kinase Reaction Mixtures

This methodology is based on studies of the chemical stability of phosphotyrosine to base (5 N KOH) at high temperature (155°). These conditions yield complete hydrolysis of proteins in 30–35 min.[35] The yield of
Tyr-P in base hydrolysates of the phosphoamino acid as well as four
proteins containing phosphotyrosine residues was 80 ± 2%.[18] The yield of
Tyr-P was established by fluorometric or ^{32}P radioactivity measurements,
using an amino acid analyzer (Fig. 3). This methodology was reported in
an earlier volume.[10] Subsequently, it was found the method could be used
to measure [32P]Tyr-P in cell homogenates or purified cellular fractions
incubated with [γ-32P]ATP.[36] Trichloroacetic acid precipitates of reaction
mixtures are hydrolyzed in base containing [14C]Tyr-P. The procedure
destroys [32P]phosphoserine and [32P]phosphothreonine residues, yielding
32Pi and [14C,32P]Tyr-P. Chromatography of the hydrolysate on short
columns of Dowex 50 (H+) separates 32Pi from [14C,32P]Tyr-P, which eliminates using an amino acid analyzer for quantification. The P_i is unadsorbed, and Tyr-P elutes after 7–10 column volumes. This is the method
by which chemically synthesized Tyr-P is purified from phosphoric acid.[13]
The volume of solution required to elute Tyr-P from Dowex 50 increases

[35] R. L. Levine, *J. Chromatogr.* **236**, 499 (1982).
[36] T. M. Martensen, *Fed. Proc. Fed. Am. Soc. Exp. Biol.* **42**, 2165 (1983).

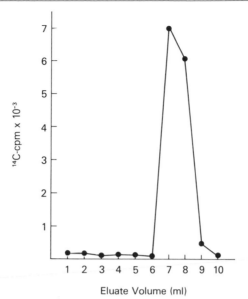

FIG. 4. Dowex 50 chromatography of tyrosine O-phosphate. One milliliter of [^{14}C]Tyr-P, 14,000 cpm (~20 nM) was applied to a 1-ml (0.8 × 2 cm) column of Dowex 50-X8 (H$^+$, 200–400 mesh) and eluted with 0.1 N formic acid. After collection of 5 ml, the column was eluted with water.

as the pH is lowered in the range 1.7–2.4. Figure 4 shows the elution profile of [^{14}C]Tyr-P using 0.1 N formic acid (pH 2.4). This procedure was used to purify Tyr-P from a base digest of phosvitin.[18] Use of 0.1 N formic/1% H$_3$PO$_4$ (pH 1.7) retards the elution of Tyr-P to ~10 column volumes and enhances the washout of ^{32}P$_i$.

Figure 5 shows the purification of [^{32}P]Tyr-P formed in a protein tyrosine kinase reaction mixture and confirmation of its identity by the amino acid analyzer. The recovery of Tyr-P is obtained from the recovery [^{14}C]Tyr-P measured by channel ratio scintillation counting; ^{14}C counts give the recovery of the internal standard, and ^{32}P counts give the quantity of phosphorylated tyrosine residues produced with [γ-^{32}P]ATP of a known specific activity.

Procedure for Quantification of Tyr-P in Proteins by Amino Acid Analysis[1]

1. The protein sample is made 5 N in base by (1) adding 5.0 N KOH to lyophilized samples or trichloroacetic acid precipitates; (2) adding

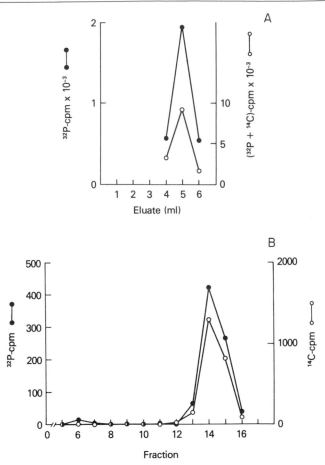

FIG. 5. Dowex 50 chromatography of base-hydrolyzed protein tyrosine kinase reaction mixture. The trichloroacetic acid precipitate was dissolved in 0.1 ml of 5 N KOH, 0.5 μM [^{14}C]tyrosine O-phosphate (32,000 cpm). The sample was hydrolyzed for 30 min at 155° prior to application to a column (0.8 × 1 cm) containing 0.5 g of Dowex 50-X8 (H$^+$, 200–400 mesh). The sample was eluted with 0.1 N formic acid, 1% H$_3$PO$_4$ (3 ml), and water (3 ml), as described. (A) Radioactivity in 0.5-ml fractions 4–6 was measured in a scintillation counter set for discriminating ^{32}P (●) counts from ^{14}C counts (○). (B) The remaining 0.5 ml in tube 5 was applied to an amino acid analyzer equilibrated with 0.2 N sodium formate, pH 2.0. The eluate was analyzed for ^{32}P and ^{14}C radioactivity as above.

10.0 N KOH to protein solutions (see below). If quantification of Tyr-P is measured by recovery of [^{14}C]Tyr-P, the base solution should contain the counts. The solubilized protein is added to a hydrolysis vial that is covered with a Teflon liner and capped.

2. Place the capped vials in a heating block (Pierce Reacti-Block C for the 1-dram vial, and Reacti-Block A for the 1- and 0.3-ml vials) seated in a standard bench-top heater set at 155°. After incubation at 155° for 30–35 min in the *covered* heating block, remove the vials with forceps and place them in an unheated block to cool.

3. Transfer the contents of each sample to conical test tubes (1–3 ml), and rinse each vial with distilled water equal to the sample volume. Place the tube containing the combined hydrolysate and wash on ice; add 5.0 N HClO$_4$ equivalent to the volume of 5.0 N KOH and mix; then leave the tube on ice for a few minutes. Add 2 N sodium formate, pH 2.0, to 0.2 N (0.11 the total sample volume) and mix; centrifuge the samples at −5–0° for 5 min. Remove the supernatant for amino acid analysis. The total dilution of the hydrolysate is 1 : 3.3.

Protein hydrolyses are carried out at 1–10 mg/ml (bovine serum albumin is added as carrier, if necessary). The KOH solution is stored in a tightly sealed plastic bottle. Hydrolyses with volumes greater than 1 ml are carried out in Wheaton 1-dram borosilicate vials. Hydrolyses with smaller volumes are carried out in 0.3- or 1.0-ml Reacti-Vials (Pierce). The sample volume should be at least 20% of the total volume of the vial. Cover the vials with unlined caps (PGC Scientific, Rockville, MD) containing a Teflon–silicone or Teflon–rubber cap liner (Pierce). If the usual rubber-lined caps are used, remove the liner. (A narrow metal spatula blade permits easy removal of the rubber liner. The caps should then be boiled in water to remove loosely adhering particles.) Septum caps should not be used. The cap should be seated firmly, without overtightening. Samples are normally stored at 4° prior to amino acid analysis. Samples put on ice or frozen will throw out more precipitate, which may settle out after tapping the sample tube but may necessitate recentrifugation. Amino acid analyses are done with an amino acid–peptide analyzer (Dionex) equipped with a stainless-steel microbore column (0.4 × 15 cm) packed with DC-5A resin (Dionex). The column is eluted with 67 mM sodium citrate buffer, pH 2.0, or preferably with 0.2 M sodium formate, pH 2.0 at a flow rate of 18 ml/hr. The eluate is mixed with o-phthalaldehyde in borate buffer, which enters the reaction coil at 9 ml/hr. The fluorescent amino acid derivatives are detected in a fluorometer (Gilson) connected to a recording integrator (Shimadzu). Tyr-P is measured after injection of the neutralized hydrolysate onto the column equilibrated with buffer (20–30 min). After elution of Tyr-P, the column is washed with 0.1 N NaOH, 0.1 N NaCl, 3.5 mM EDTA for 10 min prior to reequilibration. The detection system can reliably measure 50–100 pmol of unlabeled material and ~200 cpm by scintillation counting of the eluate.

Procedure for Analysis of [^{32}P]Tyr-P in Protein Tyrosine Kinase Reaction Mixtures

1. Precipitate protein in reaction mixtures (50 μl) by the addition of 10–20% trichloroacetic acid (0.5 ml). Centrifuge and remove the supernatant.

2. Add 0.1 ml of 5 N KOH containing ~20,000 cpm of [^{14}C]Tyr-P to dissolve the pellet and transfer to a 0.3-ml vial.

3. Hydrolyze the sample as described above. Transfer the hydrolysate to a (0.8 × 4 cm) column containing 0.5 g of Dowex-50 X8 (H$^+$, 200–400 mesh). [One milliliter of a 1 : 1 (w/v) stirring slurry of resin in water is added to a 2-ml polypropylene Econo column (Bio-Rad).]

4. Wash the column with two 0.5-ml volumes of 0.1 N formic acid–H$_3$PO$_4$ (1%) with minimal disturbance of the resin surface. Wash the column again with two 1-ml volumes.

5. Place the column above the new reservoir and elute [^{14}C,^{32}P]Tyr-P with three 1-ml washes of water; collect separately each 1-ml fraction and determine ^{32}P radioactivity and ^{14}C radioactivity by channel ratio counting.

This procedure normally gives about 40% recovery of the ^{14}C[Tyr-P] in the fifth milliliter of eluate which is used to determine the yield. The ^{14}C internal standard gives recovery from phosphoryl-bond hydrolysis, transfer, and chromatography. Elution by gravity gives sharp resolution during chromatography; this is achieved by following the sequence of steps listed. Measurement of ^{14}C recovery is optimal when ~8000 cpm of [^{14}C]Tyr-P is recovered in tube 5 and the [^{32}P]Tyr-P is no more than 4–5 times the amount. Since tube 5 is mostly salt free, an aliquot may be dried to confirm the identity of the ^{32}P radioactivity by a method for identification of *O*-phosphates. This method is not recommended for a one-step analysis of Tyr-P in crude cellular protein precipitated from cells incubated with ^{32}P$_i$. Further purification is necessary to separate ^{32}P-labeled phosphorylated material of unknown identity from the labeled Tyr-P. The use of the internal standard allows quantification in any case.

Preparation of carrier free [^{14}C]Tyr-P: An air stream is used to dry manufacturers vial of [^{14}C]tyrosine (250 μCi = 0.49 μmol) on a hot plate (~50°). A slurry of 2.3 g of H$_3$PO$_4$–1 g of P$_2$O$_5$ is prepared, and 0.2 ml is added to the vial, which is then sealed and incubated at 105° for 20 hr. Approximately 10 mg of P$_2$O$_5$ is added to the vial, which is reincubated for 20 hr. The sample is removed, let cool, and put on ice. One milliliter of water and 0.6 g of ice are added to the vial, which is immediately vortexed. The solution is applied to a 5-ml column of Dowex 50-X8 (H$^+$) and eluted with water (15, 2-ml washes), collecting each 2-ml fraction, which is analyzed for radioactivity. The second radioactive peak contains [^{14}C]Tyr-P, which is pooled (yield ~40%). [^{3}H]Tyrosine loses >90% of its label under these conditions.

Acknowledgment

The author wishes to thank members of the Laboratories of Cell Biology, and Biochemistry, NHLBI, for their advice and comments. To Dr. E. R. Stadtman, Chief of the Laboratory of Biochemistry, NHLBI, the author expresses special thanks for support and advice.

[2] Techniques in the Detection and Characterization of Phosphoramidate-Containing Proteins

By JAMES M. FUJITAKI and ROBERTS A. SMITH

Phosphorylation on hydroxyamino acids is an important posttranslational modification of proteins that has been extensively reviewed.[1-3] A major group of amino acids that are also phosphorylated are the basic amino acids arginine, histidine, and lysine. These phosphoramidate-containing phosphoamino acids have the property of being extremely acid-labile.[4] The majority of the P-N-containing amino acids found in proteins are catalytic intermediates of enzymes. However, protein kinases are also known that phosphorylate structural proteins on basic amino acids. In rat liver and Walker 256 carcinosarcoma, histone H1 is phosphorylated on lysine residues and phosphorylation occurs on both histidine residues of histone H4.[5,6] Acid-labile phosphates occur in rat myelin basic protein in the form of phosphoarginine and phosphohistidine.[7] Acid-labile phos-

[1] E. C. Krebs and J. A. Beavo, *Annu. Rev. Biochem.* **48,** 923 (1979).

[2] F. Wold, *Annu. Rev. Biochem.* **50,** 783 (1981).

[3] J. R. Knowles, *Annu. Rev. Biochem.* **49,** 877 (1980).

[4] R. A. Smith, R. M. Halpern, B. B. Bruegger, A. K. Dunlap, and O. Fricke, *Methods Cell Biol.* **19,** 153 (1978).

[5] C. C. Chen, B. B. Bruegger, C. W. Kern, Y. C. Lin, R. M. Halpern, and R. A. Smith, *Biochemistry* **16,** 4852 (1977).

[6] B. B. Bruegger, Ph.D. Dissertation, University of California, Los Angeles, 1977.

[7] L. S. Smith, C. W. Kern, R. M. Halpern, and R. A. Smith, *Biochem. Biophys. Res. Commun.* **71,** 459 (1976).

phates have been detected also in ribosomes.[8] The function of such phosphorylations is unknown.

The more commonly employed assay procedures and isolation methods in the investigations of phosphoproteins involve the use of acid. Acid conditions obviously are precluded when investigating P-N-containing proteins. Since acid treatment of phosphoprotein during purification and during determination of phosphoamino acids is quite routine, only phosphohydroxyamino acids (P-O) have been studied, and the existence of acid-labile phosphates has been largely overlooked. It is postulated that this type of phosphorylation may be more widespread than is generally believed. Several techniques have been devised to detect, quantify, and characterize phosphorylated basic amino acids on proteins.

Determination of Phosphorylated Basic Amino Acid

High-Performance Liquid Chromatography (HPLC)

Acidic buffers are normally used in HPLC analysis of amino acids and therefore would be unsuitable for the detection of compounds containing the P-N bond. An HPLC system has therefore been devised that can be used in the separation of N-phosphorylated basic amino acids.[9] Under isocratic conditions, the constraint of using only neutral elution buffers on an anion-exchange column (use of high pH buffers required extremely high salt concentrations for elution and resulted in poor resolution) requires the use of two HPLC systems, one for the phosphohydroxyamino acids and phosphohistidine using a high ionic strength buffer (buffer B) and the other for phosphoarginine and phospholysine using a low ionic strength buffer (buffer A). Separation was afforded on a 250×4.6 mm i.d. column of Chromex DA-X12-11 anion-exchange resin (polystyrene quaternary amine type, 12% cross-linking, Durrum). Buffer B consisted of 250 mM KH_2PO_4 (pH 6.3) containing 1 g of phenol per liter; buffer A consisted of 15 mM KH_2PO_4, (pH 7.5) containing 1 g of phenol per liter. Elution was performed at 50° (water bath) with a flow rate of 0.5 ml/min. Phosphoamino acids were detected by fluorescence after mixing with *o*-phthalaldehyde. Typical elution times and relative fluorescence of the phosphoamino acids are listed in Table I. Separation of all phosphoamino acids during a single analysis may be accomplished using a gradient of ionic strength or pH.

[8] D. P. Ringer, R. L. King, Jr., and D. E. Kizer, *Fed. Proc. Fed. Am. Soc. Exp. Biol.* **39,** 735 (1982).
[9] A. W. Steiner, E. R. Helander, J. M. Fujitaki, L. S. Smith, and R. A. Smith, *J. Chromatogr.* **202,** 263 (1980).

TABLE I
HPLC[a] Elution Times and Relative Fluorescence of
Phosphoamino Acids[9]

Phosphoamino acid	Buffer A (min)	Buffer B (min)	Relative fluorescence[b]
Phosphoserine	—	21.8	1.34
Phosphothreonine	—	16.0	1.37
τ-Phosphohistidine	—	18.0	1.63
ω-N-Phosphoarginine	22.5	—	2.61
ε-N-Phospholysine	20.0	—	1.00

[a] By the procedure described in the text.
[b] Fluorescence is expressed relative to tyrosine. A plot of relative fluorescence vs amount of phosphoamino acid was determined for each, and the slope was divided by the slope obtained for tyrosine.

Phosphorus-31 Nuclear Magnetic Resonance (^{31}P NMR)

^{31}P NMR affords a unique means of detecting specific phosphoamino acids on intact phosphoproteins in solution. Correlation between ^{31}P resonances of phosphoamino acid standards and those of denatured phosphoprotein is generally very good, as studies have indicated.[10-12] However, caution must be exercised in interpreting and quantitating NMR data because such factors as metal ion chelation, steric strain, and chemical anisotropy may affect the shape and position of phosphorus resonances. A major drawback of ^{31}P NMR is the relatively high concentration of phosphorylated sample (millimolar range) required for adequate data acquisition within a reasonable period of time. However, because it is a nondestructive method of analysis, ^{31}P NMR eliminates possible phosphoryl transfer reactions during the harsh conditions of either low or high pH often employed in the characterization and isolation of the phosphoamino acids moiety.

A chart of pH versus ^{31}P resonances of the various free phosphoamino acids is given in Fig. 1. In most instances, complete denaturation of the phosphoprotein [using urea, guanidinium hydrochloride, sodium dodecyl sulfate (SDS), etc.] is required for good correlation of the ^{31}P resonances of standard phosphoamino acid and protein-bound phosphate. When phosphorylation on basic amino acid residues is being investigated, neu-

[10] M. Gassner, D. Stehlik, O. Schrecker, W. Hengstenberg, W. Mareer, and J. Ruterjans, Eur. J. Biochem. 75, 287 (1977).
[11] G. Dooijewaard, F. F. Roosien, and G. T. Robillard, Biochemistry 18, 2996 (1979).
[12] J. M. Fujitaki, G. Fung, C. Y. Oh, and R. A. Smith, Biochemistry 20, 3658 (1981).

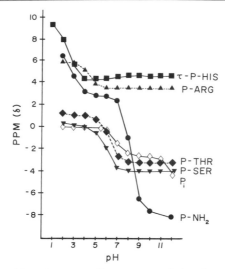

FIG. 1. Phosphorus-31 nuclear magnetic resonance of phosphoamino acids vs pH under conditions described in the text. The reference is 85% orthophosphoric acid. P-NH₂ is the dipotassium salt of phosphoramidate.

tral or slightly basic conditions must be employed. (Data accumulation under acidic pH results in the rapid hydrolysis of acid-labile phosphates.)

The utility of ^{31}P NMR is demonstrated in the investigation of chemical and enzymatic phosphorylation of histidine residues in histone H4.[12] The ^{31}P NMR chemical shifts of denatured phosphohistone and phosphohistidine standards match well (see Table II; phosphorylated HPr is included as another example). ^{31}P NMR spectra were obtained on a Bruker WP-200 NMR spectrometer operating in the Fourier transform mode at 81.02 MHz for ^{31}P resonance and equipped with a field-frequency lock on the deuterium resonance. Protein samples (5 ml) contained 0.1% SDS and at least 20% deuterium oxide. Sample tube diameter was 10 mm. Typical parameters employed were 3 kHz spectrometer width, 70° pulse angle, 166 μsec dwell time, 8000 data points in free induction decay, 1.0 sec relaxation delay. Chemical shifts were corrected with 0 ppm positioned at the corrected value of 85% orthophosphoric acid. Sample temperature was 300°K, and approximately 10,000 to 20,000 transients were performed. These NMR parameters must be varied depending upon the amount and properties of the phosphoprotein being investigated. But this example does illustrate the usefulness of ^{31}P NMR analysis, as this study was able to confirm earlier findings indicating the formation of different isomers of phosphohistidine of histone H4 catalyzed by protein kinases of nuclei from rat liver and from Walker 256 carcinosarcoma.[6]

TABLE II
^{31}P NMR CHEMICAL SHIFTS OF PHOSPHORYLATED HISTIDINE
RESIDUES AT pH 9–10 RELATIVE TO 85%
ORTHOPHOSPHORIC ACID

Species	Shift (ppm)
τ-Phosphohistidine standard	4.7
π-Phosphohistidine standard	5.4[10]
Chemically phosphorylated H4,[a] denatured[b]	4.8
Phosphorylated H4, rat regenerating liver, denatured	5.4
Phosphorylated H4, Walker-256, denatured	4.9

[a] Chemically phosphorylated H4 was prepared by mixing histone H4 (10 mg) with dipotassium phosphoramidate (10 mg) in 1 ml of distilled water. The solution was allowed to sit overnight at room temperature and was subsequently dialyzed against several changes of distilled water. τ-Phosphohistidine was isolated from base hydrolysates of phosphohistone prepared in this manner.[12]

[b] The solution contained 0.1% SDS.

Paper Chromatography and Electrophoresis

Phosphorylated basic amino acids can be analyzed by any paper chromatographic or high-voltage electrophoretic system not employing acidic buffers and not involving pyridine. Examples of acceptable procedures for paper chromatography are as follows: Whatman 3 MM paper in a solvent system of 2-propanol–ethanol–water–triethylamine (30 : 30 : 39 : 1 by volume)[13] or ethanol–1 M triethylamine–CO_2, pH 9.5 (3 : 1, v/v).[6] High-voltage electrophoresis can be performed on Whatman 3 MM paper at pH 8.25 in a buffer system of 0.1 M ammonium acetate–0.001 M EDTA for 30 min at 57 V/cm in a water-cooled apparatus[5] or at pH 9 in a buffer system of 1% ammonium bicarbonate.[14] The phosphorylated basic amino acids can be detected by ninhydrin, Pauly reagent[15] (for histidine), Sakaguchi reagent[16] (for arginine), molybdate reagent[17] (for phosphate), or by analyzing for radioactivity when ^{32}P-labeled phosphoamino acids are used.

[13] O. Zetterqvist and L. Engström, *Biochim. Biophys. Acta* **113,** 520 (1966).
[14] S. Narindrasorasak and W. A. Bridger, *J. Biol. Chem.* **252,** 3121 (1977).
[15] B. N. Ames and H. K. Mitchell, *J. Am. Chem. Soc.* **74,** 252 (1951).
[16] J. C. Bennett, this series, Vol. 11, p. 330.
[17] C. Hanes and F. Isherwood, *Nature (London)* **164,** 1107 (1949).

Column Chromatography

The hydrolysates of ^{32}P-labeled phosphoproteins can be chromatographed on a Dowex 1 column (bicarbonate form) with a linear gradient of 0.1 to 1.0 M potassium bicarbonate.[18,19] Identification of the phosphorylated basic amino acid is accomplished on the basis of coelution of ^{32}P radioactivity with carrier standard phosphoamino acid.

Discussion

Aside from ^{31}P NMR, determination of the amino acid that is phosphorylated on a protein necessitates the hydrolysis of the phosphoprotein into its individual amino acids. In the investigations of proteins containing the P-N bond, hydrolysis requires strong base or proteases. Base hydrolysis (3 N KOH, 120°, 3 hr)[20] destroys arginine and phosphoarginine as well as the phosphohydroxy amino acids.[21] Intramolecular chemical transphosphorylation may occur under basic condition.[22,23] Certain proteins are not completely hydrolyzed by proteases. Thus, isolation of a putative phosphorylated basic amino acid from hydrolysates must be confirmed by several techniques. It is also suggested that the isolated phosphorylated basic amino acid be acid treated and the resulting free amino acid be characterized to confirm the identification whenever possible.

Diagnostic Tests

Unequivocal demonstration that a certain amino acid of a protein is phosphorylated requires the isolation or detection of the corresponding phosphoamino acid. However, there are several diagnostic tests that can be initially performed to determine the nature of the phosphorylated amino acid. These tests are summarized below and include those for thiophosphates and acyl phosphates as well as those for phosphoramidates. Tests for all three are included as they all are characterized by some form of acid lability. These tests may aid in either the determination or

[18] C. C. Chen, Ph.D. Dissertation, University of California, Los Angeles, 1977.
[19] A. M. Spronk, H. Yoshida, and H. G. Wood, *Proc. Natl. Acad. Sci. U.S.A.* **73**, 4415 (1976).
[20] H. R. Mahler and E. H. Cordes, "Biological Chemistry." Harper & Row, New York, 1971.
[21] R. H. A. Plimmer and W. M. Bayliss, *J. Physiol. (London)* **33**, 439 (1906).
[22] T. I. Nazarova, N. Y. Fink, and S. M. Avaeva, *FEBS Lett.* **20**, 167 (1972).
[23] C. C. Chen, D. L. Smith, B. B. Bruegger, R. M. Halpern, and R. A. Smith, *Biochemistry* **13**, 3785 (1974).

elimination of a phosphoamino acid from possible candidates. Caution must be exercised in the interpretation of the results of some of the diagnostic tests that specifically inhibit phosphorylation, since any modification of an amino acid may also alter protein structure. The change in protein conformation rather than the alteration of the putative phosphate accepter amino acid residue may result in the inability of protein to be phosphorylated.

Thiophosphates

S-Phosphates of cysteine are labile in slightly acidic conditions but are quite stable at very high or very low pH.[24] This phosphorylation is inhibited by heavy metals[25] and sulfhydryl-directed chemical modification[26] (e.g., the alkylating agent iodoacetate). Phosphate hydrolysis is catalyzed by iodine[27] or mercuric salts.[28]

Acyl Phosphates

Acyl phosphates are labile at either pH extreme.[29] These phosphate anhydrides are cleaved by hydroxylamine[30] and are destroyed by reductive cleavage.[31]

Phosphoramidates

As mentioned previously, the phosphoramidates are extremely acid labile[32]; they are relatively base stable[4] (except for phosphoarginine in hot alkali). Histidine phosphorylation is inhibited by diethyl carbonate,[33] photooxidation by Rose Bengal[34] and other histidine-specific reagents. Pyridine and hydroxylamine catalyze the hydrolysis of these phosphoramidates.[30]

[24] V. Pigiet and R. R. Conley, *J. Biol. Chem.* **253,** 1910 (1978).
[25] F. R. N. Gurd and P. E. Wilcox, *Adv. Protein Chem.* **11,** 311 (1956).
[26] C. H. W. Hirs, this series, Vol. 11, p. 197.
[27] S. Akerfeldt, *Acta Chem. Scand.* **14,** 1980 (1960).
[28] J. F. Riordan and B. L. Vallee, this series, Vol. 11, p. 541.
[29] D. E. Koshland, Jr., *J. Am. Chem. Soc.* **74,** 2286 (1951).
[30] G. DiSabato and W. P. Jencks, *J. Am. Chem. Soc.* **83,** 4393 (1961).
[31] C. Degani and P. D. Boyer, *J. Biol. Chem.* **248,** 8222 (1973).
[32] W. P. Jencks and M. Gilchrist, *J. Am. Chem. Soc.* **87,** 3199 (1965).
[33] E. W. Miles, this series, Vol. 47, p. 431.
[34] E. W. Westhead, *Biochemistry* **4,** 2139 (1965).

Preparation of Phosphorylated Basic Amino Acid Standards

Phosphohistidine

The potassium salt of τ-phosphohistidine can be prepared from phosphoramidate and histidine by described methods[35] and purified by Dowex 1 chromatography using a 0 to 0.25 M KHCO$_3$ linear gradient. π-Phosphohistidine can be isolated during the initial incubation of the above reaction. [^{32}P]Phosphohistidine can be synthesized using [^{32}P]phosphoramidate obtained from [^{32}P]phosphorus oxychloride[36,37] or from inorganic [^{32}P]phosphate and phosphorus pentachloride.[38] Large quantities of unlabeled phosphoramidate can be derived from described procedures.[36] [^{32}P]Phosphorus oxychloride can be prepared by a high-pressure exchange method involving inorganic [^{32}P]phosphate and unlabeled phosphorus oxychloride.[39]

Pholpholysine

ε-N-Phospholysine, α-N-phospholysine, and the diphospho isomer can be isolated from the reaction of phosphorus oxychloride with lysine.[40] Purification requires several column chromatography steps. The generation of the α and diphospho isomers can be avoided by protecting the α-amino group of lysine with protecting groups that can be removed without acid treatment. Thus, reaction of phosphorus oxychloride with N-acetyl methyl esters of lysine or with polylysine (both commercially available) followed by base hydrolysis should yield only ε-N-phospholysine. However, a less expensive method involving the copper chelate of lysine has been developed that also gives ε-N-phospholysine as its sole product.[41] Depending upon the amount desired, 1–20 g of the monohydrochloride salt of lysine was dissolved in distilled water (10 ml/g), and the solution was heated in a reflux system. While refluxing, an excess of copper carbonate was gradually added until the solution was saturated. The mixture was allowed to reflux gently for 15 min, and the unreacted copper carbonate was removed by gravity filtration using Whatman No. 1 paper. The dark blue filtrate was evaporated to dryness on a rotary evaporator, and the blue residue was recrystallized from ethanol–water and

[35] D. E. Hulquist, R. W. Moyer, and P. D. Boyer, *Biochemistry* **5**, 322 (1966).
[36] H. N. Stokes, *Am. Chem. J.* **15**, 198 (1983).
[37] R. A. Smith and M. C. Theisen, this series, Vol. 4, p. 403.
[38] R. C. Sheridan, J. F. McCullough, and Z. T. Wakefield, *Inorg. Synth.* **13**, 23 (1971).
[39] R. D. O'Brien, "Toxic Phosphorus Esters." Academic Press, New York, 1960.
[40] O. Zetterqvist and L. Engström, *Biochim. Biophys. Acta* **141**, 523 (1967).
[41] J. M. Fujitaki, A. W. Steiner, S. E. Nichols, E. R. Helander, Y. C. Lin, and R. A. Smith, *Prep. Biochem.* **10**, 205 (1980).

allowed to air dry. The conversion of amino acid to its copper chelate is quantitative, but recovery yields were about 80%. The copper chelate (0.2 g) was dissolved in distilled water (3 ml), and a saturating amount of triethylamine was added. The solution was cooled to 0–5° using an ethanol–ice bath, and any excess triethylamine was decanted. Freshly distilled phosphorus oxychloride (0.32 ml) was added dropwise. The pH of the solution was maintained manually at 11 during the addition. The solution was kept on ice with occasional stirring for 30 min. The pH of the solution was then carefully adjusted to 8.0 and hydrogen sulfide was bubbled through it until the blue color disappeared. The black copper sulfide was removed by gravity filtration. A few drops of a saturated solution of barium acetate was added to the filtrate and the mixture was filtered by gravity to remove the insoluble salts. Ethanol was added to the filtrate to induce crystallization. The residue was recrystallized from ethanol and water (1 : 1, v/v). The products of the reaction of phosphorus oxychloride with protected and with unprotected lysine are shown on a high-performance liquid chromatography system in Fig. 2. The ^{32}P-radioactive compound can be synthesized from [^{32}P]phosphorus oxychloride.

Phosphoarginine

ω-N-Phosphoarginine is commercially available or can be synthesized enzymatically or chemically. Purification is accomplished using a 0 to 0.25 M KHCO$_3$ linear gradient on a Dowex 1 column. The enzymatic product obtained from arginine kinase with ATP and arginine affords ω-N-phosphoarginine. The phosphoamino acid can be synthesized chemically by the combination of arginine and phosphorus oxychloride.[42,43] Specific chemical phosphorylation is accomplished using an α-amino protecting groups such as the copper chelation described previously.[41] [^{32}P]Phosphoamino acid can be prepared using the appropriate ^{32}P-labeled reactant in the reactions.

Determination of P-N Content in Phosphoproteins

When examining for acid-labile phosphorylation, care must be taken not to perform any procedure at low pH. Several methods have been employed that can detect acid-labile phosphorylation. The methods outlined below can be used as a general screening for acid-labile phosphates; they do not, however, indicate any specific acid-labile phosphoamino acid.

[42] N. Thiem, N. Thoai, and J. Roche, *Bull. Soc. Chim. Biol.* **5**, 322 (1962).
[43] F. Marcus and J. F. Morrison, *Biochem. J.* **92**, 429 (1963).

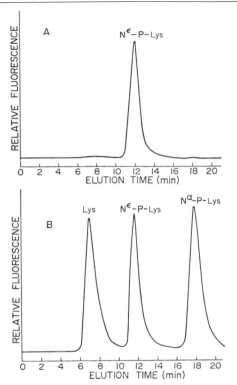

Fig. 2. Elution profile of phosphorylated lysine on HPLC under the following conditions: anion-exchange resin DA-X12-11, column temperature 50°, elution buffer 15 mM KH_2PO_4, detection by o-phthalaldehyde fluorescence.   (A) Crude product of the phospholysine synthesis by the copper chelate method.   (B) Unpurified product of the phospholysine synthesis using unprotected lysine. Phospholysine peaks were identified by reaction with 2,4-dinitrofluorobenzene yielding appropriate derivatives.

Neutral or Basic Polyacrylamide Gel Electrophoresis

There are numerous polyacrylamide gel electrophoretic systems that can be employed in the investigation of proteins containing acid-labile phosphate and in the determination of P-N and P-O content. Not only must the gel buffers be of neutral or basic pH, but the staining and destaining procedures must also adhere strictly to the absence of acid when dealing with phosphoramidate-containing proteins. Neutral and basic nondenaturing as well as SDS-containing electrophoresis systems are available.[44] To ascertain the presence of any acid-labile phosphates in a phosphorylated protein sample, the sample can be divided into two ali-

[44] R. Hardison and R. Chalkley, *Methods Cell Biol.* **17**, 235 (1978).

quots, one of which is incubated with hot acid and subsequently neutralized before application onto the gel. The two aliquots can then be subjected to electrophoresis in adjacent tracks on a neutral or basic slab gel. After neutral staining and destaining, any ^{32}P radioactivity loss due to the acid treatment can be detected by autoradiography or can be measured by counting gel slices for radioactivity. A P-O to P-N ratio can therefore be calculated.

An example of the parallel treatment with acid and neutral loading buffers for SDS gels is illustrated (see Fig. 3)[12]: the 2× concentrate of neutral loading buffer (N) consisted of 4% SDS, 20% glycerol, 10% 2-mercaptoethanol, 0.25 M Tris-HCl, pH 8; and the 2× concentrate of the acid loading buffer contained 4% SDS, 20% glycerol, 10% 2-mercaptoethanol, 0.5–1 N HCl. After mixing with identical phosphoprotein samples (the loading buffers can also serve to quench enzymatic phosphorylating reactions), both solutions were incubated for 10 min at 37°. The acid-treated sample was neutralized with concentrated Tris buffer (the volume of the neutral sample was adjusted with water). Equal volumes of both samples were applied to the gels. As an alternative to this method, neutrally quenched duplicate phosphoprotein samples can be run on gels. After electrophoresis, one sample is treated with acid (during staining and/or destaining) and the ^{32}P radioactivity from each sample can be compared. The latter methodology is less desirable, since differences in gel shrinkage, protein leakage, and dye binding to protein may affect comparison.

After electrophoresis, acetic acid is normally used to fix proteins within the polyacrylamide matrix. Obviously, the presence of P-N bond precludes staining and destaining in acid medium. Neutral staining can be accomplished in a solution of high alcohol content. For example, a solution of 0.1% Coomassie Blue in 25% isopropanol has been routinely used to stain gels.[12] Neutral destaining is accomplished in 10% isopropanol (NaHCO$_3$ is sometimes included to maintain a high pH in both staining and destaining procedures). To prevent certain proteins from washing out, formaldehyde or glutaraldehyde treatment may aid in fixing the protein. Several silver staining procedures are acceptable for P-N bond analysis of phosphoproteins.

Ion-Exchange Method

This method developed by Boyer and Bieber[45] has been modified[46] to assay for histone kinases involved in acid-labile phosphorylation. The

[45] P. D. Boyer and L. L. Bieber, this series, Vol. 10, p. 768.
[46] D. L. Smith, C. C. Chen, B. B. Bruegger, S. L. Holtz, R. M. Halpern, and R. A. Smith, *Biochemistry* **13**, 3780 (1974).

COOMASSIE BLUE   AUTORADIOGRAPHY

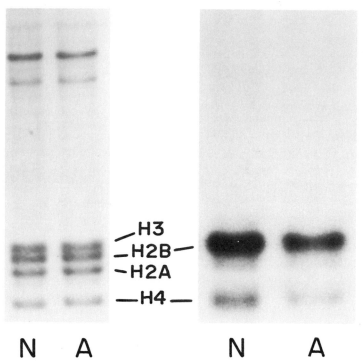

FIG. 3. Acid-lability demonstration of enzymatically phosphorylated histones on SDS–polyacrylamide gels. Phosphorylation was performed using crude nuclear preparation of regenerating rat liver.[12] Samples were treated with neutral quench (N) or acid quench (A) as described in the text and run on a neutral SDS–polyacrylamide gel. The photograph of the autoradiogram was done on a larger scale.

reactions (0.2 ml) are stopped by the addition of 1 ml of solution of 7.4 M urea and 10 mM NaOH. The mixture is heated at 90° for 15 min to destroy alkali-labile linkages of phosphohydroxyamino acids. After cooling, the mixture is subjected to ion-exchange chromatography on a Dowex 1-X8 (OH⁻ form, 100–200 mesh) column (0.8 × 4.3 cm) to remove ATP and inorganic phosphate. The column is eluted twice with 1 ml of urea–NaOH solution. An aliquot of the combined eluents is used to measure the ^{32}P-radioactivity of acid-labile phosphates.

Dialysis Techniques

Two methods[12] involving dialysis can be used to assay for acid lability.

1. A neutral solution of ^{32}P-labeled phosphoprotein is placed in a dialysis membrane tubing and permitted to equilibrate with the surrounding neutral solution. The buffer is changed as necessary. After adjusting the pH of the surrounding solution to 11–12 with concentrated NaOH, the ^{32}P content of the surrounding buffer is monitored by counting samples of the solution on a liquid scintillation counter. After equilibration, the pH is adjusted to 1–2 using HCl, and the amount of ^{32}P released into the surrounding solution is measured as before.

2. Plastic scintillation vials (Wheaton 986644) containing 0.5 ml of 0.2 N HCl are covered by individual sheets of dialysis membrane (Spectrapor 3). Gel slices from polyacrylamide gel electrophoresis of a [^{32}P]phosphoprotein mixture are placed on top of the membranes along with a drop of water. By securing the membrane in place with a screw cap, a cavity is formed containing the gel slice. The vial is then inverted and agitated for several hours. The cap, gel slice, and membrane are removed, and the ^{32}P-radioactivity in the vial is measured by Cerenkov counting. The control involves dialysis against water instead of acid.

Extraction Methods

The following two methods involving extraction procedures have been used to determine the P-O to P-N ratio occurring in nuclear proteins (see Table III).

Phenol Extraction Method.[4,45] A sample of phosphoprotein (2 ml) is mixed with 10 ml of 88% phenol (adjusted with 0.01 M sodium phosphate

TABLE III
RATIO OF ACID-LABILE PHOSPHATE TO ACID-STABLE PHOSPHATE IN
NUCLEI AND NUCLEAR PROTEIN FRACTION[5]

	Ratio of acid-labile phosphate to acid-stable phosphate[a]			
Tissue	Nuclei	Acidic protein	Deoxynucleoprotein complex	Nucleoplasm
Walker 256	0.6 ± 0.1	1.2 ± 0.5	1.0 ± 0.5	1.3 ± 0.1
Rat liver	1.7 ± 0.4	3.1 ± 0.8	1.4 ± 0.4	0.6 ± 0.1

[a] The values include the square root of the variance using determinations from each of the two methods (two-phase system and phenol extraction).

buffer). After thorough mixing with a wash buffer (10 ml) consisting of 0.01 M Na_2EDTA, 0.01 M $Na_4P_2O_7$, 0.01 M sodium phosphate buffer, pH 8.3, the mixture is centrifuged to facilitate the separation into two phases. The top aqueous layer (containing free phosphates) is discarded, and the phenol layer (containing phosphoprotein) is washed again with an equal volume of wash buffer. Proteins from an aliquot of the phenol layer are precipitated with 5 volumes of cold acetone. The resulting precipitate is treated with acid (0.3 N TCA, 0.001 M Na_2HPO_4, 60°, 5 min). The ^{32}P radioactivity released upon acid treatment represents the contribution of P-N linkages, whereas the radioactivity remaining in the precipitate represents the contribution of P-O linkages.

Two-Phase Aqueous Polymer Extractions.[5,47] An aliquot from a solution containing 30 g of 20% (w/w) dextran and 30 g of 40% (w/w) polyethylene glycol-6000 and 40 g of 0.01 M sodium phosphate buffer (pH 7) is mixed thoroughly with a phosphoprotein sample and NaCl (final salt concentration of 4.5 M). The top phase is removed, and the bottom layer is reextracted with an equal volume of fresh top layer. The combined top layers are then dialyzed against 100 volumes of distilled water to remove free phosphates. The solution is subjected to hot 5% TCA, and the released inorganic phosphate precipitated as the phosphomolybdate complex represents the contribution of P–N linkages. The supernatant contains the phosphates (P–O type) still attached to the protein.

[47] P. A. Albertson, "Partition of Cell Particles and Macromolecules." Wiley, New York, 1960.

[3] Phosphorus-31 Nuclear Magnetic Resonance Studies of Phosphorylated Proteins

By MANFRED BRAUER and BRIAN D. SYKES

A major mechanism for the regulation of cellular metabolism and function is the phosphorylation or dephosphorylation of target proteins within the cell.[1-4] The actual phosphorylation or dephosphorylation of the target protein represents, in most cases, the final step in a regulatory cascade

[1] P. Cohen, *Nature (London)* **296**, 613 (1982).
[2] O. M. Rosen and E. G. Krebs, eds., "Protein Phosphorylation," *Cold Spring Harbor Symp.,* Vols. 1 and 2 (1981).
[3] E. G. Krebs and J. A. Beavo, *Annu. Rev. Biochem.* **48**, 923 (1979).
[4] M. Bárány and K. Bárány, *Annu. Rev. Physiol.* **42**, 275 (1980).

starting with hormonal or neural stimulation, an intracellular second messenger (such as cyclic GMP, cyclic AMP, or the Ca^{2+}–calmodulin complex) and activation or inhibition of a cascade of specific protein kinases or phosphatases. The focus of this chapter is the characterization of the phosphorylated target protein by phosphorus-31 nuclear magnetic resonance (³¹P NMR). The preparation of a sample of phosphorylated protein for analysis by ³¹P NMR is described. Aspects of the methods for obtaining a ³¹P NMR spectrum of a phosphorylated protein is discussed in as "user-friendly" a way as possible; however, the authors assume some prior knowledge of NMR theory and methods or refer the reader to more general reviews on these subjects. Having described the steps needed to obtain the spectrum, the analysis of the spectrum is outlined and the types of experiments that can be done are discussed. Finally, we survey the literature to illustrate what has already been accomplished by ³¹P NMR studies of various phosphorylated protein systems.

What can we hope to learn about a phosphorylated protein by the technique of ³¹P NMR? ³¹P NMR can be used to assign what type of amino acid residue the phosphate is covalently attached to, such as the β-hydroxyl of a serine or threonine residue, or the N-1 or N-3 nitrogen of the imidazole ring of a histidine. The stoichiometry of phosphorylation can be ascertained. By varying the pH, the pK_a of the phosphoryl group can be determined. Metal ion binding to the group can be studied. The interactions of a phosphoryl group with other groups on the protein can be detected and the overall mobility of each phosphoryl group can be determined. Thus, ³¹P NMR can provide a range of information about the nature and dynamics of the phosphorylation site that cannot be obtained by other methods. The greatest strength of the NMR method relative to all the above is that if two or more phosphorylation sites exist on a protein, each phosphoryl group can be individually monitored by ³¹P NMR.

³¹P NMR has several important practical advantages. First, it is a nondestructive spectroscopic method relying on the absorption of very low radiofrequency (40 to 200 MHz) irradiation. Thus, the sample can be recovered intact after NMR analysis. The NMR experiment can be done under conditions of temperature and pH that are favorable for the stability of the phosphorylated protein. The method is not particularly sensitive to dust, strong oxidizing or reducing reagents, or small amounts of contaminants, except paramagnetic metal ions. ³¹P NMR does not require any nonphysiological or radioactive probes, except that a small amount (10–50%) of deuterium oxide (D_2O) is usually needed. The presence of D_2O does not in general affect the biological system under study.

The major disadvantage of ³¹P NMR, compared to other methods for the detection and characterization of covalently bound phosphates, is its inherent insensitivity. In spite of major advances in probe design, pulsed NMR methods, and wide-bore magnets that allow larger sample volumes, one still needs 1 ml of more than 0.1 mM phosphorylated protein to be able to detect a ^{31}P resonance. These high concentrations of protein can cause problems; because the sample must be concentrated, it must be stable at these high levels (i.e., no aggregation, precipitation, or polymerization of the protein). We discuss various strategies for dealing with the sensitivity problem in the section on concentration of the sample, but one must have more than 0.1 μmol of phosphorylated protein (5 mg of a 50,000 molecular weight protein) at >0.1 mM concentration in order to get a good ^{31}P NMR spectrum.

Preparation of a Phosphorylated Protein Sample for ^{31}P NMR

Purity of Sample

From the ^{31}P NMR point of view, sample purity is not a major requirement for a successful experiment. Since only ^{31}P resonances are detected, only phosphorus-containing impurities will show up in the spectrum. Impurities of large molecular weight usually will give very broad ^{31}P resonances and are not usually troublesome. A common high-molecular-weight contaminant is DNA, which does not show up on standard electrophoresis gels stained with Coomassie Blue. Impurities of very small molecular weight (e.g., inorganic phosphate, sugar phosphates, AMP) will yield sharp ^{31}P resonances that are readily detectable at low concentrations. This problem can be minimized by gel filtration or dialysis (assuming that the low-molecular-weight species do not bind strongly to the phosphorylated protein). A major contamination problem in NMR is the effect of paramagnetic metal ions such as Mn(II), Cu(II), Cr(III), Fe(III), and V(IV). These paramagnetic ions will dramatically broaden the observed resonances of the sample and shorten their T_1 values. A common error in NMR studies is the use of potassium dichromate cleaning solution to clean NMR tubes and other glassware; since this reagent has paramagnetic Cr(III), even trace amounts of residual cleaning solution can obliterate the spectrum. We recommend an alcoholic KOH cleaning solution that is not paramagnetic: 120 g of KOH dissolved in 120 ml of H_2O and added to 1000 ml of 95% ethanol. While sample purity is not a vital criterion from the NMR standpoint (other than being careful to exclude paramagnetic ions), purity may be important from a biochemical viewpoint. Since the phosphorylated protein concentrations needed for ^{31}P NMR are quite high, a small amount of contaminating protease or phosphatase may rapidly degrade the sample. Typically a ^{31}P NMR experiment may take from 1 to 24 hr, and the sample must be stable over this timeframe.

Concentration of the Phosphorylated Protein Sample

As we have already mentioned, the inherent insensitivity of ^{31}P NMR spectroscopy requires that the concentration of phosphorylated protein in the sample be very high. Generally speaking, the higher the protein concentration, the better (assuming no aggregation, precipitation, or polymerization). However, at very high protein concentrations (>100 mg of protein per milliliter), the viscosity of the solution increases sufficiently[5] so that a broadening of ^{31}P resonances and an increase in their T_1 values can be detected,[6] both factors contributing to a decrease in the signal-to-noise ratio (S/N) of the spectrum.

To get some idea of the minimum concentration of sample needed to obtain a good ^{31}P NMR spectrum, we must consider some of the parameters involved in determining the S/N of the spectrum. The S/N is dictated in large part by the actual NMR spectrometer and probe used, which will be discussed under instrumental techniques. The S/N of the spectrum also increases as the number of scans (n) taken increases, but S/N is proportional to $n^{1/2}$.[7] Thus, to double the S/N, one must run 4 times the number of scans. The number of scans that can be taken is usually limited by the stability of the sample, the T_1 of the resonances, the availability of the spectrometer, and the patience of the investigator. This usually means that one may acquire data on one sample for from 1 hr to 2 days (2000–100,000 scans). Considering the parameters involved, typically one would need at least a 0.1 mM phosphorylated protein solution to obtain a reasonable ^{31}P NMR spectrum, assuming a globular protein of molecular weight 50,000 with a phosphoryl group giving a linewidth of 10 Hz, with the accumulation of 20,000 scans over 11 hr on a narrow-bore 109.3 MHz ^{31}P (270 MHz ^{1}H) NMR spectrometer (10-mm NMR tube, sample volume 1.1 ml). Thus, one must have a phosphorylated protein that can be concentrated to at least 0.1 mM concentration and must be able to obtain at least 0.1 μmol or 5 mg of the protein.

With the advent of wide-bore superconducting magnets one can obtain the ^{31}P NMR spectrum of a protein that can be obtained in reasonably large amounts but cannot be concentrated up to the 0.1 mM level. Improvements in magnet and probe designs have allowed investigators to

[5] R. Simha, *J. Appl. Phys.* **23**, 1020 (1952).

[6] W. J. Ray, Jr., A. S. Mildvan, and J. B. Grutzner, *Arch. Biochem. Biophys.* **184**, 453 (1977).

[7] R. R. Ernst, *Adv. Magn. Reson.* **2**, 1 (1966).

run samples in NMR tubes that are 20–30 mm in diameter. The increase in sample volume allows one to analyze much greater amounts of a concentration-limited sample, providing a substantial increase in S/N. An increase in S/N is also possible with the advent of sideways spinning probes,[8] although these systems are somewhat less convenient in terms of sample manipulation. Thus, if the user has access to these improvements in NMR technology, he can obtain ^{31}P NMR spectra with lower concentrations of phosphorylated proteins than previously delineated.

Now that we have some idea of the degree to which a sample of phosphorylated protein must be concentrated in order to obtain a ^{31}P NMR spectrum, we must decide how to concentrate the sample. The simplest method is lyophilization to dryness followed by redissolving the sample in a smaller volume of solution. This method also results in concentration of the buffer, salts, and other small molecules in the sample. If this is a problem, one may wish to desalt the sample before lyophilization, either by dialysis or gel filtration. It is usually advisable to use a volatile buffer such as ammonium acetate, formate, or bicarbonate. (A common experimental error is to adjust the pH of a volatile buffer with a relatively nonvolatile acid or base such as HCl or KOH. Adjusting the pH with ammonium hydroxide to raise the pH and with acetic acid, formic acid, or carbon dioxide to lower the pH will ensure that the lyophilized sample does not contain appreciable salt.) Many proteins tend to aggregate or denature upon lyophilization, however, and other concentration methods must be used. Many commercial protein concentrators rely on ultrafiltration of water and small molecules through a semipermeable membrane, while retaining the protein. Depending upon the biological system involved, protein concentration can also be accomplished by ultracentrifugation, ammonium sulfate precipitation, or ion-exchange chromatography.

Virtually all modern NMR spectrometers have the ability to "lock" onto D_2O, which is usually added to the biological sample during or subsequent to the concentration step. Typically 10–50% (v/v) D_2O will be needed in the NMR sample in order to lock onto the deuterium signal. (If the temperature is low, or viscosity of the sample solution is very high, a higher volume percentage of D_2O may be needed). Since the addition of D_2O will dilute the sample, one may need to reconcentrate the sample. In some cases, a sample is dialyzed against D_2O, although this can be expensive. Most commercially available supplies of D_2O are contaminated with trace amounts of paramagnetic metal ions. These can readily be removed by equilibrating the D_2O with Chelex (Bio-Rad Laboratories), an EDTA-

[8] D. T. Hoult, *Prog. Nucl. Magn. Reson. Spectrosc.* **12,** 41 (1978).

like resin that has an affinity for metal ions. The beads of resin are readily removed from the D_2O by centrifugation or simply on standing. Other methods of removing trace metal ions, such as extraction with dithizone,[9] can also be used, but are not as convenient as the Chelex method.

Before running the NMR spectrum of the phosphorylated protein sample, often it is advisable to recheck the pH to make sure that it has not changed during concentration or addition of D_2O. If any particulate matter is present, the sample should be clarified by centrifugation or filtration. Also, the sample should be equilibrated at the temperature for which the ³¹P NMR spectrum will be taken. The air bubbles that form in aqueous solutions that are warmed up will reduce the homogeneity of the sample in the magnetic field and result in excessively broad ³¹P resonances. Particulate matter in the sample has the same effect on the ³¹P NMR spectrum.

After suitable preparation, the sample is placed in an NMR tube appropriate to the particular NMR spectrometer and probe, i.e., a 10-, 12-, 20-, or 30-mm outer diameter polished, precision bore, thin-walled NMR tube. Round-bottom or flat-bottom NMR tubes are available. If one has only a limited volume of sample, flat-bottom tubes are preferable, since the sample within the rounded part of a round-bottom tube is not usually detected by the NMR spectrometer and is thus wasted volume. The curvature at the top of the sample due to the meniscus as well as the vortexing effect upon sample spinning within the spectrometer can be eliminated by introducing a Teflon (or other non-phosphorus containing) vortex plug on top of the sample. The plug is held in place by virtue of its snug fit within the NMR tube aided by O-rings or flexible fins, or by a string from the plug to the top of the NMR tube. The sample of phosphorylated protein is now ready to be put into the NMR spectrometer for acquisition of the ³¹P NMR spectrum.

Instrumental Techniques for Obtaining the ³¹P NMR Spectrum of a Phosphorylated Protein

The detailed procedure for obtaining a ³¹P NMR spectrum from a suitably prepared phosphorylated protein sample varies with the particular model and manufacturer of the NMR spectrometer available. The canonical spectrometer relevant to these discussions is a conventional high-resolution, pulsed NMR spectrometer with a superconducting magnet and deuterium lock system, and with the signal from the sample recorded digitally using quadrature phase detection (QPD) and stored in a

⁹ E. B. Sandell, *in* "Chemical Analysis" (B. L. Clarke, P. J. Elving, and I. M. Kolthoff, eds.), Vol. III, p. 138. Wiley (Interscience), New York, 1959.

computer. In depth discussions of more general instrumental techniques for acquiring an NMR spectrum have been published elsewhere.[10-13] In this section, we emphasize those aspects of these techniques that are of particular importance in obtaining ^{31}P NMR spectra of phosphorylated proteins.

S/N is dictated in large part by the actual NMR spectrometer and probe used.[14,15] If several different pulsed NMR spectrometers are available, a major factor in deciding which one to use is the spectrometer frequency of each. Assuming that T_1 and T_2 values are not affected by the strength of the magnetic field, the S/N should increase as the spectrometer frequency, ω_0, increases (S/N is proportional to $\omega_0^{7/4}$).[14] [To keep the terminology straight, one must realize that ω_0 is in units of radians per second; there are 2π radians per cycle, so 1 Hertz (Hz) = 2π radians per second; $\omega_0 = \gamma H_0$, where γ is the gyromagnetic ratio and is a unique constant for each type of nucleus; H_0, the magnetic field strength, is in units of gauss or Tesla, where 10,000 gauss = 1 Tesla; with the result that 1 Tesla corresponds to 42.6 MHz ^1H frequency or 17.2 MHz ^{31}P frequency.] However, unlike most other commonly studied nuclei, the linewidths of the resonances of ^{31}P nuclei are significantly controlled by a relaxation mechanism called chemical shift anisotropy (CSA).[16,17] The CSA contribution to the linewidth increases with the square of the spectrometer frequency. Empirically, we have found that the spectrometers giving optimum ^{31}P NMR spectra are those ranging from 89.1 to 121.4 MHz ^{31}P frequency (220 to 300 MHz proton frequency).

Locking on the Sample

After the phosphorylated protein sample has been put into the magnet (so that the sample is positioned within the coil of the probe) and the NMR tube is spinning, the first major step is to lock onto the deuterium resonance of the D_2O in the sample. "Locking on" means that a lock circuit is activated that detects the deuterium resonance of the solvent and automatically corrects for any drifts in magnetic field, which are manifested as a change in the frequency of the resonance during the

[10] B. D. Sykes and W. E. Hull, this series, Vol. 49, p. 270.
[11] F. R. N. Gurd and P. Keim, this series, Vol. 27, p. 836.
[12] M. Bárány and T. Glonek, this series, Vol. 85, p. 624.
[13] T. C. Farrar and E. D. Becker, "Pulse and Fourier Transform NMR." Academic Press, New York, 1971.
[14] D. I. Hoult and R. E. Richards, *J. Magn. Reson.* **24**, 71 (1976).
[15] D. I. Hoult and R. E. Richards, *Proc. R. Soc. London Ser. A* **344**, 311 (1975).
[16] M. Brauer and B. D. Sykes, *Biochemistry* **20**, 6767 (1981).
[17] W. E. Hull and B. D. Sykes, *J. Mol. Biol.* **98**, 121 (1975).

course of the experiment. The deuterium signal is also used in shimming the magnet; the homogeneity of the magnetic field can be adjusted by maximizing the height (i.e., minimizing the linewidth) of the D_2O resonance.

Locking on a sample of phosphorylated protein may be difficult if the level of D_2O in the sample is low, the viscosity is high, or the spectrometer has not been adjusted to magnetic field homogeneity values roughly appropriate for the type and size of sample being studied. It is often advisable to first lock onto a sample of pure D_2O with the same volume and type of NMR tube as that of the sample. One can then easily shim this pure D_2O sample so that the spectrometer will be adjusted to give a roughly homogeneous magnetic field for the protein sample. One must usually increase both the power and sensitivity of the deuterium lock circuitry to achieve lock for a protein sample as compared to a sample of pure D_2O. The necessity of locking the sample has decreased with the advent of the superconducting magnet, since the magnetic fields in these magnets are much more stable than those of the older iron-core magnets. Thus, if the drift in the frequency of the observed resonances are appreciably smaller than the natural linewidth of these resonances, one can run spectra unlocked without compromising the quality of the spectrum. Typically, superconducting magnets drift at less than 1 Hz per hour, so samples must be locked if they are to be run for a very long time (>12 hr) and/ or if the resonances are sharp (linewidths less than 10 Hz).

Shimming the Magnet

Once the spectrometer has been locked on, one can shim the magnet to adjust for small inhomogeneities in the magnetic field induced by the particular sample. The nonspinning shims usually do not change appreciably from sample to sample, so one must usually adjust only the spinning shims (Z, Z^2, Z^3, and Z^4). It is best to adjust first the Z shim for maximum lock level, then Z^2 and Z interactively (set Z^2 and optimize Z, etc.). Then shim Z^3 and Z interactively, followed by Z^2 and Z interactively, if necessary. Similarly Z^4 and Z^2 are shimmed interactively, followed by Z^2 and Z if necessary. Z^5 (if present on the spectrometer used) is shimmed interactively with Z^3 and Z, in a similar manner. Since the S/N for most ³¹P NMR spectra of phosphorylated proteins is quite low, it is not feasible to evaluate the homogeneity by examining the spinning side-bands or the lineshape of the resonance. However, these can be assessed with a separate sample of fairly concentrated inorganic phosphate, for example.[12]

One should note that the spinning shim values vary with the height of the sample within the NMR tube. If the sample is short and does not fill up the effective volume of the transmitter–receiver coil, these shim settings will change rapidly from their normal ranges. The distortions in the field homogeneity induced by the small sample may exceed the ability of the shim circuits to compensate for these distortions, and result in broader resonances.

Setting Up the Pulse Experiment

One must now make some basic decisions about the instrumental parameters to use for data acquisition.[10,13] This discussion will apply to the operation of a standard pulsed NMR spectrometer equipped with a computer to record and store the free induction decay (FID) that results from application of the excitation pulse to the sample, and the ability to Fourier transform the FID to produce a normal spectrum. If one has not used a particular NMR spectrometer before, it is best to start with a standard, fairly concentrated ^{31}P NMR sample for which a spectrum can be obtained with one pulse (for example, 30 mM inorganic phosphate), rather than trying to run the more dilute phosphorylated protein sample immediately. Note that the probe tuning may change with the ionic strength of the sample.[18] Therefore the inorganic phosphate should be in the same buffer as the protein to be studied. As a starting point, let us use the following parameters and discuss each one in turn with the view of optimizing each parameter for the particular sample involved: excitation pulse 50°, spectral width (SW) ±2500 Hz, filter width ±2500 Hz, computer block size 8192 and QPD on.

The length of the desired excitation pulse will depend upon the length of the pulse needed to drive the nuclear spins 90° from the z axis maximally into the xy plane, where those resonances are detected (called the 90° pulse). The 90° pulse varies from one spectrometer to another and should be determined using the standard inorganic phosphate sample in the same buffer as the protein sample. One can determine the 90° pulse length by plotting the resonance intensity after one pulse versus the pulse length and looking for a maximum. One must wait at least 1 min between each trial to allow the nuclear spins to fully relax back to the z axis, since the T_1 value for inorganic phosphate is more than 10 sec. Alternatively, one can determine the 360° pulse length, after which the signal intensity should be zero, and divide by 4. This procedure reduces the need for a long waiting period between trials, since the magnetization is left near the $+z$ axis. For a 10-mm NMR sample in D_2O, typical ^{31}P 90° pulse times with conventional spectrometers (not "high power" spectrometers) are

[18] P. Daugaard, H. J. Jakobsen, A. R. Garber, and P. D. Elles, *J. Magn. Reson.* **44**, 224 (1981).

20–30 μsec, with longer times for larger sample tubes, i.e., 20- or 30-mm tubes. For the fast pulsing conditions needed for a phosphorylated protein sample, a 30–70° pulse is needed (see next section).

The center of the frequency range of the excitation pulse (called the carried frequency) is set to the center of the ^{31}P NMR spectrum if QPD is employed. QPD should be used if it is available, as it allows an increase in S/N of a factor of $2^{1/2}$ by removing foldover of noise around the carrier frequency.

The value of the SW should not be substantially greater than the range of ^{31}P resonances expected for the sample. Most phosphorus-containing compounds of biological importance have ^{31}P resonances ranging from about +10 to −25 ppm from an 85% phosphoric acid external standard. For a 300 MHz NMR spectrometer (121.4 MHz for ^{31}P nuclei), this represents a range of 4250 Hz. Thus an SW of ±2500 Hz should be adequate for most biological samples. If a resonance is present but is outside of the SW "window," it will be "folded back" into the spectrum, generally out of phase with the other resonances, and will not shift in the same way as the other resonances when the carrier frequency is shifted. The filter widths are usually set the same as the SW to prevent a similar "folding-back" of noise outside of the SW window.

The storage of the FID within the computer must be considered. The size of the data block must be large enough to define adequately the shapes of all the various resonances within a spectrum. For example, an 8K data block (8192 computer points) will define the real part of the spectrum with the first 4096 computer points and the imaginary part of the spectrum with the last 4096 computer points. With an SW of ±2500 Hz, there will be 1.22 Hz per computer point, which is sufficient to define all but the sharpest resonances. The acquisition time (AQ) or the time needed to record the FID, also depends upon the size of the data block. With QPD, two FIDs 90° out of phase with one another (along the +X and +Y axes) are recorded at the same time. Thus an FID recorded in an 8K data block with QPD actually results in two 4K FIDs (see Fig. 1A). The sampling theorem dictates that at least two computer points per cycle are needed to define a waveform digitally. If the SW is ±2500 Hz, the computer must sample every 200 μsec [1/2 × 2500, the dwell time (DW)]. The 4096 computer points for each of the two recorded FIDs would then take 4096 × 200 μsec = 0.819 sec (AQ).

The acquisition delay (DE) is the time between the end of the excitation pulse and the beginning of the acquisition of the spectrum. During this period, the excitation pulse is allowed to dissipate before the induced FID from the sample is received. It is common to make DE = DW. A DE $\simeq$ 0.6 DW can be useful for decreasing first-order phase corrections re-

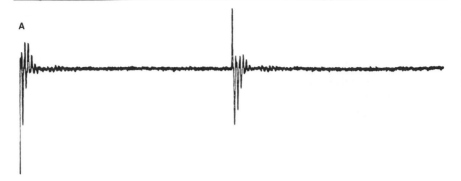

NO PROTON DECOUPLING

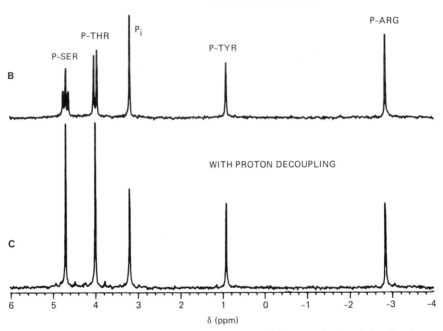

FIG. 1. ^{31}P NMR spectra of phosphorylated amino acids. (A) The free induction decay (FID) of phosphoserine (pH 9.5, 27°, 1 mM EDTA, in D_2O) collected using quadrature phase detection (QPD). Acquisition parameters: 60° pulse (17 μsec), spectral width (SW) ± 100 Hz, data block size 4096 points, hence dwell time (DW) is 5000 μsec, acquisition time (AQ), 10.24 sec. (B) The spectrum (after Fourier transformation of the FID) of a mixture of phosphorylated amino acids (pH 8.0, 27°, 1 mM EDTA, in D_2O). Acquisition parameters: 60° pulse (17 μsec), SW ± 1000, block size 4096, DW 500 μsec, AQ 1.02 sec, delay time 1.0 sec. (C) Same sample and acquisition parameters as in (B), except for broad-band decoupling during AQ and during delay time.

sulting from the ringing of standard low-pass Butterworth filters.[19] Particularly for broad resonances (linewidth > 1000 Hz) where the FID relaxes quickly, DE should be as short as possible without resulting in transmitter pulse breakthrough. (One can readily tell when one is picking up signal from the transmitter, i.e., pulse breakthrough, as the final spectrum exhibits a badly rolling baseline that cannot be phased.)

Having set up the parameters needed to obtain a single FID and having checked these parameters with a standard concentrated sample, one can now consider the repetition rate between scans appropriate for a phosphorylated protein sample. The repetition rate is usually controlled by the acquisition time plus a variable delay period after the acquisition of the FID. If the main purpose of an experiment is to determine the relative concentrations of several phosphorylated species within a sample, then one must have a repetition rate greater than three times the longest T_1 within the sample unless the T_1 values of the resonances of all the phosphorylated species under investigation are equal. Since the T_1 values of a phosphoryl group on a protein can range from 0.5 to 5.0 sec, this requirement can dramatically limit the number of scans that can be taken per unit of time. This results in a decreased S/N. If one is not worried about the relative stoichiometry of various phosphorylated groups, then one can utilize a rapid pulsing method for the most time-efficient data accumulation.[7,20] The relationship between the T_1 of a resonance, the flip angle (in degrees) of the excitation pulse (α), and the time between scans (T) is described by $\alpha = \arccos[e^{-(T/T_1)}]$. If we assume that the typical T_1 value of a phosphoryl group on a protein is 2.0 sec and the acquisition time is 0.82 sec, then the optimum flip angle is 48°.

Once the phosphorylated protein sample has been properly set up and is being scanned at a suitable repetition rate, one can (with most NMR spectrometers) analyze the accumulating FID without disturbing the data acquisition. This usually involves transferring the FID from one part of the computer memory where it is accumulating to another where it can be manipulated. An exponential multiplication of the FID is introduced to decrease the spectral noise at the cost of a 2- to 40-Hz linebroadening. This is followed by a Fourier transformation (FT) of the FID into the more familiar frequency spectrum and the phasing of the spectrum into its real (absorption mode) and imaginary (dispersion) parts. (The exact methods of data manipulation vary with the model and manufacturer of the spectrometer.) Thus, one can evaluate the spectrum as it is accumulating data and scan the sample until the S/N is suitably high.

[19] D. I. Hoult, C.-N. Chen, H. Eden, and M. Eden, *J. Magn. Reson.* **51**, 110 (1983).
[20] R. R. Ernst and W. A. Anderson, *Rev. Sci. Instrum.* **37**, 93 (1966).

One should not allow the FID to grow higher than the computer's capacity (i.e., the computer word length) to record. Before the FID exceeds the computer capacity, it will automatically scale down the FID, usually by a factor of 4 and reduce the digitizer resolution. (Typically, the computers of most NMR spectrometers have a computer word length of 20 bits (2^{20}) or more and a 12-bit digitizer.) It is desirable to use the block averaging technique for long-term signal averaging where blocks of scans are stored separately rather than allowing the accumulated FID to exceed the maximum word length. Block averaging also allows one to compare the spectrum obtained from each block to determine whether the spectrum changes with time (due to sample degradation, phosphorylation or dephosphorylation, etc.).

Proton Decoupling

Up to this point, we have described the acquisition of a ^{31}P NMR spectrum following excitation of the ^{31}P resonances. Advantages do exist, particularly with regard to S/N and spectral simplification, in simultaneously irradiating the proton resonances in the sample. This heteronuclear proton decoupling eliminates ^{1}H—^{31}P scalar (through-bond) couplings. For example, phosphoseryl and phosphothreonyl residues have a coupling between the phosphorus nucleus and the proton(s) on the β-carbon. These couplings are significant and are manifest by $^{3}J_{POCH}$ splittings of about 6 Hz for phosphoserine and 8 Hz for phosphothreonine[21] (see Table I). The effect of the collapse of ^{1}H—^{31}P scalar splittings is to sharpen and intensify the ^{31}P resonance, since the area of all the resonances in a multiplet is now in only one central peak (see Fig. 1B and C). The proton decoupling also can increase the total integrated intensity of the ^{31}P resonance due to the nuclear Overhauser enhancement (NOE). If the phosphorus is very highly mobile and its relaxation is dominated by dipolar interactions, the NOE can approach the theoretically maximum enhancement of 1.24; i.e., one can increase the relative intensity of the ^{31}P resonance from 1.0 before proton irradiation to 2.24 after proton irradiation.[22,23] If the phosphorus is fairly immobile, the NOE enhancement will drop to almost zero (0.022); i.e., the total area of the ^{31}P resonance will not change with proton irradiation. This variation of the NOE with the

[21] Chien Ho, J. A. Magnuson, J. B. Wilson, N. S. Magnuson, and R. J. Kurland, *Biochemistry* **8,** 2074 (1969).
[22] J. H. Noggle and R. Shirmer, "The Nuclear Overhauser Effect: Chemical Applications." Academic Press, New York, 1971.
[23] H. Shindo, *Biopolymers* **19,** 509 (1980).

TABLE I

CHEMICAL SHIFTS, SPIN–SPIN COUPLING CONSTANTS, AND pK_a VALUES OF
PHOSPHORYLATED AMINO ACIDS

Amino acid	$\delta_H{}^a$ (ppm)	$^3J_{PXCH}{}^b$ (Hz)	$\delta_0{}^c$ (ppm)	$^3J_{PXCH}{}^d$ (Hz)	pK_a	Reference
Phosphoserine	+0.9[e]	5.3	+4.8[e]	6.4	6.13	21[f]
	+0.7[e]	—	+4.6[e]	—	5.9	92
	+0.8	—	+4.6	—	5.8	83
	+0.7[e]	—	+4.7[e]	—·	5.85	48
	+0.7	5.5	+4.7	6.7	5.8	116
In 8 M urea	+0.7[e]	5.5	+5.0[e]	6.1	—	21
	—	—	+4.5	—	6.1	78
	+0.7	5.7	+4.8	7.3	6.2	116
Phosphothreonine	−0.1[e]	—	+4.0[e]	8.1	—	21
	−0.1[e]	—	+4.0[e]	—	—	88
	−0.2	—	+3.7	—	5.9	83
	−0.3	7.4	+4.0	7.8	6.0	116
In 8 M urea	—	—	+4.2[e]	8.0	—	21
	−0.4	8.0	+4 1	8.4	6.4	116
Phosphotyrosine	−3.3	<1	+0.9	<1	5.75	116
In 8 M urea	−3.2	<1	+0.9	<1	6.15	116
Phosphoarginine	—	—	−2.7[e]	—	—	12
	—	—	—	—	4.5	36
	−5.2	<1	−2.8	—	4.45	116
In 8 M urea	−5.3	<1	−2.8	—	4.9	116
Phosphohistidine						
1-Phosphohistidine	−4.5[e]	—	−4.9[e]	—	7.04	108
3-Phosphohistidine	−3.6[e]	—	−3.9[e]	1.6–1.8	6.2	108
Diphosphohistidine P$_1$	−4.5[e]	—	−4.1[e]	—	—	108
Diphosphohistidine P$_3$	−3.6[e]	—	−3.5[e]	—	—	108
N^ε-Phospholysine	0.0[e]	—	+4.0[e]	—	—	88
Phosphoaspartic acid	—	—	−11.6	—	—	112
Inorganic phosphate	+0.8	0	+3.3	0	6.8	36
	+0.7	0	+3.3	0	6.8	116
In 1.6 M KCl	+0.8	—	+3.4	0	6.5	36
	—	—	—	—	7.4	78
In 8 M urea	+0.7	0	+3.3	0	7.35	116

[a] Chemical shift (relative to 85% H_3PO_4 external standard) of the protonated form of phosphorylated amino acid.
[b] ^{1}H—^{31}P coupling constant for protonated form.
[c] Chemical shift of deprotonated form.
[d] ^{1}H—^{31}P coupling constant for deprotonated form.
[e] Chemical shift corrected (add 0.8 ppm) to correspond to that obtained using a superconducting magnet (see "Chemical Shift Standards and Conventions").
[f] Numbers in this column refer to text footnotes.

molecular motion has been used to determine the mobility of phosphoryl groups on proteins.

Proton decoupling does have some disadvantages. Since various phosphorylated species have different NOE values, one cannot correlate the areas under various ^{31}P resonances with the relative concentrations. Proton decoupling can result in some heating of biological samples in high ionic strength buffers. Finally, when one collapses the multiplet structure, one does lose the information that the ^{31}P—^1H scalar couplings can provide.

The main type of proton decoupling used in ^{31}P NMR is broad-band (noise modulated) decoupling.[24] The objective is to irradiate all the protons over a wide frequency range, although the range of irradiating frequencies should be centered on those protons that are involved in the scalar couplings with the phosphorus. For example, with phosphoseryl and phosphothreonyl residues, the methylene proton(s) of the β-carbon are the predominant protons involved. Thus, the maximum proton decoupling effect on the ^{31}P NMR spectrum of phosphoseryl or phosphothreonyl residues will be felt by irradiating over the methylene region of the ^1H NMR spectrum, i.e., at about 4.1 ppm from TMS.[6]

Determination of T_2, T_1, and NOE Values

The determination of the T_2 value of a ^{31}P resonance can be made directly from its linewidth ($\Delta\nu$), using the relationship $T_2 = 1/\pi\Delta\nu$. Inhomogeneity in the magnetic field can contribute to the observed $\Delta\nu$, but this contribution is usually less than 1 Hz. Since the ^{31}P resonances of most phosphorylated proteins have $\Delta\nu$'s greater than 3 Hz, field inhomogeneity is usually of only minor concern. For very sharp resonances ($\Delta\nu \leq 3$ Hz), T_2 values can be more accurately measured by spin echo-type experiments, such as the Carr–Purcell[25] technique with the Meiboom–Gill[26] modification. Another factor that will contribute to the observed $\Delta\nu$ is homonuclear or heteronuclear spin coupling. With phosphorylated proteins, one must consider ^{31}P—^1H spin coupling, which can be eliminated by broad-band proton decoupling. The FID of most ^{31}P NMR spectra is digitally filtered with an exponential multiplication function to decrease spectral noise, prior to the FT step. This process results in some artificial linebroadening, which must be taken into account when determining the natural $\Delta\nu$, and hence the true T_2, of the ^{31}P resonance.

[24] R. R. Ernst, *J. Chem. Phys.* **45**, 3845 (1966).
[25] H. Y. Carr and E. M. Purcell, *Phys. Rev.* **94**, 630 (1954).
[26] S. Meiboom and D. Gill, *Rev. Sci. Instrum.* **29**, 688 (1958).

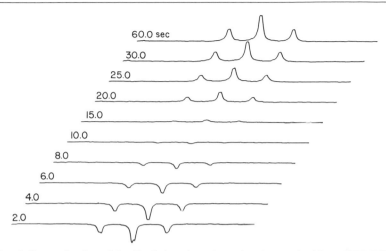

FIG. 2. Determination of the T_1 of phosphoserine using the standard inversion-recovery method (same sample and acquisition parameters as in Fig. 1A, except for delay time of 120 sec). The τ values used for each spectrum are indicated. A T_1 value of 12.5 sec was determined.

Unlike the T_2 value, the T_1 of a ^{31}P resonance cannot be determined directly from the ^{31}P NMR spectrum. The standard pulse method for T_1 determinations is the inversion recovery method[25,27] in which the nuclear spins of the sample are driven from the z axis to the $-z$ axis by a 180° pulse, a variable delay time (τ) is provided for the nuclei to relax partially back toward their equilibrium intensity along the z axis, and then a 90° pulse is given followed by the acquisition of the FID. This [180°-τ-90°] pulse sequence is repeated with the same τ value until an acceptable S/N has been built up, and then the same number of scans are used for a series of τ values (see Fig. 2). The major problem with the inversion recovery method is that a delay of about 5 T_1's is needed between each [180°-τ-90°] sequence to allow all the resonances fully to relax. This requirement makes this method very time-inefficient, especially for sensitivity-limited samples.

Several modifications of the basic inversion recovery method have been developed, such as the fast inversion recovery[28] method to increase the time efficiency. A relatively simple steady-state method of determining T_1 values that is practical for ^{31}P nuclei in phosphorylated proteins (particularly those with long T_1 values) is the progressive saturation

[27] R. L. Vold, J. S. Waugh, M. P. Klein, and D. W. Phelps, *J. Chem. Phys.* **48**, 3831 (1968).
[28] D. Canet, G. C. Levy, and I. R. Peat, *J. Magn. Reson.* **18**, 199 (1975).

method.[29] This method involves a 90° pulse followed by a delay time τ that is the sum of the AQ plus a variable delay, immediately followed by the next 90° pulse. This $(90°\text{-}\tau)_N$ sequence results in a steady-state equilibrium after a few scans, in which the heights of the resultant resonances depend upon the total time τ between excitation pulses and the T_1 of the resonance. The obvious advantage of this method is that many more scans of a sample can be taken over a specified time period than by the standard inversion recovery method, allowing better S/N of the resultant spectra. This method does require that $T_1 \gg T_2$, but this condition usually applies for ^{31}P resonances of phosphorylated proteins. (A typical phosphorylated protein may have a T_1 of 2 sec and a $\Delta\nu$ of 10 Hz or T_2 of 0.032 sec.) Also, the shortest τ value possible is limited by AQ. The AQ can be reduced to some extent by decreasing the computer block size or increasing the SW, but both eventually compromise the digital resolution of the final spectrum. Finally, the 90° pulse applied to the sample must be accurate, although composite pulse sequences that generate an almost perfect 90° pulse[30] can be employed.

Once the spectra of a sample have been taken at various τ values by any of the methods discussed, one can determine the T_1 from the increase in peak heights $M(\tau)$ as τ increases. The equation describing the return of the magnetization to its equilibrium value M_∞ is

$$M(\tau) = M_\infty[1 - \alpha e^{-(\tau/T_1)}] \tag{1}$$

where $\alpha = 1$ for the progressive saturation method, $\alpha = 2$ for the standard inversion recovery method, and $\alpha = 2 - e^{-}(T_0/T_1)$ for the fast inversion recovery method (where T_0 is the delay between the [180°-τ-90°] sequences). The best way to analyze the T_1 data is via a nonlinear least-squares fit of the data, solving for both T_1 and M_∞.[10,31] A nonlinear least-squares program, written in BASIC, has been published.[10]

Cross-relaxation is a mechanism whereby rapidly relaxing nuclei can enhance the relaxation of more slowly relaxing nuclei and has been shown to provide serious problems in ^{1}H NMR studies of proteins.[32] The possibility of cross-relaxation between ^{31}P nuclei and the abundant ^{1}H spins within a protein has been considered.[16,33] In the non-extreme narrowing limit ($\omega_0^2\tau_c^2 \gg 1$), relevant to most protein-bound ^{31}P nuclei (see section Analysis of Relaxation Times, below), cross-relaxation mechanisms play

[29] R. Freeman and H. D. W. Hill, *J. Chem. Phys.* **51**, 3140 (1969).
[30] R. Freeman, S. P. Kempsell, and M. H. Levitt, *J. Magn. Reson.* **38**, 453 (1980).
[31] H. Hanssum, *J. Magn. Reson.* **45**, 461 (1981).
[32] B. D. Sykes, W. E. Hull, and G. H. Snyder, *Biophys. J.* **21**, 137 (1978).
[33] W. E. Hull and B. D. Sykes, *J. Chem. Phys.* **63**, 867 (1975).

an insignificant role in T_1 relaxation and meaningful ^{31}P relaxation measurements can be made in the absence or the presence of ^{1}H decoupling.

Values for NOE are determined from the intensities of ^{31}P resonances in the presence of a broad-band proton decoupling field compared to their intensities in the absence of this field.[22] A complication arises in that the decoupling can cause an increase in the height (but not the total area) of the resonance by collapsing ^{31}P—^{1}H scalar couplings, so that integrated areas must be measured. The NOE determinations can be made by comparing the spectrum of a sample undergoing constant proton irradiation (inducing both the NOE and scalar decoupling) to one using gated decoupling in which protons are irradiated only during data acquisition (to collapse ^{31}P—^{1}H multiplet structures).[34] However, an appreciable delay between pulses must be employed to prevent a buildup of an NOE from the irradiation during data acquisition. The time course for the buildup (or conversely for the decay) of an NOE is related to the T_1 of the resonance. Many modern spectrometers have the computer software to allow one to store alternately one FID with proton irradiation in one block and the next FID with no irradiation (or only during acquisition) in another block. This normalizes any temperature effects that may build up owing to the proton irradiation.

Analysis of Chemical Shifts

Once a suitable ^{31}P NMR spectrum of a phosphorylated protein has been obtained, one can consider the analysis of the spectrum. The main properties of a ^{31}P resonance that can be experimentally determined are its chemical shift (the frequency at which that resonance absorbs energy), scalar splittings, linewidth, intensity, T_1, and NOE.

Chemical Shift Standards and Conventions

While each ^{31}P resonance has an absolute frequency of maximum absorption, it is customary (and technically easier) to express its chemical shift position δ in parts per million relative to some standard ^{31}P resonance normalized to the frequency of the spectrometer. Thus, we define

$$\delta = \frac{(\text{frequency of resonance} - \text{frequency of standard}) \times 10^6}{\text{frequency of spectrometer}} \quad (2)$$

If the sample is field-frequency locked with the D_2O in the sample, one might think the D_2O would be a good convenient reference frequency.

[34] D. Canet, *J. Magn. Reson.* **23**, 361 (1976).

However, this is not the case. The D_2O resonance frequency varies with temperature, pH, and salt. [31]P chemical shifts are usually determined relative to the external standard, 85% phosphoric acid, where the reference compound is run in a 1-mm sealed capillary tube mounted coaxially in the NMR tube. The frequency of the external standard is generally quite constant under the same conditions of temperature, pH, and solvent (D_2O) and does not need to be determined for each sample, provided the conditions are roughly the same. If the sample conditions are changed appreciably, the frequency of the external standard should be redetermined. For example, the frequency of 85% phosphoric acid changes by about 25 Hz (0.2 ppm for a 109.29 MHz ^{31}P NMR spectrometer) upon raising the temperature from 4 to 28°. A problem with the use of an external standard sample is the change in its frequency with bulk diamagnetic susceptibility changes caused by the geometric relationship between the sample and the magnetic field.[35] The chemical shifts, relative to coaxially oriented external 85% phosphoric acid, of compounds run on superconducting magnets (where the external standard is parallel to the magnetic field) will be 0.8 to 1.1 ppm more positive than the same compound run with iron-core magnets (where the external standard is perpendicular to the magnetic field).[35–37] Another frequently used external standard, which is convenient because its chemical shift is far from that of most ^{31}P resonances and because its diamagnetic susceptibility is the same as that of pure H_2O or D_2O, is 1.0 M methylene diphosphonate at pH 9.5 in D_2O. It is 16.35 ppm downfield of 85% phosphoric acid in an iron-core magnet or 17.45 ppm downfield of 85% phosphoric acid in a superconducting magnet.

The concept of positive and negative chemical shifts is another possible source of confusion. The older literature and much of the modern ^{31}P NMR literature has listed chemical shifts downfield (or up in frequency) of the standard as negative. However, the International Union of Pure and Applied Chemistry has recommended that downfield shifts be regarded as positive and upfield shifts as negative in keeping with the conventions of ^{1}H, ^{19}F, and ^{13}C NMR. In this chapter, all chemical shifts will be expressed by this latter convention. Chemical shift axes are usually, but not always, presented with downfield to upfield going from left to right.

[35] M. Batley and J. W. Redmond, *J. Magn. Reson.* **49**, 172 (1982).

[36] D. G. Gadian, G. K. Radda, R. E. Richards, and P. J. Seeley, *in* "Biological Applications of Magnetic Resonance" (R. G. Shulman, ed.), p. 463. Academic Press, New York, 1979.

[37] Chemical shifts determined on iron magnets are adjusted in this chapter to correspond to those obtained on superconducting magnets (add 0.8 ppm), as described in "Chemical Shift Standards and Conventions."

Chemical Shifts (Observed) of the Major Classes of Phosphorus-Containing Compounds

There are several groups of biologically important phosphorus-containing compounds that have observed ³¹P resonances grouped in similar chemical shift ranges. In the highly deshielded region (+30 to +7 ppm) are the phosphonates[38] (containing a phosphorus–carbon linkage), and thiophosphates[39] (containing a phosphorus–sulfur linkage). These compounds are very rare in nature but are being used increasingly in ³¹P NMR studies as analogs of natural phosphorus-containing compounds (ATP analogs, phosphoserine analogs, etc.). The chemical shift region from +7 to −3 ppm contains resonances from a large variety of cyclic phosphodiesters and orthophosphates, including inorganic phosphate, nucleotide monophosphates, sugar phosphates, and phosphorylated serine and threonine. The resonances of pyrophosphate diesters (e.g., NADH), nucleotide diphosphates, and end phosphates of tri- and polyphosphates are found in the −2 to −15 ppm region. Also phosphorylated histidine, arginine, and glutamic and aspartic acids have ³¹P resonances in this region. Finally, in the highly shielded region (−15 to −30 ppm) the resonances of the middle phosphates of triphosphate (for example ATP) and polyphosphates are found.

The chemical shifts of a large number of specific compounds have been compiled[36,40,41]; those of particular relevance to the study of phosphorylated proteins are given in Table I. The type of phosphorylated moiety covalently attached to a protein can often be determined from the chemical shift of the ³¹P resonance. Since the protein itself may provide a unique microenvironment for the phosphorylated moiety, it is often useful to denature the protein in 8 M urea to obtain an unperturbed chemical shift. Table I gives the chemical shifts of the commonly phosphorylated amino acids in the absence and in the presence of 8 M urea, as an aid in assigning the phosphorylated residue of a particular protein.

Changes in Chemical Shift with pH

The observed chemical shift of a ³¹P resonance under one particular set of conditions may be very useful in assigning the phosphorylated residue. Monitoring the chemical shift as a function of pH may also aid in

[38] R. Deslauriers, R. A. Byrd, H. C. Jarrell, and I. C. P. Smith, *Eur. J. Biochem.* **111,** 369 (1980).

[39] E. K. Jaffe and M. Cohn, *Biochemistry* **17,** 652 (1978).

[40] I. K. O'Neill and I. C. P. Richards, *Annu. Rep. NMR Spectrosc.* **10A,** 133 (1980).

[41] C. T. Burt, S. M. Cohen, and M. Bárány, *Annu. Rev. Biophys. Bioeng.* **8,** 1 (1979).

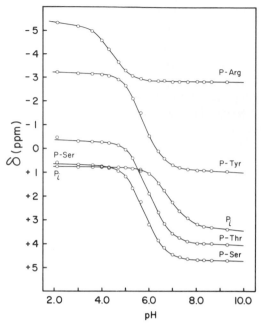

Fig. 3. Change in chemical shift with pH for a mixture of phosphorylated amino acids (same sample and acquisition parameters as in Fig. 1B). Chemical shifts, $^3J_{POCH}$ values and pK_a values (in the absence and in the presence of 8 M urea) are given in Table I.

the assignment of the phosphorylated residue and provide information about its microenvironment within the native protein (see Fig. 3).

Inorganic phosphate has three ionizable groups with pK_a values of 2.12, 7.20, and 12.40 at 25° (extrapolated to ionic strength $\mu = 0$). The ^{31}P chemical shift changes in response to the dissociation of each proton, with a chemical shift of 0.0 for H_3PO_4, +0.4 for $H_2PO_4^{-1}$, +3.0 for HPO_4^{-2}, and +5.5 ppm for PO_4^{-3}.[42] At physiological pH, it is the second transition, $H_2PO_4^{-1}$ to HPO_4^{-2}, that is the important one. Alkyl phosphate monoesters (R—O—PO_3H_2) have only two titratable groups with pK_a values of about 1.5 and 6.7, and alkyl phosphate diesters (R—O—PO_2R) have only one titratable group with a pK_a of about 1.5. For example, ethyl phosphate has pK_a values of 1.60 and 6.62, whereas diethyl phosphate has one pK_a of 1.39.[43] The fact that alkyl phosphate diesters do not titrate in the physiological pH ranges (and hence have ^{31}P chemical shifts that are

[42] M. M. Crutchfield, C. H. Dungan, and J. R. Van Wazer, *Top. Phosphorus Chem.* **5**, 1 (1967).

[43] W. D. Kumler and J. J. Eiler, *J. Am. Chem. Soc.* **65**, 2355 (1948).

constant above pH 3) has been used to differentiate them from monoesters and other titratable phosphorus-containing compounds. With the alkyl phosphate monoesters, as with inorganic phosphate, it is the second transition of $R—O—PO_3H^{-1}$ to $R—O—PO_3^{-2}$ that is important at physiological pH. The first transition is accompanied by only a very minor change in ³¹P chemical shift. The second transition is usually accompanied by a 3–4 ppm downfield shift. For example, the ³¹P resonance of phosphoserine shifts from about +0.8 ppm in its monoanionic form to about +4.7 ppm in its dianionic form, with a pK_{a2} of about 5.8 (see Table I). The pK_a for the phosphoseryl residue in the tripeptide glycyl-DL-phosphoserylglycine is 5.77, indicating that the free amino acid data are relevant to the residue in the polypeptide chain.[44]

It should be noted that the pK_a of a phosphorylated moiety can change appreciably with different experimental conditions, including temperature, ionic strength, metal ions, and solvents.[36] For phosphates, the effect of temperature on the pK_a is not great because entropy of proton ionization is not substantial. Even so, the pK_{a2} of phosphate was found to vary from 7.31 at 0° to 7.20 at 25°.[45] The pK_{a2} of phosphate was found to vary with ionic strength from 6.8 (ionic strength less than 50 mM) to 6.5 in 1.6 M KCl.[36] The chemical shifts of the protonated and unprotonated forms were not affected by the high salt concentrations. The 8 M urea did not have an appreciable effect on the chemical shifts of the protonated or unprotonated forms of a variety of phosphorylated amino acids. However, the pK_a values of phosphoserine, phosphothreonine, phosphotyrosine, phosphoarginine, and inorganic phosphate were all increased by at least 0.4 pH unit in the presence of 8 M urea (see Table I). Mg^{2+} binding was found to lower the pK_a of inorganic phosphate from 6.95 to 6.25.[46,47] The deuterium isotope effect has a small influence on the pK_a of a phosphate. The pK_{a2} of phosphoserine was found to be 5.78 in H_2O and 5.85 in D_2O.[48] Thus, one must consider the implications of the effects of experimental conditions on the pK_a values of phosphorylated moieties when pK_a values are used to assign a phosphorylated residue or to evaluate the effect of the protein on the phosphorylated group.

The pH titration of a phosphorylated protein under denaturing conditions, such as 8 M urea, can greatly aid in the assignment of the type of phosphorylated residue present by comparing the pK_a values and chemi-

[44] R. Osterberg, *Acta Chem. Scand.* **16,** 2434 (1962).
[45] H. A. Sober and R. A. Harte, "Handbook of Biochemistry, Selected Data for Molecular Biology J113." Chemical Rubber Publ. Co., Cleveland, Ohio, 1968.
[46] J. K. M. Roberts and O. Jardetzky, *Biochim. Biophys. Acta* **639,** 53 (1981).
[47] J. K. M. Roberts, N. Wade-Jardetzky, and O. Jardetzky, *Biochemistry* **20,** 5389 (1981).
[48] R. S. Humphrey and K. W. Jolley, *Biochim. Biophys. Acta* **708,** 294 (1982).

cal shifts of the protonated and deprotonated species with those of the free amino acids (Table I). The titration of a phosphorylated protein under native conditions can be very useful in characterizing the interactions of the phosphorylated group with the rest of the protein. If the ^{31}P resonance titrates normally, with the same pK_a and chemical shifts as the free amino acid, this indicates that the phosphorylated group is likely exposed to the solvent and not strongly interacting with other groups on the protein. This situation is fairly common for phosphorylation at regulatory sites of many proteins. If the ^{31}P resonance titrates with an unusual pK_a, an unusual chemical shift for its acid and/or base form, or exhibits cooperativity in its titration, this may indicate some strong interaction (hydrogen bonding, salt bridge, etc.) with another group or that the phosphate is in a hydrophobic, or highly charged pocket within the protein. If the ^{31}P resonance does not titrate at all under native conditions, it may indicate that the phosphorus is completely buried and unexposed to the solvent—a common situation for phosphorylated enzyme intermediates.

Changes in Chemical Shift with Addition of Metal Ions

In some cases, the binding of a metal ion to a phosphorylated group on a protein can be monitored by ^{31}P NMR. The binding of Mg^{2+} to the dianionic form of phosphoserine causes a 0.8 ppm upfield shift that is not due to any effect of the Mg^{2+} on the pK_a of the phosphate.[6] This binding is relatively weak, with a dissociation constant of about 40 mM at pH 7.9. Similar upfield shifts induced by Mg^{2+} have been shown to occur with the dianionic form of inorganic phosphate; these effects are independent of the effects of Mg^{2+} in lowering the pK_{a_2} of phosphate.[46,47] The binding of Ca^{2+} was also found to cause an upfield shift of about 0.4 ppm in the ^{31}P resonance of the dianionic form of phosphoserine.[49]

It is also possible to monitor the binding of paramagnetic metal ions at or near the phosphorylated group of a protein by ^{31}P NMR. Paramagnetic ions can greatly alter the chemical shifts of a ^{31}P resonance if there are appreciable interactions between the unpaired electrons of the metal and the ^{31}P nucleus (contact shifts). This usually requires the phosphate to be in the first coordination sphere of the metal. Paramagnetic ions can also change the chemical shift of the ^{31}P nucleus via through-space interactions (pseudocontact shifts). The paramagnetic ion can also dramatically affect the T_1 and T_2 values of a ^{31}P nucleus, again via both through-bond and through-space interactions. If the electron spin relaxation time for the paramagnetic metal is very long [e.g., 10^{-9} to 10^{-10} sec for Mn(II) or

[49] A. Bennick, A. C. McLaughlin, A. A. Grey, and G. Madapallimattam, *J. Biol. Chem.* **256,** 4741 (1981).

Gd(III)], the relaxation processes will be very efficient and the ^{31}P nucleus will be broadened out (since linewidth $= 1/\pi T_2$) and its T_1 greatly reduced if it is close to the metal. On the other hand, if the electron spin relaxation time is very short [e.g., 10^{-13} sec for Co(II) or Ni(II)], the ^{31}P resonance may not be appreciably broadened but show a substantial change in chemical shift. Both paramagnetic chemical shift changes[50] and relaxation enhancement effects[51,52] have been observed by ^{31}P NMR on model compounds. If the phosphorylated protein does not have a specific metal ion binding site, and the metal ion stays free in solution, paramagnetic broadening agents can be used to determine whether the phosphorylated group is close to the surface of the protein. Where a specific metal ion binding site does exist within a phosphorylated protein, it is possible to determine the distance[6,53,54] between the protein-bound paramagnetic ion and the phosphorylated group and/or correlation times for the complex.

On a more practical level, one should be aware of some of the problems involved with paramagnetic ions. The lanthanide series, from La(III) to Lu(III), irreversibly form insoluble complexes at pH ≥ 7.0, unless the metal is already firmly bound to the protein. The divalent paramagnetic ions do not suffer from this limitation in pH range.

Confirmation of Covalently Phosphorylated Protein

One potential pitfall in ^{31}P NMR studies of phosphorylated proteins is that the ^{31}P resonance of a highly mobile, covalently linked phosphoryl group, such as a phosphoseryl or phosphothreonyl residue, may have a similar chemical shift, pK_a, metal titration behavior, and linewidth to free inorganic phosphate or other small phosphate monoesters, such as 5′-AMP, glucose 6-phosphate, or fructose 6-phosphate. One can unequivocally eliminate the possibility that one is merely studying the ^{31}P resonance of inorganic phosphate by adding a small amount of inorganic phosphate to a sample whose ^{31}P NMR spectrum has just been taken. Two separate resonances should be observed (see Fig. 4). The resonance for inorganic phosphate can also be distinguished from that of a phosphoseryl or phosphothreonyl residue if proton decoupling is not used; the

[50] P. Tanswell, J. M. Thornton, A. V. Korda, and R. J. P. Williams, *Eur. J. Biochem.* **57,** 135 (1975).

[51] F. F. Brown, I. D. Campbell, R. Henson, C. W. J. Hirst, and R. E. Richards, *Eur. J. Biochem.* **38,** 54 (1973).

[52] H. Sternlicht, R. G. Shulman, and E. W. Anderson, *J. Chem. Phys.* **43,** 3123 (1965).

[53] J. E. Coleman and J. F. Chlebowski, *in* "Advances in Inorganic Chemistry" (G. L. Eichhorn and L. G. Marxilli, eds.), Vol. 1, p. 1. Elsevier/North-Holland, Amsterdam, 1979.

[54] M. Brauer and B. D. Sykes, *Biochemistry* **21,** 5934 (1982).

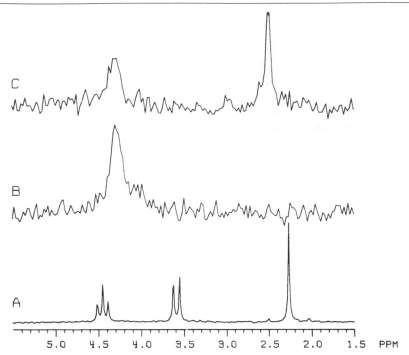

FIG. 4. ^{31}P NMR spectra of a fragment of phosphorylated troponin-T.[85]   (A) Phosphoserine, phosphothreonine, and inorganic phosphate (+4.35, 3.49, and 2.17 ppm, respectively) at pH 7.0, 27° in D_2O.   (B) Cyanogen bromide, 1.0 mM, fragment, CB-1 (residues 1
to 151) of troponin-T (pH 7.0, 4° in 0.1 M KCl, 30 mM imidazole, 20% D_2O). Acquisition
parameters: 60° pulse, SW ± 2500 Hz, block size 4096, DW 200 μsec, AQ 0.410 sec, delay
time 1.6 sec, 5 Hz linebroadening to decrease spectral noise, 4000 scans.   (C) Same as (B)
but with 0.7 mM inorganic phosphate.

covalently bound phosphate may then exhibit a ^{1}H—^{31}P splitting not possible for free inorganic phosphate. If the addition of a protein phosphatase
(acid or alkaline phosphatase) causes the disappearance of the covalent
phosphate resonance and the appearance of the resonance for free inorganic phosphate, this confirms the assignment of the original resonance.[55]
It also indicates that the phosphoryl group must be near the outside of the
protein and exposed to the phosphatase. Some commercially available
phosphatases, such as *Escherichia coli* alkaline phosphatase, have broad
specificities and can be used with many phosphorylated systems.[56] The
converse problem is preventing small amounts of contaminating phospha-

[55] A. Mak, L. B. Smillie, and M. Bárány, *Proc. Natl. Acad. Sci. U.S.A.* **75**, 3588 (1978).
[56] A. Torriani, this series, Vol. 12B, p. 212.

tases from dephosphorylating the protein of interest. Sodium phosphate, vanadate (end-product inhibitor), and EDTA (binds activating metal ions) are often added to inhibit common, nonspecific phosphatases, although the former may obscure the resonance of the protein-bound phosphate. Many other specific phosphatases are being discovered that may not be easily inhibited.[1-3]

The 31P resonance of free AMP can sometimes be mistaken for that of a covalently bound phosphate. Free AMP has a similar chemical shift, titration behavior, linewidth, and 1H—31P spin–spin coupling as that of a mobile phosphoseryl or phosphothreonyl residue. Free AMP can arise in systems containing ATP or ADP, such as a system including a kinase, ATP, and the target protein to be phosphorylated. Small contaminating amounts of adenylate kinase often found in enzyme preparations can catalyze the reaction (2 ADP ⇌ ATP + AMP), resulting in a free AMP resonance that may be mistaken for that of a phosphorylated protein.

Analysis of Scalar Coupling Constants

Protein nuclei that are covalently linked to the 31P nucleus can affect the spectrum of a phosphorylated protein. This through-bond scalar coupling between the nuclei is manifest, in the absence of proton decoupling during acquisition, as a splitting of the 31P resonance into a doublet (if the 31P nucleus studied is interacting with only one proton) or a triplet with relative intensities $1:2:1$ (if the 31P nucleus is interacting with two identical protons). The magnitude of the splitting, measured as the number of Hertz between any two lines, is called the scalar coupling constant or ^nJ, where n is the number of bonds between the nuclei. ^nJ decreases fairly dramatically as n increases. For 31P NMR studies of phosphorylated proteins, this usually involves a P—O—C—H or P—N—C—H linkage, for which there are three covalent bonds and $^3J_{PXCH}$ is usually 5–10 Hz (see Table I). We will not consider ^4J splittings; these are usually <2 Hz (0.4 Hz for phosphoserine, for example) and will be less than the natural linewidths of the 31P resonance of most phosphorylated proteins. Also, rapidly exchanging protons, such as those on primary amines, carboxylic acids, alcoholic hydroxyl protons, and amide protons, are generally exchanging too quickly to induce a splitting of the 31P nucleus.

The pattern of the J splitting can be very useful in assigning the phosphorylated group involved. For example, a 31P resonance of a phosphoseryl residue can be readily discerned from that of a phosphothreonyl residue, since the former is split by two equivalent protons on the β-carbon (giving a triplet) whereas the latter is split by only one proton on the β-carbon (giving a doublet) (see Fig. 1B). Phospholysine is split by two protons on the ε-carbon to give a triplet spectrum while 1-phospho-histidine is split by only one carbon (the ring C-2 proton) to give a doublet. Phosphoaspartic acid, phosphoglutamic acid, and phosphoarginine do not have a nonexchanging proton within three bonds of the ^{31}P nucleus, so the ^{31}P—^1H splittings are small.

The magnitude of the $^3J_{PXCH}$ splitting varies with the dihedral angle between the ^{31}P and the ^1H nuclei. The $^3J_{POCH}$ splitting for various phosphate monoesters of the form P—O—CH$_2$—R has been estimated to be about 3 Hz for the conformation where both protons on the carbon are gauche to the phosphorus.[57] A splitting of 19–24 Hz has been estimated for the two conformations where one proton is gauche and the other proton is trans. The variation in $^3J_{POCH}$ values with dihedral angle θ has been described for 3',5'-cyclic nucleotides as[58] $^3J_{POCH} = 16.3 \cos^2\theta - 4.6 \cos\theta$; this equation gives a $^3J_{POCH}$ of 1.8 Hz for the gauche form ($\theta = 60°$) and $^3J_{POCH}$ of 20.9 Hz for the trans form ($\theta = 180°$). This Karplus-like relationship gives reasonable dihedral angles for 2',3'-cyclic nucleotides and has general applicability for acyclic systems.[59] The observed $^3J_{POCH}$ coupling for any particular phosphate monoester is then a weighted average dependent upon the fractional population of the three rapidly inter-converting forms. With this approach, 5'-AMP and free phosphoserine have been found to be in the gauche form about 80% of the time and in one of its two trans forms about 20% of the time.[6,60]

A change in the $^3J_{POCH}$ value with changes in temperature or pH can reflect a change in the relative population of the various rotamers. For example, 5'-AMP[61] and trimethyl phosphate[62] show no appreciable change in $^3J_{POCH}$ with temperature or pH. However, a phosphate with bulkier substituents, tris(β-chloroethyl) phosphate, does exhibit a temper-ature dependence of its $^3J_{POCH}$ splitting[62] with the gauche conformer domi-nating at lower temperatures (as indicated by a lower $^3J_{POCH}$ value) and an equal distribution of all conformers at the higher temperatures (as indicated by a higher $^3J_{POCH}$ value). Monomethyl, monoethyl, and diethyl phosphate[62] showed a decrease of about 1 Hz in $^3J_{POCH}$ upon ionization of the first hydroxyl group of the phosphate. This probably reflects the in-creased steric requirements of the monoanionic phosphate, favoring the gauche rotamer, although this effect could also be due to a change in the Karplus relationship of $^3J_{POCH}$ versus θ due to the ionization. 3'-AMP

[57] M. Kainosho, A. Makamura, and M. Tsuboi, *Bull. Chem. Soc. Jpn.* **42**, 1713 (1969).
[58] B. J. Blackburn, R. D. Lapper, and I. C. P. Smith, *J. Am. Chem. Soc.* **95**, 2873 (1973).
[59] R. D. Lapper and I. C. P. Smith, *J. Am. Chem. Soc.* **95**, 2880 (1973).
[60] F. E. Evans and R. H. Sarma, *J. Biol. Chem.* **249**, 4754 (1974).
[61] P. J. Cozzone and O. Jardetsky, *Biochemistry* **15**, 4860 (1976).
[62] A. A. Bothner-By and W.-P. Trautwein, *J. Am. Chem. Soc.* **93**, 2189 (1971).

exhibits a decrease in $^{3}J_{POCH}$ as the pH is increased.[61] This decrease coincides with the second ionization of the phosphate. On the other hand, the $^{3}J_{POCH}$ of 3'-dAMP (lacking the 2'-hydroxyl group) does not show any pH dependence. This indicates a specific interaction between the 2'-hydroxyl and the 3'-phosphate—an interaction that more strongly favors the gauche conformer when the phosphate is dianionic. Unlike 3'-AMP, monomethyl, monoethyl, and diethyl phosphate, the $^{3}J_{POCH}$ of phosphoserine increases as the pH is raised, indicating a higher fraction of trans rotamer at high pH.[21] Dinucleotides have lower $^{3}J_{POCH}$ coupling values than do the corresponding mononucleotides[61] owing to the bulkier substituents on the phosphate, favoring the gauche rotamer. These splittings do increase at higher temperatures as the trans rotamers become more common.

Analysis of Relaxation Times

The analysis of T_2, T_1, and NOE values for the ^{31}P resonance of a phosphorylated protein can provide significant useful biochemical information obtainable only by NMR. Knowing the T_1 value of a resonance is very useful in determining the optimum repetition rate for data acquisition and for evaluating whether the relative intensities of resonances correspond to their relative concentrations under experimental conditions. Perhaps the most important information that can be obtained from relaxation times is that regarding the motions and mobility of a phosphorylated group. The theory of NMR relaxation processes is quite complex and is dealt with elsewhere.[13,63] We will present only the general theoretical equations relevant for phosphorylated proteins and solve them for the case of a phosphoseryl residue within a protein. This should give the reader a general feel for interpretation of T_2, T_1, and NOE values.

Chemical shift anisotropy (CSA) is a dominant relaxation mechanism for ^{31}P resonances,[16,17,63] especially at higher magnetic field strengths (see Table II). This mechanism involves the interaction between the ^{31}P nucleus and the magnetic field modulated by the nonspherical nature of the chemical shift tensor. The relaxation rates are given by

$$1/T_{2,CSA} = 2/90 \cdot \omega_P{}^2(\Delta\sigma)^2(1 + \eta^2/3)[4J(0) + 3J(\omega_P)] \qquad (3)$$
$$1/T_{1,CSA} = 2/15 \cdot \omega_P{}^2(\Delta\sigma)^2(1 + \eta^2/3)[J(\omega_P)] \qquad (4)$$

where $\Delta\sigma$ represents the anisotropy of the chemical shift, η is the asymmetry term,[17,63] and ω_P is the precessional frequency of the ^{31}P nucleus. J is the spectral density function defined as $\tau_c/(1 + \omega^2\tau_c{}^2)$ for isotropic

[63] A. Abragam, "Principles of Nuclear Magnetism." Oxford Univ. Press, London and New York, 1961.

TABLE II

THEORETICAL T_2 AND T_1 RELAXATION RATES FOR THE ^{31}P RESONANCE OF A PHOSPHOSERYLPROTEIN

$\tau_c{}^a$	$MW_p{}^b$	$1/T_{2,CSA}{}^c$	$1/T_{2,DD}{}^d$	$1/T_{2,total}{}^e$	Linewidth^f	$1/T_{1,CSA}{}^c$	$1/T_{1,DD}{}^d$	$1/T_{1,total}{}^e$	T_1	$\eta_0{}^g$	η^h
40.5 MHz											
10^{-11}	2×10^1	0.0007	0.0033	0.0040	0.001	0.0006	0.0033	0.0039	250	1.24	1.05
10^{-10}	2×10^2	0.007	0.033	0.040	0.013	0.006	0.033	0.039	25	1.24	1.05
10^{-9}	2×10^3	0.07	0.26	0.33	0.10	0.063	0.24	0.30	3.3	0.87	0.69
10^{-8}	2×10^4	0.42	0.78	1.21	0.38	0.09	0.18	0.27	3.7	0.04	0.03
10^{-7}	2×10^5	4.2	6.7	10.9	3.5	0.01	0.02	0.03	33	0.02	0.01
10^{-6}	2×10^6	42	67	109	35	0.001	0.002	0.003	330	0.02	0.01
161.9 MHz											
10^{-11}	2×10^1	0.012	0.003	0.015	0.005	0.010	0.003	0.013	74	1.24	0.30
10^{-10}	2×10^2	0.12	0.03	0.15	0.05	0.10	0.03	0.13	7.4	1.16	0.29
10^{-9}	2×10^3	0.92	0.12	1.0	0.33	0.50	0.07	0.57	1.7	0.15	0.02
10^{-8}	2×10^4	6.8	0.7	7.5	2.4	0.10	0.01	0.11	9.0	0.02	0.00
10^{-7}	2×10^5	68	7	75	24	0.01	0.001	0.011	90	0.02	0.00
10^{-6}	2×10^6	680	70	750	240	0.001	0.0001	0.0011	900	0.02	0.00

^{a} Rotational correlation time of the ^{31}P nucleus (seconds).

^{b} Corresponding molecular weight of the protein, assuming spherical shape, viscosity of water, and a temperature of 25°.

^{c} Based on $(\Delta\sigma)(1 + \eta^2/3)^{1/2}$ term of 86 ppm for free phosphoserine. CSA, chemical shift anisotropy.

^{d} Based on dipole–dipole (DD) interactions between ^{31}P nucleus and two nonexchangeable β-methylene protons 2.87 Å away.

^{e} $1/T$ total = $1/T_{DD} + 1/T_{CSA}$. Note that CSA dominates at 161.9 MHz, but DD dominates at 40.5 MHz.

^{f} Linewidth = $1/\pi T_{2,total}$.

^{g} Nuclear Overhauser effect (NOE) enhancement assuming 100% DD relaxation.

^{h} NOE enhancement accounting for CSA relaxation $\eta = \eta_0 [(1/T_{1DD})/(1/T_{1,total})]$.

rotation, where τ_c is the rotational correlation time. The $(\Delta\sigma)(1 + \eta^2/3)^{1/2}$ term for the dianionic form of phosphoserine in the solid state is 86 ppm.[64]

For a phosphorylated group on a protein, the ^{31}P nucleus can be relaxed also by protons on the protein or within the protein's hydration shell. This dipole–dipole (DD) relaxation tends to be the dominant relaxation mechanism at low magnetic field strengths ($\leq$80 MHz ^{31}P frequency).[16] The equations for DD relaxation between unlike spins (^{1}H and ^{31}P nuclei) are[63]

$$\frac{1}{T_{2,\text{DD}}^{\text{HP}}} = \frac{1}{20}\frac{\gamma_H^2\gamma_P^2\hbar^2}{r_{\text{HP}}^6}$$

$$[4J(0) + J(\omega_H - \omega_P) + 3J(\omega_P) + 6J(\omega_H) + 6J(\omega_H + \omega_P)] \quad (5)$$

$$\frac{1}{T_{1,\text{DD}}^{\text{HP}}} = \frac{1}{10}\frac{\gamma_H^2\gamma_P^2\hbar^2}{r_{\text{HP}}^6}[J(\omega_H - \omega_P) + 3J(\omega_P) + 6J(\omega_H + \omega_P)] \quad (6)$$

where the relaxation rates depend upon the distance between the two nuclei ($1/r_{\text{HP}}^6$), the gyromagnetic ratios γ_H and γ_P, and the spectral density functions. The values for γ_H and γ_P are 2.6735×10^4 and 1.0822×10^4 radians sec^{-1} Gauss^{-1}, and $\hbar = 1.055 \times 10^{-27}$ erg-sec. The distance, r_{HP}, between the ^{31}P nucleus and either of the protons on the β-carbon of phosphoserine can be determined from crystallographic data.[65,66] Assuming the predominant gauche-gauche conformation between the ^{31}P nucleus and the two protons, $r_{\text{HP}} = 2.87$ Å.

Closely related to the $1/T_{1,\text{DD}}$ rate is the NOE. For ^{31}P NMR studies of phosphorylated protein, where I' is the intensity of the ^{31}P resonance with the protons saturated and I_0 is the intensity in the absence of proton irradiation, the NOE enhancement is

$$\eta = \frac{I' - I_0}{I_0} = \frac{\gamma_H}{\gamma_P}\frac{6J(\omega_H + \omega_P) - J(\omega_H - \omega_P)}{3J(\omega_P) + J(\omega_H - \omega_P) + 6J(\omega_H + \omega_P)} \quad (7)$$

The NOE is due to D–D interactions. Relaxation by other mechanisms such as CSA will decrease the magnitude of the NOE. The observed NOE depends upon the fraction of the total $1/T_1$ rate due to D–D relaxation (see Table II).

Neglecting any possible cross-correlation,[67,68] the contributions of each relaxation mechanism to the overall relaxation rate are additive.

[64] S. J. Kohler and M. P. Klein, *J. Am. Chem. Soc.* **99**, 8290 (1977).
[65] E. Putkey and M. Sundaralingam, *Acta Crystallogr. Sect. B* **B26**, 782 (1970).
[66] M. Sundaralingam and E. Putkey, *Acta Crystallogr. Sect. B* **B26**, 790 (1970).
[67] L. B. Werbelow and A. G. Marshall, *Mol. Phys.* **28**, 113 (1974).
[68] L. B. Werbelow and A. G. Marshall, *J. Magn. Reson.* **11**, 299 (1973).

Thus if CSA and DD are the major relaxation mechanisms relevant to the ^{31}P NMR study of phosphorylated proteins, then $1/T^{observed} = 1/T_{CSA} + 1/T_{DD}$. Let us assume that the protein does not affect the $(\Delta\sigma)(1 + \eta^2/3)^{1/2}$ term for the CSA relaxation mechanism, introduce a proton closer to the ^{31}P nucleus than 2.87 Å, or provide more than one unique microenvironment for the phosphorylated group (i.e., introduce chemical exchange contributions to T_1 and T_2 relaxation rates). We can then calculate the total (CSA + DD) theoretical $1/T_1$ and $1/T_2$ relaxation rates for a phosphoseryl residue within a protein as a function of the mobility (τ_c) of the ^{31}P nucleus (see Fig. 5 and Table II). (In actual practice, the observed linewidth of a resonance in a sample of phosphorylated protein will generally be greater than 0.2 Hz, owing to sample or field inhomogeneity, trace amounts of paramagnetic impurity, or increased viscosity due to the pres-

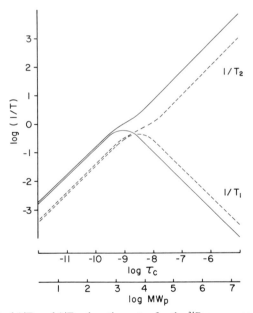

FIG. 5. Theoretical $1/T_2$ and $1/T_1$ relaxation rates for the ^{31}P resonance of a phosphoserylprotein as a function of τ_c. Rates were calculated as the sum of the theoretical CSA and dipole–dipole (DD) relaxation rates (see Table II), assuming $(\Delta\sigma)(1 + \eta^2/3)^{1/2}$ of 86 ppm for the phosphoryl group and DD contributions to the ^{31}P nucleus from only the two nonexchangeable β-methylene protons 2.87 Å away. The presence of additional protons closer to the ^{31}P nucleus or increased shielding anisotropy due to interactions with the protein would increase both $1/T_2$ and $1/T_1$ rates. A molecular weight scale for globular proteins (MW$_p$), which corresponds to the τ_c scale (at 25°, assuming the viscosity of water), is also given. -----, Relaxation rates at 40.5 MHz ^{31}P frequency; ——, Relaxation rates at 161.9 MHz ^{31}P frequency.

ence of large concentrations of protein or salt. Similarly, one will not generally be able to measure T_1 values greater than 20 sec.) For the immobilized phosphoryl group, the τ_c of the ^{31}P nucleus represents the rotation of the whole protein. For isotropic rotation of a spherical protein, we can use the Stokes–Einstein equation

$$\tau_c = \tfrac{4}{3}\pi a^3(\eta_s/kT) = V_s(\eta_s/kT) \tag{8}$$

where a is the radius of the protein (in cm), η_s the viscosity of the solvent (in g sec^{-1} cm^{-1}), T is temperature (in °K), and $k = 1.3805 \times 10^{-16}$ erg-deg^{-1} molecule^{-1}. The volume of a spherical protein ($V_s = \tfrac{4}{3}\pi a^3$) can be estimated from its molecular weight MW$_P$, assuming a partial specific volume ($\bar{v}_P$) for a protein of 0.72 cm^3 g^{-1} and hydration of 0.3–0.4 g of H$_2$O per gram of protein.[69] V_s (in cm^3) $\simeq$ (MW$_P\bar{v}_P$ + 0.35 MW$_P\bar{v}$H$_2$O)/6.023 $\times$ 10^{23} = 1.78 $\times$ 10^{-24} $\times$ MW$_P$. Thus, a protein of given MW should have a τ_c that can be calculated from the Stokes–Einstein equation, assuming spherical protein shape, known solvent viscosity, and temperature. In Fig. 5, the τ_c axis and corresponding MW axis (at 25°, assuming the viscosity is the same as pure water) are plotted together. If the τ_c value determined from the measured T_1 and T_2 values of the ^{31}P nucleus is more than one order of magnitude shorter than the theoretical τ_c value for that protein (assuming the MW$_P$ is known), then the phosphoseryl residue is probably not rigidly held within the protein but has appreciable internal mobility.

Literature Review of ^{31}P NMR Studies of Phosphoproteins

Now that we have described the basic procedures for the preparation of samples and for obtaining a ^{31}P NMR spectrum, as well as ways of analyzing δ, $^3J_{PXCH}$, T_1, T_2, and NOE values, we will describe the ^{31}P NMR studies of phosphoproteins that have been done to date. Many exciting advances have been made in ^{31}P NMR studies of phospholipids, phosphorus-containing metabolites within whole intact cells and tissues, and phosphorus-containing ligands that interact noncovalently with enzymes.[12,40,70–73] We will, however, confine ourselves to ^{31}P NMR studies

[69] C. R. Cantor and P. R. Shimmel, "Biophysical Chemistry," Vol. II. Freeman, San Francisco, California, 1980.

[70] R. G. Griffin, this series, Vol. 72, p. 108.

[71] D. G. Gadian and G. K. Radda, *Annu. Rev. Biochem.* **50**, 69 (1981).

[72] M. Cohn and B. D. Nageswara Rao, *Bull. Magn. Reson.* **1**, 38 (1979).

[73] D. G. Gadian, "Nuclear Magnetic Resonance and Its Applications to Living Systems." Oxford Univ. Press (Clarendon), London and New York, 1982.

of proteins to which phosphate is covalently attached including studies of stable phosphorylated enzyme intermediates.

^{31}P NMR Studies of Phosphoseryl Proteins

Phosphoseryl Proteins of Eggs and Milk. The most common and most extensively studied site of phosphorylation on proteins is at the hydroxyl group of serine residues.[74] The first ^{31}P NMR studies of phosphorylated proteins were of α_s-casein and phosvitin.[21,75] Both proteins have multiple phosphorylation sites (approximately 8 sites per α_s-casein and approximately 56 sites per phosvitin[76]) and are believed to play nutritional roles *in vivo*. The ^{31}P NMR resonances for both proteins in 8 M urea titrated with pH in the same manner as phosphoserine itself titrates. The linewidths, with proton decoupling, were 3–10 Hz wide. The absence or the presence of 8 M urea did not affect the chemical shifts or titration behavior of the ^{31}P resonances of phosvitin, indicating that the phosphoseryl moieties were fully exposed, not buried in unique microenvironments within the protein. The assignment of the resonances to phosphoseryl groups, rather than phosphothreonyl groups that also titrate in the same pH range, was confirmed from spectra with no proton decoupling; these showed a triplet spin–spin coupling pattern for the splitting of the phosphorus resonance by the two β-methylene protons of the serine group. The spin–spin coupling constants were about 6 Hz, consistent with the values for free phosphoserine (see Table I).

More recent ^{31}P NMR studies have revealed significant additional detail about the casein proteins.[48,77] The ^{31}P NMR spectrum of α_{s_1}-casein B showed at least four separate ^{31}P resonances, indicating distinct microenvironments for the phosphoseryl residues.[48] Partial assignments of ^{31}P resonances to individual phosphoseryl residues were made by comparing the spectra of peptide fragments (excluding some of the residues) with spectra of the intact protein. A set of four phosphoseryl residues are close together in the primary sequence of the protein, have very similar ^{31}P NMR chemical shifts, and were postulated to form a cluster that binds Ca^{2+} and to help maintain the structure of the overall casein micelle. Similar ^{31}P NMR results were obtained with β-casein A_2, a protein with a homologous phosphoseryl cluster.[48,77] Particular care was needed in eliminating paramagnetic metal ions (up to 20 mM EDTA or passage down a

[74] G. Taborsky, *Adv. Protein Chem.* **28,** 1 (1974).
[75] Chien Ho and R. J. Kurland, *J. Biol. Chem.* **241,** 3002 (1966).
[76] L. J. Banaszak and J. Seelig, *Biochemistry* **21,** 2436 (1982).
[77] R. W. Sleigh, A. G. Mackinlay, and J. M. Pope, *Biochim. Biophys. Acta* **742,** 175 (1983).

Chelex column), since the phosphoseryl cluster binds metal ions tightly.[48] T_1 values of 1.2 to 2.3 sec and linewidths of about 0.5 Hz for the ³¹P resonance of each phosphoseryl residue were consistent with a great deal of internal mobility for the phosphoseryl residues. The residues in the cluster titrated with pK_a values of 6.6 to 6.9 and showed negative cooperativity throughout the pH titration (Hill coefficient $n < 1$) indicating interactions among the phosphoseryl residues.

Ovalbumin, a protein found in abundance in egg white, has two phosphoseryl residues and two observable ³¹P resonances.[77,78] Both residues were shown to be relatively mobile and exposed to the solvent, since both ³¹P resonances had the same $^3J_{POCH}$ value and titrated with the same pK_a as phosphoserine. However, one residue was believed to be somewhat less exposed and less mobile on the basis of an increased linewidth and resistance to dephosphorylation at high pH values.[78]

The lipovitellin–phosvitin complex is a large lipoprotein (MW 456,000) with up to 170 phosphoserine residues per complex.[76] The ³¹P NMR spectrum shows one fairly narrow phosphoserine resonance and two phospholipid resonances. The T_1 of the phosphoserine resonance was about 2 sec and increased with temperature. From the Stokes–Einstein equation, one can see that τ_c becomes shorter as the temperature is raised. This indicates that the motions of the phosphoseryl residues were fast and in the extreme narrowing limit $\omega_0^2 \tau_c^2 \ll 1$ (see Fig. 5). Rotational correlation times of 0.4–0.7 nsec were estimated, clearly showing the internal mobility of the phosphoseryl residues within this very large protein complex. Virtually the same T_1 results were obtained when the complex was in the form of a wet pellet, using solid-state NMR methods with a strong proton-decoupling field. Using dry crystals of lipovitellin–phosvitin complex, a ³¹P NMR powder pattern of the phosphoseryl protein could be obtained. Principal elements of the chemical shift tensor were obtained (+43, −1, and −57 ppm) and were similar to values obtained for free phosphoserine (+46, +3, and −52 ppm).[64,79] These elements give chemical shift anisotropy values $[(\Delta\sigma)(1 + \eta^2/3)^{1/2}]$ of 87 ppm for the protein and 85 for free phosphoserine. These anisotropy values indicate that the electronic structure of the phosphoseryl groups is not appreciably altered by the protein, and they show that CSA will be a dominant relaxation mechanism for phosphoseryl groups in proteins, especially at high magnetic field strength.

[78] H. J. Vogel and W. A. Bridger, *Biochemistry* **21**, 5825 (1982).
[79] S. J. Kohler and M. P. Klein, *Biochemistry* **15**, 967 (1976).

Phosphoseryl Proteins of Bones and Teeth. ^{31}P NMR studies of demineralized dentine phosphoproteins again reflect phosphorylated serine residues that were largely exposed and mobile.[80–82] These proteins again have multiple phosphorylation sites with 53 phosphoserines per phosphoprotein from bovine dentine[82] and about 70 per phosphoprotein from rat incisor dentine.[81] These proteins, in conjunction with collagen and calcium, play an essential structural role in the formation of teeth. The ^{31}P resonances of demineralized bovine dentine phosphoprotein shifted from +5.0 ppm[37] at pH 9.5 to +1.0 ppm at pH 3.46 with an apparent pK_a of 6.80, exhibiting almost identical titration behavior as free phosphoserine.[82] Not all the phosphoserine residues were in identical microenvironments, however, as at least three component resonance bands could be observed by ^{31}P NMR. The addition of 1 mol of Ca^{2+} per mole of phosphoseryl residue resulted in very great line-broadening, indicating an interaction between the Ca^{2+} and the phosphoryl group that immobilized the phosphoryl group. Phosphoprotein from rat incisor dentine also exhibited a ^{31}P resonance that titrated like phosphoserine.[82] As with the bovine dentine phosphoprotein, Ca^{2+} binds to the phosphoprotein, increasing the resonance linewidth with a reduction in the mobility of the phosphoseryl residues. La^{3+}, Mn^{2+}, and to a lesser degree Mg^{2+} could also bind. It was suggested that —Asp—phosphoseryl— centers form high affinity Ca^{2+} binding sites that can induce the nucleation of hydroxyapatite in mineralized dentine from calcium phosphate. The two major phosphopeptides E_3 and E_4 of embryonic bovine dental enamel were also studied by ^{31}P NMR[80] to confirm that the phosphorylation is only at the level of serine residues, with no acid-labile phosphorylated amino acids or complex mixed bonds (such as between organic phosphate and carboxylic acid groups) present. Phosphoseryl residues in conjunction with aspartate residues were found by ^{31}P NMR to be involved in Ca^{2+} binding in human salivary proteins A and C.[49] Two distinct ^{31}P resonances were observed for the two phosphoseryl residues, and the pK_a's of both residues decreased by 0.6–0.7 pH units upon Ca^{2+} binding.

Phosphoseryl Proteins of Muscle. The phosphorylation of muscle proteins has become the object of ^{31}P NMR studies. The troponin complex was found to have one endogenous phosphorylation site at the N-terminal

[80] A. Roufosse, E. Strawich, E. Fossel, S. Lee, and M. J. Glimcher, *FEBS Lett.* **115**, 309 (1980).
[81] D. J. Cookson, B. A. Levine, R. J. Williams, M. Jontell, A. Linde, and B. Bernard, *Eur. J. Biochem.* **110**, 273 (1980).
[82] S. L. Lee, A. Veis, and T. Glonek, *Biochemistry* **16**, 2971 (1977).

serine hydroxyl of the TN-T subunit.[83] The ^{31}P NMR spectrum of the troponin complex included one fairly narrow resonance (linewidth of 3 Hz) which titrated with pH identically to free phosphoserine, indicating that the phosphorylation site is highly mobile. The addition of Ca^{2+} or a 1:1 molar ratio of tropomyosin to troponin complex did not have any effect on the ^{31}P resonance. Similar results were obtained with the cyanogen bromide fragment (CB1) of the purified TN-T subunit.[84,85] A narrow ^{31}P resonance at the same chemical shift as free phosphoserine was observed. This resonance was not altered in the presence of a nonpolymerizable derivative of tropomyosin or of several tropomyosin fragments, indicating that the phosphorylation site of TN-T is not immobilized upon formation of the troponin complex or on binding to tropomyosin.

A phosphorylation site at serine-238 (the penultimate residue at the C-terminal end) of α-tropomyosin was shown to have a ^{31}P resonance at the chemical shift of free phosphoserine.[55] Digestion with alkaline phosphatase resulted in the disappearance of this resonance and the appearance of a resonance corresponding to inorganic phosphate. This phosphoserine has been postulated to be involved in the head-to-tail overlap of α-tropomyosin during polymerization, with a salt bridge formed between the phosphoserine and lysine-6 at one side of the overlap region and lysine-12 on the other side.

A covalently bound phosphate has also been demonstrated at serine-14 or -15 of the "P-light chain" (LC_2) of rabbit skeletal muscle myosin.[86] The ^{31}P NMR spectrum of phosphorylated whole myosin showed a resonance having the same chemical shift and pH dependence as free phosphoserine. This ^{31}P resonance was unaffected by the absence or the presence of Ca^{2+}. The linewidth of the resonance was ≈40 Hz, considerably less than would be expected of an immobilized phosphate bound to a large molecule like myosin (MW ≈500,000). Thus, this phosphoseryl residue is proposed to have appreciable internal mobility and may prevent the interaction of the N-terminal segment of the light chain with other parts of the myosin protein. The general trend with the muscle proteins seems to involve phosphorylation near the N- or C-terminal ends resulting in a modulation of the interaction between these ends and the rest of the protein.

[83] J. E. Sperling, K. Feldman, H. Meyer, O. Jahnke, and L. M. G. Heilmeyer, *Eur. J. Biochem.* **101,** 581 (1979).
[84] J. R. Pearlstone and L. B. Smillie, *Can. J. Biochem.* **58,** 649 (1980).
[85] A. Mak, personal communication, 1982.
[86] B. Koppitz, K. Feldmann, and L. M. G. Heilmeyer, *FEBS Lett.* **117,** 199 (1980).

Phosphoglucomutase. One enzyme, for which a variety of ^{31}P NMR experiments have been performed including T_1 determinations and effects of proton decoupling and metal ion binding, is phosphoglucomutase.[6] This protein, which catalyzes the interconversion of glucose 1-phosphate and glucose 6-phosphate, has one phosphoseryl residue when it is in a stable intermediate form. Its ^{31}P resonance is at +4.6 ppm[37] at pH 7.9 with a linewidth of 14 Hz and a T_1 of about 4 sec, as compared to free phosphoserine with a chemical shift of +4.7 ppm and linewidth of 0.5 Hz. The chemical shift of the ^{31}P resonance of phosphoglucomutase did not shift upfield as the pH was decreased to pH 5.0 (below which denaturation occurred), indicating that the phosphoseryl group is likely dianionic and buried within the protein and hence does not titrate normally. Assuming that dipolar interactions were the dominant relaxation mechanism at low magnetic field strength (40.5 MHz ^{31}P frequency) and considering only the interaction between the phosphorus and the two nonexchangeable methylene protons of the serine residue, a rotational correlation time of 70 nsec was determined from the T_1 and T_2 values of the ^{31}P resonance. The theoretical tumbling time for phosphoglucomutase, assuming a spherical shape, was about 50 nsec, indicating that this phosphoserine residue is essentially immobile within the protein. More recent studies confirm this, but indicate that the phosphoseryl residue has increased internal mobility upon substrate binding.[87]

Proton decoupling experiments were also done on phosphoglucomutase. For free phosphoserine,[6] proton decoupling caused an increase in ^{31}P signal height of 400% due to the collapse of the $^3J_{POCH}$ coupling and the NOE enhancement (theoretical maximum of 1.24). The maximum effect was obtained upon proton irradiation at +4.1 ppm (from TMS), the chemical shift of the β-methylene protons. With phosphoglucomutase the maximum effect was again obtained upon proton irradiation at about +4 ppm. However, only a 30% increase in ^{31}P signal intensity was obtained and none of this increase was due to an NOE, again consistent with the premise of an immobilized phosphoserine residue within the protein. The binding of Mg^{2+} to the phosphoserine residue could be determined by ^{31}P NMR. Mg^{2+} causes a 0.8 ppm upfield shift in free phosphoserine at pH 7.9, but no shift in the ^{31}P resonance of phosphoglucomutase, indicating no divalent metal ion binding to the protein-bound phosphoserine. This was confirmed by paramagnetic T_1 relaxation studies of the effects of Ni^{2+} on the ^{31}P resonance; a Ni^{2+}—[31]P distance of 4.5 to 6.0 Å was determined,

---

[87] J. L. Markley, G. I. Rhyu, and W. J. Ray, Jr. *Fed. Proc. Fed. Am. Soc. Exp. Biol.* **41,** Abst. No. 5069 (1982).

which is too long a distance to correspond to direct coordination between the phosphate and the divalent metal ion (2.9–3.1 Å).

Alkaline Phosphatase. The diversity of experiments done on alkaline phosphatase illustrate some of the great potential of ^{31}P NMR in the study of phosphoproteins. Alkaline phosphatase is a Zn(II)-containing dimeric protein of MW 86,000 that can exist in two forms—a noncovalent protein–phosphate complex, designated as E · P, at high pH values (pH >6.5), and a covalently phosphorylated protein, designated as E-P, at low pH values (pH <5.3). (This discussion will focus on the E-P form.) Both forms could be distinguished by ^{31}P NMR, with a resonance for E · P at +4.2 ppm[37] and a resonance for E-P at +9.3 ppm.[53,88,89] In the pH range 5.3 to 6.5, both resonances could be observed with no exchange broadening, indicative of the slow exchange limit for the two forms (exchange rate <10^3 sec^{-1}). From chemical studies E-P is known to involve a phosphoserine residue (residue 99) and the ^{31}P resonance for the E-P form was shown to narrow by about 7 Hz upon proton decoupling, consistent with the proton–phosphorus coupling of a phosphoserine moiety. However, the chemical shift for the E-P form of +9.3 ppm is very atypical of a phosphoserine (+4.7 to +0.8 ppm) and was postulated to be the result of a strained phosphoserine residue.[53,88] A hydrogen bond between the phosphate and a group on the protein could mimic a strained cyclic ester. A concomitant 2° narrowing of the O—P—O bond angle might then account for the observed chemical shift of +9.3 ppm.[90,91] The resonance did not titrate with pH,[88,89] indicating that the phosphoserine residue is probably buried within the protein and likely stabilized in its dianionic form (judging from the downfield shift) by hydrogen bonding or electrostatic interactions with the protein.

The linewidth of E · P was found to vary with the ratio of enzyme to P_i,[89] indicating that the exchange between E · P and P_i was rapid enough to cause exchange broadening. The exchange rates were determined, and it was found that the dissociation of P_i from the E · P complex is the rate-limiting step for alkaline phosphatase under alkaline conditions. In contrast, the linewidth of the E-P form remained constant with changes in P_i concentration or pH. This was due to the very slow conversion of E-P to E · P and was consistent with this step being rate-limiting under acidic conditions. The observed linewidth for E-P at 40.5 MHz of 15–20 Hz

[88] J. L. Bock and B. Sheard, *Biochem. Biophys. Res. Commun.* **66**, 24 (1975).
[89] W. E. Hull, S. E. Halford, H. Gutfreund, and B. D. Sykes, *Biochemistry* **15**, 1547 (1976).
[90] D. G. Gorenstein, *J. Am. Chem. Soc.* **97**, 898 (1975).
[91] D. G. Gorenstein and K. Kar, *Biochem. Biophys. Res. Commun.* **65**, 1073 (1975).

included the $^3J_{POCH}$ splitting contribution. With proton decoupling, the linewidth for metal-free E-P was found to be 13.3 Hz at 36.4 MHz.[92]

The effects of proton decoupling on the ^{31}P NMR spectrum of E-P have been studied.[53,92,93] The effects of the NOE on the ^{31}P NMR spectrum of free phosphoserine and apo E-P (metal-ion free) were separated from those due to the collapse of the ^{31}P—^{1}H spin–spin coupling by the use of gated proton decoupling. The NOE for free phosphoserine in D_2O was found to be 0.83. For apo E-P in H_2O, the NOE enhancement was 0.22 and for apo E-P in D_2O, the NOE was 0.12.[93] The T_1 values for free phosphoserine in D_2O, apo E-P in H_2O, and apo E-P in D_2O were 19.0, 1.5, and 8.3 sec, respectively. If the phosphoryl group was firmly bound within the protein, and only the two nonexchangeable β-methylene protons of the phosphoryl residue contributed to DD relaxation, one would expect an NOE enhancement of 0.0 and T_1 of 20–30 sec. The large increase in T_1 from 1.5 to 8.3 sec in changing the solvent from H_2O to D_2O indicates that >80% of the DD relaxation in H_2O is due to the presence of exchangeable protons near the ^{31}P nucleus. Even assuming a full complement of protons hydrogen-bonded to the phosphate, internal rotation or proton exchange with a correlation time of about 1 nsec must be present to account for the rapid $1/T_1$DD and NOE of the ^{31}P nucleus in H_2O.[93]

Valuable information about the phosphoserine residue of alkaline phosphatase could be obtained from proton decoupling and NOE experiments, and additional information could be obtained from the measured $^3J_{POCH}$ values, where the proton–phosphorus spin–spin interactions are not disrupted. The $^3J_{POCH}$ coupling constants were found to be 6.1, 7.0, and 13.0 Hz for free phosphoserine, for the Cd(II) · E-P complex and for metal-free E-P, respectively.[92] The $^3J_{POCH}$ values vary with the tortional angle between the nuclei, with couplings of 3 ± 2 Hz and 25 ± 3 Hz determined for the gauche (^{31}P–^{1}H dihedral angle of 60°) and trans (^{31}P–^{1}H angle of 180°) rotamers, respectively. Thus, the $^3J_{POCH}$ of 6.1 Hz for free phosphoserine indicates that the gauche rotamer is dominant in free solution. However, the phosphoserine residue in metal-free E-P exists to a large degree in the trans rotamer, since its $^3J_{POCH}$ value is 13.0 Hz.

The role of the bound metal ion in the structure of the E-P form of alkaline phosphatase was assessed by comparing the metal-bound and apo forms of E-P. The chemical shift of the ^{31}P resonance of $(Zn)_2$ · E-P,

[92] J. F. Chlebowski, I. M. Armitage, P. P. Tusa, and J. C. Coleman, *J. Biol. Chem.* **251**, 1207 (1976).

[93] J. E. Coleman, I. M. Armitage, J. F. Chlebowski, J. D. Otvos, and A. Viterkamp, *in* "Biological Applications of Magnetic Resonance" (R. G. Shulman, ed.), p. 345. Academic Press, New York, 1979.

$(Cd^{2+})_2 \cdot$ E-P, and apo-E-P did not change with pH[92] over a pH range in which free phosphoserine titrates fully.[88,89] This indicates that the phosphoserine residue of E-P must interact with the protein to stabilize one ionization state (likely the dianionic form) over a wide pH range and these interactions do not require the presence of a divalent metal ion. However, the metal ion does seem to decrease the mobility of the phosphoserine residue, since the linewidth of the ^{31}P resonance of apo E-P (13.3 Hz) was increased upon binding of two Cd^{2+} ions per dimer [19.9 Hz for $(Cd^{2+})_2 \cdot$ E-P]. T_1 values for apo E-P were found to be 1.5 sec[92] and 2.3 sec for a Cd^{2+} and Mg^{2+} complex of E-P,[94] again consistent with the role of the metal in making the phosphoserine residue more rigid. Paramagnetic Mn^{2+} or Co^{2+} caused extensive broadening of the ^{31}P resonance of noncovalently bound P_i (E $\cdot$ P) indicating that the metal bound at the catalytic site and the phosphorus must be reasonably close together within the protein.[88,89,92,95] Direct evidence that the noncovalently bound phosphate (E $\cdot$ P) is involved in first sphere coordination with the bound metal (Cd^{2+}) was obtained from ^{31}P NMR studies of E $\cdot$ P bound with enriched $^{113}Cd^{2+}$.[96] A $^{113}Cd^{2+}$–^{31}P spin–spin coupling constant of 30 Hz could be clearly observed. No coupling could be seen for the E-P form, indicating that the covalently bound serine phosphate must be somewhat farther away. The chemical shift of the ^{31}P resonance of apo E-P was shifted 2 ppm upfield (to +7 ppm) relative to the chemical shift of metal ion-bound E-P. This shift is in the direction of the chemical shift of free phosphoserine, indicating that the phosphoserine in an apo E-P is in a somewhat strained conformation and this strain is increased upon addition of metal ion (O–P–O tetrahedral angle of 100–102°). The strain in the phosphoserine residue has been implicated in the susceptibility of this phosphoserine to hydrolysis and the role of this susceptibility in the enzymatic mechanism of alkaline phosphatase has been considered.[53,92]

Glycogen Phosphorylase. The classic example of the role of phosphorylation in metabolic control is the activation of glycogen phosphorylase *b* to phosphorylase *a* via phosphorylation of serine-14.[97] Investigation using ^{31}P NMR to study the phosphorylase system has concentrated on the role of the active-site pyridoxal 5′-phosphate (PLP) in the enzymatic mecha-

[94] J. D. Otvos, I. M. Armitage, J. F. Chlebowski, and J. E. Coleman, *J. Biol. Chem.* **254,** 4707 (1979).
[95] R. E. Weiner, J. F. Chlebowski, P. H. Haffner, and J. E. Coleman, *J. Biol. Chem.* **254,** 9739 (1979).
[96] J. D. Otvos, J. R. Alger, J. E. Coleman, and I. M. Armitage, *J. Biol. Chem.* **254,** 1778 (1979).
[97] R. J. Fletterick and N. B. Madsen, *Annu. Rev. Biochem.* **49,** 31 (1980).

nism of action of phosphorylase b[98–103]; however, some ^{31}P NMR studies of the regulatory phosphoserine residue (or derivatives of it) of phosphorylase a have been performed. It was recognized quite early that phosphorylation of the serine-14 did alter dramatically the ^{31}P NMR resonances of the PLP.[102]

A major complication in studying the phosphoserine resonance of phosphorylase a is that this ^{31}P resonance is obscured by the resonances of the active-site PLP that is necessary for enzymatic activity.[102,103] One method that allowed simultaneous study of both the phosphoserine and the PLP moieties by ^{31}P NMR involved formation of a thiophosphoserine residue at serine-14 by phosphorylation with ATPγS.[104] This derivative had its ^{31}P resonance at +42 to +44 ppm, well away from the other resonances of phosphorylase a.[103] In the presence of the activating allosteric effector AMPS (adenosine 5'-O-thiomonophosphate, a thio analog of AMP), a single thiophosphoseryl resonance at +42.2 ppm was observed. This resonance was associated with the active form of the enzyme (R) and the active form of PLP (PLP resonance III). In the absence of AMPS, two resonances for the thiophosphoserine residue were apparent, a broad one at +42.2 ppm and the other, a narrow resonance at +44.2 ppm. This latter resonance dominated under conditions of high pH or upon succinylation of the protein, both conditions being associated with the inactive form of the enzyme (T) and the inactive form of PLP (PLP resonance I).[105] Similar results were obtained by other workers in that one resonance for the thiophosphoseryl residue occurred at +42.5 ppm (linewidth 80 Hz) in the presence of the activators glucose 1,2-cyclic phosphate and AMPS, and another resonance at +43.4 ppm (linewidth 60 Hz) occurred in the presence of the inhibitor glucose.[100] It has been postulated[105] that the broader upfield resonance represents a "bound" dianionic thiophosphoserine-14 involved in a salt bridge with Arg-69 on the same subunit and Arg-43 on the other subunit, which results in fixing the N-terminal region to the main core of the protein. This resonance is in

[98] H. W. Klein and E. J. M. Helmreich, *FEBS Lett.* **108**, 209 (1979).
[99] S. G. Withers, B. D. Sykes, N. B. Madsen, and P. J. Kasvinsky, *Biochemistry* **24**, 5342 (1979).
[100] S. G. Withers, N. B. Madsen, and B. D. Sykes, *Biochemistry* **20**, 1748 (1981).
[101] S. G. Withers, N. B. Madsen, B. D. Sykes, M. Takagi, S. Shimomura, and T. Fukui, *J. Biol. Chem.* **256**, 10759 (1981).
[102] S. J. W. Busby, D. G. Gadian, G. K. Radda, R. E. Richards, and P. J. Seeley, *FEBS Lett.* **55**, 14 (1975).
[103] K. Feldmann and W. E. Hull, *Proc. Natl. Acad. Sci. U.S.A.* **74**, 856 (1977).
[104] D. Gratecos and E. H. Fischer, *Biochem. Biophys. Res. Commun.* **58**, 960 (1974).
[105] M. Hoerl, K. Feldmann, K. D. Schnackerz, and E. J. M. Helmreich, *Biochemistry* **18**, 2457 (1979).

slow exchange with the narrower, downfield resonance representing a more mobile thiophosphoserine, which does not form this salt bridge. This proposed conformational change involving the phosphoserine group and the fixing of the N-terminal segment of phosphorylase a in activating the enzyme has also been demonstrated from crystallographic studies in the presence of activators and inhibitors.[97,106]

The ³¹P resonances of unmodified phosphoserine could be observed directly by using a derivative of PLP, pyridoxal 5'-deoxymethylene phosphonate, which shifted the PLP resonance away from those of the phosphoserine resonances.[105] Here also, two separate resonances could be seen for the phosphoserine residue—one at +3.76 ppm for the narrow (presumably more mobile) resonance and one at +4.58 ppm for the broad (presumably less mobile) resonance. The addition of the inhibitor glucose appeared to increase the intensity of the narrower "free" phosphoserine resonance, and decrease that of the broader "bound" form consistent with the two forms of thiophosphoserine just described. (Glucose is known to increase the susceptibility of the phosphoserine residue to phosphorylase phosphatase degradation, presumably by making the residue more exposed.[97])

Magnetic field dependence studies of phosphorylase a have been conducted to determine the relative contributions of CSA and DD toward the linewidth of the phosphoserine resonance.[107] The CSA contribution depends on two variables, the correlation time of the nucleus and the anisotropy term $(\Delta\sigma)(1 + \eta^2/3)^{1/2}$. An anisotropy term $(\Delta\sigma)(1 + \eta^2/3)^{1/2}$ of 220 ppm was determined, assuming no internal mobility for the phosphoserine, although this is an overestimate since two phosphoserine resonances do exist for phosphorylase a.[105] The extrapolation of linewidth to zero magnetic field strength gives the dipolar contribution to the linewidth; this dipolar contribution was determined to be about 6 Hz. This value was judged to be too small to allow for the presence of one proton on the phosphate. Thus, the phosphoserine residue was presumed to be in a dianionic form in phosphorylase a.

³¹P NMR Studies of Phosphohistidyl Proteins

The phosphorylation of histidine residues on proteins has become the focus of several ³¹P NMR studies. Studies of phosphohistidyl proteins have been complicated by the acid-labile nature of the P—N bond involved.

[106] N. B. Madsen, P. J. Kasvinsky, and R. J. Fletterick, *J. Biol. Chem.* **253,** 9097 (1978).
[107] H. J. Vogel, W. A. Bridger, and B. D. Sykes, *Biochemistry* **21,** 1126 (1982).

The phosphorylation of histidine can lead to three possible derivatives, 1-phospho-, 3-phospho-, and 1,3-diphosphohistidine. The ^{31}P NMR resonances for each derivative were observed,[108] and their chemical shifts were monitored as a function of pH to determine the pK_a of each (see Table I). These model studies were applied to the ^{31}P NMR studies of the *Staphylococcus aureus* phosphocarrier protein HPr phosphorylated enzymatically (via the phosphoenolpyruvate-dependent phosphotransferase system) or chemically (via phosphoamidate).[108] From comparisons of the chemical shifts and pK_a's of denatured HPr with those of the free phosphohistidines, it was concluded that enzymatic phosphorylation produced a 1-phosphohistidyl residue and chemical phosphorylation resulted in a 3-phosphohistidyl HPr derivative. This distinction is important, as the 1-phosphohistidyl HPr was an active phosphorylated enzyme intermediate whereas the 3-phosphohistidyl HPr was inactive. From ^{1}H and ^{31}P NMR studies, the pK_a values of the histidine residue of HPr, enzymatically phosphorylated HPr, and chemically phosphorylated HPr were 6.0, 8.3, and 6.9, respectively.

Similar studies were done with *E. coli* HPr.[109] There again, enzymatic phosphorylation was shown to occur at the N-1 position of the imidazole ring, on the basis of the chemical shift of the denatured phospho-HPr. The pK_a of the histidine residue (His A) was increased from 5.6 (determined via ^{1}H NMR) to 7.8 (determined via ^{1}H NMR and ^{31}P NMR) upon phosphorylation, similar to the results with *S. aureus* HPr phosphorylated enzymatically.[108] However, this change in pK_a upon phosphorylation was interpreted to be due to a conformational change in the HPr[109] rather than simply the local effect of the phosphate covalently bound to the imidazole ring. The HPr is postulated to be in one conformation (I) in which the histidine is deprotonated at neutral pH (pK_a 5.6), making the imidazole ring a good nucleophile to accept phosphate from phosphorylated E$_I$, the first phosphoenzyme in this phosphotransferase system. The other conformation (II) of HPr has the histidine residue protonated at neutral pH (pK_a 7.8), making the positively charged imidazole a good leaving group from the phosphate, i.e., making conformation II a good phosphate donor. The phosphorylation of the histidine residue (uncharged for conformation I of HPr) favors conformation II, in which the histidine will be positively charged. Thus, the phosphorylation of HPr converts this protein from a potent phosphate acceptor (I) to a phosphate donor (II). This overall process results in the passage of the phosphate from phosphory-

[108] M. Gassner, D. Stehlik, O. Schrecker, W. Hengstenberg, W. Maurer, and H. Ruterjans, *Eur. J. Biochem.* **75**, 287 (1977).
[109] G. Dooijewaard, F. F. Roossien, and G. T. Robillard, *Biochemistry* **18**, 2996 (1979).

lated E_I to the next acceptor in this phosphotransferase system, either the F_{III} enzyme or hexose with the transport of this sugar into the cell.

The phosphorylation of histone H_4 was shown to yield different phosphohistidine derivatives, depending upon the source and type of kinase present.[110] When histone H_4 was enzymatically phosphorylated with extracts from rapidly growing cells of regenerating rat liver, a resonance at -4.5 ppm[37] was detected for the denatured protein. This is consistent with the presence of 1-phosphohistidine. If the phosphorylation was done with Walker 256 carcinosarcoma extract, a resonance at -4.1 ppm for the denatured phosphorylated protein indicated the presence of 3-phosphohistidine. Chemical phosphorylation with phosphoramidate also gave a phosphorylated histone H_4 with a prominent peak at -4.1 ppm when denatured, corresponding to 3-phosphohistidine. However, when the ³¹P NMR spectrum of phosphorylated H_4 was run under native rather than denaturing conditions, two major resonances were observed—one sharp resonance at -4.0 ppm (linewidth ~9 Hz) and one broad resonance at -7.5 ppm (linewidth ~55 Hz). Chymotrypic fragment 38–102, when similarly phosphorylated, had only a broad resonance at -6.2 ppm and fragment 1–23 exhibited only the sharp resonance at -4.0 ppm. Since the amino acid sequence for histone H_4 is known, and only two histidine residues are present, the broad resonance at -6.2 ppm could be assigned to His-75 and the sharp resonance at -4.0 ppm to His-18. The increased linewidth and unusual chemical shift of His-75 indicates that this residue is at least partially immobilized and interacting with several groups within the globular part of the spectrum, while His-18 is on the relatively mobile and exposed N-terminal region of the protein. Consistent with this concept, it was found that the ³¹P resonance of phosphorylated His-18 dephosphorylated more quickly than did that of His-75.

Recent ³¹P NMR studies of the 3-phosphohistidyl catalytic intermediate form of succinyl-CoA synthetase have been done to study the relaxation behavior of the ³¹P resonance and the effects of substrates and inhibitors on the phosphohistidine residue.[107,111] Using the CSA contribution,[107] which was dominant at high magnetic field strength, and assuming an anisotropy term $(\Delta\sigma)(1 + \eta^2/3)^{1/2}$ of 187 ppm (based on the model compound imidazole diphosphate), a rotational correlation time of 84 nsec was determined. This was almost twice the correlation time of the overall protein (44 nsec) and indicates that the phosphohistidine residue is probably immobile. From the DD contribution, it was suggested that the phos-

[110] J. M. Fujitaki, G. Fung, E. Y. Oh, and R. A. Smith, *Biochemistry* **20,** 3658 (1981). See also Fujitaki and Smith, this volume [2].
[111] H. J. Vogel and W. A. Bridger, *J. Biol. Chem.* **257,** 4834 (1982).

phate of the 3-phosphohistidine residue be monoanionic at pH 7.2. The effects of various substrates, activators, and inhibitors on the ^{31}P resonance of this active-site phosphohistidine were examined.[111]

^{31}P NMR Studies of Other Phosphorylated Proteins

One rather unusual phosphorylation site on a protein for which ^{31}P NMR spectra have been obtained is the phosphorylation of the β-carboxyl group of an active-site aspartyl residue on Na$^+$,K$^+$-ATPase and Ca^{2+}-ATPase.[112] A ^{31}P resonance with a chemical shift of about -17.5 ppm and $T_1 <0.2$ sec was observed in duck salt gland Na$^+$,K$^+$-ATPase preparations under conditions known to stabilize the phosphorylated intermediate (i.e., the presence of ATP + Mg^{2+} + Na$^+$, or P$_i$ + Mg^{2+}, or the inhibitor ouabain). This resonance decreased upon addition of K$^+$, known to lead to the hydrolysis of the enzyme-bound phosphate. A similar resonance at -17.5 ppm was found for rabbit skeletal muscle sarcoplasmic reticulum (Ca) ATPase in the presence of Ca^{2+}, Mg^{2+}, and ATP. The resonance could be eliminated by treating the enzyme with hydroxyalamine, known to hydrolyze acyl phosphates. The authors assign the resonance at -17.5 ppm to a phosphorylated β-carboxyl aspartyl group, since phosphorylation of the dipeptide L-seryl-L-aspartate yielded resonances at $+3.7$ and -17.4 ppm, presumably for phosphoserine and phosphoaspartate, respectively.

^{31}P NMR studies have shown the presence of a covalent phosphorus bound to a protein in a phosphodiester linkage.[113–115] A flavodoxin and glucose oxidase were found to have a ^{31}P resonance at $+1.5$ and -1.5 ppm,[37] respectively. Neither resonance titrated with pH either in the absence or the presence of protein denaturants such as 8 M urea or 3 M guanidine hydrochloride. The phosphodiester linkage for the flavodoxin has been shown to be between a serine and a threonine residue.[114] The ^{31}P resonance of the phosphodiester in glucose oxidase was found to be dramatically broadened by added Mn(III) free in solution, indicating that the ^{31}P nucleus was at or near the protein surface.

The ^{31}P NMR studies of phosphorylated proteins discussed here have illustrated several important uses of this method. The identity of the phosphorylated species can be readily determined from the chemical shift of the resonance and its pH titration behavior in 8 M urea, where the local

[112] E. T. Fossel, R. L. Post, D. S. O'Hara, and T. W. Smith, *Biochemistry* **20**, 7215 (1981).
[113] D. E. Edmondson and T. L. James, *Proc. Natl. Acad. Sci. U.S.A.* **76**, 3786 (1979).
[114] T. L. James, D. E. Edmondson, and M. Husain, *Biochemistry* **20**, 617 (1981).
[115] G. L. Gustafson and J. E. Gander, this volume [9].

environment within the intact protein cannot complicate the structural assignment. The chemical shift of the resonance in the native protein may indicate the steric stress on the phosphate exerted by the protein. The pH titration of the ^{31}P resonance in the native protein can give one information about whether the phosphoryl group is buried or interacting with the protein. Determination of the T_2, T_1, and NOE of the ^{31}P resonance can help determine the mobility of the group. The $^3J_{\mathrm{PXCH}}$ couplings in the absence of proton irradiation can help in the estimation of the dihedral angle between the phosphorus and the β-methylene protons of the phosphoserine. Metal binding to the phosphate can be evaluated and distances determined.

Acknowledgments

We are very grateful to Drs. L. Smillie and A. Mak for allowing us to use some of their unpublished results (Fig. 4). This work was supported by the Medical Research Council of Canada (Group on Protein Structure and Function) and the Alberta Heritage Fund for Medical Research (Postdoctoral Fellowship to M. B.).

[4] Protein Kinases

By PETER J. ROACH

General Properties

Reaction Catalyzed

Protein kinases catalyze the transfer of the γ-phosphate of a nucleoside triphosphate to form phosphomonoesters of amino acid residues of a substrate polypeptide. The modified amino acid residues are usually serine, threonine, or tyrosine but the existence of enzymes specific for other amino acids cannot be excluded.[1-4] The term "protein kinase" is not used to describe the formation of phosphorylated enzyme species that occur as catalytic intermediates. All protein kinases that have been characterized can use ATP as phosphoryl donor; K_m values of most but not all

[1] M. Weller, "Protein Phosphorylation." Pion, London, 1979.
[2] G. Taborsky, *Adv. Protein Chem.* **28,** 1 (1974).
[3] J. M. Fujitaki and R. A. Smith, this volume [2].
[4] T. M. Martensen, this volume [1].

protein kinases are below 100 μM. The only other significant phosphoryl donor known is GTP, which can be utilized by a minority of protein kinases, although it is not known if GTP is a physiologically significant substrate. As for other phosphoryl transferase enzymes, divalent metal ions are necessary for the reaction. Mg^{2+} is normally the preferred cation but Mn^{2+} can often substitute. The optimal cation concentration is usually in excess of that required for chelation to the nucleoside triphosphate. In the first instance, an assay system[5] for protein kinase activity would include ATP and Mg^{2+}, and would monitor either the transfer of the γ-phosphate of ATP to an appropriate substrate or else the modification of some measurable property of the protein substrate, often catalytic activity. Phosphorylation of sufficiently pure proteins can be determined by the incorporation of ^{32}P from $[\gamma\text{-}^{32}P]ATP$ into protein. With less pure systems, gel electrophoresis followed by autoradiography may be used to monitor the phosphorylation of specific proteins or polypeptides. It is not difficult to imagine, however, novel protein kinase-like enzymes whose activities would not register with such procedures.

Distribution and Isozymes

Protein kinases are probably present in all eukaryotic cells[1-4,6-9] although most is known of the enzymes of vertebrates. Evidence also exists for the presence of protein phosphorylation and protein kinases in bacteria.[10] Several animal viruses code for protein kinases,[11,12] as is true for at least one bacteriophage, bacteriophage T7.[13] Protein kinases are therefore widely distributed in nature. In cells of higher species, protein kinases are to be found in diverse cellular compartments. Enzymes have been described that are specifically localized in the cytoplasm, nucleus, mitochondria, chloroplast, microsomes, and the cell membrane. In addition, the same enzyme may be distributed among different compartments.

[5] Robert Roskoski, Jr., this series, Vol. 99, p. 3.

[6] C. S. Rubin and O. M. Rosen, *Annu. Rev. Biochem.* **44**, 831 (1975).

[7] E. G. Krebs and J. A. Beavo, *Annu. Rev. Biochem.* **48**, 923 (1979).

[8] P. B. Chock, S. G. Rhee, and E. R. Stadtman, *Annu. Rev. Biochem.* **49**, 813 (1980).

[9] D. A. Flockhart and J. D. Corbin, *CRC Crit. Rev. Biochem.* **12**, 133 (1982).

[10] J. Y. J. Wang and D. E. Koshland, *in* "Protein Phosphorylation" (O. M. Rosen and E. G. Krebs, eds.), p. 637. Cold Spring Harbor, 1981.

[11] T. Hunter and B. M. Sefton, *Mol. Aspects Cell. Regul.* **2**, 333 (1982).

[12] J. M. Bishop and H. Varmus, *in* "RNA Tumor Viruses" (R. Weiss, N. Teich, H. Varmus, and J. Coffin, eds.), p. 999. Cold Spring Harbor, 1982.

[13] H. J. Rahmsdorf, S. H. Pai, H. Ponta, P. Herrlich, R. Roskoski, Jr., M. Schweiger, and F. W. Studier, *Proc. Natl. Acad. Sci. U.S.A.* **71**, 586 (1974).

Many of the protein kinases that will be discussed are insufficiently characterized to judge whether differences in properties observed for the enzymes of different cellular compartments, tissues, and species reflect the presence of separate isozymes. However, it is clear that isozymes of protein kinases can exist, as exemplified by the two well-characterized enzymes cAMP-dependent protein kinase and phosphorylase kinase. Therefore, in future characterizations of other protein kinases the potential for isozymes with significantly different properties must be considered.

Regulation[6-9]

Protein kinases are subject to the same classes of control as other enzymes: ligand binding, covalent phosphorylation, and control of enzyme concentration. A number of specific examples are given later. The protein kinases for which control mechanisms are best documented nearly all involve allosteric ligand binding in some way.[9] Several protein kinases appear to be regulated by covalent phosphorylation. Several are activated by limited proteolysis but it is not clear whether this constitutes a physiologically relevant mechanism to increase protein kinase activity. By their nature, protein kinases allow for a type of control unique to enzymes that modify other proteins: effects on the substrate protein may be translated to an altered efficacy of the protein kinase.[14] Ligand binding to the substrate protein may alter protein kinase action.[15] Phosphorylation at one site on a protein can affect the phosphorylation of a separate site.[16,17] The potential also exists for regulation of the relative substrate specificity of protein kinases. However, there is no good example of radical alterations of the specificity of a protein kinase arising from modification of its activity by ligand binding, phosphorylation, or proteolysis.

Protein Substrates

Specificity

Specificity for protein substrates is central to the function of protein kinases. Assessment of the selectivity of a protein kinase, however, must

[14] J. Preiss and D. A. Walsh, in "Biology of Carbohydrates" (V. Ginsburg, ed.), Vol. 1, p. 199. Academic Press, New York, 1981.
[15] M. R. El-Maghrabi, T. H. Claus, and S. J. Pilkis, this series, Vol. 99, p. 212.
[16] C. Picton, J. Woodgett, B. Hemmings, and P. Cohen, FEBS Lett. 150, 191 (1982).
[17] A. A. DePaoli-Roach, Z. Ahmad, M. Camici, J. C. Lawrence, Jr., and P. J. Roach, J. Biol. Chem. 258, 10702 (1983).

be made at both chemical and biological levels. If one considers only the chemical determinants involved in substrate recognition, more stringent requirements would define a more specific protein kinase in a chemical sense. An enzyme judged selective by this criterion would be expected to phosphorylate few proteins out of a random selection. Recognition of physiologically pertinent substrates, however, not only requires the accessibility of enzyme and substrate in the cellular locale but may also involve evolved characteristics of the protein substrate. Thus, there is no a priori requirement that the *in vitro* and *in vivo* selectivities of a protein kinase coincide. Some protein kinases may have evolved to phosphorylate a single substrate whereas the role of others may reside precisely in the modification of multiple substrates. For several protein kinases, sequence studies of phosphorylation sites[18] and the use of synthetic peptide substrates[19,20] have allowed recognition features to be identified in terms of local amino acid sequences. Acidic or basic residues with a particular location relative to the modified residue are important determinants of the action of several protein kinases. Nonetheless, primary structure alone is not always sufficient to ensure recognition. For example, cAMP-dependent protein kinase does not phosphorylate serine-14 of muscle phosphorylase even though this residue has an apparently appropriate surrounding amino acid sequence.[21] Similarly, the same protein kinase does not recognize some serines in the native lysozyme molecule that it can phosphorylate if the lysozyme is denatured.[22]

Another aspect of protein kinase specificity is the phenomenon of autophosphorylation. Most protein kinases studied can catalyze the phosphorylation of one or more constituent subunits. Experimentally, one has to ascertain that apparent "autophosphorylation" is not due to the phosphorylation of a copurifying contaminant. Often, autophosphorylation is detected using purified enzymes at relatively high concentrations under conditions that are not physiological. In some, although not all, cases, autophosphorylation can affect protein kinase activity. It is difficult to judge whether autophosphorylation is a physiologically important occurrence or not, and of course this might vary with different protein kinases.

[18] L. Engström, P. Ekman, E. Humble, U. Ragnarsson, and Ö. Zetterqvist, this volume [6].
[19] D. B. Glass, this series, Vol. 99, p. 119.
[20] D. J. Graves, this series, Vol. 99, p. 268.
[21] G. M. Carlson, P. J. Bechtel, and D. J. Graves, *Adv. Enzymol.* **50,** 41 (1979).
[22] D. B. Bylund and E. G. Krebs, *J. Biol. Chem.* **250,** 6355 (1975).

Practical Considerations

Questions of specificity have several important practical implications. Specificity is relative and given sufficient protein kinase activity and a sensitive enough detection system (e.g., very high specific activity [γ-^{32}P]ATP), detectable but irrelevant phosphorylation of just about any protein can be obtained. Observation of a significant phosphorylation stoichiometry (i.e., modification of an appreciable proportion of substrate molecules) is an important step in judging whether a phosphorylation reaction is meaningful. It is, however, not always necessary to obtain a perfect integral stoichiometry. In addition, an acceptable rate of phosphorylation is desirable. Assessment of rates and stoichiometries of phosphorylation is obviously more difficult the less purified the system under study is. In judging rates of phosphorylation, it commonly happens that several substrates, physiological or not, can be compared (note the inherent problems if more than one protein kinase is present). Comparisons of substrate specificity are often helpful in characterizing a protein kinase. Screening of protein substrates under a fixed set of reaction conditions, and in particular at a fixed substrate concentration, can reveal gross specificity differences among substrates. However, more complete kinetic studies are necessary for definitive statements about substrate specificity.

A protein kinase assay is only as well defined as the substrate used. Where one is forced to use a relatively impure system, with assay of substrate activity or phosphoproteins separated by gel electrophoresis, numerous factors besides the protein kinase can influence phosphorylation. Even with relatively purified substrates, caution must still be exercised. Highly purified protein preparations can still contain significant amounts of contaminating protein kinase and/or phosphatase activity, for example, and appropriate controls must be made. Another important factor is the level of endogenous phosphorylation of the substrate. If relevant phosphorylation sites are already modified to any extent, phosphorylation reactions will be correspondingly attenuated. An increasing number of proteins appear to contain multiple phosphorylation sites, a fact that can further complicate definition of the substrate. Although in some instances different sites are modified specifically by distinct protein kinases, in others a single protein kinase modifies multiple sites of the same substrate. In the latter case, the different sites are essentially separate substrates for the protein kinase and evaluation of kinetic parameters will be more difficult. If a property other than phosphorylation is monitored, enzyme activity for example, it may happen that individual phosphorylation sites differentially affect the property being measured. Examples are also known where the state of phosphorylation of one site can influence the phosphorylation of other sites.

Nomenclature

No truly universal nomenclature exists. The enzyme commission number EC 2.7.1.37 for protein kinase (ATP:protein phosphotransferases) cannot accurately represent the multiplicity of enzymes now known. Where specific effectors of protein kinases are recognized, the enzymes can be usefully named for the effector (e.g., cGMP-dependent protein kinase). Sometimes, however, multiple enzymes may be covered by such nomenclatures (e.g., calmodulin-dependent protein kinases). Often, the substrate allows a useful name (e.g., phosphorylase kinase), but even then the name becomes less precise if new substrates are discovered. In this area of study, where novel protein kinases and substrates are being uncovered with regularity, it is unavoidable that multiple, trivial designations be used. A case can in fact be made for some degree of trivial nomenclature during formative studies, and I would argue that names overly suggestive of function and/or identity with other investigations be employed only when the degree of certainty is high. Generic or operational names are often helpful (e.g., a heparin-sensitive protein kinase, a casein kinase, a tyrosine protein kinase).

Classifications of Protein Kinases

By Effectors

Cyclic AMP-Dependent Protein Kinase[6–9,21–24]

Cyclic AMP-dependent protein kinase, the best characterized protein kinase, exists in two major isozymal forms, type I and type II, that were first differentiated on the basis of elution from DEAE-cellulose.[25] Both types have a structure R_2C_2, of M_r 170,000–180,000, where R and C denote regulatory and catalytic subunits, respectively. The isozymes differ in distribution and in the R-subunit (M_r of 49,000 for R_{II} and 54,000–56,000 for R_I). Both contain a catalytic subunit, of M_r 38,000–42,000, with very similar properties. The R_{II} subunit undergoes phosphorylation

[23] P. Cohen, *Curr. Top. Cell. Regul.* **14**, 117 (1978).
[24] D. B. Glass and E. G. Krebs, *Annu. Rev. Pharmacol. Toxicol.* **20**, 363 (1980).
[25] J. D. Corbin, S. L. Keely, and C. R. Park, *J. Biol. Chem.* **250**, 218 (1975).

by the C-subunit[26] and also by casein kinase II[27,28] and F_A/GSK-3.[28] The free catalytic subunit is susceptible to autophosphorylation. Evidence for further subclasses of cAMP-dependent protein kinase has been presented.[29] Methods for the purification of the holoenzyme,[30] the C-subunit,[31] and R_I and R_{II} subunits[32] have been described.

The holoenzyme is inactive. Cyclic AMP binding to the R-subunit, which contains two binding sites, results in the subsequent dissociation of the enzyme to yield active monomeric C-subunits and the complex R_2-cAMP$_4$. Several other factors, notably ionic strength, the substrate histone, and the phosphorylation state of the R_{II}-subunit may affect the dissociation.[33] Several protein inhibitors[34] of cAMP-dependent protein kinase are known, of which the most important practically is the heat-stable protein described by Walsh *et al.*[35] This inhibitor, which interacts with the C-subunit, is a useful specific probe for the presence of cAMP-dependent protein kinase. The protease inhibitor, α-N-tosyl-L-lysine chloromethyl ketone (TLCK), also inactivates the protein kinase.

Cyclic AMP-dependent protein kinase is specific for ATP, with an apparent K_m in the range of 3–15 μM. In terms of protein substrates, it has broad specificity *in vitro*. Some substrates of physiological interest include phosphorylase kinase, glycogen synthase (sites 1a, 1b, and 2), hormone sensitive lipase, pyruvate kinase (liver), fructose-6-phosphate, 2-kinase, phosphatase inhibitor-1, ATP citrate-lyase, myosin light chain kinase, and phospholamban. The enzyme effectively phosphorylates histones but not phosvitin or caseins (except for the specific genetic variant β-casein B[36]). Serine, or occasionally threonine, is the amino acid residue phosphorylated and the local amino acid sequence appears important in substrate recognition: a pair of basic amino acids two or three residues N-

[26] J. Erlichman, R. Rosenfield, and O. M. Rosen, *J. Biol. Chem.* **249,** 5000 (1974).
[27] D. F. Carmichael, R. L. Geahlen, S. M. Allen, and E. G. Krebs, *J. Biol. Chem.* **257,** 10440 (1982).
[28] B. A. Hemmings, A. Aitken, P. Cohen, M. Rymond, and F. Hofmann, *Eur. J. Biochem.* **127,** 473 (1982).
[29] D. Sarkar, J. Ehrlichman, N. Fleischer, and C. S. Rubin, this series, Vol. 99, p. 187.
[30] J. A. Beavo, P. J. Bechtel, and E. G. Krebs, *Methods Enzymol.* **38,** 299 (1974).
[31] E. M. Reimann and R. A. Beham, this series, Vol. 99, p. 51.
[32] H. D. White, S. B. Smith, and E. G. Krebs, this series, Vol. 99, p. 162.
[33] J. D. Corbin, this series, Vol. 99, p. 227.
[34] S. Whitehouse and D. A. Walsh, this series, Vol. 99, p. 80.
[35] D. A. Walsh, C. D. Ashby, C. Gonzalez, D. Calkins, E. H. Fischer, and E. G. Krebs, *J. Biol. Chem.* **246,** 1977 (1971).
[36] B. E. Kemp, D. B. Bylund, T.-S. Huang, and E. G. Krebs, *Proc. Natl. Acad. Sci. U.S.A.* **72,** 3448 (1975).

terminal to the serine residue is characteristic of sites phosphorylated by this enzyme. Cyclic AMP-dependent protein kinase is perhaps the best documented example of a protein kinase involved in cellular regulation. It mediates, at least in part, the actions of a variety of extracellularly acting agents, mostly hormones, that can regulate the cellular cAMP level.

Cyclic GMP-Dependent Protein Kinase[7,9,24,37]

Cyclic GMP-dependent protein kinase is present at much lower levels than the cyclic AMP-dependent counterpart. The enzyme has been purified to homogeneity[38–41] from heart and lung. The native enzyme, M_r 175,000–180,000, is composed of two identical subunits, of M_r 74,000–82,000, linked by interchain disulfide bridges. Each subunit contains both a catalytic site and a regulatory site. Activation, half-maximal at 0.02–0.05 μM cyclic GMP, involves an allosteric mechanism in which cyclic GMP binds to the regulatory site. The enzyme is specific for ATP with apparent K_m in the range of 10–20 μM. The specificity for protein substrates is very similar, although not identical, to that of cAMP-dependent protein kinase although rates of phosphorylation by the cGMP-dependent enzyme are usually significantly slower. There are instances in which cGMP-dependent protein kinase is the more effective as exemplified by the phosphorylation[42,43] of the R_I-subunit of cAMP-dependent protein kinase (which does not undergo autophosphorylation). The subunit of cGMP-dependent protein kinase is subject to autophosphorylation. Unequivocal physiological substrates for the protein kinase await definition and the role of the enzyme is still not clearly understood. It has been proposed that cAMP- and cGMP-dependent protein kinases are closely related structurally.[44,45] The superficially different mechanisms of cyclic nucleotide activation can be rationalized when the cGMP-dependent enzyme is viewed as containing separate catalytic and regulatory domains, analogous to the R- and C-subunits but located in a single polypeptide chain. In this context, it is interesting that an insect protein kinase[46] has

[37] G. N. Gill and R. W. McCune, *Curr. Top. Cell. Regul.* **15,** 1 (1979).
[38] T. M. Lincoln, W. L. Dills, Jr., and J. D. Corbin, *J. Biol. Chem.* **252,** 4269 (1977).
[39] G. N. Gill, K. E. Holdy, G. M. Walton, and C. B. Kanstein, *Proc. Natl. Acad. Sci. U.S.A.* **73,** 3918 (1976).
[40] V. Flockerzi, N. Speichermann, and F. Hofmann, *J. Biol. Chem.* **253,** 3397 (1978).
[41] T. M. Lincoln, this series, Vol. 99, p. 62.
[42] R. L. Geahlen and E. G. Krebs, *J. Biol. Chem.* **255,** 9375 (1980).
[43] R. L. Geahlen and E. G. Krebs, *J. Biol. Chem.* **255,** 1164 (1980).
[44] T. M. Lincoln and J. D. Corbin, *Proc. Natl. Acad. Sci. U.S.A.* **74,** 3239 (1977).
[45] G. N. Gill, *J. Cyclic Nucleotide Res.* **3,** 153 (1977).
[46] A. Vardanis, this series, Vol. 99, p. 71.

been described that can be activated, with approximately equal sensitivities, by either cAMP or cGMP.

Ca^{2+}-Activated Protein Kinases

Phosphorylase Kinase.[7-9,14,21,23,47] The enzyme from skeletal muscle[48] has M_r 1.2–1.3 × 10^6 and is composed of four distinct subunits arranged in the stoichiometry $(\alpha\beta\gamma\delta)_4$. Subunit molecular weights are approximately 145,000 (α), 128,000 (β), 42,000 (γ), and 17,000 (δ). The δ-subunit is identical to the Ca^{2+}-binding protein calmodulin.[49] Recent studies suggest a similarity of structure for cardiac[50] and liver[51] phosphorylase kinase. Isozymes,[52] characteristic of white or of red and cardiac muscle, differ in the M_r of the α-subunit. Identification of the catalytic subunit has been long debated but there is good evidence that the γ-subunit exhibits catalytic activity.[53,54] Intact phosphorylase kinase is almost inactive in the absence of Ca^{2+} ions which bind to the δ-subunit (calmodulin). In the presence of Ca^{2+}, added calmodulin can elicit a further, lesser activation via a separate interaction. Phenothiazine drugs, such as trifluoperazine, block the effects of added calmodulin but do not inhibit the basic Ca^{2+}-stimulated activity. The α- and β-subunits can be phosphorylated by cAMP-dependent protein kinase, cGMP-dependent protein kinase, and by autophosphorylation. The phosphorylation correlates with activation of the enzyme. Limited proteolysis, such as by trypsin, degrades the α- and β-subunits with concomitant activation. Proteolysis also eliminates interaction with added calmodulin. As purified, muscle phosphorylase kinase displays a low pH activity ratio of 6.8 : 8.2 in the presence of Ca^{2+} and the activation by phosphorylation or proteolysis is best observed as an increase in this ratio. Glycogen[55] and heparin[56] activate phosphorylase kinase.

Phosphorylase kinase is specific for ATP with apparent K_m of approximately 200 μM. Two substrates are glycogen phosphorylase and glycogen

[47] P. J. Roach, *Curr. Top. Cell. Regul.* **20**, 45 (1981).

[48] P. Cohen, this series, Vol. 99, p. 243.

[49] R. J. A. Grand, S. Shenolikar, and P. Cohen, *Eur. J. Biochem.* **113**, 359 (1981).

[50] H. S. Sul, B. Dirden, K. L. Angelos, P. Hallenbeck, and D. A. Walsh, this series, Vol. 99, p. 250.

[51] T. D. Chrisman, J. E. Jordan, and J. H. Exton, *J. Biol. Chem.* **257**, 10798 (1982).

[52] H. P. Jennissen and L. M. Heilmeyer, *FEBS Lett.* **42**, 77 (1974).

[53] J. R. Skuster, K. F. J. Cahn, and D. J. Graves, *J. Biol. Chem.* **255**, 2203 (1980).

[54] K.-F. Jesse Chan and D. J. Graves, this series, Vol. 99, p. 259.

[55] R. J. DeLange, R. T. Kemp, W. E. Riley, R. A. Cooper, and E. G. Krebs, *J. Biol. Chem.* **243**, 2200 (1968).

[56] T. D. Chrisman, J. E. Jordan, and J. H. Exton, *J. Biol. Chem.* **256**, 12981 (1981).

synthase (site 2). The phosphorylation of phosphorylase is almost certainly involved in Ca^{2+} and cAMP-mediated mechanisms for stimulating glycogenolysis.

Myosin Light Chain Kinase.[57,58] The enzyme contains a single catalytic subunit, M_r 90,000–155,000, that expresses activity only when bound to calmodulin by an interaction that requires the presence of Ca^{2+}. Trifluoperazine totally inhibits the enzyme. Myosin light kinase has been purified from several mammalian tissues.[59,60] The catalytic subunit of myosin light chain kinases can be phosphorylated by cAMP-dependent protein kinase.[61,62] A consequence is to weaken calmodulin binding so that the enzyme is inactivated at nonsaturating calmodulin concentrations. The enzyme has apparent K_m for ATP of 50–300 μM. The most important substrate is myosin light chain. It is important to note that the isolated smooth muscle 20,000-dalton light chain is a substrate for numerous protein kinases[63] and its phosphorylation cannot be viewed, without other information, as a reliable discriminator for myosin light chain kinase. The phosphorylation of myosin light chains may have an important role in the process of muscle contraction[57,58,64] and the case is particularly strong for smooth muscle in which phosphorylation of the 20,000-dalton light chain is necessary for the activation of myosin ATPase activity by actin.

Calmodulin-Dependent Glycogen Synthase Kinase.[65,66] The enzyme from rabbit liver has been best characterized, and has a native M_r of 275,000–500,000. It is composed of approximately equal proportions of subunits of M_r 51,000 and 53,000 that may be distinct polypeptides. An apparently analogous enzyme has been reported in skeletal muscle.[67] Activity is totally dependent on the presence of Ca^{2+} and calmodulin, like myosin light chain kinase. Similarly, total inhibition by trifluoperazine can

[57] M. P. Walsh and D. J. Hartshorne, *in* "Calcium and Cell Function" (W. Y. Cheung, ed.), Vol. 3, p. 223. Academic Press, New York, 1982.

[58] R. S. Adelstein and E. Eisenberg, *Annu. Rev. Biochem.* **49,** 921 (1980).

[59] R. S. Adelstein and C. B. Klee, *J. Biol. Chem.* **256,** 7501 (1981).

[60] M. P. Walsh, S. Hinkins, R. Dabrowska, and D. J. Hartshorne, this series, Vol. 99, p. 279.

[61] R. S. Adelstein, M. A. Conti, D. R. Hathaway, and C. B. Klee, *J. Biol. Chem.* **253,** 8347 (1978).

[62] D. R. Hathaway, C. R. Eaton, and R. S. Adelstein, *Nature (London)* **291,** 252 (1981).

[63] For example, five of five glycogen synthase kinases tested could phosphorylate the isolated, smooth muscle light chain (A. A. DePaoli-Roach, P. J. Roach, and D. R. Hathaway, unpublished).

[64] P. J. England, *Mol. Aspects Cell. Regul.* **1,** 153 (1980).

[65] M. E. Payne and T. R. Soderling, this series, Vol. 99, p. 299.

[66] Z. Ahmad, A. A. DePaoli-Roach, and P. J. Roach, *J. Biol. Chem.* **257,** 8348 (1982).

[67] J. R. Woodgett, N. K. Tonks, and P. Cohen, *FEBS Lett.* **148,** 5 (1982).

be achieved. The enzyme undergoes autophosphorylation. ATP is the preferred phosphoryl donor with apparent K_m of 27 μM. Besides muscle glycogen synthase (sites 1b and 2), isolated smooth muscle, but not cardiac muscle, myosin light chain is an effective substrate. The enzyme is, however, quite distinct from myosin light chain kinase. Caseins are moderate substrates. The calmodulin-dependent glycogen synthase kinase is a possible candidate to mediate Ca^{2+}-dependent inactivation of liver glycogen synthase.

Ca²⁺ and Phospholipid-Dependent Protein Kinase (Protein Kinase C).[68] The enzyme, although widely distributed, is best characterized from rat brain[69,70] primarily through the work of Nishizuka and colleagues. The protein kinase, which is distributed between soluble and membrane fractions of the cell, is composed of a single subunit of M_r 82,000. The enzyme requires the presence of both phospholipid (such as phosphatidylserine) and Ca^{2+} for activity. The additional presence of diacylglycerol markedly increases the sensitivity to both Ca^{2+} and phospholipid. The enzyme is unaffected by calmodulin even though it can be inhibited by trifluoperazine, probably via interaction with phospholipid. Diacylglycerol can be effectively substituted by tumor-promoting phorbol esters such as 12-*O*-tetradecanoylphorbol 13-acetate (TPA).[71] Proteolysis, such as by a Ca^{2+}-dependent protease, causes irreversible activation, generating a subunit of M_r 51,000 that no longer requires Ca^{2+} or phospholipid for activity. Histones or protamine can be used to assay for the protein kinase. The full range of physiological substrates has yet to be established. However, from studies of thrombin stimulation of platelets, a 40,000-dalton protein of unknown function has been identified as a probable substrate for protein kinase C both *in vitro* and *in vivo*. Because of its activation by diacyl glycerol, a product of phosphatidyl inositol breakdown, protein kinase C may be involved in mediating cellular stimulation by a number of agents that affect phospholipid turnover in the plasma membrane.[68]

Others. Several other, less extensively characterized, Ca^{2+}-activated protein kinases have been reported. Some may turn out to coincide with enzymes discussed above. Notable examples are calmodulin-dependent protein kinases that phosphorylate phospholamban,[72] tryptophan and

[68] Y. Nishizuka, *Trends Biochem. Sci.* **8**, 13 (1983).
[69] U. Kikkawa, R. Minakuchi, Y. Takai, and Y. Nishizuka, this series, Vol. 99, p. 288.
[70] U. Kikkawa, Y. Takai, R. Minakuchi, S. Inohara, and Y. Nishizuka, *J. Biol. Chem.* **257**, 13341 (1982).
[71] M. Castagna, Y. Takai, K. Kaibuchi, S. Sano, U. Kikkawa, and Y. Nishizuka, *J. Biol. Chem.* **257**, 7847 (1982).
[72] C. J. Le Peuch, J. Haiech, and J. G. Demaille, *Biochemistry* **18**, 5150 (1979).

tyrosine monooxygenases,[73] and protein I from brain membranes[74] and neurotubulin.[75]

By Substrates

Casein and Phosvitin Kinases[76]

Caseins and phosvitin have been used as nonphysiological monitors of protein kinase activity in many studies. It should be appreciated that many protein kinases display some casein and/or phosvitin kinase activity, the extent of which may be difficult to evaluate in crude systems. Note also that whole casein, frequently used as a substrate, actually consists of several distinct proteins. Two major casein/phosvitin kinase activities, however, have emerged as distinct, well-defined enzymes and will be denoted here casein kinase I and casein kinase II.[76] Both enzymes are widely distributed in higher organisms.

Casein kinase I[77] is a monomer with a subunit of M_r 30,000–40,000 that undergoes autophosphorylation. The enzyme has an apparent K_m for ATP of 10–20 μM and is ineffective in utilizing GTP as a phosphoryl donor. The enzyme has been considered to be unaffected by heparin although in one study of the rabbit liver enzyme significant heparin inhibition was observed.[78] Glycogen synthase phosphorylation is inhibited by glycogen.[78] Casein kinase I modifies only serine residues in caseins and phosvitin. A glutamic acid two residues N-terminal to the serine may be involved in site recognition. Proteins phosphorylated *in vitro* also include eukaryotic initiation factors eIF-4B and eIF-5, mRNP particles, spectrin, band 3 protein, nonhistone nuclear proteins, RNA polymerase, glycogen synthase, and phosphorylase kinase. It is unclear which are the important physiological substrates and the function of the enzyme is unknown.

Casein kinase II[79] probably has a structure ($\alpha_2\beta_2$) with M_r 130,000–180,000. The α-subunit has M_r 43,000 and the β-subunit 25,000. Species of M_r 35,000–40,000 are sometimes present and may represent proteolytic

[73] T. Yamauchi, H. Nakata, and H. Fujisawa, *J. Biol. Chem.* **256,** 5404 (1981).
[74] M. B. Kennedy and P. Greengaard, *Proc. Natl. Acad. Sci. U.S.A.* **78,** 1293 (1981).
[75] B. E. Burke and R. J. De Lorenzo, *Proc. Natl. Acad. Sci. U.S.A.* **78,** 991 (1981).
[76] G. M. Hathaway and J. A. Traugh, *Curr. Top. Cell. Regul.* **21,** 107 (1982).
[77] G. M. Hathaway, P. T. Tuazon, and J. A. Traugh, this series, Vol. 99, p. 308.
[78] Z. Ahmad, M. Camici, A. A. DePaoli-Roach, and P. J. Roach, *J. Biol. Chem.,* in press (1984).
[79] G. M. Hathaway and J. A. Traugh, this series, Vol. 99, p. 317.

degradation of the α-subunit. The β-subunit undergoes autophosphorylation. Casein kinase II is totally inhibited by low concentrations of heparin. Polyamines, especially spermine and spermidine, can cause activation of the enzyme under some conditions. Inhibition by 2,3-diphosphoglycerate has also been reported. ATP (apparent K_m 4–30 μM) as well as GTP (apparent K_m 7–40 μM) are both effective phosphoryl donors, a characteristic of this enzyme. Casein kinase II is able to modify both serine and threonine residues in caseins and phosvitin. From studies of casein phosphorylation, it appears that the enzyme recognizes highly acidic regions of a polypeptide, with glutamic or aspartic residues toward the C-terminus. Other substrates phosphorylated *in vitro* include eukaryotic initiation factors eIF-2, eIF-3, eIF-4B, eIF-5, mRNP particles, 60 S ribosomal proteins, spectrin, glycophorin, nonhistone nuclear proteins, RNA polymerase, troponin T, glycogen synthase (site 5), and the R_{II} subunit of cAMP-dependent protein kinase. As for casein kinase I, the significant physiological substrates for this enzyme are unknown.

The slime mold *Physarum polycephalum* contains a protein kinase[80,81] that phosphorylates ornithine decarboxylase in a reaction totally dependent on the presence of polyamines, spermine, and/or spermidine. Similar enzymes may exist in other eukaryotes. Putrescine antagonizes spermine/spermidine activation. It is suggested that protein kinase activity is associated with a polypeptide of M_r 26,000, and which can utilize ATP or GTP as phosphoryl donor. These properties, together with elution from phosphocellulose at high (0.7 M) NaCl concentration, make an interesting parallel with the casein kinase II enzyme, although the precise degree of structural similarity remains to be elucidated. Polyamine-stimulated phosphorylation of ornithine decarboxylase has been proposed as part of a negative feedback control involving alterations in both ornithine decarboxylase activity and its actual rate of synthesis.[81]

The casein kinases described above are not responsible for the physiological phosphorylation of casein, and indeed modify caseins at sites different from those phosphorylated *in vivo*. A separate casein kinase,[82,83] associated with the Golgi apparatus and which has been partially purified from mammary gland, can phosphorylate caseins at the same sites phosphorylated *in vivo,* and is likely to be the physiological enzyme. The mammary gland casein kinase does not phosphorylate phosvitin.

[80] V. J. Atmar and G. D. Kuehn, this series, Vol. 99, p. 366.
[81] G. D. Kuehn and V. J. Atmar, *Fed. Proc. Fed. Am. Soc. Exp. Biol.* **41,** 3078 (1982).
[82] E. W. Bingham, H. M. Farrell, Jr., and J. J. Basch, *J. Biol. Chem.* **247,** 8193 (1972).
[83] A. G. Mackinlay, D. W. West, and W. Manson, *Eur. J. Biochem.* **76,** 233 (1977).

Glycogen Synthase Kinases[14,47,84]

Glycogen synthase is perhaps the best example of a metabolic enzyme subject to complex, multiple phosphorylation of its subunit. More than seven sites can be phosphorylated *in vitro* by at least seven different protein kinases. Five of these enzymes, cAMP-dependent protein kinase, phosphorylase kinase, calmodulin-dependent glycogen synthase kinase, casein kinase I, and casein kinase II ($PC_{0.7}$) were already discussed. Two other glycogen synthase kinases are known. One, $PC_{0.4}$ or glycogen synthase kinase-4, is relatively specific for glycogen synthase and phosphorylates site 2. It is not extensively characterized. The other enzyme, denoted here $F_A/GSK-3$[85] (also called F_A or glycogen synthase kinase-3), has been identified in liver and muscle, and is probably a monomer of M_r ~51,000. The enzyme has an apparent K_m for ATP of 25 μM and can also utilize, albeit rather poorly, GTP as a phosphoryl donor (apparent K_m of 0.4 mM). The enzyme phosphorylates a group of three serine residues (sites 3a, 3b, and 3c) located in a nine-amino-acid stretch of the glycogen synthase subunit. These phosphorylations are the most potent known as regards inactivation of glycogen synthase. Phosphorylation at site 3 is facilitated if glycogen synthase has already been phosphorylated by casein kinase II ($PC_{0.7}$) at site 5.[16,17] A distinguishing characteristic of $F_A/GSK-3$ is its ability to act as an activator (F_A activity) of an ATP–Mg^{2+}-dependent protein phosphatase.[86,87] The mechanism of activation may involve phosphorylation of a component (identifiable as phosphatase inhibitor-2) of the phosphatase.[88] $F_A/GSK-3$ also phosphorylates the R_{II} subunit of cAMP-dependent protein kinase.

Pyruvate Dehydrogenase and Branched Chain α-Ketoacid[89–91] Dehydrogenase Kinases

Pyruvate dehydrogenase and branched chain α-ketoacid dehydrogenase are both mitochondrial enzymes that exist as large multienzyme

[84] P. Cohen, *Nature (London)* **296,** 613 (1982).
[85] B. A. Hemmings and P. Cohen, this series, Vol. 99, p. 337.
[86] J. R. Vandenheede, S. D. Yang, J. Goris, and W. Merlevede, *J. Biol. Chem.* **255,** 11768 (1980).
[87] S. Shenolikar and T. S. Ingebritsen, this volume [5].
[88] B. A. Hemmings, T. J. Resink, and P. Cohen, *FEBS Lett.* **150,** 319 (1982).
[89] P. J. Randle, G. J. Sale, A. L. Kerbey, and A. Kearns, *in* "Protein Phosphorylation" (O. M. Rosen and E. G. Krebs, eds.), p. 687. Cold Spring Harbor, 1981.
[90] L. J. Reed and F. H. Pettit, *in* "Protein Phosphorylation" (O. M. Rosen and E. G. Krebs, eds.), p. 701. Cold Spring Harbor, 1981.
[91] M. S. Olsen, R. Scholz, C. Buffington, S. C. Dennis, A. Padma, T. B. Patel, P. Waymack, and M. S. DeBuysere, *in* "The Regulation of Carbohydrate Formation and Utilization in Mammals" (C. M. Veneziale, ed.), p. 153. University Park Press, Baltimore, 1981.

complexes, with M_r 7×10^6 and $\geq 2 \times 10^6$, respectively. Each contains three enzyme activities that together catalyze the oxidative decarboxylation of α-keto acids yielding the corresponding thioesters of CoA. Both complexes can undergo phosphorylation of the α-subunit of the decarboxylase component, leading to inactivation. The protein kinase copurifies with the complexes and these are interesting examples of a very close physical association of a protein kinase and its physiological substrate.

Pyruvate dehydrogenase kinase[92] from bovine heart has an M_r of 100,000 and is composed of two subunits of M_r 48,000 and 45,000. It modifies three serine residues in the α-subunit of the complex. The apparent K_m for ATP is 25 μM. Phosphorylation of pyruvate dehydrogenase is stimulated by acetyl CoA and NADH, and inhibited by pyruvate and ADP. The protein kinase is inhibited by dichloroacetate.[93] Less is known of the branched chain α-ketoacid dehydrogenase kinase[89,94,95] although certain properties parallel those of the pyruvate dehydrogenase kinase. The apparent K_m for ATP is, for example, 25 μM. The phosphorylation is inhibited by substrates for the decarboxylation, namely α-ketoisocaproate, α-ketoisovalerate, and α-keto-β-methylvalerate. The pyruvate dehydrogenase kinase and branched chain α-ketoacid dehydrogenase kinase are probably not identical, however. The ketoacid analog, α-chloroisocaproate, strongly inhibits the branched chain α-ketoacid dehydrogenase kinase, but is less effective in inhibiting pyruvate dehydrogenase phosphorylation.[96] Dichloroacetate, in contrast, is a more effective inhibitor of pyruvate dehydrogenase kinase. Phosphorylation of these two multienzyme complexes is likely to be involved in metabolic and hormonal control mechanisms.

3-Hydroxy-3-methylglutaryl-CoA (HMG–CoA) Reductase Kinase[97,98]

HMG–CoA reductase, the enzyme that regulates the pathway of cholesterol biosynthesis, was proposed some years ago as an enzyme subject to a covalent phosphorylation–dephosphorylation cycle that controls enzyme activity. Phosphorylation, which causes inactivation, is catalyzed by a specific HMG–CoA reductase kinase. Separate forms of this protein kinase have been purified from microsomes[98] and cytosol.[99] The microsomal enzyme, of M_r 360,000, contains a subunit of M_r 58,000. This

[92] F. H. Pettit, S. J. Yeaman, and L. J. Reed, this series, Vol. 99, p. 331.
[93] S. Whitehouse, R. H. Cooper, and P. J. Randle, *Biochem. J.* **141**, 761 (1974).
[94] R. Odessey, *Biochem. J.* **204**, 353 (1982).
[95] R. Paxton and R. A. Harris, *J. Biol. Chem.* **257**, 14433 (1982).
[96] R. A. Harris, R. Paxton, and A. A. DePaoli-Roach, *J. Biol. Chem.* **257**, 13915 (1982).
[97] T. S. Ingebritsen and D. M. Gibson, *Molec. Aspects Cell. Regul.* **1**, 63 (1980).
[98] Z. H. Beg and H. B. Brewer, Jr., *Curr. Top. Cell. Regul.* **20**, 139 (1980).
[99] H. J. Harwood, K. G. Brandt, and V. W. Rodwell, *J. Biol. Chem.*, in press (1984).

HMG–CoA reductase kinase itself is subject to phosphorylation, leading to activation, by a separate protein kinase. The "reductase kinase kinase" activity has been identified in both soluble and microsomal fractions of liver but its characteristics are not well defined. The phosphorylation cascade for HMG–CoA reductase may mediate short-term effects on the rate of cholesterol biosynthesis by hormones such as insulin and glucagon.

By Other Classifications

Protein Synthesis

A number of ribosomal proteins have been reported to undergo phosphorylation.[100] In general, unequivocal functional correlates are lacking. Protein kinases in the casein kinase category and cAMP-dependent protein kinase have been implicated in such phosphorylation. Several initiation factors of eukaryotic protein synthesis can be phosphorylated (see Casein and Phosvitin Kinases). Most is known of eukaryotic initiation factor 2 (eIF-2), which is composed of three distinct subunits, α, β, and γ, and which functions to promote binding of Met-tRNA$_f$ to the 40 S ribosomal subunit. Two eIF-2α kinases are known.[101-104] One, termed HCR (heme-controlled repressor) or HRI (heme-regulated inhibitor), may be linked to the well-established property of heme to overcome inhibition of protein synthesis in reticulocyte lysates. Phosphorylation of the α-subunit by HCR inactivates the eIF-2. Heme inhibits HCR, by an unresolved mechanism, thereby decreasing phosphorylation and inactivation of eIF-2. The structure of HCR is incompletely understood and the protein may exist in several forms. An autophosphorylated polypeptide of M_r approximately 90,000 may be associated with the protein kinase activity. The second eIF-2α kinase is an enzyme stimulated by low concentrations of double-stranded RNA.[105] Phosphorylation of eIF-2 by the double-stranded RNA-dependent protein kinase activity occurs at the same site(s) in the α-subunit modified by HCR and likewise inactivates eIF-2. The protein kinase activity may be associated with autophosphorylatable polypeptides of M_r 67,000. Particular interest in the enzyme rests in the suggestion that it can be induced by interferon and is involved in the

[100] D. P. Leader, *Mol. Aspects Cell. Regul.* **1,** 203 (1980).
[101] G. Kramer and B. Hardesty, *Curr. Top. Cell. Regul.* **20,** 185 (1981).
[102] R. Jagus, D. Crouch, A. Konieczny, and B. Safer, *Curr. Top. Cell. Regul.* **21,** 35 (1982).
[103] T. Hunt, *Mol. Aspects Cell. Regul.* **1,** 175 (1980).
[104] S. Ochoa and C. de Haro, *Ann. Rev. Biochem.* **48,** 549 (1979).
[105] R. Petryshyn, D. H. Levin, and I. M. London, this series, Vol. 99, p. 346.

antiviral action of interferon. Phosphorylation of the β-subunit (nomenclature as in Ref. 101) of eIF-2 has been observed and the protein kinase involved can be identified as casein kinase II. Effects of phosphorylation at the β-subunit on eIF-2 activity have been harder to demonstrate but one report[101] does document an increased activity under certain conditions.

Nuclear Protein Kinases

Phosphorylation of a number of nuclear proteins has been observed and both cAMP-dependent and -independent protein kinase activities have been identified. Two protein kinases, with preferences for acidic protein substrates, have been distinguished,[106-108] and have sometimes been designated nuclear kinase I and nuclear kinase II.[106] By several criteria, these enzymes are identifiable with casein kinase I and casein kinase II, respectively, and phosphorylate certain of the nonhistone chromatin proteins.[109] Other potential substrates include RNA polymerases I[110] and II,[111] and poly(A) polymerase.[112] Histones also undergo phosphorylation and much interest has been directed at histone H1, which contains multiple phosphorylation sites.[113] One serine residue is specifically modified by cAMP-dependent protein kinase and, *in vivo*, its phosphorylation is promoted by hormonal stimulation such as by glucagon.[114] Phosphorylation of a different set of five or six serine and threonine residues correlates with cell growth and is likely to result from the action of a particular histone H1 protein kinase found only in growing cells. The enzyme has been partially characterized and has an M_r of 90,000.[115,116] A specific histone H3 protein kinase has also been described.[117] Despite the many appealing possibilities, precise mechanisms for the role of protein phosphorylation in regulating nuclear processes are lacking.

[106] P. R. Desjardins, P. F. Lue, C. C. Liew, and A. G. Gornall, *Can. J. Biochem.* **50,** 1249 (1972).

[107] W. Thornburg, A. F. O'Malley, and T. J. Lindell, *J. Biol. Chem.* **253,** 4638 (1978).

[108] W. Thornburg and T. J. Lindell, *J. Biol. Chem.* **252,** 6660 (1977).

[109] J. L. Christmann and M. E. Dahmus, *J. Biol. Chem.* **256,** 3326 (1981).

[110] B. W. Duceman, K. M. Rose, and S. T. Jacob, *J. Biol. Chem.* **256,** 10755 (1981).

[111] M. E. Dahmus, *J. Biol. Chem.* **256,** 3332 (1981).

[112] K. M. Rose and S. T. Jacob, *Biochemistry* **19,** 1472 (1980).

[113] T. A. Langan, C. Zeilig, and B. Leichtling, *in* "Protein Phosphorylation" (O. M. Rosen and E. G. Krebs, eds.), p. 1039. Cold Spring Harbor, 1981.

[114] T. A. Langan, *Proc. Natl. Acad. Sci. U.S.A.* **64,** 1276 (1969).

[115] R. S. Lake, *J. Cell Biol.* **58,** 317 (1973).

[116] J. Schlepper and R. Knippers, *Eur. J. Biochem.* **60,** 209 (1975).

[117] C. B. Shoemaker and R. Chalkley, *J. Biol. Chem.* **255,** 11048 (1980).

Tyrosine-Specific Protein Kinases (Tyrosine Protein Kinases)

A class of protein kinases, distinguished by an exclusive specificity for tyrosine residues in substrate proteins, has been recognized only recently, and enzymological definition of its members is incomplete. Two broad categories will be formed: tyrosine protein kinases associated with RNA tumor viruses and tyrosine protein kinases stimulated by growth factors.

Some half-dozen different retroviruses encode tyrosine protein kinases.[11,12] Most is known of the *src* gene product, pp60^{src}, of Rous sarcoma virus (RSV).[118] Expression of the *src* gene is essential for cell transformation by the virus. The protein pp60^{src} is a tyrosine-specific protein kinase of M_r 60,000 and is itself a phosphoprotein. It is phosphorylated at both a serine and a tyrosine residue, the latter being a potential site of autophosphorylation. At least a portion of pp60^{src} molecules is membrane associated and the membrane-binding domain may reside in the N-terminal region of the polypeptide. A C-terminal segment of the molecule bears the catalytic site. Normal cells also contain a gene closely related to the viral *src* gene and do express tyrosine protein kinase activity, albeit at low levels. Transformation correlates with a significant increase in tyrosine phosphorylation in proteins, presumably due to increased expression of the viral *src* gene. As a protein kinase, pp60^{src} is effective in utilizing both ATP and GTP, and Mn^{2+} can replace Mg^{2+}. Much current effort addresses identification of physiologically important substrates for pp60^{src} (and other tyrosine protein kinases).[119] One prominent substrate, exhibiting elevated tyrosine phosphorylation in RSV-transformed cells, is a 36,000-dalton membrane-associated polypeptide (p36) of unknown function. Named substrates for pp60^{src} include vinculin (a cytoskeletal protein located at adhesion plaques) and the glycolytic enzymes enolase, phosphoglycerate mutase, and lactate dehydrogenase. The functional importance of these phosphorylations is not clear.

The protein kinases encoded by the other sarcoma viruses[120] and by the Abelson murine leukemia virus[121] share several properties, besides tyrosine specificity, with pp60^{src}. All are phosphoproteins, containing at

---

[118] R. L. Erikson, *Curr. Top. Microbiol. Immunol.* **91**, 25 (1981).

[119] T. Hunter and J. A. Cooper, *Prog. Nucl. Acid Res. Mol. Biol.* **29**, 221 (1983).

[120] These viruses include Y73 (p90$^{gag-yes}$), Esh (p80$^{gag-yes}$), Fujinami (p140$^{gag-fps}$), PRCII (p105$^{gag-yes}$), UR-1 (p150$^{gag-fps}$), Snyder-Theilen feline (p85$^{gag-fes}$), Gardner-Arnstein feline (p95$^{gag-fes}$), and UR-2 (p68$^{gag-ros}$) sarcoma viruses. In parentheses is the designation for the corresponding tyrosine protein kinase which specifies also the polypeptide molecular weight.

[121] J. Y. J. Wang and D. Baltimore, this series, Vol. 99, p. 373.

least one phosphotyrosine and usually also phosphoserine. All can utilize Mn^{2+} as divalent cation. All contain sequences derived, at least in part, from cellular genes. In some cases, there is evidence for common substrates (e.g., p36, enolase, phosphoglycerate mutase, and lactate dehydrogenase). The amino acid sequences surrounding the phosphorylated tyrosine residue of the protein kinases themselves show considerable homology, and are characterized by acidic residues on the N-terminal side. Nonetheless, some of the protein kinase activities can be distinguished by biochemical criteria (e.g., ATP versus GTP use, Mg^{2+} versus Mn^{2+} use) and definitive classification of this group of enzymes should perhaps await further characterization.

The second group of tyrosine protein kinases share (1) an association with a membrane receptor and (2) increased activity upon binding of the corresponding agonist. The hypothesis is that the receptor itself displays protein kinase activity. Three examples are the receptors for epidermal growth factor (EGF),[122] insulin[123,124] (which has significant growth-promoting activity), and platelet-derived growth factor (PDGF).[125,126] In all cases, binding of the growth factor leads to increased phosphorylation, at tyrosine residues, of the receptor. EGF and PDGF action, in certain cell types, also leads to increased tyrosine phosphorylation in the same p36 protein affected by several of the RNA tumor viruses. Enzymological characterization of these receptor kinases is ongoing, and will allow the degree of similarity to the viral enzymes to be evaluated. The conceptual link between tyrosine phosphorylation and growth control is, of course, striking.

Evaluation of a Phosphorylation Reaction

The following brief outline contains some suggestions for characterizing a newly encountered phosphorylation reaction. (1) Establish that phosphorylation is indeed involved: incorporation of ^{32}P from $[\gamma\text{-}^{32}P]ATP$ into protein; presence of alkali-labile, acid-stable phosphate in "phosphorylated" substrate; presence of phosphoserine, phosphothreonine, or phosphotyrosine in acid hydrolysate of protein; removal of phosphate by

[122] S. Cohen, this series, Vol. 99, p. 379.
[123] M. Kasuga, F. A. Karlsson, and C. R. Kahn, *Science* **215**, 185 (1982).
[124] L. M. Petruzelli, S. Cianguli, C. J. Smith, M. H. Cobb, C. S. Rubin, and O. M. Rosen, *Proc. Natl. Acad. Sci. U.S.A.* **79**, 6792 (1982).
[125] B. Ek, B. Westermark, A. Wasteson, and C.-H. Heldin, *Nature (London)* **295**, 419 (1982).
[126] J. Nishimura, J. S. Huang, and T. F. Deuel, *Proc. Natl. Acad. Sci. U.S.A.* **79**, 4303 (1982).

SOME EXPERIMENTALLY USEFUL EFFECTORS OF PROTEIN KINASES[a]

Compound	Ancillary conditions	Effect[b]	Enzyme(s) affected
cAMP	—	↑	cAMP-dependent protein kinase (R_2C_2 not free C subunit)
Heat-stable inhibitor protein[c]	—	↓	cAMP-dependent protein kinase (free C subunit)
cGMP	—	↑	cGMP-dependent protein kinase
Ca^{2+}	No calmodulin[d] No phospholipid	↑	Phosphorylase kinase
Calmodulin	Ca^{2+}	↑	Phosphorylase kinase Myosin light chain kinase Calmodulin-dependent glycogen synthase kinase
Phospholipid (e.g., phosphatidylserine)	Ca^{2+}	↑	Protein kinase C
Diacylglycerol or phorbol esters (e.g., TPA)	Ca^{2+} and phospholipid[d]	↑	Protein kinase C
Trifluoperazine	Ca^{2+} and calmodulin	↓	All calmodulin-stimulated activities
Glycogen	—	↑ ↓	Phosphorylase kinase Casein kinase I
Heparin	—	↑ ↓ ↓	Phosphorylase kinase Casein kinase II ? Casein kinase I
Spermine and/or spermidine	—	↑[e]	Casein kinase II Ornithine decarboxylase kinase
Dichloroacetate	—	↓	PDH kinase[f] > BCKADH kinase
α-Chloroisocaproate	—	↓	BCKADH kinase > PDH kinase

[a] Note that with impure systems, compounds may have indirect effects on the phosphorylation of a substrate protein (e.g., effects on phosphatases; presence of protein kinase cascade).

[b] ↑, activation; ↓, inhibition.

[c] D. A. Walsh, C. D. Ashby, C. Gonsalez, D. Calkins, E. H. Fischer, and E. G. Krebs, J. Biol. Chem. **246,** 1977 (1971).

[d] Consider always the possible presence of endogenous levels of effector, in this case calmodulin or phospholipid.

[e] Polyamine effects on casein kinase II can be dependent on a variety of reaction conditions, including divalent and monovalent cation concentrations.

[f] PDH, pyruvate dehydrogenase; BCKADH, branched chain α-ketoacid dehydrogenase.

phosphatase. (2) Characterize protein kinase: even at a stage of impurity, test for possible effectors (see the table); note that enzyme could be desensitized to an effector by proteolysis (e.g., phosphorylase kinase to calmodulin) or dissociation (e.g., free catalytic subunit of cyclic AMP-dependent protein kinase is not stimulated by cyclic AMP); test for ATP versus GTP use; undertake further study involving usual characterizations of an enzyme, including molecular weight, subunit structure, and in this case also protein substrate specificity using available substrates; look for similarities to known protein kinases. (3) Characterize the phosphorylation reaction: stoichiometry of phosphorylation and number of sites modified; mapping of phosphopeptides; effects of phosphorylation on protein properties, such as enzyme kinetic properties; ultimate determination of amino acid sequence surrounding phosphorylation site(s). (4) Evaluate the possible physiological role: Can the phosphorylation occur rapidly enough? Does the phosphorylation occur *in vivo* or in whole cells? Does the phosphorylation state vary with relevant physiological stimuli or conditions? Is there correlation with expected changes in properties of the protein (e.g., enzyme activity)? The above list is neither complete nor feasible in every case, but it includes many of the immediate questions to be asked of a novel phosphorylation reaction.

Conclusion

The present article must not be considered an exhaustive survey of protein kinases, a virtually impossible task in this fast-expanding area. The intent has been to provide some feeling for the diversity of properties and function of the better characterized protein kinases that might aid, conceptually and practically, entry into the study of protein phosphorylation. The general references[1,2,6-9] may provide starting points for subjects not treated specifically. Also, in attempting to curtail the bibliography, I have omitted many primary articles in favor of reviews, and my apologies for any uneven coverage.

Acknowledgments

Thanks go to David Hathaway, Tony Hunter, Larry Jones, Tom Langan, and Ralph Paxton for discussions or for providing access to manuscripts. I am grateful to Peggy Smith for typing the manuscript. Work from the author's laboratory was supported in part by NIH Grants AM27221 and AM27240, and by the Showalter Foundation. The author is recipient of Research Career Development Award AM01089 from the NIH.

[5] Protein (Serine and Threonine) Phosphate Phosphatases

By S. Shenolikar and T. S. Ingebritsen

As early as 1938, it was known that the enzyme glycogen phosphorylase could exist in two forms, *a* and *b*, distinguishable by their requirement for the nucleotide 5-AMP for full expression of their activity.[1] Subsequently, an activity that converted phosphorylase *a* to *b* was discovered in skeletal muscle extracts.[2] Since this conversion was thought to involve the removal of a prosthetic group (i.e., pyridoxal phosphate) from phosphorylase *a*, it was termed the "prosthetic-group-removing enzyme" (PR-enzyme). In the mid-1950s, Sutherland and Wosilait made the important observation that the PR-enzyme was actually a protein phosphatase that catalyzed the release of inorganic phosphate from phosphorylase *a* to yield phosphorylase *b*.[3,4] Hence, ironically, phosphorylase phosphatase was discovered before the enzyme, phosphorylase kinase, that catalyzes the activation of phosphorylase *b* to *a*. Yet, although much is known about the structure and mechanism of action of phosphorylase kinase,[5–8] neither the structure of the native form of phosphorylase phosphatase nor the nature of its substrate recognition has yet been elucidated.

Friedman and Larner demonstrated a protein phosphatase converting glycogen synthase *b* to *a* form.[9] In 1968, the phosphorylase kinase phosphatase activity was discovered.[10] Subsequently, studies designed to fractionate protein phosphatases by ion-exchange chromatography using several phosphoprotein substrates identified two protein phosphatases in skeletal muscle extracts. One phosphatase dephosphorylated the β subunit of phosphorylase kinase faster than the α subunit whereas the second

[1] G. T. Cori, S. P. Colowick, and C. F. Cori, *J. Biol. Chem.* **123**, 381 (1938).

[2] G. T. Cori and C. F. Cori, *J. Biol. Chem.* **158**, 321 (1945).

[3] E. W. Sutherland and W. D. Wosilait, *Nature (London)* **175**, 169 (1955).

[4] W. D. Wosilait and E. W. Sutherland, *J. Biol. Chem.* **218**, 469 (1956).

[5] T. Hayakawa, J. P. Perkins, and E. G. Krebs, *Biochemistry* **12**, 574 (1973).

[6] P. Cohen, *Eur. J. Biochem.* **34**, 1 (1973).

[7] P. Cohen, A. Burchell, J. G. Foulkes, P. T. W. Cohen, T. C. Vanaman, and A. C. Nairn, *FEBS Lett.* **92**, 287 (1978).

[8] S. Shenolikar, P. T. W. Cohen, P. Cohen, A. C. Nairn, and S. V. Perry, *Eur. J. Biochem.* **100**, 329 (1979).

[9] D. G. Friedman and J. Larner, *Biochemistry* **2**, 669 (1963).

[10] W. D. Riley, R. J. Delange, G. E. Bratuold, and E. G. Krebs, *J. Biol. Chem.* **243**, 2209 (1968).

TABLE I
CLASSIFICATION OF PROTEIN PHOSPHATASES[14]

Protein phosphatase	Inhibition by inhibitors 1 and 2	Specificity for phosphorylase kinase	Substrate specificity	Phosphorylase phosphatase activity	Regulators
1a	Yes	β Subunit	Broad	High	Inhibitor-1, inhibitor-2
1b	Yes	β Subunit	Broad	High	Glycogen synthase kinase-3 and Mg-ATP, inhibitor-1, inhibitor-2
2A	No	α Subunit	Broad	High	Unknown
2B	No	α Subunit	Narrow	Very low	Calcium ions, calmodulin
2C	No	α Subunit	Broad	Very low	Magnesium ions

enzyme preferentially dephosphorylated the α subunit.[11] Additional experiments showed that β-phosphorylase kinase phosphatase copurified with the major phosphorylase phosphatase and glycogen synthase phosphatase activities. α Phosphorylase kinase phosphatase also had some glycogen synthase and slight phosphorylase phosphatase activity. Both enzymes are active against phosphohistones.[12] The finding that these enzymes demonstrated quite different substrate specificities and sensitivity to a variety of regulators led to the classification of protein phosphatases into two categories.[13] More recent studies[14–20] indicate that four enzymes, termed protein phosphatases 1, 2A, 2B, and 2C, (Table I) account for virtually all the protein phosphatase activity in cellular extracts acting on phosphorylated proteins involved in the control of glycogen metabolism, glycolysis/gluconeogenesis, fatty acid synthesis, cholesterol synthesis, and protein synthesis (Table II). In addition, these enzymes appear to

[11] J. F. Antoniw and P. Cohen, Eur. J. Biochem. 68, 45 (1976).
[12] J. F. Antoniw, H. G. Nimmo, S. J. Yeaman, and P. Cohen, Biochem. J. 162, 423 (1977).
[13] P. Cohen, J. G. Foulkes, J. Goris, B. A. Hemmings, T. S. Ingebritsen, A. A. Stewart, and S. J. Strada, in "Metabolic Interconversion of Enzymes" (H. Holzer, ed.), p. 28. Springer-Verlag, Berlin and Heidelberg, 1981.
[14] T. S. Ingebritsen and P. Cohen, Science 221, 331 (1983).
[15] T. S. Ingebritsen and P. Cohen, Eur. J. Biochem. 132, 255 (1983).
[16] T. S. Ingebritsen, J. G. Foulkes, and P. Cohen, Eur. J. Biochem. 132, 263 (1983).
[17] T. S. Ingebritsen, J. Blair, P. Guy, L. Witters, and D. G. Hardie, Eur. J. Biochem. 132, 275 (1983).
[18] M. D. Pato, R. A. Adelstein, D. Crouch, B. Safer, T. S. Ingebritsen, and P. Cohen, Eur. J. Biochem. 132, 283 (1983).
[19] A. A. Stewart, T. S. Ingebritsen, and P. Cohen, Eur. J. Biochem. 132, 289 (1983).
[20] T. S. Ingebritsen, A. A. Stewart, and P. Cohen, Eur. J. Biochem. 132, 297 (1983).

TABLE II
Substrate Specificity of Type 1 and Type 2 Protein Phosphatases[a]

Metabolic pathway	Substrates	Protein kinase	Concentration		Moles phosphate per mole	Relative activities			
			mg/ml	μM		PrP-1	PrP-2A	PrP-2B	PrP-2C
Glycogen metabolism	Phosphorylase kinase	cAMP-PK	0.8	2.4	1.8	100 (β)	100 (α)	100 (α)	100 (α)
	Phosphorylase a	PhK	1.0	10	1.0	27	17	1	2
	Glycogen synthase (site 1a)	cAMP-PK	0.11	1.2	0.4	57	92	3	21
	Glycogen synthase (site 2)	PhK, cAMP-PK, GSK-4, CaM-dependent GSK	0.11	1.2	0.6	103	92	<1	86
	Glycogen synthase (sites 3a, 3b, 3c)	GSK-3	0.11	1.2	0.9	28	63	2	8
	Inhibitor-1	cAMP-PK	0.04	2	0.9	54	100	192	19
Glycolysis and gluconeogenesis	Pyruvate kinase	cAMP-PK	0.06	1.0	0.7	20	39	3	58

Fatty acid synthesis	ATP-citrate lyase	cAMP-PK	0.12	1.0	0.8	18	79	2	15
	Acetyl-CoA carboxylase	cAMP-PK	0.24	1.0	0.7	21	28	1	18
Cholesterol synthesis	HMG-CoA reductase	RK	3.0 U/ml	—	—	21	17	<1	186
	HMG-CoA reductase kinase	RKK	1.0 U/ml	—	—	80	124	4	1814
Protein synthesis	Initiation factor eIF-2	eIF-2α kinase	0.014	0.1	0.9	16	44	<1	5
Muscle concentration	Myosin light chain	MLCK	0.07	4	0.5	40	449	74	19
	Histone H1	cAMP-PK	0.04	2	0.6	523	231	2	9
	Histone H2B	cAMP-PK	0.030	2	0.8	228	340	8	224

[a] Activity is expressed as percentage of phosphate released in the assays relative to the dephosphorylation of either the α or β subunit of phosphorylase kinase for all substrates except HMG-CoA reductase and HMG-CoA reductase kinase. With the latter enzymes, activity is expressed as percentage of conversion to the active or inactive form, respectively. The concentration of HMG-CoA reductase is expressed as the enzyme activity after complete activation by protein phosphatase.[15] Assay conditions are described by Ingebritsen and Cohen.[15] Values are an indication of the substrate that each protein phosphatase is capable of acting on, but should not necessarily be interpreted as being representative of rates that occur *in vivo*. The abbreviations are: cAMP-PK, cAMP-dependent protein kinase; PhK, phosphorylase kinase; GSK, glycogen synthase kinase; CaM. calmodulin: RK, HMG-CoA reductase kinase; RKK, HMG-CoA reductase kinase kinase; HMG-CoA, 3-hydroxy-3-methylglutaryl CoA; eIF, eukaryotic initiation factor (protein synthesis); MLCK, myosin light-chain kinase. See Ingebritsen *et al.*[16] for a complete description of the substrates.

explain most if not all of the protein phosphatase activities that have been reported in the literature.

The protein phosphatases can be grouped into two classes. The type 1 protein phosphatase (protein phosphatase 1) selectively dephosphorylates the β subunit of phorphorylase kinase and is inhibited by nanomolar concentrations of two thermostable regulatory proteins, termed inhibitor-1 and inhibitor-2. Type 2 protein phosphatases (protein phosphatases 2A, 2B, and 2C) selectively dephosphorylate the α subunit of phosphorylase kinase and are insensitive to inhibitor-1 and inhibitor-2.

Two mechanisms for regulating protein phosphatase 1 have been identified. One involves the phosphorylation of inhibitor-1 on a specific threonine residue by cyclic AMP-dependent protein kinase.[21,22] The phospho form of the protein is a potent inhibitor of protein phosphatase 1, and the dephospho form is inactive. In skeletal muscle, the state of phosphorylation on inhibitor-1 is increased in response to epinephrine[23,24] and decreased by insulin,[25,26] which appears to act by antagonizing the effects of low concentration of β agonists. Interestingly, protein phosphatase 2B appears to be the major inhibitor-1 phosphatase activity in skeletal muscle or liver extracts when assays are carried out in the presence of micromolar calcium ions.[20] Although it is not yet known whether the state of phosphorylation of inhibitor-1 is decreased in response to physiological stimuli that elevate cytosolic calcium, these results suggest that the activity of protein phosphatase 1 may be regulated both by cyclic AMP and calcium ions.

A second major mechanism for regulating protein phosphatase 1 was reported by Merlevede and Riley in 1966.[27] These workers noted that phosphorylase phosphatase activity in extracts from bovine adrenal cortex was stimulated after incubation in the presence of MgATP. Subsequently, this MgATP-dependent phosphorylase phosphatase activity was found to be present in a number of other tissues,[28] and the activity was resolved into two protein components, termed factor Fa and Fc enzyme.[29] Fc enzyme was an inactive protein phosphatase whose activity was ex-

[21] P. Cohen, D. B. Rylatt, and G. A. Nimmo, *FEBS Lett.* **76**, 182 (1977).

[22] P. Cohen, G. A. Nimmo, S. Shenolikar, and J. G. Foulkes, *FEBS Symp.* **54**, 161 (1978).

[23] J. G. Foulkes and P. Cohen, *Eur. J. Biochem.* **97**, 251 (1979).

[24] B. S. Khatra, J. L. Chisson, H. Shikama, J. H. Exton, and T. R. Soderling, *FEBS Lett.* **114**, 253 (1980).

[25] J. G. Foulkes, L. S. Jefferson, and P. Cohen, *FEBS Lett.* **112**, 21 (1980).

[26] J. G. Foulkes, P. Cohen, S. J. Strada, W. V. Everson, and L. S. Jefferson, *J. Biol. Chem.* **257**, 12493 (1982).

[27] W. Merlevede and G. A. Riley, *J. Biol. Chem.* **241**, 3527 (1966).

[28] S. D. Yang, J. R. Vandenheede, J. Goris, and W. Merlevede, *FEBS Lett.* **111**, 201 (1980).

[29] J. Goris, G. Defreyn, and W. Merlevede, *FEBS Lett.* **99**, 279 (1979).

pressed only after incubation with Fa and MgATP. More recently, it has been shown that the enzymatic properties of active Fc enzyme are indistinguishable from those of protein phosphatase 1,[30] suggesting that the two enzymes may share the same catalytic subunit. This idea has been substantiated by the observation that an enzyme, composed of 1 : 1 complex of an M_r 33,000 form of protein phosphatase 1 (see below) and inhibitor-2, has properties that are indistinguishable from those of Fc enzyme.[31] Incubation of this complex with Fa and MgATP resulted in the activation of the protein phosphatase as a consequence of the phosphorylation of inhibitor-2 on a threonine residue. The activating factor, Fa, has been purified to homogeneity[32] and shown to copurify with an enzyme, termed glycogen synthase kinase-3, that phosphorylates glycogen synthase on three residues termed sites 3a, 3b, and 3c.[33,34] Thus protein phosphatase 1 and Fc enzyme appear to be interconvertible forms of the same enzyme. In view of this result and the involvement of protein phosphorylation in the interconversion, Fc enzyme has been renamed protein phosphatase 1b and the active form of the enzyme is termed protein phosphatase 1a.[14] It is not presently known whether the ratio of protein phosphatase 1a to protein phosphatase 1b is altered *in vivo* in response to physiological stimuli.

Type 2 protein phosphatases have been separated into three enzymatic species, differing in their relative specificity to phosphoprotein substrates and their regulation by divalent cations.[14,15] These enzyme activities, termed protein phosphatase 2A, 2B, and 2C, constitute only a small proportion of the protein phosphatase activity regulating the enzymes of glycogen metabolism in skeletal muscle. However, protein phosphatases 2A and 2C are present in higher concentration in liver and are more likely to be important in the regulation of metabolic pathways in that tissue. Protein phosphatase 2B is the dominant α-phosphorylase kinase phosphatase activity in skeletal muscle. Protein phosphatases 2A and 2C have broad but distinct substrate specificities, whereas protein phosphatase 2B has significant activity on only three known substrates, namely the α subunit of phosphorylase kinase, inhibitor-1, and the isolated P-light chain of myosin (Table II). Protein phosphatase 2B is a Ca^{2+}-dependent

---

[30] A. A. Stewart, B. A. Hemmings, P. Cohen, J. Goris, and W. Merlevede, *Eur. J. Biochem.* **115,** 197 (1981).

[31] B. A. Hemmings, T. J. Resink, and P. Cohen, *FEBS Lett.* **152,** 319 (1982).

[32] B. A. Hemmings, D. Yellowees, J. C. Kernohan, and P. Cohen, *Eur. J. Biochem.* **119,** 443 (1981).

[33] D. B. Rylatt, A. Aitken, T. Bilham, G. D. Condon, N. Embi, and P. Cohen, *Eur. J. Biochem.* **107,** 529 (1980).

[34] N. Embi, P. J. Parker, and P. Cohen, *Eur. J. Biochem.* **115,** 405 (1982).

enzyme ($A_{0.5}$ = 0.5–1 μM) whose activity is stimulated 10-fold by the calcium binding protein, termed calmodulin ($A_{0.5}$ = 6.0 nM). This protein phosphatase has been shown to be identical[35] to a major calmodulin binding protein in brain termed calcineurin[36] or CaM-BP$_{80}$.[37] Its immunocytochemical localization in postsynaptic densities and microtubules of postsynaptic dendrites may suggest a possible role in neurotransmitter action. Protein phosphatase 2C is absolutely dependent on Mg^{2+} ($A_{0.5}$ = 1.0 mM) for activity with all known substrates, and protein phosphatase 2A is active in the absence of divalent cations.

Preparation of Phosphoprotein Substrates

Phosphorylase a

[^{32}P]Phosphorylase a is prepared by incubating phosphorylase b with phosphorylase kinase and [γ-^{32}P]ATP Mg.[6] One mole of phosphate per subunit is incorporated into a distinct serine residue.[14]

Reagents

Solution A: 50 mM sodium glycerophosphate, pH 7.0–2.0 mM EDTA–0.1% (v/v) 2-mercaptoethanol

Solution B: 5.0 mM Tris-HCl, pH 7.0 (25°)–1.0 mM EDTA–0.3% (v/v) 2-mercaptoethanol

Solution C: 50 mM Tris-HCl, pH 7.0 (25°)–250 mM NaCl–0.3% (v/v) 2-mercaptoethanol

Phosphorylase b: This protein is purified by the method of Fischer and Krebs[38] and recystallized three times. The crystals, redissolved in solution A, are shaken with Dowex AG-1-X4 resin to remove residual 5′-AMP,[39] which is utilized in the crystallization procedure.[28] The supernatant is dialyzed against solution A containing 50% (v/v) glycerol and stored at −20° at a concentration of 50–100 mg of protein per milliliter (the $E_{260/280}$ ratio of AMP free enzyme was 0.54, and E_{280} of phosphorylase b = 13.1). Prior to use, the protein is dialyzed overnight against solution A containing 0.2% (v/v) 2-mercaptoethanol to remove glycerol.

[35] A. A. Stewart, T. S. Ingebritsen, A. Manalan, C. B. Klee, and P. Cohen, *FEBS Lett.* **137**, 80 (1982).
[36] C. B. Klee, T. H. Crouch, and M. A. Krinks, *Proc. Natl. Acad. Sci. U.S.A.* **87**, 6260 (1979).
[37] R. W. Wallace, E. A. Tallant, and W. Y. Cheung, *Biochemistry* **19**, 1831 (1980).
[38] E. H. Fischer and E. G. Krebs, *J. Biol. Chem.* **231**, 65 (1958).
[39] D. P. Gilboe, K. L. Larson, and F. Q. Nuttall, *Anal. Biochem.* **47**, 20 (1972).

Phosphorylase kinase b: This protein is prepared as described by Cohen.[6]

Procedure. The incubation is composed of phosphorylase b (12.5 mg/ml), phosphorylase kinase (0.2 mg/ml), $CaCl_2$ (125 μM), [γ-^{32}P]ATP (300 μM), magnesium acetate (3.0 mM), 50 mM Tris-HCl, 50 mM sodium glycerophosphate, pH 8.2. After incubation for 60 min at 30°, the reaction is terminated by the addition of an equal volume of 90%-saturated ammonium sulfate, pH 7.0. After 1 hr at 4°, the precipitate is centrifuged at 15,000 g for 10 min and resuspended in 1 ml of solution B. The [^{32}P]phosphorylase a is dialyzed against solution B (500 volumes) for 16 hr with one change of dialysis buffer. Nucleotides, specifically the remaining [γ-^{32}P]ATP, are more efficiently removed by the inclusion of a second dialysis bag containing Dowex AG-1-X4 resin throughout the dialysis. The crystals of phosphorylase a formed during dialysis are centrifuged at 15,000 g for 10 min. The supernatant, containing unreacted phosphorylase b and phosphorylase kinase, is discarded. The crystals are redissolved in solution C at 30° and stored at a protein concentration of 30–60 mg/ml at 4°.

Phosphorylase Kinase a

[32]P-labeled phosphorylase kinase a containing approximately 1 mol of phosphate in each α and β subunit is prepared by incubating phosphorylase kinase b with the catalytic subunit of cyclic AMP-dependent protein kinase and [γ-^{32}P]ATP Mg.[30] Two serines surrounded by distinct amino acid sequences are phosphorylated.[14]

Reagents

Solution A: 50 mM Tris-HCl, pH 7.0 (25°)–0.3% (v/v) 2-mercaptoethanol

Phosphorylase kinase b: See preparation of phosphorylase a.

Cyclic AMP-dependent protein kinase: The catalytic subunit of this enzyme is prepared as described by Beavo *et al.*[40] One unit of activity catalyzes the incorporation of 1 nmol of phosphate into 0.2 mg of mixed histones in 1 min at 30°.[41]

Procedure. The incubation, containing phosphorylase kinase b (2 mg/ml), cyclic AMP-dependent protein kinase (0.5 U/ml), 0.4 mM EDTA, 0.2 mM EGTA, 2.0 mM magnesium acetate, 0.2 mM [γ-^{32}P]ATP, 0.1% (v/v) 2-mercaptoethanol, 50 mM sodium glycerophosphate, pH 7.0, is carried

[40] J. A. Beavo, P. J. Bechtel, and E. G. Krebs, this series, Vol. 38, p. 299.
[41] P. Cohen, G. A. Nimmo, and J. F. Antoniw, *Biochem. J.* **162,** 435 (1977).

out at 30°.[25] The incubation is continued until the amount of phosphate in the α subunit reaches the level of the β subunit (approximately 1.8–2.0 mol of phosphate per $\alpha\beta\gamma\delta$ unit) and then terminated by the addition of 0.1 volume of 100 mM EDTA, pH 7.0, and 0.1 volume of 500 mM NaF. Then 90%-saturated ammonium sulfate solution (0.5 volume) is added; after 5 min at 4°, the precipitate is centrifuged at 15,000 g for 5 min. The precipitate is redissolved in 1.0 ml of solution A containing 10 mM NaF and 1.0 mM EDTA and dialyzed overnight against 1 liter of the same buffer. This procedure inactivates endogenous protein phosphatases and removes ammonium sulfate and some unreacted [γ-^{32}P]ATP. The solution is centrifuged at 15,000 g for 2 min and gel-filtered on Sephadex G-50 superfine (1 × 10 cm) equilibrated in solution A to remove all residual [γ-^{32}P]ATP and NaF. The protein is stored at 4° and used within 2 weeks.

In order to determine the incubation time required to give equal phosphorylation of the α and β subunits, a pilot incubation is carried out under the conditions described above. At appropriate times (1–10 min), aliquots are removed and the extent of phosphorylation of the α and β subunits of phosphorylase kinase is estimated after separation of the subunits by electrophoresis on 5% (w/v) polyacrylamide gels in the presence of sodium dodecyl sulfate (SDS). ^{32}P radioactivity in the α and β subunits is determined by Cerenkov counting after cutting these components from the gel. Phosphorylase kinase has a higher K_m for ATP than cyclic AMP-dependent protein kinase and is totally dependent on Ca^{2+} for activity.[42] Hence, the use of a low concentration of ATP and the presence of EGTA in the incubation mixture prevents significant autophosphorylation of phosphorylase kinase. The concentration of phosphorylase kinase is determined using $E_{280} = 12.4$.[6]

Preparation of Other ^{32}P-Labeled Substrate Proteins

Methods for purifying and phosphorylating other protein substrates listed in Table II are given by Ingebritsen and Cohen[15] and Stewart et al.[30]

Specific Assays for Type 1 and Type 2 Protein Phosphatases

Protein Phosphatase 1a

Protein phosphatase 1a is assayed by following the time-dependent release of trichloroacetic acid-soluble radioactivity from ^{32}P-labeled phosphorylase a. [The activity of protein phosphatase 2A (assayed as below) is subtracted.]

[42] P. Cohen, *Curr. Top. Cell Regul.* **14**, 117 (1978).

Reagents

Solution A: 50 mM Tris-HCl, pH 7.0 (25°)–0.01% Brij 35

Solution B: 50 mM Tris-HCl, pH 7.0 (25°)–3.0 mM EGTA–6.0 mM
  MnCl$_2$–1 mg of bovine serum albumin per milliliter–0.3% (v/v) 2-
  mercaptoethanol

Solution C: 50 mM Tris-HCl, pH 7.0 (25°)–0.3% (w/v) 2-mercapto-
  ethanol

^{32}P-Labeled phosphorylase a: This substrate is diluted in solution C
  to a final concentration of 3.0 mg/ml (30 μM) and warmed for a few
  minutes at 30° to dissolve the crystals.

Assay Procedure. Protein phosphatase 1a (0.02 ml) appropriately diluted in solution B is mixed with 0.02 ml of solution A and preincubated
for 5 min at 30°. The reaction is started by addition of 0.02 ml of phosphorylase a. Incubations are carried out for 1–30 min at 30°. The assay is
terminated by the addition of 0.1 ml of ice-cold 20% (w/v) trichloroacetic
acid and 0.1 ml of bovine serum albumin (6 mg/ml), acting as carrier. After
10 min on ice, the solutions are centrifuged at 15,000 g for 2 min. ^{32}P
radioactivity in the supernatant (0.2 ml) may be determined either directly by Cerenkov counting or by liquid scintillation counting after addition of 1.0 ml of scintillation fluid (Aquasol, New England Nuclear Company). The assay is linear with respect to time and concentration of
protein phosphatase up to 20–30% release of inorganic [^{32}P]phosphate. A
control incubation is carried out in the absence of protein phosphatase 1a
at appropriate times to correct for any endogenous protein phosphatase
activity in the phosphorylase a preparation. Since protein phosphatase 2A
also acts on phosphorylase a, the activity due to this enzyme (see below)
must be subtracted from the total activity in order to assess specifically
the activity due to protein phosphatase 1a. Protein phosphatase 2B and
2C do not dephosphorylate phosphorylase a at significant rates. Protein
phosphatase 1a is gradually converted during purification and storage to a
form that is completely dependent on Mn^{2+} for activity. Hence MnCl$_2$ is
routinely included in the assays in order to optimize activity. It should be
noted, however, that in freshly prepared tissue extracts, the dephosphorylation of phosphorylase a by protein phosphatase 1a is decreased by 40–
50% in the presence of Mn^{2+}. Crude tissue fractions contain low molecular weight inhibitors of protein phosphatase 1a, which can be removed by
gel filtration on Sephadex G-50 equilibrated with buffer C.

Other Substrates. Protein phosphatase 1a is active on a variety of
phosphorylated proteins involved in the control of glycogen metabolism,
glycolysis/gluconeogenesis, fatty acid synthesis, cholesterol synthesis,
protein synthesis, and muscle concentration (Table II). Since many of
these phosphorylated proteins are also substrates for protein phosphatases

2B and 2C, background activity in the presence of inhibitor-2 will be somewhat higher when using these proteins, rather than phosphorylase *a*, as substrates. For proteins that are substrates for protein phosphatase 2B, background activity due to this enzyme can be blocked by carrying out assays in the presence of 100 μM trifluoperazine. Background activity due to protein phosphatase 2C can be eliminated by omitting Mn^{2+} from the assays, although this may also decrease the activity of protein phosphatase 1a (see above).

Units. One unit of protein phosphatase 1a releases 1 nmol of inorganic phosphate from phosphorylase in 1 min under standard assay conditions.

Protein Phosphatase 1b

Protein phosphatase 1b is assayed by following the time-dependent release of trichloroacetic acid-soluble radioactivity from ^{32}P-labeled phosphorylase *a* after complete activation of the phosphatase by incubation with MgATP and glycogen synthase kinase-3.

Reagents

Solution A: 50 mM Tris HCl, pH 7.0 (25°)–1 mg of bovine serum albumin per milliliter–0.3% (v/v) 2-mercaptoethanol
Solution B: 50 mM Tris-HCl, pH 7.0 (25°)–0.3% (v/v) 2-mercapto-ethanol
^{32}P-Labeled phosphorylase *a*: See assay for protein phosphatase 1a.

Assay Procedure. Protein phosphatase 1b (0.02 ml) appropriately diluted in solution A, 0.01 ml of glycogen synthase kinase-3 (diluted in solution A), and 0.01 ml of 0.2 mM ATP plus 0.4 mM magnesium acetate in solution B are mixed and preincubated at 30° for 15 min. The assay is started by addition of 0.02 ml of phosphorylase *a*, and incubations are continued for 1–30 min. The reaction is stopped, and the samples are further processed as described for the assay of protein phosphatase 1a. Control incubations are carried out in the absence of MgATP to correct for activity due to protein phosphatases 1a and 2A. Pilot assays must be carried out in order to determine the amount of glycogen synthase kinase-3 required to give maximal activation of protein phosphatase 1b during the 15-min preincubation.

Other Substrates. The substrate specificity of protein phosphatase 1b is identical to that of protein phosphatase 1a.

Units. See assay for protein phosphatase 1a.

Protein Phosphatase 2A

This enzyme is assayed by following the time-dependent release of trichloroacetic acid-soluble radioactivity from ^{32}P-labeled phosphorylase *a* in the presence of sufficient inhibitor-2 to block completely the activity of protein phosphatase 1a.

Reagents. See assay for protein phosphatase 1a.

Assay Procedure. The assay is identical to that described for protein phosphatase 1a except that solution B contains 10,000 U of inhibitor-2 per milliliter (see assay of inhibitor-2 for definition of units) to block completely any activity due to protein phosphatase 1a. In freshly prepared tissue extracts, the activity of protein phosphatase 2A is stimulated 1.5- to 2-fold when Mn^{2+} is included in the assays. Furthermore, protein phosphatase 2A, like protein phosphatase 1a, is gradually converted during purification and storage to a form that is completely dependent on Mn^{2+} for activity. When assaying crude tissue fractions, they should be subjected to gel filtration on Sephadex G-50 (see protein phosphatase 1a assay) prior to assay to remove low molecular weight substances that inhibit protein phosphatase 2A.

Other Substrates. Protein phosphatase 2A, like protein phosphatase 1, has a very broad substrate specificity. When using phosphorylated proteins that are also substrates for protein phosphatases 2B and 2C, background activity due to protein phosphatase 2B can be eliminated by carrying out the assays in the presence of 100 μM trifluoperazine. Background activity due to protein phosphatase 2C can be blocked by carrying out the assays in the absence of Mn^{2+}, although this will also decrease the activity of protein phosphatase 2A (see above). Alternatively, the activity due to protein phosphatase 2C may be assessed separately (see below) and subtracted from the total activity observed in the presence of Mn^{2+} to give the net activity due to protein phosphatase 2A. This correction is usually relatively small, since protein phosphatase 2C accounts for only a minor proportion of the total activity toward most substrates.[20]

Units. See protein phosphatase 1a assay.

Protein Phosphatase 2B

This enzyme is assayed by following the time-dependent release of trichloroacetic acid-soluble radioactivity from ^{32}P-labeled phosphorylase kinase in the presence of 3 μM free Ca^{2+}.

Reagents

Solutions A and C: See assay for protein phosphatase 1a.
Solution B: 50 mM Tris-HCl, pH 7.0 (25°)–1 mg of bovine serum albumin per milliliter–0.1% (v/v) 2-mercaptoethanol
Solution D: 50 mM Tris-HCl, pH 7.0 (25°)–3.0 mM EGTA–2.64 mM $CaCl_2$–1 mg of bovine serum albumin per milliliter–0.3% (v/v) 2-mercaptoethanol

[32]P-Labeled phosphorylase kinase: This substrate contains approximately 1 mol of phosphate in each α subunit and β subunit; it is diluted in solution C to a concentration of 2.4 mg/ml (7.2 μM).

Assay Procedure. Solution A (0.01 ml) containing 10,000 U of inhibitor-2 per milliliter, 0.1 ml of solution D, and 0.02 ml of phosphorylase kinase is mixed and preincubated for 5 min at 30°. The reaction is started by addition of 0.02 ml of protein phosphatase 2B appropriately diluted in solution B. After 1 min, the assay is terminated, and samples are further processed as described in the assay for protein phosphatase 1a. By starting the assay with the protein phosphatase rather than the substrate and by using a short incubation time, problems with the rapid inactivation of protein phosphatase 2B by Ca^{2+}-dependent proteinases are minimized.[19,20] Problems are particularly apparent in tissue extracts and during the early stages of the purification of the enzyme. Protein phosphatases 1, 2A, and 2C also act on phosphorylase kinase. The presence of inhibitor-2 and the absence of Mg^{2+} block the activities of protein phosphatases 1 and 2C, respectively. In order to estimate the activity due to protein phosphatase 2A, a control assay is carried out in the presence of 100 μM trifluoperazine (added in solution A with inhibitor-2), which specifically inhibits protein phosphatase 2B. The activity due to protein phosphatase 2A is subtracted from the total activity to give the net activity specifically due to protein phosphatase 2B. Control incubations in the absence of protein phosphatase 2B (with and without trifluoperazine) are also carried out to correct for endogenous protein phosphatase activity in substrate.

Mn^{2+} is capable of substituting for Ca^{2+} in the activation of protein phosphatase 2B. In the presence of 1.0 mM $MnCl_2$ the activity is similar to that observed in the presence of 3 μM Ca^{2+}. Mn^{2+} is capable of interacting with calmodulin at low ionic strength; however, it is not yet clear whether the activity observed in the presence of Mn^{2+} is a consequence of this interaction or results from trace contamination of Ca^{2+} in the $MnCl_2$ solutions.

It should be noted that after chromatography of protein phosphatase 2B on calmodulin–Sepharose (the final step in the purification procedure) the enzyme is converted to a form that has a low specific activity in the presence of Ca^{2+}. At this stage, reactivation can be better achieved by carrying out the assays in the presence of 1.0 mM $MnCl_2$ than with Ca^{2+} (3 μM). Other properties of the enzyme, e.g., its substrate specificity and its stimulation by calmodulin (see below), are unaltered.[19]

Other Substrates. Protein phosphatase 2B also has significant activity toward inhibitor-1 and the P-light chain of myosin. The assay described above can also be used with these substrates, however, the activity is stimulated up to 10-fold by calmodulin (6.0 nM), depending upon the degree of purification of the protein phosphatase. Activity toward phosphorylase kinase is not stimulated by calmodulin owing to the presence of this protein as an integral subunit of phosphorylase kinase.[7,8]

Units. One unit of protein phosphatase 2B releases 1 nmol of inorganic phosphate from phosphorylase kinase in 1 min under standard assay conditions.

Protein Phosphatase 2C

This enzyme is assayed by following the time-dependent release of trichloroacetic acid-soluble radioactivity from ^{32}P-labeled phosphorylase kinase in the presence of 10 mM Mg^{2+}.

Reagents

Solutions A and C: See assay for protein phosphatase 1a.
Solution B: 50 mM Tris-HCl, pH 7.0 (25°)–3.0 mM EGTA–30 mM magnesium acetate–1 mg of bovine serum albumin per milliliter–0.3% (v/v) 2-mercaptoethanol
[32]P-Labeled phosphorylase kinase: See assay for protein phosphatase 2B.

Assay Procedure. Solution A (0.02 ml) containing 10,000 U of inhibitor-2 per milliliter is mixed with 0.02 ml of protein phosphatase 2C appropriately diluted in solution C and preincubated for 5 min at 30°. The reaction is started by addition of 0.02 ml of phosphorylase kinase and allowed to continue for 1–30 min. The assay is terminated, and samples are further processed as described in the assay for protein phosphatase 1a. Protein phosphatases 1, 2A, and 2B are also active on phosphorylase kinase. The presence of inhibitor-2 and EGTA block the activity due to protein phosphatases 1 and 2B, respectively. In order to eliminate activity due to protein phosphatase 2A, samples are incubated in the presence of 10 mM EDTA and 50 mM NaF for 30 min at 30° and then gel filtered prior to assay on a small column of Sephadex G-50 equilibrated with solution C to remove NaF and EDTA. This procedure inactivates protein phosphatase 2A, but has no effect on the activity of protein phosphatase 2C. In order to correct for any remaining protein phosphatase 2A activity, control incubations are carried in the absence of Mg^{2+}. This activity (usually less than 10% of the total) is subtracted from the total activity observed in the presence of Mg^{2+}. Protein phosphatase 2C can also be activated by Mn^{2+}, giving a maximal activity (observed at 1.0 mM Mn^{2+}) of about 80% of that in the presence of saturating Mg^{2+} (10 mM). This cation cannot be used, however, when assaying fractions that also contain protein phosphatase 2A, since the NaF/EDTA inactivation of this protein phosphatase is partially reversed by Mn^{2+}.

Other Substrates. Protein phosphatase 2C has a rather broad substrate specificity, although it is distinct from that of protein phosphatases 1 and 2A. The assay described above can be used with each of these substrates.

Units. One unit of protein phosphatase 2C releases 1 nmol of inorganic phosphate from phosphorylase kinase in 1 min under standard assay conditions.

Assay of Regulators of Protein Phosphatase 1

Inhibitor-1

The activity of this protein is assessed by its ability to inhibit protein phosphatase 1a. Prior to assay, the inhibitor is fully activated by incubation with cyclic AMP-dependent protein kinase. The activation of inhibitor-1 correlates with the incorporation of 1 mol of phosphate per mole of protein into a threonine residue.

Reagents

Solution A: See assay for protein phosphatase 1a.
Cyclic AMP-dependent protein kinase: See preparation of phosphorylase kinase for method of preparation of the catalytic subunit and definition of units.
Protein phosphatase 1a: See Purification of Type 1 and Type 2 Protein Phosphatases (below). For definition of units see assay of protein phosphatase 1a.

Procedure. Inhibitor-1 is first fully phosphorylated by incubation with cyclic AMP-dependent protein kinase (10–20 U/ml), 2.0 mM magnesium acetate, 0.2 mM ATP, 20 mM Tris-HCl, pH 7.0 (25°). The high concentration of protein kinase is necessary to overcome the effects of the heat-stable inhibitor of cyclic AMP-dependent protein kinase that is present in heat-treated extracts and copurifies with inhibitor-1 through several steps of its purification.[43] Inhibitor-1 is fully phosphorylated (1.0 mol of phosphate per mole) after 30–60 min of incubation at 30°. The reaction is terminated by incubation in a boiling water bath for 2 min, and the mixture is appropriately diluted in solution A. The activity of the inhibitor-1 is assessed by adding a 0.02-ml aliquot to a standard assay for protein phosphatase 1a, containing 0.02 unit of protein phosphatase 1a and omitting $MnCl_2$. Protein phosphatase 1a is active on inhibitor-1, but, this reaction

[43] G. A. Nimmo and P. Cohen, *Eur. J. Biochem.* **87,** 341 (1978).

is unusual in that it requires Mn^{2+} whereas the dephosphorylation of other substrates by this protein phosphatase is not Mn^{2+}-dependent. The omission of $MnCl_2$ from the assay thus prevents the dephosphorylation of inhibitor-1. However, if the purification of protein phosphatase 1a has been carried out in the presence of $MnCl_2$, protein phosphatase 1a must be preincubated with 1.0 mM EDTA for 90 min at 30° prior to use in the assays. The percentage inhibition of protein phosphatase 1a plotted against the inhibitor-1 concentration is linear up to 30% inhibition.[43]

Units. One unit of inhibitor-1 is the amount that inhibits 0.02 U of protein phosphatase 1 by 50% in the standard assay.

Inhibitor-2

The activity of this protein is assessed by its ability to inhibit protein phosphatase 1a. Prior to assay, any contaminating inhibitor-1 is inactivated by treatment with a protein phosphatase.

Reagents. See assay for inhibitor-1.

Procedure. The assay for inhibitor-2 is identical to that described for inhibitor-1 except that the preincubation with cyclic AMP-dependent protein kinase is omitted. If inhibitor-1 is also present, it can be inactivated by including 1.0 mM $MnCl_2$ during the preincubation with protein phosphatase 1a and in the assay.

Activating Factor (Fa) of the MgATP-Dependent Protein Phosphatase (Protein Phosphatase 1b)

The activity of Fa (or glycogen synthase kinase-3) is evaluated by its capacity to activate protein phosphatase 1b in the presence of ATP-Mg.[32]

Reagents

Solutions A and B: See assay for protein phosphatase 1b.
Glycogen synthase kinase-3: For preparation, see below.
^{32}P-Labeled phosphorylase *a*: See assay for protein phosphatase 1a.

Procedure. Protein phosphatase 1b (maximum potential activity of 0.04 U; for definition of units, see above) in solution A (0.02 ml) is preincubated with glycogen synthase kinase-3 (0.01 ml appropriately diluted in solution A) at 30° for 5 min. The reaction is initiated by the addition of 0.01 ml of 0.2 mM ATP and 0.4 mM magnesium acetate in solution B. After 5 min at 30°, protein phosphatase 1b is assayed by the addition of 0.02 ml of ^{32}P-labeled phosphorylase *a* as described above. The activation of protein phosphatase 1b by glycogen synthase kinase-3 is linear to 0.001 U of the activator, corresponding to 10% conversion of protein phosphatase 1b to its active form.[32]

Other Substrates. The activating factor (or glycogen synthase kinase-3) may also be assayed by its ability to phosphorylate glycogen synthase with $[\gamma\text{-}^{32}P]ATP\text{-}Mg^{32}$ at three distinct phosphorylation sites, termed 3a, 3b, and 3c.

Units. One unit of Fa is defined as the amount that increases the phosphorylase phosphatase activity of protein phosphatase 1b by 1.0 U/min (i.e., catalyzing the release of 1.0 nmol of inorganic phosphate from phosphorylase in 1 min).

Identification of Protein Phosphatases

1. In the protein phosphatase assays described above, activity is assessed by measuring the appearance of trichloroacetic acid-soluble radioactivity. A potential artifact in this procedure is that many low molecular weight peptides are also soluble in trichloroacetic acid. Thus, when initially characterizing a protein phosphatase, it is important to confirm that the trichloroacetic acid-soluble radioactivity released is actually inorganic [^{32}P]phosphate. This can be specifically measured[44] by mixing 0.1 ml of the trichloroacetic acid supernatant with 0.2 ml of 1.25 mM KH_2PO_4 in 1.0 N H_2SO_4 and 0.5 ml of isobutanol–toluene (1 : 1). Then 0.1 ml of 5% (w/v) ammonium molybdate is added, the sample is mixed again, and the organic and aqueous phases are separated by centrifugation for 2 min at 15,000 g. A 0.3 ml aliquot of the upper organic phase (0.5 ml total volume) is removed and counted as described in the assay for protein phosphatase 1a.

2. The next step in the procedure is to decide whether the enzyme is a type 1 or type 2 protein phosphatase. This determination is based on the sensitivity of the enzyme to inhibitor-1 and -2 and the specificity of the enzyme for the α or β subunit of phosphorylase kinase (Table I). Since, in the presence of Mn^{2+}, each of the type 1 and type 2 protein phosphatases is active on phosphorylase kinase, this substrate can be used conveniently for this procedure. Incubations are carried out as described in the assay for protein phosphatase 1a in the presence or the absence of 100 units of inhibitor-1 or inhibitor-2. In order to minimize the dephosphorylation of inhibitor-1 under these conditions, short assay times (1 min) are used and the reaction is started with the protein phosphatase rather than the substrate. If the purification of the protein phosphatase has been monitored with another substrate, a second set of assays should be carried out using this substrate instead of phosphorylase kinase.

[44] J. F. Antoniw and P. Cohen, *Eur. J. Biochem.* **68**, 45 (1976).

In order to assess the dephosphorylation of the α and β subunits of phosphorylase kinase, a time course is carried out and at each time point, duplicate aliquots are removed. One aliquot is used to estimate the release of trichloroacetic acid-soluble radioactivity, and the second is subjected to SDS–polyacrylamide gel electrophoresis. The amount of radioactivity in the protein bands corresponding to the α and β subunits is estimated as described for the preparation of ^{32}P-labeled phosphorylase kinase. The amount of ^{32}P-radioactivity remaining in the α and β subunits at each time point is estimated from the total ^{32}P-radioactivity released and the ratio of the counts in the bands corresponding to the α and β subunits.

3. If the protein phosphatase dephosphorylates the β subunit of phosphorylase kinase and is inhibited by inhibitor-1 and -2 with all substrates, it is clearly a form of protein phosphatase 1a (see above). Conversely, if the activity of the preparation is not blocked by the inhibitor proteins and the preparation dephosphorylates the subunit of phosphorylase kinase, it contains one or more of the type 2 protein phosphatases. However, if another substrate was used to monitor the activity of the protein phosphatase during its purification, it is also possible that the activity toward this substrate is due to a unique enzyme. Thus, if the enzyme appears to be a type 2 protein phosphatase, it is necessary to use additional criteria to decide whether it is a unique enzyme or corresponds to one of the type 2 protein phosphatases. These include a comparison of the elution positions of the protein phosphatase after chromatography on DEAE-cellulose and Sephadex G-200 (see below), the effects of divalent cations on its activity, its substrate specificity, and its subunit structure with those of protein phosphatases 2A, 2B, and 2C.

4. All procedures for the identification of the protein phosphatases utilizing the substrates described above, are intended to select for phosphoseryl and phosphothreonyl protein phosphatases. Phosphotyrosyl protein phosphatases have been demonstrated to be distinct and separable protein phosphatase species.[45] Hence the use of uncharacterized phosphoprotein substrates for the isolation of protein phosphatases necessitates the identification of the ^{32}P-labeled phosphoamino acids present in the substrates and the analysis of their specific dephosphorylation. Despite these precautions, distinct phosphoseryl protein phosphatases, which do not possess the properties of the protein phosphatases described here, may exist. For example, in the mitochondria, pyruvate dehydrogenase is activated by a specific protein phosphatase that demonstrates

[45] J. G. Foulkes, E. Erickson, and R. L. Erickson, *J. Biol. Chem.* **258**, 431 (1983).

Ca^{2+}-dependency but clearly differs from protein phosphatase 2B in its subunit composition and its Ca^{2+} binding properties.[46,47]

Isolation of Type 1 and Type 2 Protein Phosphatases

Reagents

Solution A: 4.0 mM EDTA, pH 7.0–250 mM sucrose–0.1% (v/v) 2-mercaptoethanol

Solution B: 10 mM Tris-HCl, pH 7.0 (25°)–0.1 mM EGTA–0.1% (v/v) 2-mercaptoethanol

Solution C: 50 mM Tris-HCl, pH 7.0 (25°)–0.1 mM EGTA–10% ethanediol–0.1% (v/v) 2-mercaptoethanol

Solution D: 50 mM Tris-HCl, pH 7.0 (25°)–0.1 mM EGTA–2% (v/v) glycerol–1.0 mg of bovine serum albumin per milliliter–1.6% (v/v) 2-mercaptoethanol

Solution E: 20 mM Tris-HCl, pH 7.0 (25°)–10% glycerol–0.1% (v/v) 2-mercaptoethanol

Extraction of Protein Phosphatases

This is carried out by homogenizing the appropriate tissue in 3.0 volumes of solution A and then centrifuging at 12,000 g for 10 min. This and all subsequent operations are carried out at 4°. The supernatant (tissue extract) is decanted and further centrifuged for 60 min at 100,000 g. The supernatant (cytosol) is decanted, and the pellet is resuspended in solution A to one-fifth of the extract volume with the aid of a Potter–Elvehjem homogenizer.

Protein phosphatases 1, 2A, 2B, and 2C are present in all tissues so far examined. However, the concentration of each enzyme varies considerably from tissue to tissue (Table III). Consequently, the choice of starting material depends on the protein phosphatase that one desires to obtain. The richest source of protein phosphatase 1 is the skeletal muscle, whereas the highest concentration of protein phosphatases 2A and 2C is in the liver. The best sources of protein phosphatase 2B are skeletal muscle and brain.

In liver and skeletal muscle, protein phosphatases 2A, 2B, and 2C are localized almost exclusively in the cytosolic fraction. In the case of protein phosphatase 1, however, 50–60% of the activity in skeletal muscle

[46] L. J. Reed, F. A. Pettit, D. H. Bleile, and T. L. Wa, "Metabolic Interconversions of Enzymes." (H. Holzer, ed.), p. 124. Springer-Verlag, Berlin and Heidelberg, 1981.

[47] M. L. Pratt, J. F. Maher, and T. E. Roche, *Eur. J. Biochem.* **125,** 349 (1982).

TABLE III
TISSUE DISTRIBUTION OF TYPE 1 AND TYPE 2 PROTEIN PHOSPHATASE (PrP)[20]

Species	Tissues	Specific activity in tissue extracts (U/mg protein)[a]				Total activity (U/g wet weight)[a]			
		PrP-1	PrP-2A	PrP-2B	PrP-2C	PrP-1	PrP-2A	PrP-2B	PrP-2C
Rabbit	Skeletal muscle	1.13 (9) ±0.09	0.35 (9) ±0.04	0.35 (2) ±0.08	0.022 (5) ±0.002	24.9 (9) ± 2.4	7.5 (9) ± 0.8	8.1 (5) ±1.7	0.5 (5) ±0.05
	Heart muscle	0.41 (9) ±0.06	1.17 (3) ±0.07	0.09 (3) ±0.02	0.085 (3) ±0.007	5.4 (3) ± 0.7	15.5 (3) ± 1.4	1.2 (3) ±0.2	1.1 (3) ±0.1
	Brain	0.58 (3) ±0.14	1.93 (3) ±0.27	0.56 (3) ±0.05	0.089 (3) ±0.005	6.2 (3) ± 1.0	20.9 (3) ± 1.3	6.1 (3) ±0.6	1.0 (3) ±0.2
	Adipose tissue	0.15 (3) ±0.03	0.60 (3) ±0.10	0.22 (2) (0.18, 0.27)	0.037 (3) ±0.005	0.3 (3) ± 0.04	1.1 (3) ± 0.2	0.4 (2) (0.3, 0.4)	0.1 (3) ±0.01
	Liver	0.32 (13) ±0.03	0.67 (13) ±0.04	0.10 (4) ±0.02	0.041 (5) ±0.003	13.2 (11) ± 1.8	26.1 (11) ± 1.5	3.4 (4) ±0.5	1.4 (5) ±0.1
Rat	Liver	0.32 (4) ±0.04	0.58 (4) ±0.08	0.06 (1)	0.040 (4) ±0.004	12.6 (4) ± 2.1	22.2 (4) ± 4.1	2.6 (1)	1.6 (4) ±0.2

[a] Numbers in parentheses refer to number of animals used for protein phosphatase determinations.

extracts and 30% of the activity in liver extracts is recovered in the 100,000 g pellet fraction. In skeletal muscle, the activity in this fraction is primarily associated with the protein–glycogen complex and may be solubilized by incubation of the fraction at 30° for 60 min in the presence of α-amylase (0.1 mg/ml). In the liver, the activity is associated with both microsomal membranes and the protein–glycogen complex. Protein phosphatase 1b is difficult to quantitate in tissue extracts and crude subcellular fractions for reasons discussed by Yang et al.[28] Consequently, the relative amounts of protein phosphatase 1a and 1b in these fractions are unclear. Protein phosphatase 1b activity is first detected after chromatography of the cytosolic fraction on DEAE-cellulose (see below).

Separation of Cytosolic Protein Phosphatases

Typically, 300 ml of the cytosolic fraction is applied to a column of DEAE-cellulose (4 × 16 cm) equilibrated with solution B. The column is washed with solution B containing 50 mM NaCl until the A_{280} approaches zero (approximately 1000 ml) and then developed with a linear gradient (1200 ml) from 50 to 500 mM NaCl in solution B. The elution positions of protein phosphatases 1, 2A, 2B, and 2C can be detected by using the specific assays for these protein phosphatases described above.

Protein phosphatases 1a and 1b elute in the same position (0.16 M NaCl) from DEAE-cellulose, but can be resolved by a further chromatography on Affi-Gel Blue (Bio-Rad)[31] or Blue Sepharose CL-6B (Pharmacia)[48] equilibrated with solution E. Protein phosphatase 1a binds to the column under these conditions, but protein phosphatase 1b is not retained. Protein phosphatase 1a can be eluted from the column with solution E containing 0.4 M NaCl.[48] Protein phosphatase 2A is resolved into three species, termed protein phosphatases $2A_0$ ($M_{r_{app}} = 210,000$), $2A_1$ ($M_{r_{app}} = 210,000$), and $2A_2$ ($M_{r_{app}} = 150,000$), which elute at 0.15 M, 0.20 M, and 0.28 M NaCl, respectively. Each of these species contains the same catalytic subunit ($M_{r_{app}} = 38,000$) but differs in the number and type of other subunits present in the higher molecular weight complexes. The activity of protein phosphatase $2A_0$ is observed only after dissociation of the $M_r = 38,000$ catalytic subunit. Aliquots (0.01 ml) of appropriate column fractions are diluted with 0.19 ml of solution D and frozen at $-20°$ for 60 min. The solution is allowed to thaw at 4° and any insoluble material is removed by centrifugation at 15,000 g for 2 min. An aliquot of the mixture (0.02 ml) is then assayed using the standard assay for protein phosphatase 2A. Protein phosphatase 2B ($M_{r_{app}} = 98,000$) elutes at the

[48] J. R. Vandenheede, S. D. Yang, and W. Merlevede, J. Biol. Chem. 256, 5894 (1981).

same position as protein phosphatase $2A_0$ on DEAE-cellulose, but the two enzymes can be resolved by gel filtration on Sephadex G-200 (2.5 × 45 cm) equilibrated with solution C. Protein phosphatase 2C ($M_{r_{app}}$ = 43,000) elutes at the leading edge of protein phosphatase $2A_1$ peak from the ion-exchange column. These two enzymes are completely separated by gel filtration on Sephadex G-200 as described above.

Purification of Type 1 and Type 2 Protein Phosphatases

Protein Phosphatase 1

Several groups have reported that the major phosphorylase phosphatase in skeletal muscle extracts (protein phosphatase 1a) has an $M_{r_{app}}$ of 250,000–260,000. Purification by a variety of procedures results in its conversion to a number of lower M_r forms in range of 120,000–140,000, 60,000–80,000, and 30,000–45,000 (for discussion see Ingebritsen and Cohen[15]). The only form of protein phosphatase 1a that has been purified to homogeneity is a species that migrates with an $M_{r_{app}}$ of 32,000–33,000 on SDS–polyacrylamide gels that may represent the catalytic subunit of protein phosphatase 1a.[49,50] The molecular organization of the larger species is unknown.

Yang *et al.*[51] have purified protein phosphatase 1b to apparent homogeneity from rabbit skeletal muscle. The purified enzyme had an $M_{r_{app}}$ of 70,000 on SDS–polyacrylamide gels. Hemmings *et al.*[31] have prepared an enzyme consisting of a 1 : 1 complex of the $M_r - 33,000$ form of protein phosphatase 1a and inhibitor-2, which has properties indistinguishable from those of protein phosphatase 1b (see above).

Protein Phosphatase 2A

Protein phosphatase $2A_1$ and $2A_2$ have been purified to homogeneity from rat liver,[52,53] turkey gizzard,[54] rabbit reticulocytes,[55] and arterial

[49] M. K. Ganapathi, S. R. Silberman, H. Paris, and E. Y. C. Lee, *J. Biol. Chem.* **256**, 3213 (1981).

[50] E. Y. C. Lee, S. R. Silberman, M. K. Ganapathi, S. Petrovic, and H. Paris, *Adv. Cyclic Nucleotide Res.* **13**, 95 (1980).

[51] S. D. Yang, J. R. Vandenheede, J. Goris, and W. Merlevede, *J. Biol. Chem.* **255**, 11759 (1980).

[52] S. Tamura, H. Kikuchi, K. Kikuchi, A. Hiraga, and S. Tsuiki, *Eur. J. Biochem.* **104**, 347 (1980).

[53] S. Tamura and S. Tsuiki, *Eur. J. Biochem.* **114**, 217 (1980).

[54] M. D. Pato and R. S. Adelstein, *J. Biol. Chem.* **255**, 6535 (1980).

[55] D. Crouch and B. Safer, *J. Biol. Chem.* **255**, 7918 (1980).

smooth muscle.[56] In addition to the catalytic subunit, protein phosphatase $2A_1$ contains two other subunits with apparent M_r 60,000 and 55,000, while protein phosphatase $2A_2$ contains only the $M_r = 60,000$ subunit complexed with the catalytic subunit.[14] Protein phosphatase $2A_0$ has not yet been obtained in homogeneous form, and its subunit composition is presently not known. The $M_{rapp} = 38,000$ catalytic subunit of protein phosphatase 2A has been obtained in homogeneous form from rabbit skeletal muscle.[49,50] The stoichiometry of the association of the α (M_r 38,000), β (M_r 60,000), and γ (M_r 55,000) subunits in protein phosphatases $2A_0$, $2A_1$, and $2A_2$ is discussed by Cohen et al.[13]

Protein Phosphatase 2B

This enzyme has been purified to homogeneity from rabbit skeletal muscle[19] and consists of two subunits $M_{rapp} = 60,000$ and 15,000. The smaller subunit binds 4 mol of Ca^{2+} per mole with affinities in the micromolar range, and the larger subunit may possess the catalytic domain. Calcineurin[36] or CaM-BP$_{80}$,[37] which appears to be identical to protein phosphatase 2B (see above), has been purified to homogeneity from brain.

Protein Phosphatase 2C

This enzyme, which consists of a single subunit ($M_{rapp} = 43,000$), has been purified to homogeneity from rat liver[57] and turkey gizzard.[54]

Purification of Protein Modulators of Protein Phosphatase-1

Inhibitor-1

The procedure for purifying inhibitor-1 was originally developed by Nimmo and Cohen[43] and subsequently modified by Foulkes and Cohen[23] and Aitken et al.[58]

Reagents

Solution A: 0.5 M Tris-HCl, pH 8.0 (25°)
Solution B: 5.0 mM Tris-HCl, pH 8.0 (25°)
Solution C: 5.0 mM Tris-HCl, pH 8.5 (25°)

[56] D. K. Werth, J. R. Haeberle, and D. R. Hathaway, J. Biol. Chem. 257, 7306 (1982).
[57] A. Hiraga, K. Kikuchi, S. Tamura, and S. Tsuiki, Eur. J. Biochem. 119, 503 (1981).
[58] A. Aitken, T. Bilham, and P. Cohen, Eur. J. Biochem. 126, 235 (1982).

Solution D: 20 mM Tris-HCl, pH 8.5 (25°)–1.0 mM EDTA–80% (v/v) ethanol

Solution E: 5.0 mM Tris-HCl, pH 8.5 (25°)–0.1 mM EDTA

Procedure. Rabbits are injected with a lethal dose of Nembutal and exsanguinated; the hind limb and back muscle is rapidly removed and chilled on ice. The muscle (10–20 kg in a typical preparation) is minced and homogenized for 45 sec at 4° in a Waring blender in 4 volumes of ice-cold 2% (w/v) trichloroacetic acid–4.0 mM EDTA. The homogenate is centrifuged at 6000 g for 45 min, and the supernatant is decanted through glass wool. The mixture is then adjusted to 15% (w/v) trichloroacetic acid by the slow addition of 100% (w/v) trichloroacetic acid. After standing at 4° for at least 3 hr, the suspension is centrifuged at 20,000 g for 10 min. The supernatant is discarded, and the pellet is resuspended in solution A and the pH adjusted to 7.4 with 10 M ammonium hydroxide. The suspension is dialyzed at 4° for 20 hr against solution B with one change of dialysis buffer. The heavy precipitate is removed by centrifugation at 20,000 g for 10 min. The supernatant is heated in 200-ml aliquots by immersion in a boiling water bath for 15 min and is cooled to 4° on ice. The precipitated protein is removed by centrifugation at 20,000 g for 10 min. Solid ammonium sulfate is added to the supernatant to bring the solution to 60% saturation. After standing at 4° for 4 hr, the suspension is centrifuged at 20,000 g for 20 min. The precipitate is redissolved in solution A containing 1.0 mM EDTA and dialyzed for 24 hr against the same buffer containing 200 mM NaCl and 45% (v/v) ethanol at 4°. The suspension is centrifuged at 20,000 g for 10 min and the supernatant is dialyzed against solution D at 4° for 24 hr. The suspension is centrifuged at 20,000 g for 10 min and the supernatant is discarded. The precipitate is redissolved in solution C and subjected to gel filtration on Sephadex G-100 superfine (3.5 × 55 cm) equilibrated in the same buffer. Fractions containing inhibitor-1 are pooled and converted to the active form using [γ-^{32}P]ATP and the catalytic subunit of cyclic AMP-dependent protein kinase (see assay of inhibitor-1). The mixture is then heated at 90° for 5 min to denature the protein kinase and dialyzed overnight against 100 volumes of solution E. The solution is then applied to a DEAE-cellulose column (1.5 × 5 cm) equilibrated in the same buffer and washed with buffer until the absorbance at 280 nm approaches zero. The column is then developed with a linear gradient (400 ml) of 0 to 0.25 M NaCl in the same buffer. The ^{32}P-labeled fractions are pooled and precipitated with 10% (w/v) trichloroacetic acid. The precipitate is washed once with 10% (w/v) trichloroacetic acid and three times with ether to remove the trichloroacetic acid and then redissolved in water. Inhibitor-1 is obtained in essentially homogeneous form in an overall yield of about 20%. Approximately 2 mg of inhibitor-1 is obtained per kilogram of skeletal muscle.

Inhibitor-2

The procedure for preparing homogeneous inhibitor-2 from rabbit skeletal muscle was originally developed by Foulkes and Cohen.[59] The simplified procedure outlined below was subsequently devised by Yang et al.[60]

Reagents

Solution A: 20 mM Tris-HCl, pH 8.0 (25°)–0.5 mM dithiothreitol–2 mM EDTA–2 mM EGTA–0.5 mM benzamidine–0.1 mM phenyl-methanesulfonyl fluoride–0.1 mM 1-chloro-3-tosylamido-7-amino-L-2-heptanone–0.1 mM L-1-tosylamido-2-phenylethyl chloro-methyl ketone

Solution B: 20 mM Tris-HCl, pH 7.0 (25°)–0.5 mM dithiothreitol

Procedure. Rabbits are killed and exsanguinated, and muscle is removed from the back and hind limbs as described for the purification of inhibitor-1. The muscle (1.5 kg) is homogenized in solution A and centrifuged for 20 min at 10,000 g. The supernatant (pH 7.0) is filtered through glass wool and directly absorbed onto DEAE-Sephadex A-50 (one-half the supernatant volume) equilibrated with solution B. The resin is washed with 4 volumes of solution B containing 0.2 M NaCl, packed into a column, and further washed with the same buffer until the absorbance at 280 nM approaches zero. The column is then eluted with 0.4 M NaCl in solution B. The peak protein fractions are combined, concentrated by precipitation with ammonium sulfate (35–50%), and dialyzed for 1 hr against solution B. The dialyzed fraction (5–10 ml) is heat-treated by dropwise addition of the fraction to 5 ml of boiling solution B, and the complete mixture is further heated for 5 min in a boiling water bath. After cooling on ice, the mixture is centrifuged for 10 min at 10,000 g and the supernatant is saved. The pellet is extracted with 2 × 5-ml aliquots of solution B, and the extracts are combined with the original supernatant. The resulting fraction is dialyzed extensively against solution B and applied to a column (1 × 10 cm) of Blue Sepharose CL-6B (Pharmacia) equilibrated with solution B. The column is washed extensively with solution B containing 0.1 M NaCl and eluted with solution B containing 0.6 M NaCl. The eluate is concentrated by vacuum dialysis, dialyzed exten-

[59] J. G. Foulkes and P. Cohen, *Eur. J. Biochem.* **105,** 195 (1980).
[60] S. D. Yang, T. R. Vandenheede, and W. Merlevede, *FEBS Lett.* **132,** 293 (1981).

sively against solution B, and stored at $-20°$ in small aliquots. Inhibitor-2 (M_r 30,500) is obtained in essentially homogeneous form in an overall yield of 40%. Approximately 0.7 mg of protein is obtained from 1 kg of skeletal muscle. The presence of other proteins, which may be present in trace amounts, may be entirely eliminated by further ion-exchange chromatography on DEAE-cellulose at pH 5.0, according to Foulkes and Cohen.[59]

Glycogen Synthase Kinase-3

The procedure outlined below was developed by Hemmings *et al.*[32]

Reagents

Solution A: 10 mM Tris-HCl, pH 7.0 (0°)–1.0 mM EDTA–0.1% (v/v) 2-mercaptoethanol

Solution B: 50 mM Tris-HCl, pH 7.0 (0°)–1.0 mM EDTA–5% glycerol–0.1% (v/v) 2-mercaptoethanol

Preparation of (Glycogen Synthase)–Agarose. Affi-Gel-15 (Bio-Rad, 12.5 ml) is washed with water and resuspended in 5 ml of 100 mM sodium bicarbonate, pH 8.0, containing 15 mg of glycogen synthase. After mixing for 16 hr at 4°, 10.0 ml of 1.0 M glycine, pH 8.0, is added. After a further 1 hr at 4°, the gel is washed with 50 mM Tris-HCl, pH 7.0 (4°), containing 0.5 M NaCl, and then equilibrated with solution B. Ninety-five percent of the glycogen synthase is covalently attached to the agarose using this procedure.

Procedure. Skeletal muscle (2.4 kg), obtained from rabbits as described for the purification of inhibitor-1, is minced and homogenized with 2.5 volumes of 4 mM EDTA, pH 7.0–0.1% (v/v) 2-mercaptoethanol in a Waring blender for 40 sec at low speed. The resulting homogenate is centrifuged at 6000 g for 45 min, and the supernatant (pH 6.6–6.8) is decanted through glass wool. This step and all subsequent ones are carried out at 4°. The pH of the supernatant is adjusted to 6.1 with 1.0 N acetic acid. After 20 min, the suspension is centrifuged at 6000 g for 45 min. The supernatant is adjusted to pH 7.0 with 10 M ammonium hydroxide, and then solid ammonium sulfate is added to 33% saturation. After stirring for 45 min the suspension is centrifuged at 6000 g for 45 min. The supernatant is adjusted to 55% saturation by addition of further solid ammonium sulfate, stirred for 60 min, and then centrifuged at 6000 g for 60 min. The precipitate is resuspended in solution A, dialyzed exhaustively against the same buffer, and centrifuged at 80,000 g for 30 min. The supernatant is applied to a DEAE-cellulose column (9 × 15 cm) equilibrated with solution A. Most of the glycogen synthase kinase-3 is not retained by the column, but it is necessary to wash the column with 3 volumes of equilibration buffer to elute the remainder of the enzyme. The combined fractions are adjusted to 50 mM Tris-HCl, pH 7.0 (0°)–1.0 mM EDTA–5% (v/v) glycerol–0.1% (v/v) 2-mercaptoethanol (solution B). The material from two such preparations is combined and applied to a phosphocellulose column (3.4 × 14 cm) equilibrated in solution B. The column is washed with solution B containing 50 mM NaCl until the absorbance at 280 nm is below 0.05 and then eluted with the solution B containing 0.2 M NaCl. The most active fractions, which elute after the bulk of the protein, are combined, dialyzed against solution C, and applied to an Affi-Gel Blue column (1.5 × 3.5 cm) equilibrated with the same buffer. The column is washed with solution C containing 0.15 M NaCl and eluted with a linear gradient (1000 ml) from 0.15 to 1.0 M NaCl in solution C. The most active fractions (eluting at 0.45 M NaCl) are combined, concentrated to 10–15 ml of vacuum dialysis, and dialyzed against solution C. The fraction is then applied to a column (1.5 × 4.5 cm) of (glycogen synthase)–agarose equilibrated with solution B. The column is washed with solution B and eluted with a linear gradient (200 ml) from 0 to 0.3 M NaCl in solution B. The active fractions are pooled, concentrated by vacuum dialysis to about 5 ml, dialyzed against solution C, and rechromatographed on (glycogen synthase)–agarose as described above. Approximately 0.08 mg of protein is recovered from 5.0 kg of skeletal muscle of which 50% of the protein is glycogen synthase kinase-3. The final specific activity on the preparation is 2.1 nmol of phosphate incorporated min^{-1} mg protein^{-1}. The enzyme, purified through the Affi-Gel Blue chromatography step, is stable for at least 3 months when stored in solution B containing 50% (v/v) glycerol at −20°, but becomes much less stable after the final affinity chromatography step on glycogen synthase–agarose.

The purified enzyme maintains both Fa and glycogen synthase kinase-3 activities and is represented by a major protein on SDS–polyacrylamide gels of M_r 50,000–51,000.

Concluding Remarks

Since the discovery of cyclic AMP[61] and the subsequent demonstration of its role as an intracellular "second messenger" of β-adrenergic agents via the activation of the cyclic AMP-dependent protein kinase, considerable interest has been focused on the nature of protein kinases

[61] G. A. Robison, R. W. Butcher, and E. W. Sutherland, "Cyclic AMP." Academic Press, New York.

and their regulatory mechanisms.[62,63] In contrast, little progress was made in characterizing protein phosphatases until the mid 1970s. In the last few years, it has become increasingly apparent that protein phosphatases, as well as protein kinases, are important targets for cellular regulation, and some progress has now been made in elucidating the structure and mechanisms of regulating these enzymes.[64]

Type 1 protein phosphatases show the capacity to be regulated by the phosphorylation state of inhibitor-1 in response to β-adrenergic agonists and insulin.[26] In addition, type 1 protein phosphatases may demonstrate two interconvertible forms, differing in their association with inhibitor-2. This activation–inactivation process appears to be under the control of glycogen synthase kinase-3 (which phosphorylates inhibitor-2 to 1 mol of phosphate per mole of protein on a threonine residue[31]). The phosphorylation of glycogen synthase by this kinase may be influenced by insulin.[64] However, the precise nature of the control of protein phosphatase-1 activity by this cyclic nucleotide-independent kinase in response to hormonal stimuli remains unclear.

Type 2 protein phosphatases demonstrate differing regulation by cations. Protein phosphatase-2A may exist in a number of enzymatic forms distinguished by their association with a number of protein components that may modulate the activity of this broad-specificity protein phosphatase. The question of the potential hormonal regulation of type 2 protein phosphatases is currently unanswered.

The final issue as yet unresolved relates to both type 1 and type 2 protein phosphatases, where the molecular organization of the multiple catalytic and regulatory components in the native enzyme and the nature of their substrate recognition is not yet understood.

Acknowledgments

The authors would like to acknowledge the considerable part played by Professor Philip Cohen and his colleagues in our current understanding of the regulation of protein phosphatases, their data being a major part of this chapter. We would also like to express our gratitude to Professor Samuel J. Strada for helpful discussions during the preparation of the manuscript.

[62] D. A. Walsh, J. P. Perkins, and E. G. Krebs, *J. Biol. Chem.* **243**, 3763 (1968).
[63] P. Greengard, *Science* **199**, 146 (1978).
[64] P. Cohen, *Nature (London)* **296**, 613 (1982).

[6] Detection and Identification of Substrates for Protein Kinases: Use of Proteins and Synthetic Peptides

By Lorentz Engström, Pia Ekman, Elisabet Humble, Ulf Ragnarsson, and Örjan Zetterqvist

Two types of intracellular phosphoproteins are known: enzymes that are intermediately phosphorylated at their active sites and proteins that are phosphorylated by protein kinases on serine, threonine, or tyrosine residues. In both cases the proteins are phosphorylated after their synthesis.

Intermediate phosphoryl enzymes are formed rapidly, usually within milliseconds, on incubation of the enzymes with their substrates. At the active sites of phosphoglucomutase and alkaline phosphatase, a serine residue is phosphorylated, whereas in all other known cases a histidine, a glutamic acid, or an aspartic acid residue is the acceptor of the phosphoryl groups.[1]

When phosphoproteins are formed through the action of protein kinases, steady-state levels are usually reached only after several minutes. This type of phosphorylation mainly takes place on serine residues, while phosphorylation on threonine residues represents a minor fraction. Fairly recently, the phosphorylation of tyrosine residues by specific protein kinases has been described, both in normal cells and after infection with tumor virus. The amount of tyrosine-bound phosphate in normal cells is very low[2] compared with that of serine- and threonine-bound phosphate, which in several mammalian tissues is about 0.5–1.0 μmol per gram of wet tissue.[3,4]

Since the detection of a protein kinase in 1954,[5] several different protein kinases and their substrates have been found in a number of tissues and cell fractions. The enzymes differ with regard to substrate specificity and factors that influence their activity, as seen in the table. In the present chapter, the detection and use of substrates for protein kinases active on serine and threonine residues will be discussed, with emphasis on substrates and kinases present in mammalian tissues.

---

[1] M. Weller, *in* "Protein Phosphorylation," p. 163. Pion, London, 1979.

[2] T. Hunter and B. M. Sefton, *Proc. Natl. Acad. Sci. U.S.A.* **77,** 1311 (1980).

[3] T. A. Langan, *in* "Regulation of Nucleic Acid and Protein Biosynthesis," BBA Library, Vol. 10, p. 233. Elsevier, Amsterdam, 1967.

[4] H. Forsberg, Ö. Zetterqvist, and L. Engström, *Biochim. Biophys. Acta* **181,** 171 (1969).

[5] G. Burnett and E. P. Kennedy, *J. Biol. Chem.* **211,** 969 (1954).

Detection of Protein Phosphorylation in Crude Tissue Extracts

The search for a particular protein phosphorylation can be made with use of whole animals, perfused tissues, intact cells, cell fractions, or crude tissue extracts. Although with intact cells there may be a lower risk of artifactual phosphorylation, crude tissue extracts usually offer a more convenient system for an initial scanning. In this section, experiments on crude extracts will be described. Phosphorylation in intact cells is discussed in a later section.

The crude tissue extract may consist in isolated cell supernatant or extracts of the particulate cell fractions. When the extracts are incubated with [γ-^{32}P]ATP and Mg^{2+}, considerable incorporation of [^{32}P]phosphate into proteins occurs, owing to the presence of protein kinases and endogenous substrates. This has been illustrated in experiments with rat liver cell sap.[6] A fairly large incorporation of [^{32}P]phosphate, i.e., near 1 nmol per milligram of cell sap protein, has been observed after incubation with 5 mM [^{32}P]ATP for 60 min at 30°. By incubation of the same material for a short period at low temperature, e.g., 15 sec at 0°, the incorporation of [^{32}P]phosphate into intermediate phosphoryl enzymes can be estimated, as was demonstrated for rat liver cell sap. This incorporation is of a much lower degree than that due to the protein kinase reactions and amounts to about 0.04 nmol per milligram of cell sap protein.[6,7]

A significant part of the ^{32}P-labeling is accounted for by a few, relatively abundant proteins. However, the detection of phosphorylation of a particular, minor component may be of equal interest. This usually requires a highly efficient separation method, such as the two-dimensional polyacrylamide gel electrophoresis of O'Farrell.[8] In the few cases in which the phosphoprotein is identified and specific antibodies are available, immunoprecipitation followed by one-dimensional polyacrylamide gel electrophoresis in sodium dodecyl sulfate may be used.

To achieve detectable ^{32}P-labeling of as many phosphoproteins of the extract as possible, several aspects have to be considered.

1. Although the apparent K_m with respect to ATP is rather low (10^{-5} to 10^{-4} M) for a number of protein kinases, the concentration of [^{32}P]ATP should initially be in the millimolar range to allow extensive ^{32}P labeling before the [^{32}P]ATP is hydrolyzed by ATP hydrolases.

2. To detect phosphorylation of a minor protein, the specific radioactivity of the [^{32}P]ATP should be high enough to give a detectable spot on

[6] O. Ljungström and L. Engström, *Biochim. Biophys. Acta* **336**, 140 (1974).
[7] Ö. Zetterqvist, *Biochim. Biophys. Acta* **141**, 533 (1967).
[8] P. H. O'Farrell, *J. Biol. Chem.* **250**, 4007 (1975).

an autoradiogram of a polyacrylamide gel electrophoresis. With enforcing screens and preexposure of the film with a photo flash, about 50 dpm of ^{32}P per square centimeter is detectable in 24 hr.[9] A specific radioactivity of 20 dpm/pmol would then permit the detection of a compound representing 1/1000 of the ^{32}P-labeled phosphoproteins of cell sap from 1 mg of liver tissue (i.e., from about 50 μg of cell sap protein).

3. Allowance should be made for the endogenous ATP present in the extract, if this compound has not been removed on a Sephadex G-50 column.

4. The specific radioactivity of the [^{32}P]ATP will decrease during the incubation, owing to the action of ATP hydrolases and adenylate kinase. To permit true estimates of the phosphorylation, [^{32}P]ATP therefore has to be isolated and the specific radioactivity of the γ-phosphorus determined.[6]

5. Phosphorylated or phosphorylatable sites may be more sensitive to proteolysis than the rest of the protein.[10] Inhibitors of proteolytic enzymes should therefore be added to the incubation mixture to reduce this risk.

6. Phosphorylatable proteins may have been partially phosphorylated in the cell. Thus a small or slow incorporation of [^{32}P]phosphate might be due to the presence of bound, nonradioactive phosphate at the beginning of the incubation, and the observed ^{32}P labeling will rather be a consequence of the turnover of the phosphate caused by the activity of phosphoprotein phosphatases present in the system.

The method of interruption of the phosphorylation greatly influences the amount and type of [^{32}P]phosphoproteins obtained. When further studies on native phosphoprotein are to be made, the reaction may be interrupted by the removal of free Mg^{2+} with EDTA. Phosphoprotein phosphatases may be inhibited by sodium fluoride or orthophosphate.[11] The ^{32}P-labeled phosphoproteins are separated from most of the low molecular weight, ^{32}P-labeled compounds by rapid gel filtration. It is important to keep in mind, however, that considerable amounts of these compounds may be adsorbed to the protein of crude extracts after the gel filtration. To obtain reliable estimates of the true protein phosphorylation,

[9] R. A. Laskey and A. D. Mills, *FEBS Lett.* **82,** 314 (1977).
[10] G. Bergström, P. Ekman, E. Humble, and L. Engström, *Biochim. Biophys. Acta* **532,** 259 (1978).
[11] T. S. Ingebritsen and D. M. Gibson, *in* "Recently Discovered Systems of Enzyme Regulation by Reversible Phosphorylation" (P. Cohen, ed.), p. 63. Elsevier/North-Holland, Amsterdam, 1982.

the adsorbed material must be removed by denaturation procedures (see below).

To prove definitely that an apparent protein phosphorylation is due to covalent binding to serine and threonine residues, it is necessary to isolate the phosphoamino acid after degradation of the protein, generally by partial acid hydrolysis in 2 M HCl for 10–20 hr at 100°, a method originally introduced by Lipmann.[12] After lyophilization of the acid hydrolysates, the phosphoamino acids are generally isolated by chromatography[13] or electrophoresis.[14] In this context, the importance of using more than one system at the identification of a particular phosphoamino acid is emphasized.

Identification of Protein Kinase Activities

If phosphorylation of an unknown protein substrate has been found in a crude tissue extract, attempts to identify the protein kinase activity and the substrate can be performed in parallel. In some cases, the substrate protein and the protein kinase activity may be enriched in the same fraction during a preliminary purification. Thus, rat liver pyruvate kinase and cyclic AMP-dependent protein kinase copurified during an acid precipitation (pH 5.5) and an ammonium sulfate fractionation step, which facilitated the identification of liver pyruvate kinase as substrate of cyclic AMP-dependent protein kinase.[15]

Use of Activators and Inhibitors of Protein Kinases

Addition of different activators of protein kinases (see the table), e.g., cyclic AMP, cyclic GMP, calcium ions with or without calmodulin, or phosphatidylserine and diolein, to crude systems may increase the rate of the phosphorylation reaction and thus point to the type of protein kinase active in the system.

The cyclic AMP-dependent protein kinase activity of a crude fraction can be inhibited by adding the protein inhibitor described by Walsh.[16] Inhibition of calcium-stimulated protein kinases can be achieved by the

---

[12] F. Lipmann, *Biochem. Z.* **262**, 3 (1933).

[13] L. Engström, *Biochim. Biophys. Acta* **52**, 49 (1961).

[14] J. A. Cooper, N. A. Reiss, R. J. Schwartz, and T. Hunter, *Nature (London)* **302**, 218 (1983).

[15] O. Ljungström, G. Hjelmquist, and L. Engström, *Biochim. Biophys. Acta* **358**, 289 (1974).

[16] D. A. Walsh, C. D. Ashby, C. Gonzalez, D. Calkins, E. H. Fischer, and E. G. Krebs, *J. Biol. Chem.* **246**, 1977 (1971).

addition of EGTA. The concomitant inhibition of the phosphorylation of the unidentified protein will give an indication as to which protein kinase is active.

Addition of Purified Protein Kinases

If the phosphorylation of an endogenous protein is enhanced by the addition of a purified kinase preparation, this indicates that the corresponding protein kinase is responsible for the phosphorylation of that protein when studied in the crude extract. However, it should be kept in mind that many proteins are phosphorylated by more than one protein kinase, as has been most clearly demonstrated for rabbit muscle glycogen synthase (see the table). In such cases, it seems to be common that different residues are phosphorylated by the different protein kinases. One way to establish that different residues are involved is to compare the patterns of the [^{32}P]phosphopeptides obtained after degradation by proteolytic enzymes and by partial acid hydrolysis.[17]

Tissue Distribution of a Protein Kinase Active on an Identified Substrate

If the substrate of a protein kinase has been identified and is available for enzyme assays, while the protein kinase itself is unidentified, investigation of the tissue distribution of the protein kinase activity[18–21] may facilitate the latter identification. This approach was used to identify the calcium-activated, phospholipid-dependent protein kinase described by Nishizuka and collaborators[22] as one of the enzymes that phosphorylates human fibrinogen.[23]

Identification of Natural Substrates for Protein Kinases

Preliminary Characterization of a Phosphorylatable Component

Some guidance for the identification of phosphorylatable proteins may be obtained from the estimation of their native and subunit molecular

[17] C. Milstein and F. Sanger, *Biochem. J.* **79**, 456 (1961).
[18] J. F. Kuo and P. Greengard, *Proc. Natl. Acad. Sci. U.S.A.* **64**, 1349 (1969).
[19] J. F. Kuo, *Proc. Natl. Acad. Sci. U.S.A.* **71**, 4037 (1974).
[20] J. F. Kuo, R. G. G. Andersson, B. C. Wise, L. Mackerlova, I. Salomonsson, N. L. Brackett, N. Katoh, M. Shoji, and R. W. Wrenn, *Proc. Natl. Acad. Sci. U.S.A.* **77**, 7039 (1980).
[21] R. Minakuchi, Y. Takai, B. Yu, and Y. Nishizuka, *J. Biochem. (Tokyo)* **89**, 1651 (1981).
[22] Y. Takai, A. Kishimoto, Y. Iwasa, Y. Kawahara, T. Mori, and Y. Nishizuka, *J. Biol. Chem.* **254**, 3692 (1979).
[23] P. Papanikolaou, E. Humble, and L. Engström, *FEBS Lett.* **143**, 199 (1982).

weights, e.g., by gel chromatography and by polyacrylamide gel electrophoresis in sodium dodecyl sulfate under reducing conditions, respectively. Other properties may also be investigated, e.g., their behavior in ion-exchange chromatography in different systems or the isoelectric point as determined by isoelectric focusing or chromatofocusing.

The amount of the phosphorylatable component is also useful for its identification. This may be estimated from the maximal [^{32}P]phosphate incorporation obtained, under the assumption that only one phosphate group is incorporated per subunit.

If the phosphorylatable protein is an adaptive enzyme, some clues to its identity may be provided by feeding the experimental animals with different diets and then determining the amounts of phosphorylatable proteins. The tissue distribution and subcellar location of the phosphorylatable component may also be helpful in the identification.

Identification from Known Hormonal Effects

The effect of many hormones at the molecular level is to increase the intracellular concentration of cyclic AMP, whose main, or perhaps sole, effect in mammals is to activate cyclic AMP-dependent protein kinase. The physiological actions of these hormones are thus exerted through phosphorylation of specific proteins. An unknown substrate of cyclic AMP-dependent protein kinase in a tissue should thus be related to some process that is regulated by hormones acting on the tissue via cyclic AMP. The identification of liver pyruvate kinase type I, as a substrate of cyclic AMP-dependent protein kinase was facilitated by this approach.[15] Glucagon is the main hormone that increases the concentration of cyclic AMP in the liver. An important physiological effect of glucagon is to raise the level of blood glucose. This is accomplished partly by increasing the mobilization of glycogen and partly by stimulating gluconeogenesis with an increased synthesis of phosphoenolpyruvate from pyruvate.[24] On this basis, a protein in rat liver cell sap that was found to be phosphorylated under the influence of cyclic AMP and had the same subunit molecular weight and approximately the same concentration as pyruvate kinase could be identified as this enzyme.

In a similar way, Avruch and collaborators have identified a protein of fat and liver tissues, whose phosphorylation is stimulated by insulin and is identical to ATP citrate-lyase.[25]

[24] G. A. Robison, R. W. Butcher, and E. W. Sutherland, "Cyclic AMP," p. 232. Academic Press, New York, 1971.
[25] M. C. Alexander, E. M. Kowaloff, L. A. Witters, D. T. Dennily, and J. Avruch, J. Biol. Chem. **254,** 8052 (1979).

Studies of Amino Acid Sequences as a Means of Identifying Protein Substrates

A number of substrates of cyclic AMP-dependent protein kinase, e.g., pyruvate kinase, glycogen synthase, and phosphorylase kinase show similarities between their phosphorylated sites (see the table). Concerning liver pyruvate kinase, it has been found that the native conformation of the enzyme is not a prerequisite for phosphorylation, since both alkali-inactivated enzyme and a cyanogen bromide fragment from the same protein are phosphorylated at a greater rate than the native enzyme, although the final degree of phosphorylation is the same.[26] These observations initiated phosphorylation experiments with synthetic peptides representing part of the phosphorylatable site of pyruvate kinase, in which the essential role of the amino acid sequence around the phosphorylatable amino acid residue was demonstrated. Thus, by studying known primary structures of proteins, it should be possible to find substrates of cyclic AMP-dependent protein kinase. This method led to the finding that fibrinogen is a substrate *in vitro* of cyclic AMP-dependent protein kinase.[27]

However, the mere existence of a favorable amino acid sequence in a protein does not make it a protein kinase substrate. Another prerequisite is that the phosphorylation site be accessible to the protein kinase. In fact, a protein may be made phosphorylatable to a certain extent by denaturation.[28] Obviously, caution is needed in the interpretation of phosphorylation *in vitro*. Eventually, its occurrence *in vivo* always has to be considered.

Synthetic Peptides as Substrates of Protein Kinases

[^{32}P]Phosphopeptides were derived from phosphorylated pig and rat liver pyruvate kinase[29,30] and shown to exhibit one distinct feature, namely, two vicinal arginine residues. This feature, together with results from experiments with denatured pyruvate kinase,[26] induced our group[31]

[26] E. Humble, L. Berglund, V. Titanji, O. Ljungström, B. Edlund, Ö. Zetterqvist, and L. Engström, *Biochem. Biophys. Res. Commun.* **66**, 614 (1975).

[27] L. Engström, B. Edlund, U. Ragnarsson, U. Dahlqvist-Edberg, and E. Humble, *Biochem. Biophys. Res. Commun.* **96**, 1503 (1980).

[28] D. B. Bylund and E. G. Krebs, *J. Biol. Chem.* **250**, 6355 (1975).

[29] G. Hjelmquist, J. Andersson, B. Edlund, and L. Engström, *Biochem. Biophys. Res. Commun.* **61**, 559 (1974).

[30] B. Edlund, J. Andersson, V. Titanji, U. Dahlqvist, P. Ekman, Ö. Zetterqvist, and L. Engström, *Biochem. Biophys. Res. Commun.* **67**, 1516 (1975).

[31] Ö. Zetterqvist, U. Ragnarsson, E. Humble, L. Berglund, and L. Engström, *Biochem. Biophys. Res. Commun.* **70**, 696 (1976).

and other investigators[32] to explore synthetic peptides based on the phosphorylatable site of liver pyruvate kinase.

Peptide Synthesis

Synthesis of peptides consisting of up to about 10 amino acid residues can conveniently be carried out by the solid-phase method.[33] The early and simple manual version[34] is quite satisfactory as long as only a moderate number of peptides are required, and it is very suitable for gaining some practical experience of solid-phase peptide synthesis.[35] If large series of peptides are required, however, the synthetic manipulations can be speeded up considerably by using an automatic synthesizer.

Synthesis of Leu-Arg-Arg-Ala-Ser-Val-Ala. As an example, the synthesis of a heptapeptide, representing the phosphorylatable site of rat liver pyruvate kinase, will be described in some detail.

Chloromethylated, 1% cross-linked polystyrene (Biobeads, Bio-Rad) is esterified with Boc-alanine.[34,35] This resin, 1.75 g, containing 0.400 mmol of Boc-Ala, is loaded into the reaction vessel of a Beckman Model 990 peptide synthesizer. The Boc group is removed by exposure to a solution of 33% trifluoroacetic acid in dichloromethane for 30 min. Free amino groups are liberated by neutralization with 10% triethylamine in the same solvent for a total of 10 min. After further careful washing with CH_2Cl_2, the next amino acid, valine, is attached as its Boc derivative (2.5 equivalents) in CH_2Cl_2 with the aid of dicyclohexylcarbodiimide (also 2.5 equivalents) for 2 hr. To ensure the highest possible yield in this step, the coupling is repeated once with the same amounts of fresh reagents before proceeding.

The protecting Boc group is removed from the valine residue as described above and the next amino acid, serine, is attached, using the conditions described above. The serine derivative used is Boc-Ser (Bzl). Similarly, Boc-Ala, Boc-Arg(NO$_2$) (twice), and Boc-Leu are added. As Boc-Arg(NO$_2$) does not dissolve in CH_2Cl_2, a mixture of dimethylformamide and CH_2Cl_2 (2 and 9 ml, respectively) is used in these coupling steps.

The peptide is cleaved from the resin with liquid hydrogen fluoride.[35] Simultaneously, the groups protecting the side chains are removed from the peptide. After extraction of the polymer with 10% acetic acid, the peptide solution is lyophilized.

[32] B. E. Kemp, D. J. Graves, E. Benjamini, and E. G. Krebs, *J. Biol. Chem.* **252,** 4888 (1977).

[33] R. B. Merrifield, *J. Am. Chem. Soc.* **85,** 2149 (1963).

[34] R. B. Merrifield, *Biochemistry* **3,** 1385 (1964).

[35] J. M. Stewart and J. D. Young, "Solid Phase Peptide Synthesis." Freeman, San Francisco, California, 1969.

An aliquot (about 100 mg of the product) is subjected to chromatography on a 1.4×16 cm column of carboxymethylcellulose. After equilibration with an ammonium acetate buffer, pH 5.5, 0.025 M in ammonium ions, elution is performed with a 600-ml linear gradient of 0.025 to 0.25 M ammonium acetate, at a flow rate of about 20 ml/hr. The peptide, which appears in the middle of the gradient, is traced by its absorbance at 235 nm, appropriate fractions are lyophilized, the residue is dissolved in water, and the solution is again lyophilized to give about 50 mg of a crystalline material. This preparation initially contains insignificant amounts of ammonium ions as seen at amino acid analysis. A typical analysis gave Arg 2.03, Ser 0.96 (after correcting for 11% decomposition during the acid hydrolysis), Ala 2.00, Val 0.99, Leu 1.01, and a peptide content of 62%.

The purity of this and similar peptides can be further determined by HPLC. The basicity and hydrophilicity of the peptide require a specially designed ion-pair chromatographic system.[36] Incidentally, the same system can be applied to separate the peptide from the corresponding phosphopeptide and also to separate different phosphopeptides from each other.[37]

Minimum Peptide Substrates for Cyclic AMP-Dependent Protein Kinase

Smaller peptides, lacking N-terminal Leu, Leu-Arg, and Leu-Arg-Arg or C-terminal Ala and Val-Ala, can be similarly prepared and assayed as substrates for the cyclic AMP-dependent protein kinase. With use of a set of these peptides, the peptide Arg-Arg-Ala-Ser-Val was shown to represent the minimum substrate phosphorylated at a significant rate.[31]

By a similar approach, a second type of substrate of cyclic AMP-dependent protein kinase, based on the sequence of a fragment from the β subunit of phosphorylase kinase,[38] has been explored.[39] In this case, the minimum substrate proved to be Arg-Thr-Lys-Arg-Ser-Gly-Ser-Val (Ser at the seventh position being phosphorylated). Although the two arginine residues of this peptide were not located next to each other, both were apparently essential to the phosphorylation, as shown by the substitution of glycine for either arginine residue. In addition, substitution of glycine for the lysine residue showed that the lysine was also of significance to the phosphorylation, although the influence of this exchange on the apparent K_m was less than with the arginine residues. The second hydroxyamino

[36] B. Fransson, U. Ragnarsson, and Ö. Zetterqvist, J. Chromatogr. 240, 165 (1982).
[37] B. Fransson, U. Ragnarsson, and Ö. Zetterqvist, Anal. Biochem. 126, 174 (1982).
[38] S. J. Yeaman, P. Cohen, D. C. Watson, and G. H. Dixon, Biochem. J. 162, 411 (1977).
[39] Ö. Zetterqvist and U. Ragnarsson, FEBS Lett. 139, 287 (1982).

acid, threonine, on the other hand, seemed to be replaceable without significant changes of the kinetic parameters.

Cyclic GMP-Dependent Protein Kinase

The specificity of cyclic GMP-dependent protein kinase, obtained from pig lung,[40] has also been investigated with use of synthetic peptides.[41,42] This enzyme seems to have structural requirements that are similar, although not identical, to those of cyclic AMP-dependent protein kinase. One observed difference is that, for a significant rate of phosphorylation, the cyclic GMP-dependent protein kinase seems to require two amino acid residues C-terminal to the phosphorylatable serine, compared with only one in the two minimum peptides for cyclic AMP-dependent protein kinase.

Techniques for Quantification of Phosphorylation of Proteins and Synthetic Peptides

Proteins

Incubation Mixtures. The concentration of protein in crude extracts may amount to about 10–15 mg/ml. When purified protein substrate is used, a practical concentration is around 2 mg/ml, or 50 μM. A specific radioactivity of [^{32}P]ATP of about 20 dpm/pmol for the γ-phosphorus and a 5 mM concentration of free Mg^{2+} are generally adequate. Several types of buffers have been used,[43] although in some cases orthophosphate has been shown to inhibit protein kinase activity.[44] However, in crude extracts this inhibition may only be apparent, since orthophosphate inhibits phosphoprotein phosphatase and will retard the removal of unlabeled phosphate present in the protein *in vivo*.

For experiments with the Ca^{2+}, calmodulin-activated protein kinase, 50 μM Ca^{2+}, and 0.57 μM calmodulin (10 μg/ml) may be used. The cyclic nucleotide-dependent protein kinases need about 1 μM cAMP or cGMP in a system with purified substrate and protein kinase. This concentration should be increased to 100 μM if the incubation mixture is suspected to contain diesterase activity, which hydrolyzes the cyclic nucleotide. The

[40] K. Nakazawa and M. Sauo, *J. Biol. Chem.* **250,** 7415 (1975).

[41] T. M. Lincoln and J. D. Corbin, *Proc. Natl. Acad. Sci. U.S.A.* **74,** 3239 (1977).

[42] B. Edlund, Ö. Zetterqvist, U. Ragnarsson, and L. Engström, *Biochem. Biophys. Res. Commun.* **79,** 139 (1977).

[43] S. A. Rudolph and B. K. Krueger, *Adv. Cyclic Nucleotide Res.* **10,** 107 (1979).

[44] J. H. Wang, J. T. Stull, T.-S. Huang, and E. G. Krebs, *J. Biol. Chem.* **251,** 4521 (1976).

diesterase may also be inhibited by the addition of 1 mM methylxanthine. The Ca^{2+} concentration required for phospholipid-activated protein kinases is in the order of 10^{-7} to 10^{-4} M, and concentrations exceeding 0.5 mM are usually inhibitory. A commonly used concentration of phosphatidylserine is 10–25 μg/ml.

In systems that are not extensively purified, the phosphorylation should be preferably performed at 30° for a rather short time, i.e., 5–10 min, since contaminating proteases and phosphoprotein phosphatases may decrease the recovery of the protein-bound [^{32}P]phosphate. Experiments with pure systems may be continued for about 2 hr when the maximal incorporation is to be estimated.

Termination of the Reaction and Analysis of ^{32}P Labeling. Two methods of interrupting the reaction and determining the phosphorylation of the total protein of the sample will be described. In the first method, the incubation is interrupted by applying a 25-μl aliquot of the incubation mixture to a Whatman 3 MM paper disk, 11 × 11 mm. The paper is then immediately placed in a stainless, wire-netting support with separate compartments for each paper, which stands in a beaker containing ice-cold 10% trichloroacetic acid and 50 mM phosphoric acid. The presence of 50 mM phosphoric acid in the trichloroacetic acid is important for preventing the adsorption of a possible trace amount of [^{32}P]orthophosphate present in the [^{32}P]ATP preparation. Free [^{32}P]ATP is removed by washing the papers for 10 min in the acid under magnetic stirring. This washing is repeated twice with fresh acid. The paper disks are then dried in ethanol for 5 min and, finally, in ethyl ether for 5 min. The papers are left in the wire-netting support throughout the washing and drying procedure. This procedure is similar to that described by Corbin and Reimann.[45] However, with the wire-netting support the washing is more efficient and more reproducible. The dry filter paper disk is placed on the bottom of an empty scintillation vial, and the Čerenkov radiation is measured in a liquid scintillation counter.

When it is suspected that a protein is not quantitatively adsorbed to the paper (which is the case for instance with fructose-1,6-bisphosphatase) or when the incubation volume has to be increased to facilitate detection of the incorporated [^{32}P]phosphate, precipitation with trichloroacetic acid in a test tube is preferred. It should be noted, however, that the concentration of acid required for quantitative precipitation can vary between different proteins. In most cases, a final trichloroacetic acid concentration of 10% is sufficient. Some basic proteins, however, require

---

[45] J. D. Corbin and E. M. Reimann, this series, Vol. 38 [41], p. 287.

the addition of tungstate to be precipitated.[46] The trichloroacetic acid may be added as solid acid or, preferably, as an aliquot of a concentrated stock solution. If the protein concentration is below 1 mg/ml, a carrier protein, for instance bovine serum albumin, can be added just before the addition of trichloroacetic acid. After immersion of the mixture in an ice-water bath for 10 min, the precipitate is collected by centrifugation at 1000 g for 5 min, dissolved in 0.5 M NaOH, and immediately reprecipitated by trichloroacetic acid at the appropriate final concentration. The protein should be dissolved and reprecipitated totally 4 times. After the last centrifugation, the protein is dissolved in 0.5 M NaOH and the Čerenkov radiation is measured.[47] Serine- and threonine-bound phosphate is fairly easily removed from phosphoproteins in strong alkaline solutions by β-elimination, even at room temperature.[48] Therefore, any treatment of this type of phosphoprotein with strong alkaline solutions should be short and preferentially performed in an ice-water bath. On the other hand, the bound phosphate remains very stable when exposed to strong acid at room temperature. Treatment with 1 M HCl for 30 min at room temperature will therefore remove histidine- and aspartic acid-bound phosphate without losses of serine- or threonine-bound phosphate.[7]

Preparation of Native Phosphoprotein. When it is desirable that a purified protein should remain in a native state after phosphorylation, the phosphoprotein and ATP may be separated by Sephadex G-50 gel chromatography. The phosphorylation reaction can be stopped by the addition of EDTA before the chromatography if the reaction time is critical. Otherwise, the reaction will be stopped upon separation of protein from ATP on the column. The volume of the sample should not exceed 10% of the Sephadex G-50 column volume, in order to get a good separation. The use of a phosphate buffer with a concentration of at least 50 mM is of advantage to decrease the adsorption of [^{32}P]ATP and trace amounts of [^{32}P]orthophosphate that are usually present in the [^{32}P]ATP preparation. Rechromatography usually reduces the adsorption even further. The degree of adsorption may be determined by treating an aliquot as described in the preceding paragraph.

Synthetic Peptides

Incubation Mixtures. Synthetic peptides have been used as substrates of various protein kinases, such as cyclic AMP-dependent protein kinase,[31,32] cyclic GMP-dependent protein kinase,[41,42] phosphorylase ki-

[46] G. N. Gill and G. M. Walton, *Adv. Cyclic Nucleotide Res.* **10,** 93 (1979).
[47] S. Mårdh, *Anal. Biochem.* **63,** 1 (1975).
[48] G. Taborsky, *Adv. Protein Chem.* **28,** 1 (1974).

nase,[32] and casein kinase.[49] In experiments with cyclic AMP-dependent protein kinase, the presence of bovine serum albumin keeps the activity of the usually highly diluted protein kinase reasonably constant during the incubation.[32] This is particularly important in kinetic studies. Another difficulty with kinetic experiments arises during measurement of the rate of phosphorylation of peptides with low K_m values. In order to measure the rate at sub-K_m concentrations of the peptide, the radiochemical purity of [^{32}P]ATP is particularly important. Even trace amounts of labeled impurities may give unacceptably high blank values.

Termination of the Reaction. The method for terminating the reaction depends on the further processing of the material. In kinetic experiments trichloroacetic acid[31] has been added to a final concentration of 10% or acetic acid[32] to a final concentration of 30%, and in the preparation of larger amounts of phosphopeptide, boiling for 3 min has been used.[50]

Analysis of Phosphorylation. Separation of [^{32}P]phosphopeptide from [^{32}P]ATP in kinetic experiments on purified protein kinase is achieved by electrophoresis[31] or by ion-exchange chromatography, e.g., on AG-1-X8, equilibrated with 30% acetic acid and packed in Pasteur pipettes.[32] In the former case, the trichloroacetic acid-containing sample is neutralized by five extractions with water-saturated ethyl ether. Electrophoresis is performed on Whatman 3 MM paper or on a thin layer of cellulose powder on a plastic base, e.g., Polygram Cel 300 (Machery-Nagel, Düren, Germany). Radioactive spots are located by autoradiography, cut out, and counted in empty scintillation vials (Čerenkov radiation). When isolated by an anion exchanger, the [^{32}P]phosphopeptide is eluted directly into an empty scintillation vial. The eluate is made alkaline with sodium hydroxide before counting, to prevent the evaporation of acetic acid in the scintillation counter.

Preparation of Phosphopeptides. For preparation purposes, the phosphopeptide is easily separated from unphosphorylated peptide and ATP by ion-exchange chromatography, preferably on CM-cellulose.[50] The phosphopeptide is easily located in the chromatogram by the addition of trace amounts of ^{32}P-labeled phosphopeptide to the sample loaded onto the column. More recently, HPLC systems with a capacity to separate a number of phosphopeptides from each other have been developed.[37]

Phosphorylation of Proteins *in Vivo* and in Intact Cells

Ideally, in an unprejudiced search for a physiologically relevant protein phosphorylation, intact animals injected with [^{32}P]orthophosphate

[49] B. E. Kemp, E. Benjamini, and E. G. Krebs, *Proc. Natl. Acad. Sci. U.S.A.* **73,** 1038 (1976).
[50] V. P. K. Titanji, Ö. Zetterqvist, and U. Ragnarsson, *FEBS Lett.* **78,** 86 (1977).

should be used. In practice, however, this approach presents a number of difficulties, one being the limits as to the specific radioactivity that can be used in such a system.

An alternative approach is to investigate isolated cells, obtained by collagenase treatment of an organ, for instance, the liver,[51] heart,[52] or fat pads.[53] If 4 MBq of [^{32}P]orthophosphate is added to a suspension of 3×10^6 hepatocytes in 1 ml of Krebs–Ringer solution, containing 1.18 μmol of orthophosphate, the specific radioactivity of the γ-phosphorus of [^{32}P]ATP in the cells will increase to a steady state of about 10 dpm/pmol in 30–40 min at 37°. Thus, the total amount of radioactivity handled may be considerably lower than that required in experiments on whole animals. Nevertheless, the specific radioactivity is sufficiently high to permit detection even of minor phosphoproteins.

There is a need for rapid isolation of [^{32}P]phosphoprotein to avoid losses of the [^{32}P]phosphate bound and for removal of other phosphoproteins in order to quantitate the ones under investigation. One method is to precipitate the protein from a homogenate of the ^{32}P-labeled cells by means of ammonium sulfate. The phosphorylation–dephosphorylation will then stop, owing to the lack of Mg^{2+} and ATP. A particular [^{32}P]phosphoprotein can also be isolated by immunoprecipitation, and the [^{32}P]phosphate incorporated can be measured directly in the immunoprecipitate. Another procedure is to lyse the cells in denaturing media that are used to prepare samples for one- or two-dimensional polyacrylamide gel electrophoresis.[8]

After the preincubation of the cells with radioactive phosphate, the excess can be removed by centrifugation and the incubation of the cells can continue in the presence or the absence of hormones that are known or presumed to affect the phosphorylation of the protein in question. In addition to samples for measurement of the incorporation of [^{32}P]phosphate, samples for determination of the specific radioactivity of the γ-phosphorus of the [^{32}P]ATP should be collected.[6]

Concluding Remarks

For a protein phosphorylation to be physiologically important, certain criteria have to be fulfilled.[54–56] A main criterion is that the phosphory-

[51] M. N. Berry and D. S. Friend, *J. Cell Biol.* **43**, 506 (1969).
[52] T. Powell and V. W. Twist, *Biochem. Biophys. Res. Commun.* **72**, 327 (1976).
[53] M. Rodbell, *J. Biol. Chem.* **239**, 375 (1964).
[54] E. G. Krebs, *in* "Endocrinology, Proceedings of the Fourth International Congress of Endocrinology," p. 17. Excerpta Medica, Amsterdam, 1973.
[55] E. G. Krebs and J. A. Beavo, *Annu. Rev. Biochem.* **48**, 923 (1979).
[56] H. G. Nimmo and P. Cohen, *Adv. Cyclic Nucleotide Res.* **8**, 145 (1977).

lation occurs *in vivo*. In order to study the *in vivo* phosphorylation of minor components, extremely high specific radioactivity of the endogenous ATP is required. This may be impossible to achieve with whole animals, in which case experiments on isolated cells will have to be accepted.

One way to establish that the phosphorylation detected in an *in vitro* system is the same as that observed *in vivo* or in intact cells is to determine the amino acid sequence near the phosphorylated amino acid residue. Alternatively, the pattern of [^{32}P]phosphopeptides obtained after enzymatic digestion of partial acid hydrolysis[17] may be investigated.

For many studies on protein phosphorylation, however, broken-cell preparations are employed. If a protein in a crude extract is phosphorylated by a purified protein kinase, it may be reasonable to assume that the phosphorylation is of physiological significance. However, artifactual phosphorylation may be obtained upon denaturation. It has been shown that a short segment of the peptide chain around the phosphorylatable amino acid contains the information that is sufficient for a significant rate of phosphorylation.[31] This points to the risk that partial denaturation of a protein will unmask a phosphorylatable site that is not exposed in the native state. It has been demonstrated, in fact, that denatured lysozyme, in contrast to the native enzyme, is phosphorylated by cyclic AMP-dependent protein kinase.[28]

Synthetic peptides, representing the phosphorylatable site of a protein, have proved to be invaluable tools in the elucidation of the mechanism of, and structural requirements in, protein phosphorylation. However, in the experience of the authors, peptides that are phosphorylatable by cyclic AMP-dependent protein kinase are extremely sensitive to peptidases of crude extracts. This is true also for the corresponding phosphopeptides.[57] Since the cleavage may separate the arginine residues from the remainder of the phosphopeptide, the initially basic phosphopeptide turns into an acid compound that may not be detected in an assay procedure designed to detect basic phosphopeptides. The use of peptide substrates in the determination of cyclic AMP-dependent protein kinase activity in crude extracts may therefore be subject to errors.

The number of enzymes and other identified proteins that are phosphorylated by protein kinases has been increasing rapidly in recent years. The table represents an attempt at a comprehensive list with respect to enzymes. With a few exceptions, only mammalian protein kinases and substrates are listed. Phosphorylations in which the substrate has been less well defined are mostly excluded. Despite the length of the list, a number of intracellular substrates of protein kinases are still unidentified.

[57] Ö. Zetterqvist and R.-M. Bålöw, *Biochem. Soc. Trans.* **9**, 233 P (1981).

PROTEIN SUBSTRATES OF PROTEIN KINASES[a]

Protein kinase	Inhibitors and activators of the protein kinase	Protein substrate	Effect of the phosphorylation on the biological activity of the substrate	Amino acid sequence at the phosphorylated site	References[b]
cAMP-dependent protein kinase = glycogen synthase kinase I	cAMP (+)	Acetyl-CoA carboxylase	−		1
	Walsh inhibitor (−)	Actin			2
		ATP citrate-lyase			3
		Cholesterol esterase	+	ThrAlaSerPPheSerGlu	4, 5
		Cyclic nucleotide phosphodiesterase			6
		Fibrinogen		ThrThrArgArgSerPCysSerLys	7
		Filamin			8
		Fructose-1,6-bis-phosphatase	+	SerArgTyrSerPLeuProLeu	9
		Fructose-2,6-bis-phosphatase/fructose-6-P,2-kinase	+		10
			−		11
		Fructose-6-P kinase	−	SerArgLysArgSerPGlyGluAla	12, 13
		Glycogen synthase	−	ProArgArgAlaSerPCysThrSer	14
				LysArgSerAsnSerPValAspThr	14
				SerArgThrLeuSerPValSerSer	14
				ArgArgLysAspThrPProAlaLeu	15
		G-substrate			16
		Guanylate cyclase	+		
		Histone		ArgLysAlaSerPGlyProPro (H1)	17, 18
				ThrArgSerSerPArg (H2A)	17, 18
				LysArgSerPArgLysGluSerPTyr (H2B)	17, 18

(continued)

PROTEIN SUBSTRATES OF PROTEIN KINASES (*continued*)

Protein kinase	Inhibitors and activators of the protein kinase	Protein substrate	Effect of the phosphorylation on the biological activity of the substrate	Amino acid sequence at the phosphorylated site	References[b]
		HMG 14	+	LysArgLysValSerPSerAlaGlu	19
		Hormone-sensitive lipase/diglyceride lipase			20, 21
		Lipomodulin	−		22
		Myelin basic protein		GlyArgGlyLeuSerPLeuSerArg	23
				ArgHisArgAspThrPGlyIleLeu	23
				GlnArgHisGlySerPLysTyrLeu	23
		Myosin light-chain kinase	−		24, 25
		Na$^+$,K$^+$-ATPase			26
		Phenylalanine hydroxylase	+	SerArgLysLeuSerPAsxPheGly	27
		Phosphatase inhibitor 1	+	ArgArgArgProThrPProAlaThr	14
		Phosphorylase kinase	(+)	PheArgArgLeuSerPIleSerThr (α-chain)	28
			+	LysArgSerGlySerP$^{Val}_{Ile}$TyrGlu (β-chain)	14
		Protamine		ArgArgSerSerPArgProIle	29
		Pyruvate kinase	−	LeuArgArgAlaSerP$^{Val}_{Leu}$AlaGlx	30
		RNA polymerase	+		31
		Self: Catalytic subunit		GluIleAlaValSerPIleAsnGlu	32, 33
		Regulatory subunit (R_{II})	−	AspArgArgValSerPValCysAla	34

Kinase	Modulator	Substrate		Sequence	Ref.
		Troponin I	+	ArgValArgMetSerPAlaAspAla	35
			−	ArgArgSerPAspArgAla	35, 36
		Tyrosine hydroxylase	+		37
cGMP-dependent protein kinase	cGMP (+)	ATP-citrate lyase	−		38
	Protein modulator (+)	cAMP-dependent protein kinase (regulatory subunit R_1)	−	ArgGlyAlaIleSerPAlaGluVal	34
		Fructose-1,6-bisphosphatase			39
		Glycogen synthase	−		39
		G-substrate		ArgArgLysAspThrPProAlaLeu	39
		Histone		ArgLysAlaSerPGlyProPro (H1)	15
				LysArgSerPArgLysGluSerPTyr (H2B)	18, 40
				LysArgLysValSerPSerAlaGlu	18, 40
		HMG 14			19
		Hormone-sensitive lipase	+		41
		Phosphorylase kinase	+		42
		Pyruvate kinase	−		39
		Self	−		43
		Troponin I			44, 45
Casein kinase I = casein kinase S		Casein		ValAsnGluLeuSerPLysAspIle (αs_1)	46
				SerPGluGluAsnSerPLysLysThr (αs_2)	46
				SerPSerPGluGluSerPIleIleSerP (αs_2)	46
				SerPSerPGluGluSerPIleThrArg (β)	46

(continued)

PROTEIN SUBSTRATES OF PROTEIN KINASES (*continued*)

Protein kinase	Inhibitors and activators of the protein kinase	Effect of the phosphorylation on the biological activity of the substrate	Protein substrate	Amino acid sequence at the phosphorylated site	References[b]
		−	Glycogen synthase		46
		+	Phosphorylase kinase		28
			Phosvitin		47
			RNA-polymerase		48
			Self		46
			Spectrin		46
Casein kinase II = casein kinase TS = troponin T kinase = glycogen synthase kinase V = casein kinase G = casein kinase NII	Heparin (−) 2,3-Diphosphoglycerate (−) Polyamines (+)		Acetyl CoA carboxylase		49
			Calsequestrin		50
			cAMP-dependent protein kinase (regulatory subunit R_{II})	AlaAspSerPGluSerPGluAspGlu	51
			Casein	GlySerPGluSerPThrPGluAspGln (αs_1)	46
				GluGlnLeuSerPThrPSerPGluGlu (αs_2)	46
				GluGlnGlnGlnThrPGluAspGlu (β)	46
			eIF-2β		46
		−	Glycogen synthase	SerProHisGlnSerPGluAspGlu	47, 52
			Phosphatase inhibitor 1		50
			Phosvitin		47
		+	RNA polymerase		53
			Self		46
			Troponin T	AcSerPAspGluValGlu	54

148

Kinase	Effectors	Substrate		Sequence	Ref
Mammary gland casein kinase		Casein			46, 55
Calcium- and phospholipid-activated protein kinase	Ca²⁺ (+) μM (−) mM	Fibrinogen			56
	Phospholipids (+)	Histone			57, 58
	Diolein (+)	Myelin basic protein			59
	Phorbol esters (+)	Protamine			58
	Heparin (−)				
	Melittin (−)	Self			58
	Phenothiazines (−)	Troponin I			59
	Polyamines (−)	Troponin T			59
Phosphorylase kinase = glycogen synthase kinase II	Ca²⁺ (+)	Calcium-transport ATPase	−		60
	Calmodulin (+)	Fructose-6-P,2 kinase			11
	Phenothiazines (−)	Glycogen synthase	−	SerArgThrLeuSerPValSerSer ArgLysGlnIleSerPValArgGly	61
		Phosphorylase	+		62
		Self	+		63
		Troponin I		AlaIleThrPAlaArgArg	35
				AlaLeuSerPThrArgCys	35
				SerPAsnGluGluValGlu	35
		Troponin T		AlaLeu(SerP,Ser)MetGlyAla	35
				AsnTyr(SerP,Ser)Tyr	35
Glycogen synthase kinase III		cAMP-dependent protein kinase (regulatory subunit R_{II})		AlaArgSerPArgAlaSerPThrPro	51
		Glycogen synthase	−	ProArgProAlaSerPValPro- -ProSerPProSerLeuSerPArg	51
		Self		SerArgThrLeuSerPValSerSer	64
Glycogen synthase kinase IV		Glycogen synthase	−	SerArgThrLeuSerPValSerSer	49

149

(continued)

Protein Substrates of Protein Kinases (*continued*)

Protein kinase	Inhibitors and activators of the protein kinase	Protein substrate	Effect of the phosphorylation on the biological activity of the substrate	Amino acid sequence at the phosphorylated site	References[b]
Glycogen synthase kinase	Ca²⁺ (+) Calmodulin (+) Phenothiazines (−)	Glycogen synthase Myosin light chain Self	−	SerArgThrLeuSerPValSerSer	65 66 66
Myosin light-chain kinase	Ca²⁺ (+) Calmodulin (+) Phenothiazines (−)	Myosin light chain	+	ArgAlaThrSerPAsnValPhe GlyGlySerSerPAsnValPhe	67 67
Hemin-dependent eIF-2α-kinase	Hemin (−) N-Ethylmaleimide (+)	eIF-2α Self	−		68 69
Double stranded RNA-dependent protein kinase	Double-stranded RNA (+) [high concentration (−)] N-Ethylmaleimide (−)	eIF-2α Histone Self	−		70 69 70
Pyruvate dehydrogenase kinase	Acetyl-CoA (+) NADH (+) Pyruvate (−) ADP (−)	Pyruvate dehydrogenase	−	TyrHisGlyHisSerPMetSerAsn-ProGlyValSerPTyrArg	71 71
Acetyl-CoA carboxylase kinase		Acetyl-CoA carboxylase	+	GlyMetGlyThrSerPValGluArg	72
Acetyl-CoA carboxylase kinase		Acetyl-CoA carboxylase	−		73

Protein kinase	Substrate	Effector	+/−	Reference
Not identified	Histone			73
	Protamine			73
Not identified	Aminoacyl-tRNA synthetase		−	74
ATP citrate-lyase kinase	ATP citrate-lyase			75
Branched chain α-ketoacid dehydrogenase kinase	Branched chain α-ketoacid dehydrogenase		−	76, 77
Calcium and calmodulin-stimulated protein kinase	Tubulin	Ca^{2+} (+)		78
	Tryptophan hydroxylase	Calmodulin (+)	+	79
	Tyrosine hydroxylase	Phenothiazines (−)	+	79
Not identified	Glycerophosphate acyltransferase		−	80
Hydroxymethylglutaryl-CoA reductase kinase	Hydroxymethylglutaryl-CoA reductase		−	81
	Self			81
Hydroxymethylglutaryl-CoA reductase kinase kinase	Hydroxymethylglutaryl-CoA reductase kinase		+	81
Polyamine-dependent protein kinase	Ornithine decarboxylase	Polyamines (+)	−	82

151

(continued)

PROTEIN SUBSTRATES OF PROTEIN KINASES (*continued*)

Protein kinase	Inhibitors and activators of the protein kinase	Protein substrate	Effect of the phosphorylation on the biological activity of the substrate	Amino acid sequence at the phosphorylated site	References[b]
Rhodopsin kinase		Rhodopsin			83
		Self			83
Not identified		Tyrosine amino-transferase			84

[a] The protein kinases mentioned toward the end of the table have as yet been less extensively investigated. Autophosphorylation ("self") is included even if it has not been proved that the kinase preparation was free of other protein kinases. When known, the effect of the phosphorylation on the biological activity of the substrate is indicated (+, increase; −, decrease). The amino acid sequence at the phosphorylated site is given only if the sequence of an isolated phosphopeptide has been determined or if the amino acid composition of the phosphopeptide and amino acid sequence of the protein was known. Some minor phosphorylation sites have been omitted. Owing to space limitations, not all the reported phosphorylation sites in nonenzyme substrates, such as casein and protamine, are listed.

The references consist mainly of recent and comprehensive articles containing either original or reviewed data.

[b] Key to references: (*1*) R. W. Brownsey and D. G. Hardie, *FEBS Lett.* **120**, 67 (1980); (*2*) M. P. Walsh, S. Hinkins, and D. J. Hartshorne, *Biochem. Biophys. Res. Commun.* **102**, 149 (1981); (*3*) M. W. Pierce, J. L. Palmer, H. T. Keutmann, T. A. Hall, and J. Avruch, *J. Biol. Chem.* **257**, 10681 (1982); (*4*) G. S. Boyd and A. M. S. Gorban, *Mol. Aspects Cell. Regul.* **1**, 95 (1980); (*5*) J. C. Khoo, E. M. Mahoney, and D. Steinberg, *J. Biol. Chem.* **256**, 12659 (1981); (*6*) R. K. Sharma, T. H. Wang, E. Wirch, and J. H. Wang, *J. Biol. Chem.* **255**, 5916 (1980); (*7*) L. Engström, B. Edlund, U. Ragnarsson, U. Dahlqvist-Edberg, and E. Humble, *Biochem. Biophys. Res. Commun.* **96**, 1503 (1980); (*8*) D. Wallach, P. J. A. Davies, and I. Pastan, *J. Biol. Chem.* **253**, 4739 (1978); (*9*) P. Ekman and U. Dahlqvist-Edberg, *Biochim. Biophys. Acta* **662**, 265 (1981); (*10*) M. R. El-Maghrabi, E. Fox, J. Pilkis, and S. J. Pilkis, *Biochem. Biophys. Res. Commun.* **106**, 794 (1982); (*11*) C. S. Richards and K. Uyeda, *J. Biol. Chem.* **257**, 8854 (1982); (*12*) E. Furuya and K. Uyeda, *J. Biol. Chem.* **255**, 11656 (1980); (*13*) D.-H. Söling and I. A. Brand, *Curr. Top. Cell. Regul.* **20**, 107 (1981); (*14*) P. Cohen, *Nature (London)* **296**, 613 (1982); (*15*) A. Aitken, T. Bilham, P. Cohen, D. Aswad, and P. Greengard, *J. Biol. Chem.* **256**, 3501 (1981); (*16*) J. Zwiller, M.-O. Revel, and P. Basset, *Biochem. Biophys. Res. Commun.* **101**, 1381 (1981); (*17*) S. V. Shlyapnikov, A. A. Arutyunyan, S. N. Kurochkin, L. V. Memelova, M. V. Nesterova, L. P. Sashchenko, and E. S. Severin, *FEBS Lett.* **53**, 316 (1975); (*18*) D. B. Glass and E. G. Krebs, *Annu. Rev. Pharmacol. Toxicol.* **20**, 363 (1980); (*19*) G. M. Walton, J. Spiess, and G. N. Gill, *J. Biol. Chem.* **257**, 4661 (1982); (*20*) L. Berglund, J. C. Khoo, D. Jensen, and D. Steinberg, *J. Biol. Chem.* **255**, 5420 (1980); (*21*) N. Ö. Nilsson, P. Strålfors, G. Fredrikson, and P. Belfrage, *FEBS Lett.* **111**, 125 (1980); (*22*) F. Hirata, *J. Biol. Chem.* **256**, 7730 (1981); (*23*) P. Daile, P. R. Carnegie, and I. D. Young, *Nature (London)* **257**, 416 (1975); (*24*) M. A. Conti and R. S. Adelstein, *J. Biol. Chem.* **256**, 3178 (1981); (*25*) A. M. Edelman and E. G.

Krebs, *FEBS Lett.* **138**, 293 (1982); (26) S. Mårdh, *Curr. Top. Membr. Trans.* **19**, 999 (1983); (27) M. Wretborn, E. Humble, U. Ragnarsson, and L. Engström, *Biochem. Biophys. Res. Commun.* **93**, 403 (1980); (28) T. J. Singh, A. Akatsuka, and K.-P. Huang, *J. Biol. Chem.* **257**, 13379 (1982); (29) S. Shenolikar and P. Cohen, *FEBS Lett.* **86**, 92 (1978); (30) L. Engström, *Mol. Aspects Cell. Regul.* **1**, 11 (1980); (31) E. G. Kranias, J. S. Schweppe, and R. A. Jungman, *J. Biol. Chem.* **252**, 6750 (1977); (32) Y. S. Chiu and M. Tao, *J. Biol. Chem.* **253**, 7145 (1978); (33) S. Shoji, D. C. Parmelee, R. D. Wade, S. Kumar, L. H. Ericsson, K. A. Walsh, H. Neurath, G. L. Long, J. G. Demaille, E. H. Fischer, and K. Titani, *Proc. Natl. Acad. Sci. U.S.A.* **78**, 848 (1981); (34) R. L. Geahlen, D. F. Carmichael, E. Hashimoto, and E. G. Krebs, *Adv. Enzyme Regul.* **20**, 195 (1981); (35) S. V. Perry, *Biochem. Soc. Trans.* **7**, 593 (1979); (36) K. Yamamoto and I. Ohtsuki, *J. Biochem. (Tokyo)* **91**, 1669 (1982); (37) A. M. Edelman, J. D. Raese, M. A. Lazar, and J. D. Barchas, *J. Pharmacol. Exp. Ther.* **216**, 647 (1981); (38) G. D. Swergold, O. M. Rosen, and C. S. Rubin, *J. Biol. Chem.* **257**, 4207 (1982); (39) T. M. Lincoln and J. D. Corbin, *Proc. Natl. Acad. Sci. U.S.A.* **74**, 3239 (1977); (40) E. Hashimoto, M. Takeda, Y. Nishizuka, K. Hamana, and K. Iwai, *J. Biol. Chem.* **251**, 6287 (1976); (41) J. C. Khoo, P. J. Sperry, G. N. Gill, and D. Steinberg, *Proc. Natl. Acad. Sci. U.S.A.* **74**, 4843 (1977); (42) P. Cohen, *FEBS Lett.* **119**, 301 (1980); (43) J. L. Foster, J. Guttman, and O. M. Rosen, *J. Biol. Chem.* **256**, 5029 (1981); (44) D. K. Blumenthal. J. T. Stull. and G. N. Gill, *J. Biol. Chem.* **253**, 334 (1978); (45) T. M. Lincoln and J. D. Corbin. *J. Biol. Chem.* **253**, 337 (1978); (46) G. M. Hathaway and J. A. Traugh, *Curr. Top. Cell. Regul.* **21**, 101 (1982); (47) E. Itarte, M. A. Mor, A. Salavert, J. M. Pena, J. F. Bertomeu, and J. J. Guinovart, *Biochim. Biophys. Acta* **658**, 334 (1981); (48) M. E. Dahmus. *J. Biol. Chem.* **256**, 3332 (1981); (49) P. Cohen, D. Yellowless, A. Aitken, A. Donella-Deana, B. A. Hemmings, and P. J. Parker, *Eur. J. Biochem.* **124**, 21 (1982); (50) F. Meggio, A. Donella-Deana, and L. A. Pinna. *J. Biol. Chem.* **256**, 11958 (1981); (51) B. A. Hemmings, A. Aitken, P. Cohen, M. Rymond, and F. Hofmann, *Eur. J. Biochem.* **127**, 473 (1982); (52) C. Picton, J. Woodgett, B. A. Hemmings, and P. Cohen, *FEBS Lett.* **150**, 191 (1982); (53) D. A. Stetler and K. M. Rose, *Biochemistry* **21**, 3721 (1982); (54) L. A. Pinna, F. Meggio, and M. M. Dediukina, *Biochem. Biophys. Res. Commun.* **100**, 449 (1981); (55) J. C. Pascall, A. P. Boulton, and R. K. Craig, *Eur. J. Biochem.* **119**, 91 (1981); (56) P. Papanikolaou, E. Humble, and L. Engström, *FEBS Lett.* **143**, 199 (1982); (57) M. Castagna, Y. Takai, K. Kaibuchi, K. Sano, U. Kikkawa, and Y. Nishizuka. *J. Biol. Chem.* **257**, 7847 (1982); (58) U. Kikkawa, Y. Takai, R. Minakuchi, S. Inohara, and Y. Nishizuka, and L. M. G. Heilmeyer, Jr., *FEBS Lett.* **131**, 223 (1981); (61) N. Embi, D. B. Rylatt, and P. Cohen, *Biochem. J.* **209**, 189 (1983); (60) M. Varsanyi and L. M. G. Heilmeyer, Jr., *FEBS Lett.* **131**, 223 (1981); (61) N. Embi, D. B. Rylatt, and P. Cohen, *Eur. J. Biochem.* **100**, 339 (1979); (62) C. Nolan, W. B. Novoa, E. G. Krebs, and E. H. Fischer, *Biochemistry* **3**, 542 (1964); (63) G. M. Carlson, P. J. Bechtel, and D. J. Graves, *Adv. Enzymol. Relat. Areas Mol. Biol.* **50**, 41 (1979); (64) B. A. Hemmings, D. Yellowlees, J. C. Kernohan, and P. Cohen, *Eur. J. Biochem.* **119**, 443 (1981); (65) J. R. Woodgett, N. K. Tonks, and P. Cohen, *FEBS Lett.* **148**, 5 (1982); (66) Z. Ahmad, A. A. DePaoli-Roach, and P. J. Roach, *J. Biol. Chem.* **257**, 8348 (1982); (67) Symposium Oct. 1981: Coordination of Metabolism and Contractibility by phosphorylation in cardiac, skeletal and smooth muscle, *Fed. Proc. Fed. Am. Soc. Exp. Biol.* **42**, 7 (1983); (68) R. Jagus, D. Chrouch. A. Konieczny, and B. Safer, *Curr. Top. Cell. Regul.* **21**, 35 (1982); (69) H. Grosfeld and S. Ochoa, *Proc. Natl. Acad. Sci. U.S.A.* **77**, 6526 (1980); (70) R. Petryshyn, D. H. Levin, and I. M. London, *Proc. Natl. Acad. Sci. U.S.A.* **79**, 6512 (1982); (71) L. J. Reed, *Curr. Top. Cell. Regul.* **18**, 95 (1981); (72) R. W. Brownsey, G. J. Belsham, and R. M. Denton, *FEBS Lett.* **124**, 145 (1981); (73) B. Lent and K.-H. Kim, *J. Biol. Chem.* **257**, 1897 (1982); (74) Z. Damuni, F. B. Caudwell, and P. Cohen, *Eur. J. Biochem.* **129**, 57 (1982); (75) S. Ramakrishna, D. L. Pucci, and W. B. Benjamin. *J. Biol. Chem.* **256**, 10213 (1981); (76) K. S. Lau, H. R. Fatania, and P. J. Randle, *FEBS Lett.* **144**, 57 (1982); (77) R. Paxton and R. A. Harris, *J. Biol. Chem.* **257**, 14433 (1982); (78) B. E. Burke and R. J. Delorenzo, *Brain Res.* **236**, 393 (1982); (79) T. Yamauchi, H. Nakata, and H. Fujisawa, *J. Biol. Chem.* **256**, 5404 (1981); (82) G. D. Kuehn and V. J. Atmar, *Fed. Proc., Fed. Am. Soc. Exp. Biol.* **41**, 3078 (1982); (83) R. H. Lee, B. M. Brown, and R. N. Lolley, *Biochemistry* **21**, 3303 (1982); (84) R. H. Stellwagen and K. K. Kohli, *Biochem. Biophys. Res. Commun.* **102**, 1209 (1981).

153

One reason for this is the fact that unequivocal identification of a substrate requires its extensive purification. When the substrate is an enzyme, such a purification generally calls for methods of stabilizing the enzyme. Since most phosphorylatable enzymes are labile, the identification of a substrate of a protein kinase may be a fairly difficult problem. In addition, the identification may be hampered by the fact that several particle-bound protein kinase activities are not very well characterized. It is even possible that there are protein kinases that still remain to be discovered.

[7] Assays for Regulatory Properties of Polyamine-Dependent Protein Kinase

By VAITHILINGAM SEKAR, VALERIE J. ATMAR, and GLENN D. KUEHN

[EC 2.7.1.37, ATP: protein phosphotransferase]
$$ATP^{4-} + \text{ornithine decarboxylase enzyme} \rightarrow ADP^{3-} + H^+$$
$$+ \text{phosphoornithine decarboxylase}^{2-}$$

The naturally occurring polyamines putrescine, spermidine, and spermine are synthesized in appreciable amounts in all living cells.

Putrescine: $H_2N\text{-}CH_2CH_2CH_2CH_2\text{-}NH_2$
Spermidine: $H_2N\text{-}CH_2CH_2CH_2CH_2\text{-}NH\text{-}CH_2CH_2CH_2\text{-}NH_2$
Spermine: $H_2N\text{-}CH_2CH_2CH_2\text{-}NH\text{-}CH_2CH_2CH_2CH_2\text{-}NH\text{-}CH_2CH_2CH_2\text{-}NH_2$

The combined charged cationic and aliphatic properties of the polyamines, under physiological conditions, suggest a wide variety of interactions that may occur between them and various cellular components. They can link through ionic forces to anionic nucleic acids, proteins, and phospholipids. The aliphatic character of their methylene group clusters indicates possible interactions with hydrophobic environments such as those occurring in membranes. Their amino-linked protons can hydrogen-bond to electronegative atoms. Thus, the polyamines have been implicated to act in a wide variety of cellular processes. Their participation has been invoked in virtually every phase of macromolecular biosynthesis.

The polyamines are synthesized in eukaryotes by a pathway that begins with the decarboxylation of the amino acid ornithine. The enzyme ornithine decarboxylase [L-ornithine carboxy-lyase, EC 4.1.1.17] catalyzes putrescine formation through pyridoxal phosphate-dependent decarboxylation of L-ornithine. Ornithine decarboxylase is ubiquitous. It can be assayed in virtually every type of biological cell grown under appropriate conditions. The catalytic activity of the enzyme was first described by Gale as an inducible activity in bacteria.[1] Later Morris and Pardee[2] reported an isozyme in *Escherichia coli* that they termed a "biosynthetic ornithine decarboxylase" to distinguish it from the inducible enzyme. Ornithine decarboxylase activity was detected simultaneously in animal tissues by three different laboratories in 1968.[3-5] Interest in this enzyme was immediately stimulated by the finding of two of its unique properties. First, ornithine decarboxylase demonstrated the shortest apparent half-life of any known eukaryotic enzyme, i.e., approximately 10 min.[6] Second, the enzyme demonstrated striking inducibility by a variety of chemical, hormonal, and physical stimuli. This result suggested an important regulatory mechanism operative in the control of ornithine decarboxylase.

Early attempts to identify an efficient regulatory mechanism for ornithine decarboxylase centered on testing potential, classical feedback modifiers. *In vitro* inhibition of purified, or partially purified, preparations of ornithine decarboxylase was demonstrated by putrescine, spermidine, and spermine. The biosynthetic ornithine decarboxylase from *Escherichia coli* was inhibited 60% by 2 mM spermine, 8 mM spermidine, and 20 mM putrescine. These inhibitors appeared to act in a competitive manner with respect to ornithine.[7] In addition, certain nucleoside phosphates, including GTP, GMP, ATP, UTP, and CTP, were found to be positive effectors for the enzyme. GTP was the most effective activator with a K_m for activation of less than 1 μM.[8] In contrast, control by low molecular weight metabolite effectors was not found to be of major importance with ornithine decarboxylase from eukaryotes. An early report that putrescine, spermidine, and spermine were weak competitive inhibitors (K_i values being 1, 2.7, and 9 mM, respectively) of ornithine decarboxylase from rat prostatic tissue did not prove to be universal.[4,9] No enzyme isolated

[1] G. F. Gale, *Adv. Enzymol. Relat. Subj. Biochem.* **6,** 1 (1946).
[2] D. R. Morris and A. B. Pardee, *Biochem. Biophys. Res. Commun.* **20,** 697 (1965).
[3] J. Jänne and A. Raina, *Acta Chem. Scand.* **22,** 1349 (1968).
[4] A. E. Pegg and H. G. Williams-Ashman, *Biochem. J.* **108,** 533 (1968).
[5] D. H. Russell and S. H. Synder, *Proc. Natl. Acad. Sci. U.S.A.* **60,** 1420 (1968).
[6] D. H. Russell and S. H. Snyder, *Mol. Pharmacol.* **5,** 253 (1969).
[7] D. R. Morris, W. H. Wu, D. Applebaum, and K. L. Koffron, *Ann. N. Y. Acad. Sci.* **171,** 968 (1970).
[8] D. M. Applebaum, J. C. Dunlap, and D. R. Morris, *Biochemistry* **16,** 1580 (1977).
[9] A. E. Pegg, *Ann. N. Y. Acad. Sci.* **171,** 977 (1970).

from other eukaryotic sources has been found to be susceptible to feedback inhibition. Other compounds have been tested, including nucleotides and intermediates from neighboring metabolic pathways, but no low molecular weight effectors have been found.

The failure to identify any classical feedback modifiers for ornithine decarboxylase, which might have explained the capacity of the enzyme to demonstrate rapid and large fluctuations in its activity in eukaryotes, spawned numerous alternative proposals. These proposals have been discussed in detail in several recent reviews.[10-13] Proposals have included (1) control at the level of mRNA synthesis; (2) translational control by the polyamines; (3) antizyme induction or release by polyamines; (4) unspecified transitions between active and less active forms; and, (5) posttranslational covalent modifications such as transglutamination,[14] proteolysis,[15] and phosphorylation.[16] Reversible phosphorylation has emerged as a strong candidate for explaining rapid control of ornithine decarboxylase because of our discovery of a polyamine-dependent protein kinase enzyme that phosphorylatively inactivates the enzyme.[16] The protein kinase reaction is absolutely dependent on spermidine and spermine. Activation of the protein kinase by spermidine and spermine is antagonized by putrescine. These properties of the protein kinase provide a cogent molecular explanation consistent with the observations that the terminal end products of polyamine biosynthesis exert a rapid, negative feedback control on ornithine decarboxylase that is reversed by putrescine.[17,18] None of the other proposals for explaining the regulation of ornithine decarboxylase accommodate these properties. Moreover, polyamine-dependent protein kinase autophosphorylates in the absence of ornithine decarboxylase.[19] Autophosphorylation is stimulated by the calcium ion–calmodulin

[10] P. P. McCann, in "Polyamines in Biomedical Research" (J. M. Gaugas, ed.), p. 109. Wiley, New York, 1980.
[11] E. S. Canellakis, J. S. Heller, D. A. Kyriakidis, and D. Viceps-Madore, in "Polyamines in Biology and Medicine" (D. R. Morris and L. J. Marton, eds.), p. 83. Dekker, New York, 1981.
[12] E. S. Canellakis, D. Viceps-Madore, D. A. Kyriakidis, and J. S. Heller, Curr. Top. Cell. Regul. **15**, 155 (1979).
[13] D. H. Russell and B. G. M. Durie, Progr. Cancer Res. Ther. **8**, 59 (1978).
[14] A. K. Tyagi, H. Tabor, and C. W. Tabor, Biochem. Biophys. Res. Commun. **109**, 533 (1982).
[15] J. E. Seely, H. Pösö, and A. Pegg, J. Biol. Chem. **257**, 7549 (1982).
[16] V. J. Atmar and G. D. Kuehn, Proc. Natl. Acad. Sci. U.S.A. **78**, 5518 (1981).
[17] T. J. Paulus and R. H. Davis, J. Bacteriol. **145**, 14 (1981).
[18] P. S. Mamont, M. N. Joder-Ohlenbusch, and J. Grove, Biochem. J. **196**, 411 (1981).
[19] G. D. Kuehn and V. J. Atmar, Adv. Polyamine Res. **4**, 615 (1982).

complex.[19,20] Thus, autophosphorylation of polyamine-dependent protein kinase may eventually explain the role of calcium ion in the induction of ornithine decarboxylase activity in many different cells.[20] In procedures presented below, methods are described for demonstrating calcium–calmodulin-dependent autophosphorylation of polyamine-dependent protein kinase.

Treatment of mammalian cells with interferon is accompanied by a rapid loss of intracellular ornithine decarboxylase activity.[21,22] Recent studies suggest that this inactivation may be related to the induction by interferon of a protein kinase activity that is activated by the polyamines. Early studies demonstrated that pretreatment of mammalian cells with interferon results in the appearance of a new protein kinase activity that requires double-stranded RNA (dsRNA) for maximum catalytic activity. This protein kinase phosphorylates two endogenous polypeptides of apparent molecular weight (M_r) 35,000 and 67,000. The polypeptide of 35,000 M_r is proposed to be the smallest of the subunits of one of the initiation factors eIF-2, involved in protein synthesis.[23] The identity of the 67,000 M_r phosphopeptide is unknown. We have shown that polyamine-dependent protein kinase activity is also markedly induced in Ehrlich ascites tumor cells that were pretreated with interferon.[24] The same endogenous phosphate-acceptor protein appears to serve as the substrate for both polyamine-dependent protein kinase and dsRNA-dependent protein kinase. The phosphate-acceptor protein is a protein of 68,000–70,000 M_r. Both protein kinases also demonstrate the same elution properties from phosphocellulose chromatography or from poly(I) · poly(C)–agarose chromatography. The two protein kinases, however, may not be identical,[24] since polyamine-dependent protein kinase does not demonstrate the capacity of dsRNA-dependent protein kinase to phosphorylate protein initiation factor eIF-2.

Since interferon induction of polyamine-dependent protein kinase is a facile method of inducing (30- to 100-fold) polyamine-dependent protein kinase in Ehrlich ascites tumor cells, methods are described below that achieve this large enhancement of activity.

[20] G. D. Kuehn, in "Recently Discovered Systems of Enzyme Regulation by Reversible Phosphorylation" (P. Cohen, ed.), p. 185. Elsevier/North-Holland, Amsterdam, 1983.
[21] T. Sreevalsan, J. Taylor-Papadimitriou, and E. Rozengurt, Biochem. Biophys. Res. Commun. 87, 679 (1979).
[22] Y. Gazitt, Cancer Res. 41, 2959 (1981).
[23] W. K. Roberts, A. G. Hovanessian, R. E. Brown, M. J. Clemens, and I. M. Kerr, Nature (London) 264, 477 (1976).
[24] V. Sekar, V. J. Atmar, M. Krim, and G. D. Kuehn, Biochem. Biophys. Res. Commun. 106, 305 (1982).

Assay Method for Autophosphorylation of Polyamine-Dependent Protein Kinase

Principle. The assay for autophosphorylation of polyamine-dependent protein kinase requires the demonstration that transfer of $[^{32}P]HOPO_3^{2-}$ from $[\gamma-^{32}P]ATP$ to the protein kinase is markedly stimulated by calcium ion and calmodulin. The procedure involves measurement of acid-stable $[^{32}P]HOPO_3^{2-}$ incorporated into the protein kinase after termination of the protein kinase reaction by a cold solution of EDTA containing carrier bovine serum albumin.

Reagents

Buffer reagent containing 0.375 M β-glycerol phosphate disodium salt, 37.5 mM $Mg(CH_3CO_2)_2 \cdot 4H_2O$. The final pH is adjusted to 7.5 with HCl at 25°

$[\gamma-^{32}P]ATP \cdot$ Tris reagent, 100 mM, with specific radioactivity of approximately 5×10^9 cpm/μmol or 2250 mCi/mmol, prepared by a published procedure[25]

$CaCl_2$ solution containing 0.6 mM $CaCl_2$

Spermidine/spermine solution containing 6 mM spermidine (free base) and 6 mM spermine (free base)

Calmodulin solution containing 10.8 mg calmodulin per milliliter dissolved in water

Termination reagent containing 100 mM disodium EDTA and 50 mg of bovine serum albumin per milliliter

$HClO_4$, solutions of 5 and 10%

Scintillation cocktail containing 1 liter of toluene, 1 liter of 2-ethoxyethanol, and 8 g of Omnifluor

Procedure. The procedure described below has successfully measured the activity of calcium ion–calmodulin-stimulated autophosphorylation of polyamine-dependent protein kinase from the acellular slime mold *Physarum polycephalum*. Incubation mixtures at pH 7.5 contain 40 μl of buffer reagent, 10 μl of $[\gamma-^{32}P]ATP \cdot$ Tris reagent, 10 μl of calmodulin solution, 10 μl of $CaCl_2$ solution, 40 μl of purified protein kinase preparation, and 10 μl of H_2O. Polyamine-dependent protein kinase from *P. polycephalum* is prepared as previously described.[26] Calmodulin is prepared from *P. polycephalum*[27] or from bovine brain,[28] which also stimulates autophosphorylation of polyamine-dependent protein kinase.

[25] I. M. Glynn and J. B. Chappell, *Biochem. J.* **90,** 147 (1964).
[26] G. D. Kuehn, V. J. Atmar, and G. R. Daniels, this series, Vol. 94, p. 147.
[27] J. Kuznicki, L. Kuznicki, and W. Drabikowski, *Cell Biol. Int. Rep.* **3,** 17 (1979).
[28] J. H. Wang and R. K. Sharma, *Adv. Cyclic Nucleotide Res.* **10,** 187 (1979).

Calmodulin is assayed by its capacity to stimulate hydrolysis of cyclic AMP to 5'-AMP catalyzed by bovine brain cyclic nucleotide phosphodiesterase.[29] All components are preincubated at 30° for 5 min. [γ-^{32}P]ATP · Tris reagent is added to initiate the reaction at zero time. After 20 min, the reaction is stopped by addition of 15 μl of termination reagent. The reaction tube is placed on ice. Zero-time control reactions are treated similarly except that [γ-^{32}P]ATP is added to the reaction mixture after addition of the termination reagent. From the reaction tube, a sample of 0.1 ml is immediately spotted uniformly over the area of a disk, 2.5 cm diameter, of Whatman 3 MM paper. Disks are prerinsed in distilled water before use to remove loose fibers and are dried at room temperature. A maximum of 30 disks are collectively washed successively in a beaker for 30 min in 10% HClO$_4$ at 5°, again for 30 min in 10% HClO$_4$ at 5°, and last in 5% HClO$_4$ for 30 min at room temperature. One liter of each HClO$_4$ solution is used. When these washings are complete, the disk is blotted on absorbent paper to remove excess HClO$_4$ and dried at room temperature or under a heat gun. Each dried disk is suspended in 5 ml of nonaqueous scintillation cocktail and is counted for [^{32}P]phosphate content. The specific radioactivity of the stock [γ-^{32}P]ATP solution is determined by counting a sample of the stock concurrently with the dried disk.

The polyamines spermidine and spermine inhibit the autophosphorylation reaction by polyamine-dependent protein kinase in the absence of calcium ion–calmodulin.[30] This property can be demonstrated in enzyme assay reactions containing the 40 μl of buffer reagent, 10 μl of [γ-^{32}P]ATP · Tris reagent, 10 μl of spermidine/spermine solution, 40 μl of purified protein kinase preparation, and 20 μl of H$_2$O. The other assay details are those described above.

Calcium ion–calmodulin complex reverses the inhibition caused by spermidine/spermine. This property can be demonstrated by comparing results from the assay mixtures described in the preceding paragraph with enzyme assay reactions containing 40 μl of buffer reagent, 10 μl of [γ-^{32}P]ATP · Tris reagent, 10 μl of calmodulin solution, 10 μl of CaCl$_2$ solution, 10 μl of spermidine/spermine solution, and 40 μl of purified protein kinase preparation. The other assay details are those described in the first paragraph of this section.

Representative data that demonstrate these regulatory effects on polyamine-dependent protein kinase are shown in the table.

[29] J. N. Wells, C. E. Baird, Y. J. Wyl, and J. G. Hardman, *Biochim. Biophys. Acta* **384**, 430 (1975).
[30] G. R. Daniels, V. J. Atmar, and G. D. Kuehn, *Biochemistry* **20**, 2525 (1981).

STIMULATION OF AUTOPHOSPHORYLATION OF POLYAMINE-DEPENDENT PROTEIN KINASE
BY CALCIUM ION-CALMODULIN COMPLEX

Reaction mixture	Specific activity [nmol min^{-1} (mg enzyme)$^{-1}$]
Complete[a]	19.0
Complete plus 50 μM CaCl$_2$, 50 μM calmodulin	51.5
Complete plus 0.5 mM spermidine, 0.5 mM spermine	6.2
Complete plus 50 μM CaCl$_2$, 50 μM calmodulin, 0.5 mM spermidine, 0.5 mM spermine	44.1
Complete plus 50 μM CaCl$_2$	10.0
Complete plus 50 μM calmodulin	17.3

[a] The complete reaction mixture contained 40 μl of buffer reagent, 10 μl of [γ-^{32}P]ATP · Tris reagent (specific radioactivity 8.40 × 10^8 cpm/μmol), 40 μl of purified protein kinase preparation (0.02 mg of protein), and 30 μl of H$_2$O. Other details are in the text.

Notes on Regulatory Properties of Polyamine-Dependent Protein Kinase. No low molecular weight modifiers that have been previously recognized to influence other special protein kinases have been found to affect polyamine-dependent protein kinase.[30] As yet only the polyamines and calcium ion-calmodulin have been found to modulate its activity.

The natural substrate for polyamine-dependent protein kinase is the enzyme ornithine decarboxylase.[16] Calcium ion, calmodulin, or calcium ion–calmodulin complex do not affect the direct phosphorylation of ornithine decarboxylase by polyamine-dependent protein kinase (data not shown).[19,20] Only the autophosphorylation reaction by polyamine-dependent protein kinase is modulated by calcium ion–calmodulin complex. Autophosphorylation correlates with inhibition of the capacity of the protein kinase to phosphorylate ornithine decarboxylase.[19]

Induction of Polyamine-Dependent Protein Kinase by Interferon in Ehrlich Ascites Tumor Cells

Principle. Polyamine-dependent protein kinase activity is induced over 30-fold in nuclei of Ehrlich ascites tumor cells pretreated with interferon. The simple procedures employed to induce and demonstrate the activity in partially fractionated preparations are well suited to small volumes of extract readily available from interferon-treated cell cultures.

Reagents for Cell Growth and Fractionation

Ehrlich ascites tumor cell culture, American Type Culture Collection No. CCL-77

Eagle's minimum essential spinner medium supplemented with 10% fetal calf serum (Irvine Scientific Co., Santa Ana, CA)

Mouse fibroblast β-interferon

Poly(I) · Poly(C)–agarose (P-L Biochemicals, Milwaukee, WI)

KCl buffer: 90 mM or 1 M KCl in 10 mM N-2-hydroxyethylpiperazine-N'-2-ethanesulfonic acid (HEPES), 1.5 mM Mg(CH$_3$CO$_2$)$_2$ · 4H$_2$O, 7 mM 2-mercaptoethanol containing 20% glycerol (v/v) (referred to as 90 mM or 1 M KCl buffer). After preparation, the pH is adjusted to 7.5 with NaOH

Tris-saline buffer: 0.35 mM Tris-HCl (pH 7.6 at 25°) in 146 mM NaCl

Extraction buffer, 10 mM Tris-HCl (pH 7.5 at 25°), 7 mM 2-mercaptoethanol, 10 mM KCl, 1.5 mM Mg(CH$_3$CO$_2$)$_2$ · H$_2$O, 20% glycerol (v/v)

Dialysis buffer: 50 mM Tris-HCl (pH 7.5 at 25°), 0.3 M NaCl

Procedure. Ehrlich ascites tumor cells are grown under 5% CO$_2$ at 37° in Eagle's minimum essential spinner medium supplemented with 10% fetal calf serum. Cells are inoculated at an initial cell density of 1×10^5 cells/ml into 200 ml of medium contained in a 500-ml Erlenmeyer flask. Cultures are very gently agitated with a magnetic stirring bar. After 24 hr, the cell density reaches 4×10^5 cells/ml, at which time 100 ml of culture is removed and 100 ml of fresh growth medium is added. After another 24 hr, the cell density increases to 8×10^5 cells/ml. Again, 75 ml of culture is removed and 75 ml of fresh medium is added. After 24 hr the cell density reaches 2×10^6 cells/ml and 400 units/ml of highly purified mouse fibroblast β-interferon (specific activity: 3×10^5 reference units/ml based on the NIH Standard) dissolved in 0.1 M sodium phosphate (pH 7.4)–20 mM NaCl is added. A control culture receives an equal volume of phosphate-saline buffer (0.27 ml) without interferon. After 21 hr the cell density increases to 3.5×10^6 cells/ml and the cultures are harvested by sedimentation at 2500 g for 10 min at 4°. The pelleted cells are washed twice in cold Tris-saline buffer. After each wash, the cells are recovered by centrifugation at 2500 g. The final wet-packed cell pellet is suspended in 1.5 volumes of cold extraction buffer. The suspended cells are ruptured by treatment with 20 strokes in a glass Dounce homogenizer. The concentration of KCl in the homogenate is increased to 90 mM by addition of solid KCl, and the homogenate is centrifuged at 2500 g for 10 min at 4°. Subsequent treatment of the pelleted nuclear fraction is described in the next paragraph. The resulting cytosolic fraction is centrifuged at 200,000 g for 2 hr at 4°. The sedimented pellet is discarded. The supernatant fraction, hereafter referred to as the S-200 cytosolic fraction, is fractionated by chromatography on phosphocellulose or on poly(I) · poly(C)–agarose.

The nuclear pellet fraction derived as in the preceding paragraph is lysed by suspension in cold 90 mM KCl buffer supplemented with 0.5%

Nonidet P-40 detergent. The suspension is vortexed for 1 min and incubated for 5 min before centrifuging at 10,000 g for 10 min at 4°. The resulting supernatant fraction, hereafter referred to as the S-10 nuclear fraction, is next fractionated by chromatography on phosphocellulose or on poly(I) · poly(C)–agarose.

Phosphocellulose Chromatography. The S-200 cytosolic fraction and the S-10 nuclear fraction from control and interferon-treated cultures are dialyzed to equivalent conductance against 50 mM Tris-HCl (pH 7.5)–0.3 M NaCl. Each respective fraction is next treated with Bio-Rex 70 cation exchange resin as previously described.[26] All subsequent procedures beyond this step are exactly like those described for fractionation of polyamine-dependent protein kinase from *P. polycephalum.*[26]

Poly(I) · Poly(C)–Agarose Chromatography. Poly(I) · poly(C)– · agarose is equilibrated with 90 mM KCl buffer in a 4 × 21 mm column constructed from a 1-ml polypropylene plastic syringe. The S-200 cytosolic fraction and the S-10 nuclear fractions from control and interferon-treated cultures are each applied without further treatment to respective columns of poly(I) · poly(C)–agarose. The column effluent is collected during the addition of the S-200 or S-10 fractions to the column. This effluent is reapplied and cycled through the column a second time. This recycling procedure is repeated two additional times such that the original S-200 or S-10 fraction is passed over the poly(I) · poly(C)–agarose bed a total of four times. Quantitative absorption of polyamine-dependent protein kinase to the dsRNA–agarose column occurs during these procedures. After applying the protein fraction, the column is washed extensively with 90 mM KCl buffer to remove unbound protein. Next, the poly(I) · poly(C)–agarose-bound enzymes are eluted from the column with 1 M KCl buffer. Fractions of 1 ml are collected. Methods of enzyme assay needed to detect polyamine-dependent protein kinase activity in column fractions are those described earlier.[26]

Notes on Procedures and Isolated Fractions. The procedures described permit comparisons of the occurrence of polyamine-dependent protein kinase in control and interferon-treated cytosolic and nuclear compartments of Ehrlich ascites tumor cells. In general, the interferon-treated cells will demonstrate over 4 times higher polyamine-dependent protein kinase activity than control cultures in the cytosolic fraction and 30 times more activity in the nuclear fraction.[24] Clearly, the induction of polyamine-dependent protein kinase activity is most pronounced in nuclei.

Chromatography of cytoplasmic and nuclear fractions on phosphocellulose or poly(I) · poly(C)–agarose both yield polyamine-dependent protein kinase. However, poly(I) · poly(C)–agarose chromatography is preferred, since it yields polyamine-dependent protein kinase preparations with 3 times greater specific activity than phosphocellulose fractionation.[24]

Other pertinent comments regarding the stability and properties of polyamine-dependent protein kinase can be found in Vol. 94 of this series.[26]

[8] UDP-N-acetylglucosamine : Lysosomal Enzyme N-Acetylglucosamine-1-phosphotransferase

By MARC L. REITMAN, LES LANG, and STUART KORNFELD

$$\text{UDP-GlcNAc} + \text{Man}\alpha1 \rightarrow \text{X} \longrightarrow \text{UMP} + \text{GlcNAc}\alpha1 \rightarrow \text{P-6Man}\alpha1 \rightarrow \text{X}$$

UDP-N-acetylglucosamine : lysosomal enzyme N-acetylglucosamine-1-phosphotransferase (N-acetylglucosaminylphosphotransferase) transfers N-acetylglucosamine 1-phosphate en bloc to the C-6 oxygen of certain mannose residues in high mannose-type oligosaccharides of lysosomal enzymes.[1,2] N-Acetylglucosaminylphosphotransferase is selective for lysosomal enzymes, as opposed to other newly synthesized glycoproteins, and thus is believed to be responsible for the initial and determining step in the segregation of lysosomal proteins into lysosomes.[3]

The N-acetylglucosaminylphosphotransferase has been studied using [32P]UDP-GlcNAc as the donor substrate and either α-methyl-D-mannoside or purified lysosomal enzymes as acceptor substrates. Synthesis of the [β-32P]UDP-GlcNAc, assay procedures using both types of acceptors, and a partial purification of the transferase from rat liver are described.

Assay

Principle. The transfer of ^{32}P-labeled GlcNAc 1-phosphate from [β-^{32}P]UDP-GlcNAc to various acceptor substrates is quantitated by one of two methods. In Method I, α-methylmannoside (or other neutral monosaccharides or oligosaccharides) is the acceptor. Ion-exchange chroma-

[1] M. L. Reitman and S. Kornfeld, *J. Biol. Chem.* **256,** 4275 (1981).
[2] A. Hasilik, A. Waheed, and K. von Figura, *Biochem. Biophys. Res. Commun.* **98,** 761 (1981).
[3] M. L. Reitman and S. Kornfeld, *J. Biol. Chem.* **256,** 11977 (1981).

METHODS IN ENZYMOLOGY, VOL. 107

tography is used to separate the phosphodiester product (GlcNAcα1 → ^{32}P-6Manα1 → methyl, when using α-methylmannoside), which has only one negative charge at neutral pH, from free phosphate, phosphomonoesters, and nucleotide sugars.

Method II is used to measure GlcNAc 1-phosphate transfer to glycoprotein and glycopeptide acceptors. After incubation, the acceptor glycoproteins are digested with Pronase, producing glycopeptides that are then subjected to affinity chromatography on concanavalin A (Con A)–Sepharose. The ^{32}P radioactivity bound to the Con A–Sepharose is the product.

UDP-[^{3}H]GlcNAc can also be used as the sugar phosphate donor. The ^{3}H-labeled substrate is commercially available and has a long half-life. Unfortunately, it is also a substrate for side reactions whose products interfere with the assays, and we therefore do not routinely use this substrate.

Reagents

UDP-GlcNAc, 1 mM, pH 6.5

[β-^{32}P]UDP-GlcNAc, prepared as described below

ATP, 100 mM, pH 6.5

Bovine serum albumin (BSA), 100 mg/ml

Dithiothreitol, 25 mM, freshly prepared

Lubrol PX, 10% (w/v) (Sigma)

Tris-HCl, 1 M, pH 7.5

GlcNAc, 500 mM

α-Methylmannoside, 1 M

α-Methylgalactoside, 1 M

MgCl$_2$, 1 M

MnCl$_2$, 1 M (in 1 mM 2-mercaptoethanol to prevent oxidation of the Mn^{2+})

QAE-Sephadex, (Q-25-120, Sigma) equilibrated in 100 mM Tris base and then washed and equilibrated in 2 mM Tris base

Pronase buffer: 100 mM Tris-HCl (pH 8.0)–100 mM β-glycerol phosphate–50 mM GlcNAc–20 mM CaCl$_2$–0.02% (w/v) NaN$_3$

Pronase (grade B, Calbiochem) made fresh to 40 mg/ml in Pronase buffer and preincubated (56°, 10 min) to digest contaminating phosphatases

Con A-Sepharose (Pharmacia)

Phosphate-buffered saline (PBS): 150 mM NaCl–10 mM sodium phosphate, pH 7.5

Synthesis of [β-^{32}P]UDP-GlcNAc. Dialyze 28 units (0.2 mg) of yeast hexokinase (Boehringer-Mannheim #127-809) and 160 units (0.8 mg) of rabbit muscle phosphoglucomutase (Boehringer-Mannheim #108-383) together against 500 ml of 10 mM Tris-HCl (pH 8.0)/1 mM 2-mercapto-ethanol/1 mM MgCl$_2$/0.1 mM EDTA (4°, 2 hr). Resuspend 25 units of UDP-glucose pyrophosphorylase (Sigma U-8501) and 100 units of inorganic pyrophosphatase (Sigma I-4503) together in 80 μl of the same dialysate and store on ice (dialysis is not necessary). Simultaneously, dry the [γ-^{32}P]ATP into a 1.5-ml polypropylene tube, using a centrifuge evaporator. Add the appropriate reagents to the [γ-^{32}P]ATP so that the reaction mixture contains 100 mM Tris-HCl (pH 8.0)/5 mM MgCl$_2$/5 mM 2-mercaptoethanol/0.2 mg/ml BSA/10 mM glucosamine/2.5 mM ATP/4 μM glucose-1,6-diphosphate/0.5 mM EDTA and 64.7 μl of the hexokinase/phosphoglucomutase mix in a final volume of 100 μl. Care must be taken (use neutralized stock solutions) so that the pH of the final reaction is 8.0. After incubation at 37° for 60 min, add 44 μl of the pyrophosphorylase/pyrophosphatase mix and bring the volume to 160 μl with UTP to a final concentration of 10 mM. Incubate again at 37° for 2 hr, add remaining pyrophosphorylase/pyrophosphatase mix, and incubate an additional 2 hr. Water (320 μl) and saturated NaHCO$_3$ (53 μl) are added, followed by 70 μl of a freshly prepared acetic anhydride/water mix (3 : 67). After vigorous vortexing, this mixture is incubated at room temperature for 5 min and subsequently heated for 3 min in a boiling water bath.

The [β-^{32}P]UDP-GlcNAc is purified by preparative thin layer chromatography using silica gel 60F-254 plates (cut to 5-cm width) containing a concentrating zone (MCB Manufacturing Chemists, Merck #13792-7). The entire reaction mixture is streaked across the upper two-thirds of the concentrating zone; the plate is developed for 5 hr in ethanol/1 M ammonium acetate, pH 3.8 (7 : 3 by volume), and air dried. UDP-GlcNAc migrates with an R_f = 0.61 relative to the solvent front, and the radioactive product can be identified by its UV absorption as well as from a short (15-sec) exposure autoradiograph of the plate. The region containing the product is scraped into two 1.5-ml polypropylene tubes and eluted by vortexing in 1.0-ml aliquots of water (~8.0 ml total) until most of the radioactivity is recovered. The eluate is then filtered through 0.8- and 0.22-μm Millex filters to remove residual silica, made to 50% ethanol/0.1 mM cold UDP-GlcNAc, and stored at −20°. In experiments in which a more highly purified product is required, the sample is rechromatographed on paper in the same solvent as described above (see below).

Yields >65% are typically achieved, and the product is 95–99% radiochemically pure after elution from the TLC plate. No increase in the assay background is seen during a 2-month storage of the labeled sugar nucleotide at −20°.

Assay Procedure: Method I. A mix containing ATP, UDP-GlcNAc, and [β-^{32}P]UDP-GlcNAc is prepared and aliquoted into 1.5-ml polypropylene test tubes such that each tube contains 100 nmol of ATP and 7.5

nmol of UDP-GlcNAc with enough ^{32}P (usually 300,000 cpm) so that the expected activity will be easily detected. The solvent is removed using a centrifuge evaporator. A stock reaction mixture is prepared (fresh daily), containing 50 μl of MgCl$_2$, 50 μl of dithiothreitol, 250 μl of Tris-HCl, 500 μl of GlcNAc, 100 μl of BSA, 500 μl of α-methylmannoside, and 50 μl of MnCl$_2$ (which is added last, just before use, to minimize oxidation). Each reaction tube then receives 15 μl of the stock mix, as well as enzyme and enzyme buffer to a final volume of 50 μl. Final concentrations during reaction are as follows: 150 μM [β-^{32}P]UDP-GlcNAc, 100 mM α-methyl-mannoside, 2 mM ATP, 10 mM MgCl$_2$, 10 mM MnCl$_2$, 0.25 mM dithiothreitol, 2 mg of BSA per milliliter, 50 mM GlcNAc, and 50 mM Tris-HCl. No detergent is added, since the enzyme buffer contains 0.3% Lubrol.

Samples are incubated at 37°; the reaction is terminated by adding 1 ml of 5 mM NaEDTA, pH 7.8, and vortexing. The preparation is applied directly to a QAE-Sephadex column prepared as follows: 1.0 ml of QAE-Sephadex in a polypropylene Econocolumn is washed with 2 ml of 2 mM Tris base, the sample is applied, and the column is again washed with 2 ml of 2 mM Tris base. The reaction product is eluted directly into scintillation vials using 4 ml of 30 mM NaCl–2 mM Tris base and counted after the addition of 12 ml of scintillation fluid. One unit of enzyme activity is defined as the transfer of 1 pmol of N-acetylglucosamine 1-[^{32}P]phosphate per hour at 37° using 100 mM α-methylmannoside as acceptor.

The exact conditions for the QAE-Sephadex chromatography are important. Scrupulous care should be taken to pour columns of constant size and to add buffer such that the resin is not disturbed. No ^{32}P radioactivity should run through the column, and blanks (i.e., without enzyme, zero time, or with α-methylgalactoside as acceptor) should contain <0.03% of the input radioactivity. If the background is high or if duplicates do not agree, one should vary the elution conditions (use less or more 30 mM NaCl–2 mM Tris base) while monitoring the radioactivity in each 1-ml fraction. Because method I measures only the phosphodiester product, N-acetylglucosamine is included to inhibit any phosphodiester glycosidase activity. Both N-acetylglucosamine and ATP inhibit breakdown of the [β-^{32}P]UDP-GlcNAc.

Although described for α-methylmannoside, this assay method can be used for other neutral monosaccharide or oligosaccharide acceptors. When using larger oligosaccharide acceptors, the eluting NaCl concentration is lowered to 20 mM.

When cultured fibroblasts are to be assayed, the cells are mechanically harvested at confluence (about 7–10 days after passage, as the specific activity decreases after this point) and washed with 10 ml of 20 mM Tris-

HCl (pH 7.5)–155 mM NaCl. The cell pellet is suspended in 2 volumes of 50 mM Tris-HCl (pH 7.5)–0.25 mM dithiothreitol–0.75% (w/v) Triton X-100 (or Lubrol PX) with 10 strokes of a motor-driven Teflon homogenizer or by sonication using a microprobe (60 W, 0°, 2 × 15 sec). The homogenate is assayed as described above except for the following modifications: the final ATP concentration is 5 mM and the final detergent concentration is 1% (add 500 μl of 10% Lubrol to the 1.5 ml of reaction mix and add 20 μl of reaction mix per tube). The Lubrol concentration should be maintained at 1% whenever a crude extract is assayed. [β-^{32}P]UDP-GlcNAc, which has been rechromatographed on paper, must be used for these assays. The TLC eluate contains a minor contaminant which coelutes from the QAE columns with the assay product, making low levels of activity difficult to detect. When enzyme activities are very low, e.g., in fibroblasts from I-cell patients, the only control that adequately measures background radioactivity is a parallel assay tube using a nonacceptor such as α-methylgalactoside. Normal control fibroblasts are also assayed in parallel, and typically 250–500 μg of protein is used.

Assay Procedure: Method II. The ATP, UDP-GlcNAc, and [β-^{32}P]UDP-GlcNAc are mixed, aliquoted, and dried down as in Method I, except that the specific activity of the [β-^{32}P]UDP-GlcNAc is increased such that each tube contains 5.0 nmol of UDP-GlcNAc (100 μM final concentration) with 1.5 × 10^6 cpm of ^{32}P; 10 μl of reaction mix (made exactly as for Method I but without α-methylmannoside) is then added to each tube. The appropriate acceptor (i.e., lysosomal enzyme), N-acetylglucosaminylphosphotransferase, and water to bring the volume to 50 μl are added to start the reaction. Samples are incubated at 37°, and the reaction is terminated (typically after ~1 hr) by adding 100 μl of ice-cold 2.5 mM UDP-GlcNAc–1 mg of BSA per milliliter plus 300 μl of ice-cold 1.5% (w/v) phosphotungstic acid–0.75 N HCl, vortexing, and incubating on ice for 10 min. Precipitates are collected by centrifugation (5 min, 12,000 g, 4°), and the supernatants are carefully removed and discarded. Pellets are resuspended by sonication in 0.9 ml of ice-cold 1% phosphotungstic acid–0.5 N HCl and centrifuged as above, and supernatants are removed and discarded. Washed pellets are resuspended by sonication in 450 μl of Pronase buffer, after which 50 μl of preincubated Pronase (40 mg/ml) is added, bringing the final concentration of Pronase to 4 mg/ml. After 30 min at 56° (making sure that all visible protein clumps are gone), the samples are heated in a boiling water bath for 10 min, 1 ml of PBS is added, and the tubes are centrifuged (5 min, 12,000 g, room temperature). The supernatant is applied to a 0.6-ml Econocolumn of Con A–Sepharose and washed with 2 ml of PBS. Care should be taken not to exceed the binding capacity of the Con A–Sepharose (~50 nmol of high mannose oligosaccharide per milliliter of resin). If necessary, load only a fraction of the Pronase digest. The resin is extruded directly into a scintillation vial using a 1-ml syringe containing 1 ml of 500 mM α-methylmannoside, and then counted after addition of 12 ml of scintillation fluid. Blanks incubated for zero time or without acceptor are used to correct for background radioactivity and endogenous acceptor, if present. One unit of enzyme activity is defined as the transfer of 1 pmol of N-acetylglucosamine 1-[^{32}P]phosphate per hour at 37°.

When this method is used with nonphosphotungstic acid-precipitable glycopeptide acceptors, a number of modifications are made. After the incubation, the reactions are stopped by adding 100 μl of 2.5 mM UDP-GlcNAc and then heating (100°, 5 min). Next, 0.5 ml of 4.8 mg of Pronase per milliliter in Pronase buffer is added, and the samples are treated as described for glycoprotein acceptors after Pronase treatment, except that the Con A–Sepharose columns are washed extensively (100–200 ml per column over 1 hr).

Purification from Rat Liver

All procedures are performed at 4° except the adjustment of buffer pH, which is done at room temperature. Lubrol PX was chosen for its low absorbance at 280 nm; in preliminary experiments Triton X-100 was also used. All buffers after preparation of the Golgi membranes contained 0.17 unit of Aprotinin per milliliter. The inclusion of glycerol in the purification buffers resulted in increased enzyme stability and yield.

Preparation of Golgi Membranes. Liver Golgi membranes are prepared from 200 g male Sprague–Dawley rats by a modification of the method of Leelevathi *et al.*[4] All sucrose solutions contain 50 mM Tris-HCl (pH 7.5)–5 mM MgCl$_2$. Rats are individually etherized (just long enough to make them manageable) and killed by decapitation. Livers are excised, minced, and immediately placed in 0.5 M sucrose on ice. The finely minced liver is homogenized in 0.5 M sucrose (1 g of tissue per 3 ml of sucrose) for 30 sec with a Polytron homogenizer (type PT 10 OD, Brinkmann Instruments, Inc., Westbury, NY) at a low speed to minimize frothing. Nuclei and intact cells are removed from the homogenate by centrifugation at 600 g for 10 min in a Sorvall-refrigerated centrifuge. The postnuclear supernatant is removed by aspiration, and Aprotinin is added to 0.17 unit/ml. Nineteen-milliliter portions are layered on top of 10 ml of 1.3 M sucrose in Spinco No. 30 rotor centrifuge tubes and then centri-

---

[4] I. Tabas and S. Kornfeld, *J. Biol. Chem.* **254,** 11655 (1979).

fuged at 27,000 rpm (63,000 g_{av}) for 2 hr. The crude smooth membrane fraction appears as a white band above the 1.3 M sucrose layer and is generously pooled and adjusted to 1.1 M sucrose with 2.0 M sucrose. Twenty-seven-milliliter portions of this crude membrane fraction are loaded into cellulose nitrate tubes (Beckman 302237). Seven milliliters of 1.0 M sucrose are layered over the membrane fraction, and 5 ml of 0.5 M sucrose are layered over the 1.0 M sucrose. The sample is centrifuged in a SW 27 rotor at 25,000 rpm (83,000 g_{av}) for 90 min. The Golgi membranes appear as a band of small white pieces between the 0.5 M sucrose and the 1.0 M sucrose. This band is collected, adjusted to 0.25 M sucrose–0.4 M NaCl (add 2 volumes of ice-cold distilled water and 0.333 volume ice-cold 4 M NaCl), and pelleted by centrifugation in a No. 60 Ti rotor at 45,000 rpm (143,000 g_{av}) for 45 min, yielding salt-washed Golgi membranes.

Differential Protein Extraction and Removal of Endogenous Acceptors. The salt-washed Golgi membranes are suspended to 10–40 mg of protein per milliliter in 10 mM mannose 6-phosphate–0.1% Lubrol PX–5 mM MgCl$_2$–0.25 mM dithiothreitol–10 mM Tris HCl, pH 7.4 (buffer A plus mannose 6-phosphate), homogenized with 5 strokes of a motor-driven Teflon homogenizer, and incubated at 0° for 30 min. The membranes are then diluted to 3.5 mg of protein per milliliter with buffer A and pelleted at 45,000 rpm (133,000 g_{av}) for 45 min at 4°.

Detergent Solubilization. The membranes are solubilized by sonication (60 W, 0° in 15-sec bursts) in 2% Lubrol PX–0.5% sodium deoxycholate–20% glycerol–5 mM MgCl$_2$–3 mM 2-mercaptoethanol–25 mM Tris-HCl (pH 7.4). The volume of solubilization buffer is half the volume of buffer A used in the preceding step. The extract is centrifuged (133,000 g_{av}, 45 min, 4°), and the pellet is extracted twice more with the same buffer, each time using one-tenth of the original buffer A volume. All three supernatants are pooled.

DEAE-Cellulose Chromatography. The solubilized membranes are applied to a DE-52 column (about 3 mg of protein per milliliter of resin) equilibrated with 0.3% Lubrol PX–20% glycerol–5 mM MgCl$_2$–25 mM Tris-HCl (pH 7.4) (buffer B). The column is washed with 1–2 column volumes of buffer B and then with 55 mM NaCl in buffer B until the A_{280} is <0.2. The *N*-acetylglucosaminylphosphotransferase is batch-eluted with 250 mM NaCl in buffer B, and fractions are assayed for activity using α-methylmannoside as acceptor. Peak fractions are pooled and dialyzed twice against 1 liter of 50 mM Tris-HCl (pH 7.5)–10 mM MgCl$_2$–10 mM MnCl$_2$–0.3% Lubrol–20% glycerol. The partially purified enzyme is stored at 0°.

Purification, Yield, and Stability. Following the above protocol, the best preparation of *N*-acetylglucosaminylphosphotransferase was 950-

PURIFICATION OF RAT LIVER N-ACETYLGLUCOSAMINYLPHOSPHOTRANSFERASE[a]

Fraction	Volume (ml)	Protein (mg)	N-Acetylglucosaminyl-phosphotransferase			
			Total activity (units)	Specific activity (units/mg)	Yield (%)	Purification (fold)
Homogenate	282	14,100	831,900	59	100	1
Postnuclear super-natant	232	9280	742,400	80	89	1.4
Crude smooth mem-branes	81	1296	672,624	519	81	9
Golgi membranes	19	95	323,950	3410	39	58
Salt-washed Golgi membranes	14.2	30	314,700	10,490	38	178
Detergent extract	8.4	15	168,420	11,228	20	190
DEAE-cellulose eluate	7.4	3.6	159,664	44,351	19	752

[a] Golgi membranes were prepared from a total of 10 rats (74 g of liver). Details of the purification are in the text.

fold purified with a 22% yield. This enzyme activity remained stable (<10% loss of activity) over 2 months when stored at 0°. A typical enzyme preparation is shown in the table.

Contaminating Activities. The partially purified N-acetylglucosaminyl-phosphotransferase contained other enzyme activities capable of transferring N-acetylglucosamine from UDP-GlcNAc to appropriate acceptors. N-Acetylglucosaminyltransferase I is one such activity, and although over 98% of this contaminant can be removed by selective binding to UDP-hexanolamine Sepharose 4B,[5] the residual contaminating activity is sufficient to prevent the use of commercially available UDP-[³H]GlcNAc as the donor in assays using lysosomal enzyme acceptors.

Properties of Rat Liver N-Acetylglucosaminylphosphotransferase

General. Rat liver N-acetylglucosaminylphosphotransferase is an integral membrane glycoprotein.[6] Unless otherwise noted, the activity has been characterized using α-methylmannoside as the acceptor.

Metal and pH Requirements. Enzyme activity requires the presence of a divalent cation. This requirement is satisfied by Mn^{2+} better than

[5] C. L. Oppenheimer, A. E. Eckart, and R. L. Hill, *J. Biol. Chem.* **256,** 11477 (1981).
[6] M. L. Reitman, unpublished data, 1981.

Mg^{2+}, but not by Ca^{2+}, Cu^{2+}, Cd^{2+}, Co^{2+}, Zn^{2+}, or Hg^{2+}.[6] The pH optimum is 6.5 to 7.5, with Tris-HCl giving 50% more activity than sodium cacodylate.[6]

Kinetic Parameters. The apparent K_m for UDP-GlcNAc is 38 μM (at 100 mM α-methylmannoside and 20 mM Mn^{2+}). The apparent K_m for α-methylmannoside is 183 mM (at 150 μM UDP-GlcNAc and 20 mM Mn^{2+}).

Inhibitors. No significant inhibition is observed with the following compounds (at 5 mM): AMP, ADP, ATP, UMP, UTP, GDP, CDP, UDP-Gal, UDPglucuronic acid, mannose 6-phosphate, glucose 6-phosphate, N-acetylglucosamine 1-phosphate, glucose 1-phosphate, and phosphate or (at 100 mM) GlcNAc, GalNAc, N-acetylmannosamine, xylose, and lyxose. At 5 mM, UDP inhibits 63%. Surprisingly, UDP-Glc is a competitive inhibitor with respect to UDP-GlcNAc and is also a substrate for the enzyme, with a catalytic efficiency about 12-fold worse than UDP-GlcNAc.[6]

Acceptor Specificity. Studies of the ability of various monosaccharides to accept GlcNAc 1-phosphate have demonstrated that the transferase is highly selective for α-linked mannose residues.[6] The nonreducing terminal mannose residues of the disaccharide, Manα1 → 2Man, and the trisaccharide, Manα1 → 6Manα1 → 6Man, are equally as efficient substrates for the enzyme as α-methylmannoside. High-mannose oligosaccharides have been shown to accept GlcNAc 1-phosphate, but with a catalytic efficiency essentially equal to that of α-methylmannoside.[3] More recent studies have probed the selectivity of the phosphotransferase toward different size high-mannose oligosaccharides still bound to lysosomal enzymes. Results suggest that there is no selectivity for one oligosaccharide over another.[7]

The most striking property of N-acetylglucosaminylphosphotransferase is its glycoprotein acceptor specificity. This enzyme phosphorylates lysosomal enzymes in an *in vitro* assay at least 100-fold more efficiently than either other glycoproteins with similar carbohydrate chains or free oligosaccharides.[3] Using deglycosylated lysosomal enzymes as competitive inhibitors we have found that this specificity is encoded solely by the protein portion of the acceptor.[7]

Other Properties

In the rat liver fractionation used, N-acetylglucosaminylphosphotransferase activity largely copurifies with the Golgi marker galacto-

---

[7] L. Lang, unpublished data, 1983.

syltransferase. However, in both the BW5147 and P388D₁ murine cell lines, N-acetylglucosaminylphosphotransferase can be separated from galactosyltransferase by sucrose density gradient centrifugation and thus may reside in either the "early" or "cis" portion of the Golgi or in the transitional endoplasmic reticulum.[9,10]

In addition to rat liver, N-acetylglucosaminylphosphotransferase has been demonstrated in human fibroblasts,[2,10] placenta,[2] brain,[11] spleen,[11] kidney,[11] liver,[11,12] and peripheral blood leukocytes,[13] as well as Chinese hamster ovary cells,[1] sheep liver,[6] BW5147 murine lymphoma cells,[9] J774.2,[6] and P388D₁ murine macrophage-like cells.[9] The enzyme is probably present in all normal cells synthesizing lysosomal enzymes. However, N-acetylglucosaminylphosphotransferase activity is deficient in patients with the autosomal recessive errors of metabolism I-cell disease (mucolipidosis II) and pseudo-Hurler polydystrophy (mucolipidosis III) and is believed to be the primary genetic lesion in these patients.[2,10,13]

---

[8] R. Pohlmann, A. Waheed, A. Hasilik, and K. von Figura, *J. Biol. Chem.* **257,** 5323 (1982).

[9] D. E. Goldberg and S. Kornfeld, *J. Biol. Chem.* **258,** 3159 (1983).

[10] A. Varki, M. L. Reitman, and S. Kornfeld, *Proc. Natl. Acad. Sci. U.S.A.* **78,** 7773 (1981).

[11] A. Waheed, R. Pohlmann, A. Hasilik, K. von Figura, A. van Elsen, and J. G. Leroy, *Biochem. Biophys. Res. Commun.* **105,** 1052 (1982).

[12] M. Owada and E. Neufeld, *Biochem. Biophys. Res. Commun.* **105,** 814 (1982).

[13] A. Varki, M. L. Reitman, A. Vannier, S. Kornfeld, J. H. Grubb, and W. S. Sly, *Am. J. Hum. Genet.* **34,** 717 (1982).

[9] O^β-(N-Acetyl-α-glucosamine-1-phosphoryl)serine in Proteinase I from *Dictyostelium discoideum*

By GARY L. GUSTAFSON and JOHN E. GANDER

Proteinase I is a lysosomal thiol endopeptidase isolated from stationary-phase amoebae of the cellular slime mold, *Dictyostelium discoideum*.[1] It is distinguished from other thiol proteinases in this organism by its ability to catalyze an inactivation of the enzyme UDPglucose pyrophosphorylase. Proteinase I was shown to be a phosphoprotein based on the

---

[1] G. L. Gustafson and L. A. Thon, *J. Biol. Chem.* **254,** 12471 (1979).

following criteria[2]: (1) the purification of ^{32}P-labeled enzyme to constant specific radioactivity; (2) the coimmunoprecipitation of ^{32}P isotopic activity and proteinase enzymatic activity with proteinase-specific antisera; and (3) the coelution of ^{32}P isotopic activity with purified, denatured enzyme protein during dissociation chromatography. The initial identification of O^β-(N-acetyl-α-glucosamine-1-phosphoryl)serine in proteinase I was based on the demonstrated releases of phosphate groups from the enzyme either as N-acetyl-α-glucosamine 1-phosphate (by β-elimination) or as O^β-phosphorylation (by acid hydrolysis).[3]

In this chapter we summarize the chemical characterization of the O^β-(N-acetyl-α-glucosamine-1-phosphoryl)serine group and provide the first description of the ^{31}P and ^{13}C nuclear magnetic resonance signals associated with this phosphodiester.

Chemical Characterizations

Procedures

Principle. Phosphoesters at the hemiacetal hydroxyls of sugars are all moderately stable to alkali but are labile to mild acid hydrolysis.[4] In contrast, phosphoesters at the β-hydroxyls of serine and threonine residues in proteins are moderately stable to acid hydrolysis but are readily released by an alkali-catalyzed β-elimination reaction.[5] Because of the differences in chemical properties of these two phosphoester bonds, it is possible to demonstrate that proteinase I from *D. discoideum* contains O^β-(N-acetyl-α-glucosamine-1-phosphoryl)serine by showing (1) that essentially all the phosphate in the proteinase can be isolated as N-acetyl-α-glucosamine 1-phosphate following treatment of the enzyme with alkali; and (2) that a substantial portion of proteinase phosphate can also be isolated as O^β-phosphorylserine following acid hydrolysis.

Preparation of Proteinase I. A fractionation procedure, yielding 1–2 mg of highly purified proteinase I, has been described.[1] This procedure can be readily scaled-up to provide for the isolation of much larger quantities of enzyme of comparable purity. From 75 g of lyophilized, stationary-phase amoebae, approximately 80 mg of proteinase I can be isolated. For storage, the purified enzyme is dialyzed against 5 mM ammonium bicar-

[2] G. L. Gustafson and L. A. Thon, *Biochem. Biophys. Res. Commun.* **86,** 667 (1979).
[3] G. L. Gustafson and L. A. Milner, *J. Biol. Chem.* **255,** 7208 (1980).
[4] L. F. Leloir and C. E. Cardini, this series, Vol. 3 [115].
[5] G. Taborsky, *Adv. Protein Chem.* **28,** 1 (1974).

bonate and lyophilized. The purified enzyme contains equal amounts (0.69 μmol/mg dry wt) of total phosphate and acid-labile hexosamine.

Assay Methods

Total phosphate is determined as inorganic phosphate according to the methods of Chen *et al.*[6] after samples are ashed in the presence of $Mg(NO_3)_2$.[7]

Acid-labile hexosamine, released by hydrolysis of proteinase I in 1 *N* HCl in a boiling water bath for 6 min, is quantified by the procedures of Good and Bessman.[8]

N-Acetylglucosamine, released from *N*-acetyl-α-glucosamine 1-phosphate, by *Escherichia coli* alkaline phosphatase, is assayed either by gas–liquid chromatography[9] or by a colorimetric method.[10]

Reducing sugar is assayed by the Park and Johnson[11] ferricyanide method.

Isolation of N-Acetylglucosamine 1-phosphate. Purified proteinase I, 4 mg, is dissolved in 0.5 ml of 0.4 *N* NaOH and incubated under N_2 at 23° for 20 hr. The mixture is adjusted to pH 8 by additions of 0.1 ml of 0.1 *M* ammonium bicarbonate and 0.1 ml of 2 *N* HCl. *N*-Acetylglucosamine 1-phosphate, released from the enzyme with alkali, is separated from the protein by chromatography of the neutralized reaction mixture on a column of Sephadex G-25 (1.5 × 29.5 cm). The column is developed with 0.01 *M* ammonium bicarbonate. All the phosphate in the initial reaction mixture elutes as a single peak at 1.6 column void volumes and is clearly resolved from the protein, which elutes in 1 column void volume. Fractions containing phosphate are combined and evaporated to dryness at 40° under vacuum.

Characterization of Isolated N-Acetylglucosamine 1-phosphate. Aliquots (0.5 ml) of the phosphoryl moiety isolated as described above, containing 0.38–0.42 μmol of total phosphate, are treated either with (1) *N* HCl at 100° for 6 min, or (2) 0.4 unit of purified *E. coli* alkaline phosphatase (Sigma Chemical Co.) at pH 8, 37° for 8 hr. Colorimetric analyses indicate that samples treated in either manner contain equal molar amounts of total hexosamine, reducing sugar equivalents, and inorganic

[6] P. S. Chen, T. Y. Toribara, and H. Warner, *Anal. Chem.* **28,** 1756 (1956).

[7] B. N. Ames and D. T. Dubin, *J. Biol. Chem.* **235,** 769 (1960).

[8] T. A. Good and S. P. Bessman, *Anal. Biochem.* **9,** 253 (1964).

[9] C. C. Sweeley, R. Bentley, M. Makita, and W. W. Wells, *J. Am. Chem. Soc.* **85,** 2497 (1963).

[10] J. L. Reissig, J. L. Strominger, and L. F. Leloir, *J. Biol. Chem.* **217,** 959 (1955).

[11] J. T. Park and M. J. Johnson, *J. Biol. Chem.* **181,** 149 (1949).

phosphate.[3] In contrast, untreated samples of the isolated phosphoryl moiety contain no detectable free hexosamine or reducing sugar. Although a fraction (approximately 10%) of the total phosphate in the untreated, isolated material is detected as inorganic phosphate, this is attributable to the partial hydrolysis of N-acetylglucosamine 1-phosphate by the acidic conditions used in the assays for inorganic phosphate. The hexosamine released by acid or phosphatase treatments of the isolated phosphoryl moiety is identified as N-acetylglucosamine by gas–liquid chromatographic analysis.[3] The rate of hydrolysis of the isolated N-acetylglucosamine 1-phosphate in 0.5 N HCl at 50° indicates that it has the [α] configuration.[3]

Isolation of O^β-Phosphorylserine. Purified proteinase I, 7 mg, is dissolved in 1 ml of 2 N HCl and incubated under vacuum at 100° for 18 hr. The HCl is removed from the resulting hydrolysate by repeated evaporation at 40° under vacuum. The dry residue is dissolved in 0.1 ml of distilled water and applied to a thin-layer plate of Avicel (Analtech Inc.). The plate is developed by electrophoresis at 350 V for 3.5 hr, with 10% formic acid as the electrode buffer. The apparatus described by Gracy[12] is used for electrophoresis.

The developed plate is dried at room temperature and sprayed lightly with solutions of 10% triethylamine in methylene chloride and 0.1% fluorescamine in acetone. After this treatment, compounds containing α-amino groups are detected as fluorescent bands when the plate is viewed under ultraviolet light.[13] The single fluorescent band, displaced from the origin in the direction of the anode, is scraped from the glass plate, and this material is extracted five times with distilled water. In each extraction the Avicel–water mixture is briefly mixed and centrifuged at 1000 g for 30 min. The supernatant fluids recovered from the five extractions are combined and evaporated to dryness at 40° under vacuum.

Characterization of O^β-Phosphorylserine. Approximately 1 mmol of total phosphate is recovered in the material extracted from the Avicel thin layer. The isolated phosphoryl moiety exhibits an R_f ($R_f = 0.81$) identical to that of O^β-phosphorylserine when chromatographed on a thin layer with 40% ethanol : 28% NH_4OH (9 : 1).

Identification of the isolated phosphoryl moiety as O^β-phosphorylserine is confirmed by hydrolyzing an aliquot of the material under vacuum, in 6 N HCl at 110° for 22 hr, and subjecting the hydrolysate to quantitative amino acid analysis in an amino acid analyzer.[3]

[12] R. W. Gracy, this series, Vol. 47 [22].
[13] A. M. Felix and M. H. Jimminez, *J. Chromatogr.* **89,** 361 (1974).

Comments

The conditions of acid hydrolysis required to release O^β-phosphorylseryl residues from peptide linkages (i.e., 2 N HCl, 100°, 18 hr) cause hydrolysis of 76% of the phosphoester bonds in this moiety. Because of this significant destruction of the phosphoester during its isolation, it is not possible to demonstrate directly that all the N-acetylglucosamine 1-phosphate residues recovered from alkali-treated proteinase are linked through phosphodiester bonds to serines. However, correcting for the predicted loss in O^β-phosphorylserine incurred during hydrolysis, it is estimated that at least 95% of the enzyme phosphate is esterified to peptidylserines. Furthermore, no phosphorylthreonine is detected in the phosphorylated material that is isolated from acid hydrolysates of proteinase I.

Planter and Carlson[14] have described methods for following the progress of an alkali-catalyzed, β-elimination reaction and for quantifying the olefinic amino acyl residues generated in the reaction. In proteins containing a mixture of phosphodiesters, these methods can aid in evaluating the stoichiometric relationships between the products of β-elimination reactions.

In situations where very limited amounts of purified phosphoglycoprotein are available, the general approach used by Varki and Kornfeld[15] in detecting the N-acetyl-α-glucosamine-(1,6)-phosphomannose linkage may be adapted for identifying the O^β-(α-N-acetylglucosamine-1-phosphoryl)serine modification. This method employs pig liver α-N-acetylglucosaminidase to release free sugar residues from an isotopically labeled phosphodiester. The strategy would be to demonstrate (1) that a [^{32}P]phosphoprotein contains phosphorylserine residues that are protected from hydrolysis by alkaline phosphomonoesterase, and (2) that treatment of the protein with purified α-N-acetylglucosaminidase provides for releasing the block protecting these moieties. In using this approach, it would be best to identify the labeled, phosphorylated amino acid by paper electrophoresis under conditions that resolve phosphorylserine from phosphorylthreonine.[16]

^{31}P and ^{13}C NMR Spectroscopy of Proteinase I

Procedures

Principle. Nuclear magnetic resonance (NMR) spectroscopy provides a means of characterizing covalently linked glycosyl and phosphoryl moi-

---

[14] J. J. Planter and D. M. Carlson, this series, Vol. 28 [3].
[15] A. Varki and S. Kornfeld, *J. Biol. Chem.* **255**, 8398 (1980).
[16] L. Bitte and D. Kabat, this series, Vol. 30 [54].

eties of proteins *in situ* under conditions that do not destroy the material being analyzed. Thus, artifacts, as sometimes occur in wet chemical analyses, can be avoided. When used in conjunction with enzymatic and other chemical procedures, NMR spectroscopy provides a powerful analytical tool. ^{13}C and ^{31}P NMR spectroscopy have been used to characterize poly- and oligosaccharides attached to glycoproteins and glycopeptides.[17–19] The use of ^{13}C NMR spectroscopy has considerable potential as an analytical tool, because most of the resonances of the carbon atoms in saccharides are well dispersed and occur in a portion of the spectrum (55–110 ppm) distinct from those of the carbon atoms of amino acyl residues. The resonance frequency of the β-carbon of seryl and threonyl residues are exceptions to this generalization.[17]

Natural abundance ^{13}C NMR spectroscopy in a high-field superconducting spectrometer with Fourier transform capability requires of the order of 10–20 μmol of each residue to be characterized. Since the natural abundance of ^{13}C is approximately 1.1%, 100-fold smaller quantities of sample will provide adequate signals for the enriched carbons if the substance being analyzed contains ^{13}C atoms, enriched with 90–100 atom % [13]C, in nonadjacent positions, compared with a sample containing natural abundance [13]C.

The resonance spectra of monosaccharide residues are well spread, so that it is usually possible to identify the sugars if there are relatively few species or if polymers contain repeating units of relatively small oligosaccharides. Furthermore, it is also possible to ascertain the position(s) of linkage of one sugar residue to its neighbor and the spin–spin coupling constant, J_{C-H}, between the anomeric carbon of the glycosyl residue and its H-1 atom; J_{C-H} values for α- and β-linked glycosyl residues are of the order of 170 and 160 Hz, respectively.[20]

A 400-fold greater sensitivity is obtained using ^{31}P NMR compared to ^{13}C NMR spectroscopy of substances containing natural abundance of [13]C. Although ^{31}P NMR spectroscopy has not been widely applied to the characterization of conjugated proteins, it undoubtedly will be used to a greater extent in the future on phosphoglycoproteins and phosphoproteins. A minimum of 0.05 mmol of phosphoprotein is required if it contains one phosphorous atom per molecule.

Preparation and Modification of Proteinase I. Proteinase I is purified by the large-scale fractionation procedure described above. Spectroscopy is carried out on 30 mg of the freeze-dried preparation dissolved in 3.9 ml of 99.8% ^2H_2O.

[17] K. Dill and A. Allerhand, *J. Biol. Chem.* **254,** 4524 (1979).
[18] C. J. Unkefer and J. E. Gander, *J. Biol. Chem.* **254,** 12131 (1979).
[19] C. J. Unkefer, C. Jackson, and J. E. Gander, *J. Biol. Chem.* **257,** 2491 (1982).
[20] K. Bock and C. Pederson, *J. Chem. Soc. Perkin Trans.* **2,** 293 (1974).

After the spectroscopic analyses of unmodified proteinase I, the same sample of enzyme is treated with 10 ml of reagent grade, aqueous 48% hydrofluoric acid (Mallinckrodt) for 48 hr at 0°. This treatment is known to cleave phosphodiester linkages without cleaving either polypeptide or glycopyranosyl linkages.[21] The reaction mixture is neutralized with ammonium hydroxide, and the low molecular weight substances are removed by dialysis against 50 mM ammonium bicarbonate buffer, pH 8.0. Dialysis is conducted in 3500 molecular weight cutoff dialysis tubing (Spectrapore II). After dialysis, the sample is filtered through a Millipore (22 μm) filter and freeze dried. The sample is dissolved in ^2H_2O as described above, and its NMR spectra are again examined.

Methods of NMR Spectroscopic Analyses. Proton-coupled and protein-decoupled ^{13}C NMR spectra are obtained at 25° in a 70.5 kG field on a Nicolet spectrometer operated in the Fourier transform mode at a frequency of 75.46 MHz. The field is locked on the deuterium signal. Decoupled spectra are obtained with broad band proton decoupling; coupled spectra are obtained with the decoupler gated off to observe ^1H–^{13}C spin–spin coupling. Spectra are collected using a 90° pulse and recycle times greater than 5 T_1 (3 sec). At least 15,000 transients are collected, and signal averaging is carried out. Chemical shifts are reported in parts per million (ppm) from internal (trimethylsilane)-1-propane sulfonate (TSP) added to the sample. The anomeric carbon of β-D-glucopyranose resonates at 98.8 ppm in this system.

Proton-decoupled ^{31}P NMR spectroscopy is carried out with broadband proton decoupling with the instrument described above operating in the Fourier transform mode at a frequency of 121.49 MHz with the field locked on the deuterium signal. Spectra are collected using a 90° pulse. At least 1000 transients are collected and signal-averaged on the sample containing unmodified proteinase I; 15,000 transients are collected on the protein isolated after treatment with aqueous 48% HF. Chemical shifts of the phosphorous in proteinase I are reported in parts per million from an external reference of uridine-5'-monophosphoric acid, whose resonance occurs at 0.47 ppm with reference to phosphoric acid.

^{31}P NMR Spectroscopy

Preliminary ^{31}P NMR spectroscopy of proteinase I shows only one signal in the general region where phosphodiesters and orthophosphate resonate. An external reference of uridine-5'-monophosphoric acid is added to the sample, and the spectrum obtained shows a single resonance at -1.31 ppm (Fig. 1). This signal is similar to that observed for O^β-

---

[21] C. J. Unkefer, C. Jackson, and J. E. Gander, unpublished observations, 1983.

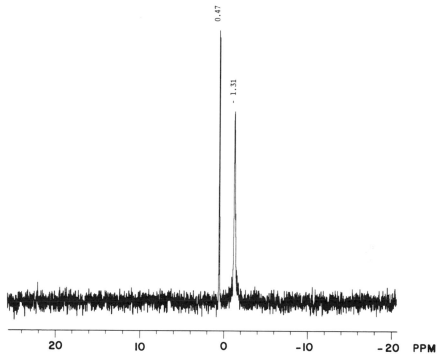

Fig. 1. Proton-decoupled ^{31}P NMR spectroscopy of proteinase I at 121.49 MHz with a recycle time of 3 sec, 1200 accumulations. Spectroscopy is conducted on 30 mg of protein at 25°. An external standard of uridine-5′-monophosphoric acid referenced to phosphoric acid is used as the reference. The reference resonates at 0.47 ppm.

(ethanolamine-1-phosphoryl)serine[22] and dissimilar to those of phospho-monoesters or 2′,3′-cyclic phosphodiesters, which all have resonances considerably farther downfield. There is no evidence for the occurrence of phosphomonoester in proteinase I.

Treatment of proteinase I with aqueous 48% HF removes all detect-able phosphorus from the enzyme as low molecular weight substances that can be removed by dialysis. No ^{31}P resonances are observed after 15,000 acqusitions with the modified enzyme (data not shown).

^{13}C NMR Spectroscopy

Proton-decoupled ^{13}C NMR spectroscopy of proteinase I shows major signals at 96.9, 56.6, 73.3, 72.4, 76.0, and 63.2 ppm (Fig. 2). These signals

[22] J. M. Chalovich, C. T. Burt, S. M. Cohen, T. Glonek, and M. Barany, *Arch. Biochem. Biophys.* **182,** 683 (1977).

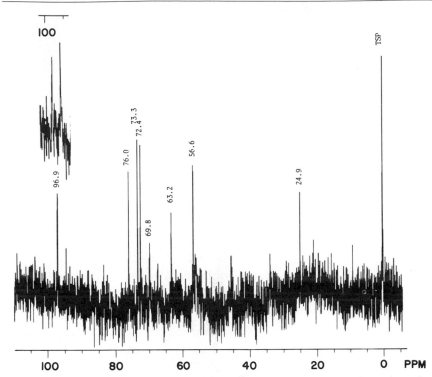

FIG. 2. Proton-decoupled natural abundance ^{13}C NMR spectroscopy of proteinase I at 75.46 MHz with a recycle time of 3 sec, after 15,000 accumulations. Spectroscopy is conducted on 30 mg of protein at 25°. An internal standard of (trimethylsilane)-1-propane (TSP) is used as the reference. The inset shows a portion of the spectrum around 100 ppm of the protein analyzed with proton coupling using the same parameters with the exception that 123,000 accumulations are obtained.

are those associated with C-1 through C-6 of the glucosamine residue of O^β-(N-acetylglucosamine-1-phosphoryl)serine residues. Except for C-1, these resonance frequencies compare favorably with the respective carbons of N-acetylglucosamine 1-phosphate.[23] The phosphodiester apparently causes greater deshielding of C-1 than that of the phosphomonoester derivative. The signals at 96.9 and 56.6 ppm are split because of spin–spin coupling between C—O—P and C—N bonds, which occur at C-1 and C-2 positions, respectively. The signal at 24.9 ppm is that of the methyl carbon of the acetyl residue. A resonance is obtained at 177 ppm for the carbonyl carbon of the acetyl residue (not shown in Fig. 2). Several sig-

[23] D. R. Bundle, H. J. Jennings, and I. C. P. Smith. *Can. J. Chem.* **51,** 3812 (1973).

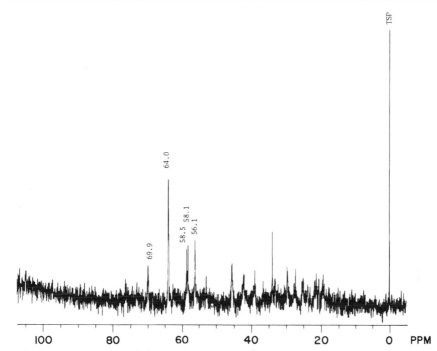

FIG. 3. Proton-decoupled natural abundance ^{13}C NMR spectroscopy of proteinase I modified by treatment with 48% aqueous HF. Preparation of the sample is described in the text. The spectrum is obtained with the same parameters as given in Fig. 2, except that 46,000 accumulations are obtained.

nals for the carbonyl and carboxyl groups of amino acyl residues in the protein are observed around 175 ppm. The unsubstituted β-carbons of seryl and threonyl residues have resonances at 63.4 and 69.3, respectively. Substitution with α-mannosyl residues shifts these signals to 68.5 and 76.8 ppm, respectively.[24] The signal at 69.8 and the lack of a signal around 77 ppm support the conclusion that phosphorus is attached only to seryl residues. A proton-coupled spectrum of proteinase I is obtained, and the coupling constant between ^{1}H–^{13}C at the anomeric carbon is measured. The value of 172.5 Hz determined for this constant (Fig. 2, inset) confirms that the anomeric carbon is attached to the aglycon by an α linkage.[20]

As shown in Fig. 3, aqueous HF treatment of proteinase I results in the loss of the signals associated with N-acetylglucosamine residues and

[24] A. Allerhand, K. Dill, E. Berman, J. M. Lacombe, and A. A. Pavia, *Carbohydr. Res.* **97**, 331 (1981).

the appearance of two new major signals at 64.0 and 56.1 ppm. The signal at 64.0 is that expected of an unsubstituted β-carbon of seryl residue; whereas the signal at 56.1 is at the frequency expected of the α-carbon of a seryl residue. These data show that treatment with 48% aqueous HF, under the conditions used, does not significantly degrade the polypeptide. This same treatment of peptidophosphogalactomannan from *Penicillium charlesii* does not cleave O-glycosidic bonds between mannopyranosyl residues,[21] and it is unlikely that it would cleave N-glycosidic linkages between *N*-acetylglucosamine residues attached to asparagine. Therefore, all of the *N*-acetyl-α-glucosamine residues are apparently present as phosphodiesters.

Comments

The results of ^{31}P and ^{13}C NMR spectroscopy provide corroborative evidence that essentially all the *N*-acetyl-α-glucosamine 1-phosphate residues in proteinase I are esterified to peptidylserines through phosphodiester bonds. This conclusion could be strengthened by examining the [1]H–[31]P spin–spin coupling of the phosphodiester and demonstrating that it is coupled to one proton on the glycosidic side of the phosphorous and to two protons on the seryl side. Preliminary studies indicate that [1]H–[31]P spin–spin coupling cannot be resolved with intact proteinase I—presumably because of the high molecular weight of the polymer. Therefore, such an analysis would require the isolation of low molecular weight polypeptide fragments of the proteinase, which contain the phosphodiester.

Concluding Remarks

In addition to the chemical and spectroscopic methods described above, it appears that immunological approaches may also be useful for detecting sugar 1-phosphate residues occurring in diester linkage in proteins. Studies by Gustafson and Milner[25] have shown that the phosphodiester group in proteinase I is a determinant for antibodies elicited against the enzyme. Similarly, others have shown that sugar 1-phosphates occurring in glycans of bacteria and yeast also function as immunological determinants.[26–28] Presumably, monoclonal antibodies directed against

[25] G. L. Gustafson and L. A. Milner, *Biochem. Biophys. Res. Commun.* **94**, 1439 (1980).
[26] J. H. Pazur, A. Cepure, J. A. Kane, and W. W. Karakawa, *Biochem. Biophys. Res. Commun.* **43**, 1421 (1971).
[27] P. N. Lipke, W. C. Raschke, and C. E. Ballou, *Carbohydr. Res.* **37**, 23 (1974).
[28] J. H. Pazur, *J. Biol. Chem.* **257**, 589 (1982).

these moieties could serve as highly specific probes for phosphodiester-linked sugars in proteins.

Finally, with regard to the physiological significance of O^β-(N-acetyl-α-glucosamine-1-phosphoryl)serine in proteinase I, a likely possibility is that these acidic moieties influence ionic interactions of the proteinase with substrate proteins and thus direct the specificity of the proteinase toward the preferential degradation of basic protein substrates. Consistent with this view are the observations[1] (1) that proteinase I has an unusually high affinity for the basic polypeptide, protamine sulfate; and (2) that, among several enzymes isolated from *D. discoideum,* the proteinase has a marked specificity for degrading UDPglucose pyrophosphorylase, an enzyme with a very basic isoelectric point ($pI = 8.4$–8.0^{29}). An alternative possibility is that the phosphoryl moiety serves some function in directing the intracellular compartmentation of proteinase I. For example, in mammalian systems mannose 6-phosphate residues in acid hydrolases serve in targeting newly synthesized hydrolases to lysosomes.[30]

Acknowledgment

NMR spectroscopy was conducted at the University of Minnesota NMR facility, Gortner Laboratory, St. Paul Campus; Thomas P. Krick, operator.

[29] R. E. Manrow and R. P. Dottin, *Proc. Natl. Acad. Sci. U.S.A.* **77,** 730 (1980).
[30] W. S. Sly, M. Natowicz, A. Gozalez-Noriega, J. H. Grubb, and H. D. Fischer, *in* "Lysosomes and Lysosomal Storage Diseases" (J. W. Callahan and J. A. Lowden, eds.), p. 131. Raven, New York, 1981.

[10] 5'-Nucleotidyl-*O*-tyrosine Bond in Glutamine Synthetase

By SUE GOO RHEE

Adenylyl and uridylyl groups attached in phosphodiester linkage to the phenolic hydroxyl group of tyrosine residues in protein were discovered during the study on the regulation of glutamine synthetase (GS) in

METHODS IN ENZYMOLOGY, VOL. 107

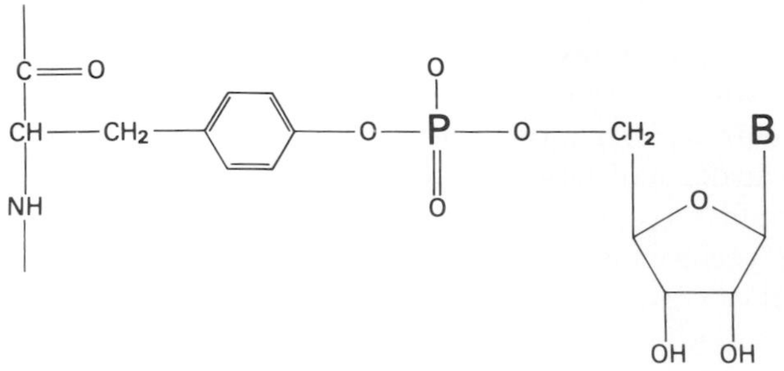

FIG. 1. 5'-Nucleotydyl-*O*-tyrosyl bond. B is adenine or uridine.

Escherichia coli.[1-3] It has since been shown that many other organisms utilize the same adenylylation and uridylylation mechanisms to regulate the activity of glutamine synthetase. In addition, the modification of several other proteins has been shown to involve the nucleotidylation of tyrosine residues. For example, the single-stranded RNA genome of poliovirus is ligated to a virus-specified protein through a phosphodiester linkage between the 5'-phosphate of UMP and the hydroxyl group of tyrosine,[4,5] and DNA is ligated to topoisomerase through a phosphodiester bond between the 3'-phosphate of DNA and the hydroxyl group of tyrosine.[6,7]

It has also been reported that activity of the lysine-sensitive aspartokinase in *E. coli* is regulated by adenylylation.[8] However, this result has not been confirmed, nor has the AMP attachment site been identified. This chapter will be limited to the 5'-adenylyl-*O*-tyrosine and 5'-uridylyl-*O*-tyrosine linkage (Fig. 1) involved in the regulation of glutamine synthetase.

[1] E. R. Stadtman and A. Ginsburg, *in* "The Enzymes" (P. D. Boyer, ed.), 3rd ed., Vol. 10, p. 755. Academic Press, New York, 1974.

[2] B. M. Shapiro, H. S. Kingdon, and E. R. Stadtman, *Proc. Natl. Acad. Sci. U.S.A.* **58**, 642 (1967).

[3] M. S. Brown, A. Segal, and E. R. Stadtman, *Proc. Natl. Acad. Sci. U.S.A.* **68**, 2949 (1971).

[4] P. G. Rothberg, T. J. R. Harris, A. Nomoto, and E. Wimmer, *Proc. Natl. Acad. Sci. U.S.A.* **75**, 4868 (1978).

[5] V. Ambros and D. Baltimore, *J. Biol. Chem.* **253**, 5263 (1978).

[6] Y. C. Tse, K. Kirkegaard, and J. C. Wang, *J. Biol. Chem.* **255**, 5560 (1980).

[7] J. J. Champoux, *J. Biol. Chem.* **256**, 4805 (1981).

[8] E. G. Niles and E. W. Westhead, *Biochemistry* **12**, 1723 (1973).

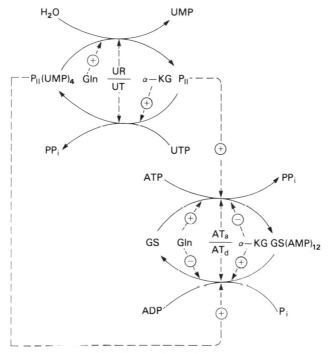

FIG. 2. The bicyclic cascade of glutamine synthetase.

Glutamine Synthetase Regulation in *E. coli*

Most of the facts regarding the adenylylated and uridylylated proteins were obtained from *E. coli* and have been the subjects of earlier reviews.[1] Here, we review briefly the information pertinent to detecting nucleotidylated proteins involved in the regulation of glutamine synthetase.

Glutamine synthetase is composed of 12 identical subunits and plays a central role in the assimilation of ammonia in *E. coli*. Cellular glutamine synthetase is regulated in response to the quality and abundance of the nitrogen source.[1,9,10] One way to regulate glutamine synthetase activity in *E. coli* involves the cyclic adenylylation and deadenylylation of a unique tyrosine group in each subunit. As illustrated in Fig. 2, the adenylylation reaction is catalyzed at the adenylylation site (AT_a) of an adenylyltrans-

[9] D. Mecke and H. Holzer, *Biochim. Biophys. Acta* **122,** 341 (1965).
[10] B. Magasanik, *Annu. Rev. Genet.* **16,** 135 (1982).

ferase (AT)[11] and involves transfer of adenylyl groups from ATP to the enzyme with the concomitant formation of PP_i. Removal of the adenylyl group from glutamine synthetase (deadenylylation) is achieved by phosphorolysis of the adenylyl-O-tyrosyl bond to yield ADP and unmodified glutamine synthetase.[12] This reaction is catalyzed at the deadenylylation site (AT_d) of AT.[13-15] Indiscriminate coupling of the adenylylation and deadenylylation reaction is prevented by the action of a tetrameric protein P_{II} (M_r 44,000), which also exists in an unmodified form, P_{IIA}, and a uridylylated form, P_{II} $(UMP)_4$ $(=P_{IID})$.[3,16,17] The capacity of AT to catalyze the adenylylation of glutamine synthetase is dependent on the concentration of P_{IIA}, whereas the capacity to catalyze the deadenylylation reaction is determined by the concentration of P_{IID}. Interconversion between P_{IIA} and P_{IID} is catalyzed by uridylyltransferase (UT) and uridylyl-removing enzyme (UR) activities.[3,17] These activities are contained in a single polypeptide (molecular weight 95,000).[18] Because glutamine synthetase and P_{II} are multimeric proteins composed of several identical proteins, the final number of nucleotides attached to the protein can be varied from 0 to 12 for glutamine synthetase and from 0 to 4 for P_{II}. The final states of nucleotidylation of these two proteins are dependent on various metabolites that affect the catalytic activities of one or more of the regulatory enzymes AT_a–AT_d, UT–UR. By far the most important of these metabolites are α-ketoglutarate and glutamine, which exhibit opposite and reciprocal effects on the adenylylation–deadenylylation pairs of reactions.[15,19,20] The sites of covalent modification are known. The amino acid sequence of tryptic peptides containing adenylylated tyrosine[21] and uridylylated tyrosine[22] show no apparent homology, as shown in Fig. 3. When the adenylylated tryptic peptide was incubated under the deadenylylating reaction conditions, the peptide was not deadenylylated.[23]

[11] D. Mecke, K. Wulff, K. Liess, and H. Holzer, *Biochem. Biophys. Res. Commun.* **24**, 452 (1966).
[12] W. B. Anderson and E. R. Stadtman, *Biochem. Biophys. Res. Commun.* **41**, 704 (1970).
[13] W. B. Anderson, S. B. Hennig, A. Ginsburg, and E. R. Stadtman, *Proc. Natl. Acad. Sci. U.S.A.* **67**, 1417 (1970).
[14] C. E. Caban and A. Ginsburg, *Biochemistry* **15**, 1569 (1976).
[15] S. G. Rhee, R. Park, P. B. Chock, and E. R. Stadtman, *Proc. Natl. Acad. Sci. U.S.A.* **75**, 3138 (1978).
[16] B. M. Shapiro, *Biochemistry* **8**, 659 (1969).
[17] S. P. Adler, D. Purich, and E. R. Stadtman, *J. Biol. Chem.* **250**, 6264 (1975).
[18] E. Garcia and S. G. Rhee, *J. Biol. Chem.* **258**, 2246 (1983).
[19] P. J. Senior, *J. Bacteriol.* **123**, 407 (1975).
[20] E. R. Stadtman and P. B. Chock, *Curr. Top. Cell. Regul.* **13**, 53 (1978).
[21] R. L. Heinrikson and H. S. Kingdon, *J. Biol. Chem.* **246**, 1099 (1971).
[22] S. G. Rhee, unpublished results.
[23] S. R. Jurgensen and P. B. Chock, unpublished results.

AMP
|
Glutamine Synthetase: Ile–His–Pro–Gly–Glu–Ala–Met–Asp–Lys–Asn–Leu–Tyr–
                               Asp–Leu–Pro–Pro–Glu–Gln–Ala–Lys

UMP
|
P$_{II}$:   Gly–Ala–Glu–Tyr–Met–Val–Asp–Phe–Leu–Pro–Lys

FIG. 3. Sequence of tryptic peptides containing covalently bound nucleotide.

Method to Detect the Adenylylation of Glutamine Synthetase

Several methods are available to prove the presence of adenylylated glutamine synthetase in *E. coli*. Since glutamine synthetase and the enzymes involved in the covalent modification of glutamine synthetase in some bacteria are different from those in *E. coli*, the procedures developed for the assay of adenylylation subunits in *E. coli* had to be modified before being applied to other organisms. Most of the methods described here are to demonstrate the principles rather than detailed procedures.

Enzymatic Procedures

It is possible to detect the presence of adenylylated glutamine synthetase as well as to quantitate the average state of adenylylation in crude extracts. Exploiting several different properties between adenylylated and unadenylylated glutamine synthetase from *E. coli*, Stadtman and co-workers[24] developed six different enzyme procedures for determining the average state of adenylylation ($\bar{n}$). All six methods depend on the fact that, when measured in the presence of Mn^{2+}, the pH profile for γ-glutamyltransfer activity of glutamine synthetase varies with the value of $\bar{n}$; however, for any specific condition, the pH activity profiles of all preparations cross one another at a common point (Fig. 4). Activity measurements at the unique crossover point (isoactivity point) is therefore a measure of the total enzyme concentration, independent of the value of $\bar{n}$. To determine the average state of adenylylation of a given crude extract, it is therefore necessary only to measure glutamine synthetase activity at the isoactivity pH (total activity) and additionally under conditions in which the adenylylated subunits of GS are either completely or nearly completely inactive. From these two measurements, it is possible to calculate the fraction of adenylylated and unadenylyated subunits and therefore the value of $\bar{n}$.

[24] E. R. Stadtman, P. Z. Smyrniotic, J. N. Davis, and M. E. Wittenberger, *Anal. Biochem.* **95,** 275 (1979).

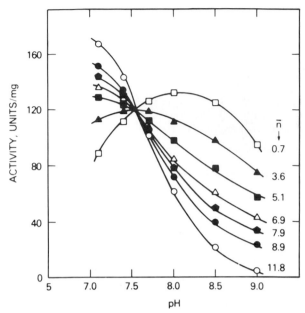

FIG. 4. Effect of pH and state of adenylylation on the γ-glutamyltransferase activities of glutamine synthetase measured in the triethanolamine-dimethylglutaric acid (TEA–DMG) buffer system containing 0.4 mM MnCl$_2$. Values of $\bar{n}$, as indicated, refer to number of adenylylated subunits per dodecamer.

Variations in the methods for measuring the value of $\bar{n}$ are due to the fact that the isoactivity pH is very much dependent upon the exact conditions of the assay system (such as the buffer composition, the ionic strength, the degrees of substrate saturation, and the kind of salt) and that the selective suppression of γ-glutamyl transfer activity of an adenylylated form can be achieved in several different ways. When a saturating level of substrates and 0.4 mM MnCl$_2$ is used, the isoactivity pH in the imidazole buffer is at 7.38; and in the triethanolamine dimethylglutaric acid (TEA–DMG) buffer system, it is at 7.57. When the Mn^{2+} assay mixtures are supplemented with 60 mM MgCl$_2$, the activity of adenylylated subunits is completely suppressed, whereas unadenylylated subunits are still active, but their pH activity profiles are altered. In addition, in the Mn^{2+}–TEA–DMG buffer system at pH 7.57, the activity of adenylylated subunit can be selectively suppressed either by lowering ADP concentration from 0.4 mM to 0.005 mM or by adding an inhibitor AMP, which inhibits the activity of adenylylated subunit but has no effect on unadenylylated glutamine synthetase.

All six methods developed in Stadtman's laboratory can be used to

measure the average adenylylation of glutamine synthetase from *E. coli.*
A procedure used most widely is described here. The procedure requires
measuring glutamine synthetase activity in the Mn^{2+}–TEA–DMG buffer
with and without $MgCl_2$.

Assay Solutions. The *Mn^{2+}–TEA–DMG (pH 7.57) assay solution,* 24
ml, is prepared by mixing 1.25 ml of the TEA–DMG buffer, 0.55 g of L-
glutamine, 0.5 ml of 1.0 *M* neutral NH_2OH, 0.5 ml of 0.02 *M* Na^+-ADP, 5
ml of 1.5 *M* KCl, and 0.5 ml of 1 *M* $KAsO_4$ (pH 7.0). After adjusting the
pH to 7.57 with 2 *M* TEA and the volume to 23 ml with water, 1.0 ml of
0.01 *M* $MnCl_2$ is added. To make the *Mg^{2+}–TEA–DMG assay (pH 7.57)
mixture,* the above procedure is followed, but before addition of Mn^{2+} and
after pH adjustment to 7.57 with TEA, the pH is raised to 7.77 with 1 *M*
KOH and the volume is adjusted to 20 ml with water, 1.5 ml of 1.0 *M*
$MgCl_2$ is added, and the pH is finally adjusted to 7.57 with 2 *M* HCl;
then 1.0 ml of 0.01 *M* $MnCl_2$ is added, and the final volume is adjusted to
24 ml with water. The *stop mixture* is prepared by mixing 40 ml of 10%
$FeCl_3$, 10 ml 24% trichloroacetic acid, 5 ml of 6 *N* HCl, and 65 ml of
water.

Assay Procedure. To each test tube containing 0.48 ml of the appropriate assay mixture, 20 μl of enzyme preparation is added. After 30 min of
incubation at 37°, the reaction is stopped by adding 2.0 ml of stop mixture.
Then the absorbance of the colored iron-hydroxamate complex is measured at 540 nm. Under these conditions, 1 μmol of hydroxamic acid will
yield an absorbancy of 0.360. The concentration of enzyme added to the
assay mixtures is adjusted so that the absorbance at 540 nm does not
exceed 1.5 OD.

To correct for the small amount of γ-glutamyl hydroxamate produced
nonenzymatically, the absorbancies of parallel incubation mixtures containing no enzymes are subtracted from the corresponding enzymatic
incubation mixtures. With crude enzyme preparations containing significant glutaminase activity, assay should also be carried out in similar
mixtures lacking ADP and arsenate to correct for the production of γ-
glutamyl hydroxamate catalyzed by glutaminase.

Calculation of $\bar{n}$. The activity at pH 7.57 in the presence of 60 m*M*
Mg^{2+} is inversely proportional to the state of adenylylation. And the
activity of fully unadenylyated GS in the presence of Mg^{2+} is 86% of that
observed in Mn^{2+}, whereas the fully adenylylated GS in inactive. From
this relationship it can be shown that in the TEA–DMG buffer system at
pH 7.75 the state of adenylylation is given by the expression

$$\bar{n} = 12 - 13.95 \left(\frac{\text{activity in 60 m}M \text{ Mg}^{2+} + 0.4 \text{ m}M \text{ Mn}^{2+}}{\text{activity in 0.4 m}M \text{ Mn}^{2+}} \right)$$

For the remaining five enzymatic procedures, it is recommended that readers refer to the original paper by Stadtman *et al.*[24] It was emphasized by Stadtman *et al.* that the procedures described by them need suitable modification before they are applied to other organisms. This is because the pH activity profiles of glutamine synthetase in other bacteria are not identical to that of the *E. coli* enzyme. Nevertheless, the principles established have been used in the development of valid $\bar{n}$ measurements in organisms other than *E. coli*.[25–30] To develop such a procedure, it is essential to determine the isoactivity pH value in the pH activity profile of enzyme preparations with different states of adenylylation and, additionally, to find a condition that will measure only either unadenylylated or adenylylated subunit activity. To obtain a pH activity profile, it is not necessary to use purified glutamine synthetase; crude extracts containing equal amounts of glutamine synthetase with different states of adenylylation are sufficient. These extracts can be obtained by growing cells under NH_4^+ excess and limiting conditions[2,9,25] or by treating the crude extract containing highly adenylylated glutamine synthetase with snake venom phosphodiesterase,[2,24,29,30] or by incubating the crude extract under either the adenylylating (ATP, glutamine) or deadenylylating (P_i, α-ketoglutarate) conditions.[17]

The isoactivity pH values measured for various organisms are pH 7.55 for *Klebsiella aerogenes*[25] in the imidazole buffer, pH 7.9 for *K. pneumoniae*[26] in HEPES buffer, pH 8.45 for *Pseudomonas fluorescens*[27] in TEA–DMG buffer, and pH 7.6 for *Azotobacter vinelandii*[28] in the TEA–DMG buffer. However, in the pH activity profile of glutamine synthetase from *Streptomyces cattleya*,[29] there is no crossover point because adenylylated enzyme is always inactive. Thus, the total amount of glutamine synthetase in a crude extract is determined by treating a sample with snake venom phosphodiesterase and assaying for γ-glutamyltransferase activity after complete digestion. *Rhizobium*[31,32] has two forms of glutamine synthetase, GS I and GS II, and only GS I is adenylylated. Therefore, the

[25] R. A. Bender, K. A. Janssen, A. D. Resnick, M. Blumenberg, F. Foor, and B. Magasanik, *J. Bacteriol.* **129,** 1001 (1977).
[26] R. B. Goldberg and R. Hanau, *J. Bacteriol.* **137,** 1282 (1979).
[27] J. M. Meyer and E. R. Stadtman, *J. Bacteriol.* **146,** 705 (1981).
[28] J. Siedel and E. Shelton, *Arch. Biochem. Biophys.* **192,** 214 (1979).
[29] S. L. Streicher and B. Tyler, *Proc. Natl. Acad. Sci. U.S.A.* **78,** 229 (1981).
[30] P. E. Bishop, J. G. Guevara, J. A. Engelke, and H. J. Evans, *Plant Physiol.* **57,** 542 (1976).
[31] R. A. Darrow and R. R. Knotts, *Biochem. Biophys. Res. Commun.* **78,** 554 (1977).
[32] R. A. Ludwig, *J. Bacteriol.* **135,** 114 (1978).

estimation of the state of adenylylation of GS I by the γ-glutamyl transfer assay is not easily attainable.

Gel Electrophoresis

Although there is no significant difference in the mobilities of adenylylated and nonadenylylated glutamine synthetase in native gels, the adenylylated and unadenylylated subunits from *E. coli, K. aerogenes,* and *K. pneumoniae* are separable on the Laemmli SDS–polyacrylamide gel electrophoresis.[33] SDS-gel electrophoresis thus provides a means of estimating the adenylylation state and the quantity of glutamine synthetase present, independent of enzymatic activity measurements. This method is particularly useful when adenylylated and unadenylyated subunits from strains carrying the glnA (the structure gene for glutamine synthetase) allele are to be quantitated. As shown in *S. cattleya* extract, analysis of adenylylated and unadenylylated subunits in crude extracts can also be achieved using the O'Farrell two-dimensional polyacrylamide gel electrophoresis.[29]

Precipitation of Multiple Adenylylated Forms with Anti-AMP Antibody

Hohman *et al.*[34] showed that sheep antibodies directed against AMP–bovine serum albumin conjugate bind to the AMP moiety of adenylylated glutamine synthetase from *E. coli* and to no other antigenic determinant on the protein. At each state of adenylylation, the fraction of glutamine synthetase precipitated increases with increasing concentration of antiserum to a maximal value, and in each case, antibody excess leads to the formation of soluble immune complexes. The maximum amount of glutamine synthetase that can be precipitated is proportional to the average stage of adenylylation over the range of $\bar{n} = 4$ to 10, and GS_{12} is completely precipitated. These precipitation experiments have been useful in establishing that the glutamine synthetases from *A. vinelandii*[28] and *P. fluorescens*[27] undergo an adenylylation–deadenylylation reaction and in estimating their state of adenylylation.

Four stable murine hybridoma monoclones that produce homogeneous antibodies against BSA–AMP have been established.[35] Antibody produced from a clone does bind to the adenylylated glutamine synthe-

[33] R. Bender and S. L. Streicher, *J. Bacteriol.* **137,** 1000 (1979).
[34] R. J. Hohman, S. G. Rhee, and E. R. Stadtman, *Proc. Natl. Acad. Sci. U.S.A.* **77,** 7410 (1980).
[35] H. K. Chung and S. G. Rhee, unpublished results.

tase, but forms soluble complexes even when glutamine synthetase is at a fully adenylylated state. The remaining three clones produce antibodies that are capable of precipitating glutamine synthetase in proportion to the value of $\bar{n}$.

Exploiting the interaction with anti-AMP antibodies offers unique methods for studying the macromolecules containing a 5'-AMP moiety covalently attached through a phosphodiester linkage. For example, free tRNA, but not amino acyl-tRNA, binds to an anti-AMP monoclonal antibody Sepharose matrix, probably because only free 5'-adenylyl moiety at the 3' end of tRNA can react with the antibody.[36]

Separation of Glutamine Synthetase Species with Different $\bar{n}$ Values by Affinity Chromatography

Random processes of adenylylation and adenylylation generate multimolecular forms of glutamine synthetase that differ from one another, not only by the number of adenylylated subunits, but also by the distribution of adenylylated and unadenylylated subunits within single molecules.[37,38] Theoretical calculation shows that 382 uniquely different forms of glutamine synthetases are possible. Several lines of experimental evidence also indicate that indeed naturally occurring glutamine synthetase preparations are mixtures of various hybrids that differ from one another by number of adenylylated subunits per dodecamer. Until recently, efforts to resolve these mixtures into more or less uniquely defined molecular entities by means of conventional chromatographic, electrophoretic, or electrofocusing techniques have failed. A partial solution to this problem has been achieved by the use of the affinity chromatography of Affi-Blue Sepharose or on anti-AMP antibody immunoadsorbant.

It is known that Cibacron Blue binds to the nucleotide substrate binding site on the enzyme and can be displaced by ADP.[39] In addition, as mentioned above, the affinity of the unadenylylated subunits for ADP is several orders of magnitude greater than the affinity of adenylylated subunits for this ligand under certain conditions. Accordingly, as shown in Fig. 5, it is feasible to separate adenylylated glutamine synthetase from unadenylylated glutamine synthetase by adsorption on Sepharose columns to which Cibacron Blue was covalently attached, followed by

[36] A. J. Wittwer and T. C. Stadtman, *J. Biol. Chem.* **258,** 8637 (1983).

[37] J. E. Ciardi, F. Cimino, and E. R. Stadtman, *Biochemistry* **12,** 4321 (1973).

[38] E. R. Stadtman, R. J. Hohman, J. N. Davis, M. E. Wittenberger, P. B. Chock, and S. G. Rhee, *Mol. Biol. Biochem. Biophys.* **32,** 144 (1980).

[39] E. R. Stadtman and M. Federici, unpublished results.

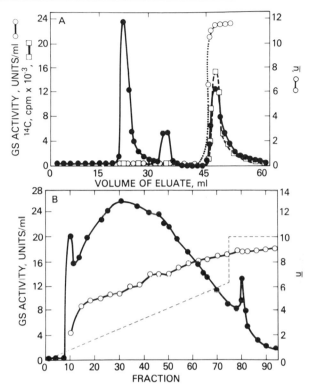

FIG. 5. Affinity chromatography of glutamine synthetase (GS) on Affi-Blue Sepharose columns. (A) A 0.19-ml sample containing 1.0 mg each of $GS_{\bar{1}}$ and [adenylyl-^{14}C]$GS_{\overline{12}}$ was placed on an Affi-Blue Sepharose column (0.9 × 10 cm) that had been preequilibrated with buffer containing 10 mM imidazole (pH 7.0), 2 mM MnCl$_2$, 100 mM KCl, and 10 mM glutamate. The column was eluted as follows: first, with 15 ml of the same buffer; second, with 15 ml of the same buffer containing 10^{-5} M ADP; third, with 15 ml of the buffer containing 10^{-4} M ADP; fourth, with 15 ml of the buffer containing 10^{-3} M ADP. Fractions (1.5 ml) were collected and assayed for GS activity, $\bar{n}$, and ^{14}C. (B) A 2.0-ml sample containing 24 mg of native $GS_{\bar{6}}$ was placed on an Affi-Blue column as above, except that the buffer contained 10 mM HEPES (pH 7.5) in place of imidazole. The sample was washed through with the same buffer. The GS was eluted with a linear gradient of ADP in the same buffer ranging from 10^{-5} to $6.3 × 10^{-4}$ M ADP (fractions to 0–75), then with buffer containing 10^{-3} M ADP (fractions 76–95), as indicated by the dashed line. Each 1.5-ml fraction was assayed for GS activity (●) and for the state of adenylylation, $\bar{n}$ (○).

selective elution with buffer containing an appropriate concentration of ADP (as described in Fig. 5 legend).[38] When a native $GS_{\bar{6}}$ preparation[40] is adsorbed on an Affi-Blue Sepharose column and then eluted with an ADP gradient ranging from 10^{-5} M to 10^{-3} M (as described in Fig. 5 legend), a

[40] The subscript indicates the average number of adenylylated subunits per dodecamer.

broad peak of enzyme activity is observed, and the state of adenylylation across this peak increases progressively from 2 to 9 (Fig. 5).

In another experiment, Hohman and Stadtman[41] prepared an affinity matrix by covalent attachment of purified polyclonal anti-AMP antibodies from sheep to Affi-Gel 10 (Bio-Rad Laboratories). After absorbing a $GS_{\bar{6}}$ preparation to the immunoadsorbant, glutamine synthetase fractions with $\bar{n}$ values ranging from 1 to 8 were eluted with a series of stepwise AMP gradients. However, each step resulted in the elution of only a single broad peak across which the values of $\bar{n}$ generally increase without clear separation of adenylylated forms of glutamine synthetase.

The prospect of separating the mixture of glutamine synthetase hybrid into the fractions containing molecule with the same $\bar{n}$ value is further investigated using the immunoadsorbents prepared by cross-linking murine anti-AMP hybridoma proteins to Affi-Gel 10.[35] Two monoclonal antibodies from clones 37-2-1 and 72-76-1, which exhibit distinctively different properties, are used. The antibody from clone 37-2-1 belongs to the IgM subclass and binds weakly to free 5'-AMP (K_d = 7.5 μM). Another antibody from 72-76-1 is IgG and binds tightly to free 5'-AMP (K_d = 0.6 μM). In addition, the maximum amount of glutamine synthetase that can be precipitated with the hybridoma protein 37-2-1 is proportional to the average state of adenylylation, while the antibody from 72-76-1 forms soluble complexes with glutamine synthetase independent of the $\bar{n}$ values.

The elution profile obtained with a $GS_{\overline{7.8}}$ preparation is shown in Fig. 6. When a linear AMP gradient is applied to the IgM column as described in the figure legend, fractions with $\bar{n}$ values ranging from 1 to 10 are obtained; a small peak (fractions 5–10) with an $\bar{n}$ value from 1 to 3, another peak (fractions 35–45) with an $\bar{n}$ value increasing from 4 to 6, and a broad trailing edge (fractions 46–110) encompassing $\bar{n}$ values of 6 to 10. The IgG column renders two small peaks (fractions 5–8 and 14–16) representing glutamine synthetase with $\bar{n}$ = 0 and 1.5, and a large peak across which $\bar{n}$ increases rapidly from 4 to 10.

Similar experiments are repeated with a mixture of glutamine synthetase adenylylated in the presence of [^{3}H]ATP (Fig. 7). The IgG column clearly separates GS_0 (fractions 5–8) from the remaining glutamine synthetase and, in addition, gives three peaks; fractions 11–20 for $GS_{\overline{1.3}}$, fractions 50–80 for $GS_{\overline{2.5}}$, and fractions 100–112 for $\bar{n}$ values ranging from 2.5 to 11. These results indicate that certain populations of glutamine synthetase with two adenylylated subunits per dodecamer, probably with two subunits far removed from each other, elute together with $GS_{\overline{1.0}}$, and that dodecameric glutamine synthetase with 2–3 adenylylated subunits

[41] R. J. Hohman and E. R. Stadtman, *Arch. Biochem. Biophys.* **218,** 548 (1982).

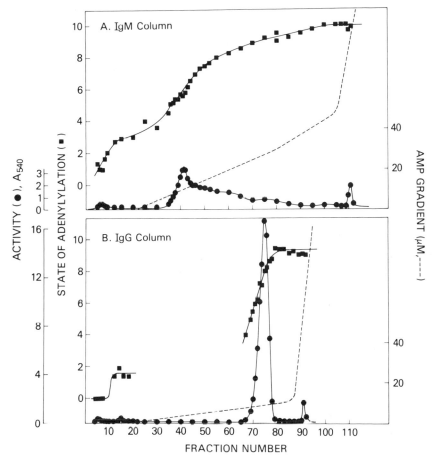

FIG. 6. Chromatography of GS$\overline{7.8}$ on anti-AMP immunoadsorbent column derived from IgM (A) and IgG (B). The column (1.5 × 20 cm) containing 20 ml of Affi-Gel 10 to which purified monoclonal antibody (124 mg of IgM from clone 37-2-1 or 197 mg of IgG from clone 72-76-1) was covalently attached, was equilibrated with a standard buffer (10 mM Tris-HCl, pH 7.3, 100 mM KCl, 1 mM MgCl$_2$, 0.02% NaN$_3$). Immediately after loading 6 mg of GS$\overline{7.8}$ in 0.2 ml of the same buffer to the column, column A was washed with a linear gradient generated from 250 ml each of 0 to 60 μM AMP (fractions 0–80), with a linear gradient generated from 120 ml each of 30 and 100 μM AMP (fractions 81–100) and finally with 10 mM AMP (fractions 101–115); column B was washed with a linear gradient generated from 250 ml each of 0 and 20 μM AMP (fractions 0–80) and with 10 mM AMP (fractions 81–95). All AMP solutions were prepared in the standard buffer. Fractions of 3.1 ml were collected at a flow rate of 9.3 ml/hr. All steps were performed at room temperature. Total glutamine synthetase activities independent of the state of adenylylation were measured at 37° in the Mn^{2+}–TEA–DMG buffer system at pH 7.57. The activities (●) are expressed as the absorbance at 540 nm per 50-μl sample per 20 min. The state of adenylylation (■) was calculated from the ratio of activities measured in the Mn^{2+}–TEA–DMG buffer with and without 60 mM MgCl$_2$ added.

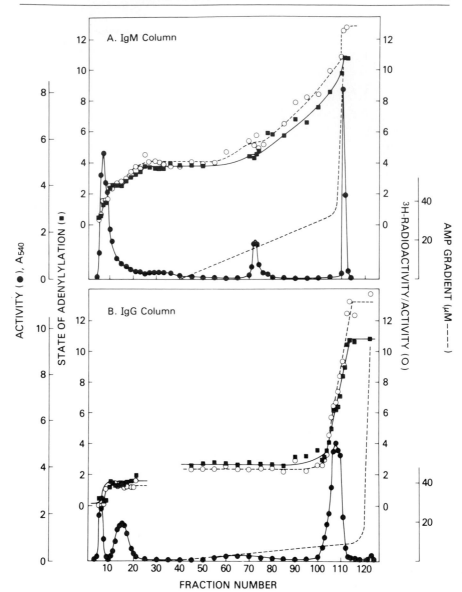

FIG. 7. Chromatography of a mixture of $[^3H]GS_{\overline{1.8}}$ + $[^3H]GS_{\overline{3.3}}$ + $[^3H]GS_{\overline{10.8}}$ on anti-AMP immunoadsorbent column derived from IgM (A) and IgG (B). Tritium-labeled glutamine synthetase preparations with $\bar{n}$ values of 1.8, 3.3, and 10.8 were prepared by adenylylating GS_0 in the presence of $[2,8-^3H]ATP$ (0.2 $\mu Ci/nmol$). The immunoadsorbent columns were identical to those in Fig. 6. A mixture of 1.6 mg of $[^3H]GS_{\overline{1.8}}$, 2.0 mg of $[^3H]GS_{\overline{3.3}}$, and 1.2 mg of $[^3H]GS_{\overline{10.8}}$ in 0.3 ml of the standard buffer was loaded onto the column. Column A was washed with the standard buffer (fractions 0–30), with a linear gradient generated from 250

are widely spread, depending on the subunit arrangement in the dodeca-mer. The IgM column gives a peak (fractions 6–14) across which $\bar{n}$ in-creases from 0 to 2.5, and a broad trailing edge (fractions 15–50) for $\bar{n} =$ 2.5 to 4.0. For glutamine synthetase with higher $\bar{n}$ values ($\bar{n} > 4$), the IgM column renders better defined fractions.

The results obtained with immunoadsorbent gels indicate that separation of glutamine synthetase into uniquely defined molecular entities is possible if various gels are prepared from monoclonal antibodies different from one another in the valency, in the flexibility at the hinge of Fab arms, and in the affinities for adenylylated moiety.

Assay of Adenylyltransferase

Although adenylyltransferase catalyzes both adenylylation and deadenylylation reaction, the adenylylation activity (AT_a) is routinely fol-lowed because the deadenylylation reaction requires another protein P_{IID}, whereas the adenylylation of glutamine synthetase proceeds well in the absence of P_{IIA} if a sufficient amount of L-glutamine is present.

Incorporation of [^{14}C]AMP to glutamine synthetase is used to monitor the formation of adenylylated GS.[14] This method requires a large amount of unadenylylated GS, [^{14}C]ATP with high specific activity, and labori-ous processes. An alternative method is colorimetric assay, which in-volves two incubation steps.[42] The first step involves prior incubation of a suitable amount of adenylyltransferase preparation with unadenylylated GS in the presence of Mg-ATP and L-glutamine. The second step is to measure the amount of adenylylated GS that was produced in the prior incubations. This is achieved by taking advantage of the facts that under the pH of 6.4 the γ-glutamyl transfer activity of adenylylated GS is much greater than that of unadenylylated GS and that L-alanine inhibits the γ-glutamyl transfer reaction of the unadenylylated enzyme more than that of the adenylylated enzyme.

---

[42] S. G. Rhee, R. Park, and M. Wittenberger, *Anal. Biochem.* **88,** 174 (1978).

---

ml each of 0 and 60 μM AMP (fractions 31–100), and finally with 10 mM AMP (fractions 101–115). Column B was washed with the standard buffer (fractions 0–30), with a linear gradient generated from 250 ml each of 0 and 20 μM AMP (fractions 31–115), and finally with 10 mM AMP (fractions 116–125). In addition to the state of adenylylation (■) measured by γ-glutamyltransferase assay, the number of AMP groups attached per dodecamer (●) was calculated from (tritium radioactivity in 0.1 ml of sample from each fraction) divided by (glutamine synthetase activity independent of the state of adenylylation). Procedures for chromatography and enzyme activity measurements were the same as described in Fig. 6 unless specifically mentioned.

Assay Solutions. The *adenylylation assay mixture* contains 40 mM 2-methylimidazole (pH 7.6), 10 mM ATP, 80 mM L-glutamine, and 60 mM MgCl$_2$. The *γ-glutamyltransferase assay mixture* contains 50 mM TEA–DMG (pH 6.4), 20 mM KAsO$_4$, 40 mM NH$_2$OH, 0.8 mM ADP, 150 mM L-glutamine, 100 mM KCl, 0.4 mM MnCl$_2$, and 60 mM L-alanine. The *stop mixture* is prepared by mixing 80 ml of 10% FeCl, 20 ml of 24% trichloroacetic acid, 10 ml of 6 N HCl, and 10 ml of water.

Assay Procedure. To the test tube containing 30 μl of the adenylylation reaction mixture, 10 μl of adenylyltransferase preparation is added. After equilibration at 26° for a few minutes, the reaction is initiated by adding 10 μl of unadenylylated *E. coli* GS (12 mg/ml) in a buffer containing 10 mM imidazole (pH 7.2), 100 mM KCl, and 20 mM MgCl$_2$. After prior incubation for a suitable period of time, 1.0 ml of the γ-glutamyltransferase assay mixture (preequilibrated at 26°) is added to determine the amount of adenylylated GS formed during the prior incubation. The reaction mixture is incubated for another 5 min, and 1.0 ml of stop mixture is added to stop the reaction and to develop the color; this resulting solution is centrifuged, and its absorbance at 540 nm is measured. The absorbance observed at 540 nm for the control tube (which is prepared by a procedure similar to that used for the sample tube except no adenylyltransferase is added) is subtracted from that of the sample tube to obtain AT$_a$ activity.

This calorimetric assay, in which unadenylylated GS from *E. coli* is used as a substrate, was utilized to measure AT$_a$ activity in the crude extracts prepared not only from *E. coli* but also from *Salmonella typhimurium*,[43] *P. putida*,[44] and *K. aerogenes*.[45]

Assay of P$_{\text{IID}}$ and Uridylyltransferase

P$_{\text{IID}}$ is required for the deadenylylation of GS catalyzed by AT$_d$. Therefore, in the presence of high adenylyltransferase concentration and other appropriate assay conditions, the deadenylylation of GS is a linear function with respect to P$_{\text{IID}}$ concentration. This linear relationship is the basis of assays described here for determining the activity of P$_{\text{IID}}$ and UT activity. P$_{\text{IID}}$ assay,[46] a procedure also used in determining UT activity, consists of two incubation steps. In the first step the P$_{\text{IID}}$ sample is incu-

[43] S. Bancroft, S. G. Rhee, C. Neumann, and S. Kustu, *J. Bacteriol.* **134,** 1046 (1978).
[44] S. G. Rhee, unpublished results.
[45] F. Foor, R. J. Cedergren, S. L. Streicher, S. G. Rhee, and B. Magasanik, *J. Bacteriol.* **134,** 562 (1978).
[46] S. G. Rhee, C. Y. Huang, P. B. Chock, and E. R. Stadtman, *Anal. Biochem.* **90,** 752 (1978).

bated in a deadenylylation mixture containing a relatively large quantity of adenylyltransferase and adenylylated GS. This is followed by a second incubation of the sample in a γ-glutamyltransferase mixture to determine the extent of deadenylylation that occurred during the first step. The differentiation between the adenylylated and unadenylylated forms of GS is accomplished by measuring the γ-glutamyltransferase activity under conditions where the unadenylylated enzyme is significantly more active than the adenylylated enzyme; i.e., at pH 8.0 in the presence of Mn^{2+}.

The first step in the UT assay[46] involves incubation of the UT sample with P_{IIA} in uridylyltransferase mixture for the uridylylation of P_{IIA}. After the first incubation step, the change in P_{IID} concentration is determined by the P_{IID} assay described above. All incubations are performed at 26°.

Assay Solution. The *stock solution A* is prepared and stored at $-20°$ for P_{IID} and UT assays. It contains 100 mM 2-methylimidazole, pH 7.6, 40 mM α-ketoglutarate, 2 mM ATP, 30 mM MgCl$_2$, 40 mM KP$_i$, 2 mM dithiothreitol, and 0.6 mM EDTA. The P_{IID} *assay mixture* should be freshly prepared and kept in an ice bath. It consists of one part of adenylyltransferase (2400 units/ml) and five parts of stock solution A. One unit of adenylyltransferase is defined as the amount of adenylyltransferase required to form 1 nmol of adenylylated GS subunit per minute at 26°. The *UT assay mixture* contains 300 mM 2-methylimidazole at pH 7.6, 500 mM KCl, 60 mM MgCl$_2$, 6 mM UTP, 30 mM α-ketoglutarate, and 1.5 mM ATP. The *UT assay blank mixture* contains all components of the UT mixture except UTP. The *deadenylylation reaction mixture* contains one part of adenylylated GS from *E. coli* (3.4 mg/ml, $n > 10$), one part of adenylyltransferase (1200 units/ml) and five parts of stock solution A. The *γ-glutamyltransferase assay mixture* contains 50 mM 2-methylimidazole at pH 8.0, 40 mM KAsO$_4$, 40 mM NH$_2$OH, 0.1 mM ADP, 150 mM L-glutamine, 100 mM KCl, and 0.4 mM MnCl$_2$. The *stop mixture* is the same as described in the assay of adenylyltransferase.

P_{IID} Assay Procedure. To the test tubes containing 30 μl of the P_{IID} mixture, 10 μl of P_{IID} is added. After equilibration at 26° for a few minutes, the deadenylylation is initiated by adding 10 μl of adenylylated GS from *E. coli* (2.3 mg/ml in 10 mM imidazole, pH 7.2, 100 mM KCl, and 20 mM MgCl$_2$ and $n > 10$) to the assay mixtures. After a suitable incubation period, 1.0 ml of the γ-glutamyltransferase assay mixture, preequilibrated at 26°, is added to measure the increase in the unadenylylated GS activity. The mixture is incubated for another 5 min followed by the addition of 1 ml of stop mixture to stop the enzymatic reactions and to develop the color. The sample is clarified by centrifugation, and its optical density at 540 nm is recorded. The blank absorbance is obtained from an assay tube containing all components except P_{IID}.

UT Assay Procedure. The mixture of 5 μl of UT sample is allowed to incubate for a few minutes at 26°, then 5 μl of *E. coli* P_{IIA} (0.4 unit/ml) is added to initiate the uridylylation reaction. After a suitable incubation period, 70 μl of the deadenylylation reaction mixture preequilibrated at 26° is added to determine the amount of P_{IID} formed. After 5 min of incubation, 1 ml of the γ-glutamyltransferase mixture is added, and after another 4 min, the reaction is stopped by adding 1 ml of the stop mixture. The resulting solution is centrifuged, and its absorbance at 540 nm is recorded. The control tube is prepared with the UT sample by a similar procedure as the sample tube with the exception that the UT assay mixture is substituted by the UT assay blank mixture.

The procedures described above can be used to measure P_{IID} and UT activity in *S. typhimurium*,[43] *P. putida*,[44] and *K. aerogenes*.[45] This is possible because P_{IID} from those organisms stimulates the deadenylylation reaction catalyzed by *E. coli* AT_d and because *E. coli* P_{IIA} is uridylylated by UT activities from various organisms.

[11] Determination and Occurrence of Tyrosine *O*-Sulfate in Proteins

By WIELAND B. HUTTNER

The term "protein sulfation" describes the modification of proteins by covalent attachment of sulfate. One can distinguish between two principal types of protein sulfation. The first type is the covalent linkage of sulfate to amino acid residues of proteins, i.e., the primary modification of the polypeptide chain itself. Only one amino acid, tyrosine, has so far been shown to undergo this modification. The second type is the covalent linkage of sulfate to carbohydrate moieties of glycoproteins and proteoglycans. In structural terms this type of sulfation is a secondary modification of the polypeptide chain, the primary one being protein glycosylation. Both sulfated tyrosine and sulfated carbohydrate residues can apparently occur in the same protein.

The sulfate linked to tyrosine is present as an O^4-sulfate ester. Such a tyrosine O^4-sulfate residue was first identified in 1954 in bovine fibrinopeptide B.[1] In the 28 years following this discovery, the presence of

[1] F. R. Bettelheim, *J. Am. Chem. Soc.* **76**, 2838 (1954).

tyrosine sulfate, initially believed to be restricted to fibrinogens and fibrins,[2] was detected in a few biological peptides, such as gastrin II,[3] phyllokinin,[4] caerulein,[5] cholecystokinin,[6] and Leu-enkephalin,[7] as well as in the small polypeptide hirudin.[8] Except for these peptides (and their precursors), however, the occurrence of tyrosine sulfate in proteins remained virtually unnoticed until recently. In fact, when proteins were found to be sulfated, this was, with a single exception,[9] generally taken to be indicative of the presence of sulfated carbohydrate residues.

In a more recent study,[10] the possible widespread occurrence of tyrosine sulfate in proteins was investigated. Tyrosine sulfate was detected in proteins from a wide variety of vertebrate tissues and cell cultures in many molecular weight ranges. Since this recognition of tyrosine sulfation as a widespread modification of proteins, at least 10 specific tyrosine-sulfated proteins have been newly identified and partially characterized. These are (1) the four major sulfated proteins of a rat pheochromocytoma cell line (PC12), designated as p113, p105, p86, and p84, which were also found to be phosphorylated on serine[11]; (2) the major soluble sulfated protein of the rat brain, designated as p120, which was found to be developmentally regulated[12]; (3) three prominent sulfated proteins of the chick retina, designated as p110, p102, and p93, which were found to move by fast axonal transport down the optic nerve[12], (4) an acidic secretory protein of the bovine anterior pituitary,[13] originally described by Rosa and Zanini[14]; (5) immunoglobulin G of some hybridoma cell lines[15]; and (6) vitellogenin of *Drosophila*.[15]

[2] F. R. Jevons, *Biochem. J.* **89,** 621 (1963).

[3] H. Gregory, P. M. Hardy, D. S. Jones, G. W. Kenner, and R. C. Sheppard, *Nature (London)* **204,** 931 (1964).

[4] A. Anastasi, G. Bertaccini, and V. Erspamer, *Br. J. Pharmacol. Chemother.* **27,** 479 (1966).

[5] A. Anastasi, V. Erspamer, and R. Endean, *Arch. Biochem. Biophys.* **125,** 57 (1968).

[6] V. Mutt and J. E. Jorpes, *Eur. J. Biochem.* **6,** 156 (1968).

[7] C. D. Unsworth, J. Hughes, and J. S. Morley, *Nature (London)* **295,** 519 (1982).

[8] T. E. Petersen, H. R. Roberts, L. Sottrup-Jensen, S. Magnusson, and D. Bagdy, *in* "Protides of Biological Fluids" (H. Peeters, ed.), Vol. 23, p. 145. Pergamon, Oxford, 1976.

[9] G. Scheele, D. Bartelt, and W. Bieger, *Gastroenterology* **80,** 461 (1981).

[10] W. B. Huttner, *Nature (London)* **299,** 273 (1982).

[11] R. W. H. Lee and W. B. Huttner, *J. Biol. Chem.* **258,** 11326 (1983); R. W. H. Lee, A. Hille, and W. B. Huttner, unpublished observations.

[12] S. B. Por and W. B. Huttner, manuscripts in preparation.

[13] P. Rosa, G. Fumagalli, A. Zanini, and W. B. Huttner, manuscript submitted for publication.

[14] P. Rosa and A. Zanini, *Mol. Cell. Endocrinol.* **24,** 181 (1981).

[15] P. A. Baeuerle and W. B. Huttner, manuscripts submitted for publication.

The role of tyrosine sulfation in the function of these proteins is the subject of current investigations. A possible common denominator may be the observation that all polypeptides so far known to contain tyrosine sulfate either are secretory proteins or show properties consistent with them being secretory proteins. It is therefore possible that tyrosine sulfation is involved in the processing, sorting, or functioning of some secretory proteins. In the case of the known proteins, tyrosine sulfation may occur during their passage through the Golgi complex; it appears to be only slowly reversible or even irreversible.[11]

It is, however, too early to draw general conclusions about the subcellular compartmentation and the degree of reversibility of tyrosine sulfation. The proteins mentioned above are only some of the tyrosine-sulfated proteins existing in the respective cell systems, and only a few cell systems have so far been studied. Clearly, many more tyrosine-sulfated proteins are yet to be discovered, and many more known proteins need to be tested for the presence of tyrosine sulfate. Hopefully, by learning more about tyrosine-sulfated proteins, we will achieve an understanding of the biological role(s) of this modification. In the following sections, some of the current procedures used to detect tyrosine-sulfated proteins and to study this modification are described.[16]

Chemical Synthesis and Properties of Tyrosine Sulfate

Synthesis. L-Tyrosine O^4-sulfate is synthesized according to the method of Reitz *et al.*[17] by the reaction of L-tyrosine with concentrated sulfuric acid at low temperature. After the reaction, the sulfuric acid is neutralized and precipitated by addition of barium hydroxide. The tyrosine sulfate is freed from unreacted tyrosine and Ba^{2+} by passage through a cation exchanger. The final product may contain variable amounts of tyrosine 3'-sulfonate, which can be distinguished from tyrosine sulfate by its slightly different electrophoretic mobility at pH 3.5, its distinct ultraviolet absorption spectrum, and its stability to acid at elevated temperature. The appearance of tyrosine 3'-sulfonate can be minimized (less than 1% of the tyrosine sulfate) by limiting the reaction time of tyrosine with sulfuric acid to 15 min and by keeping the reaction temperature low (between

---

[16] For recent overviews of sulfation, the reader is referred to two excellent books: G. J. Mulder, "Sulfation of Drugs and Related Compounds," CRC Press, Boca Raton, Florida, 1981; G. J. Mulder, J. Caldwell, G. M. J. Van Kempen, and R. J. Vonk, "Sulfate Metabolism and Sulfate Conjugation," Taylor & Francis, London, 1982.

[17] H. C. Reitz, R. E. Ferrel, H. Fraenkel-Conrat, and H. S. Olcott, *J. Am. Chem. Soc.* **68**, 1024 (1946).

−25 and −10°). Tyrosine O^4-[^{35}S]sulfate is synthesized by the same protocol, using [^{35}S]H$_2$SO$_4$.

Properties. One of the most remarkable properties of tyrosine sulfate is the lability of the ester bond in acid and its stability in alkali. More than 95% of the ester is hydrolyzed after 5 min in 1 M HCl at 100°, making it impossible to detect tyrosine sulfate after acid hydrolysis of proteins. The acid lability of the ester presumably explains why tyrosine sulfate is not observed when peptides known to contain this modified amino acid are sequenced by chemical methods. Fortunately, more than 90% of the ester remains after 24 hr in 0.2 M Ba(OH)$_2$ at 110°, making alkaline hydrolysis the method of choice to detect tyrosine sulfate in proteins. Tyrosine sulfate (Ba^{2+}, K$^+$, or Na$^+$ salt) is stable in neutral aqueous solution at 4° for weeks, and standard solutions can be kept frozen at −20° for at least a year. The ultraviolet absorption spectrum of tyrosine sulfate is different from that of tyrosine, showing a peak at 260.5 nm with a molar extinction coefficient of 283 (pH 7.0).[18]

Detection of Tyrosine Sulfate in Proteins

This chapter focuses on the detection of tyrosine-sulfated proteins and on the determination of tyrosine sulfate by methods that are based on labeling proteins with ^{35}SO$_4$. This approach has several advantages over the chemical determination of tyrosine sulfate in unlabeled proteins. It can be much more easily used to search for tyrosine-sulfated proteins and is considerably more sensitive and therefore requires less protein material for analysis. An immunological approach to detect tyrosine-sulfated proteins has been developed in our laboratory.[18] Antisera were raised against a synthetic antigen containing a large number of tyrosine sulfate residues, and antibodies were purified from the antisera by affinity chromatography. These antibodies appeared to recognize tyrosine sulfate-containing proteins in "Western" blots and in solid-phase radioimmunoassay. The general usefulness of these antibodies to screen for, immunoprecipitate, or purify tyrosine-sulfated proteins is currently being tested.

The standard method to detect tyrosine sulfate in proteins involves four main steps described in Sections 1–4 below: (1) *in vivo* labeling of tissues *in situ* or of tissue explants, tissue slices, and cells in culture with inorganic [^{35}S]sulfate; (2) separation of proteins by polyacrylamide gel electrophoresis (PAGE); (3) elution from gels and hydrolysis of individual ^{35}SO$_4$-labeled proteins; (4) separation of tyrosine [^{35}S]sulfate by thin-layer electrophoresis. Major modifications of and additions to the standard

---

[18] P. A. Baeuerle, Diploma Thesis, University of Konstanz, 1983.

steps are described in separate paragraphs at the end of each section. A general scheme of the various procedures is shown in Fig. 1.

1. Labeling with $^{35}SO_4$

When whole animals are labeled with $^{35}SO_4$, one should bear in mind that the isotope becomes distributed throughout the body and that its specific activity is reduced by the endogenous unlabeled sulfate. Thus, in order to achieve sufficient $^{35}SO_4$ incorporation into proteins of the tissue of interest, relatively large quantities of $^{35}SO_4$ of high specific activity (>900 Ci/mmol) should be used. For example, a single intraperitoneal injection of 20 mCi of $^{35}SO_4$ into a 100-g rat was sufficient for the detection of sulfated proteins in various tissues and in the plasma 18 hr after the injection by SDS–PAGE and fluorography of the gels for 30 hr.[10] More efficient labeling is found if the $^{35}SO_4$ is administered to the tissue of

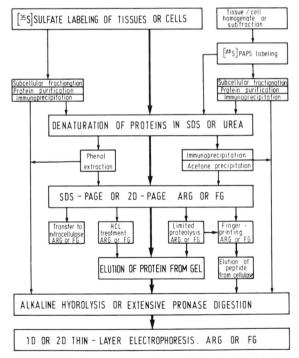

FIG. 1. Schematic outline of the sequence of procedures used to detect tyrosine sulfate in proteins. Thick arrows indicate the sequence of the standard method; thin arrows indicate modifications and additions. ARG, autoradiography; FG, fluorography.

interest directly. For example, sulfated proteins could easily be detected in the chick retina or the rat brain 1 hr after injection of 2 mCi of $^{35}SO_4$ into the chick eye or the third ventricle of the rat brain, respectively, by SDS-PAGE and fluorography of the gels for 15 hr.[12]

Mammalian tissues are not capable of reducing $^{35}SO_4$ for the synthesis of [^{35}S]methionine and [^{35}S]cysteine. Nevertheless, one encounters [^{35}S]methionine and [^{35}S]cysteine incorporation into proteins after labeling whole animals with $^{35}SO_4$. This is presumably due to the synthesis of [^{35}S]methionine and [^{35}S]cysteine by bacteria present in the gastrointestinal tract of the animals and can be prevented by using germfree animals. Although this phenomenon makes the interpretation of fluorograms of SDS–polyacrylamide gels less straightforward, it poses no problem for the identification of tyrosine [^{35}S]sulfate in protein hydrolysates, using thin-layer electrophoresis (see Section 4).

When tissue explants and cells are labeled in culture with $^{35}SO_4$, efficient $^{35}SO_4$ incorporation into proteins is achieved by adding carrier-free $^{35}SO_4$ to sulfate-free medium. Many culture systems use medium supplemented with some sort of serum, and the presence of low concentrations of unlabeled sulfate (up to about $10^{-4} M$), as is the case when sulfate-free medium supplemented with undialyzed serum is used, still allows satisfactory radioactive sulfate incorporation into proteins. In this case, the reduction in the specific activity of $^{35}SO_4$ is presumably compensated to some extent by the increase in cellular sulfate uptake. The advantage of using the isotope at high specific activity should be balanced against the potential disadvantage that may result from the use of dialyzed serum and from starving cells of sulfate. It is therefore recommended to ascertain that the use of sulfate-free medium and (when serum is required for the culture) the use of dialyzed serum during $^{35}SO_4$ labeling does not reduce the viability of the cells under investigation. In particular, the capacity for protein synthesis should not be impaired, since protein synthesis appears to be required for sulfate incorporation into some proteins.[11] In our experience, the use of carrier-free sulfate (about 0.5 mCi/ml final concentration) in sulfate-free medium supplemented with a reduced concentration (1–5%) of undialyzed serum has resulted in very efficient incorporation of radioactive sulfate into PC12 cell proteins after labeling periods of up to 18 hr.

Since incorporation of $^{35}SO_4$ into proteins may occur at both tyrosine residues and carbohydrate residues, it can be informative to perform $^{35}SO_4$ labeling in the absence and in the presence of inhibitors of N-glycosylation, e.g., tunicamycin. For example, in the slime mold *Dictyostelium discoideum*, protein sulfation was virtually abolished by tunicamycin, and analysis of sulfated proteins for the presence of tyrosine sulfate gave essentially negative results.[19] In PC12 cells, the four proteins designated p113, p105, p86, and p84, known to contain most of the incorporated sulfate as tyrosine sulfate,[11] were labeled similarly with $^{35}SO_4$ in the absence and in the presence of tunicamycin.

Labeling with 3'-Phosphoadenosine 5'-Phospho[^{35}S]sulfate (PAPS). The study of tyrosine sulfation of proteins has been extended to cell-free systems.[11] Using the radiolabeled "activated sulfate" [^{35}S]PAPS as sulfate donor, transfer of $^{35}SO_4$ to tyrosine residues of endogenous protein acceptors, catalyzed by an endogenous tyrosylprotein sulfotransferase, can be observed in cell lysates and some subcellular fractions. At present, the labeling efficiency of proteins in cell lysates is less than that obtained in intact cells, for several reasons. (1) The specific activity of commercially available [^{35}S]PAPS is relatively low (~2 Ci/mmol), compared with that of inorganic [^{35}S]sulfate (>900 Ci/mmol). (2) The proportion of the protein of interest that is in the unsulfated form, and thus a substrate for labeling, may be small at any given time point including that of cell lysis, whereas in intact cells ongoing protein synthesis continuously supplies unsulfated substrate protein. There can be little doubt, however, that, as the components of cell-free tyrosine sulfation of proteins are elucidated, sulfation of defined proteins by tyrosylprotein sulfotransferase will become more efficient, and the results obtained will increasingly contribute to our understanding of the role of tyrosine sulfation.

2. Separation of $^{35}SO_4$-Labeled Proteins

After labeling of tissues or cell cultures with $^{35}SO_4$, reactions are terminated and proteins are solubilized for separation on polyacrylamide gels. For these purposes, the use of SDS together with boiling at neutral pH appears to be the most suitable method, since it fulfills all the following requirements.

1. Rapid inactivation of enzymes. Although at present little is known about the possible regulation of the sulfation of specific proteins by extracellular and intracellular signals, only the preservation of the *in vivo* state of protein sulfation by rapid enzyme inactivation may allow the observation of such regulatory phenomena.

2. Avoidance of low pH. The tyrosine sulfate ester is labile in acidic conditions at elevated temperatures. Although the ester bond may be stable in acidic conditions in the cold, it appears safer to avoid low pH

---

[19] J. Stadler, G. Gerisch, G. Bauer, C. Suchanek, and W. B. Huttner, *EMBO J.* **2**, 1137 (1983).

whenever possible. It is for this reason that we avoid the use of trichloroacetic acid for terminating sulfation reactions.

3. Complete solubilization of proteins. Proteins solubilized in SDS can be subjected to phenol extraction, immunoprecipitation, SDS–PAGE and, after acetone precipitation, two-dimensional PAGE (see Fig. 1).

An SDS-containing, neutral solution that we have found to be suitable for most purposes is the sample buffer according to Laemmli,[20] referred to as "stop solution." Cells attached to culture dishes and cell pellets are directly dissolved in stop solution [3% (w/v) SDS, 10% (w/v) glycerol, 3.3% (v/v) 2-mercaptoethanol, a trace of bromophenol blue, and 62.5 mM Tris-HCl, pH 6.8], whereas cells in suspension are mixed with 0.5 volume of three times concentrated stop solution, followed in both cases by immediate boiling of the samples for 3–5 min. Tissues are rapidly frozen in liquid nitrogen, crushed into a fine powder under liquid nitrogen, and then dissolved in stop solution followed by boiling.

After $^{35}SO_4$ labeling, tissues and cells may be subjected to subcellular fractionation, and the subcellular fractions of interest can then be dissolved in stop solution. It should be borne in mind, however, that rapid regulatory phenomena in the sulfation of proteins, should they exist, might be lost during the fractionation process.

In an attempt specifically to remove sulfated carbohydrate residues from proteins after $^{35}SO_4$ labeling and at the same time preserve tyrosine sulfate residues, we have subjected PC12 cell proteins, known to contain tyrosine sulfate,[11] to treatment with anhydrous hydrogen fluoride according to Mort and Lamport.[21] In our experience, however, all the sulfate was removed from these proteins by this treatment. We have not used enzymatic deglycosylation to distinguish between the presence of sulfated carbohydrate residues and tyrosine sulfate residues, but this may be useful in some cases.

If specific proteins are to be subjected to immunoprecipitation prior to electrophoresis, cells and tissues can be dissolved in a neutral buffer without 2-mercaptoethanol containing 1–3% (w/v) SDS followed by boiling. The SDS is then diluted by addition of the nonionic detergent Nonidet P-40, and immunoprecipitation is performed according to standard procedures.[22] Alternatively, cells can be dissolved in RIPA buffer and subjected to immunoprecipitation as described.[23] Immunoprecipitates can be dis-

[20] U. K. Laemmli, *Nature* (*London*) **227**, 680 (1970).
[21] A. J. Mort and D. T. A. Lamport, *Anal. Biochem.* **82**, 289 (1977).
[22] S. E. Goelz, E. J. Nestler, B. Chehrazi, and P. Greengard, *Proc. Natl. Acad. Sci. U.S.A.* **78**, 2130 (1981).
[23] B. M. Sefton, K. Beemon, and T. Hunter, *J. Virol.* **28**, 957 (1978).

solved in stop solution (for SDS–PAGE) or in O'Farrell lysis buffer[24] (for two-dimensional PAGE).

Cells and tissues dissolved in stop solution can be subjected to any of the following protocols.

1. A phenol extraction protocol as described in Procedure I, designed to separate the proteins from sulfated glycosaminoglycans, followed by PAGE (see below). This procedure is similar to that introduced by Hunter and Sefton[25] for the study of tyrosine phosphorylation of proteins, and used in that case to separate proteins from nucleic acids and phospholipids. The effect of phenol extraction is illustrated in Fig. 2.

2. SDS–PAGE according to Laemmli.[20]

3. Two-dimensional PAGE using either isoelectric focusing or non-equilibrium pH gradient electrophoresis in the first dimension, as described by O'Farrell.[24] For two-dimensional PAGE, samples in stop solution are mixed with 5 volumes of acetone, kept at $-20°$ until precipitation occurs, and centrifuged. The pellets are washed in 80% (v/v) acetone, dried, and dissolved in O'Farrell lysis buffer containing 5% (w/v) Nonidet P-40. Alternatively, cells attached to culture dishes or cell pellets can also be directly dissolved into lysis buffer. In our experience, both types of sample preparation give rise to similar separations upon two-dimensional PAGE.

Procedure I. Phenol Extraction of Sulfated Proteins

1. Dissolve sample, e.g., $^{35}SO_4$-labeled cells in culture, in stop solution (0.5–5 mg of protein per milliliter of stop solution). Boil immediately for 3–5 min.

2. Unless otherwise indicated, the following steps are performed at room temperature. Prepare phenol solution: dissolve phenol in an equal amount (w/v) of HEN buffer (50 mM HEPES-NaOH, pH 7.4; 5 mM EDTA; 100 mM NaCl), mix vigorously for 5 min, let stand or centrifuge until phases are separated. The lower phase is HEN buffer-saturated phenol (referred to as phenol solution), the upper phase is phenol-saturated HEN buffer (referred to as HEN solution).

3. Mix the sample with an equal volume of phenol solution, vortex vigorously for at least 30 sec, centrifuge for 10 min at 15,000 rpm in a Sorvall SS34 rotor.

[24] P. H. O'Farrell, *J. Biol. Chem.* **250,** 4007 (1975); P. Z. O'Farrell, H. M. Goodman, and P. H. O'Farrell, *Cell* **12,** 1133 (1977).
[25] T. Hunter and B. M. Sefton, *Proc. Natl. Acad. Sci. U.S.A.* **77,** 1311 (1980).

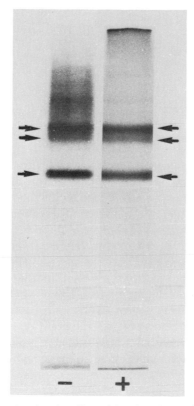

Fig. 2. Autoradiograms showing the effect of the phenol extraction protocol described in Procedure I. PC12 cells were labeled with $^{35}SO_4$ and subjected to SDS–PAGE, without ($-$) or with ($+$) prior phenol extraction. Arrows indicate major tyrosine-sulfated proteins.[11] Some of the proteins can be better seen after phenol extraction.

4. Collect the aqueous (upper) phase. If flocculent material at the interface is present, avoid disturbing it. Keep phenol (lower) phase plus interface.

5. Mix the aqueous phase with an equal volume of phenol solution; vortex and centrifuge as before. Discard the aqueous phase; keep the phenol phase plus interface.

6. Pool the phenol phases plus interfaces from first and second extraction; reextract 1–3 times with equal volume of HEN solution. Discard aqueous phases each time. The efficiency of extraction can be monitored by spotting aliquots of the aqueous phases on filter paper and observing the decline in radioactivity with a portable β-radiation monitor or by liquid scintillation counting.

7. Add five volumes of cold (−20°) ethanol to final phenol phase plus interface. If necessary, transfer to larger centrifuge tube. Mix well and keep at −20° until precipitation has occurred (2 hr or longer). Centrifuge for 10 min at 10,000 g in a Sorvall SS34 rotor. Discard the supernatant. Wash the precipitate once with chloroform–methanol (2 : 1) and collect it by centrifugation as above. Allow the precipitate to dry.

8. Dissolve the precipitate in stop solution (for SDS–PAGE), in lysis buffer (for two-dimensional PAGE), or in 0.2 M Ba(OH)$_2$ (for alkaline hydrolysis and determination of tyrosine sulfate, see Procedure III).

After electrophoresis, gels are fixed and (if desired) stained and destained by conventional procedures, using acetic acid rather than trichloroacetic acid. Fixed gels can also be subjected to an acid treatment that is described in a subsection at the end of this section. For the detection of $^{35}SO_4$-labeled proteins, gels are prepared for fluorography, dried, and fluorographed at −70°. We use exclusively the sodium salicylate method[26] for fluorography, primarily because the salicylate is water-soluble and can therefore easily be removed from gels after fluorography. This renders the proteins present in the gel suitable for further biochemical analysis, e.g., peptide mapping by limited proteolysis, tryptic fingerprinting, tyrosine sulfate analysis, (see below). If the labeling of proteins is sufficiently intense, the salicylate treatment of gels can be omitted, and the dried gels are subjected to autoradiography at room temperature. The X-ray film used in our laboratory for autoradiography and fluorography is Kodak XAR-5.

An alternative, rapid procedure to obtain an autoradiogram after SDS–PAGE or two-dimensional PAGE is to transfer the proteins from the gel onto nitrocellulose filter paper using the "Western" blotting technique. After the transfer, the nitrocellulose filter paper is either air-dried and autoradiographed or is dipped in 20% PPO in toluene, dried, and fluorographed at −70°. This procedure has two advantages.

1. The time needed for autoradiography after transfer to nitrocellulose filter paper is shorter than with dried polyacrylamide gels.

2. The sulfated material that is often found as a diffuse smear in the high molecular weight regions of SDS–polyacrylamide gels (presumably sulfated proteoglycans or glycosaminoglycans) does not appear to transfer very well from the gel to the nitrocellulose filter paper. Thus, autoradiograms obtained after transfer of $^{35}SO_4$-labeled proteins to nitrocellulose filter paper tend to be "cleaner" than those of the corresponding gels

[26] J. P. Chamberlain, *Anal. Biochem.* **98,** 132 (1979).

and can be compared to autoradiograms obtained after SDS–PAGE of phenol-extracted samples (Fig. 2).

$^{35}SO_4$-labeled proteins separated in polyacrylamide gels can be used for hydrolysis (see Section 3 below) and tyrosine sulfate analysis (Section 4). If desired (see Fig. 1), individual sulfated protein bands can first be subjected to limited proteolysis in SDS and peptide mapping in SDS polyacrylamide gels,[27] or to extensive proteolysis and two-dimensional fingerprinting by thin-layer electrophoresis/chromatography. In these cases, the resulting peptide fragments can be used for hydrolysis and tyrosine sulfate analysis.

HCl Treatment of Proteins in Polyacrylamide Gels. The details of the method are described in Procedure II, below. The rationale of this treatment is to screen, among the variety of sulfated proteins present in a sample, for those likely to contain tyrosine sulfate. Since the tyrosine sulfate ester is remarkably acid-labile and appears to be hydrolyzed faster than most carbohydrate sulfate esters, a short acid treatment will lead to a preferential loss of sulfate from tyrosine residues. When such an acid treatment is performed on $^{35}SO_4$-labeled proteins fixed in polyacrylamide gels, the resulting autoradiographic pattern may show some labeled bands that disappear quite specifically after the acid treatment (i.e., their reduction is greater than the small overall reduction in labeled bands). These specifically acid-sensitive bands may contain a large proportion of their sulfate label as tyrosine sulfate. In the examples illustrated in Fig. 3, there appears to be a good correlation between the acid sensitivity of bands and their tyrosine sulfate content.

It may, however, be too early to assume generally that a correlation between these two parameters will be found in every case because one cannot rule out false-positive and false-negative results. Possible explanations for false-positive acid sensitivity of bands include the following. (1) Sulfated residues other than tyrosine sulfate, e.g., certain carbohydrate sulfate esters, may exist that are as acid-sensitive as tyrosine sulfate ester. (2) The acid treatment may lead to hydrolysis of some peptide bonds. This may result in the formation of peptide fragments small enough to remain no longer fixed in the gel but to diffuse out. If sulfated residues other than tyrosine sulfate, e.g., sulfated carbohydrates, were located in such peptide fragments, an acid treatment-induced loss of $^{35}SO_4$ label would be observed without the actual hydrolysis of a tyrosine sulfate ester. False-negative acid sensitivity, i.e., the apparent lack of acid sensi-

---

[27] D. W. Cleveland, S. G. Fischer, M. W. Kirschner, and U. K. Laemmli, *J. Biol. Chem.* **252**, 1102 (1977).

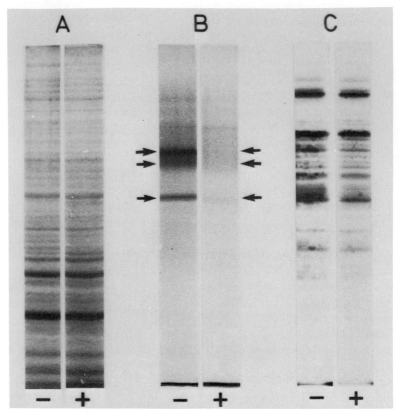

FIG. 3. Autoradiograms showing the effect of the HCl treatment described in Procedure II on proteins separated by SDS–PAGE: (A) proteins of rat pheochromocytoma cells (PC12), labeled with [35S]methionine; (B) same, labeled with [35S]sulfate; (C) proteins of *Dictyostelium discoideum*, labeled with [35S]sulfate. Compared with the control (−), the HCl treatment (+) does not lead to a loss of protein from the gel (A). The HCl treatment markedly reduces the 35SO4 label of PC12 cell proteins known to contain tyrosine sulfate[11] (arrows) (B), but only slightly the 35SO4 label of *D. discoideum* proteins known to contain sulfate on carbohydrate residues[19] (C).

tivity of bands despite the presence of tyrosine sulfate, may also be observed. For example, if a glycoprotein contained 10 sulfated carbohydrate residues and 1 tyrosine sulfate residue, the acid treatment-induced specific loss of 35SO4 label of about 10% would not readily be noticed upon comparative autoradiography. Despite these considerations, the acid treatment of gels is a useful and simple tool in searching for tyrosine-sulfated proteins, particularly when this method is combined with the procedure to determine tyrosine sulfate in individual proteins obtained from polyacrylamide gels (Sections 3 and 4).

Procedure II. Acid Treatment of Sulfated Proteins after Separation in Polyacrylamide Gels

1. This method can be applied to one- or two-dimensional wet polyacrylamide gels after electrophoresis, fixation, and (optional) Coomassie Blue staining, as well as to dried gels after autoradiography or fluorography using sodium salicylate. Wet gels are fixed after electrophoresis in 10% acetic acid–50% methanol for at least 1 hr. The same solution containing Coomassie Blue can be used simultaneously to fix and to stain the gel, if staining is required to cut a selected portion from the gel for acid treatment. Dried gels or portions of dried gels, after autoradiography or fluorography using sodium salicylate, are reswollen in 10% acetic acid–30% methanol for at least 1 hr. If the dried gel contained sodium salicylate, the solution should be changed at least once.

2. Equilibrate the gel in cold 1 M HCl for 30 min with constant shaking on ice. Use 10 volumes of HCl per original gel volume. While the gel is equilibrating, heat another 10 volumes of 1 M HCl in a separate, covered container that is placed in a boiling water bath. Make sure that this HCl solution has reached the final temperature (~90°) at the end of the equilibrating period.

3. Transfer the gel from the cold HCl into the container with the hot (~90°) HCl. Gently shake the container with the gel in the boiling water bath for 5 min. After 5 min, remove the container with the gel from the water bath and carefully pour the hot HCl and the gel into a large (10 liter) tray with ice water to stop the hydrolysis.

4. After a few minutes in ice water, place gel either into 10% acetic acid–50% methanol (if staining–destaining is not required), or into the same solution containing Coomassie Blue followed by destaining in 10% acetic acid–10% isopropanol (if staining–destaining is required), or directly into destaining solution (if gel had been stained prior to acid treatment). A minimum of 3 hr with constant shaking should be allowed for diffusion of hydrolyzed $^{35}SO_4$ from the gel.

5. Rinse the gel in water for 15 min and dry (if fluorography is not desired), or rinse in water, then shake in 1 M sodium salicylate for 30 min, and then dry (if fluorography is desired).

3. Hydrolysis of $^{35}SO_4$-Labeled Proteins Obtained from Polyacrylamide Gels

The method for the hydrolysis of proteins from polyacrylamide gels to liberate tyrosine sulfate is described in detail in Procedure III, below. Individual proteins contained in polyacrylamide gels are eluted by pro-

RECOVERY OF TYROSINE SULFATE AND SULFATED CARBOHYDRATES
AFTER HYDROLYSIS WITH BARIUM HYDROXIDE

$^{35}SO_4$-labeled sample	Eluate (Procedure III, step 3)	Neutralized supernatant (Procedure III, step 6)
Tyrosine sulfate[a]	611 cpm	576 cpm
	100%	94%
Contact site A glycoprotein	1526 cpm	3 cpm
of *Dictyostelium discoideum*[b]	100%	<1%
Hemagglutinin	553 cpm	4 cpm
of influenza virus[c]	100%	<1%

[a] Lee and Huttner.[11]
[b] Stadler *et al.*[19]
[c] Nakemura and Compans, and Huttner and Matlin.[28]

teolytic digestion. Since the proteolytic digestion serves only to facilitate the elution of the protein, the extent of digestion and the sites of cleavage in the protein are irrelevant, as long as the labeled peptide fragments are sufficiently small to diffuse out from the gel. We have used both trypsin and Pronase, with qualitatively similar results. The efficiency of elution may be slightly greater with Pronase. We routinely preincubate the Pronase solution to inactivate possible contaminating enzymes (e.g., a hypothetical sulfatase) by proteolytic digestion and have not observed any loss of tyrosine sulfate during Pronase digestion. The eluate obtained after proteolytic digestion is subjected to exhaustive alkaline hydrolysis. As mentioned above, the tyrosine sulfate ester is acid-labile and alkali-stable, making alkaline hydrolysis the method of choice to complete the liberation of tyrosine sulfate from the polypeptide chain. For alkaline hydrolysis barium hydroxide is used, followed by neutralization with sulfuric acid. The advantage of using barium hydroxide is that inorganic radioactive sulfate, generated by hydrolysis of alkali-labile sulfated carbohydrate residues of glycoproteins and proteoglycans, is precipitated as barium sulfate, whereas tyrosine sulfate remains in solution. This is exemplified by the observation (see the table) that from the 80-kilodalton contact site A glycoprotein of *Dictyostelium discoideum* and the hemagglutinin of influenza virus, proteins known to contain sulfate on carbohydrate residues,[19,28] virtually none of the radioactive $^{35}SO_4$ originally present is re-

[28] K. Nakamura and R. W. Compans, *Virology* **86**, 432 (1978); W. B. Huttner and K. Matlin, unpublished observations.

covered in the supernatant obtained after neutralization of the alkaline hydrolysate. In contrast, approximately 90% of tyrosine [^{35}S]sulfate standard is recovered in the neutralized supernatant. Thus most, if not all, of the ^{35}S radioactivity that is recovered in the neutralized supernatant after hydrolysis of ^{35}SO$_4$-labeled proteins with barium hydroxide is in the form of tyrosine sulfate. The exact proportion of ^{35}S in the neutralized supernatant that is present as tyrosine sulfate is determined after thin-layer electrophoresis, which is described in Section 4, below.

Procedure III. Liberation of Tyrosine Sulfate from Proteins Separated in Polyacrylamide Gels

1. The starting material is protein contained in one- or two-dimensional polyacrylamide gels. The gels may be wet or dried, stained or unstained, may contain the proteins noncovalently fixed (use acetic acid) or unfixed, and may or may not have been treated with sodium salicylate.

2. Locate the protein of interest according to the Coomassie Blue staining or the autoradiogram. Excise the portion of the gel that contains the desired protein and equilibrate (wet gel piece) or reswell (dried gel piece) in 10–20 ml of 10% acetic acid–30% methanol. (This and the following volume specifications are valid for gel pieces that contain up to 200 μl of fluid in the wet state. If larger gel pieces are used, increase volumes in steps 2 and 3 accordingly.) Remove filter paper, if present. Shake for 3 hr with a change of solution every hour. This serves to remove sodium salicylate, SDS, running buffer, excess Coomassie Blue, etc., while keeping the protein fixed in the gel. Wash gel piece in H$_2$O for 30 min to remove most of the acetic acid and methanol, and lyophilize the washed gel piece. If the gel piece is excised from a wet gel that has been stained and destained, it can be put directly in H$_2$O and then dried. If the gel piece is excised from a dried gel that has been stained, destained, and does not contain salicylate, it can be taken directly to step 3.

3. Make fresh Pronase solution: 50 mM NH$_4$HCO$_3$, pH 8, 50 μg/ml Pronase (Boehringer), and a trace of phenol red. Preincubate Pronase solution for 60 min at 37° in a capped tube. Reswell the dried gel piece in 1 ml of preincubated Pronase solution and incubate in capped tube with shaking for 12–24 hr at 37°. (The phenol red indicator can be used to monitor that no significant amount of acid remained in the gel piece.) Collect the eluate, thereby measuring its volume, into disposable tubes holding 3–4 ml. (These disposable tubes must tolerate the subsequent steps, i.e., centrifugation at 10,000 g and pH 14 at 110°; they must also be suitable for gastight sealing. We use uncapped 3-ml glass centrifuge tubes.) *Add another 1 ml of fresh, preincubated Pronase solution to gel piece and incubate for another 6–12 hr as before. Collect the second*

eluate and pool with first eluate. Of the pooled eluate, take a 5% aliquot and determine the radioactivity by liquid scintillation counting (use Pronase solution as blank). Lyophilize pooled eluate.

4. Make fresh 0.2 M Ba(OH)$_2$ (use degassed H$_2$O to reduce formation of BaCO$_3$). Usually, the barium hydroxide does not dissolve completely. While it is being stirred, take 1 ml of the barium hydroxide solution and add it to residue of lyophilized eluate. Seal the tube under N$_2$ so that it is gastight (for uncapped glass tubes: freeze sample, evacuate, seal by melting in flame). Place in an oven at 110° for 20–24 hr.

5. Cool the sample to 4° and break the seal. Centrifuge for 10 min at 10,000 g. Collect the supernatant and transfer it to another centrifuge tube. *Add 0.4 ml of H$_2$O to pellet, vortex, centrifuge as above, pool the second supernatant with first one.* Discard the tube with the pellet.

6. Neutralize the pooled supernatant with sulfuric acid, using the phenol red in the sample as indicator. Carefully add 1 M sulfuric acid with frequent vortexing until the phenol red turns transiently yellow at the site of sulfuric acid addition. Then continue the neutralization with 0.1 M sulfuric acid until the sample is homogeneously red-orange (pH ~7). If too much sulfuric acid has been added accidentally (yellow color), back-titrate with barium hydroxide. Centrifuge the sample for 10 min at 10,000 g. Collect the supernatant, thereby measuring its volume, and transfer to a 2-ml Eppendorf tube. *Add 0.4 ml of H$_2$O to pellet, vortex, centrifuge as above, pool second supernatant with first one.* Discard the tube with the pellet. The pooled supernatant is referred to as neutralized supernatant. Take a 5% aliquot of neutralized supernatant and determine the radioactivity by liquid scintillation counting. Lyophilize neutralized supernatant.

7. The residue of the neutralized supernatant is used for the identification of tyrosine sulfate by thin-layer electrophoresis as described in Procedure IV and Fig. 4.

8. In the above procedure, the steps given in *italics* can be omitted if only qualitative results are required.

Since of the ^{35}SO$_4$ originally present in proteins mainly tyrosine [^{35}S]sulfate appears to be recovered in the neutralized supernatant, the proportion of ^{35}S in the neutralized supernatant to that in the eluate can be taken as a rough estimate of what proportion of the total sulfate in a protein is tyrosine sulfate. If proteins contain ^{35}S not only as sulfate, but also as methionine and cysteine (see Section 1, above), such an estimation is not valid. If it can be established for a given protein that most or all of the incorporated ^{35}SO$_4$ is present as tyrosine sulfate, it is obvious that hydrolysis and tyrosine sulfate analysis need not be done routinely.

The alkaline hydrolysis of ^{35}SO$_4$-labeled protein with barium hydroxide is not restricted to proteins contained in polyacrylamide gels, but is

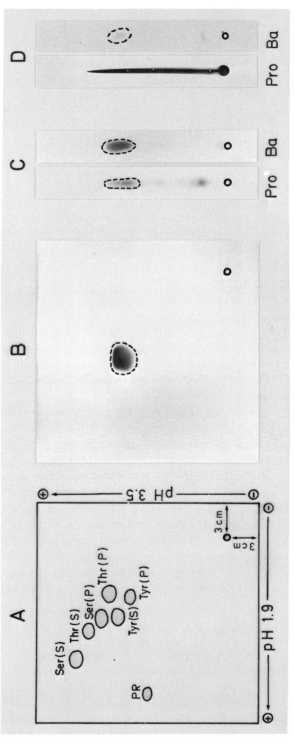

FIG. 4. Thin-layer electrophoresis of tyrosine sulfate. (A) Schematic diagram of the two-dimensional separation, giving the positions of sulfated and phosphorylated hydroxyamino acids. PR, Phenol red. (B) Autoradiogram of a two-dimensional separation of an alkaline hydrolysate (see Procedure III) of sulfated PC12 cell proteins. (C and D) One-dimensional separation and autoradiography comparing alkaline hydrolysates (Ba) with extensive Pronase digests (Pro). (C) A protein that contains sulfate predominantly as tyrosine sulfate[13]; (D) sulfated proteoglycan. Small circles in A, B, C, and D indicate the origin; the dotted lines in B, C, and D indicate the position of tyrosine sulfate.

applicable to any type of protein sample (immunoprecipitates, purified proteins, etc.). In the case of whole tissue or cells, it is advisable to carry the sample through the phenol extraction protocol (see Procedure I) to remove most of the nonprotein material prior to hydrolysis and thus to prevent overloading of the cellulose thin-layer sheet (see Section 4, below). Other protein samples can be prepared for hydrolysis by precipitation with acetone (85% final concentration).

Extensive Pronase Digestion of Proteins. The presence of tyrosine sulfate in proteins can also be detected after extensive proteolytic digestion, avoiding the strongly alkaline conditions described in Procedure III. The lyophilized pooled eluate of the desired $^{35}SO_4$-labeled protein, as obtained after step 3 of Procedure III, is dissolved in 50 μl of preincubated "high Pronase solution" (1 mg of Pronase per milliliter in 50 mM ammonium bicarbonate) and incubated for 12–24 hr at 37°. Thereafter, five volumes of acetone are added to precipitate excess Pronase, the precipitate is removed by centrifugation, the supernatant is collected, the acetone is evaporated, and the remainder is lyophilized. The residue is dissolved in pH 1.9 or pH 3.5 electrophoresis buffer and subjected to two-dimensional or one-dimensional thin-layer electrophoresis (described in step 4, Procedure IV), respectively, followed by autoradiography. Because of its mild nature, an extensive Pronase digestion is useful if the overall spectrum of sulfated residues of a protein is to be studied. This is further discussed in Section 4.

In some cases, a comparison of the result obtained by extensive Pronase digestion with that of the standard alkaline hydrolysis (see Procedure III) can be informative. For example, if a large proportion of the radioactive sulfate incorporated into a protein is lost during the hydrolysis with barium hydroxide and neutralization with sulfuric acid, this apparently indicates the presence of sulfated carbohydrate residues in excess over sulfated tyrosine residues. Such a loss would, however, be observed also if tyrosine sulfate residues of proteins, in contrast to tyrosine sulfate standard, were for some reason unstable during alkaline hydrolysis. In this case, the extensive Pronase digestion should give greater recovery of tyrosine sulfate than the alkaline hydrolysis. (We have not observed such a case so far, but cannot rule out its existence.)

4. Thin-Layer Electrophoresis of Tyrosine Sulfate

Thin-layer electrophoresis of tyrosine sulfate is performed under similar conditions as that of tyrosine phosphate[25] and is described in Procedure IV and Fig. 4. As discussed in Section 3, the hydrolysis and neutralization conditions used to liberate tyrosine sulfate from proteins render tyrosine sulfate the major, if not the only, ^{35}S-labeled substance in the sample. One-dimensional thin-layer electrophoresis therefore suffices in most instances to demonstrate the presence of tyrosine sulfate. In fact, once it has been established for the protein of interest that most or all of the ^{35}S radioactivity in the neutralized supernatant is in the form of tyrosine sulfate, the electrophoretic separation may be omitted for routine analysis. If total cell protein rather than individual proteins have been used for alkaline hydrolysis, ^{35}S-labeled material other than tyrosine sulfate can be present in the neutralized supernatant, in which case the two-dimensional thin-layer electrophoresis can be used to identify unequivocally tyrosine sulfate.

Procedure IV. Thin-Layer Electrophoresis of Tyrosine Sulfate
A. One-Dimensional Electrophoresis

1. Dissolve the sample (the residue after lyophilization of the neutralized supernatant, see Procedure III) in 20 μl of pH 3.5 electrophoresis buffer (5% acetic acid–0.5% pyridine). If the residue is small and dissolves readily, the sample will usually run well in the thin-layer electrophoresis without prior acetone precipitation; proceed to step 3. If the residue appears to be too big to be loaded on a thin-layer sheet or does not dissolve readily, do an acetone precipitation as described in step 2.

2. Add 100 μl of acetone to a 20-μl sample. Sediment the precipitate by centrifugation (Microfuge, 5 min), collect the supernatant, and evaporate the acetone.

3. Spot the sample on 20 × 20 cm plastic-backed 100-μm cellulose thin-layer sheets (e.g., Schleicher & Schüll F 1440), 3 cm from the cathodic edge of the sheet (see Fig. 4); let dry. We have run up to 10 samples in parallel per sheet. Each sample spot should contain 3 μg of unlabeled tyrosine sulfate as marker, mixed to the sample prior to spotting, and a trace of phenol red (usually already present from the hydrolysis).

4. Wet the cellulose sheet with pH 3.5 electrophoresis buffer. This can be done either by spraying (if the spot is small) or by using a "mask" of Whatman 3 MM filter paper (if the spot is to be concentrated). In the latter case, take a piece of the filter paper that is slightly larger than the cellulose sheet, cut a circular hole, slightly larger than the sample spot, at the position corresponding to that of the sample on the cellulose sheet, wet the filter paper with electrophoresis buffer and place it on the cellulose sheet such that the sample spot is not touched by the filter paper. Allow the electrophoresis buffer to concentrically migrate toward the center of the spot, thereby concentrating the sample (watch the phenol red present in the sample). The cellulose sheet should be homogeneously wet, appearing gray without "shiny" areas that are indicative of excess buffer. Place wet cellulose sheet in electrophoresis apparatus.

5. Perform electrophoresis at pH 3.5 until the phenol red marker has migrated 7–8 cm, i.e., just past the middle of the cellulose sheet. In our electrophoresis system, this takes about 40,000 V × minutes, e.g., 80 min at 500 V. Tyrosine sulfate migrates approximately 1.5 times as fast as the phenol red under these conditions and is then usually found 5–7 cm from the anodic edge of the cellulose sheet.

6. Dry the cellulose sheet. The unlabeled tyrosine sulfate marker is then detected by spraying the cellulose sheet with 1% ninhydrin in acetone and developing the color in an oven at 100° for a few minutes. (If the radioactive tyrosine sulfate of the sample is to be preserved for further biochemical analysis, omit the ninhydrin staining.) Perform autoradiography or fluorography of the cellulose sheet to detect the radioactive tyrosine sulfate. For fluorography, method 1 of Bonner and Stedman[29] is used.

B. Two-Dimensional Electrophoresis

1. As in A1, except that pH 1.9 electrophoresis buffer (7.8% acetic acid–2.2% formic acid) is used.

2. As in A2, if necessary.

3. Spot on cellulose sheet 3 cm from the "bottom" edge (the cathodic edge in the second dimension) and 3 cm from the "right" edge (the cathodic edge in the first dimension (see Fig. 4A). Otherwise as in A3.

4. As in A4, using pH 1.9 electrophoresis buffer and a mask of filter paper to wet the sample.

5. Perform electrophoresis at pH 1.9 until the phenol red marker has reached a position approximately 2 cm from the anodic "left" edge of the cellulose sheet. In our electrophoresis system, this takes about 82,500–90,000 V × minutes, e.g., 110–120 min at 750 V. Tyrosine sulfate migrates approximately 0.5 times as fast as the phenol red under these conditions and is then usually found in the middle between the anodic and the cathodic edge of the cellulose sheet. Dry the cellulose sheet.

6. Wet the cellulose sheet with pH 3.5 buffer by spraying. Then perform electrophoresis at pH 3.5 as in A5 (the cathodic and anodic edges being "bottom" and "top" of the cellulose sheet, respectively).

7. Detect tyrosine sulfate as in A6.

The extensive Pronase digestion protocol described in Section 3 usually liberates at least 50%, but rarely more than 80%, of the tyrosine sulfate residues from proteins, the rest remaining in the form of small acidic peptides that are seen as discrete spots between the tyrosine sulfate spot and the origin after one-dimensional thin-layer electrophoresis and autoradiography (see Fig. 4C). In contrast to the alkaline hydrolysis, the extensive Pronase digestion does not destroy sulfated carbohydrate resi-

[29] W. M. Bonner and J. D. Stedman, *Anal. Biochem.* **89**, 247 (1978).

dues of glycoproteins and proteoglycans. When extensive Pronase digests of $^{35}SO_4$-labeled glycoproteins and proteoglycans are analyzed by one-dimensional thin-layer electrophoresis and autoradiography, one sometimes observes some discrete spots between the tyrosine sulfate spot and the origin that may have similar electrophoretic mobilities as small tyrosine sulfate-containing peptides. In addition, one usually finds, in particular with proteoglycans, a fairly continuous streak that reaches from the origin to and beyond the tyrosine sulfate spot. In extensive Pronase digests of protein samples containing both tyrosine sulfate and sulfated carbohydrate residues, this latter $^{35}SO_4$-labeled material can fog the tyrosine sulfate spot or, if present in excess over tyrosine sulfate, hide its presence (see Fig. 4D).

For reasons outlined above (see Section 1), [^{35}S]methionine and [^{35}S]cysteine may be present in proteins, along with sulfated residues, after $^{35}SO_4$ labeling in animals. Their presence poses no problem for the detection of tyrosine sulfate, since in the one-dimensional thin-layer electrophoresis tyrosine sulfate is well separated from methionine, cysteine, and also cystine, all of which remain near the origin. Cysteic acid migrates about 1.5 times as fast as tyrosine sulfate toward the anode at pH 3.5, being very close to serine sulfate (see below).

Determination of the Stoichiometry of Tyrosine Sulfation of Proteins

In determining the stoichiometry of tyrosine sulfation of proteins, the possibility of sulfation at multiple sites must be taken into account. For example, if a stoichiometry of 1 mol of tyrosine sulfate per mole of protein is found, this would be consistent with a single tyrosine residue being sulfated in all of the protein molecules, or with two tyrosine residues being sulfated in half of the protein molecules, and so on. Evidence for multiple-site sulfation of $^{35}SO_4$-labeled proteins separated in polyacrylamide gels can be obtained by "fingerprinting." The procedures used for fingerprinting are similar to those used in the study of phosphorylated proteins.[30,31] Gel pieces from one- or two-dimensional gels containing the protein of interest are incubated with trypsin or any other specific protease until this digestion is complete. The eluted sulfated peptides are then separated in two dimensions on cellulose thin-layer sheets by electrophoresis and ascending chromatography and detected by autoradiography (the conditions of electrophoresis and chromatography may have to be adapted for the sulfated protein under study). Individual sulfated peptides

[30] W. B. Huttner and P. Greengard, *Proc. Natl. Acad. Sci. U.S.A.* **76,** 5402 (1979).
[31] W. B. Huttner, L. J. DeGennaro, and P. Greengard, *J. Biol. Chem.* **256,** 1482 (1981).

can be eluted from the cellulose with 50% pyridine and then used for hydrolysis and tyrosine sulfate analysis.

There are a number of possible approaches to determine the stoichiometry of tyrosine sulfation in individual sulfated proteins or in sulfated peptide fragments of these proteins. These include approaches based on the following considerations.

1. The direct chemical determination of tyrosine sulfate. This has the advantage that labeling of the protein of interest with radioactive isotopes is not required. However, the protein must be available in sufficient quantities to yield, after exhaustive alkaline hydrolysis, an amount of tyrosine sulfate that can be detected by amino acid analysis (about 10 pmol).[31a] With the advances made in the sensitivity of amino acid analysis, it should be feasible to develop methods for the determination of the stoichiometry of tyrosine sulfate in protein spots from two-dimensional polyacrylamide gels.

2. The determination of the proportion of tyrosine present as tyrosine sulfate. If the number of tyrosine residues in the protein under study is known, the stoichiometry of tyrosine sulfation can be determined as follows. Cells containing the protein are labeled with [³H]tyrosine or double-labeled with $^{35}SO_4$ and [³H]tyrosine (if the sulfate labeling is necessary to detect the protein in the subsequent steps). The protein is then obtained by immunoprecipitation or from two-dimensional gels (see Section 2) and subjected to exhaustive alkaline hydrolysis as described in Section 3. The residue of the neutralized supernatant (after step 6, Procedure III) is subjected to electrophoresis at pH 1.9 on cellulose thin-layer sheets (origin at 7 cm from the anodic edge of the sheet; 750 V for 40 min). Under this condition, tyrosine (migrating toward the cathode) is separated from tyrosine sulfate (migrating toward the anode). The tyrosine sulfate and the tyrosine spots, detected by ninhydrin staining of standard markers, are then counted for ³H radioactivity. The values obtained (after correction for the recovery of internal standards) can be used to calculate the proportion of tyrosine that is present as tyrosine sulfate and, taking into account the moles of tyrosine per mole of protein, the moles of tyrosine sulfate per mole of protein. We have used this method to estimate the proportion of tyrosine sulfate to tyrosine in the total protein of PC12 cells and chicken embryo fibroblast, and have obtained values ranging from 0.2 to 0.5%. This suggests that tyrosine sulfate in proteins is far more abundant than tyrosine phosphate.

3. The determination of the specific activity of the $^{35}SO_4$ transferred to proteins. After prolonged labeling of cells with $^{35}SO_4$, including several changes of the medium containing the isotope, it is reasonable to assume

---

[31a] J. Dodt, H. P. Mueller, U. Seemueller, and J.-Y. Chang, *FEBS Lett.* **165**, 180 (1984).

that the specific activity of the $^{35}SO_4$ in the medium (which can be calculated) is similar to that of the intracellular sulfate donor and also to that of the sulfate bound to tyrosine residues of proteins. The ^{35}S radioactivity in the tyrosine sulfate spot generated from a known amount of the protein (see Sections 3 and 4) can in this case be used to estimate the stoichiometry of sulfation.

4. The use of purified tyrosylprotein sulfotransferase *in vitro*. An enzymatic activity catalyzing the sulfation of proteins on tyrosine residues, designated as tyrosylprotein sulfotransferase, has been described in cell lysates.[11] Once purified preparations of this enzyme are available, it should be possible to sulfate the protein of interest *in vitro* using 3'-phosphoadenosine-5'-phospho [^{35}S]sulfate of known specific activity, and thus to determine the stoichiometry of sulfation.

5. The determination of charge shifts in two-dimensional gels. If the unsulfated form of the protein of interest can be identified in two-dimensional polyacrylamide gels (e.g., by immunoblotting), it is possible to estimate the number of single charge shifts between the unsulfated form and the sulfated form(s) by the carbamoylation of these proteins.[32,33]

Concluding Remarks

Tyrosine sulfation of proteins is emerging as a widespread posttranslational modification in animals. The methods described in this chapter can be used to identify tyrosine-sulfated proteins and may help in studies on the role of these proteins in cell function. Is tyrosine sulfation the only case or just the first example of protein sulfation on amino acid residues? Do proteins undergo sulfation of serine and threonine residues? In preliminary experiments, in which extensive Pronase digests of $^{35}SO_4$-labeled rat brain proteins were analyzed by two-dimensional thin-layer electrophoresis and autoradiography, a radioactive substance with a mobility similar to that of serine O-sulfate was observed. However, it is too early at present for a definitive answer to the questions raised above. The study of protein sulfation on tyrosine and, possibly, on other amino acids is an area of research where little is known at present, but where new insights into cellular control mechanisms are likely to be gained in the future.

Acknowledgments

I wish to thank my colleagues P. Baeuerle, A. Hille, Dr. R. W. H. Lee, Dr. S. B. Por, Dr. P. Rosa, and C. Suchanek for their collaboration. This work was supported in part by DFG Grant Hu 275/3-1.

[32] D. Bobb and B. H. J. Hofstee, *Anal. Biochem.* **40**, 209 (1971).
[33] R. A. Steinberg, P. H. O'Farrell, U. Friedrich, and P. Coffino, *Cell* **10**, 381 (1977).

[12] Protein Side-Chain Acetylations

By VINCENT G. ALLFREY, EUGENE A. DI PAOLA, and
RICHARD STERNER

The acetylation of amino acids within the polypeptide chain, as distinct from acetylation of free NH_2-terminal groups, was first detected in cell nuclei as a rapid and reversible incorporation of radioactively labeled acetate into histones H3 and H4.[1,2] The acetyl groups were subsequently shown to be localized on lysine residues as ε-*N*-acetyllysine,[3,4] and the acetyl group donor was identified as acetyl-CoA.[2,5] The acetylation of histone lysyl residues is a postsynthetic modification, unaffected by inhibitors of protein synthesis such as puromycin[1,6] and readily demonstrable in intact histone molecules using the appropriate acetyltransferase (EC 2.3.1.48). The general reaction is illustrated (Fig. 1).

Acetylation of lysine ε-amino groups is not limited to histones, but takes place also on other DNA-binding proteins such as the high-mobility group (HMG) proteins.[7,8] In all cases, the neutralization of positive charge upon acetylation of the lysine side chains would be expected to weaken the electrostatic interactions between the protein and the associated DNA. This effect is amplified when more than one lysine is involved, as is usually the case. Amino acid sequence studies have established that a limited and specific set of lysine residues is subject to acetylation in each histone or HMG class. In histone H4, the potential sites of acetylation have been identified as lysine residues at positions 5, 8, 12, and 16 of the polypeptide chain. In histone H3, the modifications occur at positions 9, 14, 18, and 23. Histone H2A has a major acetylation site at Lys-5 with a minor substitution at Lys-9. Histone H2B can be acetylated at four positions (lysine residues 12, 15, 20, and 23).[9–11] The modification of HMG

[1] V. G. Allfrey, R. Faulkner, and A. E. Mirsky, *Proc. Natl. Acad. Sci. U.S.A.* **51,** 786 (1964).
[2] V. G. Allfrey, *Can. Cancer Conf.* **6,** 313 (1964).
[3] E. L. Gershey, G. Vidali, and V. G. Allfrey, *J. Biol. Chem.* **243,** 5018 (1968).
[4] R. J. DeLange, D. M. Fambrough, E. L. Smith, and J. Bonner, *J. Biol. Chem.* **244,** 319 (1969).
[5] H. Nohara, T. Takahashi, and K. Ogata, *Biochim. Biophys. Acta* **127,** 282 (1966).
[6] B. G. T. Pogo, V. G. Allfrey, and A. E. Mirsky, *Proc. Natl. Acad. Sci. U.S.A.* **55,** 805 (1966).
[7] R. Sterner, G. Vidali, and V. G. Allfrey, *J. Biol. Chem.* **254,** 11577 (1979).
[8] R. Sterner, G. Vidali, and V. G. Allfrey, *J. Biol. Chem.* **256,** 8892 (1981).
[9] G. H. Dixon, E. P. M. Candido, B. M. Honda, A. J. Louie, A. R. MacLeod, and M. T. Sung, *in* "The Structure and Function of Chromatin" (D. W. Fitzsimons and G. E. W. Wolstenholme, eds.), p. 229. Associated Scientific Publ., Amsterdam, 1975.

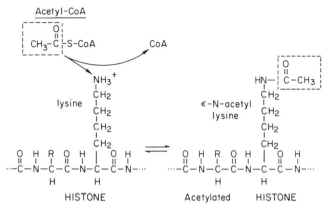

FIG. 1. Formation of ε N-acetyllysine in the polypeptide chain by acetyl group transfer from acetyl-CoA. Note the neutralization of positive charge resulting from acetylation of the ε-amino group.

proteins occurs also in the very basic NH_2-terminal domains believed to be involved in DNA binding; at lysine residues 2 and 11 in HMG-1, at positions 2 and 4 in HMG-14, and at positions 2, 4, and 10 in HMG-17.[7,8]

It is important to note that the potential for acetylation at multiple sites is not realized in every histone or HMG molecule; i.e., particular lysine residues may or may not be acetylated because this type of modification is a rapid and reversible process that can proceed in the absence of histone or HMG synthesis or degradation.[6,12,13] Acetate "turnover" generates a changing population of protein derivatives, each comprising a mixture of polypeptide chains of identical amino acid sequence, some of which are internally acetylated to different degrees while others are not acetylated at all. (Each of these subfractions may then differ with regard to other forms of substitution, such as phosphorylation, methylation, ADP-ribosylation, ubiquitination, etc.[14–17]). Because the charge neutralization associated with acetylation of the lysine ε-amino groups affects the electropho-

[10] B. H. Thwaits, W. F. Brandt, and C. von Holt, *FEBS Lett.* **71**, 193 (1976).

[11] I. Sures and D. Gallwitz, *Biochemistry* **19**, 943 (1980).

[12] L. A. Sanders, N. M. Schechter, and K. S. McCarty, *Biochemistry* **12**, 783 (1973).

[13] V. Jackson, A. Shires, R. Chalkley, and D. Granner, *J. Biol. Chem.* **250**, 4856 (1975).

[14] V. G. Allfrey, *in* "Chromatin and Chromosome Structure" (H. J. Li and R. A. Eckhardt, eds.), p. 167. Academic Press, New York, 1977.

[15] V. G. Allfrey, *in* "Cell Biology" (L. Goldstein and D. M. Prescott, eds.), Vol. 3, p. 347. Academic Press, New York, 1980.

[16] D. Mathis, P. Oudet, and P. Chambon, *Prog. Nucleic Acid Res. Mol. Biol.* **24**, 1 (1980).

[17] V. G. Allfrey, *in* "The HMG Chromosomal Proteins" (E. W. Johns, ed.), p. 123. Academic Press, New York, 1982.

retic and chromatographic behavior of the parent polypeptide chain, it is possible to separate unmodified histone molecules from those containing from one to four ε-N-acetyllysine residues. Protein subfractions differing in their degree of acetylation migrate at different rates in acid–urea–polyacrylamide gels[18] and in starch–urea–aluminum lactate gels,[19] and the identity of the various histone bands, as separated electrophoretically, has been confirmed by chromatographic separations,[20] isotopic labeling with radioactive acetate, and direct analysis. Using preparative gel electrophoresis, it is now possible to prepare the well-defined histone substrates necessary for studies of the kinetics and mode of action of isolated transacetylases or deacetylases, as described in detail below.

Procedures

Formation of ε-N-Acetyllysine in Vivo and in Vitro

The uptake of radioactive acetate[1,3,6] or acetate analogs[21] by living cells leads to the formation of the correspondingly modified histones or HMG proteins. It should be emphasized, however, that isotope incorporation is not, in itself, a sufficient proof of protein side-chain acetylation because some of the isotopic acetate rapidly enters the free amino acid pools (mainly as aspartic and glutamic acids) and is thus incorporated into nascent polypeptide chains, while other acetyl groups are transferred from acetyl-CoA to the free α-NH_2 groups at the amino termini of a variety of newly synthesized protein molecules. (Histone H1, for example, is not subject to acetylation of its lysine residues, but does contain acetylserine at position 1.) For these reasons, attempts to monitor side-chain acetylations in complex systems (intact cells, isolated nuclei, chromatin fractions, or cellular extracts containing multiple forms of acetyltransferases) may be quite misleading, and a rigorous chemical identification of the reaction product is recommended.

Detection and Quantitation of ε-N-Acetyllysine

The amide linkage between the acetyl group and the ε-NH_2 function of lysine is not stable to acid hydrolysis, and therefore standard procedures of hydrolysis and amino acid analysis cannot be used to detect ε-N-

[18] S. Panyim and R. Chalkley, Arch. Biochem. Biophys. **130,** 337 (1969).
[19] M. T. Sung and G. H. Dixon, Proc. Natl. Acad. Sci. U.S.A. **67,** 1616 (1970).
[20] L. J. Wangh, A. Ruiz-Carrillo, and V. G. Allfrey, Arch. Biochem. Biophys. **150,** 44 (1972).
[21] R. Sterner and V. G. Allfrey, J. Biol. Chem. **257,** 13872 (1982).

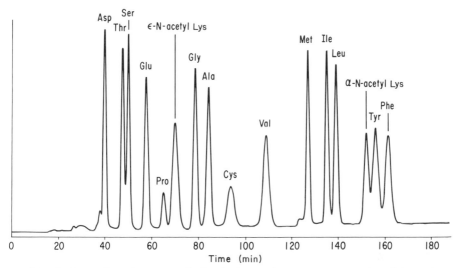

Fig. 2. Separation of ε N-acetyllysine by ion-exchange chromatography of the amino acids generated by extensive protease digestion of acetylated histones, as described under Procedures.

acetyllysine. The presence of this derivative can be determined, however, by enzymatic degradation of the modified protein and chromatographic separation of the component amino acids. This can be accomplished with some modification[22] of the procedure of Gershey et al.[3]

Protein (0.2–2 mg) is dissolved in 1 ml of 0.2 M N-ethylmorpholine acetate, pH 8.1. Trypsin (1% of the weight of the substrate) is added, and the solution is incubated for 2 hr at 37°. At this time an aliquot of fresh trypsin (1% by weight) is added, and the hydrolysis is continued for another 2 hr. The solution is then placed in a boiling water bath for 5 min and allowed to cool. Pronase (10% by weight) is added, and the reaction mixture is incubated for 18 hr at 40°. At this point it is advisable, particularly if a high-sensitivity, microbore amino acid analyzer system is to be used, to remove the proteolytic enzymes from the reaction mixture by addition of an equal volume of 0.29 M perchloric acid. The precipitated proteins are removed by centrifugation, and the supernatant is vacuum dried. In subsequent amino acid analysis, ε-N-acetyllysine emerges as a peak immediately preceding that of glycine (Fig. 2). For more rigorous identification, the putative ε-N-acetyllysine peak may be diverted and collected prior to its reaction with ninhydrin in the analyzer reaction coil and hydrolyzed in 6 N HCl at 110°. Subsequent amino acid analysis will reveal a shift in elution position to that of free lysine.

Quantitation of the extent of protein acetylation may be achieved by the use of fluorodinitrobenzene (FDNB), which reacts with free amino groups to form acid-stable dinitrophenyl (DNP) derivatives. In proteins such as histone H1, which are known not to undergo posttranslational acetylation, it has been demonstrated that reaction with FDNB results in virtually quantitative conversion of lysine residues to ε-N-DNP-lysine.[22] However, acetylated ε-amino groups do not react with FDNB and after acid hydrolysis appear as free lysine, thus allowing measurement of the original extent of acetylation in a given protein. Dinitrophenylation and amino acid analysis of the five forms of histone H4 separated by preparative gel electrophoresis has confirmed the presence of 0–4 ε-N-acetyllysine residues in the separated bands and established the validity of the method for quantitating the extent of acetylation.[23]

Proteins may be dinitrophenylated according to the following protocol.[22,24–26] The protein sample (0.2–2.0 mg) is dissolved in 1 ml of NaHCO$_3$ solution (20 mg/ml). After the addition of 2 ml of absolute ethanol, the denatured protein is shaken for 1 hr. At this time 30 µl of FDNB is added, and the mixture is shaken for 2 hr at 25°. The insoluble yellow product is washed twice with water, ethanol, and ether, successively, and then hydrolyzed for 24 hr in 4 N HCl at 110°. The hydrolysate may then be subjected to standard techniques for amino acid analysis. The ε-N-DNP-lysine is seen as a broad peak eluting after arginine.

Other direct chemical procedures for the quantitation of acetyl groups in proteins include steam distillation of protein hydrolysates and titration of the volatile acetic acid,[1,27] hydrazinolysis[28] followed by conversion of the acetyl hydrazide to a dinitrophenyl derivative,[29] and extraction of acid hydrolysates with *tert*-butyl ethyl ether followed by gas chromatographic separation of the extracted acetic acid.[30] A microenzymatic method requiring only 0.1–0.9 µmol of protein or equivalent amount of acetate has also been described.[31] These methods do not discriminate between ace-

[23] E. A. Di Paola, R. Sterner, and V. G. Allfrey, unpublished experiments.
[24] F. Sanger, *Biochem. J.* **39,** 507 (1945).
[25] E. Wofsy and S. J. Singer, *Biochemistry* **2,** 104 (1963).
[26] W. A. Schroeder and J. LeGette, *J. Am. Chem. Soc.* **75,** 4612 (1953).
[27] J. Z. Augustyniak, and W. G. Martin, *Can. J. Biochem.* **43,** 291 (1965).
[28] K. Narita, *Biochim. Biophys. Acta* **28,** 184 (1958).
[29] D. M. P. Phillips, *Biochem. J.* **86,** 397 (1963).
[30] D. N. Ward, J. A. Coffey, D. B. Ray, and W. M. Lamkin, *Anal. Biochem.* **14,** 243 (1966).
[31] L. Stegink, *Anal. Biochem.* **20,** 502 (1967).

tates present on lysine residues and those present on the free NH_2-terminal groups.

The highest sensitivities are provided by radioisotopic labeling procedures. Proteins may be monitored for possible posttranslational acetylations *in vivo* or *in vitro* by incubation with the appropriate precursors: [³H]- or [¹⁴C]acetate in intact cells (and some types of isolated nuclei), and [³H]- or [¹⁴C]acetyl-CoA in cellular homogenates, extracts, and purified enzyme preparations. The uptake of isotopic acetate can be quantitated by scintillation spectrometry of the protein samples and by fluorography of the electrophoretically separated histone derivatives,[32] but there are several caveats that require discussion. First, metabolic conversions of acetate may permit the entry of the isotopic label into the cellular amino acid pools and subsequent incorporation into nascent polypeptide chains. Inhibitors of protein synthesis, such as puromycin and cycloheximide, which do not block the acetylation of lysine residues,[1,6] can be used to confirm the posttranslational nature of the isotopic labeling. (However, these inhibitors may be expected to reduce the NH_2-terminal acetylations that are frequently coordinated with translation. When acetylation and *de novo* protein synthesis take place concurrently, as in avian erythroblasts,[33] both types of acetylation are observed.) The distinction between isotope uptake as acetate or as amino acids derived from acetate can be made by acid hydrolysis of the protein followed by steam distillation of the volatile acetic acid. Both the volatile and nonvolatile fractions can then be assayed for radioactivity.[1]

Acetate labeling experiments in complex systems are also complicated by the action of histone deacetylases, which rapidly remove the acetyl groups from the modified lysine residues. The loss of acetyl groups can be blocked almost completely by the addition of 5–10 mM sodium butyrate,[34] which has been shown to inhibit histone deacetylase activity *in vivo* and *in vitro*.[32,35] Butyrate does not inhibit acetyltransferase activity, and cells cultured in butyrate progressively accumulate the multiacetylated forms of the histones[34] and HMG proteins.[8] (This makes possible the purification of the multiacetylated forms of histone H3 and H4, as well-defined substrates for histone deacetylase assays, as described below.) The butyrate effect is rapidly reversible and most of the acetylated forms of histones H3 and H4 are converted back to the unmodified form within

[32] L. S. Cousens, D. Gallwitz, and B. M. Alberts, *J. Biol. Chem.* **254**, 1716 (1979).

[33] A. Ruiz-Carrillo, L. J. Wangh, and V. G. Allfrey, *Science* **190**, 117 (1975).

[34] M. G. Riggs, R. G. Whittaker, J. R. Neumann, and V. M. Ingram, *Nature (London)* **268**, 462 (1977).

[35] L. C. Boffa, G. Vidali, and V. G. Allfrey, *J. Biol. Chem.* **253**, 3364 (1978).

15 min after removal of the butyrate.[36,37] These changes are associated with alterations in transcription, cell morphology, and growth characteristics.[15]

For studies of the complex relationships between acetylation of nuclear proteins and cell function, rigorous characterizations of the modified proteins and identification of the sites of modification within each protein may be necessary. Even when dealing with a particular protein class, such as histone H2A, additional complications arise because different structural variants of H2A show different patterns of postsynthetic acetylation.[38] Purification of each variant should precede attempts to identify the position of the modified lysine residues. The acetylation sites can be identified by electrophoretic separation and analysis of the tryptic peptides[38,39] or by sequential Edman degradation of the radioactive protein.[7,19,40]

In assessing the purification of protein acetylating enzymes, simpler assay procedures are usually employed. After incubations in the presence of radioactive acetyl-CoA, the labeled histones can be precipitated and collected on glass-fiber filters[11,41] or transferred to phosphocellulose-paper disks[42–44] for counting, as described below.

Selective Recovery of Newly Acetylated Protein Molecules

A novel approach to the study of protein acetylation employs the acetate analog 2-mercaptoacetate, which has been shown to be incorporated into the histones and HMG proteins of living cells. The modification permits the recovery of the thio-derivatized proteins by affinity chromatography on organomercurial agarose columns.[21] For proteins that lack cysteinyl residues, including all of the nucleosomal "core" histones (except H3) and the high-mobility group proteins HMG-14 and HMG-17, the method provides a unique opportunity to recover only those molecules that were modified during the incubation with mercaptoacetate. Peptide

---

[36] G. Vidali, L. C. Boffa, R. S. Mann, and V. G. Allfrey, *Biochem. Biophys. Res. Commun.* **82,** 223 (1978).

[37] J. Covault, L. Sealy, R. Schnell, A. Shires, and R. Chalkley, *J. Biol. Chem.* **257,** 5809 (1982).

[38] P. Pantazis and W. M. Bonner, *J. Biol. Chem.* **256,** 4669 (1981).

[39] L. O. Tack and R. T. Simpson, *Biochemistry* **18,** 3110 (1979).

[40] H. D. Niall, this series, Vol. 27, p. 942.

[41] S. R. Harvey and P. R. Libby, *Biochim. Biophys. Acta* **429,** 742 (1978).

[42] K. Horiuchi and D. Fujimoto, *Anal. Biochem.* **69,** 491 (1975).

[43] E. Belikoff, L.-J. Wong, and B. M. Alberts, *J. Biol. Chem.* **255,** 11448 (1980).

[44] J. E. Wiktorowicz and J. Bonner, *J. Biol. Chem.* **257,** 12893 (1982).

analyses of the Hg-affinity-purified proteins confirm that the sites of mercaptoacetylation are the same as those modified when [³H]acetate is employed as the precursor.[21] An important advantage of this technique for selective recovery of the newly acetylated proteins is that it allows the study of coordinate or simultaneous modifications of the same polypeptide chain. A recent application showing simultaneous mercaptoacetylation and phosphorylation of histones H2A, HMG-14, and HMG-17 provides incontrovertible evidence that the same protein molecule can carry both types of modification at the same time.[45]

Assay of N-Acetyltransferases

The acetylation of lysine residues in histones and other DNA-associated proteins is catalyzed by acetyltransferases that utilize acetyl-CoA as the acetyl group donor. Multiple forms of ε-N-acetyltransferase exist in the cell nucleus, and at least one is known to be present in the cytoplasm, where it modifies nascent histone H4 chains.[33] Earlier studies on acetyltransferases have been reviewed.[14] More recent characterizations of highly purified enzyme preparations confirm that considerable complexity exists with regard to substrate specificities, kinetics of activity, pH optima, sensitivity to inhibitors, and molecular characteristics.[11,43,44,46,47] A detailed description of the various enzyme preparations is beyond the scope of this chapter, but a few examples will serve to illustrate the current state of the field. Although none of the acetyltransferases has been purified to homogeneity, several preparations approach that goal and provide useful indications of the general properties and substrate requirements of their active components.

A histone acetylase specific for histone H4 has been purified about 1200-fold from *Drosophila* embryo extracts, utilizing successive chromatography on histone-Sepharose, hydroxyapatite, and Bio-Rex 70.[47] The enzyme activity migrates on nondenaturing gels as a single peak with an approximate molecular weight (M_r) of 160,000. It has a broad pH-activity range, with an optimum at pH 9, but it is irreversibly inactivated at pH values below pH 6.5. It is not inhibited by butyrate, thus permitting the use of butyrate as a deacetylase inhibitor during studies of its function. The purified enzyme is active toward nonacetylated and monoacetylated H4 molecules. Under the assay conditions employed, the K_m for free H4 is 3.4 μM and the K_m for acetyl-CoA is 0.22 μM. The enzyme does not act

[45] R. Sterner and V. G. Allfrey, *J. Biol. Chem.* **258,** 12135 (1983).
[46] P. R. Libby, *J. Biol. Chem.* **253,** 233 (1978).
[47] R. C. Wiegand and D. L. Brutlag, *J. Biol. Chem.* **256,** 4578 (1981).

on H4 in chromatin, and it is likely to be a cytoplasmic acetyltransferase similar to the type B enzymes in mammalian cells.[11] If so, its primary function is the acetylation of nascent H4 chains.[33,48]

Several purifications of nuclear acetyltransferases have been described.[11,43,44,46] A 6000-fold purification of a calf thymus nuclear enzyme was achieved by polyethyleneglycol precipitation and subsequent chromatography on DNA-cellulose, norleucine-Sepharose, DEAE-Sephacel, and phosphocellulose.[43] The molecular weight of the enzyme, as judged by its mobility on SDS–polyacrylamide gels, was estimated at 65,000–70,000. At pH 7.8, the K_m for a mixture of the four nucleosomal histones (H2A, H2B, H3, H4) was about 20 μM, and the K_m for acetyl-CoA was 0.5–1.0 μM. All four histones were acetylated in their free forms, but the enzyme failed to acetylate H3 in chromatin or in nucleosomes.

A rat liver nuclear acetyltransferase that preferentially acetylates histone H3 has been described.[44] The enzyme was purified from nuclear extracts by hydrophobic affinity chromatography on phenyl-Sepharose, followed by gel filtration, ion-exchange chromatography on CM-cellulose, and acetyllysine affinity chromatography. The purified enzyme had optimal activity at pH 7.4–7.5. Its molecular weight, as judged by gel filtration, was 96,000. Isoelectric focusing gave a major activity peak corresponding to a pI of 7.8–8.2. Reaction and product inhibition kinetics suggested an ordered bireactant mechanism in which histone adds first to the enzyme and acetyl-CoA concentration is rate-determining.[49] The K_m for acetyl-CoA is 21 μM. The enzyme is completely inactivated by sulfhydryl reagents such as N-ethylmaleimide and Hg^{2+}.

A calf liver nuclear enzyme that preferentially acetylates histone H4, but also modifies other nucleosomal histones, has been purified about 5000-fold.[46] Its molecular weight was estimated as 150,000, and its pH optimum is 7.8. Surprisingly, the purified enzyme also acetylates polyamines, such as spermine and spermidine. This is relevant to the independent observation that a multisubstrate analog in which CoA is linked to spermidine through acetic acid is a very potent inhibitor of histone acetylation.[50] Both histone- and polyamine-acetylating activities are strongly inhibited by p-chloromercuribenzoate.[46]

Given this background, some general comments bearing on the assay of acetyltransferases are in order. First, many of the enzymes are unstable, and yields during purifications have been very low despite the addi-

---

[48] V. Jackson, A. Shires, N. Tanphaichitr, and R. Chalkley, *J. Mol. Biol.* **104,** 471 (1976).

[49] J. E. Wiktorowicz, K. L. Campos, and J. Bonner, *Biochemistry* **20,** 1464 (1981).

[50] P. M. Cullis, R. Wolfenden, L. S. Cousens, and B. M. Alberts, *J. Biol. Chem.* **257,** 12165 (1982).

tion of protease inhibitors, glycerol, carrier proteins, and other precautionary measures. At least some of the purified acetyltransferases are inhibited by divalent cations,[43,44] and the purified calf thymus nuclear acetyltransferase A is sensitive to monovalent cations, with inhibitions exceeding 90% at salt concentrations higher than 0.1 M.[51] Inhibitions by monovalent cations appear to be reversible,[51] but it is clear that precautions must be taken to assure reproducible ionic concentrations and to avoid high ionic strengths in the media employed for acetyltransferase assays. Heavy metals and sulfhydryl-reactive compounds should be avoided.

The measurement of ε-N-acetyltransferase activities in nuclei, chromatin, or crude nuclear extracts is complicated by the presence of natural inhibitors[43,46,52] and by the concurrent activity of the deacetylases, which rapidly remove the newly incorporated acetyl groups. In systems where deacetylases are likely to be present, the addition of 5–50 mM sodium butyrate is recommended. (Preparations of the cytoplasmic acetyltransferases do not appear to be contaminated in this way, and butyrate has no effect on the assay.[47])

Acetyltransferase assays, as carried out with purified histone substrates, frequently show preferences for histone classes that are known to be preferentially acetylated *in vivo;* but this is not always the case. Artifactual acetylations involving the wrong histone, or the wrong acetylation sites, are commonplace. For example, highly purified nuclear acetyltransferase will utilize histone H1 as a substrate,[11,46] although H1 is not subject to acetylation of its lysine residues *in vivo;* and careful sequence analyses of histones H3 and H4 following *in vitro* acetylation have shown the presence of ε-N-acetyllysine residues at positions that are not acetylated *in vivo.*[10,11] Even with purified histone substrates, such as H3 and H4, the combination of wrong-site acetylations and multiple acetylations involving combinations of one to four acetylation sites limits the interpretation of kinetic studies of acetate uptake. Studies of histone acetylation *in vivo* strongly suggest that acetyl group transfer is progressive, favoring multiple modifications of the same enzyme-bound histone molecule.[53] Such preferential hypermodification of a bound substrate would invalidate the simple kinetic models and K_m calculations based on total histone concentrations. It is also clear that patterns of acetylation observed for histones in solution are not the same as those for histones organized into nucleo-

[51] L.-J. Wong, unpublished experiments.

[52] L.-J. Wong, *Biochem. Biophys. Res. Commun.* **97,** 1362 (1980).

[53] A. Ruiz-Carrillo, L. J. Wangh, and V. G. Allfrey, *Arch. Biochem. Biophys.* **174,** 273 (1976).

somes. As noted earlier, some enzymes fail to acetylate nucleosomal histones at all.[47] In other cases the reaction of acetyltransferases with isolated nucleosomes was found not to be an accurate model for events in intact cells or in isolated nuclei.[54]

The isolated acetyltransferases have pH optima above neutrality (pH 7–9). Assays at the higher pH values are complicated by nonenzymatic acetylations in which acetyl-CoA also serves as the group donor.[55,56] Corrections for what may be a significant background of nonenzymatic acetylation should be made by measuring acetate uptake in enzyme-free controls.

Method

The simplest acetyltransferase assays are those based on the transfer of radioactive acetate from acetyl-CoA to protein substrates, which are then precipitated or adsorbed for counting.

Substrates. The choice of substrate is important, not only with regard to the analysis of enzyme specificity, but also in a practical sense; substrates with the maximum number of free acetylatable lysine residues are preferable to those that already have been partially or wholly acetylated. The substrates are usually histones, but all the major HMG proteins are also acetylatable.[17] Histones prepared from transcriptionally inert systems, such as echinoderm sperm[20] and *Tetrahymena* micronuclei,[57] occur exclusively in their unacetylated forms. In other cells a variable proportion of acetylated molecules is present in all the histone fractions (except H1). The pure nonacetylated forms may be obtained by preparative gel electrophoresis, as described below.

Incubation Conditions for Acetylase A.[43] A dilute enzyme sample (10 μl) is added to 110 μl of a solution at 0° containing 0.1 M KCl, 75 mM Tris-HCl, pH 7.8, 0.25 mM Na$_2$EDTA, 5 mM 2-mercaptoethanol, 1 mg of bovine serum albumin per milliliter, and 1 mg of histone substrate per milliliter. The assay is begun by the addition of 0.6 μM [^{3}H]acetyl-CoA (specific activity 0.6 Ci/mol), followed by incubation at 37° for periods of 1–20 min. (It is important to confirm that the acetyl-CoA is not rapidly

[54] R. L. Garcea and B. M. Alberts, *J. Biol. Chem.* **255,** 11454 (1980).
[55] D. Gallwitz and C. E. Sekeris, *Hoppe-Seyler's Z. Physiol. Chem.* **350,** 150 (1969).
[56] W. K. Paik, D. Pearson, H. W. Lee, and S. Kim, *Biochim. Biophys. Acta* **213,** 543 (1970).
[57] M. A. Gorovsky, G. L. Pleger, J. B. Keevert, and C. A. Johmann, *J. Cell. Biol.* **57,** 773 (1973).

degraded under the conditions of the assay; and, if deacetylases are present, it will be necessary to supplement the reaction mixture with 5–50 mM sodium butyrate to prevent the loss of newly incorporated [^{3}H]acetyl groups.)

Incubation Conditions for Acetylase B.[47] The standard assay mixture (100 μl) contains 75 mM Tris-HCl (pH 8.5), 5 mM 2-mercaptoethanol, 0.5 mM EDTA, 10% glycerol, 0.1 mg of bovine serum albumin per milliliter, 50 μg of histone substrate (preferably unacetylated H4), 1 μCi of [^{3}H]acetyl-CoA (10^8 cpm/nmol), and 0.1 M KCl. Incubations with the *Drosophila* enzyme are carried out at 17°,[47] but calf thymus enzymes are assayed at 37°.[11] The reaction is stopped by the addition of 200 μl of cold 0.1 M sodium acetate (pH 4.5) containing 0.5 mg of salmon sperm DNA (added as a carrier) per milliliter.[47] In an alternative procedure,[11] aliquots of the reaction mixture are pipetted onto filter paper disks that are immediately immersed in ice-cold 25% trichloroacetic acid.

Collection of Acetylated Proteins. A modification of the method of Horiuchi and Fujimoto[42] is described. Aliquots (100 μl) of the acetylase A reaction mixture are spotted onto 2.4-cm squares of Whatman P-81 phosphocellulose paper. The filters are batch-washed with gentle swirling three times for 5 min at 37° in 100 ml of 50 mM NaHCO$_3$–Na$_2$CO$_3$ (pH 9), washed once in acetone at 25°, dried, and then counted by scintillation spectrometry. The radioactivity of samples run without enzyme is subtracted to correct for nonenzymatic acetylation.

The reaction products of the *Drosophila* acetylase B assay[47] are precipitated by the addition of 1 ml of ethanol, transferred to GF/C glass-fiber filters, washed 4 times with 3 ml of cold 70% ethanol, dried, and counted in a toluene-based scintillation fluid. Alternatively,[11] the products are transferred to filter paper disks (Schleicher & Schuell 2043b), immersed in cold 25% trichloroacetic acid for 20 min, heated at 95° for 5 min, washed once with 25% trichloroacetic acid, ethanol, ethanol–ether (1 : 1, v/v), and ether, then dried and counted.

Identification of Modified Proteins. The transfer of acetyl groups to the targeted lysine residues eliminates the positive charges on their ε-amino groups. Consequently, the electrophoretic mobility of the protein is reduced in proportion to the number of acetyl groups transferred. Histone H4, for example, can be resolved in acid–urea–polyacrylamide gels[18] into five bands, one unmodified and four slower bands containing 1–4 ε-N-acetyllysine residues. When the proteins have been labeled with radioactive acetate, the positions of the bands and their relative intensities can be determined by fluorography of the gels. An excellent example of the approach is provided by Cousens *et al.*[32]

Deacetylation of ε-N-Acetyllysine in Proteins

Purification and Properties of Deacetylating Enzymes

The deacetylases catalyze the removal of acetyl groups from ε-NH_2 groups of lysine residues, thus converting ε-N-acetyllysine to lysine in the polypeptide chain. This reverses the modifications introduced by the acetyltransferases, and the two enzyme families act together to maintain or alter the posttranslational acetylation states of nuclear proteins.[14,15]

Early reports of deacetylase extractions[58-60] described crude preparations in which tissues or nuclei were homogenized in low ionic strength buffers and the supernatants from subsequent centrifugation served as the enzyme preparation. Subsequent reports by Vidali et al.[61] and Kaneta and Fujimoto[62] provided more extensive protocols for deacetylase purification, although the final preparations were still far from homogeneous. Vidali et al. reported an enzyme with an isoelectric point of 4.5, optimal activity at pH 8, and a molecular weight of about 150,000. The enzyme had particular affinity for acetylated histones H3 and H4.[61] Kaneta and Fujimoto[62] reported an enzyme preparation with optimal activity at pH 7.5 that apparently contained two separate activities; one acted on chromatin-bound as well as free histones, and the other released acetate from free histones only.

The most comprehensive protocol to date for the separation and partial purification of deacetylase activities is that of Cousens et al.,[32] using calf thymus as the source. The procedure involves precipitation with $(NH_4)_2SO_4$ and further fractionation on DEAE-cellulose and hydroxyapatite columns. Two deacetylase activities were separated on the ion-exchange columns, with estimated purifications of 60- and 150-fold, respectively. The procedure works well for purification of deactylase activities from either calf thymus[32] or duck erythrocytes.[23] The latter system is simpler because of the lower complexity of the protein extract (most of which is hemoglobin and is easily removed[23]). As judged by mobility in SDS–polyacrylamide gels, the putative erythrocyte deacetylases have a molecular weight of 60,000–70,000.[23]

Other recent studies have indicated that most of the chromatin-bound deacetylase activity of HeLa cells occurs in association with a high molecular weight DNA–protein complex, and it is not extractable in salt

---

[58] A. Inoue and D. Fujimoto, Biochim. Biophys. Acta 220, 307 (1970).
[59] P. R. Libby, Biochim. Biophys. Acta 213, 234 (1970).
[60] G. S. Edwards and V. G. Allfrey, Biochim. Biophys. Acta 299, 354 (1973).
[61] G. Vidali, L. C. Boffa, and V. G. Allfrey, J. Biol. Chem. 247, 7365 (1972).
[62] H. Kaneta and D. Fujimoto, J. Biochem. (Tokyo) 76, 905 (1974).

solutions (including 2 M NaCl).[63] Therefore, it is likely that isolations based on salt extractions represent only a fraction of the activity present, and they may exclude particular enzyme species. An additional complication arises in the coextraction of deacetylase inhibitors.[61]

A most significant and exploitable property of the deacetylases is their inhibition by millimolar concentrations of sodium butyrate.[32,34-36] This makes possible the recovery of multiacetylated forms of the histones as substrates for the deacetylase assay.

Assay for Deacetylase Activity

Preparation of Substrates. Defined substrates for the deacetylase reaction can be prepared by chemical[64-67] and enzymatic[11,23,32,34,38,54,66] means. In one chemical approach, solid-phase peptide synthesis was used to insert [^{3}H]- and [^{14}C]acetyllysine residues at specific sites in the H4 sequence (1–37). This substrate was used in the study of the mechanism of enzyme action on multiacetylated domains of the H4 molecule.[65] Alternatively, synthetic peptides have been completely acetylated by reaction with [^{3}H]acetic anhydride.[66,68]

Chemical acetylations have also been used to modify naturally occurring histones and derivative peptides.[1,39,67,69] Such reactions lack the specificity of the acetyltransferases and modify lysine residues that are not acetylated *in vivo*.[39] However, the reaction of [^{3}H]acetic anhydride with certain peptides, such as the (1–23) fragment of histone H4, generates a useful substrate containing all four natural acetylation sites but no other sites.[67] Complete acetylation of this fragment would yield a homogeneous population of tetraacetylated molecules, thus eliminating ambiguities in the analysis of deacetylation kinetics due to heterogeneity of the substrate and uncertainty about where its modifications occur. Chemical acetylation affords substrates of much higher specific radioactivity than would be obtained through *in vivo* labeling, and hence the sensitivity of the deacetylase assay is correspondingly heightened.

The acetylation of histones in cell cultures provides a natural set of substrates for deacetylase assays. The yield of multiacetylated forms of

[63] C. W. Hay and E. P. M. Candido, *J. Biol. Chem.* **258,** 3726 (1983).
[64] D. E. Krieger, R. Levine, R. B. Merrifield, G. Vidali, and V. G. Allfrey, *J. Biol. Chem.* **249,** 332 (1974).
[65] D. E. Kreiger, G. Vidali, B. W. Erickson, V. G. Allfrey, and R. B. Merrifield, *Bioorg. Chem.* **8,** 409 (1979).
[66] A. Kerbavon, J. Mery, and J. Parello, *FEBS Lett.* **106,** 93 (1979).
[67] J. H. Waterborg and H. R. Matthews, *Anal. Biochem.* **122,** 313 (1982).
[68] J. F. Riordan and B. L. Vallee, this series, Vol. 25, p. 494.
[69] G. Ramponi, G. Manao, and G. Camici, *Biochemistry* **14,** 2681 (1975).

the histones is ordinarily low, but it is enhanced greatly in the presence of sodium butyrate in the culture medium. Since butyrate inhibits deacetylase activity *in vivo* and *in vitro,* without inhibiting the acetyltransferases, the ensuing accumulation of hyperacetylated histone molecules permits their isolation in amounts sufficient for routine deacetylase assays. Methods for the preparation of radioactive and nonradioactive histone substrates with known levels of acetylation follow.

Log-phase HeLa cell cultures containing 5×10^5 cells/ml are exposed to 5 mM sodium butyrate for 21 hr and harvested. The cells are lysed and nuclei are isolated by method B of Krause[70] for maximizing yields of histones H3 and H4. Sodium butyrate (50 mM) is present throughout the isolation to inhibit deacetylase activity. The histones are extracted from the nuclear pellets in 0.4 M H$_2$SO$_4$, precipitated in 10 volumes of acetone, washed with acetone and ether, and dried.

In the preparation of histones containing radioactive acetyl groups, isotope incorporation as amino acids derived from [^{3}H]- or [^{14}C]acetate is an undesirable complication. For this reason we recommend preparation of [^{3}H]acetate-labeled histones from mature avian erythrocytes (which are incapable of protein synthesis but continue to acetylate histones[53] and HMG proteins[8]). Incubation conditions for labeling duck erythrocyte nuclear proteins with [^{3}H]acetate have been described.[8] Sodium butyrate should be present in the incubation medium and during isolation of the nuclei.[8]

Different classes of histones can be separated by chromatography on molecular-sieve matrices, such as BioGel P60.[71] Radioactive erythrocyte histone H4 prepared in this way is a useful substrate for the rapid assay of deacetylase activity, as described below.

For more detailed studies of the interconversions of modified histone forms by deacetylases, electrophoretic separations of the various acetylated species provide both defined substrates and the analytical means of assessing changes in the proportions of the individual acetylated forms.

The acetylated and nonacetylated forms of histones H3 and H4 are readily separated on a preparative scale[23] by electrophoresis in acid–urea–polyacrylamide gels.[18,20,72] Cylindrical gels (17 mm in diameter × 39 cm long), containing 63 ml of 15% polyacrylamide, 2.5 M urea, 0.9 M acetic acid, are prepared and cleared by electrophoresis at 3 mA/gel for 24 hr. Histone samples are dissolved at 10 mg/ml in 9 M urea, 0.9 M acetic acid, 1% 2-mercaptoethanol, and 200-μl aliquots (2 mg) are applied to each

[70] M. O. Krause, *Methods Cell Biol.* **17,** 51 (1978).
[71] C. von Holt and W. Brandt, *Methods Cell Biol.* **16,** 205 (1977).
[72] R. Hardison and R. Chalkley, *Methods Cell Biol.* **17,** 235 (1978).

preparative gel. Electrophoresis at 4.5 mA/gel separates all the acetylated forms of histone H4 in 60 hr. The various acetylated forms of histone H3 separate in 72 hr. For identification and recovery of the modified and unmodified proteins, the gels are removed from their tubes and sliced transversely into 13-cm sections. The cathodic section, containing the unacetylated H4 band and the more slowly moving acetylated forms, is cut lengthwise to yield a slice about one-sixth of the volume of the cylinder. This is stained with 0.1% Amido Black in 40% methanol–10% acetic acid, while the remaining gel is covered with glycerin and stored at −20° to minimize diffusion of the bands. After destaining in 40% methanol– 10% acetic acid, the minor slice is aligned with the major unstained slice and used as a template for excision of the individual histone bands. The slices from multiple gels are combined and placed in an Isenberg eluter (Mechanical Design Co., Corvallis, OR 47330) for electroelution at 10 mA for 48 hr. The proteins are collected in 5.5 ml of the running buffer (0.9 M acetic acid), dialyzed against water, and lyophilized. The purity of the isolated histone subfractions can be monitored by analytical gel electrophoresis and verified by the chemical procedures described above in the section Detection and Quantitation of ε-N-Acetyllysine.

Assay Procedures

The most efficient and quantitative assay procedure for deacetylase activity is that described by Cousens et al.,[32] which measures the release of [³H]acetate from histones acetylated in vivo.

Method. The standard assay contains in a 20-μl final volume, 9 μg of [³H]acetate-labeled histone substrate (ca. 10^3 cpm) dissolved in 15 mM potassium phosphate (pH 7) containing 4% (v/v) glycerol and 0.2 mM Na₃EDTA. We have modified the incubation medium by the inclusion of the protease inhibitors phenylmethylsulfonyl fluoride (PMSF) and 1,2-epoxy-3-(p-nitrophenoxy)propane (EPNP), both present at 0.1 mM.[23] After addition of the enzyme, the mixture is incubated at 37°, usually for periods of 1.5–15 min, at which time the reaction is stopped by the addition of a cold solution containing 0.5 N HCL, 0.1 M acetic acid, and 1 mg/ml of unlabeled histone (added as a carrier). The histones are precipitated by the addition of 40 μl of a saturated (at 0°) solution of ammonium tetrathiocyanodiammonochromate (ammonium Reinecke salt, Sigma). After 30 min at 0°, the histone precipitate is removed by centrifugation for 5 min in an Eppendorf 3200 centrifuge, and a 50-μl aliquot of the supernatant is counted by scintillation spectrometry.

An alternative procedure for the recovery of the released [³H]acetic acid employs selective extraction in ethyl acetate and direct counting of the organic phase.[63] An advantage of this procedure is that it precludes possible artifacts due to proteolytic degradation of the histone substrate and failure of the ammonium Reineckate to precipitate quantitatively small acetylated peptides.

In our experience, the assay procedures for deacetylases are subject to another source of error, namely, the ability of other enzymes, such as acetylcholinesterases, to remove [^{3}H]acetyl groups from the labeled histone substrates. The hydrolytic activity of the cholinesterases can be distinguished from protein deacetylase activities by its insensitivity to 100 mM sodium butyrate.[23]

In the absence of radioactively labeled substrates, the release of acetyl groups can be monitored by a rapid colorimetric procedure employed for acetylcholinesterase assays.[73] We have employed this procedure to measure the release of acetate from free acetyllysine and from acetylated histones by both deacetylases and acetylcholinesterase.[23] Direct chemical[27,39,30] or enzymatic[31] methods for acetate determination may also be employed.

None of the above assays can provide a full picture of deacetylase function, since acetate release from histones or HMG proteins is a multistep process for molecules containing more than one ε-N-acetyllysine residue and a single event for monoacetylated molecules. Measurements of total [^{3}H]acetate release, however useful for monitoring the purification of deacetylase activities, do not provide the resolution needed for the analysis of enzyme mechanisms or for following the interconversions of a complex population of fully acetylated, partially acetylated, and nonacetylated substrate molecules. The electrophoretic separation of the modified forms, combined with accurate densitometry[60,74,75] and fluorography,[32,76–78] can provide many of the insights necessary for understanding deacetylase kinetics and mode of action.

[73] F. Rappaport, J. Fischl, and N. Pinto, *Clin. Chim. Acta* **4,** 227 (1959), and Sigma Technical Bulletin 420 (1980).
[74] S. R. Grimes and H. Henderson, *Arch. Biochem. Biophys.* **221,** 108 (1983).
[75] J. Bode, M. M. Gomez-Lira, and H. Schroter, *Eur. J. Biochem.* **130,** 437 (1983).
[76] D. A. Nelson, *J. Biol. Chem.* **257,** 1565 (1982).
[77] W. M. Bonner and R. A. Laskey, *Eur. J. Biochem.* **46,** 83 (1974).
[78] R. A. Laskey and A. D. Mills, *Eur. J. Biochem.* **56,** 335 (1975).

[13] The N^ε-(γ-Glutamic)lysine Cross-Link: Method of Analysis, Occurrence in Extracellular and Cellular Proteins

By Ariel G. Loewy

The N^ε-(γ-glutamic)lysine[1] cross-link is a covalent bond formed between protein chains. It is catalyzed by a Ca^{2+}-dependent acyltransferase (EC 2.3.2.13, glutaminyl-peptide γ-glutamylyltransferase)[2] in which the γ-carboxamide groups of peptide-bound glutamine residues are the acyl donors and the ε-amino groups of peptide-bound lysine residues are the acyl acceptors (Fig. 1). The formation of this cross-link is a unique interaction between protein chains in that it does not occur in the absence of enzyme catalysis. Other examples of cross-link formation between peptide chains, such as sulfhydryl oxidation, sulfhydryl exchange, and cross-links involving aldol and aldimide groups, can occur spontaneously even though in some cases enzyme-mediated increases in reaction velocities have been observed.

The Glu–Lys cross-link has been identified in both extracellular and intracellular proteins. The frequency of protein chains cross-linked by Glu–Lys bonds is generally low. However, it appears that this isopeptide bond occurs in a large variety of cell types.

The basic strategy for the analysis of the Glu–Lys bond involves the liberation of N^ε-(γ-glutamyl)lysine from a protein preparation by exhaustive enzyme digestion followed by the chromatographic purification and measurement of the isodipeptide. The particular choices regarding the preparative procedures, the use of radioactive isotopes, the chromatographic methods of purification, and the techniques of measurement will depend on the material being studied. The role of this review is therefore seen as suggesting a variety of alternatives rather than recommending a single procedure for the measurement of the Glu–Lys cross-link.

Preparation of Material for Analysis

Before the Glu–Lys isodipeptide is purified from a protein digest and measured, a number of choices must be made regarding the fractionation

---

[1] This nomenclature is derived from recommendations of the IUPAC Commission on the Nomenclature of Organic Chemistry and the IUPAC–IUB Commission on Biochemical Nomenclature reprinted in *Biochemistry* **14**, 449 (¶1.4.3 on p. 450) (1975). For purposes of brevity, we shall refer to the Glu–Lys cross-link or bond.

[2] Systematic name: *R*-glutaminyl-peptide : amine γ-glutamylyltransferase. For purposes of brevity, we shall refer to the enzyme as transglutaminase.

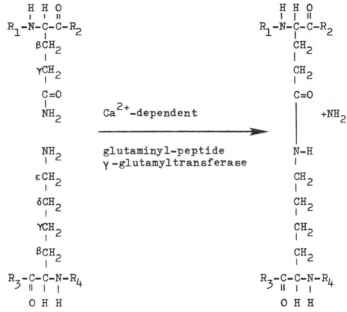

Fig. 1. Enzyme-catalyzed formation of Glu–Lys cross-link between peptide chains.

of the material being studied, the measurement of the protein sample, and the exhaustive enzyme digestion of the preparation.

Fractionation

Fractionation of a protein sample from a general cell extract not only serves to identify cell components that are Glu–Lys cross-linked, but also increases the Glu–Lys cross-link content of the material being studied. This facilitates greatly the purification and measurement procedures employed. It has been possible, however, to measure the Glu–Lys cross-link content of total cell extracts in order to determine how widely this form of interaction between protein chains is distributed in cell types of varying antiquity.[3]

A useful approach to fractionation depends on the fact that cross-linking brings about the formation of guanidine-HCl, urea, or sodium dodecyl sulfate (SDS)-insoluble matrices that can be purified by washing with these solvents.[4,5] Another method is to purify cross-linked chains by

[3] S. S. Matacic and A. G. Loewy, *Biochim. Biophys. Acta* **576**, 263 (1979).
[4] R. H. Rice and H. Green, *Cell* **11**, 417 (1977).
[5] J. L. Abernethy, R. L. Hill, and L. A. Goldsmith, *J. Biol. Chem.* **252**, 1837 (1977).

polyacrylamide gel electrophoresis,[6] and yet another method involves, as in the fibrin system, the purification of the substrate before the cross-linking reaction takes place.[7,8]

No general method for the purification of cross-linked protein chains can be described here; suffice it to point out that in general fractionation methods will depend on the relatively large size of the covalently cross-linked protein complexes

Measurement of Protein Sample

Because of the relative insolubility of cross-linked proteins, a convenient way of measuring them and preparing them for enzyme digestion is to precipitate them in 7% trichloroacetic acid, incubating at 4° for 30 min, centrifuging and resuspending in 5% trichloroacetic acid, heating at 90° for 10 min, extracting three times with ethanol–ether (1 : 1, v/v), three times with ether, and drying overnight in a vacuum desiccator in the presence of P_2O_5. The amount of protein can then be determined by direct weighing.

In the instances in which proteins have been labeled by the incorporation of radioactive lysine, the amount of protein can be measured by determining the radioactivity of an aliquot of enzyme-digested protein after the appropriate calibration measurement of the specific activity of the protein preparation has been made.[3]

Exhaustive Enzyme Digestion of Protein Sample

Methods for exhaustive enzyme digestion of the cross-linked protein sample must be adjusted to the tractability of the material being studied. It is possible to obtain close to 100% hydrolysis of fibrin with sequential enzyme digestion using Pronase twice, followed by leucine aminopeptidase.[7] On the other hand, some authors have utilized more elaborate digestion procedures involving trypsin, Pronase twice, leucine aminopeptidase, and prolidase for fibrin[8,9]; trypsin, pepsin, chymotrypsin, Pronase, leucine aminopeptidase, and carboxypeptidases A and B for hair and wool[10,11,11a]; pepsin, Pronase, aminopeptidase M, and prolidase for wool

[6] A. G. Loewy, S. S. Matacic, P. Rice, and J. Stern, Biochim. Biophys. Acta 668, 177 (1981).

[7] S. S. Matacic and A. G. Loewy, Biochem. Biophys. Res. Commun. 30, 356 (1968).

[8] J. J. Pisano, J. S. Finlayson, and M. P. Peyton, Science 160, 892 (1968).

[9] J. J. Pisano, J. S. Finlayson, and M. P. Peyton, Biochemistry 8, 871 (1969).

[10] R. S. Asquith, M. S. Otterburn, J. H. Buchanan, M. Cole, J. C. Fletcher, and K. L. Gardner, Biochim. Biophys. Acta 221, 342 (1970).

[11] H. W. J. Harding and G. E. Rogers, Biochemistry 10, 624 (1971).

[11a] K. Röper, J. Föhles, and H. Klostermeyer, this series, Vol. 106 [5].

medullary protein[12]; Pronase, leucine aminopeptidase, and carboxypeptidase A for coagulated protein of semen[13]; pepsin, Pronase, aminopeptidase M, and prolidase for the chorion of the egg of the rainbow trout[14]; Pronase, leucine aminopeptidase, and carboxypeptidases A and B for the insoluble residue of cultured human epidermal keratinocytes[15]; Pronase, cobaltous chloride-activated aminopeptidase, carboxypeptidase A, prolidase, and aminopeptidase M for the horny cell membrane of the epidermis of the newborn rat.[16]

In the above procedures, certain enzymes may be used more than once. Thymol was generally used as a bacteriostatic agent. Mixtures of enzymes without substrate were the control designed to determine whether or not the enzymes contributed to the presence of N^{ε}-(γ-glutamyl)lysine in the digest. Experiments designed to test whether or not (1) significant amounts of the Glu–Lys isodipeptide were liberated by the enzyme digestion and (2) new Glu–Lys cross-links were generated during the digestion process, will be discussed later.

Analytic Procedures

The choice of the analytic procedure to be used will depend on (1) the cross-link content of the protein being studied, (2) the sensitivity of the method of detection available, and (3) the suitability of the material for *in vitro* incorporation of [^{14}C]lysine into the protein being studied.

Sensitivity Considerations for the Direct Measurement of the Glu–Lys Isodipeptide

If the material being studied cannot be labeled by *in vitro* incorporation of [^{14}C]lysine, then the sensitivity requirement for the dipeptide measurement will depend on the amount of material available for a given analysis and its cross-link content as illustrated in Fig. 2. This nomogram shows, for instance, that if one has available only 10 μg of material with a cross-link content as low as one cross-link in 2×10^5 daltons of protein, one requires an analytical detecting system capable of measuring as little as 50 pmol of the Glu–Lys isodipeptide.

[12] H. W. J. Harding and G. E. Rogers, *Biochim. Biophys. Acta* **257**, 37 (1972).
[13] H. G. Williams-Ashman, A. C. Notides, S. S. Pabalan, and L. Lorand, *Proc. Natl. Acad. Sci. U.S.A.* **69**, 2322 (1972).
[14] H. E. Hagenmaier, I. Schmitz, and J. Foehles, *Hoppe-Seyler's Z. Physiol. Chem.* **357**, 1435 (1976).
[15] R. H. Rice and H. Green, *J. Cell Biol.* **76**, 705 (1978).
[16] K. Sugawara, *Agric. Biol. Chem.* **43**, 2543 (1979).

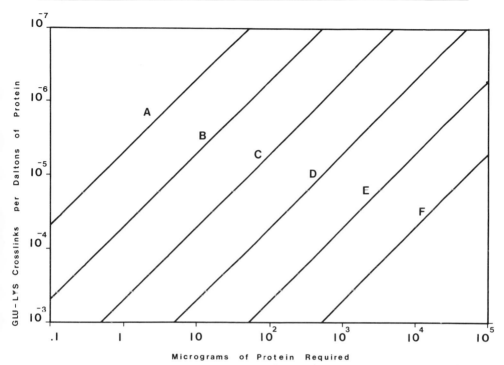

Fig. 2. Relationship between the sensitivity requirement for the dipeptide measurement, the amount of material needed for a given analysis, and the Glu–Lys cross-link content of the protein being studied. Sensitivity requirements are as follows: A = 5 pmol, B = 50 pmol, C = 500 pmol, D = 5 nmol, E = 50 nmol, F = 500 nmol.

If the material being studied can be readily labeled with [^{14}C]lysine, then analytical procedures of much lower sensitivity can be used.[3]

The Use of a ^{3}H-Labeled Glu–Lys Isodipeptide Marker

Whether or not the option of either high-sensitivity analysis or [^{14}C]lysine incorporation is to be used, it is desirable to employ a ^{3}H-labeled Glu–Lys isodipeptide marker. This facilitates both the identification and the quantification of the Glu–Lys isodipeptide liberated from the protein being studied.[7] Such a marker can be synthesized biochemically by incorporating ^{3}H lysine into casein with plasma or liver transglutaminase, enzymatically hydrolyzing the derivatized casein, and purifying the ^{3}H-labeled Glu-Lys isodipeptide. Since the α-amino group of lysine does not react in this incorporation, one obtains in this manner a pure preparation of the ε-amino derivative. The incorporation of free lysine is very slow when compared with that of an esterified amino acid, and therefore high levels of enzyme and long periods of incubation must be used.

The incubation mixture contains 20 mg of casein, 4 mg of factor XIII (fraction 5)[17] or an equivalent amount of liver transglutaminase,[18] 5 mCi of L-[^{3}H]lysine (1–5 Ci/mmol), 15 μm of $CaCl_2$, 0.1 mg of thrombin (200 Iowa units per milligram of nitrogen), 15 μmol of mercaptoethanol, in 3 ml of 0.3 Tris buffer, pH 7.5.

This mixture is incubated for 24 hr at 30°, after which 2 mg of additional factor XIII is added and the incubation is continued for another 24 hr. The labeled casein is then purified by precipitation with 10% trichloroacetic acid for 20 min at 4° and then repeatedly washed with 5% TCA until the supernatant is free of [^{14}C]lysine. The precipitate is then washed 5 times with 5 ml of ethanol–ether (1 : 1) followed by two washes with 5 ml of ether, and finally dried under vacuum over P_2O_5. Activities of 10^6 to 10^7 cpm per milligram of casein are obtained.

The ^{3}H-labeled Glu–Lys isodipeptide is released from the casein by enzyme hydrolysis as follows: Pronase (Sigma Chemical Company) is added to a 0.01 M Tris buffer, pH 7.5, solution of casein in a substrate-to-enzyme ratio of 10 : 1 and incubated for 24 hr at 37°. The pH is then readjusted to 7.5, and the same quantity of Pronase is added again. The Pronase is inactivated by boiling for 1 min, the pH is adjusted to 8.0, and $MgCl_2$ is added to 5 mM final concentration, then leucine aminopeptidase (Worthington Biochemical Corp.) is added to a substrate-to-enzyme ratio of 5 : 1 and digestion is allowed to proceed for 48 hr at 37°. The mixture is then boiled for 1 min and centrifuged. Chromatographic purification of the Glu–Lys isodipeptide from the supernatant yields a recovery of 85% of the original radioactivity. The purified product can be shown to comigrate chromatographically with authentic N^ε-(γ-glutamyl)lysine (Accurate Chemical and Scientific Corp.).

Chromatographic Purification and Measurement of the N^ε-(γ-Glutamic)lysine Isodipeptide

The Glu–Lys isodipeptide can be purified from the enzyme-hydrolyzed amino acid mixture by (1) paper[4] or thin-layer[15] chromatography, (2) ion-exchange column chromatography,[7–9,16,19,20] and (3) reverse-phase, high-performance liquid chromatography (HPLC).[21]

[17] A. G. Loewy, K. Dunathan, R. Kriel, and H. L. Wolfinger, Jr., *J. Biol. Chem.* **236**, 2625 (1961).
[18] T. Abe, S. Chung, R. P. DiAugustine, and J. E. Folk, *Biochemistry* **16**, 5495 (1977).
[19] M. S. Otterburn and W. J. Sinclair, *J. Sci. Food Agric.* **27**, 1071 (1976).
[20] K. Sugawara and T. Ouchi, *Agric. Biol. Chem.* **46**, 1085 (1982).
[21] M. Griffin, J. Wilson, and L. Lorand, *Anal. Biochem.* **124**, 406 (1982).

1. Paper or thin-layer chromatography can be of value for detecting the Glu–Lys isodipeptide when prior fractionation of the protein preparation results in a relatively high Glu–Lys content[4] or if the lysines in the proteins have been radioactively labeled.[15]

2. By far the largest number of separations of the Glu–Lys isodipeptide have been made with column ion-exchange chromatography. It is not possible here to describe the large variety of columns and elution procedures that have been used. The advantage of column ion-exchange chromatography is that automated apparatus for the measurement of amino acids and small peptides is available in many laboratories. The limitation of the method is that the resolution of the Glu–Lys isodipeptide from neighboring amino acid peaks is such that at low dipeptide : amino acid ratios two column chromatographic passages are required[7,9] to remove significant amounts of amino acid impurities.

The sensitivity of ion-exchange column chromatographic methods depends in great part on the detection system. With the ninhydrin reaction, most contemporary automatic systems can measure 5 nmol of isodipeptide. By using a fluorogenic reagent such as o-phthalaldehyde (OPA), 50 pmol can be measured.

3. Perhaps the most effective method for the purification and detection of the Glu–Lys isodipeptide is the use of HPLC equipment, reverse-phase chromatography, and precolumn derivatization with OPA.[21] By using more recent equipment than that described by Griffin et al.,[21] 5–10 pmol sensitivity is possible.[21a] This development in the technology of amino acid and peptide chromatography now permits us to work with microgram quantities of protein chains purified by acrylamide gel electrophoresis. Because of this, one may expect in the near future a great deal of work in cell biologically relevant aspects of Glu–Lys cross-link formation.

In the absence of modern HPLC equipment, picomole quantities of Glu–Lys isodipeptide can also be measured by incorporating [^{14}C]lysine into the protein under study.[3,4,6,22,23] By the use of a ^{3}H-labeled Glu–Lys isodipeptide marker added to the protein solution during enzyme digestion and by performing accurate double-label counts, it is possible to enhance greatly the precision of measurement. This double-layer method has made it possible to elute protein chains from acrylamide gels and measure the cross-link content of an 80 kilodalton polypeptide chain (2 nmol of Glu–Lys per milligram of protein).[6] The intensity and amount of

[21a] A. G. Loewy and N. M. Taggart, unpublished.

[22] P. J. Birckbichler, R. M. Dowben, S. S. Matacic, and A. G. Loewy, Biochim. Biophys. Acta 291, 149 (1973).

[23] A. G. Loewy and S. S. Matacic, Biochim. Biophys. Acta 668, 167 (1981).

radioactive label that must be incorporated into a given tissue depends on a variety of considerations relating to the culture conditions and metabolic properties of the material being studied, its cross-link content, the chromatographic equipment used, and the scintillation counting apparatus available. It is therefore not possible to specify in advance the precise experimental design one must utilize to carry out a double-label analysis such as the one we describe above.

The Unambiguous Characterization of the N^{ε}-(γ-Glutamyl)lysine Isodipeptide

The mobility of the isodipeptide in a given chromatographic separation is insufficient evidence as to the identity of the molecule. The following procedures can and have been employed that together result in an unambiguous identification of the compound.

1. Acid hydrolysis of the presumptive N^{ε}-(γ-glutamyl)lysine peak and demonstration that the product is lysine and glutamic acid in equimolar quantities.[4,5,7]

2. Comigration of the presumptive isodipeptide in ion-exchange column chromatography with biochemically synthesized ^{3}H-labeled N^{ε}-(γ-glutamyl)lysine, which itself had been shown to comigrate with authentic N^{ε}-(γ-glutamyl)lysine.[5,7,8]

3. Comigration of ^{3}H-labeled N^{ε}-(γ-glutamyl)lysine marker and ^{14}C-labeled isodipeptide from [^{14}C]lysine-labeled protein digests.[3,6,23]

4. Comigration in thin-layer chromatography of ion-exchange chromatographically purified Glu–Lys isodipeptide with authentic N^{ε}-(γ-glutamyl)lysine of the presumptive isodipeptide.

5. Dansylation of the presumptive Glu–Lys isodipeptide,[24] followed by acid hydrolysis and thin-layer chromatography (or paper electrophoresis), showing only dansylglutamic acid and α-monodansyl lysine in the hydrolysate.[2,16]

6. Demonstration that authentic N^{α}-(α-glutamyl)lysine and N^{α}-(γ-glutamyl)lysine has very different chromatographic properties from N^{ε}-(γ-glutamyl)lysine.[4,24]

Estimating the Percentage of Hydrolysis

Exhaustive enzyme hydrolysis has been reported in a variety of studies to have accounted for 95–100% of the protein sample. As we shall see presently, a precise estimate of the amount of hydrolysis is important because it has bearing on the issue of measurement artifacts.

[24] E. Squires, L. Feltham, and A. Woodraw, *Biochem. J.* **185**, 761 (1980).

A convenient strategy for measuring the degree of enzyme hydrolysis of a protein is to determine total protein of two equivalent samples and then compare the amino acid content of the enzyme-hydrolyzed sample with that of an acid-hydrolyzed control sample. It is useful to measure total protein by direct weighing of samples that have been dehydrated by washing with anhydrous ether and storing under vacuum over P_2O_5. The amino acid content of the enzyme hydrolysate can be measured during the Glu–Lys isodipeptide purification procedure and can then be compared with that of an equivalent chromatographic separation of the acid hydrolysate. A convenient alternative procedure to measuring each amino acid is to inject a known aliquot of the digest into the analyzer and bypassing the chromatographic column, thereby measuring the total amino acid concentration.

An important source of uncertainty in the above approach is the contribution of self and mutual digestion of the proteolytic enzymes. This effect can be evaluated by a control experiment consisting of enzymes without protein substrates. We found that when only Pronase and leucine aminopeptidase are used, the levels of amino acid contribution from these enzymes are insignificantly low. In experiments using a greater variety of enzymes, it is likely that precise measurements of percentage of hydrolysis become more difficult.

A way of circumventing the problem of amino acid contamination derived from the autodigestion of the proteolytic enzymes is to incorporate [^{14}C]lysine into the protein being studied. It is then necessary only to measure the specific activity of the lysine in a sample of the protein, after which the measurement of the radioactivity of the lysine peak will yield an accurate estimate of the amount of protein digested irrespective of the contamination contributed by the proteolytic enzymes.

Possibilities of Artifacts in the Measurement of Glu–Lys Cross-Links

There are two classes of possible artifacts that may attend the above procedures of Glu–Lys measurement.

1. *Incomplete hydrolysis of the protein sample* may mask the presence of some Glu–Lys cross-links remaining in the unhydrolyzed residue. If this effect is random so that the Glu–Lys cross-link frequency in the unhydrolyzed residue is no larger than that of the 95–98% hydrolyzed moiety, we have an artifact that is well within the experimental error of the method of analysis. If, however, Glu–Lys cross-linking of protein chains interferes with complete hydrolysis, then one can imagine that the 2–5% unhydrolyzed residue is in fact enriched with Glu–Lys bonds. A great deal of analytic work is required to eliminate such a possibility. The

most direct approach is to dansylate or dinitrophenylate the unhydrolyzed moiety of the protein, hydrolyze it, and then check whether any free lysine appears in the hydrolysate. As an alternative to dansylation or dinitrophenylation, it is possible to use cyanoethylation with acrylonitrile.[8,9] The above methods for derivatizing amino groups may not be sufficiently precise for detecting one cross-linked lysine in 100 free lysines in an unhydrolyzed protein. However, in a situation in which one tests for the possible presence of a relatively high amount of cross-linked lysine, as possibly in the 2–5% unhydrolyzed remains of a protein, the above methods of derivatizing free lysine ε-amino groups are likely to be effective. This is expecially so if one is concerned with eliminating the possibility of an artifact involving an enrichment of Glu–Lys cross-links in the unhydrolyzed remains of a protein preparation.

2. *The generation of Glu–Lys cross-links* during enzyme hydrolysis is the other possible source of artifact inherent in the method of analysis. The possibility of this occurring can be eliminated only by demonstrating the absence of significant levels of Glu–Lys cross-links in control proteins known not to contain Glu–Lys cross-links.[3] It is important to carry out such a control experiment whenever new enzymes, or even new enzyme preparations, are used.

On the whole, no evidence of substantial amounts of artifactually formed Glu–Lys cross-links has so far been reported. Further evidence that these cross-links do not form during enzyme hydrolysis comes from studies of known protein sequences. For instance, in the use of the γ-dimer of fibrin, in which the primary structure of the chain has been determined, two previously defined cross-links have been identified irrespective of conditions of hydrolysis.[25]

Another category of evidence showing that Glu–Lys cross-links are not generated artifactually in significant amounts during the exhaustive enzyme digestion of the protein sample comes from cyanoethylation studies with acrylonitrile, a reagent that has been used very effectively in proteins such as fibrin containing relatively high amounts of cross-links.[8,9,26] In this procedure, free amino groups, including the ε-amino groups of lysine, are allowed to react with acrylonitrile. The derivatized protein is then acid-hydrolyzed, and this releases the lysines involved in isopeptide cross-links as free lysines. The limitation of this method is that it is not specific for the Glu–Lys bond, since it includes in its measurement any inaccessible lysine ε-amino group. It also gives relatively high

[25] J. E. Foïk and J. S. Finlayson, *Adv. Protein Chem.* **31**, 1 (1977).

[26] J. J. Pisano, T. J. Bronzert, M. P. Peyton, and J. S. Finlayson, *Ann. N. Y. Acad. Sci.* **202**, 98 (1972).

values with proteins containing low levels of Glu–Lys cross-links.[22] However, good correspondence between cross-linked lysines measured by the cyanoethylation technique and direct measurement of Glu–Lys cross-links have been reported,[8,9,26] and this can be taken as evidence that the latter are not generated in significant amounts during exhaustive enzyme digestion.

Occurrence in Extracellular Proteins

The Glu–Lys cross-link was first discovered in vertebrate fibrin,[7,8] an extracellular polymer involved in blood clotting. We now know a great deal about the identity of the polypeptide chains and the precise Glu–Lys bonds involved in this polymerization process.[25] Later, it was demonstrated in the lobster that clotting occurs when fibrinogen is cross-linked with Glu–Lys bonds.[27] This interesting observation suggests that the Glu–Lys-mediated polymerization is an evolutionarily more ancient phenomenon in the blood-clotting system than the aggregation reaction induced by thrombin. Another clotting reaction involving a basic protein in the seminal fluid of guinea pigs, rats, and mice also appears to be related to Glu–Lys cross-link formation.[13] It is likely that other extracellular gelation reactions will be found to involve the formation of transglutaminase-catalyzed Glu–Lys cross-links.

A second category of extracellular proteins polymerized, at least in part, by the formation of Glu–Lys cross-links is the tough and insoluble medulla of hair and quill. Glu–Lys cross-link measurements have been made in a number of species (rat, guinea pig, rabbit, camel, porcupine) and found to be extremely high (one cross-link per 4–50 kilodaltons of protein).[12] Lower, but significant, levels of Glu–Lys cross-links have been found in keratin fractions of merino sheep wool[10] (approximately 1 cross-link per 60,000 of protein).

Finally, a third category of Glu–Lys cross-linked proteins is found among the protein polymers endowing skin with its protective qualities. Here we find that high molecular weight fractions of α-keratin isolated from human stratum corneum contain approximately one cross-link per 110,000 daltons of protein.[10]

In the case of blood and semen coagulation, the process of cross-link formation is entirely extracellular and reasonably well understood. On the other hand, in the formation of hair and the protective layer of skin, the process of enzyme-catalyzed polymerization begins in living cells or in

[27] G. M. Fuller and R. F. Doolittle, *Biochemistry* **10,** 1311 (1971).

close juxtaposition to them. In these latter cases, we do not as yet have a very clear understanding of the processes by which these important protective structures are assembled.

Occurrence in Intracellular Proteins

In the ensuing survey of Glu–Lys cross-links in intracellular proteins, we shall consider only reports in which the presence of these bonds have been ascertained by direct measurement.

Total Intracellular Protein

While concern with biological function and operational convenience would dictate that the search for Glu–Lys cross-links be carried out with large, polymerized protein fractions, it is also of interest to determine the distribution of this cross-link in the total protein of the cell in organisms of varying evolutionary antiquity.[3] By incorporating radioactive lysine into cellular proteins, it was possible to measure low cross-link concentrations ranging from one cross-link in 10^6 daltons to one cross-link in 10^7 daltons of protein.[3] Assuming an average protein chain size of 40,000 daltons, one concludes that only one in every 25–250 protein chains is cross-linked with Glu–Lys bonds. Even though the cross-link content of cell proteins is relatively low, Glu–Lys bonds seem to be widely distributed in the world of life. They are present in the prokaryote *Escherichia coli,* in the protozoan *Paramecium aurelia,* in the slime mold *Physarum polycephalum,* in tissue-cultured embryonic chick skeletal muscle cells, and in mouse skeletal muscle cells.[3]

Modulation in Muscle Cells[6,23]

The Glu–Lys cross-links of embryonic chick muscle cells are entirely localized in the glycerol-extracted residue. Unlike the cross-links of extracellular proteins, the cross-link content of muscle can be rapidly modulated: (1) Lowering the temperature of tissue-cultured skeletal cells from 37° to 4° significantly lowers the cross-link content. (2) Treatment with Mg^{2+} ATP of glycerol-extracted tissue-cultured skeletal cells also lowers the cross-link content. (3) Treatment with Mg^{2+} ATP followed by Ca^{2+} of glycerol–extracted tissue-cultured heart cells raises the cross-link content. (4) Ca^{2+} alone, without prior Mg^{2+} ATP-activation, has no effect on cross-link content.

The rapid modulation of Glu–Lys cross-links in intracellular proteins suggests that they participate actively in the physiology of the cell. Thus, the single free energy-lowering reaction suggested for the unidirectional polymerization of extracellular proteins (Fig. 1) cannot accommodate the rapid modulation of the intracellular systems observed in muscle. To explain these rapid modulations, a reaction cycle has been proposed in which an Mg^{2+} ATP activation of the γ-carboxyl group of the glutaminyl side chain is followed by a Ca^{2+}-activated cross-linking reaction that in turn is followed by a hydrolytic or phosphorylitic cleavage of the isopeptide bond.[6] The enzyme catalyzing the cross-linking reaction involving a presumptive high-energy intermediate derivative of the glutaminyl carboxyl group may not be the intracellular transglutaminase generally assayed by standard Ca^{2+}-activated amine incorporation assays. The criterion for distinguishing between such a transglutaminase and the one hypothesized above is the requirement for Mg^{2+} ATP-activation. Note that, in the intracellular studies reported below, it is not generally possible to distinguish between enzyme reactions that do require and those that do not require activation with Mg^{2+} ATP.

Fractionation studies of the protein chains in glycerol-extracted embryonic chick heart cells show that most of the Glu–Lys cross-link modulation occurs in very large, highly polymerized fractions or matrices.[6,28] Recent work shows that these matrices seem to be continuous and appear to occupy the entire volume of mature skeletal muscle cells. Thus, when glycerinated chicken breast muscle is extracted exhaustively with 6 M guanidine HCl and 5% 2-mercaptoethanol, a "tissue ghost" consisting of 1% of the total protein remains. This structure, when treated with high-purity collagenase, falls apart into "muscle fiber ghosts" containing 0.2% of the total protein (Figs. 3 and 4).[29] These muscle fiber ghosts consist of a Glu–Lys cross-linked intracellular matrix, the properties of which have yet to be explored in detail.

Membranes and Membrane Proteins

One of the earliest measurements of Glu–Lys bonds in cellular proteins was carried out with preparations of plasma membranes and of endoplasmic reticulum membranes of mouse fibroblast L cells.[22] Cross-link contents of one cross-link in 600 kilodaltons of protein were reported for plasma membrane preparations. Much lower cross-link concentrations (1500–2300 kilodaltons of protein) were measured in endoplasmic reticulum preparations.

[28] A. G. Loewy, in "Biological Structures and Coupled Flows," (A. Oplatka and M. Balaban, eds.). Academic Press, New York, and Balaban Press, Philadelphia, 1983.

[29] A. G. Loewy, F. J. Wilson, N. M. Taggart, E. A. Greene, P. Frasca, H. S. Kaufman, and M. J. Sorrell, J. Cell Motility 3, 463 (1983).

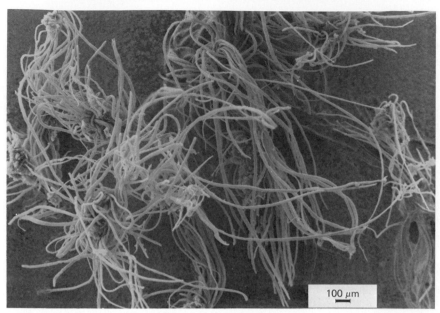

FIG. 3. Scanning electron micrograph of muscle fiber ghosts prepared by exhaustive extraction of glycerinated adult chicken pectoral muscle with 6 M guanidine–HCl, 5% 2-mercaptoethanol, and 0.1 M sodium carbonate, pH 8.5. The tissue ghost containing 1% of the total protein is then treated with high-purity collagenase. The muscle fibers, thereby released as separate entities, contain 0.2% of the total tissue protein. The Glu–Lys cross-linked intracellular matrix of the muscle fiber determines the shape of the muscle fiber ghost even though, in its dehydrated state, it occupies only 1/50 of the volume.

In erythrocyte ghosts, frozen and thawed several times before use, guinea pig liver transglutaminase brings about the cross-linking of a number of membrane proteins.[30] This contrasts with sarcoplasmic reticulum preparations, in which the cross-linking is minimal under similar conditions.

It appears that Glu–Lys cross-linking reactions of membrane proteins can occur also in the living state. In human erythrocytes,[31,32] a rise of internal Ca^{2+} brings about a polymerization reaction in membrane proteins with a concomitant disappearance of band 4.1 and a reduction in spectrin and band 3 materials.[33] The isolated protein polymer is rich in

[30] A. Dutton and S. J. Singer, *Proc. Natl. Acad. Sci. U.S.A.* **72,** 2568 (1975).
[31] G. E. Siefring, Jr., A. B. Apostol, P. T. Velasco, and L. Lorand, *Biochemistry* **17,** 2598 (1978).
[32] L. Lorand, G. E. Siefring, Jr., and L. Lowe-Krentz, *J. Supramol. Struct.* **9,** 427 (1978).
[33] For nomenclature of erythrocyte membrane protein chains, see G. Fairbanks, T. L. Steck, and D. F. H. Wallach, *Biochemistry* **10,** 2606 (1971).

10 µm

FIG. 4. Higher magnification image of muscle fiber ghosts shown in Fig. 3.

Glu–Lys cross-links containing as much as one Glu–Lys cross-link in 16 kilodaltons of protein. The formation of the polymer is inhibited by water-soluble primary amines that also inhibit the Ca^{2+}-induced stiffening often observed in erythrocytes. The authors therefore conclude that the loss of deformability in erythrocytes may be brought about by an *in vivo* transglutaminase-catalyzed cross-linking reaction of membrane proteins. Since the cross-linking reaction observed in erythrocytes occurs on the cytoplasmic side of the membrane, we do not as yet know whether or not this is a Mg^{2+} ATP-dependent reaction.

Eye Lens Tissue

Another example of what may well be a polymerization reaction related to aging has been observed in cataracts of human eye lens tissue. Here, a Glu–Lys cross-link content of one cross-link in 200 kilodaltons of protein has been measured.[34]

[34] L. Lorand, L. K. H. Hsu, G. E. Siefring, Jr., and N. S. Rafferty, *Proc. Natl. Acad. Sci. U.S.A.* **78**, 1356 (1981).

Chorion of Fertilized Rainbow Trout Eggs

An interesting involvement of Glu–Lys cross-linking in bringing about increased mechanical stability of the egg chorion of the rainbow trout has been observed. Before fertilization, no Glu–Lys cross-links can be detected, whereas after fertilization a Glu–Lys cross-link content of one cross-link per 22 kilodaltons of protein can be measured.[35]

Epidermal Differentiation

One of the most interesting and well-studied areas of intracellular Glu–Lys cross-link formation relates to the differentiation of epidermis.

Located below the plasma membrane of cells in the stratum corneum is the cornified envelope layer, which is insoluble in protein solvents containing a disulfide reducing reagent.[36] Proteolytic enzymes solubilize the cornified envelope, showing that the coherence of this structure is protein dependent.[36] Analysis of cornified envelopes for Glu–Lys cross-links revealed that approximately 18% of the total lysine is involved in cross-link formation.[36] Another series of measurements of insoluble structures (horny membranes) in the skin of newborn rats, in cow snouts, and in human stratum corneum, give values of one Glu–Lys cross-link in 32, 48, and 24 kilodaltons, respectively.[16,20] Two studies have attempted to identify soluble cytoplasmic precursors of the cross-linked cornified envelope. In one of these, convincing immunological evidence is presented that a 92-kilodalton protein chain accumulating below the plasma membrane is the precursor of the insoluble matrix in the cornified envelope.[37] In the other study, a 36-kilodalton protein has been identified immunologically and shown to be a substrate for an *in vitro* transglutaminase reaction.[38]

The developmental processes involved in the terminal differentiation of epidermis can be studied in cultured keratinocytes. It is possible to show in such cultured cells the development of Glu–Lys cross-linked cornified envelopes. It can also be demonstrated that cross-linking begins after protein synthesis has sharply declined.[39] Furthermore, cross-linking can be induced by inhibiting protein synthesis[39] as well as by raising the internal Ca^{2+} concentrations of the cells in culture.[40] Reagents known to

---

[35] H. E. Hagenmaier and J. Föhles, *Verh. Anat. Ges.* **72,** 319 (1978).
[36] R. H. Rice and H. Green, *Cell* **11,** 417 (1977).
[37] R. H. Rice and H. Green, *Cell* **18,** 681 (1979).
[38] M. M. Buxman, C. J. Lobitz, and K. D. Wuepper, *J. Biol. Chem.* **255,** 1200 (1980).
[39] R. H. Rice and H. Green, *J. Cell Biol.* **76,** 705 (1978).
[40] H. Hennings, P. Steinert, and M. M. Buxman, *Biochem. Biophys. Res. Commun.* **102,** 739 (1981).

inhibit transglutaminase activity inhibit the cross-linking reaction.[40] Keratin filaments found abundantly in the cells of the stratum corneum were not found to be Glu–Lys cross-linked.[41] It is likely that the study of keratinocytes in culture will prove to be a productive approach to the study of assembly processes of covalent cellular matrices.

Normal and Transformed Cells

Another developmental approach to the assembly of Glu–Lys cross-linked matrices is the comparison of normal and virus-transformed cells. Normal human lung cells (WI-30) have been shown to possess 20–40 times more Glu–Lys cross-links than their simian virus-transformed counterparts.[42] Furthermore, normal cells arrested in a nondividing state have more Glu–Lys bonds than do rapidly dividing cells. It may well be that during division there is a disassembly of the presumptive Glu–Lys cross-linked matrix.

Conclusions

Procedures, involving exhaustive enzyme digestion, for the analysis of Glu–Lys cross-links in proteins have been found to be free of significant levels of artifact. By labeling the protein under study, or by using modern HPLC equipment, it is now possible to measure the cross-link content of protein chains purified by acrylamide gel electrophoresis.

A number of extracellular proteins with biological functions related to their inertness or their mechanical strength form Glu–Lys bonds by a unidirectional polymerization process. Such a mechanism may be involved also in the differentiation of skin and in cell aging. However, there appear to be other cell processes in which Glu–Lys cross-link formation and breakdown are modulated rapidly. These mechanisms, distinguished by their requirement for Mg^{2+} ATP activation prior to polymerization, are likely to play an active role in the physiology of the cell.

A Glu–Lys cross-linked intracellular matrix extending throughout the entire skeletal muscle fiber has been described. It is possible that such covalently cross-linked matrices are present also in other cell types.

[41] T.-T. Sun and H. Green, *J. Biol. Chem.* **253**, 2053 (1978).
[42] P. J. Birckbichler, H. A. Carter, G. R. Orr, E. Conway, and M. K. Patterson, Jr., *Biochem. Biophys. Res. Commun.* **84**, 232 (1978).

[14] N^ε-(β-Aspartyl)lysine

By H. Klostermeyer

Whereas posttranslational cross-linking of peptide chains by N^ε-(γ-glutamyl)lysine bonds is widely distributed in nature, N^ε-(β-aspartyl)lysine cross-linking is generally found only in heated proteins (for reviews, see footnotes 1 and 2). Unlike the glutamyllysyl isopeptide bond, the N^ε-(β-aspartyl)lysine link is not hydrolyzed by homogenates of mucosa, liver, or kidney of rat,[3] and it has been concluded that the aspartyllysine isopeptide is not nutritionally available to the animal and therefore should be secreted.[4] Among the large number of ninhydrin-positive constituents of normal human urine, several laboratories have indeed identified β-aspartyl peptides,[5-7] including N^α-(L-β-aspartyl)L-lysine.[8]

N^ε-(β-Aspartyl)lysine has also been found in urine of children with various pathological conditions,[9] but its presence does not appear to be related to specific symptoms. The occurrence of β-Asp-ε-Lys links in native wool is doubtful.[10] At this time there is no definitive proof that the Asp Lys isopeptide is a product of a posttranslational enzyme-catalyzed reaction.

It is known that proteins from blood serum as well as from membranes, e.g., enzymes like γ-glutamyltransferase,[11] may accumulate in colostrum, and it is natural to predict that products of the reactions catalyzed by these enzymes might also be detected in colostrum. When colostrum was screened for the presence of isopeptide bonds not only the γ-

[1] R. S. Asquith, M. S. Otterburn, and W. J. Sinclair, *Angew. Chem.* **86,** 580 (1974).
[2] M. Otterburn, M. Healy, and W. Sinclair, *Adv. Exp. Med. Biol.* **86B,** 239 (1977).
[3] P. F. Finot, F. Mottu, E. Bujard, and J. Mauron, *Adv. Exp. Med. Biol.* **105,** 549 (1978).
[4] P. F. Finot, E. Maquenat, F. Mottu, E. Bujard, R. Deutsch, C. Dormond, A. Isely, L. Pignat, and R. Madelaine, *Ann. Nutr. Aliment.* **32,** 325 (1978).
[5] Y. Kakimoto and M. D. Armstrong, *J. Biol. Chem.* **236,** 3280 (1961).
[6] D. L. Buchanan, E. E. Haley, and R. T. Markiw, *Biochemistry* **1,** 612 (1962).
[7] J. J. Pisano, E. Prado, and J. Freedman, *Arch. Biochem. Biophys.* **117,** 394 (1966).
[8] M. F. Lou, *Biochemistry* **14,** 3503 (1975).
[9] J. P. Kamerling, G. Aarsen, P. K. de Bree, M. Duran, S. K. Wadman, J. F. G. Kliegenhart, G. I. Tesser, and K. M. van de Waarde, *Clin. Chim. Acta* **832,** 85 (1978).
[10] K. Röper, J. Föhles, and H. Klostermeyer, this series, Vol. 106, p. 58.
[11] K. Yasumoto, K. Iwami, T. Fushiki, and H. Mitsuda, *J. Biochem. (Tokyo)* **84,** 1227 (1978).

Glu-ε-Lys link, but also the β-Asp-ε-Lys link, could be detected,[12] however, only in a limited number of samples (two of seven). It must be kept in mind that the methods used to detect isopeptides include complete enzymatic digestion of proteins, requiring long-time treatments at pH 2.0 and 8.2.[10] Under these conditions the rather labile β-Asp-amide bond may well be hydrolyzed to a certain extent, and it is thus likely that small amounts of this derivative are overlooked.

Identification in Colostrum

Procedure. The proteins are completely digested by proteases,[10] and the hydrolysates are checked for Asp Lys (and Glu Lys) with an amino acid analyzer. Samples with peaks in the positions of the isopeptides are fractionated by ion-exchange chromatography. Fractions collected between the elution peaks of Leu and Tyr are hydrolyzed with hydrochloric acid and analyzed for Asp and Lys (or Glu and Lys, respectively) for identification.

Preparation of Samples. Fresh colostrum up to the third day of lactation, defatted by centrifugation, was used. One milliliter (about 50 mg of protein) was diluted with 4 ml of water and adjusted to pH 2.0 with 0.5 M HCl (total volume about 8 ml). One milligram of pepsin (crystalline, ~100 mU/mg) and a crystal of thymol were added, and the digestion mixture was incubated at 40° for 24 hr, at which time the pH was adjusted with about 1 ml of 2 M Tris to 8.2, and 1 mg of Pronase E (~70 PUK/mg) was added. After an additional 24 hr at 40°, 1 mg of aminopeptidase M (~10,000 mU/mg) and a suspension of 1 mg of prolidase (~200 U/mg) were added, and the digest was again incubated for 24 hr at 40°. The lypholized hydrolysate was finally dissolved for analysis in 0.35 M lithium citrate buffer, pH 2.2.

Ion-Exchange Chromatography.[13] To detect the presence of isopeptides in proteolytic hydrolysates, we used an amino acid analyzer with an elution program for complete separation of the common amino acids and the isopeptides: column, 58 × 0.9 cm; resin, Aminex A-6; elution buffer flow, 80 ml/hr; ninhydrin reagent flow, 40 ml/hr. The elution program was 120 min for buffer A (pH 3.26–3.27) at 40°, 100 min for buffer B (pH about 3.52) at 40° and 60 min at 55°. The buffers were prepared (2 liters) from 12.94 g of LiOH (98%), 32 g of citric acid, 5 ml of thiodiglycol, 9 ml of a

[12] H. Klostermeyer, K. Rabbel, and E. H. Reimerdes, *Hoppe-Seyler's Z. Physiol. Chem.* **357,** 1197 (1976).
[13] I. Schmitz, H. Zahn, H. Klostermeyer, K. Rabbel, and K. Watanabe, *Z. Lebensm.-Unters.-Forsch.* **160,** 377 (1976).

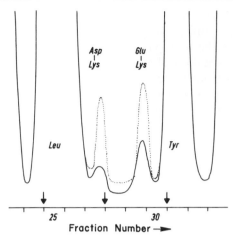

FIG. 1. Part of a chromatogram from an enzymatic digest of colostrum. ——, Original hydrolysate; ····, after addition of the synthetic isopeptides. Scale: mode of fractionation.

Brij-35 solution (20%), 0.2 ml of caprylic acid, and about 28–30 ml of HCl (37%) to adjust the pH. Final, precise pH adjustments were made with the aid of actual test runs, using standard amino acid calibration mixtures with addition of two isopeptides. Buffer B was standardized to a pH at which Asp Lys is eluted after Leu and Glu Lys before Tyr (about 120 and 180 min elution time). The column was regenerated and equilibrated with 0.2 M NaOH and buffer A. For samples with elution peaks in the position of Asp Lys a second chromotogram was run, connecting the ion-exchange column with a fraction collector instead of to the ninhydrin line, and fractions of about 7.5 ml were collected. Fractions 25–27 contained Asp Lys plus some Leu; fractions 28–30 contained Glu Lys plus some Tyr (see Fig. 1). The fractions of interest were dried on a rotatory evaporatory and hydrolyzed with 5 ml of 6.2 M HCl for 24 hr at 110°. After evaporation, the samples were dissolved in buffer, pH 2.2, and analyzed for amino acids.

If Asp and Lys (contaminated only by Leu) are found in a ratio of 1 : 1, this is considered to be strong evidence for the presence of N^ε-(β-aspartyl)lysine in the unknown samples; the other possible peptides, Asp-Lys and Lys-Asp (also Glu-Lys and Lys-Glu) are eluted in positions after Phe on this chromotographic system.

Reference Substances. For precise calibration of the ion-exchange system, synthetic isopeptides are very useful. A well-established method of synthesis is the coupling of the benzyloxycarbonyl-α-benzyl esters of aspartic acid or glutamic acid via mixed anhydrides to the copper complex of lysine and splitting off the protecting groups by catalytic hydrogenation or with HBr in acetic acid.[14,15] Purification of the isopeptides can be done by ion-exchange chromotography.[3]

[14] H. Zahn and W. Pätzold, *Chem. Ber.* **96**, 2566 (1963).
[15] M. S. Otterburn and W. J. Sinclair, *J. Sci. Food Agric.* **27**, 1071 (1976).

[15] Formation of N^ε-(Biotinyl)lysine in Biotin Enzymes

By NEIL H. GOSS and HARLAND G. WOOD

Biotin-Dependent Enzymes

The role of biotin in carboxyl transfer reactions has been the subject of considerable investigation for a number of years, and the mechanisms of the enzymes that catalyze these reactions are known in some detail.[1,2] These reactions occur at two spatially distinct subsites, the biotin being carboxylated at the $1'N$-nitrogen of the ureido ring of biotin in the first site, then moving to the second site, where the carboxyl group is transferred to an appropriate acceptor molecule. The biotin is covalently attached to the enzyme through an amide bond that links the carboxyl group of the valeric acid side chain of biotin to the ε-amino group of a lysine residue in the enzyme, yielding a biocytin residue. Of the nine known biotin enzymes, six use ATP and bicarbonate to carboxylate the biotin in the first partial reaction, as shown in Eq. (1).

$$\text{E-biotin} + \text{HCO}_3^- + \text{ATP} \rightleftharpoons \text{E-biotin}\text{\small$\sim\!\!\sim$}\text{CO}_2^- + \text{ADP} + \text{P}_i \tag{1}$$

In the case of the decarboxylases and transcarboxylase, a carboxyl donor replaces the ATP and HCO_3^-, as shown in Eq. (2).

$$\text{E-biotin} + {}_2^-\text{OC}\text{\small$\sim\!\!\sim$}\text{donor} \rightleftharpoons \text{E-biotin}\text{\small$\sim\!\!\sim$}\text{CO}_2^- + \text{donor} \tag{2}$$

The second partial reaction, in which the acceptor molecule is carboxylated by the biotin$\sim\!\!\sim$$\text{CO}_2^-$ complex, occurs as shown in Eq. (3).

$$\text{E-biotin}\text{\small$\sim\!\!\sim$}\text{CO}_2^- + \text{acceptor} \rightleftharpoons \text{E-biotin} + \text{acceptor}\text{\small$\sim\!\!\sim$}\text{CO}_2^- \tag{3}$$

In the case of the decarboxylases, the "acceptor" molecule is water.

[1] J. Moss and M. D. Lane, *Adv. Enzymol. Relat. Areas Mol. Biol.* **35**, 431 (1971).
[2] H. G. Wood and R. E. Barden, *Annu. Rev. Biochem.* **46**, 385 (1977).

Although the mechanisms of catalysis of these enzymes have much in common, their structures show considerable diversity. The enzymes vary from structures having the donor site, acceptor site, and biotin attachment site on a single polypeptide chain, as in pyruvate carboxylase of animals, to ones that have each of these three functional sites on separate polypeptides, as in transcarboxylase of *Propionibacterium shermanii* or acetyl-CoA carboxylase of *Escherichia coli*. The most common structural form for biotin enzymes appears to be between these two extremes, with the biotin and carboxylation site on one polypeptide and the acceptor site apparently located on a second polypeptide. This type of structure has been established for β-methylcrotonyl-CoA carboxylase from *Achromobacter* IVS,[3] propionyl-CoA carboxylase from *Mycobacterium smegmatis*[4] and from human livers,[5] and pyruvate carboxylase from *Pseudomonas citronellolis*.[6] A more detailed discussion of these structural variations is given by Wood and Barden.[2] Despite this great structural diversity, each of these enzymes must have the biotin covalently attached to it in a very specific manner, and, as will be seen below, the biotination occurs in an essentially identical fashion irrespective of the final holoenzyme formed.

Holoenzyme Synthetase

The covalent attachment of biotin to a specific lysine(s) in apoenzymes is catalyzed by holoenzyme synthetase in a two-step reaction [Eqs. (4) and (5)].

$$\text{ATP} + \text{biotin} \rightleftharpoons 5'\text{-adenylate-biotin} + \text{PP}_i \qquad (4)$$
$$5'\text{-Adenylate-biotin} + \text{apoenzyme} \rightleftharpoons \text{holoenzyme} + \text{AMP} \qquad (5)$$

Synthetases have been shown to be responsible for the formation of holoforms of propionyl-CoA carboxylase,[7] β-methylcrotonyl-CoA carboxylase,[8] acetyl-CoA carboxylase,[9] transcarboxylase,[10] and pyruvate carboxylase[11] from their respective apoenzymes.

---

[3] U. Schiele, R. Niedermeier, M. Stüzer, and F. Lynen, *Eur. J. Biochem.* **60,** 259 (1975).

[4] K. P. Henrikson and S. H. G. Allen, *J. Biol. Chem.* **254,** 5888 (1979).

[5] F. Kalousek, M. D. Darigo, and L. E. Rosenberg, *J. Biol. Chem.* **255,** 60 (1980).

[6] N. D. Cohen, J. A. Duc, H. Beegen, and M. F. Utter, *J. Biol. Chem.* **254,** 9262 (1979).

[7] L. Seigel, J. L. Foote, J. E. Christner, and M. J. Coon, *Biochem. Biophys. Res. Commun.* **13,** 307 (1963).

[8] T. Höpner and J. Knappe, *Biochem. Z.* **342,** 190 (1965).

[9] F. Lynen and K. L. Rominger, *Fed. Proc. Fed. Am. Soc. Exp. Biol.* **22,** 537 (1963).

[10] M. D. Lane, D. L. Young, and F. Lynen, *J. Biol. Chem.* **239,** 2858 (1964),

[11] J. J. Cazzulo, T. K. Sundaram, S. N. Silks, and H. L. Kornberg,. *Biochem. J.* **122,** 653 (1971).

The various biotin synthetases have broad specificity and cross-react with numerous apoenzymes. For example, Kosow et al.[12] demonstrated that transcarboxylase synthetase from propionibacteria catalyzes the biotination of apopropionyl-CoA carboxylase from rat liver, and, in a more comprehensive study, McAllister and Coon[13] showed that holoenzyme synthetases from rabbit liver, yeast, and P. shermanii cross-react forming active enzymes from apopropionyl-CoA carboxylase from rat liver, from apo-3-methylcrotonyl-CoA carboxylase from Achromobacter and from apotranscarboxylase from P. shermanii. The only exception was that rat liver synthetase did not biotinate apotranscarboxylase. Such a broad specificity led to a general consideration that the apobiotin enzymes have similar features that provide a recognition site for the synthetase and permit the precise attachment of the cofactor into the active sites of these enzymes.

There have been no direct studies to determine the changes that may occur following the synthesis of the polypeptides making up biotin enzymes. It is quite certain, however, that the biotination occurs as a discrete posttranslational event. Certain organisms, including rats, yeast, and some bacteria, cannot synthesize biotin, and when biotin is limiting there is a decrease in the level of the biotinated enzymes and an increased amount of nonbiotinated enzymes, and the total enzyme (apo-plus holoenzyme) remains nearly constant. After the administration of biotin, the amount of holoenzyme increases rapidly and apparently without synthesis of new protein, since actinomycin D and puromycin do not block this process.[14] Similarly, Freytag and Utter[15] have shown with avidin-treated 3T3-L1 cells that the pyruvate carboxylase level is low and is rapidly restored upon the addition of biotin. However, the amount of immunoprecipitable pyruvate carboxylase in the presence or the absence of avidin is essentially the same. Clearly, apoenzyme is synthesized in the absence of biotin, and if biotin is added, there is biotination of the preexisting apoenzyme.

The question of whether or not assembly of the intact apoenzyme is required prior to biotination was addressed by Wood et al.[16] using synthetase from P. shermanii and both assembled apotranscarboxylase and the isolated aposubunits. The results from these studies showed that the synthetase biotinates the unassembled apopolypeptide just as rapidly as it

[12] D. B. Kosow, S. C. Huang, and M. D. Lane, J. Biol. Chem. 237, 3633 (1962).
[13] H. C. McAllister and M. J. Coon, J. Biol. Chem. 241, 2855 (1966).
[14] A. D. Deodhar and S. P. Mistry, Arch. Biochem. Biophys. 131, 507 (1969).
[15] S. O. Freytag and M. F. Utter, Proc. Natl. Acad. Sci. U.S.A. 77, 1321 (1980).
[16] H. G. Wood, F. R. Harmon, B. Wühr, K. Hübner, and F. Lynen, J. Biol. Chem. 255, 7397 (1980).

	80	81	82	83	84	85	86	87	88	89	90	91	92	93	94	95	96	97	98
Pyruvate carboxylase (sheep)	Gly	Gln	Pro	Leu	Val	Leu	Ser	Ala	Met	Bct	Met	Glu	Thr	Val	Val	Thr	Ser	Pro	Val
(chicken)	Gly	Ala	Pro	Leu	Val	Leu	Ser	Ala	Met	Bct	Met	Glu	Thr	Val	Val	Thr	Ala	Pro	Arg
(turkey)	Gly	Ala	Pro	Leu	Val	Leu	Ser	Ala	Met	Bct	Met	Glu	Thr	Val	Val	Thr	Ala	Pro	Arg
Transcarboxylase (P. shermanii)	Gln	Thr	Val	Leu	Val	Leu	Glu	Ala	Met	Bct	Met	Glu	Thr	Glu	Ile	Asn	Ala	Pro	Thr
Acetyl-CoA carboxylase (E. coli)	Asn	Thr	Leu	Cys	Ile	Val	Glu	Ala	Met	Bct	Met	Met	Asn	Gln	Ile	Glu	Ala	Asn	Lys

FIG. 1. The amino acid sequences around the biocytin residues in several biotin-containing enzymes. The boxed region indicates the totally conserved tetrapeptide present in each of these sequences. The numbering is that of the residues of the biotinyl subunit of transcarboxylase.

does the intact apoenzyme, and so illustrated that, in this case there is no requirement for assembly prior to biotination.

Homology of Amino Acid Sequences near the Biotin. The biotinyl polypeptides vary greatly in size, ranging from M_r 12,000 in transcarboxylase to ~220,000 in the multifunctional subunit of acetyl-CoA carboxylase from livers of rat and chicken and the mammary gland of rat (see Table I of Wood and Barden[2] for the size of other biotinyl polypeptides). Thus far, only the biotinyl polypeptide from transcarboxylase has been completely sequenced.[17] However, others have been sequenced in the region near the biotinylated lysine,[18,19] and these are shown in Fig. 1.

The peptides are aligned at the biocytin (Bct). There is complete homology in the peptides from the pyruvate carboxylases of chicken and turkey livers, and only 3 positions out of 19 of the avian biotinyl peptide differ from that of pyruvate carboxylase from sheep. Two of these substitutions, glutamine-81 for alanine and serine-96 for alanine, are quite conservative. The substitution of valine-98 for an arginine is likely to generate a more significant difference in the structures of the respective peptides.

[17] W. L. Maloy, B. U. Bowien, G. K. Zwolinski, G. K. Kumar, H. G. Wood, L. H. Ericcson, and K. A. Walsh, *J. Biol. Chem.* **254,** 11615 (1979).
[18] D. B. Rylatt, D. B. Keech, and J. C. Wallace, *Arch. Biochem. Biophys.* **183,** 113 (1977).
[19] M. R. Sutton, R. R. Fall, A. M. Nervi, A. W. Alberts, P. R. Vagelos, and R. A. Bradshaw, *J. Biol. Chem.* **252,** 3934 (1977).

The sequence from *P. shermanii* transcarboxylase is closely homologous to the pyruvate carboxylase, 11 of the 19 residues being aligned with the avian sequence, while the acetyl-CoA carboxylase sequence from *E. coli* shows 5 of 19 residues in the same relative positions. The most striking feature of these sequence comparisons is the total conservation of the tetrapeptide from residues 87 through 90, which contains the biocytin moiety. That this sequence may have a special significance is indicated by the fact that of the many proteins so far catalogued, only the biotin enzymes contain a Met-Lys-Met grouping. Perhaps this is the sequence that specifies to the synthetase that this is the lysine, among the many present in the polypeptides, that is to be biotinated. It has been estimated that bacteria and eukaryotes diverged 3000 million years ago. It has been proposed[2] that this conservation provides evidence that these biotin enzymes may have evolved from a common ancestor. Furthermore, the cross-reactivity of the various biotin synthetases with various biotin enzymes may be a consequence of the occurrence of identical sequences in the enzyme. This subject will be considered in greater detail below.

Purification and Properties of Holocarboxylase Synthetase

Methods of Assay

There are two general methods that have been used to determine the activity of holocarboxylase synthetase. The first procedure follows the incorporation of radioactive biotin into the appropriate apoenzyme directly, and the second procedure monitors the appearance of the resultant active holoenzyme. Both are based on conditions described by Lane *et al.*[10] and each is presented below.

Method a. Incorporation of Radioactive Biotin. Incorporation of [^{14}C]biotin into apoenzyme is followed in a reaction mixture containing final concentrations (millimolar) of the following components: Tris-Cl, pH 7.6, 83; ATP, 0.75; MgCl$_2$, 0.75; glutathione, 0.125; biotin, 0.02; and serum albumin, 0.05 mg in a final volume of 0.2 ml. The concentration of apoenzyme in the assay is generally adjusted to about 2 nmol of available biotination sites per reaction mix, although lower levels of this substrate can be used. In the control experiment, the apoenzyme is replaced with buffer to correct for any endogenous apoenzyme substrate present in the holoenzyme synthetase preparation. In the studies conducted by Wood *et al.*[16] on the biotination of apotranscarboxylase, the Tris-Cl buffer was replaced with 240 mM potassium phosphate, pH 6.8–7.0, since under these conditions the dissociation of apotranscarboxylase to its subunits is reduced.[20]

The reaction mixtures are equilibrated at 30°, and the reaction is initiated by the addition of [^{14}C]biotin (final specific activity = 8.5×10^4 cpm/nmol) followed immediately by the synthetase. At appropriate time intervals, aliquots of the reaction mixture (5 to 50 μl) are removed into a 10-fold excess of cold 10% trichloroacetic acid (TCA) containing 1 mM unlabeled biotin. These aliquots are kept on ice for 5–10 min, and the precipitated material is then collected on 25 mm Amicon HAWP-45 filters (Millipore Corp.) that have been prewashed with 2 ml of 10% TCA containing 1 mM biotin. The filters are then washed with 3 ml of 10% TCA containing 1 mM biotin, dried under a 250-W heat lamp for 15 min, and counted in 20-ml scintillation vials containing 10 ml of New England Nuclear 963 cocktail scintillation mix.

The nanomoles of biotin incorporated are calculated by dividing the total radioactivity fixed, after correction for the radioactivity fixed by the synthetase alone, by the specific activity of the biotin (8.5×10^4 cpm/nmol). A typical result, using *P. shermanii* synthetase and apotranscarboxylase, is shown in Fig. 2.[16]

Method b. Determination of Active Holoenzyme Carboxylase. In this assay, the same reaction mixture is used as described in Method a above, except that nonradioactive biotin is used. At specified time intervals, aliquots of the reaction mixture are removed and the reaction is terminated either by dilution into 0.1 M EDTA[10,16] or direct addition into the appropriate holoenzyme carboxylase assay mix.[7] The assays for the specific holoenzyme carboxylases rely either on spectrophotometric detection of the product of the carboxylase reaction in a coupled assay or the direct incorporation of H^{14}CO$_3^-$ into an acid-stable product. Conditions for assay of individual enzymes differ somewhat and can be found in earlier volumes of this series.

Figure 3 illustrates the results of an experiment obtained using apotranscarboxylase and *P. shermanii* synthetase as described above. The holotranscarboxylase is assayed as described by Wood *et al.*[16] after terminating the synthetase reaction by dilution of aliquots into 0.1 M EDTA. In this case, two controls are necessary: (1) apotranscarboxylase without biotin or synthetase, as a correction for transcarboxylase present in the apotranscarboxylase preparation; and (2) synthetase with biotin, as a correction for any transcarboxylase plus apotranscarboxylase that might be present in the synthetase preparation. The sum of these two values is subtracted from that obtained with the complete system.

[20] H. G. Wood, J.-P. Chiao, and E. M. Poto, *J. Biol. Chem.* **252,** 1490 (1977).

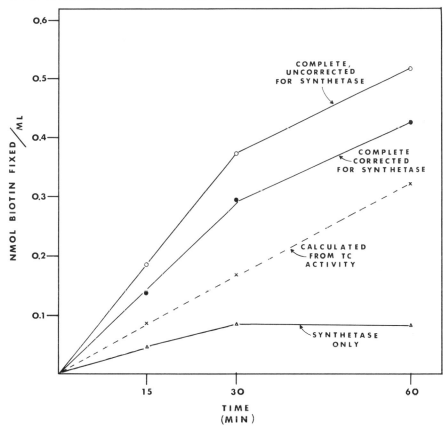

FIG. 2. Biotination of apotranscarboxylase (APO TC) using the [^{14}C]biotin assay. The conditions of the assay were as described in assay method a. Synthetase (specific activity 0.65 U/mg; 21 μg) was incubated with apotranscarboxylase (54 μg; 0.4 nmol of biotin sites) and 1.65 nmol of [^{14}C]biotin in the standard assay volume of 0.2 ml. The final incorporation of [^{14}C]biotin has been calculated on the basis of an assay volume of 1.0 ml. As seen from the dashed line, significantly more biotination occurs than can be accounted for by enzyme activity. Apparently, inactive holoenzyme is formed.[16]

Alternative Assay Procedures

Eisenberg *et al.*[21,22] have reported that the *E. coli* holoenzyme synthetase, when combined with biotinyl-5'-adenylate, forms a repressor protein–biotin complex that binds to the operator site of the biotin operon. These authors utilized several other assays for the *E. coli* repressor/

[21] O. Prakash and M. A. Eisenberg, *Proc. Natl. Acad. Sci. U.S.A.* **76**, 5592 (1979).
[22] M. A. Eisenberg, O. Prakash, and S.-C. Hsiung, *J. Biol. Chem.* **257**, 15167 (1982).

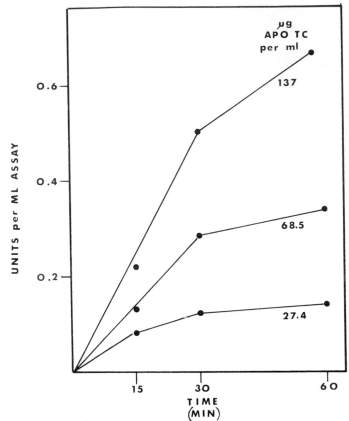

FIG. 3. Formation of transcarboxylase with different concentration of apotranscarboxylase and a constant amount of synthetase. The reaction mixture contained, in 1.0 ml, synthetase (specific activity 0.65; 53 μg), apotranscarboxylase, at the concentrations shown in the figure, and the standard components as described for the biotination of apotranscarboxylase. The pH of the reaction mix was 6.8. The transcarboxylase activity is expressed in units per milliliter of reaction mix. Units equal micromoles of oxaloacetate formed per minute.

synthetase in addition to those based on enzymatic activity, and these are presented below.

Assay a. Biotin Binding Assay. This assay relies on the strong noncovalent binding of [^{14}C]biotin to the synthetase. The assay contains, in a final volume of 0.2 ml, the following components (mM): Tris-Cl, pH 7.4, 100; potassium chloride, 200; magnesium acetate, 10; EDTA, 1; 2-mercaptoethanol, 6; [^{14}C]biotin, 8.5×10^{-7}; and appropriate amounts of the synthetase. When more highly purified synthetase preparations were used, 1.0 mg of γ-globulin was also included in the assay to aid in the subsequent protein precipitation. In the control, unlabeled biotin at a final concentration of 8.2×10^{-3} M was included to correct for nonspecific binding.

The reaction mixture was incubated at room temperature for 5 min, and the protein was then precipitated by the addition of 0.8 ml of cold saturated ammonium sulfate in the Tris-Cl buffer, pH 7.6. After standing for 10 min on ice, the precipitated protein was collected by centrifuging and washed once with 1.0 ml of cold 80% saturated ammonium sulfate in Tris buffer. The precipitate was again collected by centrifuging and then dissolved in 0.2 ml of 0.1 N NaOH. The centrifuge tube was washed twice with 0.2 ml of 0.1 N NaOH, and the radioactivity in the combined solutions was determined in Aquasol 2 scintillant (New England Nuclear). The biotin binding activity is expressed as picomoles of biotin bound per milliliter of enzyme. The apparent dissociation constant for this reaction, calculated from the Scatchard plot, is 1.3×10^{-7} M.[22]

Assay b. DNA Filter-Binding Assay. This assay relies on the specific interaction between the ^{32}P-labeled λ *bio* DNA and the synthetase–biotinyl-5'-adenylate complex. The reaction mixture contains, in a final volume of 0.1 ml, the following components (mM): Tris-Cl, pH 7.5, 1.0; EDTA, 0.1; magnesium acetate, 5; KCl, 10; dithiothreitol, 0.2; 5 μg of bovine serum albumin; 0.1 μg of λ *bio* [^{32}P]DNA; 10 μg of unlabeled λ DNA, and appropriate concentrations of repressor (synthetase) and corepressor (biotin or biotinyl-5'-adenylate).

The reaction mixture (0.4 ml) is incubated for 10 min at 25°, and aliquots (0.1 ml) are removed and filtered through Schleicher & Schuell 13 mm BA 85 membrane filters presoaked for 1 hr in 1.0 mM Tris-Cl, pH 7.5, containing 0.1 mM EDTA, 5 mM magnesium acetate, and 10 mM KCl. Triplicate samples were filtered simultaneously, and each was washed with 50 μl of the above buffer. The filters were dried, and the radioactivity was determined using Aquasol 2 in a liquid scintillation counter. The average values of the triplicates were corrected for nonspecific binding (without biotin), which amounted to approximately 10% of the total input counts.

Assay c. Repression of a Coupled Transcription–Translation System. This assay measures the ability of the synthetase–biotinyl-5'-adenylate complex (the repressor–corepressor complex) to inhibit the appearance of two enzymes coded for by the *E. coli* biotin operon. These two enzymes are 7,8-diaminopelargonic acid (DAPA) aminotransferase and dethiobiotin (DTB) synthetase. For this purpose, Prakash and Eisenberg[21] have utilized DNA carrying the biotin gene cluster isolated from a biotin transducing, plaque-forming phage, λ *biot* 124. The preparation of the S-30 extracts and the *in vitro* protein synthesis were performed essentially by the method of Zubay[23] except that the synthesizing mixture (final volume 0.25 ml) contained no cyclic AMP and had 8–10 μmol of magnesium acetate and 4–5 μmol of calcium acetate per milliliter of reaction mix. Routinely, 60 μg of λ *biot* 124 DNA per milliliter was used. The synthesis was initiated by addition of 0.4 ml of the S-30 extract per milliliter and was at 34° for 20 min on a gyrating water bath at 240 rpm in both the unrepressed and repressed systems. In the repressed system, to optimize the effectiveness of the interaction between the operator and the repressor–corepressor complex, all the components of the protein synthesis mix except the S-30 and nucleotide triphosphates were preincubated for 3 min at 0° in a final volume of 0.15 ml. The reaction then was initiated by the addition of 0.08 ml of S-30 and 0.02 ml of nucleoside triphosphates. The reactions were terminated by the addition of chloramphenicol to a concentration of 100 μg/ml. As a control, DNA was omitted from the reaction mixture.

The DAPA aminotransferase and DTB synthetase produced in the *in vitro* system described above were then assayed by the procedures of Eisenberg and Stoner[24] and Eisenberg and Krell.[25] Prior to this determination, the repressed mixtures were dialyzed exhaustively against 2 mM Tris-Cl, pH 7.4, containing 6 mM 2-mercaptoethanol and 0.01 mM pyridoxal 5′-phosphate to remove biotin. These bioassays allowed the detection of as little as 1 pmol of reaction products.

The assay mixture for DAPA aminotransferase contained (mM) in a final volume of 0.2 ml: dithiothreitol, 12.5; pyridoxal 5′-phosphate, 0.5; S-adenosylmethionine, 6.25; 7-keto-8-aminopelargonic acid, 0.025. For the DTB synthetase assay, the reaction mixture contained (mM) in a final volume of 0.1 ml: ATP, 40; NaHCO$_3$, 200; DAPA, 0.1. The appropriate assay mixes were added to the above-described transcription–translation systems, which were then incubated at 37° for 1 hr. The reactions were terminated by the addition of 0.1 ml of 0.06 N ZnSO$_4$ and heating in a boiling water bath for 1 min. The protein was then precipitated by the addition of an equivalent amount of Ba(OH$_2$) and removed by centrifuging. The supernatants were dried under vacuum over P$_2$O$_5$ and resuspended in 0.04 ml of water; 0.02 ml was applied to 6-mm filter disks for microbiological assay, using the appropriate Rif^r and Cml^r assay organism. One unit is defined as the amount of synthesized enzyme that produces 1 nmol of DAPA or DTB in 60 min under the conditions described above. The percentage of repression was estimated from the values obtained with the repressed system compared to the nonrepressed system.

[23] G. Zubay, *Annu. Rev. Genet.* **7,** 267 (1973).
[24] M. A. Eisenberg and G. L. Stoner, *J. Bacteriol.* **108,** 1135 (1971).
[25] M. A. Eisenberg and K. Krell, *J. Biol. Chem.* **244,** 5503 (1969).

Purification of Holoenzyme Synthetase

Partially purified synthetase preparations have been obtained from a variety of organisms.[10,16,26] Only recently, however, has the enzyme been purified to homogeneity, and this procedure, developed by Eisenberg *et al.* using *E. coli*,[22] is presented below.

Escherichia coli K12 cells (200–260 g), strain H 105 (λc 1857s7, λc 1857*rif*[d]18) were harvested in mid-log phase and ruptured by sonic disruption, as described by Prakash and Eisenberg,[27] in 1.3 volumes of 0.2 M Tris-Cl, pH 7.6, containing 0.2 M KCl, 0.01 M magnesium acetate, 0.3 mM dithiothreitol, 0.12 mM phenylmethylsulfonyl fluoride, and 5% glycerol. The resulting cell preparation was stirred at 0° for 1–2 hr to reduce the viscosity and then centrifuged at 30,000 rpm for 3 hr in a Beckman type 30 rotor. The supernatant was collected and diluted to ca. 20 mg/ml with the same buffer. To this solution was slowly added a 2% solution of protamine sulfate, while stirring, to give a final concentration of 0.4%. A heavy precipitate developed and was removed by centrifuging. To the supernatant from the protamine sulfate precipitation step was added enough solid ammonium sulfate to bring the solution to 30% saturation. The solution was centrifuged to remove the precipitated protein, and to the supernatant was added more solid ammonium sulfate to increase the final saturation to 60%. The precipitated protein was collected by centrifuging at 29,000 rpm for 30 min in a Beckman type 30 rotor, redissolved in a small volume of buffer I (0.1 M sodium phosphate, pH 7.6, 10% glycerol, and 10^{-4} M dithiothreitol), and dialyzed at 4° against 2 liters of the same buffer with three changes.

The dialyzed enzyme was applied to a DEAE-cellulose column (9 × 30 cm), equilibrated with buffer I and washed with 2 liters of the same buffer to remove unbound protein. Most of the synthetase activity remained bound to the column and could be eluted with buffer I containing 0.25 M NaCl. The fractions containing synthetase activity were pooled and precipitated with 70% saturated ammonium sulfate. The precipitate was collected by centrifuging, redissolved in a minimal volume of buffer II (0.01 M sodium phosphate, pH 6.5, containing 5% glycerol and 10^{-4} M dithiothreitol), and dialyzed extensively against the same buffer.

After dialysis, the enzyme was applied to a phosphocellulose column (Whatman P-11) (9 × 30 cm) equilibrated in buffer II at a flow rate of 80 ml/hr, and unbound protein was eluted with a wash of 1400 ml of the same buffer. The column was then washed in a stepwise fashion, first with

[26] L. Siegel, J. L. Foote, and M. J. Coon, *J. Biol. Chem.* **240**, 1025 (1965).
[27] O. Prakash and M. A. Eisenberg, *J. Bacteriol.* **134**, 1002 (1978).

buffer II containing 0.25 M NaCl, which elutes some protein but little synthetase activity, and second with buffer II containing 0.5 M NaCl. Most of the synthetase activity elutes from the column under these conditions. The fractions containing most of the synthetase activity were pooled precipitated at 70% ammonium sulfate saturation and collected by centrifuging. The precipitate was redissolved in buffer III (0.01 M sodium phosphate, pH 7.2, containing 10% glucose, 10^{-4} M DTT, and 2×10^{-3} M EDTA) and dialyzed against three changes of 1 liter each of the above buffer.

The dialyzed enzyme solution (ca. 25 ml) was then divided into two portions, and one half was applied to a 25-ml column (1.8 × 12 cm) of calf thymus DNA-cellulose and eluted with the equilibrating buffer III at a flow rate of 0.5 ml/min. The column was then washed with buffer III until the absorbance at 280 nm was below 0.03, and the enzyme was then eluted with 0.2 M NaCl. After elution of the enzyme, the column was regenerated using 1.0 M NaCl and reequilibrated with buffer III; the procedure was repeated on the second half of the enzyme solution. Fractions containing synthetase activity were pooled, precipitated with 70% saturated ammonium sulfate, and collected by centrifugation. The precipitated protein was redissolved in a minimal volume of buffer III and dialyzed against three changes of 500 ml each of the same buffer.

The pooled, dialyzed enzyme (9.0 ml) was reapplied to the same DNA-cellulose column and equilibrated as described above, and the column was washed with buffer III containing 0.5 M NaCl until the absorbance at 280 nm was below 0.10. The enzyme was then eluted from the column with a linear gradient of 25 ml of buffer III containing 0.05 M NaCl and 25 ml of buffer III containing 0.3 M NaCl. The synthetase eluted at 0.15 M NaCl. The fractions containing the enzyme were pooled and dialyzed against 500 ml of buffer II for 2 hr.

The enzyme obtained from the second DNA-cellulose column was applied to a DEAE-cellulose column (1.5 × 24 cm) equilibrated in buffer II, and the enzyme was eluted with a linear gradient of 50 ml of buffer II and 50 ml of buffer II containing 0.35 M NaCl. The enzyme eluted at approximately 0.25 M NaCl. This enzyme showed three bands on SDS–gel electrophoresis, corresponding to 98,000, 26,000, and 34,000 in order of increasing staining intensity. A summary of this purification is presented in the table.

To purify this synthetase preparation to homogeneity, 1 ml of the enzyme (290 μg) purified as described above was applied to a 5-ml chromofocusing column and eluted with a polybuffer 96-acetate pH gradient from pH 8.0 to pH 6.0. The results of this elution are shown in Fig. 4. Notice that the holoenzyme synthetase activity, biotin-binding activity and tran-

PURIFICATION OF THE HOLOENZYME SYNTHETASE

Preparation	Volume	Protein (mg/ml)	Biotin bound (pmol/ml)	Biotin-binding activity				Holoenzyme synthetase activity		
				Sp. Ac.[a]	Total $\times 10^6$	% Recovery	Purification (fold)	Biotin incorporated (pmol/min ml^{-1})	Sp. Ac.[b]	Purification (fold)
Crude	1784	37.8	113	3	20.9	—	—	129	3	—
Protamine sulfate	2280	26.5	80	3	18.2	87	—	110	4	—
(NH$_4$)$_2$SO$_4$, 0.6 sat.	579	58.0	317	5.5	18.4	87	1.8	447	7.8	2.4
DEAE-cellulose I, pH 7.6	586	20.5	215	10.5	12.6	60	3.5	356	17.4	5.4
Phospho-cellulose	26	21.0	3038	145	7.9	38	48	4970	237	74
DNA-cellulose I	9.6	4.1	8303	2025	8.0	38	675	8004	1952	610
DNA-cellulose II	13	0.66	3503	5307	4.6	22	1769	3516	5327	1665
DEAE-cellulose II pH 6.5, tube No. 16	3.9	0.35	3891	11,117	1.5	7	3705	3176	9074	2836

[a] Specific activity, picomoles of biotin bound per milligram of protein.
[b] Specific activity, picomoles of biotin incorporated per minute per milligram of protein.

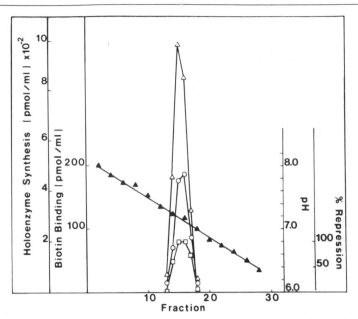

FIG. 4. Chromofocusing of *Escherichia coli* holoenzyme synthetase. The synthetase (290 μg) was applied to a 5-ml chromofocusing column and eluted as described in the text. Biotin-binding activity (○——○); holoenzyme synthetase activity (▲——▲); repressor activity (□——□); pH gradient (△——△).

scription–translation repression activity all comigrate. Since this preparation is homogeneous on SDS-gel electrophoresis, showing a single protein band at M_r 34,000, these findings strongly support the view that the *E. coli* holoenzyme synthetase is also the biotin repressor protein.

Properties of the Holoenzyme Synthetase

The *E. coli* enzyme prepared as described above has a native M_r 37,000–44,000 as determined by gel permeation chromatography in 10 mM sodium phosphate, pH 6.5, 0.5 M NaCl, 5% glycerol, and 1 mM dithiothreitol. Since the homogeneous enzyme shows only one band on SDS–polyacrylamide gel electrophoresis with a molecular weight of 34,000, the enzyme appears to be a monomer. In other studies, Rominger reported M_r 39,000–40,000 for the enzyme from *P. shermanii*,[28] although

[28] K. L. Rominger, Ph.D. dissertation, University of Munich, Munich, West Germany, 1964.

Wood *et al.*[16] found that the native enzyme from the same source had an M_r 29,000 ± 5000 on Sephadex G-75 chromatography. Rominger[28] reported an M_r ca. 90,000 for the enzyme from both brewers' yeast and bakers' yeast. This comparison is interesting in the light of the observation by Eisenberg *et al.* of a minor component of M_r 90,000, which occasionally appeared in the gel filtration chromatography experiments. Barker *et al.*[29] have also suggested an M_r of 80,000–100,000 for the *E. coli* synthetase based on similar experiments with crude extracts. Thus, the synthetase may in fact have an oligomeric structure *in vivo*.

Reaction Mechanisms

The reactions catalyzed by all synthetases so far examined occur via two partial reactions, as shown in Eqs. (4) and (5) above. The evidence for this comes from the observation that the enzyme will catalyze an exchange reaction between ATP and $[^{32}P]PP_i$ in the absence of apoenzyme and that biotinyl-5′-adenylate can replace ATP and biotin in the formation of holoenzyme from apoenzyme.[26,30] The first partial reaction requires Mg^{2+}, but the second partial reaction does not. In the overall reaction catalyzed by transcarboxylase synthetase from *P. shermanii*, the K_m for biotin was found to be 4.2×10^{-7} M, and the K_m for biotinyl-5′-adenylate was 2.2×10^{-7} M.[30] The K_m for ATP determined by Siegel *et al.* using rabbit liver propionyl-CoA carboxylase synthetase was 1.3×10^{-6} M. Other studies conducted by Lynen and Rominger[9] and Rominger[28] using acetyl-CoA carboxylase synthetase from yeast, Höpner and Knappe[8] using β-methylcrotonyl-CoA carboxylase from *Achromobacter*, and Cazzulo *et al.*[11] using pyruvate carboxylase synthetase from *Saccharomyces cerevisiae* and *Bacillus stearothermophilus*, have revealed similar low K_m values for these substrates. Other nucleotide triphosphates substitute for ATP only weakly.[10,26] (±)-*o*-Heterobiotin, (+)-homobiotin, and (+)-biotin methyl ester were inhibitors of the synthetase reaction, and (±)-*o*-heterobiotin is utilized as a substrate by transcarboxylase synthetase forming enzymatically inactive (±)-*o*-heterobiotin transcarboxylase.[10]

Using the DNA filter binding assay, the concentration for half-maximal binding ($K_{0.5}$) of biotin to the *E. coli* synthetase was 1×10^{-6} M. However, with biotinyl 5′-adenylate, this value decreased to 1.1×10^{-9} M, indicating that the activated form of biotin binds much more tightly to the enzyme.[21]

[29] D. F. Barker, J. Kuhn, and A. M. Campbell, *Gene* **13**, 89 (1981).
[30] M. D. Lane, K. L. Rominger, D. L. Young, and F. Lynen, *J. Biol. Chem.* **239**, 2865 (1964).

Possible Roles for Sequence Homology

It was noted above that a remarkable feature of the amino acid sequences of the biotin-containing enzymes so far determined is the strict conservation of the tetrapeptide that contained the lysine to which biotin is attached in the active site. Furthermore, it appeared likely that the reason for this strict conservation might be to act as a recognition site for the attachment of the biotin to a specific lysine by holoenzyme synthetase. In order to test this hypothesis, we have obtained a small peptide containing this sequence by cleaving the biotin carboxyl carrier protein $(1.3S_E)$ of transcarboxylase with *Staphylococcus aureus* V_8 protease, yielding the pentapeptide Ala-Met-Bct-Met-Glu. This peptide was isolated by gel chromatography on Sephadex G-25 in 5% acetic acid and by high-pressure liquid chromatography on a Synchropak RP-P column using a linear gradient of 0.1% TFA and 0.1% TFA containing 40% acetonitrile. The holopeptide elutes at 27.3% acetonitrile (Fig. 5A). To remove the biotin from this peptide, and so provide a substrate for the holoenzyme synthetase, 300 nmol of the peptide was incubated at 37° for 18 hr in 100 m*M* potassium phosphate, pH 7.0, with 30 μg of biotinidase. This enzyme has been purified to homogeneity from human serum, and a detailed description of this purification and characterization of the action of this enzyme of biotin-containing peptides will be presented elsewhere.[31] Under these conditions, the biotin is quantitatively removed from the pentapeptide. The apopentapeptide was separated from the reaction mixture by rechromatography (HPLC) under the same conditions described above. The apopeptide elutes at 21% acetonitrile (Fig. 5B).

When this apopeptide (40 nmol) was incubated with *P. shermanii* synthetase and [^{14}C]biotin (93,000–130,000 cpm/nmol), as described above, there was no biotination of the peptide in 24 hr, as determined by rechromatography on HPLC and radioactivity determination. It therefore appears either that this highly conserved sequence alone is not enough to act as a recognition site for the synthetase or that some chemical alteration has occurred during the cleavage or purification of the peptide that has destroyed the ability of this peptide to be recognized by the synthetase. One possible chemical alteration would be the oxidation of the two methionines that flank the biotinated lysine. We have reduced the peptide in 5% aqueous 2-mercaptoethanol at pH 8.0 for 24 hr under nitrogen to convert any methionine sulfoxide to methionine but have still failed to observe biotination of the peptide. Perhaps exposure of the carboxyl and amino groups at the ends of the peptide after cleavage may affect the

[31] D. V. Craft, N. H. Goss, and H. G. Wood, manuscript in preparation (1984).

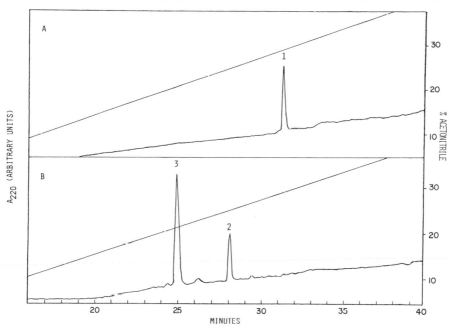

FIG. 5. Cleavage of holopentapeptide by biotinidase. (A) The holopentapeptide (1), prepared and purified as described in the text, chromatographs as a single peptide on a Synchropak RP-P column using a linear gradient of 0.1% TFA and 0.1% TFA containing 40% acetonitrile. (B) After cleavage by biotinidase, the apopeptide (3) elutes ahead of biotin (2) under the same chromatography conditions as in (A).

binding of the peptide to the synthetase. Further examination of the rates of biotination of larger apopeptides by the synthetase should provide an answer to this question and thus provide a minimal recognition site for the posttranslational addition of biotin to apobiotin enzymes.

Concluding Remarks

The studies conducted to date on biotin holoenzyme synthetases from a variety of sources make several general features of this posttranslational modification apparent. First, the ability of the enzyme to discriminate between a large number of available lysine residues and specifically biotinate the correct lysine in the active site of a variety of apoenzymes of grossly different quaternary structure strongly suggests a common structural feature or recognition site in all of these apoenzymes. Second, at least in the case of transcarboxylase, it is clear that the entire recognition site for the synthetase is contained within the 123 residues of the $1.3S_E$

biotin subunit, since this polypeptide is biotinated at the same rate as the intact apoenzyme. It is not possible at this point to assign a role in the biotination reaction to the highly conserved region immediately surrounding the biotination site, although this sequence forms a part of the recognition site for the synthetase.

In recent years, the clinical importance of both holoenzyme carboxylase synthetase and biotinidase has become apparent in relation to the multiple carboxylase deficiency disease originally described by Gompertz *et al.*[32,33] This disease manifests itself in either a neonatal or infantile form and a late-onset or juvenile form. In the neonatal form, the primary enzyme deficiency appears to be in the holoenzyme synthetase,[34] whereas recent findings by Wolf *et al.*[35] implicate a biotinidase deficiency in the late-onset form of the disease. Most patients suffering from either form of the disease show dramatic clinical improvement following administration of pharmacological doses of biotin. This topic has been reviewed by Wolf and Feldman,[36] and this reference should be consulted for further details.

Acknowledgment

The authors wish to acknowledge support from Grant GM 22579 from the National Institutes of Health.

[32] D. Gompertz, G. H. Draffan, J. L. Watts, and D. Hull, *Lancet* **2**, 22 (1971).

[33] K. Bartlett and D. Gompertz, *Lancet* **2**, 804 (1976).

[34] B. J. Burri, L. Sweetman, and W. L. Nyhan, *J. Clin. Invest.* **68**, 1491 (1981).

[35] B. Wolf, R. E. Grier, R. J. Allen, S. I. Goodman, and C. L. Kien, *Clin. Chim. Acta* **131**, 273 (1983).

[36] B. Wolf and G. L. Feldman, *Am. J. Hum. Genet.* **34**, 699 (1982).

Section II

Oxidations, Hydroxylations, and Halogenations

[16] Formation and Isomerization of Disulfide Bonds in Proteins: Protein Disulfide-Isomerase

By DAVID A. HILLSON, NIGEL LAMBERT, and ROBERT B. FREEDMAN

Introduction: Discovery

The discovery in the 1960s that reduced denatured proteins would spontaneously reoxidize and refold to form their native conformation led to the search for a physiological catalyst of this process, as the refolding and reoxidation reactions were seen as models for folding of proteins *de novo* and formation of their disulfide bonds. An enzymatic activity was found that catalyzed the formation of native proteins from the reduced denatured state, and this has been termed, successively, reduced ribonuclease reactivating enzyme, disulfide interchange enzyme, "rearrangease," and, more systematically, protein disulfide-isomerase (PDI, EC 5.3.4.1).

It was shown that the renaturation process, both nonenzymatic and enzymatic, involved initial rapid reoxidation to give a mixture of disulfide-bonded intermediates, followed by slower thiol–disulfide interchanges leading to regain of native structure.[1] The reactions catalyzed by PDI were shown to be of the latter type; formally they are thiol : disulfide oxidoreductions. Oxidoreduction between thiols and protein disulfides can give rise to a number of distinct overall reactions, each of which has physiological counterparts (see Freedman[2] for review). Several of these thiol : disulfide oxidoreduction reactions may be considered as post-translational modifications of proteins: the initial formation of native disulfides (inter- and intrachain) during protein synthesis (possibly preceded by formation of nonnative disulfides with subsequent isomerization); later rearrangement of internal thiols and disulfides; formation or reduction of mixed disulfides; cleavage of metastable disulfides, either inter- or intrachain.

Although several thiol : protein-disulfide oxidoreductases may exist, at present the best characterized is protein disulfide-isomerase. Initially this was defined as the enzymatic activity that catalyzed the reoxidation and reactivation of reduced denatured ribonuclease, but this was an unstable substrate, undergoing considerable spontaneous reactivation. Thus an assay was developed using ribonuclease that had been reduced then reox-

[1] C. B. Anfinsen, *Science* **181**, 223 (1973).
[2] R. B. Freedman, *FEBS Lett.* **97**, 201 (1979).

idized in denaturing conditions (''scrambled'') as substrate; this contains a complex undefined mixture of the 105 possible disulfide-bonded monomer structures, together with various oligomeric derivatives. When incubated in the presence of a thiol compound, protein disulfide-isomerase catalyzes the regain of native ribonuclease structure from this scrambled ribonuclease, with concomitant return of activity toward RNA.[3-5] This assay is inconvenient and is based on a patently nonphysiological substrate; nevertheless it is very sensitive and has permitted the study of PDI activity in a number of contexts, making it possible to propose a physiological role for this activity (see below).

Assay Method

Preparation of Substrate

Scrambled ribonuclease is a fully oxidized mixture containing randomly formed disulfide bonds. It is easily prepared from beef pancreatic ribonuclease A by the following method.

Incubate ribonuclease at 30 mg/ml (~2.2 mM) in 50 mM Tris-HCl buffer, pH 8.6–9 M urea–130 mM dithiothreitol (approximately 15-fold molar excess of dithiothreitol over reducible disulfide bonds) at ambient temperature for 18–20 hr, or at 35° for 1 hr.

Isolate reduced protein by acidification of the reaction mixture to pH 4 with glacial acetic acid, followed by immediate elution from a column of Sephadex G-25 with degassed 0.1 M acetic acid. Monitor eluted fractions at 280 nm, pool protein-containing fractions, and estimate protein concentration either spectrophotometrically or chemically, using native ribonuclease A as standard.

Dilute the sample of reduced ribonuclease to ~0.5 mg/ml with 0.1 M acetic acid. Add solid urea to give a final concentration of 10 M, and sarcosine hydrochloride to 0.1 M (sarcosine is included to react with cyanate ions that are present in concentrated solutions of urea and can inactivate ribonuclease by carbamylation[6]). Adjust to pH 8.5 with 1 M Tris, and incubate at ambient temperature for 2–3 days in the dark, during which time the protein is randomly reoxidized by atmospheric O_2. After this incubation, determination of free thiol groups using 5,5'-dithiobis(2-

[3] A. L. Ibbetson and R. B. Freedman, *Biochem. J.* **159,** 377 (1976).
[4] M. Drazić and R. C. Cottrell, *Biochim. Biophys. Acta* **484,** 476 (1977).
[5] N. Lambert and R. B. Freedman, *Biochem. J.* **213,** 235 (1983).
[6] G. R. Stark, W. H. Stein, and S. Moore, *J. Biol. Chem.* **235,** 3177 (1960).

nitrobenzoic acid)[7] shows reoxidation to be complete (<0.1 free thiol per ribonuclease molecule).

Recover the scrambled product by acidification to pH 4 with glacial acetic acid and elution from Sephadex G-25 in 0.1 M acetic acid. Pool fractions containing protein, adjust to pH 8 with 1 M Tris, and store at 4°.

The yield of scrambled ribonuclease through this procedure is typically 90–100%. The product is stable at 4° in solution for up to 6 months or, alternatively, may be dialyzed into 50 mM NH$_4$ · HCO$_3$, pH 7.8, and then lyophilized, yielding a white fluffy solid that may be stored indefinitely at −20°.

Scrambled ribonuclease prepared by this procedure is known to be a complex mixture of various different molecular weight components. Preliminary analysis of commercially available scrambled ribonuclease indicates that it comprises 60% monomer; our evidence suggests that this scrambled monomeric material is preferred as substrate to aggregates containing interchain disulfide bonds.[7a]

Assay Procedure

The substrate, scrambled ribonuclease, is essentially inactive in the hydrolytic cleavage of high-molecular-weight RNA, having about 2% the activity of native ribonuclease. The action of PDI in catalyzing interchange of inter- and intramolecular disulfides in scrambled ribonuclease results in regain of the native disulfide pairing, native conformation and concomitant return of ribonuclease activity against RNA. Thus, the activity of protein disulfide-isomerase is assayed by a time-course incubation during which aliquots are removed and ribonuclease activity toward RNA is measured. The principle of the assay, and a typical reactivation time course, are shown in Fig. 1. The assay procedure described here is a modification[5] of the original method,[3,4] which has been found to be most reliable in our hands.

The sample of protein disulfide-isomerase is added to 50 mM sodium phosphate buffer, pH 7.5, to a final volume of 900 μl and preincubated with 10^{-5} M dithiothreitol (10 μl of 1 mM stock solution, made fresh daily) for 2–3 min at 30°. Tris-HCl buffer is also acceptable but gives ~25% lower activities. The assay is then started by addition of a 100-μl aliquot of scrambled ribonuclease (0.5 mg/ml stock solution in 10 mM acetic acid, made fresh daily), and the incubation mixture is maintained at 30°. For work on a smaller scale, the volumes above can be reduced 10-fold, to

[7] G. L. Ellman, *Arch. Biochem. Biophys.* **82,** 70 (1959).
[7a] R. B. Freedman, unpublished observations, 1983.

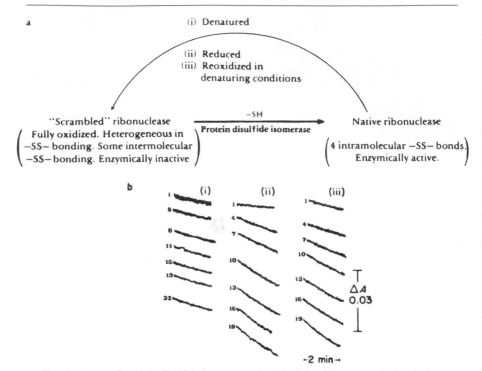

FIG. 1. Assay of protein disulfide-isomerase. (a) Principle of the assay. (b) Typical assay data. Each spectrophotometer trace is a ribonuclease assay performed at the time indicated to the left of each trace (1 min to 32 min); times are measured from the addition of protein disulfide-isomerase to the incubation. Assays from a single incubation form a vertical series. (i) Control; (ii) plus embryo chick tendon microsomes; (iii) plus embryo chick liver microsomes.

give a final assay volume of 100 μl. Aliquots of 10 μl are removed at 0.5 min and then at 2- to 3-min intervals for up to 18 min, to assay for the reactivation of scrambled ribonuclease. Each aliquot is immediately added to an assay mixture of 3 ml of TKM buffer (50 mM Tris-HCl buffer, pH 7.5–25 mM KCl–5 mM MgCl$_2$) containing 0.25 mg of highly polymerized yeast RNA (50 μl of 5 mg/ml stock solution), in a quartz cuvette previously equilibrated at 30°. Ribonuclease activity is monitored at 30° using the dual-wavelength mode of a Perkin–Elmer 356 spectrophotometer (bandwidth 2.5 nm), and measuring change in A_{260} relative to A_{280} (ΔA). The rate of RNA hydrolysis (ΔA min^{-1}) is constant over 1.5–2 min; a plot of this rate versus time of withdrawal of the aliquot from the incubation is linear for up to 15 min. The gradient of this linear portion of the time course (ΔA min^{-1} min^{-1}) is calculated by linear regression analysis of triplicate assays (with correlation coefficient routinely ≥ 0.99) and taken as a measure of protein disulfide-isomerase activity.

Control incubations are performed omitting enzyme sample to measure the rate of nonenzymatic reactivation of scrambled ribonuclease by dithiothreitol alone. These rates are usually less than $0.2 \times 10^{-3} \Delta A$ min^{-1} min^{-1} and are subtracted in the calculation of the protein disulfide-isomerase activities of enzyme samples.

One unit of protein disulfide-isomerase activity is defined[3] as the amount catalyzing reactivation of scrambled ribonuclease at a rate of one ribonuclease unit per minute; one ribonuclease unit is defined as the amount producing a change in A_{260} relative to A_{280} of 1 absorbance unit per minute.

For purified protein disulfide-isomerase, amounts of enzyme ranging from 0.5 to 5 μg have given activities in the range 0.5×10^{-3} to 3.0×10^{-3} ΔA min^{-1} min^{-1} in the standard assay procedure.[5]

Assay of Crude Tissue Preparations; Interference by Side Reactions

The above assay procedure, or the original method from which it was derived,[3] has been used to measure protein disulfide-isomerase activity in a variety of tissue preparations ranging from whole homogenates to purified homogeneous enzyme. The presence of sucrose has no significant direct effect on enzyme activity, but the enzyme shows latency in membranes (see below), so activities measured in homogenates or membrane preparations must be interpreted cautiously. Sonication of such samples before assay should allow determination of total enzyme activity. Use of the dual-wavelength mode of the Perkin–Elmer 356 spectrophotometer facilitates the assay of crude or turbid tissue samples, owing to both the inherent sensitivity of this mode and the physical proximity of the sample to the photomultiplier. Thus this assay procedure is applicable to a wide variety of preparations, from crude whole-cell homogenates to purified homogeneous enzyme. For nonturbid samples, use of the dual-wavelength mode is not essential and the ribonuclease assay can be performed in any sensitive spectrophotometer at 260 nm.

The specificity of the reactivation assay precludes any significant interference by side reactions; the only exceptions are in assay of samples containing endogenous ribonuclease or ribonuclease inhibitors or of samples containing high levels of thiols and/or disulfides. The former would interfere with estimation of the rate of hydrolysis of RNA by reactivated scrambled ribonuclease; the latter could produce significant rates of nonenzymatic thiol-disulfide interchange. However, these conditions are only rarely encountered and do not constitute general interference in the assay of protein disulfide-isomerase activity. Incubation of samples containing endogenous ribonuclease activity with the inhibitor diethyl pyrocarbonate may circumvent difficulties in assaying protein disulfide-isomerase activities in such samples.[8]

Distribution

Protein disulfide-isomerase is very widely distributed and has been detected in most vertebrate tissues, although detailed studies have been confined to the enzyme from liver. Several studies of tissue distribution have been reported, and these show that the highest activities of PDI in bovine, sheep, or rat tissues are in tissues active in the synthesis and secretion of disulfide-bonded proteins, namely liver, pancreas, lymph gland, spleen, and tendon.[8–10] Immunodeterminations show that the enzyme is present in significant quantities in mammalian liver, comprising 0.35–0.4% of total liver protein in the mouse and rat.[11,12] Further indications of this correlation come from studies on cells and tissues in which the synthesis of a single class of disulfide-bonded proteins predominates. Thus in the embryonic chick, around day 17 of development, the major disulfide-linked protein synthesized is procollagen, which contains intra- and interchain disulfide bonds in the N- and C-terminal globular domains removed extracellularly before the triple-helical collagen molecules are laid down. In the developing chick, protein disulfide-isomerase is most active in connective tissues such as tendon and cartilage, and its tissue and temporal distribution matches those of the specific hydroxylases active on procollagen Lys and Pro residues[13] (see Vol. 82 of this series). In a series of mouse lymphoid cells in culture, it has been shown that those cell lines actively synthesizing and secreting immunoglobulin have levels of protein disulfide-isomerase 10 to 100-fold greater than nonsecreting lines; in the active cells protein disulfide-isomerase can comprise approximately 1% of the extractable cell protein.[12]

Comparatively little work has been done on protein disulfide-isomerase in plant sources and almost none on microbial sources. The enzyme has been studied in wheat germ and endosperm,[14,15] where the synthesis

[8] D. A. Hillson and J. Anderson, *Biosc. Rep.* **2,** 343 (1982).

[9] F. De Lorenzo and G. Molea, *Biochim. Biophys. Acta.* **146,** 593 (1967).

[10] B. E. Brockway, S. J. Forster, R. B. Freedman, and D. A. Hillson, *Biochem. Soc. Trans.* **10,** 115 (1982).

[11] H. Ohba, T. Harano, and T. Omura, *J. Biochem.* **89,** 889 (1981).

[12] R. A. Roth and M. E. Koshland, *J. Biol. Chem.* **256,** 4633 (1981).

[13] B. E. Brockway, S. J. Forster, and R. B. Freedman, *Biochem. J.* **191,** 873 (1980).

[14] A. Grynberg, J. Nicholas, and R. Drapron, *Biochimie* **60,** 547 (1978).

[15] L. T. Roden, B. J. Miflin, and R. B. Freedman, *FEBS Lett.* **138,** 121 (1982).

and properties of disulfide-linked storage proteins are technically important, as they contribute to the properties of gluten in wheat doughs. Studies of the wheat endosperm enzyme have shown that it resembles the mammalian liver enzyme in subcellular location and molecular structure (see below). The enzyme has been detected in extracts of peas, cabbage, yeast and *Escherichia coli*,[16] but no serious detailed studies have been reported.

Developmental changes in the enzyme's activity are of interest. Both in chick bone during embryonic development and in wheat endosperm during ripening, the enzyme-specific activity passes through a maximum, corresponding to the phase of maximum specific rate of synthesis of procollagen and of disulfide-linked storage proteins, respectively.[13,17]

Thus it is evident that protein disulfide-isomerase is distributed in tissues where disulfide-bonded proteins are synthesized and secreted and is at maximal activity in developmental systems when such proteins are being produced. This is consistent with the postulated general role of protein disulfide-isomerase in biosynthesis and posttranslational modifications of disulfide-containing proteins.

Subcellular Location

In mammalian liver homogenates, protein disulfide-isomerase is found in crude microsomal membrane fractions.[4,11,18–20] Both in rat liver and wheat endosperm membranes the enzyme cosediments with markers of the endoplasmic reticulum.[3,15]

The enzyme's lateral and transverse disposition in the endoplasmic reticulum membrane is of interest, but has not been definitively established. Although some work suggested that the enzyme was active in smooth microsomes, but latent in rough microsomes owing to masking by bound ribosomes,[16] it is now clear that the enzyme is fully latent in all liver microsomes when they are carefully prepared and not exposed to osmotic stresses or vigorous sedimentation and resuspension. The enzyme can then be activated by treatment with low concentrations of detergents,[21,22] repeated centrifugation and resuspensions,[4] extraction with

[16] D. J. Williams, D. Gurari, and B. R. Rabin, *FEBS Lett.* **2,** 133 (1968).

[17] R. B. Freedman, B. E. Brockway, S. J. Forster, N. Lambert, E. N. C. Mills, and L. T. Roden, *in* "Functions of Glutathione: Biochemical, Physiological, Toxicological and Clinical Aspects" (A. Larsson *et al.,* eds.), p. 271. Raven, New York, 1983.

[18] R. F. Goldberger, C. J. Epstein, and C. B. Anfinsen, *J. Biol. Chem.* **238,** 628 (1963).

[19] D. Givol, R. F. Goldberger, and C. B. Anfinsen, *J. Biol. Chem.* **239,** 3114 (1964).

[20] H. C. Hawkins and R. B. Freedman, *Biochem. J.* **159,** 385 (1976).

[21] H. Ohba, T. Harano, and T. Omura, *Biochem. Biophys. Res. Commun.* **77,** 830 (1977).

[22] R. B. Freedman, A. Newell, and C. M. Walklin, *FEBS Lett.* **88,** 49 (1978).

organic solvents[20] or by freeze-thawing, sonication or exposure to hypotonic conditions. Since all these treatments might be expected to break down the permeability barrier of the endoplasmic reticulum membrane, it is tempting to conclude that the latency is due to the enzyme being located facing the lumen of the endoplasmic reticulum, which is known to be the site at which most disulfide bonds are formed in nascent secretory proteins.[23,24] However, there is insufficient direct structural information available to establish this conclusion. The enzyme is readily solubilized from membranes, and the pure enzyme is freely soluble in the absence of detergents; this suggests that it is a peripheral loosely bound protein and is unlikely to have a transmembrane disposition.

We have identified the position of protein disulfide-isomerase in two-dimensional electrophoretograms of rat liver microsomal proteins.[25] This confirms that protein disulfide-isomerase is a prominent acidic polypeptide comprising ~2% of total microsomal protein. Pretreatments of intact microsomes with various agents show that this spot is very resistant to proteolysis by trypsin, but is readily labeled by putatively impermeant compounds reactive toward sulfhydryl groups.[25a] Ohba et al.[11] showed that the membrane-bound enzyme bound less anti-protein disulfide-isomerase antibody and had less accessible Gln residues than did the solubilized enzyme.

It is not possible at present to reconcile all these findings; the enzyme may be located at the luminal surface of the endoplasmic reticulum membrane, or, alternatively, it may be located near the cytoplasmic surface, but with a large portion, including the active site, inaccessible, being obscured by some other membrane component. The resolution of this problem is clearly important in providing a picture of how the enzyme could operate in the posttranslational modification of nascent proteins.

Purification

The enzyme was initially purified from microsomal membranes by extraction of lipids by acetone and then extraction of the acetone powder by aqueous buffers.[26] This method and variants of it have been quite

[23] L. W. Bergman and W. M. Kuehl, J. Biol. Chem. 254, 5690 (1979).

[24] A. Oohira, H. Nagami, A. Kusakabe, K. Kimata, and S. Suzuki, J. Biol. Chem. 254, 3576 (1979).

[25] E. N. C. Mills, N. Lambert, and R. B. Freedman, Biochem. J. 213, 245 (1983).

[25a] M. A. Kaderbhai, Ph.D. Thesis, University of Kent, Canterbury, UK, 1981.

[26] F. De Lorenzo, R. F. Goldberger, E. Steers, D. Givol, and C. B. Anfinsen, J. Biol. Chem. 241, 1562 (1966).

widely used,[11,20,27] but it is time-consuming and yields are poor. A rapid, high-yielding procedure for purification of a thiol: protein disulfide oxidoreductase was introduced by Carmichael et al.[28] and subsequently refined for the preparation of homogeneous protein disulfide-isomerase.[29] The procedure, summarized in Fig. 2, uses detergent solubilization of a whole homogenate, exploiting the enzyme's heat stability and its very low pI; it is rapid, high-yielding, and reproducible.

Procedure

Homogenization in Triton. Bovine liver from freshly slaughtered animals is freed of connective tissue and cut into small cubes. The diced liver is stored at $-20°$ in 500-g lots. On demand, 500 g is thawed overnight at $4°$, washed in ice-cold saline (0.9%, w/v, NaCl) and homogenized in 1 liter of "homogenizing buffer" [0.1 M sodium phosphate, pH 7.5–5 mM EDTA–1% (v/v) Triton X-100] using a Waring blender at full speed for 4 × 30-sec bursts. The homogenate is strained through two layers of muslin and centrifuged at 18,000 g for 30 min at $4°$; the pellet is discarded.

Heat Treatment. The supernatant is decanted through glass wool and then placed in a water bath at $70°$ with constant stirring. The temperature of the extract is allowed to reach $54°$ ($±1°$), where it is maintained for 15 min. The treated extract is then transferred to an ice bath to cool; it is then centrifuged at 18,000 g for 40 min. This centrifugation and all subsequent steps are carried out at $4°$.

Ammonium Sulfate Fractionation. Solid ammonium sulfate is slowly added to the supernatant from the heat denaturation to be finally 55% saturated. After stirring for 30 min, the material is centrifuged at 18,000 g for 30 min and the pellet is discarded. Further ammonium sulfate is added to a final saturation of 85%, and the material is centrifuged as before. The supernatant is discarded, and the pellet is taken up in approximately 20 ml of 25 mM citrate buffer, pH 5.3, then dialyzed overnight against this buffer (2 × 5 liters).

Cation-Exchange Chromatography. The dialyzed extract is applied to a column of CM-Sephadex C-50 (36 × 6 cm) previously equilibrated with citrate buffer, pH 5.3, and eluted with the same buffer at a flow rate of 3–4 ml min^{-1}. Fractions (15 ml) are collected, and the void protein fractions that contain protein disulfide-isomerase activity are pooled. Protein is precipitated by addition of ammonium sulfate at 100% saturation, and the

[27] D. A. Hillson and R. B. Freedman, *Biochem. J.* **191,** 373 (1980).
[28] D. F. Carmichael, J. E. Morin, and J. E. Dixon, *J. Biol. Chem.* **252,** 7163 (1977).
[29] N. Lambert and R. B. Freedman, *Biochem. J.* **213,** 225 (1983).

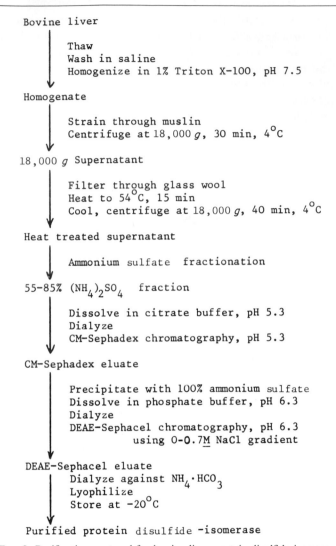

FIG. 2. Purification protocol for bovine liver protein disulfide-isomerase.

pellet is collected by centrifugation at 18,000 g for 30 min. The pellet is dissolved in about 15 ml of 20 mM sodium phosphate buffer, pH 6.3, and dialyzed overnight against the same buffer (2 × 2 liters).

Anion-Exchange Chromatography. The dialyzed extract is loaded on to a column of DEAE-Sephacel (35 × 2 cm) equilibrated with phosphate buffer, pH 6.3, and eluted with a linear gradient of 0 to 0.7 M NaCl in the same buffer at a flow rate of 20 ml hr^{-1}. Fractions (5 ml) are collected, and

PURIFICATION OF PROTEIN DISULFIDE-ISOMERASE (PDI) FROM BEEF LIVER (1 kg)

Steps	Protein (g)	PDI activity (units)	Specific activity[a] (units/g)	Yield[a] (%)
Homogenate	191	770	4.03 (3.5)	100 (100)
18,000 g Supernatant	137	750	5.47 (4.0)	97 (91)
Heat-treated supernatant	31.2	624	20.0 (14.0)	81 (79)
55–85% $(NH_4)_2SO_4$ cut	9.16	414	45.2 (41.0)	54 (81)
CM-Sephadex pH 5.3 eluate	1.45	262	181 (181)	34 (25)
DEAE-Sephacel pH 6.3 eluate	0.22	206	936 (950)	27 (18)

[a] Figures in parentheses refer to a similar preparation from rat liver.

three peaks of protein are detectable by absorbance at 280 nm. The second peak contains protein disulfide-isomerase activity; active fractions are pooled, dialyzed against $NH_4 \cdot HCO_3$ (50 mM), and freeze-dried. The freeze-dried enzyme is stored at $-20°$. After 9 months in storage the enzyme retains more than 70% of initial activity.

The table shows complete data for a purification from bovine liver and data on yields and specific activities for a comparable preparation from rat liver.[25,29]

For five preparations from bovine liver, the mean and standard deviation values for protein disulfide-isomerase specific activity and yield in the final extract were 1001 ± 136 units/g and 25 ± 6%, respectively.

The preparation contains a small proportion of bound Triton X-100 at the end of the procedure. This can be removed, if necessary, by stirring a solution of the purified enzyme (10 ml, 1 mg/ml in 50 mM sodium phosphate, pH 7.5) with 1 g of washed Amberlite XAD-2 resin, decanting the supernatant, repeating the Amberlite treatment, decanting again, and centrifugation in a bench centrifuge to remove the fines.

Molecular Properties of the Purified Enzyme

Purified protein disulfide-isomerase is homogeneous by several criteria. Determination of molecular weight by SDS–polyacrylamide gel electrophoresis in reducing and nonreducing conditions and by gel filtration show that it is a homodimer (107,000) of 57,000-dalton polypeptides. The association between subunits is noncovalent,[29] but formation of disulfide-linked aggregates can occur on storage.[30] The enzyme is extremely acidic with a pI of 4.2 (measured in 8 M urea). Analysis of the amino acid composition[29] indicates a high proportion (27%) of Glx + Asx residues,

[30] M. Pace and J. E. Dixon, *Int. J. Pept. Protein Res.* **14**, 409 (1979).

consistent with the acidic isoelectric point of the enzyme, and shows that the N-terminal residue is His. The carbohydrate content of the purified enzyme is <1%.

Work on previous preparations of the enzyme showed that it contains a catalytically essential thiol or dithiol group. Anfinsen's group[31] isolated a peptide . . . Ala · Tyr · Cys · (His,Gly) · Lys . . . claimed to contain the active-site —SH group, and they proposed that this group cooperated with a disulfide group in the catalytic mechanism of the enzyme. Further evidence for the involvement of a redox active dithiol–disulfide group in the enzyme's activity has been obtained from chemical modification studies[32] and inhibition studies.[27]

The enzymes from bovine and rat liver are closely similar in structural properties[11,25,29]; a similarity in primary structure is implied by a comparison of protease fingerprints.[25] Preliminary studies on the enzymes from chick embryo tendon and from wheat endosperm show that they resemble the liver enzymes in isoelectric point and in molecular weight, implying a high degree of conservation.

Specificity of Action

A large and unknown number of individual thiol–disulfide interchanges are involved in the conversion of a reduced or "scrambled" substrate molecule to the native state. A catalyst of this process would thus not be expected to show very restricted specificity.

A wide specificity for thiols is indeed exhibited by protein disulfide-isomerase,[27] and both crude and purified preparations show comparable activity with dithiols, such as dithiothreitol ($10^{-5} M$); simple monothiols, such as glutathione, mercaptoethanol, cysteine, cysteamine (each at $10^{-3} M$); or protein thiols, such as reduced bovine serum albumin ([SH] = $10^{-4} M$).

A similarly broad substrate specificity has been found toward the protein disulfide substrate,[33] including small single-chain polypeptides, multidomain and multichain proteins. Among proteins whose reduced or scrambled derivatives are renatured by protein disulfide-isomerase are ribonuclease, lysozyme, bovine serum albumin, soybean trypsin inhibitor, and bovine pancreatic trypsin inhibitor; the action of the enzyme on the latter has been particularly well characterized.[34] Protein disulfide-

---

[31] F. De Lorenzo, S. Fuchs, and C. B. Anfinsen, *Biochemistry* **5**, 3961 (1966).

[32] D. F. Carmichael, M. Keefe, M. Pace, and J. E. Dixon, *J. Biol. Chem.* **254**, 8386 (1979).

[33] R. B. Freedman and D. A. Hillson, *in* "The Enzymology of Post-Translational Modifications of Proteins" (R. B. Freedman and H. C. Hawkins, eds.), Vol. 1, p. 157. Academic Press, New York, 1980.

[34] T. E. Creighton, D. A. Hillson, and R. B. Freedman, *J. Mol. Biol.* **142**, 43 (1980).

isomerase also catalyzes assembly of the intact oxidized forms of the oligomeric immunoglobulins IgM and IgA from monomeric precursors.[33]

In addition to these activities leading to renaturation, isomerization of disulfide bonds leading to "scrambling" is catalyzed in proteins containing metastable disulfides, such as insulin or ribonuclease S-protein, indicating that net changes in disulfide bond arrangement catalyzed by protein disulfide-isomerase are determined by thermodynamic considerations intrinsic to the substrate and lead to the disulfide pairing consistent with the most stable tertiary structure.[33]

Despite the wide specificity of protein disulfide-isomerase toward both thiol and protein substrates, the nature of the reactions catalyzed is more limited. Detailed analysis of the enzyme's activity in the reoxidation of reduced bovine pancreatic trypsin inhibitor[34] indicates that those steps catalyzed involve intramolecular reactions between a protein-borne thiol and either a protein disulfide or a protein-reducing agent mixed disulfide. Such intramolecular thiol-disulfide interchanges are thought to be rate-determining in the formation of native disulfide bonds.[35]

In net reducing conditions (e.g., at concentrations of thiols greater than 10-fold those used in the assay of protein disulfide-isomerase), the isomerization activity of protein disulfide isomerase is diminished, presumably because the circumstances favor the formation of the fully reduced protein. For example, in these conditions the enzyme will catalyze the net reduction of insulin by reduced glutathione. More recent evidence[5,29,36] confirms earlier suggestions[37] that protein disulfide-isomerase is identical or homologous to enzymes previously identified as glutathione : insulin transhydrogenase or thiol : protein-disulfide-oxidoreductase when assayed in conditions that promote this net reduction activity.

It is not known whether protein disulfide-isomerase preparations from different sources show different specificities, for example, with preferences for source-specific proteins. For example a liver enzyme might show preference for serum albumin as the disulfide substrate, a skin enzyme may be specific for procollagen assembly, or an activity in lymphocytes may show specificity for immunoglobulin assembly and polymerization. However, Roth and Koshland[12] found no differences between enzyme preparations isolated from mouse spleen and liver.

[35] T. E. Creighton, *Prog. Biophys. Mol. Biol.* **33**, 231 (1978).

[36] S. Bjelland, K. Wallerik, J. Kroll, J. E. Dixon, J. E. Morin, R. B. Freedman, N. Lambert, P. T. Varandani, and M. A. Nafz, *Biochim. Biophys. Acta* **747**, 197 (1983).

[37] P. T. Varandani, *in* "Mechanisms of Oxidizing Enzymes" (T. P. Singer and R. N. Ondarza, eds.), p. 29. Elsevier/North-Holland, Amsterdam, 1978.

Other Enzymes

It is evident that the reactions catalyzed by protein disulfide-isomerase and other thiol : protein-disulfide oxidoreductases are wide-ranging in their effects on the thiol/disulfide status of proteins. However, their role in the *de novo* formation of disulfide bonds is far from clear. In particular, they cannot act as oxidants in any direct way, apart from participation of some low molecular weight compound, such as oxidized glutathione (GS-SG) or cystamine, or some other source of oxidizing equivalents, such as protein-S-SG mixed disulfides. These are not thought to be present in the *cytoplasm* in sufficient concentrations to enable them to act as effective oxidants for protein disulfide-bond formation, but the effective SH/SS redox potential at the site of protein disulfide formation is not known (see Freedman and Hillson[33] and elsewhere in this volume for review).

Several enzymes have been postulated to play a role in net formation of protein disulfides, none of which are thiol : disulfide oxidoreductases. Thiol or sulfhydryl oxidases[38-40] catalyze direct oxidation by molecular oxygen in glutathione, cysteine, and various proteins, thus

$$2 \text{ RSH} + \text{O}_2 \rightarrow \text{RSSR} + \text{H}_2\text{O}_2$$

None of these activities has been sufficiently well characterized to allow any firm conclusions to be drawn concerning their role in protein biosynthesis. Another enzyme postulated to act in protein disulfide formation is a microsomal "amine" oxidase.[41-43] This flavoprotein mixed-function oxidase can act on cysteamine in the presence of NADPH and O_2 to produce cystamine, which it is suggested could serve as a source of oxidizing equivalents in the formation of protein disulfides, this last step possibly being catalyzed by protein disulfide-isomerase.

Despite their promising catalytic capabilities, none of these enzymes has been firmly established as participating in the process of disulfide-bond formation in protein biosynthesis.

[38] V. G. Janolino and H. E. Swaisgood, *J. Biol. Chem.* **250,** 2532 (1975).
[39] T. S. K. Chang and B. Morton, *Biochem. Biophys. Res. Commun.* **66,** 309 (1975).
[40] M. C. Ostrowski and W. S. Kistler, *Biochemistry* **19,** 2639 (1980).
[41] D. M. Ziegler and C. H. Mitchell, *Arch. Biochem. Biophys.* **150,** 116 (1972).
[42] L. L. Poulsen, R. M. Hyslop, and D. M. Ziegler, *Biochem. Pharmacol.* **23,** 3431 (1974).
[43] L. L. Poulsen and D. M. Ziegler, *Arch. Biochem. Biophys.* **183,** 563 (1977).

[17] Enzymatic Reduction–Oxidation of Protein Disulfides by Thioredoxin

By Arne Holmgren

A protein disulfide bond may be cleaved by an enzymatic system involving the small redox protein thioredoxin.[1] Thioredoxin (M_r 12,000) and the associated enzyme thioredoxin reductase[2] constitute a thiol-dependent reduction–oxidation system that can catalyze the reduction of certain disulfides by NADPH.

Originally, thioredoxin was purified and named from *Escherichia coli* as a hydrogen donor cofactor in the biosynthesis of deoxyribonucleotides by the essential enzyme ribonucleotide reductase.[3] This enzyme contains a dithiol/disulfide that is part of the active center.[4] In its oxidized form the active center disulfide is reduced by thioredoxin. However, thioredoxin is not unique in its capacity to be hydrogen donor for ribonucleotide reductase, as shown later by the discovery of the GSH-dependent glutaredoxin system in *E. coli* mutants lacking detectable thioredoxin.[5] In mammalian species, thioredoxin is present in cells irrespective of active proliferation and DNA synthesis,[6] a finding strongly suggesting physiological functions for thioredoxin in protein disulfide reduction–oxidation and "thiol redox control" of enzyme activity.[1,2,7] The reader is referred to reviews[1,2] for an overview of the occurrence and structure–function relationships of thioredoxin and thioredoxin reductase.

Principle

The thioredoxin system is composed of three components: NADPH, thioredoxin reductase, and thioredoxin. It catalyzes thiol-dependent disulfide reduction by a combination of reactions (1) and (2).

[1] A. Holmgren, *Trends Biochem. Sci.* **6**, 26 (1981).
[2] A. Holmgren, *in* "Dehydrogenases Requiring Nicotinamide Coenzymes" (J. Jeffery, ed.), p. 149. Birkhaeuser, Basel, 1980.
[3] T. C. Laurent, E. C. Moore, and P. Reichard, *J. Biol. Chem.* **239**, 3436 (1964).
[4] L. Thelander, *J. Biol. Chem.* **249**, 4858 (1974).
[5] A. Holmgren, *Proc. Natl. Acad. Sci. U.S.A.* **73**, 2275 (1976).
[6] A. Holmgren and M. Luthman, *Biochemistry* **17**, 4071 (1978).
[7] S. Black, E. M. Harte, B. Hudson, and L. Wartofsky, *J. Biol. Chem.* **235**, 2910 (1960).

$$\text{Thioredoxin-S}_2 + \text{NADPH} + \text{H}^+ \xrightleftharpoons{\text{thioredoxin reductase}} \text{thioredoxin-(SH)}_2 + \text{NADP}^+ \quad (1)$$

$$\text{Thioredoxin-(SH)}_2 + \text{protein-S}_2 \xrightleftharpoons{} \text{thioredoxin-S}_2 + \text{protein-(SH)}_2 \quad (2)$$

$$\text{Net: NADPH} + \text{H}^+ + \text{protein-S}_2 \xrightleftharpoons{} \text{NADP}^+ + \text{protein-(SH)}_2 \quad (3)$$

Thioredoxin-S$_2$ denotes the oxidized form of thioredoxin and thioredoxin-(SH)$_2$ the reduced form. Note that the second reaction is spontaneous and that thioredoxin there may also be regarded as an enzyme.[8] The reduction of thioredoxin by NADPH at pH 7 is virtually complete; an equilibrium constant of 48 at pH 7 has been determined.[3] Since an excess of NADPH is used in reduction experiments, the reducing power of NADPH is driving the reactions (reaction 3), with thioredoxin and thioredoxin reductase in catalytic functions.

Reduction of disulfide bonds by the thioredoxin system has so far been described only for a limited number of proteins, namely; insulin,[8-12] fibrinogen and fibrin,[13] choriogonadotropin and its α and β subunits,[14] and coagulation factor VIII and the other coagulation factors in plasma.[15] The following method is used for reduction of insulin, a protein easily available and useful as a model for calibration of the reduction enzymes.

Spectrophotometric Assay of the Reduction of Protein Disulfides by Thioredoxin

Reagents

NADPH, 10 mM, in H$_2$O. Store frozen at $-20°$

Thioredoxin-S$_2$ from *Escherichia coli*,[16] 500 μM (5.8 mg/ml) dissolved in 0.06 M NH$_4$HCO$_3$. Store at $-20°$

Thioredoxin reductase from *Escherichia coli*,[17,18] 10 μM (0.7 mg/ml)

[8] A. Holmgren, *J. Biol. Chem.* **254,** 9627 (1979).

[9] N. E. Engström, A. Holmgren, A. Larsson, and S. Söderhäll, *J. Biol. Chem.* **249,** 205 (1974).

[10] A. Holmgren, *J. Biol. Chem.* **252,** 4600 (1977).

[11] A. Holmgren, *J. Biol. Chem.* **254,** 9113 (1979).

[12] M. Luthman and A. Holmgren, *Biochemistry* **21,** 6628 (1982).

[13] B. Blombäck, M. Blombäck, W. Finkbeiner, A. Holmgren, B. Kowalska-Loth, and G. Olovson, *Thromb. Res.* **4,** 55 (1974).

[14] A. Holmgren and F. J. Morgan, *Eur. J. Biochem.* **70,** 377 (1976).

[15] G. Savidge, G. Carlebjörk, L. Thorell, B. Hessel, A. Holmgren, and B. Blombäck, *Thromb. Res.* **16,** 587 (1979).

[16] A. Holmgren and P. Reichard, *Eur. J. Biochem.* **2,** 187 (1967).

[17] L. Thelander, *J. Biol. Chem.* **242,** 852 (1967).

[18] V. Pigiet and R. R. Conley, *J. Biol. Chem.* **252,** 6367 (1977).

in 50 mM potassium phosphate–1 mM EDTA, pH 7.0. Store at −20°

Insulin, bovine, 26 units/mg, 1.6 mM (10 mg/ml). This solution is prepared by suspending 10 mg of insulin in 0.5 ml of 50 mM Tris-Cl, pH 7.5, and adjusting the pH to 3 with 1.0 M HCl to dissolve the protein completely. The solution is then titrated back to pH 7.5 with 1.0 M NaOH, and the final volume is adjusted to 1.00 ml. The perfectly clear solution may be stored frozen at −20°

Procedure. Two cuvettes at 25° contained 0.10 M potassium phosphate–2 mM EDTA, pH 7.0, and 0.4 mM NADPH. To one cuvette, which was placed in the reference position in the spectrophotometer, was added insulin or any other protein (25–100 μM) and to the other an equivalent volume of buffer. To both cuvettes was added E. coli thioredoxin-S$_2$ (0.1–5 μM). The final volume was 0.50 ml. After careful determination of the initial absorbance at 340 nm, the enzymatic reaction was started by addition of 5 or 10 μl of E. coli thioredoxin reductase (10 μM) to each cuvette. The reaction was followed by recording the increase in absorbance at 340 nm at 0.5- or 1-min intervals.

Calculation of Result. The amount of disulfide reduced is calculated from the expression: $C = \Delta A_{340}/6200$ where C is the amount of disulfide reduced (molar), ΔA_{340} is the change in absorbance at 340 nm, and 6200 is the molar absorption coefficient for NADPH at 340 nm. The reduction of 1 μM of disulfide thus corresponds to a ΔA_{340} of 0.0062.

Measurements of SH Groups. Several methods may be used to determine the sulfhydryl groups formed from the disulfides reduced.

Spectrophotometric Determination with DTNB. Thiol groups are measured by 5,5′-dithiobis(2-nitrobenzoic acid) (DTNB) in 4–6 M guanidine-HCl–0.10 M Tris-Cl, pH 8.0, by Ellman's method.[19] This terminates the enzymatic reaction instantaneously. From a freshly prepared mixture of 8 M guanidine-HCl–50 mM Tris-Cl, pH 7.5–1 mM DTNB, 0.50 ml is added to the sample and reference and allowed to react for 5 min at 25°. Then the absorbance at 412 nm is determined, and the concentration of SH groups is calculated from the expression $C = \Delta A_{412}/13,600$ where C is the molar concentration of SH groups, ΔA_{412} is the net absorbance at 412 nm, and 13,600 is the molar absorption coefficient for 5′-thionitrobenzoic acid.[19] Note that 2 mol of SH are formed per mole of disulfide reduced (or NADPH oxidized). Furthermore, DTNB also measures the free SH groups of thioredoxin present in the reaction mixture.

Attempts to use DTNB to determine SH groups under nondenaturing conditions will be successful only if the activity of the thioredoxin system

---

[19] G. L. Ellman, *Arch. Biochem. Biophys.* **82,** 70 (1959).

is destroyed before addition of DTNB. This is because DTNB is an excellent disulfide substrate for the thioredoxin system; in fact it is used in assay methods.[3,16] DTNB under native conditions may be utilized if it is combined with stopping the enzymatic reaction by one of the following methods: (1) heating to 80°, which inactivates thioredoxin reductase; (2) addition of anti-thioredoxin antibodies or, better, anti-thioredoxin reductase antibodies; (3) addition of an inhibitor of thioredoxin reductase, such as N-ethylmaleimide (1–10 μM) or arsenite (1–10 μM). The conditions for these reagents may have to be tested owing to possible reactions with the newly formed protein SH groups.

Alkylation with Iodo[^{14}C]acetic Acid. A fresh solution of neutralized 0.1 M iodo[^{14}C]acetic acid (400–6000 cpm/nmol) in 2 M Tris-Cl, pH 8.0, is prepared. To the reduction reaction mixture is added the iodoacetic acid to a final concentration of 1–10 mM, depending on amount of protein. The mixture is allowed to react for 30 min at room temperature in the dark. Finally, excess reagent is separated from the protein by chromatography on a column of Sephadex G-25 in 0.06 M NH$_4$HCO$_3$, pH 8.0, or 0.2 M acetic acid. Determination of the amount of radioactivity by using a liquid scintillation counter allows the amount of carboxymethylcysteine to be determined.[20] This can also be accomplished by amino acid analysis.[20] The modified protein may be used for sequence analysis to determine the S—S bonds that have reacted. This is preferably done after complete reduction and alkylation using denaturing conditions like 6 M guanidine-HCl and dithiothreitol plus unlabeled iodoacetic acid.[14]

Comments

Reduction of disulfide bonds by the thioredoxin system has several advantages when compared with the chemical methods for reductive cleavage of disulfide bonds employing 2-mercaptoethanol or dithiothreitol (DTT).[21] However, it has also some limitations. The thioredoxin system from *E. coli* is superior to the systems from mammalian cells that are more difficult to purify and measure owing to the easy inactivation by oxidation of structural SH groups.[12]

Selectivity. It is generally assumed that only exposed protein disulfides are reduced by thioredoxin. In fibrinogen only 5 out of 29 disulfides react; all bonds cleaved are located in hydrophilic regions of the molecule. Thus, the system confers a highly selective reduction of sur-

[20] C. H. W. Hirs, this series, Vol. 11, p. 199.
[21] W. W. Cleland, *Biochemistry* **3**, 480 (1964).

REDUCTION OF PROTEIN DISULFIDE BONDS BY THIOREDOXIN-$(SH)_2$[a]

Protein	S—S bonds reduced[b]	Reactivity
Apocytochrome c[c]	1/1	Fast
Chymotrypsin	2/5	1 Fast, 1 slow
Insulin and proinsulin	3/3	3 Fast
Lysozyme	0/4	—
Ribonuclease	0/4	—
Thrombin[d]	0/4	—
Trypsin	3/6	3 Fast
Fibrinogen[e]	5/29	5 Fast
α-Subunit of choriogonadotropin[f]	5/5	5 Fast

[a] Reduction experiments were performed at 23° in 0.1 M potassium phosphate–2 mM EDTA, pH 7.0–0.4 mM NADPH with 5 μM thioredoxin and 0.1 μM thioredoxin reductase. Protein concentrations ranged from 25 to 40 μM in the different experiments. The proteins were commercial products from Sigma Chemical Co., St. Louis, MO.

[b] The first figure shows the number of S—S bonds reduced calculated from the ΔA_{340} values. The second figure denotes the total number of S—S bonds of the proteins.[23]

[c] Prepared as described by Fisher et al.[24] and oxidized in air.

[d] Highly purified. This was a kind gift from Dr. Birger Blombäck, Stockholm.

[e] Data from Blombäck et al.[13]

[f] Data from Holmgren and Morgan.[14]

face-oriented disulfides that might help to identify the structure–function relationships for individual disulfides in a protein.

Individual Reactivity. It is known that the rate of reduction of protein disulfides varies with the thioredoxin system using the homologous proteins trypsin, chymotrypsin, and thrombin as substrates.[22] So far almost nothing is known about the factors governing this reactivity. The two interchain disulfides of insulin are known to be much more efficient substrates than the low molecular weight disulfides L-cystine and GSSG.[10] By breaking the reaction at intervals, the order of reaction of disulfides may possibly be elucidated. The table summarizes results for some proteins.[13,14,23,24]

Specificity. With homogeneous preparations of thioredoxin and thioredoxin reductase, the side reactions of the reduction will be minimal, since

[22] A. Holmgren, manuscript in preparation.

[23] M. O. Dayhoff, "Atlas of Protein Sequence and Structure," Vol. 5, Suppl. I. National Biomedical Research Foundation, Silver Spring, Maryland, 1973.

[24] W. R. Fisher, H. Taniuchi, and C. B. Anfinsen, *J. Biol. Chem.* **248,** 3188 (1973).

no excess of reductant is present. The actual reductant thioredoxin-$(SH)_2$ is present in no excess, and this ought to minimize the risk of thiol–disulfide interchange reactions. Furthermore, as described above, DTNB may be used to determine SH groups; no large excess of alkylating agent is required, since there is no excess of thiol present.

General Properties of the Thioredoxin System. Thioredoxin reductase is specific for the disulfide bridge in thioredoxin. The K_m values for NADPH and thioredoxin-S_2 are 1.2 and 2.8 μM, respectively, at 25° and pH 7.6.[25] The pH optimum for reduction of insulin is 7.5. The redox potential of thioredoxin-S_2/thioredoxin-$(SH)_2$ at pH 7.0 and 25° is -0.26 V,[2] which is similar to 2 GSH/6SSG (-0.25 V) but higher than that of NADPH/NADP$^+$ (-0.31 V).

Urea up to 2 M decreases the rate of the reaction to about 50%, and considerable activity is obtained in 4 M urea. However, guanidine-HCl at low concentration (<0.1 M) is a very effective inhibitor of thioredoxin reductase.[10]

Air Oxidation. The presence of O_2 in air will reoxidize the dithiols formed and thus lead to cyclic consumption of NADPH with no corresponding net disulfide reduction. In addition, the possibilities of uncontrolled dithiol–disulfide interchange reactions will increase. The use of N_2-equilibrated buffers will eliminate this problem. However, it may also be possible to use the O_2 oxidation of dithiols to study the re-formation of disulfides by leaving the mixture in contact with air for many hours. The complete oxidation of NADPH can be measured spectrophotometrically at 340 nm, and the oxidation of SH groups can be followed by DTNB as described above.

Effect of pH. At pH 9 or above the equilibrium of reaction (2) may favor formation of disulfides. This has so far been little studied.

Use of Dithiothreitol as Reductant for Thioredoxin. Thioredoxin catalyzes dithiol–disulfide oxidoreductions with dithiothreitol or dihydrolipoamide as dithiol substrates. As a testament to the efficiency of thioredoxin-$(SH)_2$ as reductant, it should be noted that *E. coli* thioredoxin-$(SH)_2$ reduces bovine insulin at least 10,000 times more rapidly than dithiothreitol at pH 7.[11]

Acknowledgments

This investigation was supported by grants from the Swedish Medical Research Council 13X-3529, the Swedish Cancer Society (961), and Nordisk Insulinfond. The excellent secretarial assistance of Mrs. Agneta Sjövall is gratefully acknowledged.

[25] C. H. Williams, Jr., *in* "The Enzymes" (P. D. Boyer, ed.), Vol. 13, p. 89. 3rd ed.

[18] Nonenzymatic Formation and Isomerization of Protein Disulfides

By Donald B. Wetlaufer

The formation of protein disulfides can proceed either by enzymatic catalysis[1-3] or by uncatalyzed thiol–disulfide exchange.[4,5] It is not yet clear which of these mechanisms predominates *in vivo*.

It has been long known that strong acid can lead to disulfide isomerization.[6] It is also well known that disulfides can be cleaved by sulfitolysis in a mechanism that has formal similarity to thiol–disulfide exchange.[7] We will discuss neither of these further, but will deal with nonenzymatic systems that employ thiol–disulfide exchange reactions between proteins and low molecular weight thiols and/or disulfides.

In the covalent reaction (1) it is important to note that the thiolate anion, $R'''-S^-$, is the reactive species, not the protonated thiol itself, $R'''-SH$. The fact that most common alkyl thiols have proton ionization pK values in the range of 8–10 means that the exchange reaction can proceed rapidly under mildly alkaline or neutral conditions; conversely, it is inhibited by acid.

$$R'-S-S-R'' + R''S^- \rightleftharpoons \begin{cases} R'-S-S-R''' + R''S^- \\ \text{or} \\ R''-S-S-R''' + R'S^- \end{cases} \tag{1}$$

The second point to note is that the reaction is reversible; this ready reversibility proves to be very important. We can, for example, prepare the mixed disulfide between a reduced protein and cysteine (or other low molecular weight disulfide) simply by forcing the equilibrium of Eq. (1) as follows:

$$P(-S^-)_x + 1000x\,Cy-S-S-Cy \rightleftharpoons P(-S-S-Cy)_x + xCyS^- \\ + (1000x - x)Cy-S-S-Cy \tag{2}$$

[1] D. Givol, F. DeLorenzo, R. F. Goldberger, and C. B. Anfinsen, *Proc. Natl. Acad. Sci. U.S.A.* **53,** 676 (1965).

[2] H. M. Katzen and D. Stetten, Jr., *Fed. Proc. Fed. Am. Soc. Exp. Biol.* **21,** 201 (1962); H. M. Katzen and F. Tietze, *J. Biol. Chem.* **241,** 3561 (1966).

[3] R. B. Freedman and H. C. Hawkins, *Biochem. Soc. Trans. Biochem. Rev.* **5,** 348 (1977).

[4] D. B. Wetlaufer, V. P. Saxena, A. K. Ahmed, S. W. Schaffer, P. W. Pick, K.-J. Oh, and J. D. Peterson, *in* "'Protein Crosslinking," Part A (M. Friedman, ed.), p. 43. Plenum, New York, 1977.

[5] V. P. Saxena and D. B. Wetlaufer, *Biochemistry* **9,** 5015 (1970).

[6] A. P. Ryle and F. Sanger, *Biochem. J.* **60,** 535 (1955).

[7] W. W. Chan, *Biochemistry* **7,** 4247 (1968).

METHODS IN ENZYMOLOGY, VOL. 107

Bradshaw *et al.*[8] prepared the mixed disulfide of lysozyme and cysteine in just this fashion. The example in Eq. (2) makes the assumptions that the pK values and the oxidoreduction potentials of the several thiol–disulfide pairs are the same. While this is not strictly true, it is a reasonable approximation under the conditions employed. This mixed disulfide was used as the starting form of the protein in experiments, wherein lysozyme activity was regenerated[8] by the addition of mercaptoethanol, cysteine, or cysteamine to solutions buffered in the pH range 7 to 8.5.

$$
\begin{array}{c}
\text{Cy—S—S} \quad \text{S—S—Cy} \\
\diagdown \quad \diagup \\
\text{P} \\
\diagdown \\
\text{S—S—Cy}
\end{array}
+ \text{RS}^- \rightleftharpoons
\left\{
\begin{array}{l}
\begin{array}{c}
\text{Cy—S—S} \quad \text{S—S—R} \\
\diagdown \quad \diagup \\
\text{P} \\
\diagdown \\
\text{S—S—Cy}
\end{array}
+ \text{CyS}^- \\
\text{or} \\
\begin{array}{c}
\text{Cy—S—S} \quad \text{S}^- \\
\diagdown \quad \diagup \\
\text{P} \\
\diagdown \\
\text{S—S—Cy}
\end{array}
+ \text{Cy—S—S—R}
\end{array}
\right.
\tag{3}
$$

Reactions of the type indicated by Eq. (3) can occur with any one of the protein–cysteine mixed disulfides. In addition to the back reaction of Eq. (3), the lower protein product on the right-hand side can undergo intramolecular thiol–disulfide exchange [Eq. (4)].

$$
\begin{array}{c}
\text{Cy—S—S} \quad \text{S}^- \\
\diagdown \quad \diagup \\
\text{P} \\
\diagdown \\
\text{S—S—Cy}
\end{array}
\rightleftharpoons
\left\{
\begin{array}{l}
\begin{array}{c}
\text{Cy—S—S} \quad \text{S} \\
\diagdown \quad \diagup \,| \\
\text{P} \quad | \\
\diagdown \quad \text{S}
\end{array}
+ \text{CyS}^- \\
\text{or} \\
\begin{array}{c}
\text{S——S} \\
\diagdown \diagup \\
\text{P} \\
\diagdown \\
\text{S—S—Cy}
\end{array}
+ \text{CyS}^-
\end{array}
\right.
\tag{4}
$$

It should be obvious that, if one of the two protein products in Eq. (4) represents a native pairing of protein thiols, the other must be nonnative. The back-reaction of Eq. (4) provides a route for removing nonnative disulfides from reaction intermediates in an oxidative regeneration.

Often one begins with a completely reduced disulfide protein and wishes to regenerate the native disulfides rapidly and in high yield. The preparation of reduced proteins has been described elsewhere.[9,10] We can see by inspection of the above equations that the rates of some essential

[8] R. A. Bradshaw, L. Kanarek, and R. L. Hill, *J. Biol. Chem.* **242**, 3789 (1967).
[9] F. H. White, Jr., this series, Vol. 11, p. 481.
[10] S. S. Ristow and D. B. Wetlaufer, *Biochem. Biophys. Res. Commun.* **50**, 544 (1973).

reactions will depend on the concentration of R—S—S—R and others will depend on the concentration of R—S⁻. However, all the reactions are expressed by Eq. (1); that is, they are all thiol–disulfide exchange reactions.[5] Empirically we have found[5,11,12] that a concentration of R—S⁻ of approximately 1–3 mM in the presence of one-fifth to one-tenth of that concentration of R—S—S—R provides nearly optimal conditions for the oxidative re-formation of native protein disulfides. Typical conditions for such a reaction follow: Tris acetate buffer, 0.02 M, pH 8.0, 25°, 1 μM reduced protein, 3.0 mM reduced glutathione, 0.3 mM oxidized glutathione, 1 mM EDTA. If the regeneration requires more than a few hours, it may prove advantageous to exclude atmospheric oxygen.

Re-formation of protein disulfides may be monitored by a number of analytical methods.[8,10] Half-times for formation of protein disulfides have been found between a few minutes and 2 hr. It may be more relevant to monitor the appearance of a specific biological activity. The latter requires a rapid specific assay under conditions inhibitory to further thiol–disulfide exchange—for example, a few minutes at pH 5.[5,11,12] Typically, the biological activity appears more slowly than the protein disulfide.[5,8,11,12] The above regeneration system has proved to be relatively insensitive to temperature over the range 25–40°. Regenerations of hen egg white lysozyme with a glutathione system as described above showed substantially less than one protein-glutathione mixed disulfide throughout the whole course of the reaction.[13]

Means and Feeney[14] have noted, and we have also observed, that reduced proteins have a strong tendency to aggregate. This is why we recommend that oxidative renaturation be carried out at low protein concentrations. In the case of RNase A, much higher concentrations can be used, but the regeneration is very sensitive to the presence of specific ions in the buffer (sulfate and phosphate accelerate; bromide and thiocyanate inhibit).[15] In addition to glutathione, several other low molecular weight thiol–disulfide pairs can promote oxidative renaturation. These include cysteine and cystine,[8] cysteamine and cystamine,[8,12] reduced and oxidized 2-mercaptoethanol,[8,9,16] and reduced and oxidized dithiothreitol.[17]

[11] W. L. Anderson and D. B. Wetlaufer, *J. Biol. Chem.* **251,** 3147 (1976).

[12] K. Oh-Johanson, D. B. Wetlaufer, R. Reed, and T. Peters, Jr., *J. Biol. Chem.* **256,** 445 (1981).

[13] P. W. Pick, Ph.D. Thesis, University of Minnesota, 1974. Manuscript in preparation.

[14] G. E. Means and R. E. Feeney, "Chemical Modification of Proteins." Holden-Day, San Francisco, California, 1971.

[15] S. W. Schaffer, A. K. Ahmed, and D. B. Wetlaufer, *J. Biol. Chem.* **250,** 8483 (1975).

[16] P. Branca, M.S. Thesis, University of Delaware, 1982. Manuscript in preparation.

[17] T. E. Creighton, *J. Mol. Biol.* **87,** 563 (1974).

None of these systems has shown substantial advantages in terms of regeneration rate and yield.

White[9] has described the air oxidation methods early used in Anfinsen's laboratory. Air oxidation has been studied also in our laboratory.[10,16,18,19] We believe that the use of thiol–disulfide combinations is preferable because of the greater speed of the overall reactions, the often higher yields of active protein, and the greater reproducibility of the results.

Let us reexamine the regeneration of protein–cysteine and mixed disulfides with mercaptoethanol, as carried out by Bradshaw et al.[8] It seems likely that that system is rapidly transformed by thiol–disulfide exchanges, into a regeneration system very similar to what we have detailed above (where at zero time we have completely reduced protein and a mixture of low molecular weight thiols and disulfides). That is, from different beginnings, the two reaction systems converge to produce similar reaction intermediates and products. Likewise, oxidation of a reduced protein with dehydroascorbate in the presence of excess added thiol[20] will rapidly produce a mixture of low molecular weight thiol, its conjugate disulfide, and mixed disulfides. Again, this would appear to produce a convergent reaction pathway or network.

It should be apparent that the reaction formulated in Eq. (4) does not require that the role of CyS^- be played by a low molecular weight thiol; it can also be played by a protein thiol. This provides a mechanism for shuffling —S—S— bonds even in the absence of added low molecular weight thiol. The much lower reaction rates observed under these conditions probably result from the limited concentration of protein thiol and an unfavorable stereochemistry for shuffling reactions.

One final note: commercial preparations of 2-mercaptoethanol contain variable amounts of disulfide as received from the supplier and are susceptible to air oxidation once the container has been opened. Moreover, in the regeneration solution itself, even in the presence of EDTA, air oxidation is rapid enough to increase substantially the ratio of [—S—S—] to [—SH] over the course of a few hours.[16,21]

[18] J.-P. Perraudin, T. Torchia, and D. B. Wetlaufer, *J. Biol. Chem.* (in press).

[19] A. K. Ahmed, S. W. Schaffer, and D. B. Wetlaufer, *J. Biol. Chem.* **250,** 8477 (1975).

[20] D. Givol, R. F. Goldberger, and C. B. Anfinsen, *J. Biol. Chem.* **239,** 3114 (1964).

[21] P. Branca, E. Chen, and D. B. Wetlaufer, *Fed. Proc. Fed. Am. Soc. Exp. Biol.* **42,** Abstr. 1415 (1983).

[19] Disulfide Bond Formation in Proteins

By THOMAS E. CREIGHTON

Disulfide bond formation between cysteine (Cys) residues is one of the posttranslational modifications most dependent on the conformation of the protein. For a disulfide bond to be formed between them, two Cys residues must have their alpha carbon atoms within 4–9 Å of each other, and the adjoining peptide backbones must also be in the proper orientation so that all five bond rotations between the side-chain atoms can adopt favorable conformations, as is found in proteins of known crystal structure.[1,2] Which Cys residues of a protein are joined in a disulfide does not depend simply upon the number of amino acid residues between them in the polypeptide chain. The known disulfides have no characteristics to indicate that they are greatly different from the other types of interactions between amino acid side chains in proteins, in that they are determined largely by the protein conformation, while, together with all the other interactions, contributing to its stability. The covalent nature of disulfides gives them some unique and useful properties, although it has also served to inspire some unfortunate myths about their roles in proteins. A common one is that they are irreversible covalent cross-links that determine the folded structure by limiting the conformations that are possible. This obviously cannot be true at the time and under the conditions when the protein disulfides are formed; otherwise, other or no disulfides would be formed. Disulfides are also often considered to be part of the protein primary structure, probably because they are often determined in the course of a primary structure determination, but they should not be so considered.[3]

Little is known about the *in vivo* mechanism of protein disulfide bond formation. The few studies with biosynthesis of secreted proteins have indicated that disulfides are incorporated during or shortly after secretion of the nascent chain into the endoplasmic reticulum,[4-6] but not how or why.

[1] J. S. Richardson, *Adv. Protein Chem.* **34**, 167 (1981).

[2] J. M. Thornton, *J. Mol. Biol.* **151**, 261 (1981).

[3] IUPAC—IUB Commission on Biochemical Nomenclature, *J. Mol. Biol.* **52**, 1 (1970).

[4] L. W. Bergmann and W. M. Kuehl, *J. Biol. Chem.* **254**, 5690 (1979).

[5] G. Scheele and R. Jacoby, *J. Biol. Chem.* **257**, 12277 (1982).

[6] T. Peters and L. K. Davidson, *J. Biol. Chem.* **257**, 8847 (1982).

METHODS IN ENZYMOLOGY, VOL. 107

In vitro studies of disulfide bond reduction and re-formation in intact proteins have elucidated many of the details of disulfide formation, but only in a few proteins. Nevertheless, they demonstrate that whether, and which, Cys residues are linked in disulfides depends upon both the conformation of the protein and the redox potential of the environment.[7,8] A disulfide bond has no inherent stability, relative to the thiol moiety, but is dependent upon the relative concentrations of appropriate electron donors and acceptors. With a given redox potential, the stabilities of the different disulfides possible between different Cys residues depend upon the degree to which the protein conformation keeps each pair of Cys residues in proximity when no disulfide is present. This can be expressed and measured as the effective concentration of the two thiol groups, which is given by the stability of the protein disulfide relative to that of a comparable intermolecular disulfide.[9] Those with very high effective concentrations, due to the protein structure keeping the thiol groups in favorable proximity for forming a disulfide, may form a stable disulfide, even in a relatively reducing redox environment. Cys residues that might only rarely be in suitable proximity, either owing to flexibility of the protein or the need to strain it to form a disulfide, will form less stable disulfides that would be present only in more oxidizing environments. Cys residues that are kept apart by a folded protein conformation should not form a disulfide bond. Other considerations, such as the accessibilities and electrostatic environments of the Cys residues, can also affect somewhat their tendency to form disulfides. Once formed, disulfides stabilize the conformation that brought them together, so protein conformation and disulfide bond stability are linked functions and dependent upon each other.

Protein disulfide formation *in vivo* as a posttranslational modification would then be expected to depend upon both the conformation and the environment of the polypeptide chain during and after biosynthesis; very little is known about either. Extrapolations from the *in vitro* results may be informative,[7,8] but the *in vivo* process should be amenable directly to study with the appropriate protein biosynthesis system, using the techniques that have been developed *in vitro*. Those techniques and the major observations will be presented here.

[7] T. E. Creighton, *Prog. Biophys. Mol. Biol.* **33**, 231 (1978).
[8] T. E. Creighton, *Adv. Biophys.* (in press).
[9] T. E. Creighton, *in* "Functions of Glutathione: Biochemical, Physiological, Toxicological, and Clinical Aspects" (A. Larsson, S. Orrenius, A. Holmgren, and B. Mannervik, eds.), p. 203. Raven, New York, 1983.

Chemistry of Disulfide Formation

Disulfides do not form spontaneously between thiols, even when in close proximity, unless there is an appropriate electron acceptor (A) present.

$$2 -SH + A \rightleftharpoons -S-S- + AH_2 \tag{1}$$

Numerous acceptors are possible and have been used, depending upon the purpose of the study. The classic oxidant has been O_2, but the chemical reaction is not understood; consequently, interpretation of different rates of disulfide formation is not possible. The reaction is apparently dependent upon metal ions, such as Cu^{2+}, which transiently bind to the thiol group. They act catalytically in low concentrations, so the rate is difficult to control, unless large concentrations are added, when they become part of the system. The greatest practical difficulty with the oxidation reaction is that hydrogen peroxide and radicals such as superoxide anion are produced,[10,11] which can then react with other parts of the protein. Such side reactions should be minimized in all studies of disulfide bond formation and breakage by excluding air and other oxidants. Apparent covalent modifications of unknown nature have always occurred in both the oxidized and reduced forms of ribonuclease A and have seriously hindered my studies with this protein[12]; they may have arisen from only small extents of oxidation of protein and reagent thiols produced by residual oxygen.

The most pertinent electron donor and acceptor to use would be the one that is involved *in vivo*, but its identity is unknown. Glutathione (GSH) in its oxidized form, GSSG, is the most likely candidate owing to its ubiquity and relatively high concentrations in most organisms.[13] Glutathione interacts chemically with all thiols and disulfides in the thiol–disulfide exchange reaction

$$GSH + RSSR \rightleftharpoons GSSR + RSH \tag{2}$$
$$GSH + GSSR \rightleftharpoons GSSG + RSH \tag{3}$$

This reaction has the advantage of being rapid at alkaline pH (since the thiolate anion is the actual reactive species), simple, extremely specific,

[10] H. P. Misra, *J. Biol. Chem.* **249**, 2151 (1974).
[11] M. Costa, L. Pecci, B. Pensa, and C. Cannella, *Biochem. Biophys. Res. Commun.* **78**, 596 (1977).
[12] T. E. Creighton, *J. Mol. Biol.* **129**, 411 (1979).
[13] A. Larsson, S. Orrenius, A. Holmgren, and B. Mannervik, eds., "Functions of Glutathione: Biochemical, Physiological, Toxicological, and Clinical Aspects." Raven, New York, 1983.

and easy to control (see Wetlaufer, this volume [18]). The rate of the reaction depends primarily on the concentrations of the reactants, but also somewhat on the nature of the groups attached to the sulfur atoms, the ionic strength, temperature, pH, etc. Unless the thiol group is situated within a folded protein, where its environment may be drastically perturbed, these other factors affect the thiol–disulfide exchange reaction in generally straightforward ways, and different rates of forming and breaking disulfides usually may be interpreted with some degree of confidence. For example, the rate at which a thiol group reacts is predictable from its pK_a value.[14] The reactivity of a model disulfide also depends somewhat upon its electrostatic environment, but in a way that is predictable from the pK_a value of the thiol produced, and this effect is minimized at high ionic strengths (Fig. 1). Energetically strained disulfides react more rapidly, as in lipoic acid, which reacts 160-fold more rapidly than otherwise expected, presumably because the five-membered ring prevents normal disulfide geometry.[15] More stable, intramolecular disulfides react more slowly, if at all; for example, GSH is unable to reduce to any significant extent the disulfide of oxidized dithiothreitol (DTT_S^S) or that between Cys-14 and -38 in bovine pancreatic trypsin inhibitor (BPTI). In both of these molecules, the reverse reaction is extremely fast, with a half-time of about 10^{-6} sec, since the two free thiol groups generated are kept in reasonable proximity. The effective concentrations of the two thiol groups in these reduced molecules are both about $10^4 M$,[9,15-17] so that the intramolecular disulfides are highly favored.

$$GSSG + \quad DTT_{SH}^{SH} \quad \underset{K \cdot 10^4 M}{\rightleftharpoons} \quad 2\,GSH + \quad DTT_S^S \tag{4}$$

The rates of model thiol–disulfide interchange reactions may be readily measured under a variety of conditions using DTT_{SH}^{SH} and a variety of model disulfides, since DTT_S^S absorbs substantially at 260–310 nm[16,18] and because the above two-step reaction is rate-limited by only the first step, giving simple kinetics.

[14] Z. Shaked, R. P. Szajewski, and G. M. Whitesides, *Biochemistry* **19**, 4156 (1980).
[15] T. E. Creighton, *J. Mol. Biol.* **96**, 767 (1975).
[16] W. W. Cleland, *Biochemistry* **3**, 480 (1964).
[17] R. P. Szajewski and G. M. Whitesides, *J. Am. Chem. Soc.* **102**, 2011 (1980).
[18] K. S. Iyer and W. A. Klee, *J. Biol. Chem.* **248**, 707 (1973).

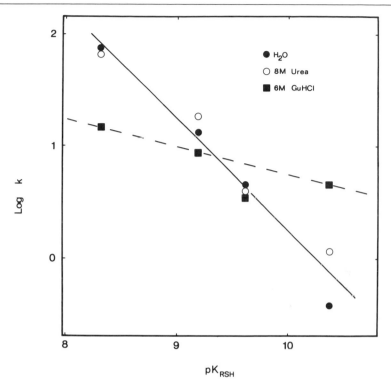

FIG. 1. Reactivities of different disulfides in the thiol–disulfide exchange reaction in three different aqueous solutions at 25°. The logarithm of the second-order rate constant for reduction by dithiothreitol, k (in units of $sec^{-1} M^{-1}$), is plotted versus the pK_a value in water of the thiol group produced. In order of increasing pK_a value, the disulfides were cystamine, glutathione disulfide, hydroxyethyl disulfide, and dithiodiglycolic acid. The rates were measured from the increase in absorbance at 283 nm resulting from the disulfide form of dithiothreitol. In each case, the buffer was 0.10 M Tris-HCl (pH 8.7), 0.20 M KCl, 1 mM EDTA, with the indicated additions. The electrostatic properties of the disulfide reagents in aqueous solution affect equally their reactivity in thiol–disulfide exchange and the ionization in water of the thiol group product of the reaction, since the solid line has a slope of unity. This electrostatic effect is apparently suppressed by the high ionic strength of 6 M guanidinium chloride, and the disulfides are more equally reactive. Taken from Creighton.[21]

Upon adding a disulfide reagent to a reduced protein, two sequential thiol–disulfide exchange reactions are required to form one protein disulfide. The first is the simple chemical reaction between the disulfide reagent and one Cys thiol group, to generate the mixed disulfide.

$$\text{(structure) SH, HS} + \text{RSSR} \underset{k_{-1}}{\overset{k_1}{\rightleftharpoons}} \text{(structure) SSR, HS} + \text{RSH} \qquad (5)$$

This reaction is bimolecular in both directions, so the rates should be dependent upon the concentrations of the disulfide and thiol forms of the reagent. The second step is the one in which a second Cys thiol group reacts with the mixed disulfide to form the protein disulfide.

$$\text{(6)}$$

This step is intramolecular in the forward direction, and its rate depends primarily upon the tendencies of the thiols of different Cys residues to come into proximity of the mixed disulfide, which is determined primarily by the protein conformation.

Adding excess disulfide reagent to a reduced protein will result in all accessible Cys residues becoming involved in either intramolecular or mixed disulfides, unless sufficient concentrations of thiol reagent are also present to make the reverse steps significant. Otherwise, the mixed-disulfide form of a Cys residue will accumulate to significant levels only if no other Cys residue reacts with it in the second step [reaction (6)] as rapidly as it was formed in the first step [reaction (5)]. On the other hand, the intramolecular step in protein disulfide formation can occur on the 10^{-6} sec time scale, so the mixed disulfide does not then accumulate. The rate of forming such a protein disulfide is then determined by the rate of the first step. Since accessible Cys thiols tend to have similar reactivities,[19] they tend to form disulfides at similar rates with such disulfide reagents, even though the second steps differ in rate. This can be an advantage, in that it ensures that intermediates in disulfide formation accumulate to significant levels, since all the disulfides tend to be formed at comparable rates.

Other possible intramolecular steps are rearrangements of protein and mixed disulfides [reactions (7) and (8)].

$$\text{(7)}$$

[19] T. E. Creighton, *J. Mol. Biol.* **96,** 777 (1975).

$$(8)$$

Other studies of protein disulfide formation have utilized as starting material the protein in which all Cys groups are present initially as mixed disulfides with a reagent.[20] Intramolecular protein disulfide formation is then initiated by adding catalytic amounts of the thiol form of the reagent [reaction (9)].

$$(9)$$

This procedure may have the practical benefit of using a more soluble starting form of the protein; reduced proteins are often notoriously insoluble, and this can be a severe technical problem. An appropriate mixed disulfide on each Cys residue can improve the solubility in some cases. However, this procedure appears to have the complication that the nature of the starting protein has been altered, in that the reagent is part of the initial protein, and that the concentrations of the thiol and disulfide forms of the reagent will fluctuate greatly during the reaction. It can also be difficult to ensure that the starting protein has all Cys residues in mixed disulfides. It is generally prepared by adding a large excess of a linear disulfide reagent to the reduced protein in an unfolding solvent; formation of mixed disulfide must then compete with intramolecular disulfide formation [reactions (6) and (9)]. However, the rate of forming intramolecular disulfides within even an unfolded protein in 8 M urea can be very rapid,[21] and it has been found to be impossible to generate only mixed disulfides with BPTI. Otherwise, the procedure using the mixed disulfide is exactly analogous to that described here using the reduced protein, except that the roles of the Cys thiol and mixed disulfides are reversed.

---

[20] R. A. Bradshaw, L. Kanarek, and R. L. Hill, *J. Biol. Chem.* **242**, 3789 (1967).
[21] T. E. Creighton, *J. Mol. Biol.* **113**, 313 (1977).

Choosing a Disulfide Reagent

Metabolites other than glutathione have been proposed as *in vivo* participants in disulfide bond formation in proteins,[22,23] often in conjunction with an appropriate enzyme. The impetus for these suggestions has usually been the mistaken belief that the GSH and GSSG levels in cells are too reducing to allow protein disulfide bond formation, since most of the glutathione is present as GSH.[13,24] However, because they are intramolecular, protein disulfides may be much more stable than those between unlinked groups, as in GSSG, when two molecules result from reduction of the disulfide. Where the folded protein conformation keeps two Cys thiol groups in appropriate proximity, they may have effective concentrations of up to 10^7 M,[9] possibly higher,[25,26] and thus a disulfide bond between them is much more stable than one between two different molecules (e.g., GSH) present at practical or physiological concentrations of about 10^{-2} M. Proposals of other means of forming protein disulfides have unnecessarily attempted to insulate the protein disulfide redox potential from that controlling the other thiol–disulfide status of the cell,[22] but the two are certain to be linked, since all thiols and disulfides would participate in the chemical thiol–disulfide exchange reaction, even in the absence of the many known enzymes or other catalysts.

Consequently, thiol–disulfide exchange involving the ubiquitous, abundant GSSG, or any other disulfide reagent, is the most favorable procedure for generating disulfides in proteins *in vitro*. Which particular reagent is used depends upon what techniques are to be used to trap, analyze, and identify the species generated (i.e., the charge and other chemical properties it will give the protein if attached via a mixed disulfide). It must be remembered also that the reagent will be part of the protein in the course of disulfide formation and thus has the potential for affecting the protein conformation and disulfide bond formation. Use of different reagents should uncover such phenomena. GSSG has been observed to have no such effects with at least some proteins and is comparable to other disulfides. It has the added advantage of possessing one net negative charge on each half, which is useful when analyzing for protein mixed disulfides.

[22] D. M. Ziegler and L. L. Poulsen, *Trends Biochem. Sci.* **2**, 79 (1977).
[23] V. G. Janolino, M. X. Sliwkowski, H. E. Swaisgood, and H. R. Horton, *Arch. Biochem. Biophys.* **191**, 269 (1978).
[24] G. Harisch, J. Eikemeyer and J. Schole, *Experientia* **35**, 719 (1979).
[25] M. I. Page and W. P. Jencks, *Proc. Natl. Acad. Sci. U.S.A.* **68**, 1678 (1971).
[26] A. J. Kirby, *Adv. Phys. Org. Chem.* **17**, 183 (1980).

Disulfide formation using a stable cyclic reagent, such as oxidized dithiothreitol (DTT_S^S),[16] is much more discriminating than with a linear reagent. The mixed disulfide is energetically unfavorable and rapidly dissociates intramolecularly, so it should not accumulate significantly. Only very favorable protein disulfides are formed with this reagent, at a rate that should be proportional to the rate of the intramolecular step of reaction (10).

$$
\begin{array}{c}
\underset{HS}{\overset{SH}{\Big\{}} + DTT_S^S \; \overset{K_{DTT}}{\rightleftharpoons} \; \left[\underset{SH}{\overset{SS\,DTT_{SH}}{\Big\{}} \right] \; \overset{k_{intra}}{\rightleftharpoons} \; \overset{S}{\underset{S}{\Big(}} + DTT_{SH}^{SH}
\end{array} \qquad (10)
$$

$$k_{obs} = k_{intra}\,K_{DTT}$$

Purification of DTT_S^S

These measurements require that the DTT_S^S be very pure, as common impurities have less stable disulfides that react much more rapidly; consequently, even relatively minor impurities can dominate the observed kinetics. Fortunately, they can be preferentially reduced by DTT_{SH}^{SH} and removed on the basis of their thiol groups, as in the following purification procedure.[27]

Commercially purchased DTT_S^S (2 g) is dissolved in 100 ml of 0.1 M NH_3 at room temperature, under an atmosphere of N_2, and approximately 200 mg of DTT_{SH}^{SH} is added. Thiol–disulfide exchange to reduce preferentially less stable disulfides is permitted for about 3 hr. Any insoluble material is then removed with a paper filter. All thiol-containing compounds are removed by passing the solution through a 1.5 cm diameter × 17 cm length column of Amberlite IRA-410 in the HO^- form. The column is washed extensively with water to recover most of the DTT_S^S, which tends to adhere to the column; DTT_S^S is monitored by its absorbance at 283 nm (molar extinction coefficient = 273 cm^{-1}).[16] The efficiency of the column is confirmed by the absence of thiol groups.[28] The DTT_S^S is recovered by lyophilization and stored at 4°. It has a melting point of between 130 and 131° and is significantly more soluble in water than the original material.

This purification procedure would not be expected to remove any impurities with more stable disulfides, but in small quantities they should

[27] T. E. Creighton, *J. Mol. Biol.* **113**, 295 (1977).
[28] G. L. Ellman, *Arch. Biochem. Biophys.* **82**, 70 (1959).

not affect significantly the kinetics of disulfide bond formation. All other disulfide and thiol reagents have been obtained commercially and used without further purification.

Choosing the Conditions

A wide variety of conditions may be used, dependent upon the individual protein and questions to be studied. The one essential requirement is that the pH be neutral or alkaline, since the rate of the thiol–disulfide interchange reaction is dependent upon ionized thiols, the pK_a values for which are generally in the region of 8 to 10. Covalent modifications of proteins become significant at very alkaline pH values, so most studies have been done at between pH 7 and 9; the values of rate and equilibrium constants given here were obtained at pH 8.7 and 25°. Ionization of protein thiol groups might also affect protein conformation; in particular, it may destabilize any folded conformation in which the thiol groups would be buried in a nonpolar environment, as they often are if involved in disulfide bonds. It is probably one factor that contributes to the tendency of reduced proteins to be unfolded.

Other than pH, the thiol–disulfide exchange reaction appears to be little affected by other variables, such as buffers and temperature.[29] High ionic strength diminishes the electrostatic difference between different thiols and disulfides (Fig. 1), in the same way that it diminishes electrostatic interactions in proteins. Different salts may be added to diminish or increase hydrophobic interactions, according to the Hofmeister series. Denaturants such as urea and guanidinium chloride have little effect on thiol–disulfide exchange (Fig. 1), but large effects on protein conformation.[21] Enzymes that catalyze thiol–disulfide exchange may also be included[30] (see Hillson et al., this volume [16]).

On the other hand, some unidentified factors can interfere with proper folding and disulfide formation. Initial experiments carried out in plastic containers gave poor yields of correctly refolded BPTI and numerous atypical intermediate species, so all folding procedures are routinely performed in glass vessels. Even so, BPTI refolding reactions have suddenly become aberrant on a few occasions; filtration through activated charcoal of all solutions involved in the refolding mixture has always alleviated this problem. The basis for these phenomena has not been investigated.

[29] T. E. Creighton, *J. Mol. Biol.* **144,** 521 (1980).
[30] T. E. Creighton, D. A. Hillson, and R. B. Freedman, *J. Mol. Biol.* **142,** 43 (1980).

Trapping Species

The order of disulfide bond formation in a protein can be determined only from the intermediate states that comprise the kinetic pathway. These intermediates are transient and will persist only if they are trapped in some way.

All protein species with free thiol groups are inherently unstable, with respect to both oxidation and participation in thiol–disulfide exchange; those encountered as transient intermediates in a multistep process are likely to be least stable. However, disulfide bond formation, breakage, and interchange can be quenched to trap such intermediates in a stable form. The trapped intermediates are sufficiently stable to be characterized in virtually any manner at one's leisure; it is necessary only to avoid conditions that might permit disulfide rearrangements.

With any trapping procedure, there is always the possibility that the quenching reaction could alter the spectrum of species present originally at the time of quenching. For example, some of the species present in a rapid equilibrium might react more rapidly than the others and thereby be trapped in elevated quantities, or the quenching reaction at one site could cause changes, such as disulfide rearrangements, at other parts of the protein. Such conceivable complications can be minimized by using a quenching reaction that is much more rapid than any other steps in the reaction being studied and has no substantial effects on the protein conformation.

Protonation of thiol groups by acidification quenches thiol–disulfide exchange, is extremely rapid, and produces a negligible change in the thiol groups, although it can affect protein conformation. The main drawback is that quenching is not complete or irreversible, since the thiol–disulfide exchange reaction is not abolished, merely slowed, decreasing by a factor of 10 for each pH unit decrease below the pK_a value. Intramolecular thiol–disulfide interchanges can occur on the 10^{-6}-sec time scale at pH 8, so they would be expected to occur on the second time scale at pH 2. Such a phenomenon has been observed with the initial two-disulfide intermediates of BPTI trapped by acid; they rearrange intramolecularly to the native-like species with the 30–51, 5–55 disulfide bonds during electrophoresis at acid pH and are not detectable in this way. Similar phenomena are almost certain to have occurred with other studies that have utilized only acid-trapping,[31–36] but would not have been detected; they

[31] L. G. Chavez and H. A. Scheraga, *Biochemistry* **16,** 1849 (1977).
[32] L. G. Chavez and H. A. Scheraga, *Biochemistry* **19,** 996 (1980).

probably account for the contradictory results obtained.[32,37] Any lengthy studies of acid-trapped species must be suspect, particularly if they are incubated with thiol- or disulfide-containing materials, such as antibodies.[31,32]

Reaction of thiols with reagents such as iodoacetic acid and iodoacetamide is irreversible under essentially all conditions and has been used extensively. Normal, accessible protein thiol groups react with these reagents with rate constants of between 2 and 16 sec^{-1} M^{-1} (Fig. 2), so the half-time for reaction with reagent added to a final concentration of 0.1 M should be 0.4–3 secs. Consequently the reaction is not instantaneous, even with such high concentrations of reagents, and the possibility that it has perturbed the spectrum of species originally present should be checked by comparing different reagents. In the case of disulfide bond formation in BPTI, it was shown that 0.1 M iodoacetamide and iodoacetate trapped the same species, in the same proportions, including the same intrinsically unstable one- and two-disulfide intermediates.[37,38] The one-disulfide intermediates were the same as those trapped with acid, which acts much more rapidly. Also, extensive variation of the conditions of refolding has demonstrated that the species trapped with iodoacetate reflect the conditions of refolding, not the trapping conditions.[21,29]

The difficulty with these reagents is that inaccessible thiol groups do not react,[39] and one two-disulfide species of BPTI was missed in the early studies because it folded into a native-like conformation with the Cys 30 and 51 thiols buried. It could be trapped only by adding denaturants after the iodoacetic acid, to make accessible the two thiol groups.[40] However, substantial rearrangements of the two disulfides occurred in this process, presumably upon unfolding and before reaction of the thiols. This is an intrinsic problem of all such studies when the protein folds up, and it also illustrates the necessity of blocking irreversibly all thiol groups to prevent intramolecular thiol–disulfide exchange. However, if a denaturant is to be added to unmask inaccessible thiol groups, it should be added after the reagent has trapped most species; if both are added at the same time,[41–43]

[33] Y. Konishi and H. A. Scheraga, *Biochemistry* **19,** 1308 (1980).
[34] Y. Konishi and H. A. Scheraga, *Biochemistry* **19,** 1316 (1980).
[35] Y. Konishi, T. Ooi, and H. A. Scheraga, *Biochemistry* **20,** 3945 (1981).
[36] Y. Konishi, T. Ooi, and H. A. Scheraga, *Biochemistry* **21,** 4734 (1982).
[37] T. E. Creighton, *J. Mol. Biol.* **87,** 579 (1974).
[38] T. E. Creighton, *J. Mol. Biol.* **87,** 603 (1974).
[39] Y. Goto and K. Hamaguchi, *J. Mol. Biol.* **146,** 321 (1981).
[40] D. J. States, C. M. Dobson, M. Karplus, and T. E. Creighton, *J. Mol. Biol.,* in press.
[41] A. S. Acharya and H. Taniuchi, *J. Biol. Chem.* **251,** 6934 (1976).
[42] W. L. Anderson and D. B. Wetlaufer, *J. Biol. Chem.* **251,** 3147 (1976).
[43] T. Dubois, R. Guillard, J.-P. Prieels, and J. P. Perraudin, *Biochemistry* **21,** 6516 (1982).

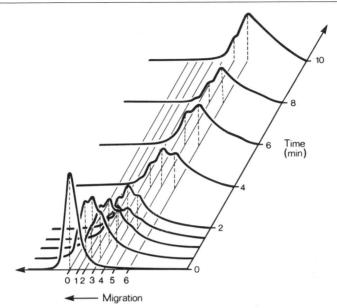

FIG. 2. Electrophoretic analysis of the kinetics of reaction of the Cys thiol groups of reduced BPTI with iodoacetate. Fully reduced BPTI (30 μM) was incubated at 25° in 0.10 M Tris-HCl (pH 8.7), 0.20 M KCl, 1 mM EDTA, and 2.5 mM iodoacetic acid. At the indicated times, the reaction was stopped by addition of iodoacetamide to 0.1 M. The protein molecules with the indicated number of acidic carboxymethyl groups were resolved by 15% polyacrylamide gel electrophoresis, using the discontinuous buffer system of R. A. Reisfield, U. J. Lewis, and D. E. Williams [*Nature* (*London*) **195,** 281 (1962)]. Densitometer traces of Coomassie Blue-stained gels are shown. Computer simulations of these kinetics indicate that the first to last thiol groups to react with iodoacetate do so with second-order rate constants varying from 5.2 to 2.2 sec^{-1} M^{-1}. The protein molecules with one carboxymethyl group were isolated by cation exchange and shown to have the carboxymethyl groups nearly uniformly distributed on the six Cys residues, indicating that their thiol groups have essentially the same reactivity in the reduced protein. The data on which this figure is based were presented by Creighton.[19]

disulfide rearrangements produced by the denaturant are almost certain to take place before trapping is complete.

High concentrations of iodoacetate or iodoacetamide can produce modifications of other protein groups, so it is often necessary also to adjust the pH to minimize such side reactions and to remove the reagent by gel filtration after reaction with thiols is complete. Nevertheless, high concentrations of reagents are required for accurate trapping and adding a slight excess of reagent[41,42,44] in the usually acceptable manner is unlikely to be satisfactory, as the rate of trapping will be too slow.

Many other thiol-blocking reagents are available and may be suitable, but the completeness and irreversibility of their reaction should be demonstrated, for even trace quantities of thiol groups can produce complete disulfide arrangements under some conditions. The other commonly used reagent, N-ethylmaleimide, does not produce stable adducts and can reregenerate thiol groups.[45]

Analyzing Trapped Species

The possible pathways of disulfide bond formation become very complex with more than just two or three Cys residues in a protein. Even formation of a single disulfide bond between two Cys residues is a multistep process with up to five different species involved, those with the two Cys residues reduced, disulfide, doubly mixed disulfide, and two single mixed disulfides (Fig. 3). Most multistep reactions produce at intermediate times complex mixtures of the original, intermediate, and final forms of the reactant (Figs. 2–5), so it is necessary to resolve the various species by fractionating the trapped mixtures. It is impractical to attempt to elucidate the pathway by following as a function of time just the average properties of the mixture, such as the average thiol, disulfide, and mixed-disulfide content or other physical or enzymatic measures. Attempts to do so have not been satisfactory,[36,42] especially if assumptions are required about the natures of the various species.[36] Also, such unsatisfactory procedures are not necessary, because the trapped species may be purified and their individual properties determined directly and unambiguously.

The various species trapped in the above ways will differ in the states of their Cys residues (whether in intramolecular disulfides, mixed disulfides with reagent, or originally free thiols that reacted with the trapping reagent) and can often be separated on this basis. For example, the disulfide reagent and the trapping reagent can be charged (e.g., negatively charged glutathione and iodoacetate), so that the net charge of the protein will depend on the number of such groups attached. Electrophoretic, ion-exchange, isoelectric focusing, etc., procedures may then be used to separate the species (Figs. 3–5). Not only the net charge, but also the clustering of charges on the molecule and its conformation, determine the ion-exchange chromatographic properties of the species, so virtually all

[44] R. R. Hantgan, G. G. Hammes, and H. A. Scheraga, *Biochemistry* **13**, 3421 (1974).
[45] E. Beutler, S. Srivastava, and C. West, *Biochem. Biophys. Res. Commun.* **38**, 341 (1970).

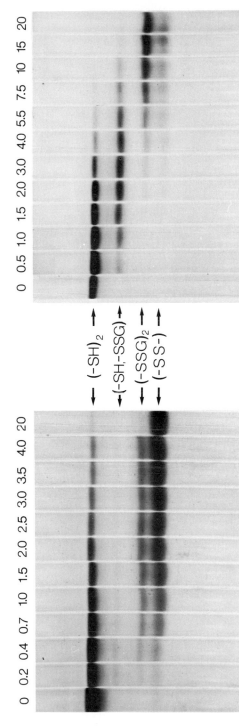

FIG. 3. Kinetics of disulfide bond formation between the two Cys residues in reduced constant fragment of an immunoglobulin light chain, using 3 mM GSSG as disulfide reagent. Reduced protein (30 μM) was incubated at 25° in 0.1 M Tris-HCl (pH 8.1), 0.15 M KCl, 1 mM EDTA, and 3 mM GSSG, either without (left) or with (right) 8 M urea. At the indicated times, the reaction was quenched by addition of iodoacetamide to 0.1 M. The trapped species were analyzed by gel electrophoresis in 15% polyacrylamide gels using the method of B. J. Davis [*Ann. N.Y. Acad. Sci.* **121**, 404 (1964)]. Photographs of Coomassie Blue-stained gels are presented; electrophoretic migration was from the top to the bottom. The protein species are identified by the states of their Cys residues at the time of trapping. The two species with mixed disulfides with glutathione on only one of the two Cys residues (—SH, —SSG), are not resolved by this procedure. The mixed-disulfide forms migrate more rapidly than the original reduced form, (—SH)₂, owing to the net negative charge of the glutathione moiety. The disulfide form, (—SS—), migrates most rapidly because it is folded into a compact conformation, whereas the other trapped species are unfolded. Taken, with permission, from Goto and Hamaguchi.[39]

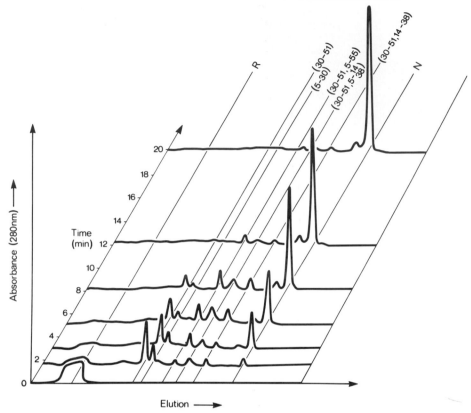

FIG. 4. Chromatographic analysis of the kinetics of disulfide bond formation in reduced BPTI using 0.15 mM GSSG as disulfide reagent. Reduced BPTI was incubated at 25° in 0.10 M Tris-HCl (pH 8.7), 0.2 M KCl, 1 mM EDTA, and 0.15 mM GSSG. At the indicated times, 0.5 M iodoacetate was added to a final concentration of 0.1 M. After 2 min, the protein was separated from reagents by gel filtration on Sephadex G-25 equilibrated with 0.02 M imidazole-HCl (pH 6.2), 1 mM EDTA. Each protein mixture was placed on a column of CM-cellulose (1.5 cm diameter by 80 cm length) equilibrated with the same buffer; elution was with a concave gradient of NaCl concentrations, from 0 to 0.6 M NaCl (1.5 liters total). The protein was monitored by its absorbance at 280 nm. Fully reduced BPTI, R, elutes first, owing to the six negatively charged carboxymethyl groups. It is followed by the one-disulfide intermediates, then the two-disulfide intermediates, and finally the refolded BPTI, N. Peak N contains native-like BPTI, with the three disulfides linking Cys 30 to 51, 5 to 55, and 14 to 38, plus native-like molecules lacking the 30 to 51 disulfide and with the 2 thiol groups buried and unreactive. The other intermediates are identified by the residue numbers of the Cys residues paired in disulfides. Only minor quantities of mixed-disulfide protein species accumulate under these conditions. The data on which this figure is based were published in Creighton.[52]

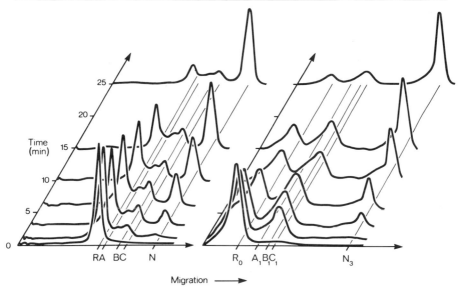

FIG. 5. Electrophoretic analysis of disulfide bond formation in reduced BPTI using 5 mM DTT$_S^S$ as disulfide reagent. Reduced BPTI (30 μM) was incubated at 25° in 0.10 M Tris·HCl (pH 8.7), 0.2 M KCl, 1 mM EDTA, and 5 mM DTT$_S^S$. At the indicated times, the reaction was quenched by the addition of either iodoacetamide (left) or iodoacetate (right) to 0.1 M. Portions of the trapped mixtures were analyzed by electrophoresis as in Fig. 2. The iodo-acetamide-trapped species (left) should all have essentially the same net charge, and their different mobilities reflect primarily their different conformations, i.e., hydrodynamic volumes. The fully reduced protein, R, is very unfolded and migrates most slowly; the refolded protein, N, most rapidly. The intermediates A, B, and C are different one-disulfide intermediates. The iodoacetate-trapped species (right) differ also in the number of negatively charged carboxymethyl groups they contain, and the mobilities of all the species, except N, are reduced accordingly. The decreased mobility of R indicates the effect of six carboxymethyl groups. The positions indicated for A_1, B_1, C_1, and N_3 are those predicted from their mobilities when trapped with iodoacetamide, assuming that their mobility is decreased in proportion to the number of carboxymethyl groups they contain and that they have the number of disulfides indicated by the subscript. The original data on which this figure is based were published in Creighton.[37]

the major intermediates trapped by iodoacetate during refolding of reduced BPTI were isolated in essentially pure form by one column chromatography separation (Fig. 4).

Using cation exchange of mixtures trapped with iodoacetate, the various species elute essentially in decreasing number of carboxymethyl (CM) groups on originally free Cys residues, i.e., increasing order of the number of disulfide bonds they contain. If one or more of these Cys residues is in a mixed disulfide with glutathione, the elution volume is only slightly altered, generally being decreased[46]; an acidic glutathione moiety is therefore nearly comparable in this case to a CM group. Neutral or basic disulfide reagents produce the expected change in elution profile, but this has been found to complicate the purification of the trapped species. It is preferable to isolate together all molecules with the same protein disulfides, irrespective of whether the other Cys residues are blocked with CM groups or mixed disulfides with the reagent. The states of these Cys residues can be determined by diagonal maps (see below) or the disulfide or trapping reagent can be radioactive, permitting their presence to be readily detected.[12,46] The number of Cys residues involved in disulfides, either intramolecular or mixed, can be determined readily in pure species by reducing the proteins, then allowing these to react with mixtures of iodoacetate and iodoacetamide, and counting the number of electrophoretic bands generated[47] (Fig. 6).

Other properties of the protein species can also depend upon the state of the Cys residues, especially their conformations, and can be used to analyze the trapped species. Electrophoretic mobility through polyacrylamide gels depends upon both the net charge and the hydrodynamic volume of the protein (Fig. 5). For example, reduced BPTI trapped by acid or iodoacetamide has essentially the same net charge as native BPTI, but only two-thirds of the electrophoretic mobility, owing to being unfolded and having a substantially greater hydrodynamic volume.[7,37] Consequently, the relative mobilities of the iodoacetamide- or acid-trapped intermediates give information about their relative compactness. Trapping the intermediates with iodoacetate introduces an acidic CM group at each free Cys residue, with consequent effects on the net charge. The charge effect of six such groups is given by the decreased mobility of reduced BPTI trapped with iodoacetate rather than iodoacetamide. There is no effect on the native protein, which has no free thiol groups. The relative mobilities of all the intermediates when trapped with iodoacetamide or iodoacetate show excellent agreement with those expected if each CM group is equivalent in charge to one of the six on RCM-BPTI and if the hydrodynamic property of the protein is the same irrespective of which trapping reagent was used. Consequently, it is possible to correlate the two electrophoretic profiles to determine the number of disulfide bonds present in each species.

Gel filtration is also sensitive to the hydrodynamic volume and has been used extensively to separate folded from unfolded species of lyso-

[46] T. E. Creighton, J. Mol. Biol. 113, 275 (1977).
[47] T. E. Creighton, Nature (London) 284, 487 (1980).

zyme generated during disulfide bond formation and breakage.[48] Unfortunately, the disulfides of the folded species were assumed to be native-like, rather than identified. Using gel filtration, a native-like three-disulfide intermediate of RNase was isolated in pure form and extensively characterized.[12,49,50]

Identifying and Characterizing the Trapped Species

Many chemical procedures are available for determining which Cys residues in a protein are paired in disulfides, which involved in mixed disulfides with the disulfide reagent, and which reacted with the trapping reagent. Nevertheless, the diagonal electrophoresis method of Brown and Hartley[51] has been found to be most useful, especially if performed at pH 3.5 with iodoacetate-trapped species. The reason for this is apparent from the diagonal maps of trapped reduced BPTI and the major one-disulfide intermediate in Fig. 7. The fully reduced form of BPTI has no disulfides, but its diagonal map has a second diagonal of six peptides that were slightly more acidic in the second dimension and contain the six CM Cys residues of the protein. These peptides do not separate from the other peptides if the electrophoresis is performed at pH 2.1 or pH 6.5 or if the Cys residues were blocked with iodoacetamide. Therefore, this "CM Cys diagonal" probably arises because the oxidation with performic acid that occurs between the two electrophoretic separations oxidizes the CM Cys residues to the sulfone (CM $CysO_2$), which lowers their pK_a values slightly, making them somewhat more acidic. Consequently, all peptides containing Cys residues lie off the diagonal of other peptides and can be identified.

Disulfide bonds between Cys residues cause their peptides to have the same mobility in the first dimension, but the performic oxidation cleaves this disulfide and converts the two Cys residues to $CysO_3H$. The two peptides are then usually more acidic than in the first dimension and lie off the major diagonal (Fig. 7, right). Which pairs of peptides were so linked originally is usually obvious from their common mobility in the first dimension. Ambiguities may often be resolved if the first-dimension mobilities of the various possible pairs are known. In Fig. 7 (right), the absence of these two peptides from the "CM Cys diagonal" indicates that the

[48] A. S. Acharya and H. Taniuchi, *Mol. Cell. Biochem.* **44**, 129 (1982).
[49] T. E. Creighton, *FEBS Lett.* **118**, 283 (1980).
[50] A. Galat, T. E. Creighton, R. C. Lord, and E. R. Blout, *Biochemistry* **20**, 594 (1981).
[51] J. R. Brown and B. S. Hartley, *Biochem. J.* **101**, 214 (1966).

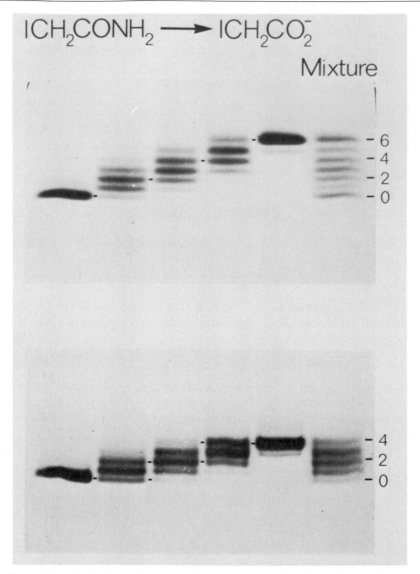

FIG. 6. Counting the integral number of Cys residues in reduced BPTI (top) and that with two Cys residues irreversibly blocked (bottom). Molecules with 0 to n acidic groups on the n Cys thiols are generated by reacting them with varying mixtures of neutral iodoacetamide and acidic iodoacetate. The proteins were reduced and unfolded by incubation at 37° for 30 min in 1.0 ml of 8 M urea, 10 mM dithiothreitol, 10 mM Tris-HCl, 1 mM EDTA, at pH 8.0. They were then alkylated by adding 200 μl of this solution to 50 μl of 0.25 M iodoacetamide, iodoacetic acid adjusted to pH 8.0 with KOH, or mixtures of the two. After 15 min at room

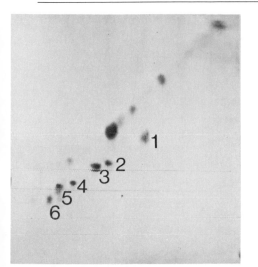 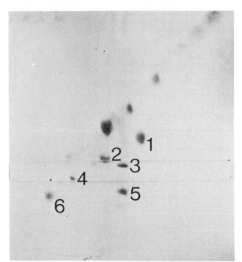

FIG. 7. Diagonal maps of (left) carboxymethylated reduced BPTI and (right) the single-disulfide intermediate with the 30–51 disulfide trapped with iodoacetate. The proteins were digested with trypsin followed by chymotrypsin. The peptides were separated by electrophoresis at pH 3.5 in the horizontal direction (anode at the right). After exposure to performic acid, the electrophoresis was repeated in the vertical direction (anode at the top). The peptides were detected by dipping through 0.4% ninhydrin, 0.1% cadmium acetate, 5% acetic acid, and 10% water in acetone. The numbered peptides lie below the diagonal of peptides with the same mobility in both dimensions. They were identified from their amino acid compositions to contain the 6 Cys residues of BPTI: peptide 1, residues 36 to 39; peptide 2, 1 to 15; peptide 3, 47 to 53; peptide 4, 5 to 15; peptide 5, 27 to 33; peptide 6, 54 to 58. The positions of peptides 3 and 5 (right) indicate that a disulfide bond was originally present between Cys 30 and 51. The absence of these peptides on the "CM Cys" diagonal, as in left, indicate that essentially all the molecules had the intramolecular 30–51 disulfide bond. Taken from Creighton.[38]

temperature, 50-μl portions were subjected to electrophoresis in 15% polyacrylamide gels containing 8 M urea and the buffer system of R. A. Reisfield, U. J. Lewis, and D. E. Williams [*Nature (London)* **195,** 281 (1962)]. The iodoacetamide-treated proteins (left) have the greatest electrophoretic mobility due to the greatest net positive charge. Molecules of BPTI (top) with 1 to 6 acidic carboxymethyl groups are generated, from left to right, by reacting with 1:1, 1:3, 1:9, and 0:1 mixtures of iodoacetamide and iodoacetate. The sample on the far right is a mixture of the five other samples. BPTI in which Cys 14 and 38 had been irreversibly blocked by reaction with iodoacetamide (bottom) gave the expected four additional bands. Taken from Creighton.[47]

protein analyzed was homogeneous and that the disulfide bond was intramolecular. The other four Cys residues were blocked with CM groups.

Mixed disulfides between protein and reagent behave analogously, depending on the nature of the reagent. Peptides that were involved in mixed disulfides with glutathione generally appear in very similar positions as those that had reacted with iodoacetate, but they may be readily distinguished by their amino acid compositions, since the former have $CysO_3H$ residues, the latter CM $CysO_2$.

The states of all the Cys residues of a homogeneous species may be determined from a single diagonal map. Mixtures of one-disulfide species may often be analyzed in the same way, even giving semiquantitative estimates of the relative proportions of the various species. However, species with two or more protein disulfides must be nearly homogeneous in order to give interpretable diagonal maps, unless they are closely related, as in differing in only one disulfide.[52]

No problems are generally encountered with disulfide interchange, even though proteolytic digestions are often carried out at pH 7.5 and 65° for 2 or 3 hr. This contrasts with earlier reports of extensive, seemingly spontaneous, interchange[53] for reasons that are unknown, but may be due to the present greater purity of reagents, irreversible blocking of all thiol groups, and care not to introduce nucleophiles that can catalyze disulfide interchange.

Besides the states of their Cys residues, the trapped intermediates can be characterized in the same way as any other protein,[50,54-56] since they are stable under all conditions except those that would permit disulfide interchange.

Preparation of Diagonal Maps

The protein to be analyzed is digested proteolytically under conditions where disulfides are stable. No thiols or other nucleophiles likely to catalyze disulfide interchange should be present, which precludes the use of thiol-containing proteases or very high pH. The ideal protease is one that produces quantitative cleavage between all the Cys residues of the protein. We routinely use thermolysin at 1 mg/ml in 0.2 M ammonium acetate

---

[52] T. E. Creighton, *J. Mol. Biol.* **95,** 167 (1975).
[53] D. H. Spackman, W. H. Stein, and S. Moore, *J. Biol. Chem.* **235,** 648 (1960).
[54] T. E. Creighton, E. Kalef, and R. Arnon, *J. Mol. Biol.* **123,** 129 (1978).
[55] P. A. Kosen, T. E. Creighton, and E. R. Blout, *Biochemistry* **19,** 4936 (1980).
[56] P. A. Kosen, T. E. Creighton, and E. R. Blout, *Biochemistry* **20,** 5744 (1981).

buffer containing 1 mM CaCl$_2$ at pH 7.0–7.5, digesting the protein at a concentration of 0.2–0.5 mM for 2–3 hr at 65°; disulfide interchange is negligible. Even very stable proteins, such as BPTI, are digested with this procedure. High concentrations of thermolysin do not interfere with diagonal maps because it contains no Cys residues and any peptides that arise from it lie on the normal diagonal. After digestion, the mixture is lyophilized.

An aliquot of the digestion mixture containing about 0.1–0.3 μmol of the original protein is applied as a 1 × 3 cm band to Whatman 3 MM paper and subjected to electrophoresis in one dimension. A mixture of the dyes xylene cyanole, acid fuchsin, and methyl orange is used to monitor the electrophoretic migration. After thorough drying of the paper at room temperature, it is placed in a closed container containing in a shallow dish a freshly prepared mixture of 19 ml of formic acid and 1 ml of 30% H$_2$O$_2$. Chloride ions should be avoided, as they are converted to chlorine by performic acid. The extent of the oxidation can be followed by the conversion of the blue tracking dye to a green color. After the 2–3 hr of exposure required for this change, the paper is removed and dried thoroughly at room temperature.

The thin strip of paper containing the peptides is then stitched across a second sheet of chromatography paper, and as much as possible of the overlapping new sheet is removed. Whatman 1 paper for the second dimension gives somewhat greater sensitivity than 3 MM paper. The paper is wetted with electrophoresis buffer on both sides of the first strip, and the merging fronts are used to concentrate the peptides into a thin, straight line down the center of its length. Electrophoresis in the second dimension is repeated in exactly the same way as in the first, but at right angles to it. The sheet of paper is then dried thoroughly at room temperature.

The peptides may be visualized by any of the available methods, but staining with fluorescamine permits recovery of the peptides for amino acid analysis. The paper is dipped once or twice through 1% triethylamine in acetone if the electrophoresis buffer was acidic. It is then dipped through a solution of 1 mg of fluorescamine per 100 ml of acetone. After drying sufficiently, the peptides are visualized by their fluorescence under an ultraviolet lamp. All fluorescent spots off the diagonal are eluted with 0.1 M NH$_3$ and subjected to amino acid analysis, for the intensity of fluorescence can be a grossly misleading guide to the amount of peptide present. The recovery of peptides is low but remarkably constant, so that the quantity of each peptide recovered can assist with interpretation of the diagonal map if the protein was not homogeneous or if the proteolytic cleavages were not complete. It should be noted that CM Cys and Met residues are converted to the sulfone by the performic acid treatment, and CM Cys sulfone is usually destroyed completely by acid hydrolysis.[57]

Reconstructing the Pathway of Disulfide Formation

Having trapped, isolated, and characterized all the intermediates that accumulate during disulfide bond formation or breakage, one must determine their kinetic roles experimentally. The identities of the species will give many clues to the pathway, but it is not sufficient to assume that the native disulfides are formed sequentially, or even that all one-disulfide species are formed directly from the reduced protein and then give rise to all the two-disulfide species, etc. The known pathways have been found to be unexpectedly complex.[7,8]

The kinetic roles of all the intermediates must be elucidated by determining experimentally their levels of accumulation as a function of time, using various concentrations of disulfide and thiol reagents; all other conditions are kept constant. The kinetics of both disulfide bond formation and breakage need to be examined. It is then possible to determine whether the rates of appearance and disappearance of an intermediate depend upon the concentration of (1) disulfide reagent, (2) thiol reagent, or (3) neither, indicating, respectively, that the rate-limiting step involves (1) protein disulfide formation, (2) disulfide breakage, or (3) intramolecular transitions, such as conformational changes or disulfide rearrangements. With sufficient kinetic data, it should be possible to elucidate a pathway with a single set of rate constants, containing where appropriate the dependence upon the concentration of thiol or disulfide reagent, which will fit all the kinetic observations. This can be demonstrated by using numerical integration to determine the time-dependence of all the relevant species present and showing that this simulates satisfactorily all the kinetic observations.[27,39,46] Still, it must be kept in mind that kinetic analysis can only exclude possible pathways, not prove any to be correct. Also remember that essential intermediates need not accumulate to detectable levels. The pathway should be reasonable in terms of the identities of the species and should also be thermodynamically sound; in particular, free energy changes around all cyclic paths should be zero, so one rate or equilibrium constant of each such cycle is determined by the values of all the others.

Complex pathways can be dissected by modifying some of the Cys thiol groups irreversibly, so that they cannot participate in disulfide bond

---

[57] C. Zervos and E. Adams, *Int. J. Peptide Protein Res.* **10**, 1 (1977).

formation.[46] Disulfide bond formation in such modified species should reflect the absence of paths involving the altered Cys residues. The trapped intermediates provide some such species, as whatever disulfides they have can be reduced, leaving the irreversible blocking groups introduced by trapping; the kinetics of disulfide bond formation in the reduced species can be determined under the normal conditions. Disulfide bond formation may be nearly normal in all such species in which the disulfides are formed directly, but will be blocked if the original disulfides arose only by rearrangements involving the other Cys residues.

The kinetic analysis should not depend upon the assumption that each kinetic species is homogeneous, for different protein molecules with the same disulfide bonds can differ in their conformation, the topology of the disulfide loops, cis–trans isomers of peptide bonds adjacent to Pro residues,[58] etc., that could affect the rates and paths with which they form disulfides. Early studies of the kinetics of protein folding[59] erred in making this assumption about the unfolded states of proteins.[60] More recently, Konishi et al.[36] concluded that there must be multiple rate-limiting steps in refolding of reduced RNase because they were unable to simulate the kinetics of appearance of native-like RNase with a single step. They assumed each of their kinetic intermediates to be homogeneous, even though multiple species were grouped together solely on the basis of the number of intramolecular and mixed disulfides they contained, irrespective of their identities. The rates at which the various one-, two-, and three-disulfide species form disulfide bonds are known to vary enormously.[8,12] Also, the types of experimental procedures they used are known to introduce presumably covalent modifications of the structure.[12] Not surprisingly, their conclusions differ from those reached by a more direct analysis of the pathway,[8,12,61] using the procedures described here.

A further check on the validity of a kinetic pathway is whether the equilibrium constants for the individual steps, and for the overall pathway, calculated from the forward and reverse rate constants agree with the experimental values at true equilibrium.[27] Equilibria can be attained by leaving reaction mixtures for suitable lengths of time in the appropriate mixtures of thiol and disulfide reagents. The approach to equilibrium can be monitored and should ideally be attained from both directions.

[58] J. F. Brandts, H. R. Halvorson, and M. Brennan, *Biochemistry* **14**, 4953 (1975).
[59] A. Ikai and C. Tanford, *J. Mol. Biol.* **73**, 145 (1973).
[60] R. L. Baldwin, *Annu. Rev. Biochem.* **44**, 453 (1975).
[61] T. E. Creighton, *J. Mol. Biol.* **113**, 329 (1977).

[20] Redox Control of Enzyme Activities by Thiol/Disulfide Exchange

By HIRAM F. GILBERT

The specific modulation of enzyme activity by posttranslational modification of specific amino acid side chains is a well-established mechanism of metabolic regulation. Essentially all the amino acids possessing chemically reactive side chains undergo some type of posttranslational modification: the amino group of lysine, the carboxyl groups of glutamate and aspartate, the imidazole side chain of histidine, the hydroxyl groups of serine, threonine, and tyrosine, and even the guanidino group of arginine. The sulfhydryl group of cysteine is potentially the most reactive nucleophile in proteins, and an extensive literature exists concerning the chemical modification of this amino acid side chain *in vitro;* however, the posttranslational modification of cysteine residues is not widely accepted as a regulatory mechanism *in vivo.*

As a functional residue in proteins, the chemistry of the sulfhydryl group is unique. The sulfhydryl group of cysteine may undergo oxidation to disulfides

$$2 \text{ RSH} \rightleftharpoons \text{RSSR} + 2 \text{ H}^+ + 2 \text{ } e^- \tag{1}$$

while the process of thiol/disulfide exchange[1-3] results in the exchange of redox status between two thiol/disulfide pairs

$$\text{RSH} + \text{R}'\text{SSR}' \rightleftharpoons \text{RSSR}' + \text{R}'\text{SH} \tag{2}$$
$$\text{RSSR}' + \text{RSH} \rightleftharpoons \text{RSSR} + \text{R}'\text{SH} \tag{3}$$

leading to the formation of mixed disulfides or symmetrical disulfides, depending on the relative redox potentials of RSH and R'SH and the relative total concentrations of oxidized and reduced species. Thiol/disulfide exchange between protein sulfhydryl groups and biologically occurring disulfides can produce mixed disulfides or intramolecular and intermolecular protein disulfides.

The process of thiol/disulfide exchange provides a mechanism for the equilibration of the sulfhydryl oxidation state of proteins with the thiol/

[1] J. M. Wilson, R. J. Bayer, and D. J. Hupe, *J. Am. Chem. Soc.* **99**, 7922 (1977).
[2] G. M. Whitesides, J. E. Lilbern, and R. P. Szajewski, *J. Org. Chem.* **42**, 335 (1977).
[3] G. H. Snyder, M. J. Cennerazzo, A. J. Karalis, and D. Field, *Biochemistry* **20**, 6509 (1981).

disulfide status of the surrounding environment. If this process occurs *in vivo* and results in changes in the activities of certain enzymes, changes in enzyme activity could be coupled to changes in the redox potential of the cell.[4-11] Originally proposed over 30 years ago by Barron,[5] the regulation of enzyme activity by thiol/disulfide exchange has received less attention than other posttranslational protein modifications. There is a large body of information from diverse sources that can be used to build an enticing circumstantial argument that thiol/disulfide exchange should be a biological regulatory mechanism. However, there is little compelling evidence in the literature requiring regulation by this mechanism.

Thiol/disulfide exchange as a biological control mechanism must exhibit at least the following six properties.

1. The intracellular thiol/disulfide status should vary *in vivo* in response to some metabolic signal.
2. Oxidation of protein sulfhydryl groups by thiol/disulfide exchange should activate some enzymes, inactivate others, and not affect the activities of others.
3. The thiol/disulfide redox potential of a regulated protein should be near the observed thiol/disulfide ratio *in vivo*.
4. Thiol/disulfide exchange reactions must be kinetically competent under physiological conditions.
5. For regulated enzymes, the oxidized and reduced forms of the enzyme should both be observable *in vivo*. The relative levels of the oxidized and reduced forms must change in response to changes in the cellular thiol/disulfide ratio.
6. The response of particular enzyme activities to oxidation by thiol/disulfide exchange should be consistent with the metabolic function of the enzyme.

This review will examine the possibility of metabolic regulation by thiol/disulfide exchange with respect to each of the six criteria outlined above.

[4] H. F. Gilbert, *J. Biol. Chem.* **257,** 12086 (1982).
[5] E. S. Barron, *Adv. Enzymol. Relat. Subj. Biochem.* **11,** 201 (1951).
[6] M. A. A. Manboodiri, J. Favilla, and D. C. Klein, *Science* **213,** 571 (1981).
[7] D. M. Ziegler and L. L. Poulson, *Trends Biochem. Sci.* **2,** 79 (1977).
[8] N. S. Kosower and E. M. Kosower, *Int. Rev. Cytol.* **54,** 110 (1978).
[9] J. Isaacs and F. Binkley, *Biochim. Biophys. Acta* **497,** 192 (1977).
[10] S. Pontremoli and B. L. Horecker, *Curr. Top. Cell Regul.* **2,** 173 (1970).
[11] H. Tschesche and H. W. McCartney, *Eur. J. Biochem.* **120,** 183 (1981).

The Thiol/Disulfide Ratio Does Vary *in Vivo*

The thiol/disulfide ratio must change in response to some metabolic signal if this mechanism of control is to be of any biological utility. The oxidized and reduced states of glutathione, the major intracellular thiol, are interrelated and presumably regulated through the glutathione cycle,[9,11,12] illustrated in Fig. 1. The elements of this cycle link the thiol/disulfide status of the cell[8] with the levels of oxidized and reduced pyrimidine nucleotides,[13] oxygen metabolism,[14] and as a possible consequence of thiol/disulfide exchange,[15] the activity of key enzymes in metabolism.

The predominant cellular thiols and disulfides are listed in Table I[9,16–27] along with estimates of their total concentrations in rat liver in the fed and fasted state. Although there is significant variation in the values reported in the literature,[8] and some uncertainty exists concerning the effects of compartmentalization, it is likely that the physiological state of the experimental animal influences the relative concentrations of thiol and disulfide species.

Isaacs and Binkley[9] have measured the GSH and Prot-SSG levels in rat liver as a function of the metabolic state of the animal. The levels of GSH and Prot-SSG undergo a reciprocal diurnal variation tied to the feeding schedule. After feeding, the levels of GSH increase as the levels of Prot-SSG decline. Between meals, the level of GSH decreases as Prot-SSG increases. Prot-SSG levels vary by a factor of 2, and GSH concentrations vary by approximately the same factor. Thus, the fed state may be

---

[12] A. Meister, *Curr. Top. Cell Regul.* **18,** 21 (1981).

[13] E. M. Scott, I. W. Duncan, and V. Ekstrand, *J. Biol. Chem.* **238,** 3928 (1963).

[14] P. W. Reed, *J. Biol. Chem.* **244,** 2459 (1969).

[15] B. States and S. Segal, *Biochem. J.* **132,** 623 (1973).

[16] J. Vina, R. Hems, and H. A. Krebs, *Biochem. J.* **170,** 627 (1978).

[17] T. P. M. Akerboom, M. Bilzer, and H. Sies, *J. Biol. Chem.* **257,** 4248 (1982).

[18] F. Tietze, *Anal. Biochem.* **27,** 502 (1969).

[19] J. T. Isaacs and F. Binkley, *Biochim. Biophys. Acta* **498,** 192 (1977).

[20] M. S. Moron, J. W. Depierre, and B. Mannervik, *Anal. Biochem.* **582,** 67 (1979).

[21] L. V. Eggleston and H. A. Krebs, *Biochem. J.* **138,** 425 (1974).

[22] N. Tateishi, T. Higashi, S. Shinya, A. Naruse, and Y. Sakamoto, *J. Biochem.* (*Tokyo*) **75,** 93 (1974).

[23] J. J. Kelley, K. A. Herrington, S. P. Ward, A. Meister, and O. M. Friedman, *Cancer Res.* **27,** 137 (1967).

[24] E. Walastys-Rode, K. E. Coll, and J. R. Williamson, *J. Biol. Chem.* **254,** 11521 (1979).

[25] B. E. Corkey, M. Brandt, R. J. Williams, and J. R. Williamson, *Anal. Biochem.* **118,** 30 (1981).

[26] P. L. Wendel, *Biochem. J.* **117,** 661 (1970).

[27] K. R. Harrap, R. C. Jackson, P. G. Riches, C. A. Smith, and B. T. Hill, *Biochim. Biophys. Acta* **310,** 104 (1973).

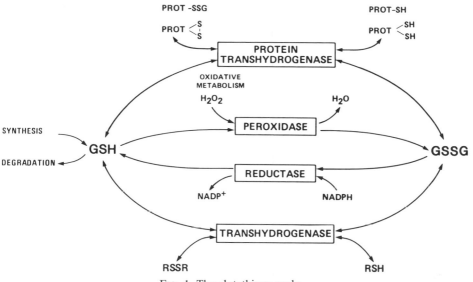

FIG. 1. The glutathione cycle.

associated with an overall thiol:disulfide ratio that may be as much as 4-fold larger (more reduced) than the fasted state. Isaacs and Binkley[19] also found hormonally induced changes in the cellular thiol/disulfide ratio. Intraperitoneal administration of glucagon or dibutyryl-cAMP results in a significant decrease in the thiol/disulfide ratio in rat liver. Increased levels of intracellular cAMP were also associated with decreased catalase activity. The decreased catalase activity could increase the levels of hydrogen peroxide.[19] Through the action of glutathione peroxidase,[14] this might result in an increased level of GSSG and, subsequently, Prot-SSG. Although a direct, causal relationship has not been established, this effect could couple the thiol/disulfide ratio to the metabolic state of the cell.

The experimental difficulties encountered in the measurement of cellular thiols and disulfides warrant a brief discussion. At neutral pH, GSH is oxidized in air to GSSG. Because of the large excess of GSH over GSSG, postmortem oxidation of a small amount of the total GSH will result in a large artifactual increase in the observed GSSG concentration. Vina and co-workers[16] have reported that the concentration of GSSG increases about 50% in 100 min in acid-soluble extracts of rat liver that have been neutralized to pH 7.4. However, in perchloric acid solution at 0 or 20° these same authors observed no increase in GSSG concentration over several hours.[16] Homogenization of tissue in the presence of perchloric acid followed by neutralization in the presence of high concentrations of

TABLE I

APPROXIMATE TOTAL CELLULAR CONCENTRATIONS OF THIOLS
AND DISULFIDES IN RAT LIVER

| Compound | Approximate concentration (mM)[a] | | References[c] |
	Fed[b]	Starved[b]	
Thiols			
GSH	7.2 (±2)	3.7 (±1)	9, 16–22
Prot-SH	14	11.5	9, 19
Cysteine	0.25	—	23
Cysteamine	0.1–0.2	—	23
CoASH	0.2	—	24, 25
Disulfides			
GSSG	0.16 (0.03–0.4)	—	17, 18, 21, 26
Prot-SSG	1.7	3.6	9, 19, 27
Prot-SS Prot	8.2	8.4	9, 19, 27
CoASSG	≤0.02	—	25

[a] Approximate millimolar concentrations were derived from reported concentrations using the conversion factors 0.1 g of protein per gram wet weight and 0.29 g dry weight per gram wet weight.[20]

[b] Values in parentheses represent the range of values reported. For levels in other tissues, see Kosower and Kosower.[8]

[c] Numbers refer to text footnotes.

N-ethylmaleimide (to trap free sulfhydryl groups and prevent oxidation) may be used to minimize oxidation of GSH to GSSG postmortem.[18] Glutathione reductase is inactivated by reasonably low levels of N-ethylmaleimide, so that this sulfhydryl reagent must be thoroughly removed from the sample before assaying GSSG.[18,28]

Another potential problem is that the concentrations of GSSG likely to occur *in vivo* are approximately 10^4 higher than expected based on the equilibrium constant for the glutathione reductase reaction, and the NADPH/NADP ratio.[16] The GSSG concentration is apparently maintained at a steady-state concentration significantly higher than the equilibrium value. If solubilization of the tissue allows the glutathione reductase reaction to come to equilibrium (or even approach it), GSSG present at high levels in the steady state could be reduced by the excess NADPH, causing artifactually low estimates of the original GSSG concentration in

---

[28] R. F. Colman and S. Black, *J. Biol. Chem.* **240,** 1796 (1965).

the cell. The relatively high levels of NADPH in cellular homogenates (0.38 mM)[29] and the favorable glutathione reductase equilibrium[16] could generate a sizable error in the observed GSSG concentration. Freeze-clamp methods,[30] in which the tissue sample is rapidly cooled immediately postmortem, followed by homogenization in perchloric acid, may minimize both kinds of artifacts. Artifactual postmortem changes in the GSSG concentration could cause underestimation or overestimation of the GSSG concentrations *in vivo*. This may explain a portion of the variability in the reported concentrations of GSSG (Table I).

There are several methods available for quantitating the levels of oxidized and reduced thiol species in cellular extracts. These have been reviewed[8] and discussed[18] previously. The concentration of sulfhydryl groups in total cellular homogenates or in acid-soluble fractions may be determined by the procedure of Ellman.[31] Protein-SH concentrations are estimated by difference between total cellular thiol and acid-soluble thiols using Ellman's reagent.[9] GSSG can be specifically quantitated by the decrease in NADPH concentration using the glutathione reductase reaction[32] or by the sensitive kinetic recycling assay of Tietze.[18] Prot-SSG and Prot-SS-Prot have been measured by following the increase in total sulfhydryl groups in the sample after denaturation and reduction of all disulfides with sodium borohydride. The increase in acid-soluble sulfhydryl groups after reduction is taken as a measure of Prot-SSG, since GSSG is presumed to be present in low levels. The increase in acid precipitable sulfhydryl groups after reduction, in excess of those expected from the increase in acid-soluble sulfhydryl groups, indicates the levels of Prot-SS-Prot.[9,19] Meister and Anderson[33a] have recently suggested that Prot-SSG concentrations may be seriously overestimated by this procedure due to entrapment of GSH and GSSG during precipitation and the reduction of thiol esters of glutathione by sodium borohydride.

Because of possible postmortem changes in the concentrations of protein and nonprotein thiols and disulfides, the measurement of *"in vivo"* levels of these species should be regarded with caution. Although the trends observed in the variation of cellular thiol and disulfide concentra-

[29] D. H. Williamson, F. Mayor, and D. Veloso, *in* "Regulation of Gluconeogenesis" (H.-D. Söling and B. Willms, eds.), p. 92. Academic Press, New York, 1971.

[30] A. Wollenberger, O. Ristau, and G. Schoffa, *Pfluegers Arch. Gesamte Physiol. Menschen Tiere* **270,** 399 (1960).

[31] G. L. Ellman, *Arch. Biochem. Biophys.* **82,** 70 (1959).

[32] H. Klotzsch and H. U. Bergmeyer, *in* "Methods of Enzymatic Analysis" (H. U. Bergmeyer, ed.), p. 363. Academic Press, New York, 1963.

[33a] A. Meister and M. E. Anderson, *Annu. Rev. Biochem.* **52,** 711 (1983).

tions with metabolic state are likely in the indicated directions, the magnitude of these variations is certainly not well established.

Oxidation of Protein Sulfhydryl Groups by Thiol/Disulfide Exchange Activates Some Enzymes and Inactivates Others

If thiol/disulfide exchange is utilized biologically as a regulatory mechanism, the oxidation of protein sulfhydryl groups should have different effects on different enzyme systems. Some sulfhydryl-containing enzymes may be unaffected by oxidation–reduction either because they lack sulfhydryl groups of appropriate redox potential or because the sulfhydryl redox changes that do occur do not affect activity. Other enzymes might display decreases or even increases in activity in response to sulfhydryl group oxidation.

Table II lists a number of enzyme activities that have been suggested to be modulated by reversible thiol/disulfide exchange.[4,11,21,33–66] All the en-

[33] F. Binkley and S. M. Richardson, *Biochem. Biophys. Res. Commun.* **106**, 175 (1982).

[34] K. Nakashima, S. Pontremoli, and B. L. Horecker, *Proc. Natl. Acad. Sci. U.S.A.* **64**, 947 (1969).

[35] U. K. Moser, M. Althaus-Salzmann, C. Van Dop, and H. A. Lardy, *J. Biol. Chem.* **257**, 4552 (1982).

[36] W. C. Carle, R. E. Cappel, E. Izbicka-Dimitrijevic, and H. F. Gilbert, unpublished results.

[37] M. J. Ernest and K.-H. Kim, *J. Biol. Chem.* **248**, 1550 (1973).

[38a] K.-H. W. Lau and J. H. Thomas, *J. Biol. Chem.* **258**, 2321 (1983).

[38] M. Magnani, V. Stocchi, P. Ninfali, M. Dacha, and G. Fornaini, *Biochim. Biophys. Acta* **615**, 113 (1980).

[39] R. Nesbakken and L. Eldjarn, *Biochem. J.* **87**, 526 (1963).

[40] A. Burchell and B. Burchell, *FEBS Lett.* **118**, 180 (1980).

[41] T. J. C. Van Berkel, J. F. Koster, G. E. J. Stall, and J. K. Kryut, *Biochim. Biophys. Acta* **321**, 496 (1973).

[42] T. J. C. Van Berkel, J. F. Koster, and W. C. Hulsman, *Biochim. Biophys. Acta* **293**, 118 (1973).

[43] M. Usami, H. Matsushita, and T. Shimazu, *J. Biol. Chem.* **255**, 1928 (1980).

[44] T. Shimazu, S. Tokutake, and M. Usami, *J. Biol. Chem.* **253**, 7376 (1978).

[45] V. Ernst, D. H. Levin, and I. M. London, *Proc. Natl. Acad. Sci. U.S.A.* **75**, 4110 (1978).

[46] G. Guillon, B. Cantu, and S. Jard, *Eur. J. Biochem.* **117**, 401 (1981).

[47] S. P. Mukherjee and C. Mukherjee, *Biochim. Biophys. Acta* **677**, 339 (1981).

[48] S. P. Mukherjee and W. S. Lynn, *Biochim. Biophys. Acta* **568**, 224 (1979).

[49] A. Baba, E. Lee, T. Matsuda, T. Kihara, and H. Iwata, *Biochem. Biophys. Res. Commun.* **85**, 1204 (1978).

[50] H. J. Brandwein, J. A. Lewicki, and F. Murad, *J. Biol. Chem.* **256**, 2958 (1981).

[51] H. F. Gilbert and M. D. Stewart, *J. Biol. Chem.* **256**, 1782 (1981).

[52] M. A. A. Namboodiri, J. T. Favilla, and D. C. Klein, *J. Biol. Chem.* **257**, 10030 (1982).

[53] M. A. A. Namboodiri, J. L. Weller, and D. C. Klein, *Biochem. Biophys. Res. Commun.* **96**, 188 (1980).

zyme activities reported in Table II undergo reversible changes in activity due to oxidation or reduction with moderate levels of biologically occurring thiols or disulfides. Many enzymes have sulfhydryl groups that can be chemically modified by a variety of sulfhydryl-specific reagents (iodoacetamide, N-ethylmaleimide, Ellman's reagent, etc.) with a resultant effect on enzyme activity. Most of these reagents are highly reactive toward sulfhydryl groups and somewhat less specific as a result. In addition, reactive disulfides such as Ellman's reagent are much better oxidizing agents than most biologically occurring disulfides. For example, the enzymes aldolase,[67] glycogen phosphorylase,[68] and thiolase[69] all have sulfhydryl groups that react with iodoacetamide, N-ethylmaleimide, or reactive disulfides; however, for all three of these enzymes, incubation with 10 mM GSSG at pH 8.0 for 24 hr (very harsh oxidizing conditions) has a minimal effect on the activity.[36] The presence of a chemically reactive sulfhydryl group is not sufficient evidence to suggest regulation of any enzyme by thiol/disulfide exchange.

Changes in enzyme activity (both increases and decreases) in response to reversible thiol/disulfide exchange may be due to changes in the overall catalytic ability (V_{max}), the steady-state affinity for substrate (K_m), the binding of allosteric effectors, or the quaternary structure of the protein. This full spectrum of effects has been observed. Therefore, it is necessary

---

[54] G. L. Cottam and P. A. Srere, Arch. Biochem. Biophys. **130,** 304 (1969).

[55] B. R. Osterlund, M. K. Packer, and W. A. Bridger, Arch. Biochem. Biophys. **205,** 489 (1980).

[56] D. W. Walters and H. F. Gilbert, unpublished results.

[57] M. G. Battelli, FEBS Lett. **113,** 47 (1980).

[58] G. Federici, D. DiCola, P. Sacchetta, C. Di Ilio, G. DelBoccio, and G. Polidoro, Biochem. Biophys. Res. Commun. **81,** 650 (1978).

[59] H. G. Pontis, J. R. Babio, and G. Salerno, Proc. Natl. Acad. Sci. U.S.A. **78,** 6667 (1981).

[60] H. W. Macartney and H. Tschesche, FEBS Lett. **119,** 327 (1980).

[61] M. A. A. Namboodiri, J. L. Weller, and D. C. Klein, J. Biol. Chem. **255,** 6032 (1980).

[62] M. J. Clemens, B. Safer, W. C. Merrick, W. F. Anderson, and I. M. London, Proc. Natl. Acad. Sci. U.S.A. **72,** 1286 (1975).

[63] R. Jagus and B. Safer, J. Biol. Chem. **256,** 1324 (1981).

[64] N. S. Kosower, G. A. Vanderhoff, and E. M. Kosower, Biochim. Biophys. Acta **272,** 623 (1972).

[65] G. L. Francis and J. F. Ballard, Biochem. J. **186,** 581 (1980).

[66] F. J. Ballard, M. F. Hoopgood, L. Reshef, and R. W. Hanson, Biochem. J. **140,** 531 (1974).

[67] B. L. Horecker, O. Tsolas, and C. Y. Lai, in "The Enzymes" (P. D. Boyer, ed.), 3rd ed., Vol. 7, p. 213. Academic Press, New York, 1972.

[68] D. J. Graves and J. H. Wang, in "The Enzymes" (P. D. Boyer, ed.). 3rd ed., Vol. 7, p. 435. Academic Press, New York, 1972.

[69] E. Izbicka-Dimitrijevic and H. F. Gilbert, Biochemistry **24,** 6112 (1982).

TABLE II

ENZYME ACTIVITIES THAT MAY BE REGULATED BY THIOL/DISULFIDE EXCHANGE

Enzyme	RSSR	Conditions		$t_{1/2}$	Effect	References[a]
		Concentration (mM)	pH			
Enzymes of carbohydrate metabolism						
Phosphofructokinase (rabbit muscle)	GSSG CoASSG CoASSCoA	0.2	8.0	50 min	Inactivation	4
Phosphofructokinase (mouse liver homogenate)	GSSG	5	7.4	<20 min	Inactivation	33
Fructose-1,6-bisphosphatase (rabbit liver)	CoASSG CoASSCoA	0.1	8.5	2.5 hr	Activation, 2- to 4-fold	34
(rat liver)	ND	—	—	—	Decreased K_i for allosteric inhibitor AMP	35
Aldolase	GSSG	10	8.0	>2 days	No effect	36
Glycogen synthase-D	GSSG	1	8.5	15 min	Inactivation	37
Glycogen synthase-I	GSSG	10	7.8	2 days at $-20°$	Inactivation	38a
Hexokinase (rabbit red blood cell)	GSSG	1	7.2	5 min	Inactivation	38
	Cystamine	1	7.5	30 min	Inactivation	39
Glucose-6-phosphatase (rabbit liver)	GSSG	100	7.0	>30 min	No inactivation observed, but absence of GSSG results in inactivation	40
Pyruvate kinase (human erythrocyte)	GSSG	2.5	8.0		K_i for activator, fructose 1,6-bisphosphate (10-fold) Increased K_m for PEP (10-fold)	41, 42
Glucose-6-P-dehydrogenase (rat liver homogenate)	GSSG	0.01–0.02	7.4	ND	Relieves product inhibition by NADPH	21

			pH	Time	Effect	Ref.
Phosphorylase phosphatase (rabbit liver, rabbit muscle)	GSSG, cystine	10	7.4	≤5 min	Inactivation	43, 44
Protein kinase (reticulocyte lysate) (crude)	GSSG	0.1–1.0	7.6	Minutes	Increased phosphorylation of eIF-2, activation of protein kinase (unidentified) activity; inactivation of eIF-2 and protein synthesis	45
Adenylate cyclase (rat adipocytes)	ND	—	—	—	Diminished response to stimulation by guanosine-5'-$\beta\gamma$-iminotriphosphate in the absence of 0.01–0.1 M mercaptoethanol	46
(rat adipocytes)	H_2O_2	0.5	—	—	Decreased activity, reversible with GSH, DTT	47, 48
(rat brain)	GSSG	0.5–2.5	7.4	20 min	Inactivation of basal and dopamine or guanosine-5'-$\beta\gamma$-iminotriphosphate-stimulated activity	49
Guanylate cyclase (rat lung)	Cystamine, cystine, GSSG, CoASSG	1	7.6	2 min	Inactivation	50
Enzymes of lipid metabolism						
Hydroxymethylglutaryl-CoA reductase (yeast)	CoASSCoA, GSSG	0.001 0.02	7.0 7.0	30 sec	Inactivation	51
Acetyl-CoA hydrolase (pineal gland)	Cystamine Various disulfide-containing peptides	10 (0.1–1 mM)	8.5 8.5	~2 min 30 min	Activation, 3-fold	52, 53
ATP citrate-lyase	GSSG	2.5	8.7	<5 min	Inactivation	54, 55
Fatty acid synthetase	CoASSG	0.08	7.5	60 min	Inactivation	56

(continued)

TABLE II (continued)

| Enzyme | Conditions | | | | Effect | References[a] |
	RSSR	Concentration (mM)	pH	$t_{1/2}$		
Miscellaneous						
Xanthine oxidase (rat liver)	GSSG	0.05	8.1	30 min	Converts NAD$^+$-dependent form (D) to an oxidase form (O)	57
Tyrosine aminotransferase	Cystine, cystamine	1	7.4	45 min	Inactivation	58
Sucrose synthase (wheat germ)	GSSG, oxidized thioredoxin	10	7.9	1 hr	Inhibits cleavage of sucrose to UDP-glucose and fructose, but not the reverse reaction	59
Collagenase (polymorphonuclear leukocytes)	GSSG, cystine insulin	0.6	7.4	3 hr	Stimulation, 10-fold; releases inhibitor, protein	11, 60
Indoleamine-N-acetyltransferase (pineal gland)	Cystamine, S—S containing peptides	5	6.5	10 min	Inactivation	61
Protein synthesis (hemin-dependent rabbit reticulocyte lysate)	GSSG	0.5		Minutes	Inhibition by oxidation of initiation factor eIF-2	45, 62–64
Protein degradation	Cystine	2	8.0	1 hr	Increased degradation of glucose-6-P-dehydrogenase and phosphoenolpyruvate carboxylase	65, 66

[a] Numbers refer to text footnotes.

to examine a number of properties of the oxidized and reduced enzyme under a variety of experimental conditions. Oxidation of human erythrocyte pyruvate kinase by GSSG results in a 10-fold increase in the K_m for the substrate phosphoenolpyruvate.[41,42] If the inactivation experiments had been performed using an enzyme assay in which the substrate concentration was held constant at $100 \times K_m$, no effect of oxidation on the enzyme activity would have been observed. Thus, the observation of enzyme inactivation or activation may depend critically on the assay procedure and other experimental variables.

Alterations in the binding of allosteric effectors may cause apparent activation or inactivation. Oxidized pyruvate kinase exhibits a 10-fold lower affinity for the allosteric activator fructose 1,6-bisphosphate than the reduced enzyme.[41,42] In contrast, oxidized fructose-1,6-bisphosphatase is less subject to allosteric inhibition by AMP because of a 2-fold increase in the K_i,[35] and the oxidized form of glucose-6-phosphate dehydrogenase has an increased inhibition constant for NADPH, a potent product inhibitor.[21] For pyruvate kinase, the increased K_m for substrates and positive effectors would appear as oxidative inactivation, whereas for glucose-6-phosphate dehydrogenase and fructose-1,6-bisphosphatase an increased dissociation constant for a negative effector would appear as enzyme activation when assayed in the presence of the inhibitor.

Phosphofructokinase is reversibly inactivated by oxidation with a number of biologically occurring disulfides.[4] Reduced phosphofructokinase is in a dynamic association equilibrium involving monomer, dimer, tetramer, and 16-mer species. In the absence of substrates, the sedimentation coefficient of the tetramer is 13.5 S. Substrate binding increases the formation of tetramers and decreases the sedimentation coefficient of the active tetramer to 12.4 S.[70–72] The oxidized enzyme still binds substrates as evidenced by a substrate-induced increase in the concentration of the tetrameric species; however, the apparent monomer–tetramer association constant is decreased approximately 4-fold compared to the reduced enzyme. The tetrameric species of the oxidized enzyme sediments at 13.5 S in the presence or the absence of substrates and does not undergo the characteristic conformation change leading to the 12.4 S species.[73] Thus, oxidation of phosphofructokinase is associated with a decrease in the V_{max} and changes in the characteristics of the tertiary and quaternary properties of the protein. Less subtle oxidation-induced changes in protein qua-

[70] L. K. Hesterberg and J. C. Lee, *Biochemistry* **19**, 2030 (1980).
[71] L. K. Hesterberg, J. C. Lee, and H. P. Erickson, *J. Biol. Chem.* **256**, 9724 (1981).
[72] L. K. Hesterberg and J. C. Lee, *Biochemistry* **21**, 216 (1982).
[73] M. A. Luther, H. F. Gilbert, and J. C. Lee, *Biophys. J.* **41**, 277 (1983).

ternary structure have been found. Activation of polymorphonuclear leukocyte collagenase by oxidation of a latent enzyme is accompanied by the release of a covalently bound inhibitory protein.[60]

In experiments with crude cellular homogenates, it is possible to observe an increase in the flux through a given metabolic step as a consequence of oxidation, even though the primary effect is enzyme inactivation. In crude cellular homogenates, the activity of glucose-6-phosphate phosphatase is stabilized by oxidation with GSSG. However, the effect is not reversible by dithiothreitol or other thiols and most likely results from a decrease in the activity of a phosphoprotein phosphatase responsible for inactivation of the active phosphorylated glucose-6-phosphate phosphatase.[40]

The multitude of mechanisms by which oxidation–reduction may affect enzyme activity requires that a number of variables must be experimentally explored in examining the potential effects of thiol/disulfide exchange on enzyme activity: the effects on substrate and effector binding, the effects on V_{max}, the effects on the tertiary and quaternary structure of the protein, and potentially the effects on other covalent posttranslational modifications.

The Thiol/Disulfide Redox Potentials of Some Proteins Are Near the Physiological Significant Range

If the thiol/disulfide status of a regulated protein is in equilibrium with the thiol/disulfide status of the cell, the equilibrium constant for the thiol/disulfide exchange reaction

$$E_{red} + RSSR \rightleftharpoons E_{ox} + RSH \tag{4}$$

$$K_{eq} = \frac{[E_{ox}][RSH]}{[E_{red}][RSSR]} \tag{5}$$

must lie near the intracellular thiol/disulfide ratio. According to this simple equilibrium model, changes in the ratio of oxidized to reduced enzyme ($[E_{ox}]/[E_{red}]$), and hence, the activity, would result from changes in the thiol/disulfide ratio ($[RSH]/[RSSR]$). We will operationally define the thiol/disulfide redox potential of a particular protein as the ratio of [GSH]/[GSSG] at which 50% of the effect on protein function (enzyme activity) is observed. It is possible that multiple pathways of oxidation and reduction of cellular thiols and disulfides could establish steady-state thiol/disulfide ratios that are not at equilibrium. Regardless, the thermodynamics of a particular redox system are still important in determining the direction and magnitude of any redox state changes.

The intracellular ratio of GSH/GSSG, the most predominant low molecular weight thiol/disulfide pair, may vary between approximately 10 and 300 (Table I), although higher ratios have been reported.[17] As described previously, the measurements of the *in vivo* GSH/GSSG are complicated by a variety of technical problems including the problem of compartmentalization. There have also been no detailed studies performed in which this specific ratio and its variation under different metabolic conditions have been determined. It has been observed that there are metabolically induced changes in the ratio Prot-SSG/Prot-SH by factors of up to 3- to 5-fold.[9,19] If these variations are translated (assuming equilibrium) into variations in the GSH/GSSG ratio, the ratio of GSH/GSSG could vary between approximately 5 and 30. Obviously, further experimental measurements of the thiol/disulfide status of cells in response to metabolic signals are necessary.

Unfortunately, there is a very limited amount of data in the literature concerning the thiol/disulfide redox potentials of individual proteins. For redox titration experiments it is necessary to ensure that equilibrium has been established with the redox buffer. This must be evidenced by no further changes in enzyme activity and the GSH/GSSG ratio. To minimize changes in GSH/GSSG ratio during equilibration due to oxidation of GSH to GSSG by oxygen, all incubations should be performed anaerobically, and the actual ratio of GSH/GSSG must be determined after equilibration. A technique devised to simplify determination of the GSH/GSSG ratio is the use of low concentrations (0.05 mM) of a reporter thiol/disulfide that equilibrates with GSH and GSSG.[4] If the thiol/disulfide ratio of the reporter can be measured conveniently, accurately, and quickly, the GSH/GSSG ratio can be determined from the known equilibrium constant between the reporter and the GSH/GSSG. The CoASSG/CoASH pair can be used as such a reporter.

$$CoASSG + GSH \rightleftharpoons CoASH + GSSG \tag{6}$$

The equilibrium constant for this "reporter reaction" (K = [CoASH][GSSGH]/[CoASSG][GSH]) is 0.34 at pH 8.0,[4] 25°, and 0.62 at pH 7.5.[56] The CoASH/CoASSG ratio can be determined by integrating the peak areas of the CoASH and CoASSG peaks from a high-performance liquid chromatographic separation of the reporter species in the equilibrated sample.[4] Since CoASSG is more difficult to reduce than GSSG by a factor of about 3, reasonably large GSH/GSSG ratios can be measured in this manner.

The oxidation state and activity of rabbit muscle phosphofructokinase equilibrates with GSH/GSSG redox buffers *in vitro*.[4] After a 16-hr equili-

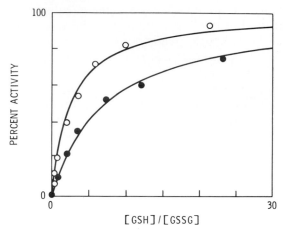

FIG. 2. Redox equilibrium titration of rabbit muscle phosphofructokinase activity. The percentage of totally reduced activity is plotted as a function of the equilibrium [GSH]/[GSSG] ratio: (●) in the absence of ATP; (○) in the presence of 0.1 mM ATP. All incubations were for 24 hr in 0.1 M Tris-HCl buffer, 25 mM MgSO₄, 25°. The total [GSH] + [GSSG] was 10 mM. Reproduced, with permission, from Gilbert.[4]

bration of a solution of phosphofructokinase with various GSH/GSSG ratios at a total concentration of GSH plus GSSG of 10.0 mM, the activity of the enzyme responds to the changing GSH/GSSG ratio according to Eq. (5) with a thiol–disulfide redox potential of 7.0 (Fig. 2). Even after 16 hr, all the initial phosphofructokinase activity could be recovered after reduction of the total reaction mixture with excess dithiothreitol regardless of the GSH/GSSG ratio. The redox potential of phosphofructokinase is sensitive to the presence of enzyme effectors. Inclusion of 1.0 mM ATP results in a shift of the thiol/disulfide redox potential from 7.0 to 2.5 (Fig. 2).

The other protein for which the thiol/disulfide redox potential has been measured is fatty acid synthetase from chicken liver. Fatty acid synthetase undergoes reversible inactivation by oxidation with a variety of disulfides including GSSG, CoASSG, and CoASSCoA.[56] Stoops and Wakil[74] have shown that Ellman's reagent inactivates chicken liver fatty acid synthetase by formation of an interchain disulfide between a cysteine residue and the sulfhydryl group of the phosphopantetheine, which cross-links the enzyme subunits into a dimer of molecular weight 500,000. Similar cross-linking is observed after oxidation with GSSG.[56] The thiol/disulfide redox potential of fatty acid synthetase (chicken liver) varies between

[74] J. K. Stoops and S. J. Wakil, *Biochem. Biophys. Res. Commun.* **104,** 1018 (1982).

approximately 1 and 11 as the total concentration of redox buffer [(GSSG] + [GSH]) is decreased from 50 mM to 6.2 mM.[56] This system obviously does not follow the simple equilibrium expression of Eq. (5), but contains higher order terms in GSH concentration, presumably due to the formation of inactive protein—SS—protein disulfides. Thus, the activity of fatty acid synthetase (and potentially other enzymes as well) would respond to metabolic changes in the GSH : GSSG ratio as well as to changes in the total glutathione concentration. Preliminary experiments in our laboratory suggest that chicken liver acetyl-CoA carboxylase is also sensitive to inactivation by oxidation with low concentrations (10–100 μM) of biological disulfides. The thiol/disulfide redox potential is not yet known with certainty, but it lies in the range 2–8.[56]

From the data of Van Berkel and co-workers,[41,42] it is possible to estimate the thiol/disulfide redox potential of pyruvate kinase. After 24 hr incubation of pyruvate kinase with 2.5 mM GSSG, 5 mM GSH was added to reactivate the enzyme. Only 20% of the lost activity was recovered even after prolonged times of incubation. Further addition of 1.0 M 2-mercaptoethanol restored complete activity. If we ignore oxidation of GSH to GSSG and assume that the 20% increase in activity accurately reflects residual activity at a GSH/GSSG ratio of 2.0, the calculated thiol/disulfide redox potential is approximately 8, similar to the value observed for phosphofructokinase.

Different enzymes will undergo different changes in activity in relation to their thiol/disulfide redox potentials. The local environment of an oxidizable sulfhydryl group would be expected to influence the rate of thiol/disulfide exchange[3] and the redox potential. Proteins with very high or very low redox potentials would stay oxidized or reduced at most physiological thiol/disulfide ratios. Thiols of intermediate redox potential (5–30) would undergo changes in redox state in response to changes in the thiol/disulfide ratio and the total glutathione concentration. It is also possible to derive models (none have been demonstrated, or even looked for) in which the formation of a mixture of Prot-SS-Prot and Prot-SSR, or any integral combination, may be required for an effect on enzyme activity. This introduces squared or higher-order terms for the [GSSG] and [GSH] into the equilibrium expression and a cooperative, sharp transition between oxidized and reduced forms of the protein.

The Kinetic Competence of the Thiol/Disulfide Exchange Process Has Not Been Demonstrated

With some notable exceptions, the thiol/disulfide exchange between protein sulfhydryl groups and biological disulfides is a reasonably slow process *in vitro* (Table II). Typical half-lives for activation or inactivation of proteins by reaction with 1–10 mM GSSG (5- to 100-fold greater than physiological) are in the range of minutes to hours at pH 8.0. Extrapolation of these rates of thiol/disulfide exchange to pH 7.4 with a concentration of GSSG of 0.2 mM, suggests that the half-life for thiol/disulfide exchange would be on the order of hours to days. For metabolic regulation by thiol/disulfide exchange to be a viable regulatory mechanism, the rate of the process *in vivo* must be at least as fast as the changes in the thiol/disulfide ratio (minutes to hours).

Catalysis of *in vivo* thiol/disulfide exchange is necessary if the process is to be a viable mechanism of regulation for most of the enzymes listed in Table II. Catalysis could be achieved by the intermediacy of either enzyme catalysts or reactive low molecular weight thiols and disulfides.[7] Enzyme activities have been isolated from a variety of sources that catalyze reversible thiol/disulfide exchange reactions. Mannervik and Axelsson[75] have isolated and characterized a rat liver cytosolic thioltransferase that catalyzes reversible thiol/disulfide exchange between proteins and low molecular weight thiols and disulfides. This enzyme accelerates the reactivation of GSSG-inactivated pyruvate kinase *in vitro*. Other enzyme activities have been observed that catalyze inactivation of glucose-6-phosphate dehydrogenase and phosphoenolpyruvate carboxykinase in the presence of cystine[65,66] presumably by thiol/disulfide exchange. Thiol/disulfide exchange reactions are required in the refolding of randomly oxidized proteins to their native form.[76] Enzyme catalysis of the thiol/disulfide exchange involved in the folding of several proteins as well as the reductive cleavage of the intrachain disulfide bonds of insulin have been reported.[77,79] The participation of these enzyme activities in a mechanism of metabolic regulation by thiol/disulfide exchange *in vivo* is unknown.

Kinetic competence is required for any regulatory system. The possibility that thiol/disulfide exchange may be enzyme catalyzed *in vivo* means that a slow rate of thiol/disulfide exchange *in vitro* cannot be used, a priori, as an argument against regulation of a particular enzyme by this process. Further characterization of the kinetics and specificity of these

[75] B. Mannervik and K. Axelsson, *Biochem. J.* **190,** 125 (1980).

[76] G. K. Smith, G. D'Alessio, and S. W. Schaffer, *Biochemistry* **17,** 2633 (1978).

[77] R. F. Steiner, F. DeLorenzo, and C. B. Anfinsen, *J. Biol. Chem.* **240,** 4648 (1965).

[78] R. A. Roth and M. E. Koshland, *Biochemistry* **20,** 6594 (1981).

[79] D. A. Wilson and R. B. Freedman, *Biochem. J.* **191,** 373 (1980).

various enzyme activities will be required to develop an overall understanding of the process *in vivo*.

Oxidized and Reduced Phosphofructokinase Has Been Observed in Mouse Liver

Enzyme regulation by thiol/disulfide exchange requires the oxidized and reduced forms of a specific enzyme to coexist *in vivo* under certain metabolic conditions. The ratio of oxidized to reduced enzyme must also change in the appropriate direction in response to changes in the cellular thiol/disulfide status. Unfortunately, there are no data available that would clearly and unambiguously address this point.

Binkley and Richardson[33] have provided some evidence that phosphofructokinase is present in both an oxidized and a reduced form in crude cellular homogenates from the livers of fasted mice, but is present only in the reduced form in fed mice. In liver homogenates from fasted mice, phosphofructokinase activity was stimulated approximately 2-fold by the addition of dithiothreitol. No significant increase in phosphofructokinase activity was observed by adding dithiothreitol to fed mouse liver homogenates. Most (but not all) of the phosphofructokinase activity in both fed and starved mouse liver homogenates could be inactivated by the addition of excess GSSG. Fasting has been associated with a decrease in the overall thiol/disulfide ratio in rat liver.[9,19] Fasting would then be expected to increase the amount of oxidized phosphofructokinase— as is observed. Although the actual cellular thiol/disulfide status was not measured in these experiments, it is reasonable to suggest that phosphofructokinase may exist in an oxidized and reduced state *in vivo*.

The conclusive demonstration of the coexistence of the oxidized and reduced states of a given protein will require physical (or chemical) separation of the oxidized and reduced enzymes and the demonstration that the only difference in the proteins is in the thiol/disulfide redox state. Moser and co-workers[35] have been able to separate oxidized and reduced fructose-1,6-bisphosphatase on the basis of differential chromatographic properties. The reduced enzyme binds strongly to phosphocellulose, but the oxidized enzyme is not retained under their conditions. The possible chromatographic resolution of oxidized and reduced enzymes should facilitate experiments designed to look for changes in the ratio of oxidized to reduced enzymes in response to metabolically induced changes in the thiol/disulfide status of the cell.

Effects of Oxidation–Reduction on the Activities of Particular Enzymes
Are Consistent with a Low Cellular Thiol/Disulfide Ratio in the
Fasted State

One of the more convincing pieces of circumstantial evidence supporting thiol/disulfide exchange as a regulatory mechanism is the consistency
of the effect of increased enzyme oxidation with the metabolic function of
the enzyme. It is frequently possible to rationalize almost any biochemical result within the general confines of metabolic regulation; however,
regulation by thiol/disulfide exchange can be rationalized on the basis of
particularly simple concepts. Experimentally, the cellular thiol/disulfide
ratio has been shown to decrease under conditions where energy demands
are greater than energy supplies.[9,19] The decrease in the thiol/disulfide
ratio can be induced by starvation or the administration of agents that
increase the intracellular concentration of cAMP, such as glucagon or
epinephrine. In contrast, when energy supplies exceed immediate energy
demands, the thiol/disulfide ratio is high. The fed state is more reduced
than the fasted state.

If changes in the thiol/disulfide ratio are coupled to changes in the
activities of some regulated enzymes, there are three general conse-
quences that might follow. (1) Enzymes that should be active only in the
fed state should lose activity on oxidation or gain activity on reduction.
(2) Enzymes that should be active only in the fasted state should either
gain activity on oxidation or lose activity on reduction. (3) Enzymes that
must function in both the fed and fasted state should either not undergo
thiol/disulfide exchange or the thiol/disulfide exchange should not affect
the enzyme activity.

Although there is scant information concerning relative redox potentials, kinetics of thiol/disulfide exchange, and the potential synergism between thiol/disulfide exchange and other posttranslational modifications,
it is instructive to consider potential metabolic effects of changing thiol/
disulfide ratios in light of the reported effects. The essential elements of
energy metabolism are outlined in Fig. 3. Pathways that should be active
in the fed (reduced) state are indicated with heavy arrows. Pathways
active primarily in the starved state are indicated by dashed arrows. Pathways active in both the starved and fasted state are indicated by solid
arrows.

Glycogen synthesis, catalyzed by glycogen synthase, is a process
characteristic of the fed state, whereas glycogen degradation catalyzed by
glycogen phosphorylase is characteristic of the fasted state. Oxidation of
glycogen synthase by GSSG results in inactivation of the enzyme.[37,38a]
Conversely oxidation of crude cellular homogenates with GSSG results in

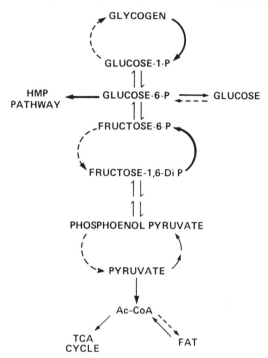

FIG. 3. The effects of thiol–disulfide exchange on energy metabolism. Enzyme activities that have been shown to increase upon oxidation are indicated by heavy, solid arrows. Enzyme activities that have been shown to decrease upon oxidation are indicated by light, dashed arrows. Other enzyme activities are indicated by light, solid arrows.

an increase in the rate of glycogen degradation. The effect of GSSG oxidation on phosphorylase activity was shown to be an indirect effect resulting from the oxidative inactivation of phosphorylase phosphatase by GSSG.[43,44] Inactivation of phosphorylase phosphatase could result in an increased level of phosphorylase *a* and a resultant stimulation of glycogenolysis. Although the effect of GSSG oxidation is apparently to inactivate both glycogen synthase and phosphorylase phosphatase, the net result of these changes is inactivation of glycogen storage and activation of glycogen degradation under conditions favoring lower thiol/disulfide ratios.

In liver, the process of gluconeogenesis (synthesis of glucose from pyruvate or lactate) is utilized in the fasted state to supply glucose to nongluconeogenic tissues such as brain and erythrocytes.[80] Glycolysis

[80] E. A. Newsholme and C. Start, "Regulation in Metabolism." Wiley, New York, 1973.

(the breakdown of glucose to pyruvate) is principally active in the fed state as a glucose catabolic pathway to supply precursors such as pyruvate for biosynthetic processes such as fatty acid synthesis.[80] Glycolysis and gluconeogenesis share many of the same chemical steps and enzymes but diverge at the irreversible and regulated steps. Glycolysis requires the activity of phosphofructokinase whereas gluconeogenesis requires the activity of fructose-1,6-bisphosphatase. The operation of both enzymes simultaneously would generate some futile cycling and ATP hydrolysis.[81] Phosphofructokinase is inactivated by oxidation[4,33] while fructose-1,6-bisphosphatase is activated (at least at moderate AMP concentrations)[34,35] by oxidation. A potential futile cycle also exists (again primarily in liver) in the interconversion of glucose and glucose 6-phosphate. Again, hexokinase activity (glycolysis) is inactivated by oxidation[38,39] while glucose-6-phosphatase activity is stimulated (indirectly) by oxidation.[40] Pyruvate kinase, another potential regulatory step in glycolysis, is also inactivated by oxidation.[41,42] The only enzyme potentially required in the fasted state that has been observed to be inactivated by oxidation is phosphoenolpyruvate carboxykinase.[65,66] However, inactivation required high concentrations of cystine (2 mM) and long periods of time (hours) in the presence of a microsomal fraction and was not totally reversible (the authors were interested in the degradation of this protein).

Fatty acid synthesis is also associated with the fed state, and in line with this expectation, acetyl-CoA carboxylase (the rate-limiting enzyme of fatty acid synthesis), fatty acid synthetase, and citrate lyase are all inactivated by oxidation.[54–56] The effect of oxidation on the rate-limiting enzymes of fatty acid β oxidation has not been examined; however, mitochondrial thiolase I (acetyl-CoA acetyltransferase), an enzyme involved in the β oxidative pathway, is highly resistant to oxidation by GSSG and a variety of other biological disulfides even though the enzyme has an essential sulfhydryl group at the active site.[69]

Protein synthesis and degradation have been reported to respond to sulfhydryl group oxidation in reciprocal fashion. Initiation factor eIF2 is inactivated by oxidation[45,62–64] or by increased phosphorylation due to increased activity of a protein kinase.[45] Both these effects would lead to decreased protein synthesis at low GSH/GSSG ratios. Ballard and co-workers[65,66] have observed that the oxidation of phosphoenolpyruvate carboxykinase and glucose-6-phosphate dehydrogenase precedes proteolysis of the enzyme during degradation and destabilizes the enzyme toward proteolysis. Similarly, the thermal stability of pyruvate kinase is

[81] H. G. Hers and L. Hue, *Curr. Top. Cell Regul.* **18,** 199 (1981).

greatly decreased on oxidation.[41,42] Both these effects are consistent with increased protein turnover in the oxidized state.

Many of the enzymes of glycolysis/gluconeogenesis are shared between the pathways and should be active in both fed and fasted state. These enzyme activities should be insensitive to oxidation–reduction. For obvious reasons, there have been no reports that oxidation of these enzymes by biological disulfides has no effect on the activity. However, fructose-1,6-bisphosphate aldolase, although it contains sulfhydryl groups that react with Ellman's reagent and other sulfhydryl reagents,[67] loses less than 10% of activity after 48 hr of incubation with 10 mM GSSG at pH 8.0.[36]

Conclusions

The coupling of cellular redox potential to enzyme activity through thiol/disulfide exchange provides an intuitively pleasing mechanism for enzyme regulation. A large amount of experimental data is consistent with the initial ideas about how such a system might function. However, it has not yet been convincingly and unequivocally demonstrated for any enzyme or enzyme system that regulation by thiol/disulfide exchange is utilized as a control mechanism *in vivo*. The crucial missing piece in this puzzle is the *in vivo* demonstration that an enzyme undergoes reversible changes in redox state in direct response to changes in the cellular thiol/disulfide status, and that these changes are reponsible for a variation in the flux through some metabolic pathway. In evaluating the potential regulation of any enzyme activity by the process of thiol/disulfide exchange, the six points outlined above should be addressed. Some of these criteria are more difficult to test experimentally than others; however, this does not obviate the necessity for doing so.

Acknowledgments

Supported by Grants GM-25347 and HL-28521 from the National Institutes of Health and Research Career Development Award HL-01020.

[21] Enzymatic Reduction of Methionine Sulfoxide Residues in Proteins and Peptides

By NATHAN BROT, HENRY FLISS,
TIMOTHY COLEMAN, and HERBERT WEISSBACH

It is firmly established that there are over a hundred different kinds of posttranslational modifications of proteins (reviewed by Uy and Wold[1]); however, only a few are considered to be reversible. These include acetylation,[2] phosphorylation,[3] methylation,[4] adenylylation,[5] and uridylylation.[5] In spite of the fact that these enzymatically catalyzed posttranslational modifications have received the most attention, nonenzymatic reactions do occur and may be of biological significance. The nonenzymatic reactions include the deamidation of asparaginyl and glutaminyl residues,[6] racemization of amino acids,[6] and the oxidation of various amino acids.[7] Among the latter reactions, the oxidation of methionine residues to methionine sulfoxide, Met(O), in proteins has only recently received special attention. It is now apparent that the nonenzymatic oxidation of methionine residues in proteins to Met(O) can occur both *in vitro* and *in vivo* and that this oxidation is associated with loss of biological activity in a wide range of proteins and peptides (reviewed by Brot and Weissbach[8]). The *in vitro* oxidation of methionine residues in proteins to Met(O) can be carried out with a variety of oxidizing agents, e.g., hydrogen peroxide, *N*-chlorosuccinimide, and chloramine-T. Although the identity of the biological oxidants that carry out this reaction *in vivo* has not been established in most instances, it appears that the hypochlorite ion, which is produced in large amounts by activated neutrophils, is one of the agents responsible. The chemical reactions by which methionine is converted into its two oxidation products, Met(O) and methionine sulfone, are shown in Fig. 1. The formation of Met(O) leads to the generation of an asymmetric center around the sulfur atom, and therefore two diaste-

[1] R. Uy and F. Wold, *Science* **198**, 890 (1977).
[2] E. L. Gershey, G. Vidali, V. G. Allfrey, *J. Biol. Chem.* **243**, 5018 (1968).
[3] E. G. Krebs and J. Beavo, *Annu. Rev. Biochem.* **48**, 923 (1979).
[4] E. N. Kort, M. F. Goy, S. H. Larsen, and J. Adler, *Proc. Natl. Acad. Sci. U.S.A.* **72**, 3939 (1975).
[5] P. B. Chuck, S. G. Rhee, and E. R. Stadtman, *Annu. Rev. Biochem.* **49**, 813 (1980).
[6] J. H. McKerrow, *Mech. Ageing Dev.* **10**, 371 (1979).
[7] R. K. Root and M. S. Cohen, *Rev. Infect. Dis.* **3**, 565 (1981).
[8] N. Brot and H. Weissbach, *Arch. Biochem. Biophys.* **223**, 271 (1983).

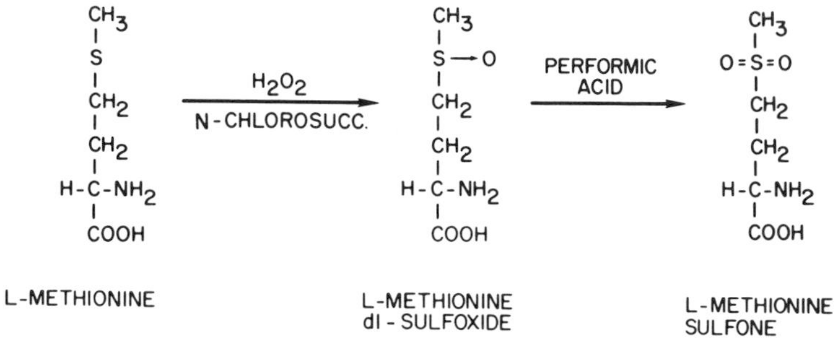

FIG. 1. Chemical conversion of methionine to methionine sulfoxide and methionine sulfone.

reoisomers can be formed from L-methionine. Met(O) can readily be reduced back to methionine in the presence of a sulfhydryl reagent; however, the oxidation to the sulfone is essentially an irreversible reaction. This chapter describes the purification and characteristics of an enzymatic system from *Escherichia coli* and human leukocytes that specifically reduces Met(O) residues in proteins and peptides to methionine. This enzymatic reduction can restore the biological activity of an inactive oxidized protein or peptide.

Materials and Methods

L-[³H]Methionine was purchased from Amersham (TRK. 583, [*methyl*-³H]methionine, 70–90 Ci/mmol), and prescored silica gel G (250 μm) thin-layer chromatography plates were obtained from Analtech, Newark, Delaware.

Synthesis of N-Acetyl-[³H]Methionine Sulfoxide

The synthesis of *N*-acetyl-[³H]methionine sulfoxide is carried out in two steps. In this procedure, methionine is first oxidized to Met(O), which is then acetylated to form *N*-AcetylMet(O) as follows. One milliliter of L-[*methyl*-³H]methionine (1 mCi) is added to 100 μl of 0.1 *M* L-methionine, 100 μl of 0.5 *N* HCl, followed by 5 μl of 30% hydrogen peroxide and is incubated at 23° for 1 hr. About 90% of the methionine is converted to Met(O). The incubation mixture is lyophilized, and the residue is dissolved in 0.5 ml of glacial acetic acid to which 0.5 ml of acetic anhydride is then added. Care is taken to exclude water from the reaction. Acetylation is carried out by incubating the reaction mixture for 90 min at 23°. The reaction is terminated by the addition of 2 ml of water, and the mixture is lyophilized. The residue is dissolved in 20 μl of water and chromatographed on a prescored silica gel G thin-layer chromatography plate (20 × 20 cm) in a solvent system consisting of n-butanol–acetic acid–water (60 : 15 : 25). Standards containing Met(O) and N-acetylMet(O) are also spotted onto one of the prescored sections of the plate. The segment containing the authentic standards is detached from the plate, sprayed with sodium hypochlorite (0.3% aqueous solution), dried, and sprayed with an aqueous solution containing 1% potassium iodide and 1% soluble starch to visualize the spots.[9] From the migration of the standard, the radioactive N-acetylMet(O) area is located, and this region of the gel is scraped from the plate, extracted with several 1-ml aliquots of water, and lyophilized. The residue is dissolved in a small quantity of water. The overall final yield of N-acetylMet(O) is about 50%.

Assay for the Reduction of N-Acetyl-[^{3}H]Met(O)

In earlier studies,[10] the assay for the presence of an enzyme, Met(O)-peptide reductase, capable of reducing Met(O) residues in proteins utilized oxidized *E. coli* ribosomal protein L12 as substrate. This assay is long and cumbersome, and the substrate is not readily available. Since it had already been established that Met(O)-enkephalin could serve as substrate,[10] other Met(O)-containing compounds were tested. One of these, N-acetylMet(O), was found to be a good substrate, and the product of the reaction, N-acetylMet, could be rapidly quantitated (Fig. 2).

Each reaction mixture contains in a final volume of 30 μl: 25 mM Tris-HCl (pH 7.4), 10 mM MgCl$_2$, 15 mM dithiothreitol, 5.0×10^{-4} M N-acetyl-[^{3}H]Met(O) (~100 cpm/pmol) and Met(O)-peptide reductase. The incubation is carried out at 37°, and the reaction is terminated by the addition of 1 ml of 0.5 M HCl. The acidified mixture is then extracted with 3 ml of ethyl acetate and centrifuged for 1 min in a table-top centrifuge and a 2-ml aliquot is assayed for radioactivity in a liquid scintillation spectrometer. It was found that under these conditions between 1 and 2% of the substrate, N-acetyl-[^{3}H]Met(O), is extracted into ethyl acetate whereas 50% of authentic N-acetyl-[^{3}H]methionine is recovered in the ethyl acetate phase. The data presented in this manuscript have been corrected for aliquot volumes and extraction efficiency. N-AcetylMet was identified as the product of the reaction by thin-layer chromatography using n-butanol–acetic acid–H$_2$O (60 : 15 : 25). The overall reaction is shown in Fig. 2.

[9] H. N. Rydon and P. W. G. Smith, *Nature (London)* **169**, 922 (1952).
[10] N. Brot, L. Weissbach, J. Werth, and H. Weissbach, *Proc. Natl. Acad. Sci. U.S.A.* **78**, 2155 (1981).

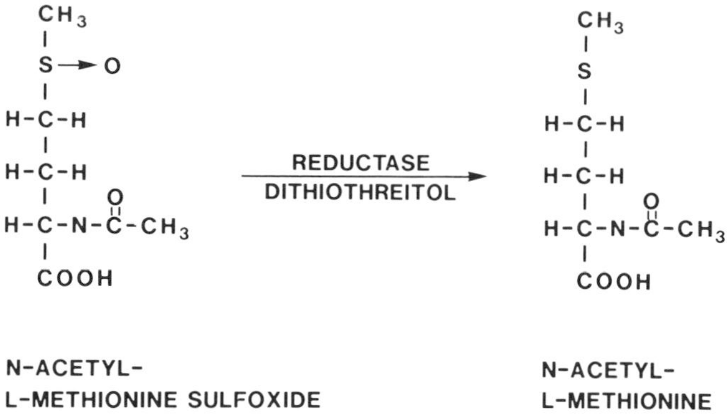

FIG. 2. Enzymatic reduction of *N*-acetylMet(O) to *N*-acetylMet.

Purification of Met(O)-Peptide Reductase from E. coli

Three hundred grams of *E. coli* B grown to 3/4 log phase in enriched medium was purchased from Grain Processing Co. (Muscatine, IA). The cell paste was suspended in about 500 ml of buffer containing 10 m*M* Tris-HCl, pH 7.5, 10 m*M* MgCl$_2$, 10 m*M* NH$_4$Cl, and 1 m*M* 2-mercapto-ethanol (buffer A). The cells were disrupted at 6000 psi in a Manton–Gualin homogenizer, and the suspension was centrifuged at 30,000 g. The pellet was discarded, and the supernatant was centrifuged at 100,000 g (S-100). The S-100 fraction (14.5 g of protein) was brought to 35% saturation with (NH$_4$)$_2$SO$_4$, and the precipitated material was removed by centrifugation and discarded. The supernatant was then brought to 60% (NH$_4$)$_2$SO$_4$ satu-ration, and the precipitate was collected by centrifugation, suspended in a buffer containing 20 m*M* Tris-HCl, pH 7.4, 25 mM KCl, 2 m*M* 2-mercap-toethanol, and 15% glycerol (buffer B) and dialyzed against this buffer. This material (8.4 g of protein) was applied to a DE-52 column (5 × 12 cm), equilibrated with buffer B, and the column was eluted with a linear gradient of 25–500 m*M* KCl in buffer B.

The fractions containing the enzymatic activity eluted at about 180 m*M* KCl. They were pooled and brought to 80% (NH$_4$)$_2$SO$_4$ saturation and the resultant precipitate (460 mg protein) was dissolved in buffer A and chromatographed on an Ultrogel AcA-44 column (2.5 × 100 cm) equilibrated with buffer A. The column was eluted with buffer A, and the fractions containing the reductase activity were pooled and concentrated by Amicon pressure filtration (YM-5). If the specific activity of the en-zyme at this stage was below 150 nmol/mg, the enzyme was rechromato-graphed on the Ultrogel column and subsequently reconcentrated as de-scribed above. The Ultrogel fraction was brought to 60% (NH$_4$)$_2$SO$_4$ by

TABLE I
PURIFICATION OF *Escherichia coli* Met(O)-PEPTIDE REDUCTASE

Step	Volume (ml)	Total protein (mg)	Total activity (nmol)	Recovery (%)	Specific activity (nmol/mg)	Purification (fold)
S-100	500	14,500	72,000	—	5.0	—
35–60% (NH$_4$)$_2$SO$_4$	280	8400	36,960	51	4.4	—
DE-52	20	460	26,840	35	60	12
Ultrogen filtration	4.9	11.4	12,102	17	1062	212
(NH$_4$)$_2$SO$_4$ extraction	2.1	1.3	2444	3.4	1880	376

the addition of a saturated (NH$_4$)$_2$SO$_4$ solution (pH 7.4), and the precipitate was collected by centrifugation. The precipitate was then extracted for 30 min sequentially with two 5-ml portions of 50, 45, 43, and 40% saturated (NH$_4$)$_2$SO$_4$. Each fraction was dialyzed against buffer A and assayed for activity. The enzyme recovered from the 40% saturated (NH$_4$)$_2$SO$_4$ extract had the highest specific activity and appeared homogeneous (see below). Some preparations, however, showed a minor higher molecular weight contaminating protein. Table I summarizes the purification steps. The final enzyme preparation was purified about 375-fold with a recovery of about 3%.

Properties and Characteristics of E. coli Met(O)-Peptide Reductase

Kinetics. The rate of reduction of *N*-acetylMet(O) by Met(O)-peptide reductase increases linearly with protein concentration up to 14 μg per assay (Fig. 3) and for at least 60 min of incubation (Table II). The dependencies for reductase and dithiothreitol (DTT) are seen in Table II.

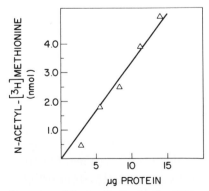

FIG. 3. Effect of protein concentration on the synthesis of *N*-acetylmethionine. Details of the incubation are described in the text. The incubations were carried out for 60 min at 37°.

TABLE II
CONVERSION OF N-ACETYLMet(O) TO
N-ACETYLMet BY Met(O)-PEPTIDE REDUCTASE[a]

System	Time (min)	N-AcetylMet (pmol)
Complete	15	660
	30	1450
	60	3270
−Enzyme	60	0
−Dithiothreitol	60	288

[a] The incubations contained 7.3 μg of a 90-fold purified Met(O)-peptide reductase from *Escherichia coli* and were carried out for the times indicated. The incubation conditions and the assay are described in the text. An extraction blank of 20 pmol has been subtracted from each value.

Although the enzyme can use DTT as a reducing agent, it appears that the biological reducing system consists of NADPH, thioredoxin, and thioredoxin reductase.[10]

Molecular Weight and Purity. The purified enzyme when electrophoresed under denaturing conditions migrates with a molecular weight of about 23,000–24,000 (Fig. 4A). One major band is observed when the electrophoresis is carried out under nondenaturing conditions (Fig. 4B), and other experiments showed that the protein extracted from the single band observed in the nondenaturing gel contained the Met(O)-peptide reductase activity.

Biological Substrates. Met(O)-peptide reductase has a broad substrate specificity and reduces Met(O) residues in both proteins and peptides. Oxidation of methionine residues can cause the inactivation of proteins or peptides that require methionine for biological activity. As examples, when the methionine residue(s) of *E. coli* ribosomal protein L12, α-1-proteinase inhibitor, and the chemotactic peptide N-formylmethionine-leucine-phenylalanine are oxidized, biological activity is lost. Enzymatic reduction of the Met(O) residues has been shown to restore activity to these polypeptides.[10–12] Met(O)-peptide reductase appears to recognize only N-blocked Met(O) residues and cannot reduce free Met(O) (unpub-

[11] W. R. Abrams, G. Weinbaum, L. Weissbach, H. Weissbach, and N. Brot, *Proc. Natl. Acad. Sci. U.S.A.* **78,** 7483 (1981).

[12] H. Fliss, G. Vasanthakumar, E. Schiffmann, H. Weissbach, and N. Brot, *Biochem. Biophys. Res. Commun.* **109,** 194 (1982).

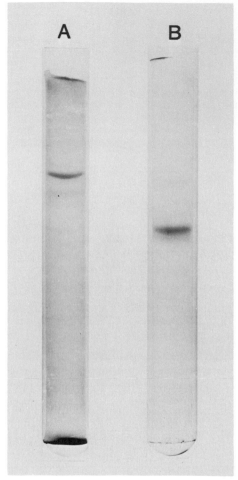

FIG. 4. Gel electrophoresis of purified *Escherichia coli* Met(O)-peptide reductase. (A) 15% polyacrylamide–SDS; (B) 13% polyacrylamide (pH 8.8), nondenaturing.

lished observation). A different enzyme, present in many tissues, has been shown to catalyze the reduction of free Met(O) to methionine.[8]

Partial Purification of Methionine Sulfoxide Reductase from Human Leukocytes

Concentrated white blood cell preparations (100 ml, 2.9×10^5 cells/ml, 60% neutrophils) were obtained by plasmaphoresis, from chronic myelocytic leukemia patients.[13] Any contaminating erythrocytes were allowed to settle at 4° for 2 hr and were discarded. The supernatant was centrifuged at 150 g for 20 min to pellet the white blood cells. The cells were washed twice with phosphate-buffered saline, suspended in a final volume of 70 ml of buffer C (10 mM Tris-HCl, pH 7.4, 10 mM MgCl$_2$, 10 mM NH$_4$Cl), and lysed by freezing and thawing. The lysate was centrifuged at 30,000 g for 30 min, and the supernatant (S-30) was retained.

Ammonium Sulfate Fractionation. The S-30 (2.1 g of protein) was made 35% saturated with ammonium sulfate and centrifuged at 1500 g for 20 min. The pellet was discarded, and the supernatant was made 80% saturated with ammonium sulfate. The precipitate was pelleted at 1500 g for 20 min, suspended in 20 ml of buffer D (10 mM Tris-HCl, pH 8.5, containing 5% glycerol) and dialyzed overnight against the same buffer.

DEAE-Cellulose Chromatography. The dialyzed (NH$_4$)$_2$SO$_4$ fraction was clarified (10,000 g, 20 min), and 1.76 g of protein was applied to a DEAE-cellulose column (Whatman, DE-52, 2.5 × 20 cm) equilibrated with buffer D. The column was washed with 300 ml of buffer D, and the protein was eluted with a 2-liter linear gradient from 20 to 200 mM KCl in buffer D. Some of the enzyme activity was not adsorbed to the column, and peaks of activity eluted in the wash and at 15, 30, and 50 mM KCl. The enzymatic activity peak eluting at 15 mM KCl showed the highest specific activity and was purified further. The fractions containing this peak were pooled and concentrated by ultrafiltration (Amicon YM-5 filter).

Ultrogel AcA-44 Chromatography. The concentrated DEAE-cellulose fraction (21.6 mg of protein) was applied to an Ultrogel AcA-44 column (5.5 × 70 cm) equilibrated with buffer C, and eluted with the same buffer at a flow rate of 200 ml/hr. A peak of activity was recovered eluting at an apparent molecular weight of about 18,000 and was concentrated by ultrafiltration (Amicon YM-5). The protein was purified approximately 20-fold with a recovery of about 5%. Table III summarizes the purification steps.

Distribution and Physiological Function

Met(O)-peptide reductase appears to be a ubiquitous enzyme. Reductase activity has been found in *Euglena gracilis, Tetrahymena pyriformis,* spinach, and various rat tissues as well as bovine and human lens.[8] The reduction of Met(O) residues in proteins may have an important physiological function. It is known that a wide variety of oxidizing agents produced in tissues can oxidize protein-bound methionine. Although these

[13] The preparations were supplied by Dr. Jean Hester of the M. D. Anderson Hospital (Houston, Texas).

TABLE III
PURIFICATION OF METHIONINE SULFOXIDE REDUCTASE FROM LEUKOCYTES

Step	Volume (ml)	Total protein (mg)	Total activity (nmol)	Recovery (%)	Specific activity (nmol/mg)	Purification (fold)
S-30	85	2083	53,312	—	25.6	—
35–80% (NH₄)₂SO₄	21	1764	31,399	59	17.8	—
DE-52 (15 mM KCl fraction)	4.6	21.6	3287	6.2	152	4.8
Ultrogel filtration	3.2	5.1	2406	5	470	18

oxidizing agents would be expected to be neutralized by various cellular enzymes (e.g., catalase, peroxidases, and superoxide dismutase), it is possible that the efficiency of these systems is not sufficient to preclude some oxidation of methionine residues in proteins. For proteins with slow turnover, this oxidation could be deleterious if the methionine residues were involved in the function of the protein. Lens proteins are an example of proteins with a slow turnover, and some lens proteins are stable for the life of the tissue. Although very little oxidation of methionine residues is observed in normal lens proteins, about two-thirds of the methionine content of cataractous lens proteins is in the form of Met(O).[14,15] It is not clear whether this is a result or a cause of cataracts.

α-1-Proteinase inhibitor (α-1-PI) is a protein that has been implicated in the etiology of diseases such as emphysema, adult respiratory distress syndrome, and rheumatoid arthritis. This protein, which functions by inhibiting elastase activity, contains methionine at its active site. The *in vitro* oxidation of a methionine residue(s) destroys the ability of α-1-PI to inactivate elastase. A similar oxidation is known to take place *in vivo*. It has been suggested that the diseases noted above may be caused in part by an imbalance between elastase and active α-1-PI. Previous studies[11] have shown that α-1-PI containing Met(O) residues (and therefore inactive) can be reactivated upon incubation with Met(O)-peptide reductase. In general, it is possible that Met(O)-peptide reductase functions as a repair enzyme to prevent the accumulation of Met(O) residues in proteins.

[14] R. J. W. Truscott and R. C. Augusteyn, *Biochim. Biophys. Acta* **492**, 43 (1977).
[15] M. H. Garner and A. Spector, *Proc. Natl. Acad. Sci. U.S.A.* **77**, 1274 (1980).

[22] Prolyl Hydroxylase in Plants

By Maarten J. Chrispeels

Hydroxyproline-containing proteins were discovered in plants in the laboratory of F. C. Steward in the 1950s (reviewed in Lamport[1]). Since then, work from many laboratories has shown that hydroxyproline occurs in at least three classes of proteins (all glycoproteins): (1) the proteins present in the cell walls of higher plants and in the extracellular matrices of algae; (2) the arabinogalactan proteins found as soluble glycoproteins in many plant tissues, on the stigma of flowers, and in the medium of cell suspension cultures; and (3) certain lectins of the Solanaceae that occur as soluble glycoproteins in the intercellular spaces. These proteins typically contain 15–25% of their amino acid residues as *trans*-4-hydroxy-L-proline. Hydroxylation occurs posttranslationally on proline residues in peptide linkage as is the case with animal protocollagen. The enzyme prolyl hydroxylase and the mechanism of hydroxylation of peptide prolyl residues have not been studied as extensively in plant systems as in animals. Detailed studies have been done with only three plant systems: aerated carrot root phloem parenchyma slices[2] and suspension-cultured cells of *Vinca rosea*,[3] and of *Lolium multiflorum*.[4] The cells in these three systems produce copious amounts of hydroxyproline-containing cell wall proteins and/or arabinogalactan proteins and should therefore be a good source of prolyl hydroxylase.

In Vivo Assay of Peptidylproline Hydroxylation

The method developed by Varner and Burton[5] to measure the activity of prolyl hydroxylase *in vivo* depends upon the removal of tritium of specifically tritiated proline, i.e., proline that has 74% of its ^{3}H in position 4-trans (as determined by NMR spectroscopy by the supplier, New England Nuclear Corp., Boston, MA). Tissue incubated in a solution of [4-^{3}H]proline takes up the labeled proline, incorporates it into protein, then hydroxylates some of the prolyl residues. Hydroxylation results in the removal of the 4-*trans*-^{3}H, and 74% of the radioactivity in the prolyl

[1] D. T. A. Lamport, *Adv. Botan. Res.* **2**, 151 (1965).
[2] D. Sadava and M. J. Chrispeels, *Biochim. Biophys. Acta* **227**, 278 (1971).
[3] M. Tanaka, H. Shibaka, and T. Uchida, *Biochim. Biophys. Acta* **616**, 188 (1980).
[4] P. B. Cohen, A. Schibeci, and G. B. Fincher, *Plant Physiol.* **72**, 754–758 (1983).
[5] J. E. Varner and J. E. Burton, *Plant Physiol.* **66**, 1044 (1980).

residues that are hydroxylated is released as $^3H^+$, which equilibrates with water. An aliquot of the medium is distilled to separate ^3H_2O from the unincorporated [^3H]proline, and radioactivity in the distillate is determined. A potential source of error is the oxidation of [4-^3H]proline in the Krebs cycle via glutamate and α-ketoglutarate. Such an oxidation would also result in ^3H_2O in the medium. To correct for this pathway, comparable tissue is incubated with [3-^3H]proline. The differential tritium loss is a measure of the hydroxylation of peptidylproline residues.

The absolute concentration of proline at the site of peptidylproline synthesis is not known. As a result, it is not possible to determine the absolute rate of hydroxylation. However, the method can be readily used to determine what proportion of the total proline incorporated into protein is hydroxylated. In addition, one can determine at what stage of development, or under what environmental conditions, peptidylproline hydroxylation becomes quantitatively important.

Extraction of Prolyl Hydroxylase for *in Vitro* Assay of Enzyme Activity

The suspension-cultured cells (150–250 g fresh weight) are washed on a sintered glass funnel, or by sedimentation and resuspension with a large volume (1.5 liters) of homogenization medium containing 50 mM Tris-HCl (pH 7.4), 0.2 M sucrose, 20 mM KCl, and 0.5 mM 2-mercaptoethanol. The cells (or tissue, if aerated carrot slices are used) are then homogenized in this medium (using 1–2 times as much medium as cells) in a Waring blender. The homogenate is centrifuged at 1500 g for 15 min, and the resulting supernatant is centrifuged for 1 hr at 100,000 g. Using carrot slices, Sadava and Chrispeels[2] observed that most of the prolyl hydroxylase was in the 40,000 g supernatant of a homogenate (the enzyme was also in the soluble fraction if the homogenate was centrifuged at 100,000 g for 1 hr). Treatment with detergent was not attempted, but apparently was not necessary to assay the activity of the enzyme when using natural protein substrates (see below). This finding has been confirmed by S. Erickson and J. E. Varner (private communication). However, in extracts of suspension-cultured cells all or most of the activity is associated with the 100,000 g sediment when synthetic substrates are used to assay the enzyme. To assay enzyme activity, the 100,000 g membrane pellet is resuspended in 10–20 ml of 50 mM Tris-HCl (pH 7.4) containing 0.1% (v/v) Triton X-100, 0.3 M NaCl, and 0.5 mM 2-mercaptoethanol, and sonicated for 1.5 min. Treatment with detergent is necessary to unmask enzyme activity. The sonicated suspension is recentrifuged to separate membranes from detergent-solubilized material. Working with *Vinca rosea* cells, Tanaka *et al.*[3] found that the prolyl hydroxylase was solubilized in this step. Using exactly the same procedure with *Lolium longiflorum* cells, Cohen et al.[4] found that all the activity remained with the membranes, and that none was solubilized.

These results indicate that the three systems so far examined show quite striking differences in the solubility of the enzyme after homogenization and/or detergent treatment. More research with different tissues is needed to clarify this problem. In evaluating these results it must be kept in mind that the different tissues used synthesize different hydroxyproline-rich molecules and may therefore have different or several prolyl hydroxylases. In addition, the various researchers measured the activity of the enzyme with different substrates. The possibility exists that the soluble enzyme hydroxylates the protein substrates more readily, while the membrane-bound enzyme hydroxylates the synthetic substrates more readily (S. Erickson and J. E. Varner, unpublished results).

Partial Purification of Prolyl Hydroxylase

A partial purification of the *Vinca rosea* enzyme can be achieved according to the procedure of Tanaka et al.[5] in which the sonicated membrane fragments are centrifuged (100,000 g for 60 min) and solid $(NH_4)_2SO_4$ is slowly added to the supernatant to a final concentration of 65% saturation. The mixture is stirred for 30 min and then centrifuged at 22,000 g for 30 min. The salted-out proteins float on top of the clear solution, which is carefully sucked off with a Pasteur pipette and discarded. The proteins are dissolved in a 50 mM Tris-HCl buffer (pH 7.8) containing 50 mM NaCl and 0.5 mM 2-mercaptoethanol, and the solution is dialyzed against the same medium containing 10% glycerol. The sample is applied to a column (1.2 × 10 cm) of DEAE-Sephadex A-50 previously equilibrated with dialysis buffer. The column is washed with dialysis buffer and eluted with solutions of increasing molarity of NaCl (0.15 M, 0.25 M, 0.35 M) in the same buffer. The enzyme elutes in the first step. The pooled fractions were dialyzed against 10 mM Tris-HCl (pH 7.4) containing 25% glycerol, 20 mM KCl, and 0.2 mM 2-mercaptoethanol. Complete purification of the plant enzyme by means of an affinity column consisting of agarose-linked poly(L-proline) as described by Tuderman et al.[6] for vertebrate prolyl hydroxylase has been achieved, but not yet described in detail.[7]

[6] L. Tuderman, E.-R. Kuutti, and K. I. Kivirikko, *Eur. J. Biochem.* **52**, 9 (1975).
[7] M. Tanaka, Annual Report of the Mitsubishi-Kasei Institute of Life Science, Tokyo, Japan **9**, 121 (1980).

Substrates

In their original work Sadava and Chrispeels[2] used unhydroxylated protein substrates isolated from carrot slices or chick embryo tibias. These tissues were incubated in the presence of [3,4-^{3}H]proline and 2,2'-dipyridyl for a brief period (5–15 min) and then extracted according to procedures that yielded protocollagen from chick embryo tibias, or a tricholoracetic acid-soluble cell wall protein precursor from carrot slices. The main disadvantages of the natural carrot substrate are its heterogeneity and instability. Recent work[3,4] has made use of synthetic substrates including poly(L-proline) polypeptides of different sizes (up to M_r 12,000) as well as polymers of (Pro-Pro-Gly). The latter are used extensively in investigations of animal prolyl hydroxylases because they are synthetic analogs of protocollagen. These substrates allow a more precise definition of the substrate specificity of the enzyme (see the table).

SUBSTRATE SPECIFICITY OF PROLYL HYDROXYLASE FROM
Vinca rosea CELLS[a,b]

Peptides	$^{14}CO_2$ released (cpm)	Relative activity (%)
Pro	0	0
Pro$_2$	0	0
Pro$_3$	56	0.9
Pro$_4$	28	0.5
Pro$_8$[c]	4979	81.8
tert-Butyloxycarbonyl-Pro$_8$	5562	91.4
Gly-Pro$_8$	5840	96.0
Poly(L-proline) (M_r 2000)	6085	100
Poly(L-proline) (M_r 6000)	5901	87.0
Poly(L-proline) (M_r 12,000)	5944	97.8
(Pro-Pro-Gly)$_5$	1150	18.9
(Pro-Pro-Gly)$_{10}$	980	16.1

[a] From Tanaka *et al.*,[3] reprinted with permission.
[b] The prolyl hydroxylase was assayed by determination of $^{14}CO_2$ released from α-[1-^{14}C]ketoglutarate for 30 min using 2.3 units of enzyme under standard conditions, except for the final concentration of α-ketoglutarate of 0.1 mM. All peptides were added to bring about a final concentration of 6.67 mM as proline residues. All values are corrected by no substrate control.
[c] Octa-L-proline was prepared from *tert*-butyloxycarbonylocta-L-proline, the *tert*-butyloxycarbonyl group of which was removed by treatment with trifluoroacetic acid at room temperature for 30 min.

Assay

The methods used to measure the activity of plant prolyl hydroxylase have all been adapted from assays used to measure the activity of the enzyme in animal cell extracts. Sadava and Chrispeels[2] adapted the method of Hutton et al.[8] The formation of hydroxyproline involves the displacement of 4-*trans* hydrogen by the incoming hydroxyl group.[9] Thus by using substrates labeled with radioactive [3,4-³H]proline with the radioactivity in the trans positions, one can measure the tritium released into water as equivalent to the [3-³H]hydroxyproline formed. That this was indeed the case was demonstrated by showing that radioactivity in tritiated water was approximately equal to radioactivity in peptidylhydroxyproline. The availability of synthetic substrates allows one to measure enzyme activity by more convenient methods. Newly formed hydroxyproline can be measured colorimetrically after hydrolysis of the partially hydroxylated substrate. Hydroxylation of peptidylproline is accompanied by the stoichiometric decarboxylation of α-ketoglutarate, making it possible to determine hydroxylase activity by measuring the amount of $^{14}CO_2$ released from α-[1-^{14}C]ketoglutarate.

Colorimetric Measurement of Hydroxyproline after Acid Hydrolysis of the Hydroxylated Substrate[4]

Final concentrations in the standard assay volume of 610 μl are 35 mM MOPS-NaOH (pH 7.5), 1 mM $FeSO_4 \cdot 7H_2O$, 0.5 mM α-ketoglutarate, and 2 mM ascorbate, with 150 μg of bovine serum albumin and 1 mg (0.2 mM) of poly(L-proline) (or other substrate). The reaction is initiated by the addition of crude enzyme (about 1.4 mg of protein) and allowed to continue with shaking for 1 hr at 20°. The reaction is stopped by the addition of an equal volume of 10% trichloroacetic acid. The supernatant is collected by centrifugation, and the pellet is washed with 5% trichloroacetic acid. Combined supernatant and washings are desalted on a column of Sephadex G-25 equilibrated with 0.1 M acetic acid. The hydroxylated substrate is evaporated to dryness and hydrolyzed with constant boiling HCl at 110° for 20 hr under nitrogen. Acid is removed by evaporation, and the hydroxyproline content is measured colorimetrically.[10]

Decarboxylation of α-Ketoglutarate

The hydroxylation of peptidylproline requires ascorbate, Fe^{2+}, and α-ketoglutarate (see below). The latter is stoichiometrically decarboxylated,

[8] J. J. Hutton, A. L. Tappel, and S. Udenfriend, *Anal. Biochem.* **16**, 384 (1966).

[9] D. T. A. Lamport, *Nature (London)* **202**, 293 (1964).

[10] G. Bondjers and S. Bjorkerud, *Anal. Biochem.* **52**, 496 (1973).

and this decarboxylation can be used to assay enzyme activity if no other decarboxylases are present. Tanaka et al.[3] adapted the method of Rhoads and Udenfriend[11] in which the released $^{14}CO_2$ is trapped and measured. Final concentrations in a standard assay volume of 300 μl are 1–10 μg of purified enzyme, 200 μg of poly(L-proline) (M_r 6000), 150 μg of bovine serum albumin, 40 mM HEPES buffer (pH 6.8), 150 mM $FeSO_4$, 2 mM ascorbate, 0.1 mM dithiothreitol, and 0.1 mM α-[1-^{14}C]ketoglutarate (7.26 $\times$ 10^6 dpm/μmol). The reaction is carried out in a 10-ml conical centrifuge tube equipped with a silicon rubber tube connected inversely to a small glass scintillation vial. The vial contains a glass filter disk with 200 μl of NCS solubilizer (Amersham) to trap the CO_2. The reaction is started by the addition of 90 μl of a mixture containing α-[1-^{14}C]ketoglutarate, $FeSO_4$, ascorbate, dithiothreitol, and buffer, to 210 μl of a mixture of poly(L-proline), bovine serum albumin, and enzyme. After incubation for 30 min at 30° with constant shaking, the reaction is stopped by injecting 800 μl of 4% trichloroacetic acid through the silicon rubber tubing. After standing for 60 min, all the released $^{14}CO_2$ has been trapped on the paper disk, and radioactivity is measured with a liquid scintillation counter.

Extent of Hydroxylation in Vitro

The extent to which the substrate is hydroxylated appears to be quite low. In their standard assay mixtures Tanaka et al.[3] use 2000 nmol of peptide substrate (as prolyl residues) and synthesize 3 nmol of peptidyl hydroxyproline in a 1-hr period. The high K_m for poly(L-proline) (0.23 mM) necessitates the use of such a large input of substrate. Using in vivo synthesized natural substrates, Sadava and Chrispeels[2] calculated that 1–2% of the prolyl residues were hydroxylated.

Cofactor Requirements of the Enzyme

Sadava and Chrispeels[2] established that the cofactor requirements of plant prolyl hydroxylase are the same as those of the enzyme isolated from animal cells. The enzyme has an absolute requirement for molecular O_2 and for ferrous ions and is almost totally inactive in the absence of ascorbate and α-ketoglutarate. These authors found that in some enzyme preparations other keto acids (pyruvate or oxaloacetate) could replace α-ketoglutarate, but this finding was not confirmed subsequently.[3] It seems likely therefore that plant prolyl hydroxylase has an absolute requirement for α-ketoglutarate, as does the enzyme from animal cells. These cofactor

---

[11] R. E. Rhoads and S. Udenfriend, *Proc. Natl. Acad. Sci. U.S.A.* **60**, 1473 (1968).

requirements are consistent with the conclusions that the enzyme is a nonheme iron mixed-function oxidase.

Properties of the Enzyme

The pH optimum of the *Vinca rosea* enzyme is pH 6.8 in PIPES buffer.[3] Potassium phosphate and sodium acetate buffer almost completely inhibit enzyme activity. The *Lolium multiflorum* enzyme has a pH optimum of pH 7.5 measured in MOPS buffer.[4] The *Vinca rosea* enzyme has a K_m for poly(L-proline) of 0.23 mM (as proline residues), of 0.09 mM for α-ketoglutarate, and of 0.06 mM for O_2 (calculated on the basis of the oxygen concentration at the half-maximum rate and the solubility of oxygen in water at 30°). Purified enzyme can be fractionated by native polyacrylamide gel electrophoresis into two abundant and two less abundant isozymes.[6] These isozymes have an approximate M_r 60,000. Mixtures of isozymes subjected to sodium dodecyl sulfate–polyacrylamide gel electrophoresis have polypeptides with M_r 32,000, 32,900, and 34,000, whereas purified isozymes have two nonidentical polypeptides.[6] Plant prolyl hydroxylase therefore is probably a dimer with two different subunits. Vertebrate prolyl hydroxylase, on the other hand, is a tetramer with a total M_r 240,000 and is composed of two different types of subunits $(\alpha_2\beta_2)$.

Subcellular Distribution of Enzyme Activity

The original work of Sadava and Chrispeels[2] showed that the activity of prolyl hydroxylase was in the 40,000 g supernatant of a carrot slice homogenate, and that little was associated with the organelles. Guzman and Cutroneo[12] showed subsequently that the activity of prolyl hydroxylase in extracts of liver, kidney, and lung was enhanced 3-fold if Triton X-100 was added to the extraction medium. Using suspension cultures of *Vinca rosea,* Tanaka *et al.*[3] found that the soluble fraction of the homogenate contained a significant amount of enzyme activity, but that nearly 8 times as much was present in the organelle fraction. This activity was latent, and treatment with Triton X-100 followed by sonication was necessary to activate and solubilize the enzyme. Fractionations of the subcellular organelles of *Lolium multiflorum* cells were carried out by Cohen *et al.*[4] The results showed that ER-rich fractions contained about 30% of the prolyl hydroxylase activity, and the Golgi-rich fraction contained the remaining 70%. The organelles were identified by the marker enzymes NADPH-cytochrome *c* reductase (EC 1.6.2.4) (ER) and inosine diphos-

[12] N. A. Guzman and K. R. Cutroneo, *Biochem. Biophys. Res. Commun.* **52,** 1263 (1973).

phatase (EC 3.6.1.6) (Golgi). By analogy with other transported proteins, hydroxyproline-rich proteins are probably synthesized by polysomes attached to the ER and pass into the lumen of the ER after synthesis. Nascent polypeptides are not yet hydroxylated,[13] but it is not known how soon after synthesis hydroxylation occurs. In aerated carrot disks newly synthesized hydroxyproline-rich proteins are associated with the endoplasmic reticulum, the Golgi apparatus, and the plasma membrane. Kinetic experiments are consistent with the interpretation that the macromolecules move from one compartment to the next in the order mentioned.[14] A large proportion of the newly synthesized hydroxyproline-rich protein is present in the Golgi apparatus prior to its secretion by the cells.[15] This finding may indicate that hydroxylation actually occurs in the Golgi as indicated by the results of Cohen et al.,[4] that the Golgi forms a much larger compartment than the ER, or that transport through the Golgi is slow because this is the site of glycosylation of the peptidylhydroxyproline with arabinosyl residues.

Substrate Specificity

Tanaka et al. studied the substrate specificity of Vinca rosea prolyl hydroxylase with synthetic peptides and polypeptides, as shown in the table. Three poly(L-prolines) prepared from N-carboxy-L-proline anhydride (average degree of polymerization of 20, 60, and 120) were suitable substrates. The enzyme did not react with free proline or prolyl peptides whose residue number was four or less, but three derivatives of octa-L-prolyl peptides were also suitable substrates. The peptides (Pro-Pro-Gly)$_5$ and (Pro-Pro-Gly)$_{10}$, which are models of protocollagen, were hydroxylated less effectively. The hydroxylation of (Pro-Pro-Gly) polymers confirmed the result of Sadava and Chrispeels[2] that the plant enzyme can hydroxylate protocollagen. Subsequent experiments by Tanaka et al.[16] indicate that the enzyme recognizes not the primary structure, but rather the secondary structure, of the substrate. They used as substrates oligo(L-proline)s and their tert-butyloxycarbonyl derivatives with a degree of polymerization of 2–8. All peptides with a residue number of 5 or greater serve as substrates if enzyme, substrate, and cofactors are preincubated at 0° and the reaction is carried out at 30°. If the preincubation at 0° is omitted, Pro$_5$ is unable to serve as substrate although t-Boc Pro$_5$ is hydroxylated quite well. The substrate Pro$_5$ has the poly(L-proline) II helix structure at 10°, but not at 30°. Furthermore, the optimum temperature for

[13] D. Sadava and M. J. Chrispeels, Biochemistry 10, 4290 (1971).
[14] D. G. Robinson and R. Glas, Plant Cell Rep. 1, 197 (1982).
[15] M. Gardiner and M. J. Chrispeels, Plant Physiol. 55, 536 (1975).
[16] M. Tanaka, K. Sato, and T. Uchida, J. Biol. Chem. 256, 11397 (1981).

Pro$_5$ hydroxylation is 15° whereas that for other substrates such as t-Boc Pro$_8$ is 30°. These and other experiments led to the conclusion that prolyl hydroxylase recognizes the poly(L-proline) II helix structure of its substrate. Because the primary structure of cell wall protein is characterized by the presence of pentapeptides of Hyp$_4$Ser,[17] Tanaka and colleagues[18] examined the effect of substitutions in Pro$_5$ peptides, and synthesized t-Boc (Pro$_m$-Gly)$_n$OH, where $m = 3$ or 4 and $n = 1$–3. All these peptides served as substrates except Boc-Pro$_3$-Gly-OH. All the peptides that served as substrate had the poly(L-proline) II helix structure. The peptides with the greater number of repeating units ($n = 3$) had a lower apparent K_m value and a higher V_{max}.

Structure of Hydroxyproline-Rich Proteins

The hydroxyproline-rich glycoproteins typically contain 15–25% of their amino acid residues as hydroxyproline. In potato lectin[19] and *Chlamydomonas* cell wall protein,[20] these residues are clustered in domains indicating that the percentage of hydroxyproline residues in these domains is even higher. The hydroxyproline-rich domain of the *Chlamydomonas* cell wall protein have the poly(L-proline) II helix structure.[21] Stuart and Varner[22] isolated from carrot slices a cell wall glycoprotein that had 50% of its residues as hydroxyproline. In this case the entire molecule seems to have the poly(L-proline) II helix formation.[23] Observations with other proteins (e.g., collagen) also indicate that pentapeptides of Pro$_5$ are not necessary for a protein to have domains with a poly(L-proline) II helix formation. Together these results may explain why prolyl hydroxylase hydroxylates certain proline residues in preference to others. The enzyme apparently requires a defined secondary structure of its substrate, rather than a defined primary structure.[16]

Acknowledgments

I am grateful to Joseph E. Varner for his careful reading of the manuscript, and to G. B. Fincher, M. Tanaka, and S. Erickson for making unpublished results available to me. I owe a debt of gratitude to David Sadava, who initiated the work on prolyl hydroxylase in plants in my laboratory.

[17] D. T. A. Lamport, *Recent Adv. Phytochem.* **11,** 79 (1977).
[18] M. Tanaka, Annual Report of the Mitsubishi-Kasei Institute of Life Science, Tokyo, Japan **10,** 110 (1981).
[19] A. K. Allen, N. M. Desai, A. Neuberger, and J. M. Creeth, *Biochem. J.* **171,** 665 (1978).
[20] K. Roberts, *Planta* **146,** 275 (1979).
[21] R. B. Homer and K. Roberts, *Planta* **146,** 217 (1979).
[22] D. A. Stuart and J. E. Varner, *Plant Physiol.* **66,** 787 (1980).
[23] G. J. Van Holst and J. E. Varner, *Plant Physiol.* **74,** 247 (1984).

[23] Mixed-Function Oxidation of Histidine Residues

By RODNEY L. LEVINE

Intracellular proteins undergo proteolytic degradation in a specific, regulated fashion.[1] Cells ranging from prokaryotic bacteria to eukaryotic mammalian cells accomplish this metabolic task with notable specificity. For example, a particular enzyme might no longer be required following a change in nutritional status. The cell can rapidly dispose of that enzyme, without affecting numerous other intracellular proteins. At present we have only a rudimentary hint of the mechanisms by which this metabolic feat is accomplished. One possibility is that cells "mark" proteins by a posttranslational modification that renders them liable to attack by intracellular proteases. Metabolic regulation of the marking reaction would provide control of proteolytic turnover of the protein.

The turnover of bacterial glutamine synthetase appears to follow this pathway.[2] The enzyme undergoes a posttranslational modification mediated by a mixed-function oxidation system. This modification leads to a loss of catalytic activity and renders the protein susceptible to proteolytic degradation. Several enzymatic, mixed-function oxidation systems mediate the modification. These include NADH oxidase (diaphorase) as well as the cytochrome *P*-450-dependent systems from bacteria or mammalian liver.[2–5] In addition to bacterial glutamine synthetase, numerous other microbial and animal enzymes are subject to modification by these mixed-function oxidation systems.[4,5] Susceptibility to modification was modulated by substrates or products, for the several proteins studied so far. This modulation provides a potential basis for metabolic control.[5,6] In the case of glutamine synthetase, susceptibility was controlled also by the extent of adenylylation of the enzyme, another posttranslational modifica-

[1] A. L. Goldberg and A. C. St. John, *Annu. Rev. Biochem.* **45,** 747 (1976).
[2] R. L. Levine, C. N. Oliver, R. M. Fulks, and E. R. Stadtman, *Proc. Natl. Acad. Sci. U.S.A.* **78,** 2120 (1981).
[3] C. N. Oliver, R. L. Levine, and E. R. Stadtman, *in* "Metabolic Interconversion of Enzymes 1980" (H. Holzer, ed.), p. 259. Springer-Verlag, Berlin, 1981.
[4] C. Oliver, L. Fucci, R. Levine, M. Wittenberger, and E. R. Stadtman, *in* "Cytochrome *P*-450: Biochemistry, Biophysics, and Environmental Implications" (E. Hietanen, M. Laitinen, and O. Hänninen, eds.), p. 531. Elsevier Biomedical Press, Amsterdam, 1982.
[5] L. Fucci, C. N. Oliver, M. J. Coon, and E. R. Stadtman, *Proc. Natl. Acad. Sci. U.S.A.* **80,** 1521 (1983).
[6] R. L. Levine, *J. Biol. Chem.* **258,** 11823 (1983).

METHODS IN ENZYMOLOGY, VOL. 107   ISBN 0-12-182007-6

tion.[2,6] Since adenylylation is under exquisite metabolic control,[7] proteolytic degradation becomes subject to regulation through control of the initial oxidative modification.

Ascorbic acid mediates some of the reactions catalyzed by the cytochrome P-450-dependent mixed-function oxidation systems. Exposure of glutamine synthetase to ascorbate in the presence of oxygen and metal cations leads to modification of the protein in a reaction that models the enzymatically mediated modification.[2,8] Ascorbate-treated proteins provide an easy source of oxidatively modified proteins. Analysis of the ascorbate-modified glutamine synthetase demonstrated that a single histidine residue was altered upon modification.[6] (Glutamine synthetase has 16 histidine residues per subunit.) Amino acid analysis of glutamine synthetase inactivated by NADH oxidase also revealed loss of a single histidine residue.[9] Thus far, we have analyzed two enzymes modified by treatment with the NADH oxidase system.[9] For 3-phosphoglycerate kinase we found a loss of 1 of 9 histidine residues, whereas yeast enolase had lost 1 of 11 histidine residues. Inactivation of superoxide dismutase by hydrogen peroxide causes loss of a single histidine residue.[10] Catechol appears to modify specifically the histidine at position 306 in tyrosinase.[11] Thus, the oxidation of specific histidine residues emerges as a common and potentially important covalent modification of proteins. The physiological role of the modification remains speculative. In addition to a possible role as a marker for proteolytic degradation, the oxidative modification of histidine may be important in host defense against microbial infection, in prevention of autolysis, in mechanisms of aging, and in oxygen toxicity.[5,6]

Each of the modifying systems studied so far requires reducing equivalents, oxygen, and a metal cation such as iron.[2,5,9] The reaction is prevented by exclusion of oxygen, or by the addition of chelating agents such as EDTA, or by the addition of catalase (implying the participation of hydrogen peroxide). The addition of 1 mM $MnCl_2$ usually provides protection, but this may not be as effective as the other three measures. We proposed a mechanism initiated by the reduction of ferric iron to the ferrous species. In the presence of oxygen (or hydrogen peroxide), the ferrous iron reoxidizes with the concomitant generation of an activated oxygen species.[4,5] If the reduced iron binds to a specific site on the pro-

[7] E. R. Stadtman, P. B. Chock, and S. G. Rhee, in "Enzyme Regulation: Mechanisms of Action" (P. Mildner and B. Ries, eds.) (*FEBS Symp.* 60), p. 57. Pergamon, Oxford, 1980.
[8] R. L. Levine, *J. Biol. Chem.* 258, 11828 (1983).
[9] R. L. Levine, unpublished observations, 1980–1982.
[10] E. K. Hodgson and I. Fridovich, *Biochemistry* 14, 5294 (1975).
[11] K. Lerch, *J. Biol. Chem.* 257, 6414 (1982).

tein, then the oxidizing species can attack nearby residues. This mechanism provides an explanation for the selectivity noted above. One would expect that enzymes possessing divalent cation binding sites and also having essential histidine residues might be susceptible to this post-translational modification. That expectation proved to be true when a number of enzymes were screened for susceptibility to oxidative modification.[5] Moreover, exposure of susceptible enzymes to oxygen and ferrous salts should lead to oxidative modification, and it does.[2,9] Finally, each mixed-function oxidation system proved to be capable of reducing ferric iron to the ferrous state.[4]

Preparation of Oxidatively Modified Proteins

Simple exposure to the various mixed-function oxidation systems provides oxidatively modified protein.[2,5] If an enzyme-mediated system is used, the modified protein can be separated from the added enzymes by the usual purification methods. However, oxidative modification does not require intimate contact between the catalytic protein and the target protein. The oxidative modification can be accomplished across dialysis membranes, at least for the cytochrome P-450 system, the NADH oxidase system, and the ascorbate system.[12] The protein to be modified may be placed inside the dialysis bag or collodion cone.[13] The mixed-function oxidation system is placed in the surrounding solution (for example, NADH, $FeCl_3$, and NADH oxidase[5]). After completion of the modification reaction, the dialysis bag or cone can be transferred to fresh buffer to remove small molecules (for this example, NADH and $FeCl_3$). We find it convenient to use the cones because they permit easy sampling of the protein for the purpose of monitoring the reaction. The course of the reaction can be followed by the loss of activity of the enzyme.

Ascorbate System

The simplest method for preparation of oxidatively modified proteins is exposure to ascorbate. One can prepare small or large quantities with ease. If large amounts are to be made, we assure adequate oxygenation by agitating the reaction flask on an oscillating water bath. Agitation is not needed for the usual test tube-scale reaction, carried out in a 37° water bath. The reaction is buffered around neutrality, but the exact pH and the buffer are not critical. One can generally utilize whichever buffer is con-

[12] L. Fucci, K. Nakamura, E. R. Stadtman, and R. L. Levine, unpublished observations, 1981–1982.
[13] Schleicher & Schuell, Inc. Keene, NH 03431.

venient for the protein being modified. However, Mn^{2+} should be avoided because it inhibits the oxidative modification. For glutamine synthetase we use 50 mM HEPES, pH 7.4, 100 mM KCl, 10 mM $MgCl_2$, 25 μM $FeCl_3$.

The KCl and $MgCl_2$ are routinely included in our laboratory to improve the stability of the glutamine synthetase (in the absence of ascorbate). They are not required for the oxidative modification. The addition of the iron assures an adequate concentration of this essential reactant, although the trace contaminants from water and buffer may suffice.

Reagent. The ascorbate is added from a stock solution of 1.28 M, prepared daily. For 10 ml, dissolve 2.25 g of ascorbic acid in 5 ml of 100 mM HEPES, pH 7.0. Readjust the pH to around neutrality with 1 M KOH. (Take care not to exceed a pH of about 7.5, as ascorbate undergoes autoxidation in alkaline solution.) Adjust the volume to 10 ml. Keep at 4°.

Procedure. Start the reaction by adding the ascorbate. Addition of 20 μl of ascorbate per milliliter of protein solution gives 25 mM ascorbate. In the case of glutamine synthetase, this leads to 50% reaction in about 10–15 min. The reaction rate can be adjusted by altering the ascorbate concentration.[8]

The reaction may be stopped by the addition of EDTA to 1 mM. This allows generation of proteins with graded modification. Addition of catalase also stops the reaction, but there is a variable lag time before the reaction ceases.

Assessment of the Extent of Modification

Loss of Enzymatic Activity

The simplest assay for the modification of susceptible enzymes is the loss of catalytic activity of the enzyme.[14] The loss of activity is directly proportional to the extent of modification of the susceptible histidine residue in glutamine synthetase[6] and in enolase,[9] the only two enzymes in which the proportionality has been examined. Loss of activity presumably provides an assessment of the extent of modification in other susceptible enzymes. This assumption can readily be confirmed by amino acid analysis.

---

[14] There is no a priori reason to expect that the oxidative modification of a histidine residue would always lead to loss of enzymatic activity. To date, that has simply been the assay used by us to screen enzymes for susceptibility to the modification. Other enzymes might demonstrate increased activity upon modification. Finally, some enzymes or proteins without catalytic activity would obviously show no effect on enzymatic activity after modification.

Amino Acid Analysis

The chemical nature of the modified histidine is not yet established. At present, loss of a histidine residue upon acid hydrolysis is the definitive assay for the oxidative modification. Glutamine synthetase has 16 histidine residues, so that loss of one residue represents only a 6% change. However, with attention to experimental detail, one can readily demonstrate the loss of a single residue by standard amino acid analysis.[6]

In addition to the loss of a histidine residue, the oxidatively modified protein gives rise to a new product that can be detected on the amino acid chromatogram. The compound has not yet been identified, but it runs between serine and glutamate under standard conditions.[6] Its presence in a protein hydrolysate provides presumptive evidence for the oxidative modification of a histidine residue. When [14C]histidine-labeled glutamine synthetase was oxidatively modified, this new compound contained 14C label.[15] Other amino acids, especially glycine and aspartate, were also labeled. This is consistent with multiple products expected from the oxidized histidine residue upon acid hydrolysis.[16]

Reaction with 2,4-Dinitrophenylhydrazine

Oxidatively modified glutamine synthetase reacts with carbonyl reagents such as phenylhydrazine and 2,4-dinitrophenylhydrazine.[17] The extent of reaction parallels the loss of catalytic activity, the loss of the histidine residue, and the gain of the unknown amino compound described above.[6] Reaction with 2,4-dinitrophenylhydrazine is the most convenient method for quantitation of the oxidative modification. Since hydrazone formation requires a low pH, we use guanidine to solubilize the protein during reaction. The extent of reaction can be monitored spectrophotometrically. Because of the high background from the reagent, we monitor the absorbance at 387 nm less that at 400 nm as a measure of the reaction ("delta absorptivity").

2,4-DINITROPHENYLHYDRAZINE REAGENT

To 9 volumes of 8 M guanidine, 26.7 mM potassium phosphate, pH 6.35, add 1 volume of 10 mM 2,4-dinitrophenylhydrazine in 2 M HCl. This diluted reagent is prepared fresh daily, but the two stock components can be stored almost indefinitely.

---

[15] J. M. Farber and R. L. Levine, unpublished observations, 1982.
[16] M. Tomita, M. Irie, and T. Ukita, *Biochemistry* **8**, 5149 (1969).
[17] R. L. Levine and S. Shaltiel, unpublished observations, 1982.

SPECTROPHOTOMETRIC METHOD

Procedure. Place 0.5 ml of reagent in the reference and sample cuvettes. If the spectrophotometer has automatic baseline correction, perform it now. Otherwise, note the baseline absorbance at 387 and 400 nm. Add 0.1 ml of buffer to the reference cell and 0.1 ml of protein sample to the sample cell. If ascorbate was used to oxidize the protein, it should be removed by dialysis or other methods. Otherwise, interfering hydrazones or osazones may form. The reaction is usually complete at room temperature in about 30 min. The cells should be capped to prevent evaporation during reaction.

The "delta molar absorptivity" (387–400 nm) for the 2,4-dinitrophenylhydrazone of glutamine synthetase is 690.[6] Thus, the test requires a fairly large amount of protein (about 0.5 mg for glutamine synthetase) for good quantitation. Smaller amounts suffice for semiquantitative detection of the modification. The "delta molar absorptivity" correlates very well with the specific activity of the enzyme,[6] providing the amusing capability to determine the specific activity of an enzyme without actually analyzing its catalytic activity.

HPLC (REVERSE PHASE) METHOD

The sensitivity of the spectrophotometric assay is relatively low because of the high background of the reagent. HPLC provides a rapid method of separating the derivatized protein from the reagent. After separation, one can take advantage of the high molar absorptivity of the 2,4-dinitrophenylhydrazones (15,000–20,000); 20 μg of protein should provide adequate sensitivity. With a dual-wavelength detector, one can monitor the protein absorbance at 210 nm (or 275 nm) and the 2,4-dinitrophenylhydrazone absorbance at 360 nm. Preliminary experience indicates that the ratio of the absorbances provides a quantitative estimate of the extent of oxidative modification of the protein. If a diode array detector is available,[18] one can also obtain a full spectrum of the derivatized protein during the chromatography (Fig. 1).

Procedure. To one volume of protein (e.g., 10 μl) add 3 volumes of 2,4-dinitrophenylhydrazone reagent. Cap the tube and incubate at 37° for 30 min. Inject the sample into an HPLC system equipped with a μBondapak C-18 column.[19,20] Elute with a gradient from 0 to 60% acetonitrile,

[18] Hewlett–Packard Model 8450A or 8451A spectrophotometers or Model 1040A HPLC spectrophotometric detector (Hewlett-Packard Company, Palo Alto, CA 94304).
[19] Waters Associates, Milford, MA 01757.
[20] Other columns may prove to be useful, but will have to be tested for their ability to separate proteins successfully. C-18 columns from other manufacturers may not be directly substitutable: L. Henderson, personal communication, 1981.

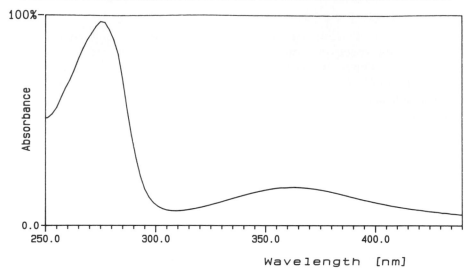

FIG. 1. Spectrum of the 2,4-dinitrophenylhydrazone of glutamine synthetase that had been oxidatively modified in the ascorbate system. The derivative was prepared and isolated by HPLC as described in the text. This spectrum was taken during chromatography by an in-line detector.[18] The absorbance at 100% is 1.67 absorbance units.

with a constant 0.05% trifluoroacetic acid. We typically run the gradient in 30 min, at a flow rate of 1 ml/min. Under these conditions, glutamine synthetase elutes at about 28 min (44% acetonitrile), well separated from the reagent at about 18 min (24% acetonitrile). The spectrum of the derivatized, oxidatively modified glutamine synthetase is shown in Fig. 1.

If an HPLC is not available, rapid separation of the protein and reagent might still be performed using cartridges filled with the same C-18 packing material as the HPLC columns (Sep-Pak[19]). We have not utilized gel filtration, but it could be useful. The columns should be run with guanidine present to assure solubility of the derivatized protein.

Reaction with Borotritide

In principle, the carbonyl group may be reduced and radioactively labeled by reaction with sodium borotritide. Preliminary experiments with oxidatively modified glutamine synthetase indicate that the reaction does occur and can be used to quantitate the oxidative modification.

[24] Dityrosine: *In Vitro* Production and Characterization

By RENATO AMADÒ, ROBERT AESCHBACH, and HANS NEUKOM

Gross and Sizer[1] were the first to describe the formation of dityrosine (3,3'-dityrosine) and trityrosine by oxidation of tyrosine in aqueous solutions with hydrogen peroxide and peroxidase. The mechanism of the reaction can be formulated as phenolic coupling of two phenoxy radicals of tyrosine.

Some years later Andersen[2] identified dityrosine in hydrolysates of resilin, the highly elastic, rubber-like structural protein found in the exoskeleton of insects. Dityrosine was subsequently also identified in hydrolysates of several other native structural proteins, where the dityrosine cross-links are partially responsible for their insoluble and elastic properties. Elastin from the aorta of chick embryos,[3] collagen of rat[4] and cow skin,[5] bovine ligamentum nuchae proteins,[6,7] uterus proteins of cats,[8] and skin proteins of arthropods[9,10] have all been shown to contain small amounts of dityrosine. Apart from these connective tissue proteins, dityrosine has been identified in tussah silk,[10] wool keratin,[11] adhesive proteins of sea mussels,[12] proteins of cataractous lens,[13,14] and bovine dental enamel matrix.[15]

Andersen[16] found that dityrosine cross-links could also be formed *in vitro* by oxidation of structural proteins. Oxidation of silk fibroin with peroxidase and hydrogen peroxide produced a fluorescent gel that contained dityrosine. The mechanism of the reaction is similar to the one

[1] A. J. Gross and I. W. Sizer, *J. Biol. Chem.* **234,** 1611 (1959).

[2] S. O. Andersen, *Biochim. Biophys. Acta* **93,** 213 (1964).

[3] F. La Bella, P. Waykole, and G. Queen, *Biochem. Biophys. Res. Commun.* **26,** 748 (1967).

[4] F. La Bella, P. Waykole, and G. Queen, *Biochem. Biophys. Res. Commun.* **30,** 333 (1968).

[5] P. Waykole and E. Heidemann, *Connect. Tissue Res.* **4,** 219 (1976).

[6] F. W. Keeley, F. La Bella, and G. Queen, *Biochem. Biophys. Res. Commun.* **34,** 156 (1969).

[7] F. W. Keeley and F. La Bella, *Biochim. Biophys. Acta* **263,** 52 (1972).

[8] J. W. Downie, F. La Bella, and M. West, *Biochim. Biophys. Acta* **263,** 604 (1972).

[9] G. Krishnan and M. H. Ravindranath, *Acta Histochem.* **44,** 348 (1972).

[10] K. Ramalingam, *Parasitology* **66,** 1 (1973).

[11] D. J. Raven, C. Earland, and M. Little, *Biochim. Biophys. Acta* **251,** 96 (1971).

[12] D. P. De Vore and R. J. Gruebel, *Biochem. Biophys. Res. Commun.* **80,** 993 (1978).

[13] S. Garcia-Castineiras, J. Dillon, and A. Spector, *Science* **199** (4331), 879 (1978).

[14] M. K. McNamara and R. C. Augusteyn, *Exp. Eye Res.* **30,** 319 (1980).

[15] J. M. Booij and J. J. Ten-Bosch, *Arch. Oral Biol.* **27,** 417 (1982).

[16] S. O. Andersen, *Acta Physiol. Scand.* **66,** Suppl. 263 (1966).

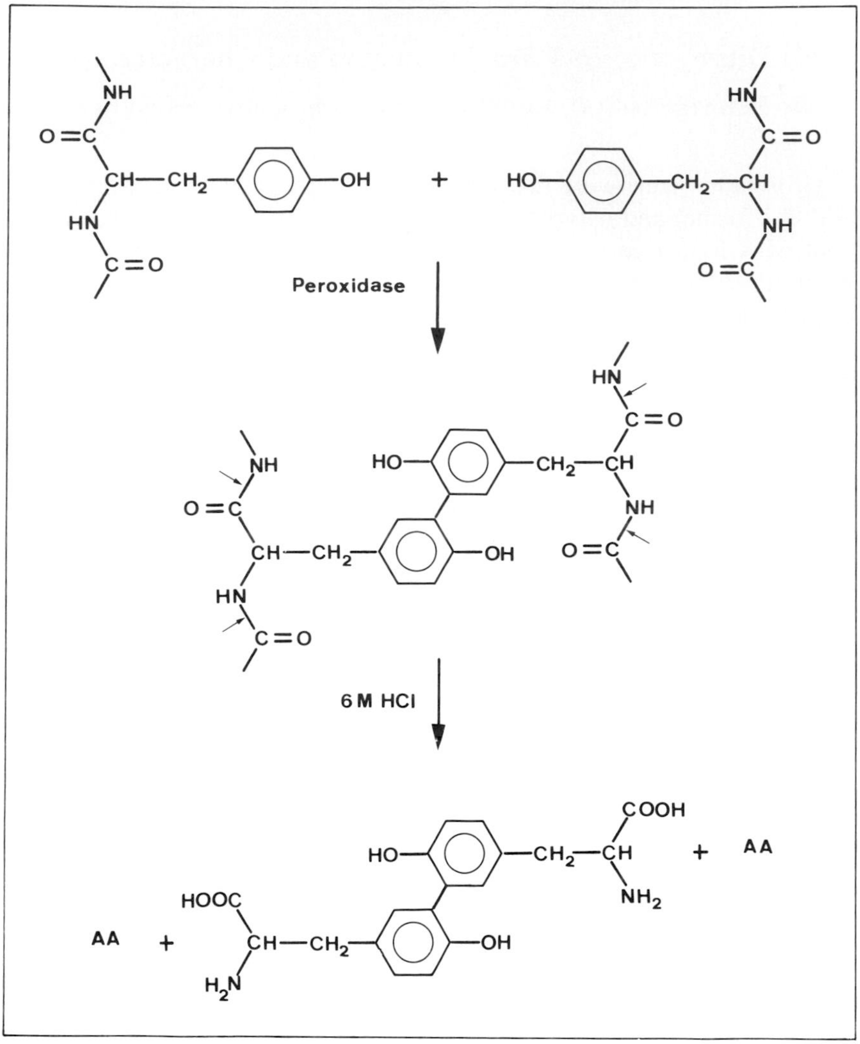

FIG. 1. Oxidative phenolic coupling of tyrosine residues in polypeptide chains with peroxidase/hydrogen peroxide and formation of dityrosine by acid hydrolysis (indicated by arrows). AA, amino acids.

formulated by Gross and Sizer,[1] however, with the oxidation taking place at the level of the protein, as indicated in Fig. 1.

Extending the work done by Andersen, Aeschbach et al.[17] were able to show that dityrosine formation could be induced by oxidation in differ-

[17] R. Aeschbach, R. Amadò, and H. Neukom, *Biochim. Biophys. Acta* **439**, 292 (1976).

ent nonstructural proteins, e.g., insulin, RNase, chymotrypsin, with peroxidase/hydrogen peroxide. Dityrosine formation from tyrosine and proteins was also observed during nitration with tetranitromethane[18,19] and more recently after ozone treatment of membrane proteins.[20-22] Finally radiation-induced dimerization of tyrosine has been shown by Boguta and Dancewicz.[23,24]

The dityrosine (native or induced by oxidation) can be detected quite easily in the intact protein or in the acid hydrolysate by its characteristic ultraviolet fluorescence or by amino acid analysis of the hydrolysate.

Dityrosine is needed occasionally as a reference compound for its identification in proteins. The present chapter describes two methods for its preparation by peroxidase oxidation of tyrosine and *N*-acetyltyrosine, respectively.

Materials

L-Tyrosine and horseradish peroxidase (EC 1.11.1.7.) were purchased from Fluka AG (Buchs, Switzerland); silica gel and cellulose thin- and thick-layer chromatography plates as well as silica gel 60 for column chromatography were obtained from Merck AG (Darmstadt, FRG). Sephadex G-10 was purchased from Pharmacia AB (Uppsala, Sweden); cellulose-phosphate CP-11 was from Whatman (Maidstone, GB). All other chemicals used were purchased from Fluka AG (Buchs, Switzerland), Siegfried AG (Zofingen, Switzerland), and Merck AG (Darmstadt, FRG) and were of highest purity available. All solvents were distilled prior to use.

General Methods

Thin-Layer Chromatography. On silica gel or cellulose plates the following solvent systems were used.
(a) 1-Butanol–acetic acid–water (55 : 15 : 30, v/v/v)
(b) 1-Butanol–acetic acid–water (4 : 1 : 1, v/v/v)

[18] J. Williams and J. M. Lowe, *Biochem. J.* **121,** 203 (1971).
[19] S. Rama and G. Chandrakasan, *Leather Sci. (Madras)* **26,** 324 (1979).
[20] H. Verweij and H. Van Steveninck, *Protides Biol. Fluids* **29,** 129 (1981).
[21] H. Verweij, K. Christianse, and J. Van Steveninck, *Chemosphere* **11,** 721 (1982).
[22] H. Verweij, K. Christianse, and J. Van Steveninck, *Biochim. Biophys. Acta* **701,** 180 (1982).
[23] G. Boguta and A. M. Dancewicz, *Nukleonika* **26,** 11 (1981).
[24] G. Boguta and A. M. Dancewicz, *Int. J. Radiat. Biol. Relat. Stud. Phys. Chem. Med.* **39,** 163 (1981).

(c) 1-Propanol–ammonia, 25% (7 : 3, v/v)

(d) Methyl ethyl ketone–2-methyl-2-propanol–water (2 : 1 : 1, v/v/v)

(e) Ethanol (96%)–water (7 : 3, v/v)

(f) 1-Butanol–ethyl acetate–acetic acid–water (1 : 1 : 1 : 1, v/v/v/v)

Ultraviolet fluorescence as well as Folin–Ciocalteau and ninhydrin reagents were used for identification of the spots.

Column Chromatography. All support materials (silica gel, Sephadex G-10, ion-exchange resins) were treated and packed into columns as recommended by the respective manufacturers. An LKB 7000 fraction collector equipped with a peristaltic pump LKB 4812 and a Uvicord II UV-detection system (LKB-Produkter, Bromma, Sweden) have been used for the chromatographic separations.

Ultraviolet and Infrared Spectroscopy. All ultraviolet spectra were performed on a Perkin-Elmer spectrophotometer, Model 402, equipped with thermostatted cuvettes (Perkin-Elmer GmbH, Ueberlingen, FRG). The infrared spectra of N-acetyltyrosine and N,N'-diacetyldityrosine were run in a Perkin-Elmer spectrophotometer, Model 337 (Perkin-Elmer, Ueberlingen, FRG) using the KBr-pellet technique.

Fluorescence Measurements. Fluorescence experiments were carried out in 0.2 M sodium borate buffer solutions at pH 9.5 with an Aminco-Bowman spectrofluorometer (silica cell of $d = 0.5$ cm).

Hydrolysis of Proteins. Acid hydrolysis of the protein samples was carried out in 6 M hydrochloric acid in sealed glass tubes under nitrogen at 110° for 20 hr. The HCl was removed by evaporation under vacuum at 40°.

Amino Acid Analyses. Amino acid analyses were performed with a Biocal BC-201 amino acid analyzer (Biocal GmbH, Munich, FRG) using a 4-buffer single-column procedure based on the method described by Spackman *et al.*[25] and modified by Werner[26] for separation of hydroxyproline-, hydroxylysine-, and hexosamine-containing protein hydrolysates. For samples where only the dityrosine content was of interest, a short program was developed by Aeschbach.[27] The details of the two programs used are given in Table I.

Preparation of Dityrosine: Method A

Step 1. Preparation of N-Acetyltyrosine. N-acetyltyrosine was prepared according to the method of Du Vigneaud and Meyer.[28] L-Tyrosine (30 g; 165 mmol) was dissolved in 85 ml of 2 N sodium hydroxide and 50

---

[25] H. D. Spackman, W. H. Stein, and S. Moore, *Anal. Chem.* **30,** 1190 (1958).

[26] G. C. Werner, Ph.D. Thesis ETH Zurich, No. 5784 (1976).

[27] R. Aeschbach, Ph.D. Thesis ETH Zurich, No. 5547 (1975).

[28] V. Du Vigneaud and C. E. Meyer, *J. Biol. Chem.* **98,** 295 (1932).

TABLE I
PROGRAMS FOR THE DETERMINATION OF
DITYROSINE BY AMINO ACID ANALYSIS[a]

Program	pH	(Na$^+$)	t (min)	T (°C)
1[b]	3.05	0.2	34	60
	3.40	0.2	56	60
	4.25	0.2	60	60
	6.45	0.73 (1.3)	126 (96)	60
	NaOH	0.4	24	60
	3.05	0.2	60	60
2[c]	4.25	0.2	20	60
	6.45	0.46	90	60
	NaOH	0.4	25	60
	4.25	0.2	35	60

[a] Amino acid analyzer, Biocal BC 201 (Biocal GmbH, Munich, FRG); resin, Aminex A 6 (Bio-Rad, Palo Alto, CA), column dimensions, 0.9 × 56 cm; buffer flow rate, 80 ml/hr; ninhydrin flow rate, 40 ml/hr. Quantitative measurements were carried out by using a calculating integrator Autolab System AA (Spectraphysics, Darmstadt, FRG).
[b] Total time (including regeneration): 360 min (330 min).
[c] Total time (including regeneration): 170 min.

ml of water. After cooling in an ice bath, 400 ml of 2 N sodium hydroxide and 40 ml of redistilled acetic acid anhydride were added in eight equal portions under vigorous shaking in the cold. The temperature in the reaction vessel was kept below 10° during the whole procedure. The reaction mixture was then allowed to stand at room temperature (22°) for 40 min; 168 ml of 6 N sulfuric acid was added, and the mixture was left overnight in the refrigerator. A small amount of nonreacting L-tyrosine precipitated and was filtered off. The filtrate was evaporated to dryness in a rotatory evaporator under vacuum and redissolved in aqueous acetone. Insoluble salts precipitated and were removed by filtration; the filtrate was again evaporated to dryness. This step was repeated several times in order to remove all salts and acetic acid. Finally, the reaction product was recrystallized from an acetone–water mixture, dried first in a desiccator over potassium hydroxide and then in high-vacuum at 0.01 torr. Finally, 28.6 g of N-acetyltyrosine (yield 82%) was obtained with a melting point of 151° (literature: 152°). Infrared data: 3350, 2580, 1720, 1600, 1560, 1260, 1230 cm^{-1}.

Step 2. Oxidation of N-Acetyltyrosine to N,N-Diacetyldityrosine. N-Acetyltyrosine, 1.115 g (5 mmol), was dissolved in 200 ml of 0.2 M sodium borate buffer (pH 9.5); 6 ml (5.3 mmol) of a 3% (v/v) hydrogen peroxide solution and 12 mg of horseradish peroxidase, dissolved in 5 ml of the above-mentioned buffer, was added, and the reaction mixture was oxidized at a constant temperature of 37° (± 0.1°) for 24 hr in an ultrathermostat. At the end of the oxidation period sodium metabisulfite was added to destroy any remaining hydrogen peroxide. The slightly brown solution was concentrated to approximately 10 ml in a rotatory evaporator under vacuum.

Step 3. Isolation and Purification of N,N'-Diacetyldityrosine. Concentrated solution (2.5-ml portions) was applied on a Sephadex G-10 column (2.5 × 85 cm) that had been equilibrated with distilled water and eluted with the same solvent. The gel filtration was performed at a flow rate of 20 ml/hr, and 3-ml fractions were collected. This preliminary purification allowed the separation of *N,N'*-diacetyldityrosine from unoxidized material, from higher oxidation products, and from the enzyme used. The peroxidase is eluted in the void volume; the unoxidized *N*-acetyltyrosine, in the total volume of the column. No clear-cut separation of the dimer could be achieved by this procedure, therefore all fractions showing fluorescence were pooled and concentrated in a rotatory evaporator under vacuum to 5 ml. The concentrate was further purified by preparative thick-layer chromatography on silica gel 60 (20 × 20 cm) plates. By using solvent system b (see General Methods) the broad zone of *N,N'*-diacetyldityrosine was very well separated from the trimer and tetramer of *N*-acetyltyrosine. The fluorescent zone corresponding to *N,N'*-diacetyldityrosine was scraped off and, after grinding the silica gel in a mortar, extracted several times with ethanol. The combined extracts were evaporated to dryness to remove all 1-butanol and acetic acid, redissolved in ethanol, and reprecipitated several times with acetone. Finally, 125 mg (yield 11.3%) of *N,N'*-diacetyldityrosine was obtained as a white powder. This pure dimer melted at 262° and showed infrared absorption bands at 3300, 2925, 1725, 1660, 1600, 1490, and 1450 cm^{-1}.

Step 4. Deacetylation of N,N'-Diacetyldityrosine and Purification of Dityrosine. *N,N'*-diacetyldityrosine (100 mg) was hydrolyzed with 50 ml of 6 M hydrochloric acid for 20 hr under reflux in the presence of nitrogen. After evaporation of the hydrochloric acid in a rotatory evaporator under vacuum, the sample was dissolved in 2 ml of solvent system b (see General Methods) and applied on a silica gel 60 column (2.5 × 25 cm) equilibrated with the same solvent system b. Chromatography was performed at a constant flow rate of 20 ml/hr; fractions of 2.5 ml were collected. The fluorescent fractions were combined, evaporated to dryness, redissolved in water, and reprecipitated several times with acetone–ethanol. Finally, 74.2 mg of dityrosine (yield 91.5%) was obtained and characterized as described below.

Preparation of Dityrosine according to Andersen[16]: Method B

L-Tyrosine (181 mg; 1 mmol) was dissolved in 200 ml of 0.2 M sodium borate buffer, pH 9.5; 1.1 ml (1.05 mmol) of a 3% (v/v) hydrogen peroxide solution, and 3 mg of horseradish peroxidase was added. The solution was oxidized at a constant temperature of 37° for 24 hr in an ultrathermostat, concentrated in a rotatory evaporator under vacuum, and acidified by adding concentrated hydrochloric acid. During the concentration step, nonreacting L-tyrosine precipitated and was removed by filtration through a glass filter funnel (G 4). The brown concentrate was applied to a cellulose-phosphate CP-11 column (2.5 × 70 cm) that was equilibrated with 0.2 M acetic acid. Elution of the column was performed with 0.2 M acetic acid containing 0.5 M sodium chloride, at a flow rate of 15 ml/hr. Fractions of 3 ml were collected. Strongly fluorescent fractions, which corresponded to a clearly separated peak in the chromatogram, were pooled and concentrated under vacuum in a rotatory evaporator. Sodium chloride that precipitated during the concentration process was filtered off; the solution was acidified to pH 2.0 and applied on a Dowex 50-X8 column (2.2 × 25 ml) (H$^+$). The column was extensively washed with water to remove all sodium chloride and acetic acid, and the dityrosine was subsequently eluted with 2 M ammonium hydroxide. The solvent was removed by several evaporations in a rotatory evaporator under vacuum. Finally, the neutral solution was freeze-dried, and 28.2 mg of chromatographically pure, slightly yellow dityrosine was obtained (yield 16.6%). Characterization of the dityrosine obtained by this method was performed in the same way as that obtained by method A (see below).

Oxidation of Proteins. Structural as well as globular proteins and different enzymes were oxidized in 0.2 M sodium borate buffer at pH 9.5 with the peroxidase–hydrogen peroxide system. To varying amounts of protein (20–50 mg of protein per milliliter of buffer), peroxidase (1 mg/20 mg of protein) and 3% (w/v) hydrogen peroxide (0.02 ml/10 mg of protein) were added. The oxidation was carried out at 37° in an ultrathermostat for 24 hr. Excess hydrogen peroxide was destroyed by the addition of sodium metabisulfite, and the oxidized proteins were dried either by freeze-drying or over P_2O_5 in a vacuum desiccator. Table II gives the results obtained and shows that different amounts of dityrosine were formed depending on the nature of the oxidized protein.

TABLE II
DITYROSINE CONTENT OF OXIDIZED STRUCTURAL AND GLOBULAR PROTEINS
AND ENZYMES[a]

Type of protein	Tyrosine content before oxidation (g/100 g protein)	Decrease of the tyrosine peak (%)	Dityrosine content (g/100 g protein)
Silk fibroin	10.86	56.8	6.17
Tussah silk[b]	9.67	trace	—
Collagen	1.0	trace	—
Casein	6.3	21.8	1.37
Soy amine	3.8	11.5	0.44
Bovine serum albumin	4.56	30.7	1.40
Gliadin	3.2	5.4	0.17
Insulin	12.1	27.2	3.29
Insulin (H_2O_2)[c]	12.1	5.1	0.62
Chymotrypsin	2.66	85.8	2.28
Chymotrypsin (H_2O_2)[c]	2.66	30.0	0.80
Trypsin	6.11	59.5	3.63
Pepsin	8.46	53.4	4.52
Lysozyme	3.58	trace	—
Urease	4.45	trace	—
Papain	11.48	28.9	3.32
Ribonuclease	6.7	17.25	1.18

[a] Calculated based on the decrease of tyrosine during oxidation; for experimental details see the text.
[b] Oxidized at pH 4.
[c] Oxidized with H_2O_2 alone.

Characterization of Dityrosine

Thin-Layer Chromatography. Thin-layer chromatography in different solvent systems was used to characterize the dityrosine obtained by the two methods described above. The R_f values in the different chromatographic systems are summarized in Table III.

UV Spectrum. The UV absorbance of dityrosine was measured in an acid (0.1 M HCl) and in an alkaline (0.1 M NaOH) solution. In 0.1 M hydrochloric acid the maximum absorbance was at 284 nm, whereas in the 0.1 M sodium hydroxide solution the maximum absorbance appeared at 317 nm. A small absorption band at 252–255 nm could be observed in neutral and alkaline solutions, indicating the presence of an *o,o'*-disubstituted biphenyl derivative.[29]

Fluorescence Measurements. The fluorescence spectrum of dityrosine in sodium borate buffer, pH 9.5, showed, for an exitation wavelength of

[29] G. Aulin-Erdtman and R. Sandén, *Acta Chem. Scand.* **17,** 1991 (1963).

TABLE III
R_f VALUES OF DITYROSINE IN DIFFERENT
SOLVENT SYSTEMS[a]

Solvent systems	Plate	R_f value
a	Silica gel	0.15
b	Silica gel	0.12
c	Silica gel	0.175
d	Silica gel	0.23
d	Cellullose	0.06
e	Silica gel	0.50
f	Cellulose	0.37

[a] See General Methods.

222 nm, an emission maximum at 400–405 nm; and for an exitation at 315 nm, an emission maximum at 410–415 nm. These results are in excellent agreement with those published by Andersen.[16] Figure 2 shows the fluorescence spectra of dityrosine and insulin (before and after oxidation) at exitation wavelengths of 220 and 276 nm, respectively.

Amino Acid Analysis. The chromatograms of a standard mixture of amino acids obtained by using programs 1 and 2 (see General Methods)

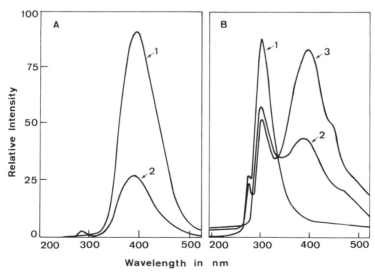

FIG. 2. Fluorescence emission spectra of dityrosine (A) and insulin (B) at pH 9.5. (A) Wavelength of exitation: 1 = 200 nm; 2 = 280 nm. (B) Wavelength of exitation: 1 = 276 nm for native insulin; 2 = 276 nm for oxidized insulin; 3 = 220 nm for oxidized insulin.

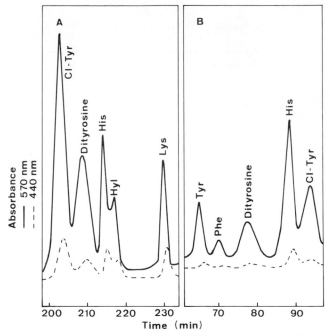

FIG. 3. Sections of chromatograms of amino acid standard mixture. (A) Program 1 (see General Methods). (B) Program 2 (see General Methods). Cl-Tyr, 3-chlorotyrosine.

are shown in Fig. 3. Sections of amino acid chromatograms of different hydrolysates of oxidized proteins are shown in Fig. 4.

Discussion

Various attempts at the chemical synthesis of dityrosine were unsuccessful.[27] Therefore, oxidation of tyrosine by different oxidizing agents was tried. Oxidation with potassium ferricyanide, sodium persulfate, and ferrichloride led to the formation of small amounts of the dimer; the portion of higher oxidation products and melanins was too high, however, and the isolation of pure dityrosine was practically impossible.[27] Oxidation of tyrosine with peroxidase and hydrogen peroxide was shown subsequently to be the method of choice for *in vitro* production of dityrosine. Gross and Sizer,[1] Andersen,[16] and Malanik and Ledvina[30] described methods for the production of dityrosine by means of the peroxidase–hydro-

[30] V. Malanik and M. Ledvina, *Prep. Biochem.* **9**, 273 (1979).

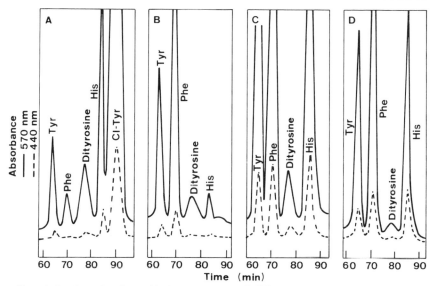

FIG. 4. Section of amino acid chromatograms of different hydrolysates, program 2 (see General Methods). (A) Standard amino acid mixture. (B) Oxidized pepsin. (C) Oxidized insulin. (D) Oxidized casein.

gen peroxide system. Using tyrosine as starting material, the yield of dityrosine was relatively small in our experience, higher oxidation products (trimer and tetramer) being the main products of the oxidation. Furthermore, the isolation of pure dityrosine from the incubation mixture proved to be difficult. Therefore *N*-acetyltyrosine (available also commercially) is suggested as the starting material for the *in vitro* production of dityrosine. This derivative has the advantage of being highly water- and ethanol-soluble, which allows the production of larger amounts of the dimer in one batch.

For comparison, dityrosine was prepared according to Andersen[16] (method B). In this procedure, the products obtained by oxidation of L-tyrosine were separated by ion-exchange chromatography on cellulose phosphate. The isolated dityrosine was identical to that obtained by using method A. Method B is easier to perform than method A, but is not suitable if larger amounts of dityrosine have to be produced in one batch.

Malanik and Ledvina[30] published a method for isolation of dityrosine after oxidation of DL-tyrosine with peroxidase and hydrogen peroxide. These authors used Sephadex G-10 gel filtration and purification on a cation-exchanger to isolate homogeneous dityrosine from the reaction mixture. The yield of dityrosine is comparable to that described here (method A), and this method represents a useful alternative to our method.

The identification of dityrosine is best done on the basis of its characteristic fluorescence (see Fig. 2). Dityrosine present in proteins can be identified after hydrolysis by amino acid analysis (see Figs. 3 and 4). The dimer behaves like a peptide and gives a relatively broad peak that elutes between phenylalanine and histidine.

Newer methods for the identification of dityrosine include chromatography on a DEAE-cellulose column[5] and high-performance liquid chromatography. The latter method is very sensitive and allows the determination of as little as 70 pmol of dityrosine per milliliter. It has been used for the determination of dityrosine in collagen from cow Achilles tendon.[31] Finally, Ohtomo et al.[32] identified dityrosine in a cell surface fraction of an encapsulated strain of *Staphylococcus aureus* by amino acid analysis with subsequent gas chromatography–mass spectrometry of the trimethylsilyl derivative of the isolated dityrosine.

[31] K. Zaitsu, S. Eto, and Y. Ohkura, *J. Chromatogr.* **206,** 621 (1981).
[32] T. Ohtomo, K. Yoshida, S. Kawamura, and Y. Suyama, *J. Appl. Biochem.* **4,** 1 (1982).

[25] Isodityrosine, a Diphenyl Ether Cross-Link in Plant Cell Wall Glycoprotein: Identification, Assay, and Chemical Synthesis

By Stephen C. Fry[1]

The highly insoluble, hydroxyproline-rich glycoprotein (extensin[1a]), characteristic of plant cell walls, has been shown to contain a new cross-linking amino acid named isodityrosine (Fig. 1). Isodityrosine is an oxidatively coupled dimer of tyrosine, with the two tyrosine units linked via a diphenyl ether bond.[2] It shares many properties with dityrosine (Amadò et al., this volume [24]), which is an oxidatively coupled dimer of tyrosine

[1] The author's work was conducted during tenure of the Royal Society Rosenheim Research Fellowship, and was partly supported by U.S. Department of Energy Contract DE-AC02-76ER0146.
[1a] See Maarten J. Chrispeels, this volume [22].
[2] S. C. Fry, *Biochem. J.* **204,** 449 (1982).

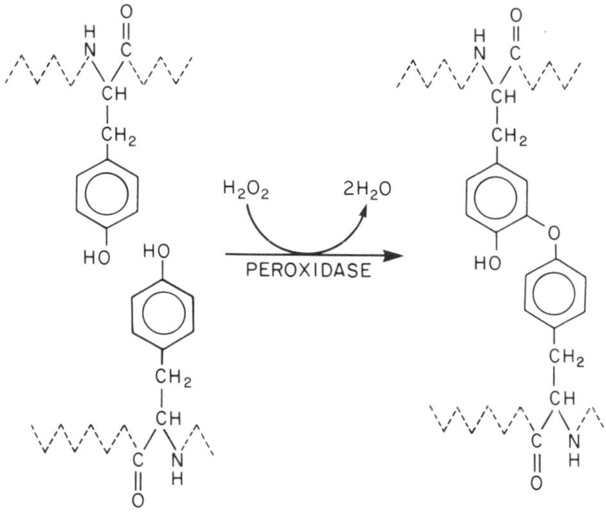

FIG. 1. The proposed mechanism whereby peptidyl tyrosine is oxidatively coupled to form peptidyl isodityrosine, cross-linking the peptides.

with the two tyrosine units linked via a C—C biphenyl bond. The biosynthesis of isodityrosine probably proceeds by the action of peroxidases on tyrosine (Fig. 1), as indicated by the fact that peroxidase inhibitors block the incorporation of [14C]tyrosine into isodityrosine.[2] It has not been fully established that formation of isodityrosine is a posttranslational modification, but experiments showing that the incorporation of [14C]tyrosine into isodityrosine in cultured plant cells is blocked by the protein synthesis inhibitor cycloheximide (unpublished), strongly suggest this. The biological function of isodityrosine is presumably to cross-link the glycoprotein molecules into a network (independent from, but perhaps entangled with, the network of cellulosic microfibrils), holding them tightly in the cell wall. This view is supported by the fact that the binding of newly synthesized glycoprotein to the cell wall is blocked by peroxidase inhibitors[2,3] and by the discovery that the glycoprotein can be solubilized with mildly acidified $NaClO_2$,[4] a reagent that splits the diphenyl ether bridge of isodityrosine.[2]

In this chapter, I strive to make isodityrosine research widely accessible by restricting myself to straightforward materials and methods avail-

[3] J. E. Varner and J. B. Cooper, in "Structure and Function of Plant Genomes" (O. Ciferri and L. Dure, eds.). Plenum, New York, 1983 (in press).
[4] M. A. O'Neill and R. R. Selvendran, Biochem. J. 187, 53 (1980).

able in practically any laboratory. Also, since particular botanical systems present particular difficulties, I have attempted to provide alternative methods where appropriate.

Isolation and Qualitative Identification

Principle. Plant cells (which may have been fed [^{14}C]tyrosine) are freed of all proteins not covalently bound within the cell wall. The remaining wall protein is hydrolyzed with 6 N HCl in the presence of phenol (to protect the phenolic amino acids from oxidation and chlorination). The isodityrosine is separated from the acidic and neutral phenols by paper electrophoresis, and then other amino acids are removed by chromatography (on the same sheet of paper) in a solvent in which isodityrosine has a remarkably low mobility. Confirmation of identity is by thin-layer chromatography (TLC) in a system that resolves isodityrosine and dityrosine.

Plant Material. Plant cells differ widely in their content of wall glycoprotein and therefore isodityrosine.[2] Cells with only primary walls are the richest sources, and suspension-cultured cells are particularly well endowed. Dicotyledons contain substantially more than monocotyledons, although dicotyledons of the order Centrospermae are poor sources, as are crown-gall tumor tissues.

Step 1 (Optional). In Vivo Labeling. L-[^{14}C]Tyrosine is an efficient *in vivo* precursor of isodityrosine in cultured cells and one that gives negligible labeling of other protein amino acids than tyrosine and isodityrosine. However, most plant cells (especially monocotyledons, but also dicotyledons) possess tyrosine ammonia-lyase activity, so that ^{14}C can enter the hydroxycinnamate pathway and cause labeling of cell wall polysaccharide-bound phenols, e.g., ferulic acid and lignin.[5] Where this occurs, electrophoresis (step 4) should definitely be included in the purification scheme.

A sample of sycamore cell suspension [ca. 50 mg fresh weight of tissue per 0.5 ml of M-6 medium[6] lacking yeast extract but supplemented with 20 mM MES [Na$^+$] buffer, pH 5.8, in a 1 × 5 cm (2-dram) vial] is fed 0.1 μCi of L-[U-^{14}C]tyrosine in the absence of nonradioactive tyrosine. The vial is plugged with cotton wool and incubated in the dark at 25° for 5 hr. During this period, ca. 50–80% of the ^{14}C is incorporated into protein. Remaining free [^{14}C]tyrosine, and any low-molecular-weight metabolites, are removed during step 2.

[5] S. C. Fry, *Biochem. J.* **203**, 493 (1982).
[6] J. G. Torrey and Y. Shigemura, *Am. J. Bot.* **44**, 334 (1957).

Step 2. Removal of Proteins Not Covalently Bound in the Cell Wall. A sample of suspension-cultured cells (ca. 0.5 g fresh weight) is added to 15 ml of 3% (w/v) sodium dodecyl sulfate containing 100 mM Na$_2$EDTA, 1% (v/v) 2-mercaptoethanol, and 100 mM NaH$_2$PO$_4$–Na$_2$HPO$_4$ (pH 7.4). The suspension is autoclaved at 120° for 20 min. The cell residue is pelleted by centrifugation (1000 g, 5 min), and the extraction is repeated several times until the supernatant no longer gives a precipitate when added to an equal volume of 30% (w/v) trichloroacetic acid[7] at room temperature (or is no longer radioactive). The final pellet is then washed several times with distilled water to remove the detergent and is freeze-dried (yield ca. 25 mg).

Step 3. Hydrolysis. The cell wall-rich residue from step 2 is suspended (at a concentration of 4 mg dry weight per milliliter) in 6 M HCl containing phenol (10 mM). Authentic dityrosine (1 μg/ml) may also be added if step 6 is to be included in the identification; it serves as a useful internal standard, since (1) it has not been found in plant proteins; (2) it comigrates with isodityrosine in steps 4 and 5 and can be detected by UV fluorescence (eliminating the need to stain guide strips); and (3) any undue losses during workup will be apparent. Hydrolysis is conducted in a sealed glass tube at 110° for 20 hr. The cooled hydrolysate is dried under vacuum or under a stream of air, and the residue is extracted with formic acid–acetic acid–water (1 : 4 : 45, v/v/v).

Steps 4, 5, and 6 are not always all necessary for the identification of isodityrosine. The choice depends on the material to be studied. The following sequences of steps have been used: (1) 4, 5, 6 (excellent); (2) 4, 6 (very good); (3) 4, 5 (good); (4) 5, 6 (fair); (5) 4 alone (fair).

Step 4. Electrophoresis. The soluble hydrolysis products are applied (alongside a spot of marker tyrosine and isodityrosine if the internal standard dityrosine has been omitted) to Whatman No. 3 paper, which is then wetted with formic acid–acetic acid–water (1 : 4 : 45, v/v/v) and electrophoresed. Ideally, this is done by high-voltage electrophoresis (5 kV for 30 min on a 56-cm sheet of paper), but it can also be done with a simple low-voltage apparatus in which the paper is trapped between glass plates as shown in Fig. 2. The isodityrosine and any dityrosine migrate toward the cathode, and show $m_{\text{tyrosine}} = 1.10$–1.15. Staining the markers is carried out, if necessary, by spraying a strip of the paper *almost* to saturation with Folin and Ciocalteu's phenol reagent (available commercially), and then hanging the moist paper in a tank saturated with NH$_3$ vapor. Tyrosine and isodityrosine show up as blue spots. [Electrophoresis is

---

[7] E. Layne, this series, Vol. 3, p. 447.

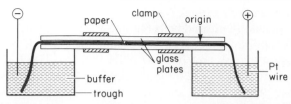

Fig. 2. A simple apparatus suitable for low-voltage paper electrophoresis of cell wall hydrolysates. The samples are loaded on a 30-cm strip of Whatman No. 3 paper at "origin." The paper is then laid on a 20×20 cm glass plate, and wetted *just* to saturation with buffer (formic acid–acetic acid–water, $1:4:45$, v/v/v). A second plate is lowered onto the paper and clamped in place with four strong stationery clips. A voltage of 150 V is maintained for 4 hr. $-$, Cathode; $+$, anode.

preferable to cation–exchange chromatography because some neutral and acidic phenols of the plant cell wall can bind strongly to cation exchangers and are eluted only under the conditions designed to elute the amino compounds. In contrast, neutral or acidic phenols do not migrate toward the cathode during electrophoresis (allowance being made for electroendoosmotic solvent flow, e.g., by reference to a marker spot of glucose).]

Step 5. Paper Chromatography. This step should preferably be used in addition to step 4 or, in preliminary work, as an alternative to step 4. In the latter case, step 5 *must* be followed by confirmation of identity (step 6). Step 5 is based on the remarkably low R_f value (ca. 0.02) of isodityrosine (and dityrosine) on paper chromatography in the solvent mixture 2-propanol–17% aqueous NH_3 (4:1, v/v). This virtual immobility is presumably related to the presence of three negative charges (two carboxy groups and the phenolic hydroxy group) on isodityrosine at high pH.

If step 4 is omitted, the samples are loaded as spots (which can be quite large, e.g., 2 cm in diameter) on Whatman No. 3 paper, which is then developed for 16–40 hr by the descending method. Since isodityrosine scarcely moves from the origin in this time, short lengths of paper (e.g., 14 cm long instead of the usual 56 cm) are adequate. Longer sheets should be used if the other amino acids are also of interest; tyrosine has an R_f of 0.28. After the paper has been dried, and any authentic standards have been stained, the isodityrosine–dityrosine spot is cut out and eluted with 2 M aqueous NH_3. The elution is also efficiently done by a miniaturized descending paper chromatography step, using 2 M NH_3 as irrigant and allowing it to drip off the end of the strip for a few hours.

When steps 4 and 5 are both to be performed, they can conveniently be carried out on the same sheet of paper (Fig. 3). After electrophoresis and the staining of a marker strip, the paper is cut (across the direction of current flow) at a point just beyond the position of isodityrosine. Chroma-

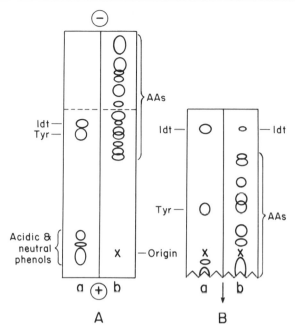

FIG. 3.   (A) Results of electrophoresis of a sycamore cell wall hydrolysate; × = origin, + = anode, − = cathode.   (B) Results of chromatography carried out subsequently on a sheet of paper identical to that shown in (A) but not stained, and cut at the dashed line. ↓ = direction of solvent flow. a, Stained with Folin and Ciocalteu's phenol reagent followed by NH₃ vapor; b, stained with ninhydrin; Idt, isodityrosine; Tyr, tyrosine; AAs, unidentified amino acids.

tography is then run in the *opposite* direction (i.e., at 180°, in contrast to the 90° usually employed in two-dimensional systems) for 16–40 hr, as above. The isodityrosine–dityrosine spot after this procedure is very pure: acidic and neutral phenols fail to move in the electrophoresis, and most amino compounds other than isodityrosine have moderate to high R_f values in the chromatography solvent. The procedure is also convenient in that many samples (approximately 20) can be processed simultaneously, on a single broad (46 cm wide) sheet of paper, thus giving maximum reproducibility as well as saving time.

Step 6. Thin-Layer Chromatography. The above methods do not resolve isodityrosine from dityrosine. This is not normally a problem because (1) dityrosine has never been detected in hydrolysates of plant protein; and (2) if a small amount of dityrosine were present, it would be detected very easily (in the presence of excess isodityrosine) by its characteristic intense blue fluorescence when the paper was exposed to NH₃ vapor and viewed in the dark under UV light. Isodityrosine does not fluoresce. The limit of detection of dityrosine by fluorescence is below that by staining either with Folin and Ciocalteu's phenol reagent or with ninhydrin. However, TLC will resolve isodityrosine and dityrosine, and this can be useful in confirmation of identity, as well as in removing the authentic dityrosine that may have been added at step 3.

The isodityrosine sample, eluted from a paper system (e.g., the isodityrosine spot in Fig. 3B), is concentrated and applied to a 20-cm silica gel TLC plate alongside authentic markers. Development (ca. 4 hr) by the ascending method in *n*-propanol–25% aqueous NH_3 (7 : 3), yields, in the marker track, spots of tyrosine (R_f 0.45), dityrosine (R_f 0.20), and isodityrosine (R_f 0.24).

Quantitative Estimation

As there are no specific assays for isodityrosine, it must be purified chromatographically before quantitation. Pure isodityrosine can then be assayed by its UV absorbance or by reaction with Folin and Ciocalteu's phenol reagent. The staining can be done either before or after elution from the chromatogram. Staining before elution is less accurate, but simpler because the position on the chromatogram of the isodityrosine to be assayed does not have to be accurately deduced before visualization.

Staining before Elution. The paper or plate is sprayed almost to saturation with Folin and Ciocalteu's phenol reagent, and then the yellow background is decolorized with NH_3 vapor. The blue spot corresponding in mobility with isodityrosine is eluted with 1 ml of 2.5 M NaOH, and the absorbance of the resulting solution is measured at 750 nm (700 nm, although below the absorption maximum of the reaction product, is suitable if this is the highest wavelength available on the spectrophotometer). Known amounts of tyrosine are used to construct a standard curve, the assumption being made that tyrosine and isodityrosine have identical molar color yields.

An alternative method consists of scanning the (unstained) TLC plate with a reflectance monitor[8] set at the absorption maximum of isodityrosine (ca. 273 nm at neutral pH).

After Elution. The unstained isodityrosine is located on the chromatogram by reference to stained markers. It is eluted from the paper or silica gel with 2 M aqueous NH_3, and assayed as follows. To 100 μl of solution is added 150 μl of Folin and Ciocalteu's phenol reagent. After 3 min at 25°,

---

[8] J. Moi and H. R. Schmutz, *Am. Lab. (Fairfield, Conn.) Jan. 1983,* p. 54 (1983).

0.75 ml of saturated aqueous Na_2CO_3 is added, and, after a further 60 min, the white precipitate due to NH_4^+ is spun down, and the A_{750} (or A_{700}) of the supernatant is read. Again, tyrosine is used for the construction of a standard curve, and the assumption is made that it has the same molar color yield as isodityrosine.

Alternatively, A_{297} of the (alkaline) eluate is measured. The molar extinction coefficient of isodityrosine in 0.1 M NaOH at 297 nm has been determined[8a] to be 4300.

Chemical Synthesis

Analysis of biological material for isodityrosine and dityrosine is greatly facilitated by the availability of authentic standards. Dityrosine can be synthesized enzymatically by treatment of tyrosine at pH 9.4 with peroxidase and H_2O_2.[2] All attempts to synthesize isodityrosine by this method, at a wide range of pH values, have failed (unpublished observations). However, isodityrosine (and dityrosine) can be synthesized by the nonenzymatic oxidation of tyrosine with ferricyanide.

Preparation. L-Tyrosine (4 g) is dissolved in 17% (w/v) aqueous ammonia (100 ml), and 6.67 ml of 0.75 M $K_3Fe(CN)_6$ is added with stirring. [Oxidation with tetranitromethane instead of ferricyanide yields the biphenyl dityrosine[9] but very little of the diphenyl ether isodityrosine (unpublished observations). In this respect the oxidation of tyrosine is like that of naphthalene.[10]] The reaction is allowed to proceed at room temperature in a sealed vessel for 3 hr. The products include low but useful yields of isodityrosine (1.8% by weight of the starting tyrosine) and dityrosine (3.4%). In addition, some trityrosine and higher homologs, and probably also isotrityrosine,[11] are formed. The products can be detected if a sample (2 μl) of the crude reaction mixture is subjected to TLC (see step 6, above) and examined (1) under UV in ammonia vapor, and (2) by staining with Folin and Ciocalteu's phenol reagent. A typical TLC separation is illustrated in Fig. 4. The crude reaction mixture is an adequate marker mixture for comparison with TLC of material obtained from plant cell wall hydrolysates, but a purification scheme is presented in the next section and should be useful for physicochemical studies of isodityrosine.

[8a] L. Epstein and D. T. A. Lamport, *Phytochemistry* (in press).

[9] J. Williams and J. M. Lowe, *Biochem. J.* **212**, 203 (1971).

[10] P. D. McDonald and G. A. Hamilton, *in* "Oxidation in Organic Chemistry, Part B" (W. S. Trahanovsky, ed.), p. 97. Academic Press, New York, 1973.

[11] D. Fujimoto, K. Horiuchi, and M. Hirama, *Biochem. Biophys. Res. Commun.* **99**, 637 (1981).

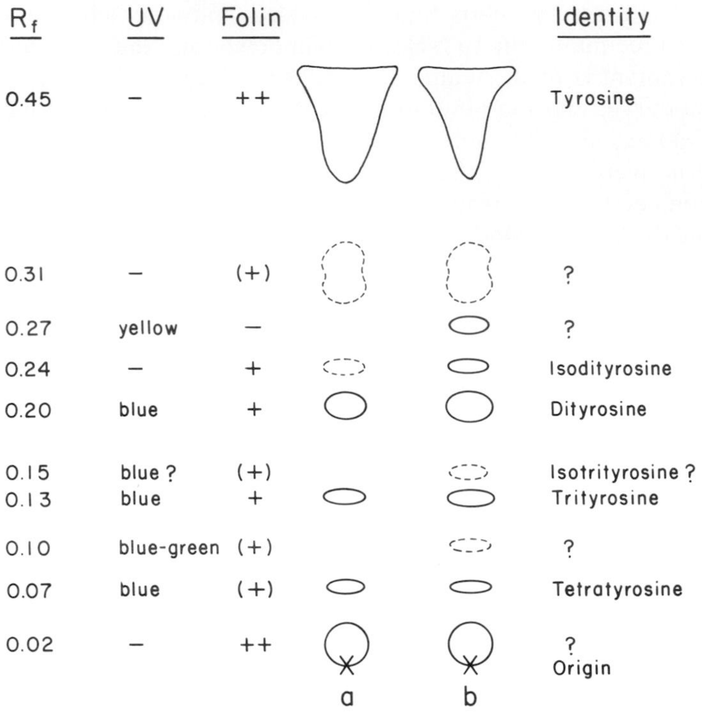

R_f	UV	Folin			Identity
0.45	—	++			Tyrosine
0.31	—	(+)			?
0.27	yellow	—			?
0.24	—	+			Isodityrosine
0.20	blue	+			Dityrosine
0.15	blue ?	(+)			Isotrityrosine ?
0.13	blue	+			Trityrosine
0.10	blue-green	(+)			?
0.07	blue	(+)			Tetratyrosine
0.02	—	++			?
			a	b	Origin

FIG. 4. Tracing of a thin-layer chromatogram of the reaction products obtained by incubation of 0.22 M L-tyrosine together with 0.048 M $K_3Fe(CN)_6$ in 17% aqueous NH_3 at 25° for (a) 20 min or (b) 180 min. Fluorescent spots visible under long-wavelength UV in the presence of NH_3 vapor were marked; the plate was then sprayed with Folin and Ciocalteu's phenol reagent followed by NH_3 vapor, and the blue-staining spots were marked.

Purification. To 105 ml of reaction mixture is added 420 ml of propan-2-ol. On standing at room temperature overnight, most of the unreacted tyrosine precipitates and can be filtered off. (Care should be taken to keep the solvent saturated with NH_3 throughout.) The filtrate is passed down a column (3 × 20 cm) of Whatman cellulose powder (preequilibrated in 2-propanol–17% aqueous NH_3, 4 : 1, v/v) and washed with ca. 1 liter of the same solvent (flow rate ca. 50 ml/hr). The isodityrosine and dityrosine (visualized by viewing the column under UV) bind to the top few centimeters of the cellulose, whereas most other compounds are eluted. The isodityrosine and dityrosine are then eluted with 2 M aqueous NH_3, concentrated under vacuum, and partially purified by preparative TLC (on a 2-mm layer of silica gel) in the solvent system given in step 6. The isodityrosine and dityrosine zones are scraped off the plate and eluted from the silica gel with 2 M aqueous NH_3. The compounds are dried under vacuum, redissolved in water, and purified by anion-exchange chromatography on QAE-Sephadex A-25. [The QAE-Sephadex is swollen in 1 M NH_4HCO_3 (pH 7.8), washed thoroughly in 10 mM NH_4HCO_3, and poured as a column of bed dimensions 20 × 1.5 cm]. The sample is applied and washed with 200 ml of 10 mM NH_4HCO_3, pH 7.8, followed by a linear gradient of 10 to 1000 mM NH_4HCO_3, pH 7.8 (200 ml total). Isodityrosine elutes at ca. 0.5 M NH_4HCO_3, and dityrosine at ca. 0.8 M NH_4HCO_3. Excellent resolution of these two components is achieved, based on the fact that, at pH 7.8, isodityrosine has two negative charges whereas dityrosine has three. The NH_4HCO_3 is removed under vacuum by repeated drying from water, and the products are stored at $-20°$ under desiccation.

Isodityrosine and dityrosine prepared by this method are chromatographically indistinguishable from the compounds isolated from plant cell walls[2] and synthesized enzymatically,[2] respectively. The fact that isodityrosine can be synthesized by oxidation of tyrosine with ferricyanide confirms the structure proposed for it earlier.[2]

Acknowledgments

I am very grateful to Professor D. H. Northcote, F.R.S. (University of Cambridge, U.K.) and Professor P. Albersheim (University of Colorado) for permission to carry out this work in their laboratories.

[26] Assay of Dihydroxyphenylalanine (Dopa) in Invertebrate Structural Proteins

By J. Herbert Waite and Christine V. Benedict

The amino acid 3,4-L-dihydroxyphenylalanine (dopa) occurs in a variety of insoluble sclerotized structures in molluscs,[1-4] annelids,[5] coelenterates,[6] and possibly flatworms.[7] Dopa is covalently attached to proteins in

[1] E. T. Degens, D. W. Spencer, and R. H. Parker, *Comp. Biochem. Physiol.* **20,** 553 (1967).
[2] J. H. Waite, *Comp. Biochem. Physiol. B* **58B,** 157 (1977).
[3] J. H. Waite and S. O. Andersen, *Biol. Bull. (Woods Hole, Mass.)* **158,** 164 (1980).
[4] J. H. Waite and M. L. Tanzer, *Biochem. Biophys. Res. Commun.* **96,** 1554 (1980).
[5] S. Hunt, "Polysaccharide–Protein Complexes in Invertebrates." Academic Press, New York, 1970.
[6] J. G. Tidball, *Cell Tissue Res.* **222,** 635 (1982).
[7] K. Ramalingam, *Acta Histochem.* **41,** 72 (1971).

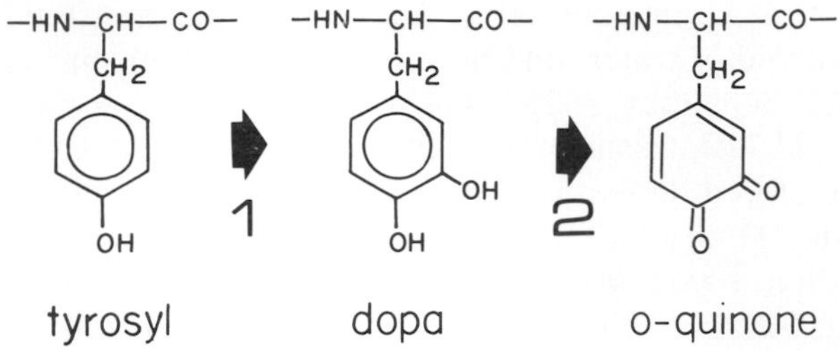

FIG. 1. The formation of dopa from tyrosine in proteins. Both reactions (1) and (2) are catalyzed by tyrosinase *in vitro* in the presence of oxygen. In invertebrate dopa-proteins, however, reactions (1) and (2) are distinctly separate. Step 1 occurs during intracellular processing of the protein, and step 2 occurs extracellularly in the presence of an *o*-diphenol oxidase.

these structures, presumably as part of the polypeptide backbone. For a number of years, it was speculated that this dopa arose as a transient intermediate in the tyrosinase- or peroxidase-catalyzed oxidation of protein tyrosines to quinones[8] during sclerotization (Fig. 1). This does indeed occur during the oxidation of many proteins *in vitro,*[9] but has yet to be demonstrated as a significant pathway in nature. The dopa in scleroproteins, however, is not simply a reaction intermediate. Dopa occurs in cellularly stored precursor proteins as well as in secreted scleroproteins and may contribute specifically to the tanning, adhesive, metal-chelating, and semiconducting properties of these materials.[10] Little to nothing is known about the posttranslational origin of dopa in these proteins. The existence of a protein-directed tyrosine 3-monooxygenase is suggested by a recent experiment.[11]

Assays for detecting dopa proteins are necessary for developing purification strategies, measuring enzymatic posttranslational tyrosine hydroxylation, and determining the interactions of dopa proteins with other substances. Owing to the importance of dopa in the biosynthesis of catecholamines, the number of available assay methods is staggering and includes colorimetric, fluorometric, polarographic, radiochemical, and electrochemical techniques. Unfortunately, most of these are applicable to dopa proteins only after prior hydrolysis of the protein to constituent amino acids.

[8] I. W. Sizer, *Adv. Enzymol. Relat. Subj. Biochem.* **14,** 129 (1953).

[9] S. Lissitzky, M. Rolland, J. Reynaud, and S. Lasry, *Biochim. Biophys. Acta* **65,** 481 (1962).

[10] J. H. Waite, *in* "The Mollusca" (P. W. Hochachka, ed.), Vol. 1, p. 467. Academic Press, New York, 1983.

[11] J. H. Waite, A. S. M. Saleuddin, and S. O. Andersen, *J. Comp. Physiol.* **130,** 301 (1979).

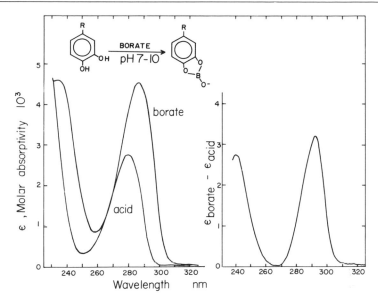

FIG. 2. Ultraviolet spectra of dopa in acid and borate buffers. Maximal absorptivity occurs at 280 nm in acid ($\varepsilon = 2800\ M^{-1}\ cm^{-1}$) and at 286.5 nm in borate ($\varepsilon = 4600\ M^{-1}\ cm^{-1}$). The maximal absorptivity difference between the borate and acid curves occurs at 292 nm for free dopa and 294 nm for peptide-bound dopa.

In the following pages, we shall describe several techniques that have proved to be useful and reliable for detecting dopa proteins. These techniques are not necessarily the most sensitive, but they are highly reproducible and for the most part require no more equipment than what is routinely available in protein chemistry laboratories.

Colorimetric Methods for the Detection of Dopa and Dopa Protein

Acid–Borate Difference Spectrum

1,2-Benzenediols, of which dopa is one, absorb maximally in aqueous acid at about 280 nm.[12] At pH 7–10 in the presence of borate, this maximum is shifted to higher wavelengths (286–287 nm) with a concomitant increase in absorptivity[13,14] (Fig. 2). This is due to the formation of a diol–

[12] T. Nagatsu, "Biochemistry of Catecholamines." Univ. Park Press, Baltimore, Maryland, 1973.

[13] U. Weser, *Struct. Bonding (Berlin)* **2,** 160 (1967).

[14] K. T. Yasunobu and E. R. Norris, *J. Biol. Chem.* **227,** 473 (1957).

borate complex (Fig. 2). Since the pH of the borate can be set well below the $pK_{\phi OH}$ of tyrosine, and since borate forms no complex with the other UV-absorbing amino acids, phenylalanine, tyrosine, or tryptophan, use of the acid–borate difference spectrum (λ_{max} 292–294 nm; Fig. 2) specifically detects the presence of dopa residues. Maximal sensitivity of the technique is between 1 and 2 μg of dopa per milliliter. Advantages are nondestruction of sample and minimal interference from other residues. One disadvantage is in the assumption that the dopa protein in question will be soluble in both buffers. Insolubility can sometimes be remedied by the addition of urea or guanidine hydrochloride to the buffers. Another disadvantage is the competition for borate by other compounds with vicinal diols (sugars[14]).

Reagents

Borate buffer: 0.1 M sodium tetraborate, prepared by dissolving 6.2 g of boric acid (Baker) in 900 ml of distilled water and adjusting the pH to 8.0 with 1 M NaOH. The volume is brought to 1 liter with distilled water

Acid buffer: 0.1 M HCl, prepared by adding 8.6 ml of concentrated hydrochloric acid (11.6 M) to 991.4 ml of distilled water

Dopa standard: 1 mM L-dopa, prepared by dissolving 197.2 mg of L-dopa (Sigma) in 1 liter of distilled water acidified with 1 ml of concentrated HCl. Pulverization of dopa crystals with mortar and pestle will expedite dissolution. Store the solution in a brown bottle at 4°

Sample solution: The sample to be tested for dopa content should be in dilute buffer, preferably acid, as this stabilizes dopa residues

Equipment: UV/vis spectrophotometer; 1-cm quartz cuvettes; pipetters

Procedure. Dispense 20, 40, 60, 80, and 100 μl of the dopa standard into two sets of five labeled tubes for the acid *and* the borate series. Adjust the final volume of each tube to 1 ml. At 292–294 nm,[14a] set to zero absorbance by adjusting the slit width with the first acid–dopa sample in the light path. Now switch to the same concentration of dopa in borate buffer and read the absorbance. Repeat for the other dopa standards and for sample dopa protein in acid and borate buffers. Determine dopa content of sample from the standard curve.

[14a] Actually the maximal absorbance difference for free dopa occurs at 292 nm, but this is shifted upward to 294 nm in peptide-bound dopa. The extinction coefficient $\Delta\varepsilon$ at each maximum, however, remains unchanged.

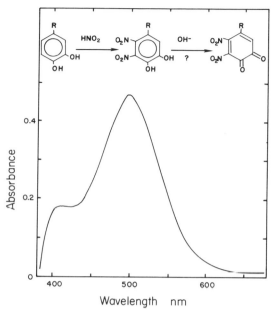

FIG. 3. The visible absorbance spectrum of the nitration product of dopa (8 μg/ml) in alkali.

Alternatively, the relationship $\Delta\varepsilon_{max} = \varepsilon_{borate} - \varepsilon_{acid} = 3200\ M^{-1}\ \text{cm}^{-1}$ (95% C.L. $\pm$ 100) has been determined for dopa in borate and acid, and using Beer's law, the concentration of dopa in the unknown sample can be calculated.

$$\Delta\varepsilon_{max} = \Delta A / Cl$$

ΔA is absorbance difference, C is dopa concentration in moles per liter, and l is cuvette light path (1 cm).

Dopa Nitration

When dopa or a variety of other 1,2-benzenediols are treated with nitrous acid, a bright yellow reaction product is generated. The yellow chromophore, possibly the dinitro derivative[15] (Fig. 3), is converted to a more stable red quinone by the addition of excess alkali. Borberg[16] was probably the first to apply the reaction to the detection of a physiological benzenediol, adrenaline, and subsequent workers extended it to other

[15] D. H. Rosenblatt, J. Epstein, and M. Levitch, *J. Am. Chem. Soc.* **75**, 3277 (1953).
[16] N. C. Borberg, *Scand. Arch. Physiol.* **27**, 350 (1912).

substances including L-dopa.[17,18,18a] This method features a high specific-ity for catechol or monosubstituted 1,2-benzenediols, including 1,2-ben-zenediol-containing macromolecules,[19–21] does not require prior hydroly-sis of dopa proteins, requires less than 0.5 hr to execute, and is sensitive to about 0.5 μg of dopa per milliliter.

Materials

Acid reagent: 0.5 M HCl, prepared by adding 4.3 ml of concentrated HCl (11.6 M) to 95.7 ml of distilled water

Nitrite reagent: 1.45 M sodium nitrite and 0.41 M sodium molybdate, prepared by dissolving 10 g of sodium nitrite (Merck) and 10 g of sodium molybdate (VI) dihydrate (99+%, Aldrich) in distilled wa-ter to 100 ml

Alkali reagent: 1 M NaOH, prepared by adding 4 g of sodium hydrox-ide pellets to a final volume of 100 ml of distilled water

Dopa standard: Described above

Sample: Dopa peptide or protein dissolved in dilute acid

Equipment: UV/visible spectrophotometer; cuvettes; pipetters

Procedure. Pipette the dopa standard in aliquots of 10, 20, 30, 40, and 50 μl into five tubes, a sixth serving as a blank. Add the acid reagent to a final volume of 0.3 ml in each case; then add nitrite (0.3 ml) to each tube, followed in rapid succession by 0.4 ml of alkali reagent. For optimal color development, it is important not to delay more than 5 min between the nitrite and alkali reagents. The yellow color in nitrite is only briefly stable. Samples are read at 500 nm (Fig. 3) upon alkalinization. The red color is stable for several hours at room temperature.

This method is selective for *o*-diphenolic residues. Tyrosine, phenyl-alanine, and tryptophan are nonreactive, as are dityrosine, hydroquinone, and disubstituted benzenediols. 3-Iodotyrosine and methyl ether dopa react, but result in rather different colors.[21] The table illustrates that a variety of synthetic and natural dopa peptides share the same maximum around 500 nm.

Some commonly encountered laboratory reagents do interfere with the assay. 2-Mercaptoethanol turns orange-red in the nitrite, whereas guanidine hydrochloride (>0.5 M) produces a white flocculent suspension in nitrite. Use of organic acids in sample buffers may compromise the

[17] B. Kisch, *Biochem. Z.* **220,** 358 (1930).
[18] L. E. Arnow, *J. Biol. Chem.* **118,** 531 (1937).
[18a] D. W. Barnum, *Anal. Chim. Acta* **89,** 157 (1977).
[19] H. Yamamoto and T. Hayakawa, *Biopolymers* **18,** 3067 (1979).
[20] H. Yamamoto and T. Hayakawa, *Biopolymers* **21,** 1137 (1982).
[21] J. H. Waite and M. L. Tanzer, *Anal. Biochem.* **111,** 131 (1981).

DETECTION OF DOPA PROTEINS BY THE NITRATION REACTION

Protein	Apparent molecular weight	Source	λ_{max} (nm) Dopa nitration	Dopa(calc)[a] (mg/mg protein)	Reference[b]
Periostracin	20,000	Periostracum (*Modiolus modiolus*)	500	0.05 (0.05)	UD
Periostracin	20,000	Periostracum (*Mytilus edulis*)	500	0.04 (0.05)	11
Polyphenolic protein	130,000	Acetic acid-extracted phenol gland (*Mytilus edulis*)	505	0.15 (0.14)	29
Tryptic dopa-peptide	1400	Polyphenolic protein (*Mytilus edulis*)	505	0.16 (0.16)	30
Poly(Lys-dopa)-HBr	22,000	Synthesized by Dr. H. Yamamoto	505	0.26 (0.25)	19
Poly(Glu-dopa)	24,400	Synthesized by Dr. H. Yamamoto	495	0.50 (0.53)	20

[a] Analyses following hydrolysis in GM HCl are given in parentheses.
[b] Numbers refer to text footnotes; UD, unpublished data.

alkalinization in the third step. This can be circumvented by removing the organic acid by prior dialysis or compensating for its presence by adding more NaOH.

Amino Acid Analysis of Dopa

Dopa is an α-amino acid and hence can be detected using the conventional ninhydrin reaction in amino acid analysis.[22] Dopa proteins must be hydrolyzed for this procedure. Since dopa is extremely labile to oxidation under hydrolysis conditions, particular care must be exercised to remove all oxygen and trace metals prior to hydrolysis.

Hydrolysis of Dopa Proteins. For the preparation of dopa proteins for hydrolysis, dialyze a known amount exhaustively against a large volume of 0.1 *M* acetic acid using dialysis tubing prepared according to McPhie.[23] Lyophilize the nondialyzable fraction in a hydrolysis tube and add hydrochloric acid reagent consisting of 6 *M* HCl and 5% (v/v) phenol. We generally add 1 ml of acid per 100 μg of protein. Evacuate the hydrolysis tubes to at least 30 μm and seal. The phenol serves as an antioxidant as

[22] S. Moore, D. H. Spackman, and W. H. Stein, *Anal. Chem.* **30,** 1185 (1958).
[23] P. McPhie, this series, Vol. 22, p. 24.

well as an indicator of seal leaks; i.e., after hydrolysis for 24 hr at 110°, the phenol turns pink with a tight seal, yellow to brown if there has been leakage. After a 48-hr hydrolysis in HCl–phenol, dopa recovery is still greater than 90%. After hydrolysis, hydrolysates are filtered through Millipore GS filters (0.22 μm) in a Swinnex-type apparatus, rinsed, and dried by vacuum evaporation at 50°. The dried film is redissolved in 0.2 M acetic acid at a protein concentration of about 1 mg/ml.

Amino Acid Analysis. There are many variations of the method for amino acid analysis described by Moore *et al.*,[22] each designed to optimize resolution between peaks corresponding to certain amino acids. We were particularly interested in separating dopa from leucine and tyrosine and hence have opted for a variation by Housley and Tanzer.[24] Conditions are briefly outlined below.

Analyses were done with a Beckman 199C Autoanalyzer using a single column (56 cm) packed with Beckman AA-15 resin. The buffers were applied in a step fashion as follows: buffer A: pH 3.20, 0.2 N Na$^+$, 0.2 N citrate, 3% n-propanol (105.5 min); buffer B: pH 4.12, 0.4 N Na$^+$, 0.2 N citrate, 2% n-propanol (59.5 min); buffer C: pH 6.40, 1.0 N Na$^+$, 0.2 N citrate (171 min), with temperature set at 45° for the first 20 min and thereafter at 55°.

Using the above program, dopa emerges at 159.9 min between leucine (155.3 min) and tyrosine (170.9 min) and has a ninhydrin color value 91% that of leucine. The method is sensitive to 0.5–1.0 μg of dopa.

Radiochemical Assay of Dopa Protein

This method was used to detect dopa residues in the protein periostracin covering the outer surface of the shell in *Mytilus edulis*.[11] It is based on the following rationale: if tyrosine, tritiated in the 3- and 5-ring positions, is taken up by tissues synthesizing dopa proteins and/or catecholamines/melanins, there will be a decrease in the specific activity of labeled tyrosine as the tritium in positions 3 or 5 is replaced by a hydroxyl group. By adding a second labeled tyrosine, this time with a uniformly distributed ^{14}C, one can calculate in trichloroacetic acid-precipitated proteins, for example, what proportion of tyrosines in the proteins are modified in position 3 or 5 simply by following the ratio ^{3}H/^{14}C in the proteins. If no modifications have occurred, the ratio will be unchanged from the initial ratio of injected labels. The assumptions made here are that the [^{3}H]- and [^{14}C]tyrosines are indiscriminately incorporated into protein,

[24] T. J. Housley and M. L. Tanzer, *Anal. Biochem.* **114,** 310 (1981).

that the [^{14}C]tyrosine is not modified so as to change its specific activity, and that the ratio of the labeled pool of amino acids does not change significantly prior to incorporation into proteins. The latter assumption does not always hold true in certain preparations.[25] The advantage of the system is high sensitivity (10–20 ng of dopa).

Materials

L-[ring-3,5-^{3}H]Tyrosine (specific activity 52 Ci/mmol)
L-[U-^{14}C]Tyrosine (specific activity 435 mCi/mmol)
Other: Scintillation counter, quench standards, Protosol (NEN), Aquasol-2 (NEN), vials

Procedure. Inject from 4 to 10 × 10^6 dpm into the animal tissue at an ^{3}H/^{14}C ratio of between 2.5 and 4. After allowing sufficient time for incorporation, extract the tissue with a buffer in which the dopa protein is soluble *and* stable. Frequently an acidic buffer (5% acetic acid) containing 0.1 mM unlabeled dopa and tyrosine will suffice. Precipitate proteins with cold 10% trichloroacetic acid. Sediment the precipitate by centrifugation 10,000 g for 30 min. Wash the pellet with 10 parts of additional trichloroacetic acid, centrifuge, wash with methanol–ether (1 : 1), and centrifuge again. Remove the organic supernatant and dissolve the pellet in 1–2 volumes of Protosol for 5 to 12 hr. Dissolved samples are added to scintillation fluid, dark-adapted for several hours to reduce spurious fluorescence, and then counted by dual-channel (^{3}H/^{14}C) counting. The double-label ratio method lends itself well to the assay of samples fractionated by electrophoresis or chromatography. In this way Waite *et al.*[11] demonstrated with polyacrylamide gel slices that about one-fifth of the tyrosine residues in periostracin were converted to dopa (Fig. 4).

Methods for Isolating Dopa and Dopa Proteins

Dopa Isolation

Dopa can easily be isolated from hydrolysates of dopa-containing proteins by using a combination of ion-exchange chromatography and gel filtration. These consequently facilitate quantitation of dopa in completely insoluble proteins.[4,26]

[25] Y. Grimm-Jørgensen, *in* "Molluscan Neuroendocrinology" (J. Lever and H. H. Baer, eds.), p. 21. North-Holland Publ., Amsterdam, 1983.
[26] J. H. Waite and S. O. Andersen, *Biochim. Biophys. Acta* **541,** 107 (1978).

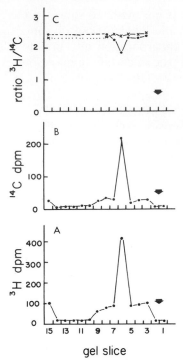

FIG. 4. Evidence for hydroxylation of tyrosyl residues in periostracin using [3,5-³H]tyrosine and [U-¹⁴C]tyrosine. Proteins were separated by electrophoresis in the presence of dodecyl sulfate. Arrows mark the position of the tracker dye. (A) Incorporation of [3,5-³H]tyrosine into periostracin; (B) incorporation of [U-¹⁴C]tyrosine into periostracin; (C) ratio of ³H/¹⁴C in gel slices (●——●). The points ×——× describe a control ratio using [2,6-³H]tyrosine instead of the 3,5 label. Dashed lines between slices 15 and 9 denote insufficient incorporation of labels for determination of ratio. From Waite *et al.*,[11] with permission.

Materials

Ion-exchange resin: Whatman cellulose phosphate P-11, precycled by sequential washing with large volumes of 0.1 *M* NaOH, distilled water, and 0.2 *M* acetic acid

Gel filtration: BioGel P-2 (200–400 mesh) preswollen in 0.2 *M* acetic acid

Buffers: 0.2 *M* acetic acid (pH 2.9) and 0.2 *M* acetic acid with 1.0 *M* NaCl

Sample: Protein hydrolysate, filtered, flash evaporated, and redissolved in 0.2 *M* acetic acid (see above for details)

Equipment: Columns; fraction collectors; peristaltic pump; gradient maker; stirrer

Procedure. After removing the fines, pour the precycled cellulose phosphate resin into a column (2 × 20 cm). Allow it to settle while pumping the starting buffer at a slow flow rate (about 25 ml/hr). Continue eluting with 0.2 *M* acetic acid until the eluent is unpigmented and the pH is about 2.9. The dopa-containing hydrolysate is then applied to the column followed by two column-volumes of 0.2 *M* acetic acid. The linear gradient of NaCl is achieved by placing 200 ml of the acetic acid into the mixing chamber and 100 ml of the acetic acid plus 1 *M* NaCl into the connecting reservoir. With the vessels linked up so that they are at the same level and with the mixing chamber stirred magnetically, the mixing chamber outlet is attached to the cellulose phosphate column by way of a peristaltic pump. Connect the fraction collector to the column and collect fractions of 2–4 ml until the mixing chamber and reservoir are nearly empty. Dopa is located by assaying 0.1 ml of alternate fractions with the nitration reaction (Fig. 5). Peak fractions are pooled, vacuum dried, and reapplied to a column (2 × 90 cm) of BioGel P-2 eluted with 0.2 *M* acetic acid. Aromatic amino acids including L-dopa elute after the internal volume markers due to their interaction with the resin (Fig. 5). Because of slight variations in the properties of cellulose phosphate P-11, the conductivity at which dopa elutes must be determined for each lot. In five different lots, the conductivity for dopa elution ranged from 27 to 35 mmhos.

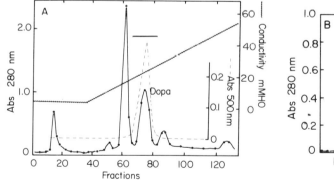 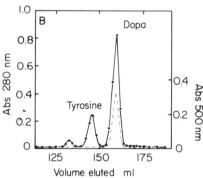

Fig. 5. (A) Ion-exchange chromatography of the byssal adhesive disk hydrolysate on cellulose phosphate. Absorbance at 280 nm measures aromatic amino acids; absorbance (40 μl of sample) at 500 nm measures the dopa nitration reaction. Conductivity was used to determine the NaCl concentration. Brackets enclose fractions pooled for concentration. (B) Gel filtration of dopa on BioGel P-2. Internal volume was at 100 ml. Assays were as described above. From Waite and Tanzer,[4] with permission.

Isolation of Dopa Proteins

There is no simple strategy for isolation that can be categorically applied to all dopa proteins. By virtue of the presence of dopa residues, however, one could expect these proteins to share a tendency to become oxidized and be adsorbed to surfaces. For this reason, it is almost always advantageous to handle dopa proteins at low pH (<5), and if higher pH levels are necessary, to include borate in the buffer. Scrupulous care must be exercised to avoid oxidizing agents at all pH values. Adsorption of the proteins may be minimized by using resins of low polarity. Along these lines, we have found it convenient to isolate dopa-proteins by polyacrylamide gel electrophoresis. A simple modification of the dopa nitration method allows for detection of dopa proteins on polyacrylamide gels. Choice of running conditions is only relevant insofar as it affects the solubility and stability of the dopa protein in question, e.g., sodium dodecyl sulfate precipitates the polyphenolic protein of *Mytilus* byssus.[4] A modified acid gel system after Panyim and Chalkley[27] has provided us with consistently good results.

Materials

> Acrylamide solution: 40 g of acrylamide, 1 g of *N,N'*-methylene bisacrylamide, distilled water to 100 ml, all filtered through activated charcoal
>
> TEMED solution: 8 ml of glacial acetic acid, 1 ml of *N,N,N',N'*-tetramethylethylenediamine, distilled water to 25 ml. Prepare fresh
>
> Persulfate solution: 1.5 g of ammonium persulfate in 100 ml of distilled water
>
> Tracker dye: 1 mg of Methyl Green in 50 ml of 5% (v/v) acetic acid with 4 *M* urea
>
> Electrode buffer: 5% (v/v) acetic acid, prepared by mixing 50 ml glacial acetic acid with 950 ml of distilled water
>
> Dopa nitration reagents: As described above
>
> Equipment: Electrophoresis unit; power supply; pipetters

Procedure. Add 1.5 parts of acrylamide solution to 5.5 parts of water, 3 parts of TEMED solution, and 1.5 parts of persulfate solution to a vacuum flask and evacuate under a strong vacuum. Pour into clean glass tubes (i.d. 3 mm) with one end stoppered. Overlay with 50% (v/v) *n*-propanol. Polymerization requires 3–5 hr. Preequilibrate the tube gels electrophoretically in the electrode buffer with the black electrode (cath-

---

[27] S. Panyim and R. Chalkley, *Arch. Biochem. Biophys.* **130,** 337 (1969).

ode) at the bottom. Recommended amperage is 2–4 mA/gel. Preequilibra-
tion requires about 5 hr. After this, the upper buffer is replaced with fresh
5% acetic acid, and samples are mixed with 0.5 part tracker dye to 1 part
sample and loaded onto the gel tops. Upon completion of electrophoresis
(1–2 hr), gels are extruded from the glass tubes and stained. Gels stained
for protein are processed in the usual manner using Brilliant Blue R-250.[28]
Parallel gels not stained for protein are rinsed off lightly with distilled
water, transferred to 2–3 parts per gel volume of nitrite reagent. Dopa
proteins appear as yellow bands in this reagent. Wait 10–15 min for the
nitrite to diffuse fully into the gel, decant excess nitrite, and add 6 or more
parts per gel of the alkali reagent. Dopa proteins will turn from yellow to a
rose-pink in a reaction lasting no more than 15–20 min (Fig. 6). The
method is sensitive to 0.1 μg of dopa per band. Nitrite-stained gels should
be photographed or scanned at 500 nm as soon as possible to avoid distor-
tion and dilution of bands caused by swelling in alkali. Inclusion of 2–4 M
urea in the acid gels results in the formation of large bubbles in the gels
during the nitration reaction.

If some other method of gel electrophoresis, e.g., sodium dodecyl
sulfate gel electrophoresis, is to be coupled with the isolation of a dopa
protein, we recommend the following steps prior to dopa staining: After
electrophoresis place the gels in perforated Tygon tubing; transfer these
to a large volume of fixative consisting of 50% (v/v) methanol and 5% (v/v)
acetic acid; and stir for 1–2 hr. This will appropriately acidify the gels as
well as fix the proteins and remove detergents, etc., in preparation for the
nitration reaction. This or similar modifications have been performed after electrophoresis in the presence of dodecyl sulfate and cetylpyridinium
bromide.[21,29]

Isolation of Dopa Peptides

Dopa peptides are derived by proteolysis from dopa proteins and
serve to elucidate the primary structure of the latter. Although the same
caveats about oxidizability and adsorption of dopa proteins apply to the
peptides, irreversible adsorption is less of a problem in dopa peptides
owing to their smaller molecular weight. The following method for the
isolation of a dopa-containing tryptic peptide from digested polyphenolic
protein[30] may be applicable to other dopa peptides.

[28] J. Segrest and R. L. Jackson, this series, Vol. 28, p. 54.
[29] J. H. Waite and M. L. Tanzer, *Science* **212**, 1038 (1981).
[30] J. H. Waite, *J. Biol. Chem.* **258**, 2911 (1983).

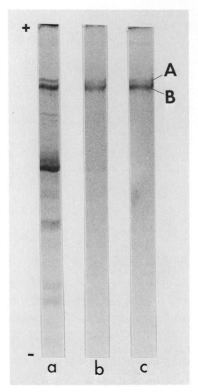

FIG. 6. Acidic polyacrylamide gel electrophoresis of polyphenolic protein-containing samples. Lane a, 0.8 *M* acetic acid extract of polyphenolic protein from phenol glands (30 μl stained with Brilliant Blue R-250); lane b, same as lane a (60 μl stained for dopa (6); lane c, polyphenolic protein after purification[30] (10 μg stained with Brilliant Blue R-250). In each case, electrophoresis was extended for 1 hr after elution of the tracker Methyl Green from the gels. From Waite,[30] with permission.

Materials

Dopa protein, purified
Trypsin (highest purity)
Buffers
    0.2 *M* acetic acid
    A. 0.2 *M* pyridine (16.1 ml/liter), 4.52 *M* acetic acid (260 ml/liter)
    B. 2.0 *M* pyridine (161 ml/liter), 2.52 *M* acetic acid (145 ml/liter)
    0.1 *M* sodium borate (pH 8.0) with 0.01 m*M* CaCl$_2$ and 4 *M* urea
    0.1 *M* sodium borate (pH 8.5)

Reagents
   Dopa nitration reagents: as above
   Ninhydrin reagent: as per Moore and Stein[31]
   Sephadex LH-20 preswollen in 0.2 M acetic acid
   Sephadex LH-60 preswollen in 0.2 M acetic acid
   SP-Sephadex C-25 precycled and swollen in buffer A
Equipment
   Cylinder of oxygen-free N_2
   Rotary vacuum evaporator
   Fraction collector
   Peristaltic pump
   Gradient maker
   UV/visible spectrophotometer
   N_2-reaction chamber
   Thin-layer electrophoresis kit
   Power supply

Procedure. Digestion of dopa proteins with trypsin or chymotrypsin poses a dilemma. On the one hand, the pH optima for these enzymes are neutral to slightly alkaline; on the other hand, many dopa proteins autoxidize or coagulate at these pH. For the digestion of the polyphenolic protein, this problem was circumvented by selecting borate for the buffer and adding 3–4 M urea. The borate stabilizes the dopa residues, and the urea retards the coagulation while not significantly inhibiting the proteases.[32] The polyphenolic protein (in 5% acetic acid) was concentrated to 1–2 ml under nitrogen, using a PM-10 (Amicon Corp.) ultrafilter, and dialyzed against 0.1 M sodium borate with $CaCl_2$ and urea for 2–3 hr. When the nondialyzable portion attained a pH of 8.0 it was transferred to a stirred reaction chamber under nitrogen, and 1 mg of trypsin per 100 mg of total protein was added. The digestion was complete in 3–4 hr but occasionally was left to proceed overnight at 25°.

Digestion was terminated by adding glacial acetic acid to a pH of 3.0. The sample was then flash-evaporated to near-dryness at 50°, redissolved in 1 ml of 0.2 M acetic acid, and fractionated on a column (90 × 1.5 cm) of Sephadex LH-60 eluted with 0.2 M acetic acid. This step functions to separate the dopa peptide from trypsin and urea–borate. Fractions giving a high absorbance reading with the dopa nitration reaction were pooled, flash-evaporated (50°) to dryness, resuspended in 1–2 ml of 0.2 M acetic acid, applied to a SP-Sephadex C-25 column (20 × 1.0 cm), and eluted

[31] S. Moore and W. H. Stein, *J. Biol. Chem.* **176,** 367 (1948).
[32] C. B. Kasper, *Mol. Biol. Biochem. Biophys.* **8,** 114 (1975).

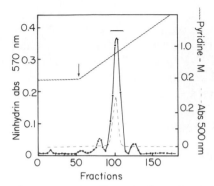

Fractions

FIG. 7. Ion-exchange chromatography of trypsin-digested polyphenolic protein on SP-Sephadex. Sample from pooled fractions off LH-60 column. The first elution buffer was 0.2 M pyridine acetate (pH 3.1); at the arrow, a linear gradient of pyridine acetate up to 2.0 M (pH 5.1) was begun. Fractions were assayed for ninhydrin color and o-diphenols. Pooled fractions were flash-evaporated. The sample applied to the column contained 2 mg of protein; 1.6 mg in the dopa-rich peak. From Waite,[30] with permission.

with a linear pyridine gradient (0.2 to 2.0 M) at room temperature. The gradient was prepared by setting up buffers A and B in the mixing and reservoir chambers, respectively, of a gradient maker, and pumping the mixture into the column (Fig. 7). Ion-exchange fractions were assayed for dopa by dopa nitration and for protein by ninhydrin or absorbance at 285 nm. The adjustment to a higher wavelength from the usual 280 nm was made to avoid strong pyridine interference. Peak fractions having a high dopa : ninhydrin (or absorbance at 285 nm) ratio were pooled and flash-evaporated to dryness three times with two 0.2 M acetic acid washes (2 ml each). The third residue was dissolved in 0.2 M acetic acid (0.5–1.0 ml) and applied to a column (90 × 1.5 cm) of Sephadex LH-20 eluted with 0.2 M acetic acid. Fractions having high absorbance at 280 nm and reacting strongly with ninhydrin and nitrite were pooled, concentrated by flash evaporation, and stored in 0.2 M acetic acid for further studies.

The final purity of the tryptic dopa peptide obtained after gel filtration on Sephadex LH-20 can be evaluated by thin-layer electrophoresis in the presence of 0.1 M borate (pH 8.5).[33] Both silica and cellulose thin-layer plates without fluorescent indicator (Merck Fertigplaten) are adequate, although silica plates pose less of a problem if peptide recovery from the resin is planned. Concentrated samples (2–10 μg of dopa/10 μl) were applied to the middle of the plates in narrow bands and air-dried. Pyro-

[33] H. Michl, *Monatsh. Chem.* **83,** 737 (1952).

catechol violet (Aldrich) was chosen as a tracker dye because of its ability to complex borate. The plates were then dampened with a fine mist of electrode buffer, and with the buffer wicks and cooling plates in place, electrophoresis was performed at 50 V/cm (12 mA) for 3–5 hr. Dopa peptides were visualized after acidification of the plates by the nitration reaction as a series of closely migrating bands on the cathodic side of the origin.[30] Peptides can be eluted from the plate with a filter funnel such as that available from Kontes Glass Co. These bands collectively had a sequence

Ala-Lys-(Pro,Hyp)-Ser-(Tyr,Dopa)-Hyp-Hyp-Thr-(Dopa,Tyr)-Lys

suggesting the only difference between them to be the degree of the posttranslational processing of Pro and Tyr.[30]

Acknowledgments

We thank Dr. Hiroyuki Yamamoto of Shinshu University, Japan, for his gift of synthetic dopa polymers, and Timothy Housley for sharing with us his expertise and time in the amino acid analysis. This work was supported by grants from NSF (PCM-8206463) and NIH (1R23-DE05956).

[27] Halogenated Tyrosine Derivatives in Invertebrate Scleroproteins: Isolation and Identification

By S. HUNT

Halogenated amino acids as natural components of animal tissues are normally associated with the function of thyroid gland in vertebrates.[1] Amino acid derivatives of chlorine, bromine, and iodine, however, are of much wider occurrence. Antibiotics from a variety of sources contain chlorinated derivatives of aliphatic and aromatic amino acids, and marine algae and some higher plants are a source of halide-substituted amino acids.[1-4] Marine invertebrates were recognized at an early date to contain protein-bound halogen as a component of skeletal material, and subsequent studies have shown this to be a common phenomenon wherever

[1] J. Roche, M. Fontaine, and J. Leloup, in "Comparative Biochemistry" (M. Florkin and H. S. Mason, eds.), Vol. 5, p. 493. Academic Press, New York, 1963.
[2] J. Berdy, "CRC Handbook of Antibiotic Compounds," Part IV. CRC Press, Boca Raton, Florida, 1980.
[3] W. Tong and I. L. Chaikoff, *J. Biol. Chem.* **215**, 473 (1955).
[4] E. A. Bell, *MTP Int. Rev. Sci. Org. Chem. Ser. One*, p. 1. (1973).

structural or scleroprotein is laid down under conditions where bromide and iodide ion concentration is relatively high.[1,5–8] More recently, the presence of chloro derivatives of amino acids in scleroproteins of marine and nonmarine species has been demonstrated, including the purely terrestrial insects.[9–13] Table I lists the halogenated derivatives of amino acids that are known to be found in proteins of vertebrates and invertebrates. Systematic searches for the common, as well as for the more unusual, halogenated amino acids now known to exist, including the more recently discovered chlorotyrosines and dityrosyl derivatives, have not been made, and it is quite likely that their distribution among phyla and tissues is more widespread than is at present appreciated. A simple technique permitting relatively rapid survey of halogenated amino acid content in proteins is therefore desirable.

In the past it was the practice when working with iodo derivatives of amino acids to use paper or thin-layer chromatographic techniques, and these were particularly valuable when radiolabeled products were available through biosynthetic routes starting from ^{131}I or ^{125}I. Such methods do not apply well to the greater diversity of halogenated derivatives found in scleroproteins. Since the presently known halogenated amino acids from these materials are confined to tyrosine and tryptophan, the known aromatic affinities of Sephadex and BioGel media provide a useful chromatographic tool for the separation of quite complex mixtures of these compounds.

Preparation of Protein Hydrolysates Containing Halogenated Amino Acids

The standard techniques of protein hydrolysis may usually be applied to release of halogenated tyrosines from scleroproteins. It should be borne in mind that concentrations of these derivatives may be low and relatively large quantities of protein may have to be hydrolyzed. Halogenated tyrosyls are in general somewhat more stable to acid hydrolysis than underivatized tyrosine, but qualitative and quantitative information

[5] J. Roche, *Experientia* **8**, 45 (1952).

[6] E. M. Low, *J. Mar. Res.* **10**, 239 (1951).

[7] C. Fromageot, M. Jutist, M. Lafon, and J. Roche, *C. R. Seances Soc. Biol. Ses Fil.* **142,** 785 (1948).

[8] J. Roche, M. Rand, and Y. Yagi, *C. R. Hebd. Seances Acad. Sci.* **232,** 570 (1951).

[9] S. O. Andersen, *Acta Chem. Scand.* **26,** 3097 (1972).

[10] S. Hunt, *FEBS Lett.* **24,** 109 (1972).

[11] S. Hunt and S. W. Breuer, *Biochim. Biophys. Acta* **252,** 401 (1971).

[12] B. S. Welinder, *Biochim. Biophys. Acta* **279,** 491 (1972).

[13] S. Hunt, unpublished studies of freshwater gastropod operculum, 1982.

TABLE I
HALOGENATED AMINO ACIDS OF ANIMAL PROTEINS

Amino acid	Sources	References[a]
3-Chlorotyrosine (MClT)	Cuticles	
	Schistocerca gregaria (Insecta)	*1*
	Limulus polyphemus (Arachnida)	*2*
	Operculae	
	Buccinum undatum	*3*
	Viviparus viviparus (Mollusca)	*4*
3-Bromotyrosine (MBrT)	Scleroproteins	
	Porifera (as spongin)	*5–7*
	Coelenterata (as gorgonin and anti-pathin)	*5–7*
	Molluscan operculum	*3*
	Cuticle	
	Cancer pagurus (Crustacea)	*8*
	Limulus polyphemus	*2*
3-Iodotyrosine (MIT)	Widely in scleroproteins of Porifera and Coelenterata	*5–7*
	In the endostyle and tunic in Ascidiacea (tunicates) and other members of the deuterostome line of invertebrates	*9–13*
	Invertebrate thyroglobulins	*5, 11*
	In *Planorbis corneus* (Mollusca)	*14*
	In *Periplaneta americana* and other insect proteins	*14–16*
	In *Nereis diversicolor* (Annelida) sclero-protein	*17*
	In byssus and periostracum of *Mytilus galloprovincialis*	*18*
3,5-Dichlorotyrosine (DClT)	Molluscan operculum	*4, 19*
	Limulus polyphemus cuticle	*2*
3,5-Dibromotyrosine (DBrT)	Molluscan operculum	*3, 19, 20*
	Limulus polyphemus cuticle	*2*
	Cancer pagurus cuticle	*8*
	In scleroproteins of sponges and coelenterates	*5–7, 21*
3,5-Diiodotyrosine (DIT)	Distribution similar to that of 3-monoiodo-tyrosine	*5–7, 9–15, 18*
3,3′-Diiodothyronine	In thyroglobulins and thyroid tissue of vertebrates	*22*
3,5,3′-Triiodothyronine	In coelenterate scleroproteins (traces)	*23, 24*
	In *Ciona intestinalis* (Ascidiacea)	*25–27*
	In nemertean mucus	*28*
	In thyroglobulins	*5, 11*

(continued)

TABLE I (continued)

Amino acid	Sources	References[a]
3,3,5'-Triiodothyronine (T₃)	In thyroglobulins	5
3,5,3',5'-Tetraiodothyronine (thyroxine, T₄)	In coelenterate scleroproteins (traces)	5–6, 23–24
	In deuterostome invertebrates	9–11, 25–27
	In insect muscle	15
	In thyroglobulins	5
3-Bromodityrosine	*Cancer pagurus* cuticle	8
3,3'-Dibromodityrosine	*Cancer pagurus* cuticle	8
3-Bromotrityrosine	*Cancer pagurus* cuticle	8
6-Bromotryptophan	*Mytilus edulis* (Mollusca) periostracum	29
2-Monoiodohistidine	In vertebrate thyroglobulins	5
	Possibly in insects	15

[a] Key to references: (*1*) S. O. Andersen, *Acta Chem. Scand.* **26,** 3097 (1972); (*2*) B. S. Welinder, *Biochim. Biophys. Acta* **279,** 491 (1972); (*3*) S. Hunt, *FEBS Lett.* **24,** 109 (1972); (*4*) S. Hunt, unpublished studies of freshwater gastropod operculae, 1982; (*5*) J. Roche, M. Fontaine, and J. Leloup, *in* "Comparative Biochemistry" (M. Florkin and H. S. Mason, eds.), p. 493. Academic Press, New York, 1963; (*6*) J. Roche, *Experientia* **81,** 45 (1952); (*7*) J. Roche, Y. Yagi, R. Michel, S. Lissitzky, and M. Eysseric-Lafon, *Bull. Soc. Chim. Biol.* **33,** 526 (1951); (*8*) B. S. Welinder, P. Roepstorff, and S. O. Andersen, *Comp. Biochem. Physiol.* **53B,** 529 (1976); (*9*) A. D. do Amaral, R. Morris, and E. J. W. Barrington, *Gen. Comp. Endocrinol.* **19,** 370 (1972); (*10*) E. J. W. Barrington and A. Thorpe, *Proc. R. Soc. London Ser. B* **163,** 136 (1965); (*11*) E. J. W. Barrington, "Hormones and Evolution," p. 67. English Universities Press, London, 1964; (*12*) E. J. W. Barrington and A. Thorpe, *Gen. Comp. Endocrinol.* **3,** 166 (1963); (*13*) G. R. Kennedy, *Gen. Comp. Endocrinol.* **7,** 500 (1966); (*14*) W. Tong and I. L. Chaikoff, *Biochim. Biophys. Acta* **48,** 347 (1961); (*15*) L. E. Limpel and J. E. Casida, *J. Exp. Zool.* **135,** 19 (1957); (*16*) B. M. Wheeler, *J. Exp. Zool.* **115,** 83 (1950); (*17*) C. R. Fletcher, *Comp. Biochem. Physiol.* **35,** 105 (1970); (*18*) J. Roche, S. Andre, and I. Covelli, *C. R. Seances Soc. Biol. Ses Fil.* **154,** 220 (1960); (*19*) S. Hunt, *Biochem. Soc. Trans.* **1,** 215 (1973); (*20*) S. Hunt and S. W. Breuer, *Biochim. Biophys. Acta* **252,** 401 (1971); (*21*) C. T. Morner, *Chem. Hoppe-Seyler's Z. Physiol.* **88,** 138 (1913); (*22*) J. Roche, R. Michel, W. Wolf, and J. Nunez, *Biochim. Biophys. Acta* **19,** 308 (1956); (*23*) J. Roche and P. Jouan, *C. R. Seances Soc. Biol. Ses Fil.* **150,** 1701 (1956); (*24*) J. Roche, S. Andre, and G. Salvatore, *Comp. Biochem. Physiol.* **1,** 286 (1960); (*25*) J. Roche, G. Salvatore, and G. Rametta, *Biochim. Biophys. Acta* **63,** 154 (1962); (*26*) G. Salvatore, *Gen. Comp. Endocrinol. Suppl.* **2,** 535 (1969); (*27*) E. J. W. Barrington, *Symp. Zool. Soc. London* **36,** 129 (1975); (*28*) C. W. Major, J. L. Hanegan, and L. Anoli, *Comp. Biochem. Physiol.* **28,** 1153 (1969); (*29*) J. H. Waite and S. O. Andersen, *Biol. Bull.* **158,** 164 (1980).

about this is lacking. It is important to avoid oxidizing conditions as much as possible.

Pretreatment of Materials

Scleroproteins frequently contain inorganic substances, and it is advisable to remove as large a proportion of these as is feasible prior to hydrolysis. For bulky materials, fragmentation or powdering of dried material may aid precleaning. Scleroproteins calcified with calcium carbonate may be decalcified by soaking in a large excess of 50% acetic acid or 12.5% trichloroacetic acid followed by extensive washing in distilled water. If other calcium salts are present, as may be the case in, for example, brachiopod or bryozoan exoskeleta, then soaking in an EDTA buffer may be necessary. Some scleroproteins, e.g., gastropod operculum, bivalve hinge ligament, and some periostraca of molluscs, contain appreciable amounts of iron, which may cause artifact formation from tyrosyls on hydrolysis.[14,15] Such iron is usually not susceptible to removal by chelators such as α,α'-dipyridyl, and drastic methods may have to be used. At 4°, 12 N hydrochloric acid will remove all the iron from gastropod operculin, in two 5-min washes, without appreciable loss of protein. The scleroprotein should then be thoroughly washed, first with 6 N hydrochloric acid and then with distilled water. For the analytical columns described here 0.1–0.5 g of protein should be taken for hydrolysis.

Hydrolysis

Hydrolysis is normally effected by 6 N hydrochloric acid at 105° for 20 hr. Traces of chlorine and bromine in hydrochloric acid may bring about formation of chloro- and bromotyrosines during hydrolysis, but this may be avoided by using only the highest grade (Aristar) acid and by including 0.1% thioglycolic acid in the 6 M hydrochloric acid as a reducing agent.[9,12,16] Hydrolysis may be achieved either by refluxing in Quickfit flasks or borosilicate tubes (soda glass hydrolysis vessels should be avoided because of the iron content[17]) that have been sealed after evacuation or gassing with nitrogen. At least a 50-fold excess of acid over protein should be used for hydrolysis. Hydrolysates are taken to dryness by rotary evaporation under vacuum at 45° with two reevaporations from

[14] S. O. Andersen, personal communication relating to the production of bismethylenedityrosine on hydrolysis of abductin.
[15] S. O. Andersen, *Nature (London)* **216,** 1029 (1967).
[16] F. Sanger and E. O. P. Thompson, *Biochim. Biophys. Acta* **71,** 468 (1963).
[17] M. Stewart and C. H. Nichols, *Aust. J. Chem.* **25,** 39 (1972).

distilled water to remove all traces of hydrochloric acid. Because most invertebrate structural proteins contain—in addition to protein—sugars, phenolic compounds, and melaninlike pigments, considerable quantities of humin usually arise during hydrolysis. The evaporated hydrolysate should therefore be taken up in a small amount of water and either clarified by centrifugation or filtered through a Millipore disk of 0.8 μm pore size. The clarified solution should then be taken to dryness once more before dissolving in the eluting medium of the appropriate column, i.e., 0.02 M acetic acid for Sephadex G-15 or 0.2 M acetic acid for BioGel P2. In general, for the columns described here a hydrolysate of 0.1–0.5 g of protein should be dissolved in no more than 0.5 ml of loading solution.

Column Chromatography

The absorption properties of dextran gels for low molecular weight substances has been recognized for some time, and this quality in Sephadexes is probably in part due to interaction of such compounds with the ether groupings of the gel cross-links.[18] The interactive behavior is, however, complex and sensitive to such factors as pH, ionic strength, temperature, and dielectric properties of the medium.[18–20] Note that pyridines and quinolines in general do not show the absorption properties under the conditions documented here. The more loosely cross-linked gels above G-25 absorb aromatic compounds relatively weakly, but G-25, G-15, and G-10 do so strongly, and elution volumes (V_e) many orders of magnitude greater than that found for a totally included noninteractive species such as glucose are typically found for benzenoid and indolyl compounds. A convenient representation of the behavior of aromatic species on gel columns is the partition coefficient (K_{av}) calculated from the equation

$$K_{av} = (V_e - V_0)/(V_t - V_0)$$

where V_0 is the void volume of the column and V_t is the total volume of the gel beads.[19] V_0 is determined with Blue Dextran 2000 (Pharmacia).

Although the Bio-Rad BioGel series of cross-linked polyacrylamide gels are chemically different to the Sephadex-dextran gels, these too absorb aromatic compounds in a behaviorally comparable but less marked manner. They may be used advantageously either to separate single mixtures or to complement separations on Sephadex gels.

---

[18] H. Determan and I. Walter, *Nature (London)* **219,** 604 (1968).

[19] J. A. Demetriou, F. M. Macias, M. J. McArthur, and J. M. Beattie, *J. Chromatogr.* **34,** 342 (1968).

[20] C. Peyron and C. Simon, *J. Chromatogr.* **92,** 309 (1974).

After some early applications of Sephadex and BioGel gels to the separation of aromatic compounds, a number of workers turned to these gels from other chromatographic techniques, for the separation of mixtures of iodoamino acids and iodopeptides of thyroidal origin.[20–25] For this purpose, Sephadex is less useful than BioGel. Sephadex G-15 and BioGel P2, however, have been found to be ideal for the separation of protein hydrolysates containing complex mixtures of halogenated amino acids including not only the simpler iodo derivatives (on Sephadex), but also the chloro and bromo compounds found in invertebrate scleroproteins.[9–12,26–28] The technique is simple and requires no complex buffer systems.

Separation of Hydrolysates on Columns of Sephadex G-15 or BioGel P2

Column dimensions are partly determined by the equipment available. In general, however, a suitable size of gel bed for qualitative analytical separation of hydrolysates containing 0.1–0.3 g of protein is in the region of 150 cm by 1.0 cm approximately for both Sephadex G-15 and BioGel P2 (Fine, 200–400 mesh). Columns of these lengths, but 2–5 cm in diameter, may be used preparatively for between 1 and 5 g of hydrolyzed protein. Glass columns are to be preferred, as aromatic compounds tend to show some adsorption behavior to the walls of plastic columns. The gels are swollen and degassed in the appropriate aqueous media according to standard procedures described by the manufacturers, and columns are poured using the now well-documented techniques.[29–31] For Sephadex G-15, the swelling and elution medium is 0.02 M acetic acid. Samples are dissolved in the appropriate eluting solution as described above and washed into the

[21] I. Radichevich and E. M. Volpert, *Endocrinol. Exp.* **11**, 105 (1966).

[22] H. J. Cahnmann, *in* "Methods in Investigative and Diagnostic Endocrinology" (S. A. Berson, ed.), p. 27. Elsevier, New York, 1972.

[23] H. Schom and C. Winkler, *J. Chromatogr.* **18**, 69 (1965).

[24] P. Thomopoulos, *Anal. Biochem.* **65**, 600 (1975).

[25] J. Michelot, D. Godeneche, J. C. Maurizis, and G. Meyniel, *J. Chromatogr.* **188**, 431 (1980).

[26] B. S. Welinder, P. Roepstorff, and S. O. Andersen, *Comp. Biochem. Physiol. B* **53B**, 529 (1976).

[27] S. Hunt, *Biochem. Soc. Trans.* **1**, 215 (1973).

[28] J. H. Waite and S. O. Andersen, *Biol. Bull.* **158**, 164 (1980).

[29] C. H. W. Hirs, this series, Vol. 47, p. 97.

[30] "Gel Filtration—Theory and Practice." Pharmacia Fine Chemicals AB, Box 175, S-75104, Uppsala 1, Sweden, 1979.

[31] Bio-Rad catalog of "Chromatography, Electrophoresis, Immunochemistry and HPLC," p. 31. Bio-Rad Laboratories, 2200 Wright Avenue, Richmond, California, 1983.

gel with two 0.5-ml aliquots of eluent before topping up with eluent. For BioGel P2, the eluent is 0.2 M acetic acid. Columns are supplied from reservoirs and pumped from the base. Flow rates of the order of 5–10 ml per hour produce quite acceptable results, and fractions of 5–10 ml may be conveniently collected. An important factor is the length of time of a column run, which may last for several days. In this case fractions of about 10 ml collected over 1-hr periods have proved to be convenient. Smaller fractions collected over shorter periods or at slower pump rates should be collected if clean preparative splitting is a requirement. A peristaltic pump draws eluent from the base of the column and passes this through a UV monitor reading at 280 nm to record the position of the peaks. Columns may be calibrated for V_0 with Dextran Blue and for V_i (volume inside the gel available to very small nonabsorbed molecules) with glucose. Phenylalanine, tyrosine, and known preparations of halogenated amino acids should also be run. It is best not to run complex mixtures of calibrants, but to observe the elution properties of more limited ranges. There is some difficulty in obtaining a full range of the natural halogenated amino acids commercially. The following are currently available: 3,5-dibromotyrosine (Sigma), 3-chlorotyrosine (Koch-Light), 3-iodotyrosine (Sigma, Koch-Light, BDH, Aldrich), 3,5-diiodotyrosine (Sigma, Koch-Light, BDH), diiodothyronine (Sigma, Koch-Light, BDH, Aldrich), triiodothyronine (Sigma, Koch-Light, BDH, Aldrich), and thyroxine (Sigma, Koch-Light, BDH, Aldrich). 3-Bromotyrosine, 3,5-dichlorotyrosine, and 3-chloro-5-bromotyrosine are not at present commercially available but may be readily synthesized.

Syntheses

Synthesis of 3,5-Dichlorotyrosine

An excess of chlorine is passed through a solution of tyrosine in acetic acid (0.5 g in 100 ml of glacial acetic acid[32]). The yellowish solution is taken to dryness on a rotary evaporator and repeatedly reevaporated from distilled water to remove traces of chlorine. The resulting white solid is dissolved in 0.2 M acetic acid (minimum amount) and applied to a column of BioGel P2 (2 × 100 cm). Unchanged tyrosine elutes well before 3-chlorotyrosine, the eluate being monitored at 280 nm and collected in 15-ml fractions. The halo derivative is recovered by fraction pooling and freeze-drying.

---

[32] R. Zeynek, *Hoppe-Seyler's Z. Physiol. Chem.* **114**, 280 (1921).

Synthesis of 3-Bromotyrosine and 3-Chloro-5-bromotyrosine

Tyrosine or 3-chlorotyrosine (100 mg) is dissolved in 100 ml of 0.5 M hydrochloric acid. A solution containing 0.28 g of potassium bromate (KBrO$_3$) and 1.19 g of potassium bromide in 100 ml of H$_2$O is added dropwise with stirring until the calculated level of bromination is achieved.[12] The brominated amino acids are concentrated on a rotary evaporator to 1.0 ml or less and fractionated on a column of BioGel P2 as described above for dichlorotyrosine. Halogenated species elute after tyrosine. 3-Chloro-5-bromotyrosine elutes after both tyrosine and 3-chlorotyrosine and may be distinguished from the latter by its distinctive ultraviolet absorption spectrum (Table II). Dibromotyrosine traces that may form under these conditions separate well from the main products and may also be distinguished by the distinctive absorption spectrum (Table II).

TABLE II

ULTRAVIOLET ABSORPTION SPECTRAL CHARACTERISTICS OF HALOGENATED AROMATIC
AMINO ACIDS AND THEIR UNSUBSTITUTED PARENT COMPOUNDS

Amino acid	$\varepsilon \times 10^{-3}$ acid[a,b]	λ_{acid} (nm)	$\varepsilon \times 10^{-3}$ alk[b]	λ_{alk} (nm)	ε alk$_{acid}$	Refer-ence
Tyrosine	1.40	274.5	2.40	293	1.71	c
	1.368	274	2.331	293	1.70	d
	1.340	274.5	2.330	293.5	1.74	e
3-Fluorotyrosine	1.525	271	2.831	288	1.86	d
3-Chlorotyrosine	1.879	278	3.329	299	1.77	d
	1.959	278.5	—	—	—	f
3-Bromotyrosine	2.045	280	—	—	—	f
3-Iodotyrosine	2.75	283	4.10	305	1.49	c
	2.488	281	3.840	303	1.54	d
3,5-Dichlorotyrosine	1.380⎱ 1.424⎰	279 ⎱ 286.5⎰	3.425	304.5	2.48	d
	1.950⎱ 1.988⎰	281⎱ 288⎰	—	—	—	f
3,5-Dibromotyrosine	2.177⎱ 2.248⎰	281.5⎱ 288.5⎰	5.059	309	2.32	d
	2.181⎱ 2.202⎰	282.5⎱ 288.5⎰	—	—	—	f
3-Chloro-5-bromo-tyrosine	1.985⎱ 1.985⎰	280⎱ 287⎰	4.711	306	2.37	d
	2.001⎱ 1.990⎰	281.5⎱ 288 ⎰	—	—	—	f
3,5-Diiodotyrosine	2.75	287	6.25	311	2.27	c
	2.459	285	5.280	309	2.15	d
	2.730	285	5.920	310	2.17	e

(continued)

TABLE II (*continued*)

Amino acid	$\varepsilon \times 10^{-3}$ acid[a,b]	λ_{acid} (nm)	$\varepsilon \times 10^{-3}$ alk[b]	λ_{alk} (nm)	ε alk$_{acid}$	Reference
3,3',5'-Triiodo-	4.09	295	4.66	322	1.14	b
thyronine	4.09	295	4.66	320	1.14	e
3,5,3',5'-Tetraiodo-	4.28	298	6.18	325	1.44	c
thyronine	4.160	295	6.210	325	1.49	e
(thyroxine)						
Dityrosine	—	283.5	—	316 (pH 10 and 13)	—	g
Monobromodityrosine	—	288	—	316	1.2	h
Dibromodityrosine	—	286	—	319	1.3	h
Trityrosine	—	286	—	322 (pH 10) 315 (pH 13)	—	g
Monobromotri- tyrosine	—	288	—	314	1.5	h
Tryptophan	5.366	271 278 286.5	5.203	272 279.5 287	—	d
	5.500 4.500	278 287.5				e
5-Bromotryptophan	4.924	278.5 287 296.5	4.756	288 298	—	d
6-Bromotryptophan	—	284–285 293				i

[a] Bracketed figures indicate discrete multiple maxima.

[b] Spectra from references *c* and *d* were recorded in 0.1 *M* HCl or 0.10 *M* NaOH except for 3,3',5'-triiodothyronine, which was recorded in 0.04 *M* HCl or 0.04 *M* NaOH. Spectra for reference *e* were recorded in 0.04 *N* HCl or 0.04 *N* KOH for 3,5-diiodotyrosine, 3,3',5'-triiodothyronine, and thyroxine and in 0.1 *N* HCl or 0.1 *N* NaOH for tyrosine and tryptophan. Spectra for reference *f* were measured in 0.2 *M* acetic acid, and for reference *h* in 0.2 *M* acetic acid or 0.1 *M* NaOH.

[c] H. Edelhoch, *J. Biol. Chem.* **237,** 2778 (1962).

[d] S. Hunt and H. Morris, previously unpublished figures, 1983.

[e] BDH Amino Acids Hand Book, published by BDH Chemicals Ltd., Poole, England, 1982.

[f] B. S. Welinder, *Biochim. Biophys. Acta* **279,** 491 (1972).

[g] S. O. Andersen, *in* "Comprehensive Biochemistry" (M. Florkin and E. Stotz, eds.), Vol. 26C, p. 633. Elsevier, Amsterdam, 1971.

[h] B. S. Welinder, P. Roepstorff, and S. O. Andersen, *Comp. Biochem. Physiol. B* **53B,** 529 (1976).

[i] S. O. Andersen, personal communication, 1981.

Hybrid tyrosine derivatives of iodine with chlorine or bromine have yet to be detected in nature; they are not commercially available, and they seem to be difficult to synthesize. 3-Fluorotyrosine (MFT) is available from Aldrich Chemical Co., Ltd., but this amino acid has so far not been identified as a natural product. 3-Bromodityrosine and 3,3'-dibromodityrosine may be synthesized by bromination of dityrosine (which is in its turn synthesized from tyrosine) by addition of saturated bromine water to a solution of dityrosine in 0.2 *M* acetic acid followed by concentration and separation on a column of BioGel P2.[26]

Separations

Figure 1 is a composite showing the separation of phenylalanine, tyrosine, MFT, MClT, MBrT, MIT, DClT, ClBrT, DBrT, DIT, glucose, and tryptophan on a Sephadex G-15 column. All these amino acids separate, although there is some overlap in the case of MIT and ClBrT. Peaks become progressively broader with increasing eluent volume. It is notable

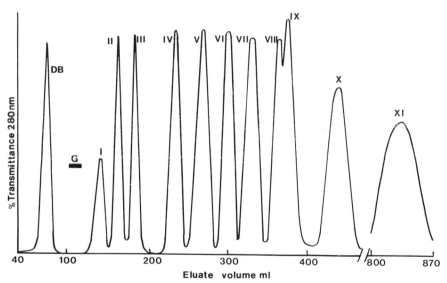

FIG. 1. Fractionation of halogenoamino acids, phenylalanine, tryptophan, glucose, and Blue Dextran 2000 on a column of Sephadex G-15 (1.20 × 148.2 cm). Elution was performed with 0.02 *M* acetic acid at a rate of 8 ml/hr. The position of glucose was detected using an orcinol sulfuric acid reagent [S. A. Barker, C. N. D. Cruickshank, and J. H. Morris, *Biochim. Biophys. Acta.* **74**, 239 (1963)]. DB, Dextran Blue; G, glucose; I, phenylalanine; II, tyrosine; III, 3-fluorotyrosine; IV, 3-chlorotyrosine; V, 3-bromotyrosine; VI, 3,5-dichlorotyrosine; VII, tryptophan; VIII, 3-iodotyrosine; IX, 3-chloro-5-bromotyrosine; X, 3,5-dibromotyrosine; XI, 3,5-diiodotyrosine. This is a composite of several individual runs.

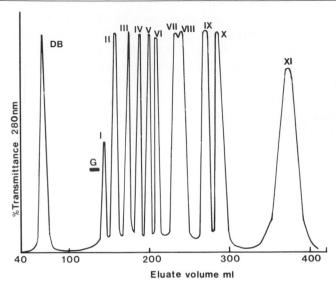

FIG. 2. Fractionation of halogenoamino acids, phenylalanine, tryptophan, glucose, and Blue Dextran 2000 on a column of BioGel P2 (1.17 × 143.5 cm). Elution was performed with 0.2 *M* acetic acid at a rate of 8 ml/hr. Glucose was detected as in Fig. 1. DB, Dextran Blue; G, glucose; I, phenylalanine; II, tyrosine; III, 3-fluorotyrosine; IV, 3-chlorotyrosine; V, tryptophan; VI, 3-bromotyrosine; VII, 3-iodotyrosine; VIII, 3,5-dichlorotyrosine; IX, 3-chloro-5-bromotyrosine; X, 3,5-dibromotyrosine; XI, 3,5-diiodotyrosine. This is a composite of several individual runs.

how the order of elution strictly follows the order of increasing halogen atomic number.

Figure 3 shows a separation, on the same column used for Fig. 1, of a hydrolysate of operculum from the large whelk *Neptunea antiqua*. Materials of this type yield hydrolysates containing a wide variety of aromatic degradation products derived from polyphenols, cross-links, and melanin-like pigments. In spite of this, separation of halogenated amino acids is good and although iodo derivatives appear to be absent all the other potential halotyrosines are clearly differentiated. The peak lying between tyrosine and 3-chlorotyrosine is oxindolylalanine, a degradation product of tryptophan.

Figures 2 and 4 show the separation of an amino acid mixture and an operculum hydrolysate, respectively, applied to a column of BioGel P2 of similar dimensions to the Sephadex G-15 column. The resolution of halogenated species is less effective, and it is clear that retardation by absorption of the aromatic residues to the column is less pronounced.

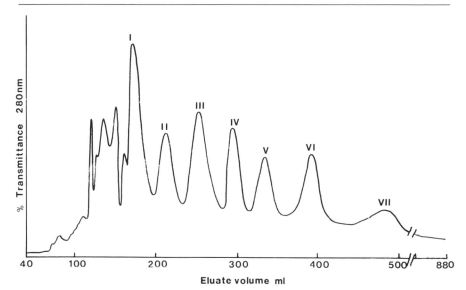

Fɪɢ. 3. Separation of a hydrolysate (6 N hydrochloric acid under nitrogen at 105° for 20 hr) of 0.2 g of *Neptunea antiqua* operculum on the same column of Sephadex G-15 used for the separation of halogenated amino acids shown in Fig. 1. I, tyrosine; II, oxindolyl-alanine; III, 3-chlorotyrosine; IV, 3-bromotyrosine; V, 3,5-dichlorotyrosine; VI, 3-chloro-5-bromotyrosine; VII, 3,5-dibromotyrosine. Note that the presence of high concentrations of other compounds (amino acids, sugars, heterocyclics) does not alter the overall spatial separation of the halogenated derivatives, but shifts their elution positions to somewhat higher V_e values than for the model mixtures.

Partition coefficients for the separation of halogenated tyrosines on these columns of Sephadex G-15 and BioGel P2 are given in Table III.

Separation of Thyroxine and Iodothyronine

The techniques described above applied to some of the invertebrate scleroproteins reputed to contain iodotyrosines, e.g., the gorgonins of coelenterates or the sponge proteins, would be likely to result in a failure to detect thyroxine and its lower iodinated homologs, as these absorb extremely strongly to Sephadex G-15 or BioGel P2 in dilute acetic acid. However, use of a shorter column of BioGel P2 (0.9 × 50 cm) equilibrated with a 0.05 M Tris-HCl buffer at pH 9 and 20° results in all halogenated derivatives of single tyrosyl residues running together at about the posi-tion of the V_i while thyroxine and triiodothyronine elute with K_{av} values of 0.92 and 1.60, respectively (Fig. 5).[24] Note that the effect of the shift to an alkaline eluting pH is to remove the effect of increasing halogen content upon elution order, so that diiodotyrosine (K_{av} = 0.37) slightly precedes

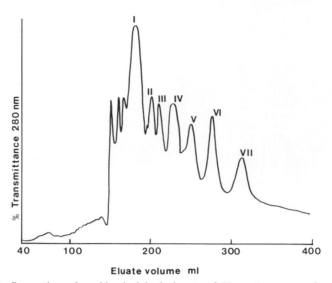

FIG. 4. Separation of an identical hydrolysate of *N. antiqua* operculum to that of Fig. 3 on the same column of BioGel P2 used for the separation of Fig. 2. I, tyrosine; II, oxindolylalanine; III, 3-chlorotyrosine; IV, 3-bromotyrosine; V, 3,5-dichlorotyrosine; VI, 3-chloro-5-bromotyrosine; VII, 3,5-dibromotyrosine. Here again, as in Fig. 3, V_e values for the halogenated amino acids are affected and increased by the presence of high concentrations of other compounds.

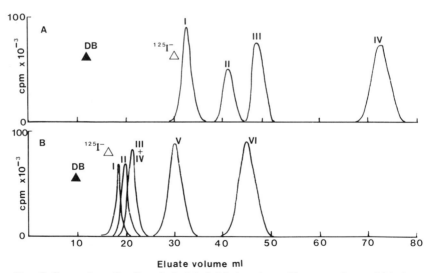

FIG. 5. Separation of radioactively labeled iodoamino acids on a column of BioGel P2 (0.9 × 50 cm). Elution in (A) is with 0.05 M Tris-maleic acid–NaOH at pH 5.3 and 20°; flow rate 6 ml/hr. DB, Dextran Blue; I, monoiodohistidine; II, diiodohistidine; III, 3-iodotyrosine; IV, 3,5-diiodotyrosine. Elution in (B) is with 0.05 M Tris-HCl at pH 9 and 20°. I, 3,5-diiodotyrosine; II, diiodohistidine; III, monoiodohistidine; IV, monoiodotyrosine; V, thyroxine; VI, triiodothyronine. Redrawn from Thomopoulos.[24]

TABLE III
PARTITION COEFFICIENTS FOR THE ELUTION OF
HALOGENATED AMINO ACIDS AND OTHER COMPOUNDS
FROM COLUMNS OF SEPHADEX G-15 AND BIOGEL P2

	K_{av}	
Compound	Sephadex G-15[a]	BioGel P2[b]
Glucose	0.44	0.71
Phenylalanine	0.76	0.85
Tyrosine	0.96	1.02
3-Fluorotyrosine	1.18	1.22
3-Chlorotyrosine	1.67	1.39
3-Bromotyrosine	2.05	1.61
3,5-Dichlorotyrosine	2.39	1.99
Tryptophan	2.60	1.53
3-Iodotyrosine	3.05	1.95
3-Chloro-5-bromotyrosine	3.17	2.37
3,5-Dibromotyrosine	3.75	2.54
3,5-Diiodotyrosine	7.89	3.52

[a] Column, 1.20 × 148.2 cm, eluted with 0.02 M acetic acid.
[b] Column, 1.17 × 143.5 cm, eluted with 0.2 M acetic acid.

monoiodotyrosine (K_{av} = 0.55) and thyroxine well precedes triiodothyronine.

Halogenated Di- and Trityrosines

Fractionation of hydrolysates of crustacean cuticle (*Cancer pagurus*) on columns of Sephadex G-15 yield chromatograms containing di- and trityrosine eluting with the tyrosine peak (Fig. 6).[26] Two peaks appearing between tyrosine and 3-bromotyrosine and a peak appearing between 3-bromotyrosine and dibromotyrosine at 1.2, 1.4, and 2.1 column volumes, respectively, are monobromodityrosine, monobromotrityrosine, and dibromodityrosine. Their ultraviolet spectral absorption characteristics are given in Table II. Clearly their position is open to confusion with monochlorotyrosine and dichlorotyrosine. Established procedures for the separation and preparation of di- and trityrosines from arthropod cuticular proteins lend themselves to separation of the halogenated derivatives of these amino acids.[33-35]

[33] S. O. Andersen, *Biochim. Biophys. Acta* **93**, 215 (1964).
[34] S. O. Andersen, *Acta Physiol. Scand.* **66**, Suppl. 263, 1 (1966).
[35] S. O. Andersen, *in* "Cuticle Techniques in Arthropods" (T. A. Miller, ed.), p. 185. Springer Publ., New York, 1980.

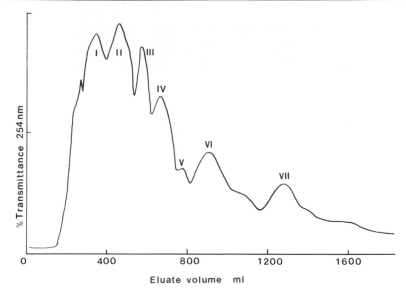

FIG. 6. Separation of a hydrolysate (6 N hydrochloric acid, 0.1% thioglycolic acid at 110°
for 20 hr) of *Cancer pagurus* exocuticle on a column of Sephadex G-15 (90 × 2.5 cm) eluted
with 0.02 M acetic acid at 80 ml/hr. Monitoring is at 254 nm. I, phenylalanine; II, tyrosine
with di- and trityrosines; III, 3-bromodityrosine; IV, 3-bromotrityrosine; V, 3-bromotyro-
sine; VI, dibromodityrosine; VII, 3,5-dibromotyrosine. Redrawn from Welinder *et al.*[26]

Separation on Cellulose Phosphate Columns

Cellulose phosphate (Whatman P11) is washed with 0.2 M NaOH until
the supernatant is colorless and then with two aliquots of 1 M acetic acid
before stirring overnight in excess 0.2 M acetic acid. Columns 1.2–1.6 cm
in diameter and 15–50 cm in height can be used for pouring the resin,
which is firmly packed by pumping 0.2 M acetic acid at a rate of 20 ml/hr.

Samples are dissolved in 5 ml of 0.2 M acetic acid and applied to the
column. Typical elution conditions for a column of 1.6 cm inside diameter
by 15 cm in length would then be 0.2 M acetic acid at 20 ml/hr for 2 hr
followed by 0.1 M NaCl in 0.2 M acetic acid until discrete ultraviolet
absorbing components, monitored at 280 nm, had ceased to be eluted. The
dityrosine group emerges at about 200 ml as a single peak containing
dityrosine, monobromotyrosine, and dibromotyrosine well separated
from both the monoamino acids and trityrosine derivatives. The single
peak pooled and concentrated can be applied to the BioGel P2 for a
complete separation of the three dityrosine derivatives.

Bromotryptophan from Molluscan Periostracum

Separation of hydrolysates of *Mytilus edulis* periostracum on a BioGel P2 column (1.5 × 50 cm) eluting with 0.2 *M* acetic acid yields a peak at 2.9 column volumes approximately (Fig. 7) which is reported to be 6-bromotryptophan.[28] This is the only published report of a protein-bound bromotryptophan, and its presence in acid hydrolysates together with tryptophan is unusual in view of the fact that tryptophan and its derivatives are unusually labile in strong, hot mineral acids. There may, however, be some special stabilizing feature of periostracal scleroproteins, as tryptophan also emerges unscathed from acid hydrolysis of *Ensis ensis* periostracum. 6-Bromotryptophan is distinguishable spectroscopically from tryptophan (Table II).

Enzymatic Release of Halogenated Tyrosine from Insect Cuticle Protein

Halogenated amino acids occur most prominently in proteins from invertebrates of marine origin. Hydrolysis of insect cuticular proteins, however, yield 3-chlorotyrosine identified after separation of the hydrolysates on columns of BioGel P2. Suspicion that the presence of the

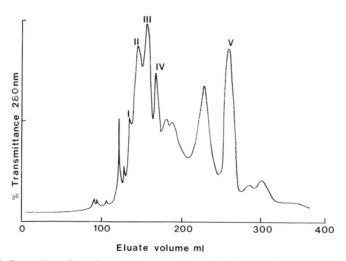

Fig. 7. Separation of a hydrolysate (6 *M* hydrochloric acid, 0.1% caprylic acid, and 1.0% thiodiglycol at 110° under vacuum for 24 hr) of *Mytilus edulis* marginal periostracum on a column of BioGel P2 (50 × 1.5 cm) eluting with 0.2 *M* acetic acid at 12.5 ml/hr. I, phenylalanine; II, tyrosine; III, 3-dihydroxyphenylalanine; IV, tryptophan; V, 6-bromotryptophan. Redrawn from Waite and Andersen.[28]

monochlorotyrosine alone might indicate artifactual origin from the hydrochloric acid is partially allayed by inclusion of a reducing agent in the hydrolysis solution but is entirely alleviated by observation that the amino acid can be released from cuticular protein by Pronase digestion.

Principle

Most scleroproteins are extremely insoluble and hence rather inaccessible to proteolytic enzymes in solution. A partial route to enzymatic digestion of such materials as are found in insect cuticle is via a preliminary partial solubilization.[9]

Reagents

Potassium tetraborate, 1%
Formic acid, 1 M
Potassium borate, 0.05 M, pH 8.0
Pronase (BDH, Sigma)

Procedure

The method describes the procedure as applied to the cuticle of fifth-instar locust (*Schistocerca gregaria*) exuviae.[9] Cuticle (2 g) is extracted with 500 ml of 1% potassium tetraborate and then washed with distilled water several times. The washed cuticle is refluxed with 500 ml of 1 M formic acid for 2 hr, and the supernatant is concentrated by rotary evaporator to 25 ml. Dialysis against 5 liters of distilled water yields a precipitate. The precipitate suspended in 10 ml of 0.05 M potassium borate, pH 8.0, is digested with 50 mg of Pronase at 40° for 48 hr. The digest is centrifuged to remove residue, and the supernatant is concentrated before fractionation on Sephadex G-15 or BioGel P2 as described above.

Pronase has been applied in the past to the release of iodinated amino acids from thyroglobulins and should be generally applicable to other iodinated proteins as well as to other halogenated species.[36]

A buffer recommended for iodoprotein is composed of 0.04 M Tris, pH 8.0, containing 0.002 M $MnSO_4$ and 0.002 M mercaptomethylimidazole (methimazole).

Mn^{2+} ions are reputed to improve hydrolysis, although this has been disputed, and methimazole is reported to reduce artifactual iodination occurring through oxidation of I^- liberated during enzymatic hydrolysis

---

[36] R. Pitt-Rivers and H. L. Schwartz, *in* "Chromatographic and Electrophoretic Techniques" (I. Smith, ed.), p. 224. Heinemann, London, 1969.

of iodoprotein.[36–40] Digests are treated as above and fractionated on BioGel P2.

Effects of Alkaline Hydrolysis

A number of early publications dealing with thyroid proteins utilized alkaline hydrolysis as a route to the release of iodinated tyrosyl and thyronyl derivatives. Alkaline hydrolysis either in sodium hydroxide or barium hydroxide cannot be considered as satisfactory, resulting as it does in extensive loss of halogen and further degradation of the aromatic species.

3,5-Diiodotyrosine has been reported to suffer oxidation on the β carbon atom during incubation at pH 14 with the formation of 3,5-diiodo-4-hydroxybenzaldehyde and oxalic acid.[41] It has also been shown that this general phenomenon applies to all the halogenated tyrosines and to tyrosine and phenylalanine, also giving rise to the appropriate halogen derivatives of 4-hydroxybenzaldehyde, the free aldehyde or benzaldehyde, respectively.[42,43]

Identification

The positions of elution of UV-absorbing species from the Sephadex G-15 and BioGel P2 columns provide a partial identification of halogeno amino acids, but in hydrolysates these may be accompanied by other components. Conventional paper and thin-layer chromatography in a range of solvents against available known standards will increase confidence in purity and identification. Ultraviolet absorption spectra of the various halogenated derivatives of tyrosine are similar, but sufficiently distinctive when examined at both acid and alkaline pH (Table II) to permit reasonably unequivocal distinction, providing that relatively pure preparations have been obtained and can be compared with authentic material.

Wherever possible, however, sufficient material should be obtained to determine mass spectra as the least equivocal route to identification.

[37] W. Tong, E. Raghupathyl, and I. L. Chaikoff, *Endocrinology* **72,** 931 (1963).
[38] K. Inoue, *Endocrinology* **79,** 601 (1966).
[39] L. L. Rosenberg and G. La Roche, *Endocrinology* **75,** 776 (1964).
[40] L. L. Rosenberg, G. La Roche, and J. M. Ehlert, *Endocrinology* **79,** 927 (1966).
[41] R. Pitt-Rivers, *Biochem. J.* **43,** 223 (1948).
[42] F. Bettzieche, *Hoppe-Seyler's Z. Physiol. Chem.* **150,** 177 (1925).
[43] S. Hunt, *in* "The Biochemistry of the Amino Acids" (G. C. Barrett, ed.). Chapman & Hall, London, in press.

The halogeno derivatives of tyrosine are sufficiently volatile for direct insertion into the mass spectrometer. Di- and trityrosine halogenates must, however, first be derivatized. Acetylation and methylation have been employed to this end.[26]

Mass spectra have been published for a number of the chloro and bromo derivatives and are reproduced in Figs. 8 and 9.[10-12,26] Isotopic abundances, either preferably at the molecule ion or in the principal aromatic fragment arising through loss of $(NH_2=CHCO_2H)^+$, give signature to the number and diversity of chlorine and bromine atoms present.

Biosynthesis

As yet there have been no serious enzymological studies, known to this author, of the halogenation process in invertebrate proteins.

The general mechanisms of tyrosine halogenation in proteins, so far as is known, seem to follow broadly similar routes. Halogenations studied appear always to be peroxidative processes with halide ion being oxidized to active halogen, hydrogen peroxide being the oxidant. Four peroxidases are known where halogenation is likely to be a specific function.

Chloroperoxidase is a bacterial enzyme capable of peroxidative halogenation with chloride as the halogen donor (although bromide and iodide are also utilizable) in the presence of hydrogen peroxide.[45,46] The general form of the reaction for this and other haloperoxidase activities is

$$X^- + H_2O_2 + \text{H-acceptor} \rightarrow \text{X-acceptor} + {}^-OH + H_2O$$

The enzyme is not particularly specific in relation to acceptor molecules; β-keto acids, β-diketones, halide ions (molecular halide being formed), and aromatic enols will all serve. Tyrosine, however, is a good acceptor. Additionally, halogen-dependent but nonhalogenative oxidations may be catalyzed and non-halogen-dependent oxidation of phenols is also mediated.

Lactoperoxidase is an enzyme occurring in raw milk and salivary, lacrimal, and Harderian glands.[47] Its general properties are similar to those of chloroperoxidase, although here iodide appears to be the preferred halide ion. In the presence of hydrogen peroxide, lactoperoxidase will form molecular iodine from iodide, iodinate tyrosine to monoiodo-

[44] J. H. Beynon, "Mass Spectrometry and Its Applications to Organic Chemistry," p. 297. Elsevier, Amsterdam, 1960.
[45] L. P. Hager, this series, Vol. 17, p. 648.
[46] P. F. Hallenberg and L. P. Hager, this series, Vol. 52, p. 521.
[47] M. Morrison, this series, Vol. 17, p. 653.

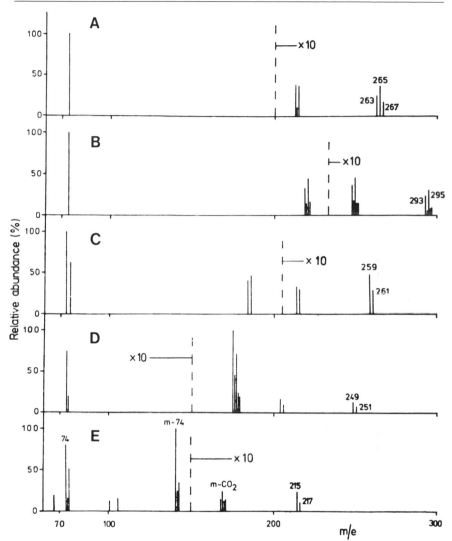

FIG. 8. Mass spectra of halogenated tyrosines. (A) 3,5-Dibromotyrosine; (B) 3-chloro-5-bromotyrosine; (C) 3,5'-dichlorotyrosine; (D) 3-bromotyrosine; and (E) 3-chlorotyrosine. The ×10 marker represents a relative abundance scale expansion above the indicated *m/e* position. 3,5-Dibromotyrosine does not give a molecular ion. Mass spectra are obtained by introduction of underivatized material. Reproduced, with permission, from Welinder.[12]

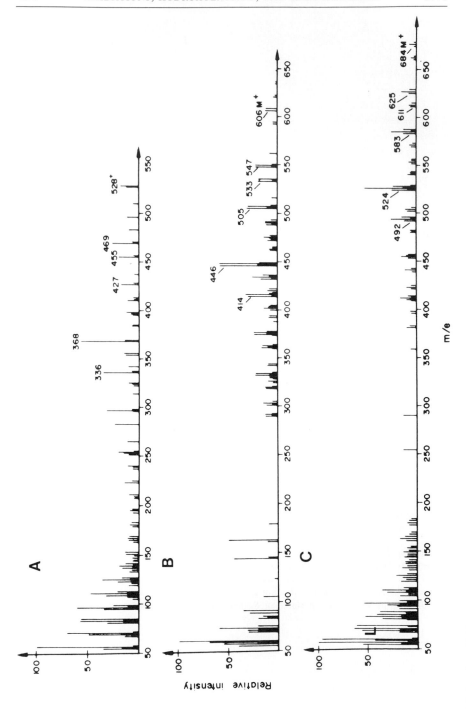

tyrosine, and oxidize monophenols to diphenols. Significantly, histidine as well as tyrosine in protein is iodinated by lactoperoxidase.[48]

The production of the iodothyronine, thyroxine, and iodohistidine residues of thyroid proteins involves the activity of thyroid peroxidase.[49] In exact analogy to chloro- and lactoperoxidase, iodide anion is oxidized to iodine or possibly iodinium (I^+) or hypoiodite (IO^-)[50] ions, in the presence of hydrogen peroxide. Tyrosine is then iodinated. The same enzyme or another peroxidase seems to be involved in the subsequent coupling reaction between two iodotyrosines to form triiodothyronine or thyroxine.[51–53] The thyroid peroxidase system is the only haloperoxidase apparently capable of going on to the production of the substituted dityrosine ether system. It is significant that natural chloro or bromo analogs of thyroxine and triiodothyronine have not been found.

Myeloperoxidase found in polymorphonuclear leukocytes again utilizes hydrogen peroxide and halide ion to generate active chlorinating agents, presumably with a bactericidal function.[54] Here hypochlorous acid is the initial product, which may perhaps then give rise to secondary chlorinating agents such as molecular chlorine or N-chloroamides and amines.[54–57]

All four enzymes are hemoproteins and subject to the action of inhibitors such as cyanide, fluoride, and azide, as might be expected for proteins containing a heme group.[45–57]

[48] T. J. Mueller and M. Morrison, *J. Biol. Chem.* **249**, 7568 (1974).
[49] M. Morrison and D. J. Danner, this series, Vol. 17, p. 658.
[50] J. M. Strum, J. Wicken, J. R. Stanbury, and M. J. Karnovsky, *J. Cell Biol.* **51**, 162 (1975).
[51] A. Taurog, M. L. Lophrop, and R. W. Estabrook, *Arch. Biochem. Biophys.* **139**, 221 (1970).
[52] J. Pommier, D. Deme, and J. Nunez, *Eur. J. Biochem.* **37**, 406 (1973).
[53] L. Lamas, M. L. Dorris, and A. Taurog, *Endocrinology* **90**, 1417 (1972).
[54] M. Morrison and G. R. Schonbaum, *Annu. Rev. Biochem.* **45**, 861 (1976).
[55] D. C. Hohn and R. I. Lehrer, *J. Clin. Invest.* **55**, 707 (1975).
[56] J. E. Harrison and J. Schultz, *J. Biol. Chem.* **251**, 1371 (1976).
[57] C. S. Foote, T. E. Goyne, and R. I. Lehrer, *Nature (London)* **301**, 715 (1983).

FIG. 9. Mass spectra of (A) dityrosine, (B) 3-monobromodityrosine, and (C) 3,3′-dibromodityrosine acetylated and methylated. Acetylation is by treatment with methanol–acetic anhydride (3 : 1, v/v), for 1 hr at room temperature. Methylation follows removal of the acetylation reagent under vacuum with solution in methanol and treatment with diazomethane in ether. Mass spectra are recorded on a Perkin-Elmer 270 mass spectrometer at 70 eV with direct sample injection and spectral recording at 100–200°. Reproduced, with permission, from Welinder *et al.*[26]

Although no specific halide peroxidases have yet been isolated from invertebrate tissues, the evidence from protochordate organs[58] producing thyroidlike hormones is suggestive of the action of a peroxidase or peroxidases similar to thyroid peroxidase. Thus, electron microscope histochemistry indicates that tunicate endostyle contains cellular peroxidase in considerable quantities.[58]

The production of thyroglobulins and related proteins is an intracellular event. Iodide must be accumulated, and the subsequent production of an activated species with consequent halogenation of protein and formation of triiodothyronine and thyroxine then takes place either within the subapical vesicles of the gland cells or at the apical plasma membrane prior to export in the usual manner.[50,58,59] In these respects vertebrate thyroidal and protochordate endostylar function are essentially similar. There is some evidence that in tunicate endostyle halogenation may continue within the vesicles after their release into the endostylar lumen.[59]

It is unfortunate that so little is known of the process of secretion of extracellular scleroprotein in the coelenterates, where traces of triiodothyronine and thyroxine are produced as protein components.[1,5] It is thus impossible to answer at present with any certainty the very important question as to whether iodination and tyrosyl coupling is an intra- or extracellular process or even if a peroxidase is involved. In *Eunicella* [131]I incorporation into gorgonin apparently takes place only in living tissue and is inhibited by thiocyanate, as in thyroid, but incorporation stops at mono- and diiodotyrosines; triiodothyronine and thyroxine are not produced.[60]

Most discussions of invertebrate scleroprotein halogenation are vague about mechanisms, but presume the process to be a secondary consequence of the cross-linking reactions.[9,11,12,26,61,62] This latter process is usually assumed (not always justified) to follow routes involving the oxidation of phenolic residues derived from tyrosine, dihydroxyphenylalanine, or some related phenol, e.g., dihydroxybenzoic acid.[63] Arthropod cuticles, for which most data exist, contain phenol oxidases and peroxidases that will convert diphenols to quinones and tyrosine to di- or trityrosyls.[26,34,63–65] It is therefore possible that these oxidases, intermediates

[58] A. Thorpe and M. C. Thorndyke, *Symp. Zool. Soc. London* **36**, 159 (1975).
[59] H. Fujita, *Acta Histochem. Cytochem.* **5**, 207 (1972).
[60] J. Roche, S. Andre, and G. Salvatore, *Comp. Biochem. Physiol.* **1**, 286 (1960).
[61] W. Tong and I. L. Chaikoff, *Biochim. Biophys. Acta* **48**, 347 (1961).
[62] E. J. W. Barrington, *Symp. Zool. Soc. London* **36**, 129 (1975).
[63] P. C. J. Brunet, *Insect Biochem.* **10**, 467 (1980).
[64] K. Bhagrat and D. Richter, *Biochem. J.* **32**, 1397 (1938).
[65] H. Hurst, *Nature (London)* **156**, 194 (1945).

formed in the oxidation process, or quinones themselves could oxidize halide to active halogen. Bivalve mollusc periostracum and byssus scleroproteins are produced by glandular tissues that also contain phenol oxidases.[66-69] Some periostraca, and certainly byssus, contain halogenated tyrosine and tryptophan.[28,70,71] The observation that thiocyanate or thiourea inhibits halogenation in these situations[71] argues against the quinone oxidation of halide theory and argues for an enzymatic involvement, but does not necessarily mean that halogenation is a cellular process. Phenol oxidase is present in periostracum.[66] No iodine is found in byssus during secretion, which, like mytilid periostracum, takes up iodine only in regions exposed to the surrounding seawater and seems to be unaffected by iodine absorbed through the tissues.[72] This situation is paralleled by the iodine binding shown by ascidian tunic[62] and by the posttanning appearance of chlorotyrosine in insect cuticle.[9]

This leads to the view that scleroproteins that are laid down and stabilized extracellularly are also halogenated extracellularly by enzymes secreted for stabilization processes and to the division of halogenation into two categories[62]: the specific and intracellular leading to the formation of thyroid-type derivatives, and the nonspecific and extracellular leading to production of various halogenated monotyrosines.

This hypothesis is partly upheld by the distribution of halogenoamino acids in proteins. Thus, bromo- followed by chlorotyrosines are certainly quantitatively most significant in scleroproteins.[9-12,26] It should be remembered that while the literature concentrates upon iodo derivatives, because of the comparative aspects of thyroidal evolution, these are often present only in traces whose significance has been accentuated through use of ^{125}I and ^{131}I incorporation as a route to detection of small quantities. A deceptive picture of relative amounts of halogenated amino acid may therefore be presented. There is a considerable need for critical quantitative study. The quantitative importance of chloro and bromo substitution is understandable in an extracellular context. Iodide concentration is environmentally low in concentrations (0.39×10^{-6} M in seawater), whereas bromide and chloride are much higher (0.8×10^{-3} M and 0.54 M, respectively, in seawater).[12] The dominance of bromotyrosines can be seen as

[66] J. H. Waite and K. M. Wilbur, *J. Exp. Zool.* **195,** 359 (1976).

[67] S. P. Kapur and M. A. Gibson, *Can. J. Zool.* **46,** 165 (1968).

[68] A. Bolognani-Fantin, J. P. Pujol, J. Bouillon, J. Bocquet, and M. V. Gervaso, *Bull. Soc. Linn. Normandie* **104,** 194 (1973).

[69] L. D. Gruffydd, *Zool. Scri.* **7,** 277 (1978).

[70] S. Hunt, unpublished studies of molluscan periostraca, 1980.

[71] J. Roche, S. Andre, and I. Covelli, *C. R. Seances Soc. Biol. Ses Fil.* **154,** 2201 (1960).

[72] G. E. Beedham and E. R. Trueman, *Q. J. Microsc. Sci.* **99,** 199 (1958).

being a function of relative ease of oxidation of the available halide ions. Thus, if halogenation is not taking place intracellularly, after membrane-mediated transport and concentration of iodide, as in thyroid, we might expect to see low levels of incorporation. And indeed in freshwater and terrestrial species, where environmentally bromide also is low, we see scleroproteins containing only chlorine.[9,13]

Aside from other factors, it may also be of some moment to restrict the possibility of random halide oxidation to the extracellular situation, since the chemical activity of oxidized halide intermediates with free access to the cytoplasm would be too great to be permitted.[73]

The nonspecific mechanism is not a wholly satisfactory explanation for the constant ratios of the different chloro and bromo derivatives that occur in, for example, operculins,[10,11,27] although this may be a function of protein conformation and tyrosyl environments.

Finally, a wider view of halogenation in the invertebrates must be taken, and protein derivatization viewed in such a context. A vast diversity of halogenated organic molecules including tyrosine-derived phenols and indoles are found in invertebrates.[74-80] Their mode of halogenation is again uncertain, although peroxidase-catalyzed bromination of p-hydroxybenzoic acid has been suggested as a starting point for the production of bromophenols.[80] Posttranslational modification of proteins produced by these same organisms may be a consequence of enzyme activities associated with the biosynthesis of secondary metabolites.

[73] A. van Zyl and H. Edelhoch, *J. Biol. Chem.* **242,** 2423 (1967).
[74] G. Cimino, S. de Stefano, L. Minale, and G. Sodano, *Comp. Biochem. Physiol. B.* **50B,** 279 (1975).
[75] G. E. Krejcarek, R. H. White, L. P. Hager, W. O. McClure, R. D. Johnson, K. L. Rhinehart, J. A. McMillan, and I. C. Paul, *Tetrahedron Lett.* **8,** 507 (1975).
[76] G. Cimino, S. de Stefano, and L. Minale, *Experientia* **31,** 756 (1975).
[77] T. Higa and P. J. Scheuer, *Tetrahedron* **31,** 2379 (1975).
[78] T. Higa, T. Fujiyama, and P. J. Scheuer, *Comp. Biochem. Physiol. B* **65B,** 525 (1980).
[79] S. L. Manley and D. J. Chapman, *FEBS Lett.* **93,** 97 (1978).
[80] Y. M. Sheikh and C. Djerassi, *Experientia* **31,** 265 (1975).

[28] Protein Bromination by Bromoperoxidase from *Penicillus capitatus*

By J. A. MANTHEY, L. P. HAGER, and K. D. McELVANY

Halogenation of protein molecules can be carried out under a variety of chemical and enzymatic reaction conditions.[1-5] The resulting differences in the labeled proteins often reflect the strengths and weaknesses of the approaches used.[6-8] The most notable difference between the systems lies with the sites of attachment of the halogen atom and the subsequent stability of the protein–halogen bond.[9] Hydrogen ion concentration is perhaps the main determining factor that affects the site of halogenation. Halogenation at acidic or basic pH values enhances the attachment of the halogen atom to the sulfur atom of cysteine residues, whereas halogenation at neutral pH occurs predominantly on the aromatic ring of tyrosine residues. Since the cysteine–halogen bond is extremely labile, even at neutral pH, cysteine residues do not represent a useful labeling site. At neutral pH values, where halogenation of tyrosine residues is favored, a relatively stable carbon–halogen bond is formed.[9,10] Therefore, optimal labeling conditions are those that minimize cysteine halogenation while enhancing halogenation of tyrosine residues.

In the field of nuclear medicine, there is considerable interest in the use of radioactively labeled proteins to study biological processes *in vivo*. Efficient techniques are needed that will readily label proteins to a high

[1] S. L. Karonen, P. Mörsky, M. Siren, and U. Seuderling, *Anal. Biochem.* **67,** 1 (1975).

[2] J. J. Marchalonis, *Biochem. J.* **113,** 299 (1969).

[3] W. M. Hunter and F. C. Greenwood, *Nature (London)* **194,** 495 (1962).

[4] A. S. MacFarlane, *Nature (London)* **182,** 53 (1958).

[5] J. F. Harwig, R. E. Coleman, S. S. L. Harwig, L. A. Sherman, B. A. Siegel, and M. J. Welch, *J. Nucl. Med.* **16,** 756 (1975).

[6] K. A. Krohn, L. C. Knight, J. F. Harwig, and M. J. Welch, *Biochim. Biophys. Acta* **490,** 497 (1977).

[7] J. Metzger, R. Secker-Walker, K. Krohn, M. Welch, and E. Potchen, *J. Lab. Clin. Med.* **82,** 267 (1973).

[8] B. L. Wajchenber, H. Pinto, I. Torresde Toledoe Souza, A. C. Lerario, and R. R. Pieroni, *J. Nucl. Med.* **19,** 900 (1978).

[9] L. C. Knight and M. J. Welch, *Biochim. Biophys. Acta* **534,** 185 (1978).

[10] K. D. McElvany, L. C. Knight, M. J. Welch, J. F. Siuda, R. F. Theiler, and L. P. Hager, "Marine Algae in Pharmaceutical Science" (H. H. Hoppe, T. Leurng, and Y. Tanaka, eds.), p. 429. de Gruyter, Berlin, 1978.

specific radioactivity and yet retain the protein's biological activity. In addition, it is often convenient to have the label detected and quantitated externally from the animal or patient. The iodine isotopes, ^{125}I and ^{131}I, are perhaps the most commonly used radiolabels attached to proteins. However, the use of the iodine isotopes poses some disadvantages for patient application. The carbon–iodine bond is subject to slow hydrolysis that results in liberation of ^{125}I and ^{131}I *in vivo*. The released iodide subsequently accumulates in the thyroid gland, and, in addition, ^{131}I emits high levels of damaging β particles during decay.

The use of ^{77}Br presents an excellent alternative to ^{125}I and ^{131}I. Not only does the carbon–bromine bond hydrolyze more slowly than the corresponding carbon–iodide bond, but also the decay of ^{77}Br produces very little particle emission and bromide ion does not accumulate in the thyroid gland. The short half-life of ^{77}Br, 2.3 days, also decreases exposure time, which serves to decrease the overall patient radiation dosage.[10] The problem that ^{77}Br presents is 2-fold. The first problem is the lack of ready availability of the isotope. Second, many of the chemical techniques used for the ^{77}Br attachment to the protein produce strong oxidizing conditions that can lead to protein denaturation.

One technique used to avoid the denaturation problem has been the indirect bromination of proteins by the attachment of the radiobrominated acylating agent, SHPP [N-succinimidyl-3-(4-hydroxyphenyl)propionate].[9,11,12] This technique produces labeled proteins that are stable against debromination. The site of attachment is predominantly at lysine residues, with smaller amounts of labeling at histidine and tyrosine residues.[9] Although the problems with this technique are not severe, the success of the indirect bromination technique depends largely on sterically accessible lysine side chains. Moreover, the attachment of such large reporter molecules to the protein may affect its biological activity.

An alternative to indirect bromination using SHPP is catalytic halogenation by peroxidases. Both horseradish peroxidase and lactoperoxidase are widely used for protein iodination. The enzymatic halogenation labeling procedure has been greatly simplified by the attachment of the peroxidases to a solid-phase support.[1,13,14] Lactoperoxidase (LPO) is more commonly used, since it is more active toward protein halogenation than horseradish peroxidase. Thyroglobulin and albumin have been suc-

[11] A. E. Bolton and W. M. Hunter, *Biochem. J.* **133,** 529 (1973).
[12] L. C. Knight, S. S. L. Harwig, and M. J. Welch, *J. Nucl. Med.* **18,** 282 (1977).
[13] K. McElvany, "Studies on the Radiohalogenation of Proteins Using the Enzymes Bromoperoxidase and Myeloperoxidase." Ph.D. Thesis, Washington University, 1980.
[14] R. G. Parsons and R. Kowal, *Anal. Biochem.* **5,** 568 (1979).

cessfully brominated at low pH using chloroperoxidase (CPO).[15] However, the low pH needed for CPO activity limits the selection of proteins which remain stable under these adverse reaction conditions. In addition, much of the label attached using CPO is readily released as free bromide owing to the predominant bromination of cysteine residues at low pH.

Myeloperoxidase (MPO) and bromoperoxidase (BPO), isolated from crude algal extracts of *P. capitatus,* provide excellent systems for enzymatic radiolabeling of proteins.[10,13,16,17] Unlike CPO, BPO and MPO are active at neutral pH. Both MPO and the BPO preparations catalyze bromination mainly at tyrosine residues in canine fibrinogen and human serum albumin (HSA).[13,16,17] The brominated canine fibrinogen retains all the clotting activity associated with the native protein, and no detectable change in biological activity occurs upon bromination of HSA. Thus it appears that these enzymes provide a means by which [77]Br may be rapidly and stably attached to protein molecules without serious deleterious effects.

Enzymatic Bromination of Proteins

Procedure. The usual test reaction mixture contains either 0.2 unit of BPO or 8 μg of MPO, 25 μg of the test protein, 20 nmol of H_2O_2, and 2 μCi of carrier-free [77]Br containing approximately 0.02 pmol of bromide in a total volume of 0.25 ml. The reaction is initiated by the addition of H_2O_2 and is incubated at 37° for 30 min. For routine assay purposes the reaction is terminated by the addition of 1.0 ml of 10% trichloroacetic acid (TCA). The solution is then centrifuged for 1 min to separate out the labeled proteins. The pellet is washed with 1 ml of 1 N NaOH and counted. The measure of percentage of labeling is determined from Eq. (1).

% Protein-bound radioactivity =

$$\frac{\text{activity (TCA ppt)}}{\text{activity (TCA ppt + supernatant + wash)}} \times 100 \quad (1)$$

The test system described above can be easily scaled up to larger volumes in order to prepare substrate amounts of a brominated protein. In the preparatory procedure the trichloroacetic acid precipitation step is re-

[15] L. Knight, K. A. Krohn, M. J. Welch, B. Spomer, and L. P. Hager, "Radiopharmaceuticals" (G. Subramanian, B. A. Rhodes, J. F. Cooper, and V. J. Sodd, eds.), p. 149, 1975.
[16] K. D. McElvany and M. J. Welch, *J. Nucl. Med.* **21,** 953 (1980).
[17] K. D. McElvany, J. W. Barnes, and M. J. Welch, *Int. J. Appl. Radiat. Isot.* **31,** 679 (1980).

placed by a suitable isolation procedure that separates the brominated protein from the other components of the reaction mixture. The separation of HSA from small molecules has been accomplished by the use of gel filtration chromatography.[13] For canine fibrinogen, separation was achieved by a 25% ammonium sulfate precipitation step.[13]

Assay Method for Bromoperoxidase

Principle. The assay for bromoperoxidase (BPO) activity parallels that previously developed for chloroperoxidase.[18] The assay utilizes the loss of absorbance at 278 nm upon the halogenation of 1,1-dimethyl-4-chloro-5,5'-cyclohexanedione [monochlorodimedon (MCD)] to the dihalo derivative. Hydrogen peroxide serves as the oxidant to form a reactive enzyme-bound brominating species that will then react with the nucleophilic acceptor to transfer bromine from the enzyme to the acceptor molecule. The time course for the assay is dependent upon the order of addition of the substrates. Initiation of the reaction with bromide ion yields a linear assay curve for the loss of absorbance as a function of time. Initiation of the reaction with H_2O_2 or BPO results in nonlinear assay curves.

Reagents

Potassium phosphate buffer, 0.10 M, pH 6.8, containing 80 μM MCD
Potassium bromide, 3 M
Hydrogen peroxide, 0.01 M

Procedure. The reaction mixture contains in a total volume of 2.7 ml, 2.5 ml of 0.10 M potassium phosphate buffer, pH 6.8, and 80 μM MCD, 0.05 ml of 0.01 M H_2O_2, 0.05 ml of BPO, and 0.10 ml of 3 M potassium bromide. All the reagents except potassium bromide are mixed in a quartz cuvette, and the initial absorbance at 278 nm is recorded. The addition of bromide ion initiates the reaction, and the rate is measured as the loss in absorbance as a function of time. The MCD extinction coefficient at 278 nm is 12,200 M^{-1} cm^{-1}. Activity units are defined as the formation of 1 μmol of monobromomonochlorodimedon per minute under standard assay conditions.

Purification of Bromoperoxidase from *P. capitatus*

The enzyme is purified in a manner similar to a previously published procedure.[19] Several changes have been introduced to increase the recov-

---

[18] D. R. Morris and L. P. Hager, *J. Biol. Chem.* **241,** 1763 (1966).
[19] J. A. Manthey and L. P. Hager, *J. Biol. Chem.* **256,** 11232 (1981).

ery of enzyme and decrease the overall time required for purification. In the revised procedure, step gradients have replaced the original continuous gradients, and the column dimensions have been changed to allow for greater flow rates. A Sephadex G-200 column has also been added as a final step in the purification procedure.

Step 1. Preparation of Crude Extract. Penicillus capitatus is collected in shallow water along the Florida Keys and is frozen and stored at −20°. A routine enzyme preparation involves the homogenization of approximately 2.5 kg of algae. Approximately 200-g aliquots of the algae are homogenized for 3 min in a stainless steel beaker at 4° using a Willem Polytron homogenizer. The resulting homogenates are pooled, filtered through cheesecloth, and centrifuged at 16,000 *g* at 4° for 30 min. The supernatant fraction (approximately 6 liters) is concentrated to a volume of 1.5 liters using an Amicon hollow-reed concentrator.

Step 2. Precipitation with Ammonium Sulfate. The 25% ammonium sulfate precipitation step formerly used is now omitted owing to the small amount of purification associated with this step. Instead, in the new procedure, ammonium sulfate is slowly added to the concentrated extract to achieve 60% saturation. This solution is stirred for 2 hr at 0°. Longer exposure to 60% ammonium sulfate decreases the recovery. The precipitated enzyme is pelleted by centrifugation at 16,000 *g* for 45 min at 0°. The pellets are redissolved in approximately 200 ml of 0.25 *M* Tris-chloride buffer, pH 7.0. The resuspended enzyme solution is centrifuged at 115,000 *g* for 45 min in order to remove undissolved particulate material. The resulting clear supernatant is dialyzed overnight against 2 liters of 0.10 *M* Tris chloride buffer, pH 7.0.

Step 3. Chromatography on DEAE-Sephadex. The dialyzed supernatant fraction is applied to a DEAE-Sephadex A-50 column, 17 × 4 cm, previously equilibrated with 0.25 *M* Tris-chloride buffer, pH 7.0. After the enzyme is applied, the column is washed with 1.0 liter of 0.30 *M* Tris-chloride buffer, pH 7.0. This washing is followed by elution of the column with 500 ml each of 0.40 *M* and 0.50 *M* Tris-chloride buffer, pH 7.0. The fractions containing bromoperoxidase activity are pooled and concentrated to approximately 200 ml with an Amicon concentrator using a PM-10 membrane. This solution is dialyzed against 2 liters of 0.25 *M* Tris-chloride buffer, pH 7.0, at 4° and is then applied to a second DEAE-Sephadex A-50 column, 8 × 3 cm, which has been preequilibrated with 0.25 *M* Tris-chloride buffer, pH 7.0. This second column is sequentially washed and eluted at a flow rate of 60 ml/hr with 500 ml each of 0.30, 0.40, and finally 0.50 *M* Tris-chloride buffer, pH 7.0. The fractions containing bromoperoxidase activity are pooled and concentrated to approximately 7 ml. This concentrated solution is dialyzed overnight at 4° against 1 liter of 0.10 *M* potassium phosphate buffer, pH 7.0.

TABLE I
PURIFICATION OF BROMOPEROXIDASE FROM *P. capitatus*

Step	Volume (mg/ml)	Protein (mg/ml)	Total activity[a] (units)	Specific activity (units/mg)	Purification (fold)
1. Crude extract	1800	0.76	1188	0.87	1
2. 60% ammonium sulfate	235	2.90	478	0.07	—
3. First DEAE-Sephadex	80	1.07	862	10.07	12
4. Second DEAE-Sephadex	130	0.16	558	26.8	31
5. Sephadex G-200	31	0.30	400	43.0	49

[a] These activity values differ from those reported earlier.[19] The previous values had been incorrectly calculated.

Step 4. Chromatography on Sephadex G-200. The dialyzed solution from step 3 is applied to a Sephadex G-200 column, 100 × 2.5 cm, that has been preequilibrated with 0.10 M potassium phosphate buffer, pH 7.0. The column is run in an ascending mode at a flow rate of 10 ml/hr, and the active fractions are collected. Depending on the enzyme purity at this step, the enzyme may be reapplied to the G-200 column. A summary of a typical purification run is listed in Table I.

Properties of Protein Bromination

Both MPO and BPO are similar in their requirements for optimal bromination conditions. For both, the optimum concentration of HSA in the radiolabeling reaction is approximately 0.20 mg/ml. On either side of this optimum concentration the percentage of labeling sharply decreases. The decreased labeling efficiency at the higher protein concentrations is probably due to the fact that tyrosine residues can serve both as a bromination and a free radical peroxidation site. At high tyrosine concentrations, the free radical reaction can effectively compete with the bromination reaction.

The percentage of labeling is also strongly dependent on the H_2O_2 concentration. As with most peroxidases, levels of H_2O_2 on either side of the optimum concentration lead to sharply decreased activity. The pH optimum with BPO is 6.5–6.8, whereas for MPO the optimum centers around 6.0. For BPO, this pH profile parallels that shown for the MCD assay. Since both of these enzymes can be used effectively at neutral pH values, bromination principally at tyrosine residues in the protein is ensured.

TABLE II
DISTRIBUTION OF BROMINATED AMINO ACID RESIDUES

	Bromoperoxidase		Myeloperoxidase	
Residue	Human serum albumin	Canine fibrinogen	Human serum albumin	Canine fibrinogen
Br⁻	5.0 ± 0.5	4.6 ± 0.5	3.5 ± 0.3	5.1 ± 0.2
Bromohistidine	10.1 ± 0.9	2.6 ± 0.4	8.7 ± 3.0	2.7 ± 1.8
Bromotyrosine	81.4 ± 0.7	89.4 ± 1.1	8.7 ± 2.0	92.2 ± 2.1
Dibromotyrosine	3.5 ± 0.6	3.4 ± 0.9	—	—

The bromination sites in both HSA and canine fibrinogen were determined by HPLC analysis of Pronase digests of each protein.[13] The relative levels of each brominated amino acid are presented in Table II. In each case, there is only a small amount of unstable [77]Br found in the Pronase digests. The free [77]Br is thought to come from the hydrolysis of labile sulfenyl halide bonds. Of the stably bound isotopes, the major portion of the bromine was attached to tyrosine residues.

These results demonstrate the effectiveness of peroxidative bromination of both HSA and canine fibrinogen by both MPO and BPO. Bromine-77 presents a desirable alternative to the commonly used iodine radionucleotides. The enzymatic bromination systems achieve both high specific radioactivity and minimal alteration of the test protein.

[29] Iodination by Thyroid Peroxidase

By JOSEPH T. NEARY, MORRIS SOODAK,* and FARAHE MALOOF

Iodination of tyrosyl moieties in thyroglobulin and other proteins is catalyzed by the thyroid peroxidase (TPO), an integral membrane, heme glycoprotein. Tyrosine and tyrosyl peptides can also be iodinated by TPO. The iodination reaction requires H_2O_2 (or a peroxide-generating source), I , an enzyme-associated iodinating intermediate (TPO-I_{oxid}),

* Deceased, September 5, 1979. We wish to dedicate this article to our beloved collaborator, who exemplified the following quotation from a poem by Robert Frost: "We dance around in a ring and suppose, but the secret sits in the middle and knows."

and an iodide acceptor, such as free or protein-bound tyrosine. The general reaction sequence is as follows:

$$TPO + H_2O_2 \rightarrow TPO_{oxid}(Cpd\ I) + H_2O \tag{1}$$
$$TPO_{oxid}(Cpd\ I) + I^- \rightarrow TPO\text{-}I_{oxid}(I^+) \tag{2}$$
$$TPO\text{-}I_{oxid}(I^+) + acceptor \rightarrow I\text{-}acceptor + TPO \tag{3}$$

The TPO-catalyzed reaction produces both free and protein-bound mono- and diiodotyrosine. TPO can catalyze also the formation of tetraiodothyronine (thyroxine) by a coupling mechanism.[1]

The enzyme is isolated from thyroid particulate material that sediments upon centrifugation for 1 hr at 105,000 g. Iodotyrosine deiodinase is also isolated from this particulate fraction[2] and is probably an integral membrane protein that may, together with TPO, regulate iodotyrosine concentrations in the thyroid. As indicated in Table I, TPO has been solubilized by proteolytic, detergent, and combined proteolytic and detergent techniques.[3-14] Solubilized forms of the enzyme have been purified to varying degrees of purity; molecular weight (M_r) estimates have ranged from 50,000 to 450,000. Preparations with the highest specific activity have been obtained by purifying the enzyme that results from combined trypsin and detergent treatment; the molecular weight of TPO prepared by these procedures ranges from 92,000 to 103,000.[5,9,11,12] However, proteolytic treatment of thyroid membranes has been shown to yield fragmented proteins,[15] and it is now generally agreed that trypsin treatment probably affects the size of TPO.[5,9,12] In order to obtain an intact TPO, we have developed a procedure for solubilizing

---

[1] J. Nunez, this volume [30].

[2] I. N. Rosenberg and A. Goswani, this volume [31].

[3] C. C. Yip, *Biochim. Biophys. Acta* **96,** 75 (1965).

[4] C. C. Yip, *Biochim. Biophys. Acta* **128,** 262 (1966).

[5] N. M. Alexander, *Endocrinology* **100,** 1610 (1977).

[6] T. Hosoya and M. Morrison, *J. Biol. Chem.* **242,** 2828 (1967).

[7] D. J. Danner and M. Morrison, *Biochim. Biophys. Acta* **235,** 44 (1971).

[8] A. Taurog, M. L. Lothrop, and R. W. Estabrook, *Arch. Biochem. Biophys.* **139,** 221 (1970).

[9] A. B. Rawitch, A. Taurog, S. B. Chernoff, and M. L. Dorris, *Arch. Biochem. Biophys.* **194,** 244 (1979).

[10] S. Ohtaki, A. Nakagawa, M. Nakamura, and I. Yamazaki, *J. Biol. Chem.* **257,** 761 (1982).

[11] J. Pommier, S. de Prailanne, and J. Nunez, *Biochimie* **54,** 483 (1972).

[12] J. Nunez, *in* "The Thyroid Gland" (M. DeVisscher, ed.), p. 45. Raven, New York, 1980.

[13] C. P. Mahoney and R. P. Igo, *Biochim. Biophys. Acta* **113,** 507 (1966).

[14] J. T. Neary, D. Koepsell, B. Davidson, A. Armstrong, H. V. Strout, M. Soodak, and F. Maloof, *J. Biol. Chem.* **252,** 1264 (1977).

[15] J. T. Neary, A. Armstrong, B. Davidson, F. Maloof, and M. Soodak, *Biochim. Biophys. Acta* **379,** 262 (1975).

TABLE I

METHODS OF SOLUBILIZATION, SPECIFIC ACTIVITIES UPON PURIFICATION, AND APPARENT MOLECULAR WEIGHTS OF THYROID PEROXIDASE (TPO) FROM VARIOUS ANIMAL SOURCES

Method of solubilization	Approximate M_r	Specific activity $\left(\dfrac{\mu\text{mol oxid. guaiacol/min}}{\text{mg protein}}\right)$	R_z (A 410/280)	Source of tissue	References[a]
Trypsin	50,000	—	—	Bovine	3, 4
Deoxycholate and trypsin	92,000	464	0.55	Bovine	5
Cholate followed by trypsin	104,000	48.9	—	Porcine	6
Cholate followed by trypsin	200,000	—	—	Porcine	7
Deoxycholate followed by trypsin	62,000	—	0.34	Porcine	8
Trypsin and deoxycholate	93,000	794	0.54	Porcine	9
Deoxycholate followed by trypsin	—	465	0.42	Porcine	10
Trypsin followed by digitonin	103,000	600–800	0.04–0.11	Porcine	11, 12
Deoxycholate	≤450,000	6.08	—	Bovine	13
Triton X-100	135,000	10.2	—	Porcine	14

[a] Numbers refer to text footnotes.

TPO with nonionic detergents.[16] When correction is made for bound detergent, TPO solubilized by Triton X-100 has an M_r of 135,000.[14] In a manner similar to that previously found for other integral membrane proteins, such as cytochrome b_5[17,18] and NADH cytochrome b_5 reductase,[19,20] it appears that proteolytic solubilization of TPO cleaves a hydrophilic, catalytically active heme and carbohydrate-containing portion of TPO from the membrane, whereas detergent solubilization yields an amphipathic protein containing the hydrophilic, catalytically active fragment and a hydrophobic portion (ca. 30,000–40,000 M_r fragment), which is responsible for the binding of TPO to thyroid membranes. Both forms of TPO catalyze the iodination of free and protein-bound tyrosine and peroxidation of chromogenic substrates such as guaiacol.

Assay Methods

During solubilization and purification, TPO activity can be conveniently monitored by a spectrophotometric guaiacol assay.[21,22] Peroxidative activity can also be determined by measuring the conversion of I^- to I_3^-.[23] Iodination activity is measured by trichloroacetic acid precipitation of an ^{125}I-labeled acceptor protein[24] or by spectrophotometric recording of the conversion of tyrosine to mono- and diiodotyrosine.[25,26] Thyroid homogenates contain ascorbic acid (10^{-3} M; 90.8% reduced), glutathione (10^{-3} M), and catalase (6×10^{-11} M), which interfere with oxidative and iodination assays.[27,28] Because of this, the particulate preparation is used as the starting material for determining recovery of TPO during purification.

[16] J. T. Neary, B. Davidson, A. Armstrong, H. V. Strout, F. Maloof, and M. Soodak, *J. Biol. Chem.* **251**, 2525 (1976).

[17] A. Ito and R. Sato, *J. Biol. Chem.* **243**, 4922 (1968).

[18] L. Spatz and P. Strittmatter, *Proc. Natl. Acad. Sci. U.S.A.* **68**, 1042 (1971).

[19] L. Spatz and P. Strittmatter, *J. Biol. Chem.* **248**, 793 (1973).

[20] K. Mihara and R. Sato, *J. Biochem. (Tokyo)* **78**, 1057 (1975).

[21] T. Hosoya, Y. Kondo, and N. Ui, *J. Biochem. (Tokyo)* **52**, 180 (1962).

[22] T. Hosoya, *J. Biochem. (Tokyo)* **53**, 381 (1963).

[23] N. M. Alexander, *Anal. Biochem.* **4**, 341 (1962).

[24] A. Taurog, *Arch. Biochem. Biophys.* **139**, 212 (1970).

[25] B. Davidson, M. Soodak, J. T. Neary, H. V. Strout, J. D. Kieffer, H. Mover, and F. Maloof, *Endocrinology* **103**, 871 (1978).

[26] B. Davidson, J. T. Neary, H. V. Strout, F. Maloof, and M. Soodak, *Proc. Int. Thyroid Conf. 7th, Int. Congr. Ser. Excerpta Med.* **378**, 143 (1976).

[27] M. Suzuki, M. Nagaskina, and K. Yamamoto, *Gen. Comp. Endocrinol.* **1**, 103 (1961).

[28] F. Maloof and M. Soodak, *Pharmacol. Rev.* **15**, 43 (1963).

Peroxidation of Guaiacol

Thyroid peroxidase, in the presence of H_2O_2, catalyzes the conversion of guaiacol to tetraguaiacol, an orange, oxidized product. This reaction can be recorded spectrophotometrically at 470 nm. The final conditions are 33 mM guaiacol, 0.3 mM H_2O_2, 47 mM Tris (pH 7.4), and a suitable amount of TPO to give a linear absorbance change for at least 30 sec. A guaiacol-Tris buffered solution is prepared daily by diluting 0.43 ml of guaiacol and 5 ml of 1 M Tris-HCl (pH 7.4) to 100 ml with distilled H_2O; this solution is stored in a brown bottle. If the commercial lot of guaiacol is not colorless, it must be distilled before use. To a 1-cm pathlength cuvette is added 2.8 ml of guaiacol–Tris solution and 0.1 ml of TPO solution. A baseline absorbance rate is recorded prior to adding H_2O_2. A change in absorbance before adding H_2O_2 has been observed with impure guaiacol or guaiacol–Tris solutions stored at room temperature for more than a day. The reaction is initiated by adding 0.1 ml of 9 mM H_2O_2 (prepared daily by diluting 0.1 ml of 30% H_2O_2 to 100 ml with H_2O). The absorbance change at 470 nm is recorded at ambient temperature. (If an excess of TPO has been used, a curved line will result during the initial reaction period.) The activity is calculated from the extinction coefficient of oxidized guaiacol (5.57 cm^{-1} mM^{-1}). One unit of guaiacol activity is defined as the amount of TPO that catalyzes the peroxidation of 1 μmol of guaiacol per minute.

Iodination of Thyroglobulin or Bovine Serum Albumin (BSA)

In the presence of I^- and H_2O_2 (either added directly or generated enzymatically or chemically), TPO catalyzes the iodination of tyrosyl moieties on proteins such as thyroglobulin and BSA.[24] For routine assays during purification, we have used BSA as an I^- acceptor.[16] The final reaction mixture contains 5 mg of BSA per milliliter, 0.1 mM KI, 50,000–100,000 cpm of $^{125}I^-$, 0.3 mM H_2O_2, 50 mM potassium phosphate buffer (pH 7.0), and a suitable amount of TPO (to give ≤5% incorporation of $^{125}I^-$ into protein) in a final volume of 1.0 ml. An assay solution is prepared from 310 mg of BSA, 0.625 ml of 10 mM KI, 6.25 ml of 0.5 M potassium phosphate buffer (pH 7.0), and 100 μl of 50 μCi/μl $^{125}I^-$ in a final volume of 50 ml. Tubes containing 800 μl of the assay solution and 100 μl of TPO solution are prepared, and iodination is initiated by adding 100 μl of 3 mM H_2O_2 (prepared by diluting 33 μl of 30% H_2O_2 to 100 ml with H_2O). Control tubes contain the complete reaction mixture minus H_2O_2. After 15 min at ambient temperature, 1 ml of ice-cold 10% trichloroacetic acid is added, and the tubes are vortexed, placed on ice for 15 min, and centrifuged in a Sorvall HL-8 rotor at 3000 rpm for 10 min at 5°.

The pellets are washed twice by resuspension in 1 ml of ice-cold 5% trichloroacetic acid and counted in an autogamma spectrometer. Total $^{125}I^-$ cpm are determined by counting 800 μl of the assay solution. A unit of iodination activity is defined as the amount of TPO that catalyzes the incorporation of 1 μmol of I^- into BSA per minute.

Identification of Iodinated Products

The protein-bound, iodinated tyrosines and thyronines can be identified by paper chromatography after Pronase digestion in the presence of 4 mM 1-methyl-2-mercaptoimidazole (MMI).[29] The tripeptide Glu-Tyr-Glu (apparent $K_m = 3 \times 10^{-6} M$) has also been used in TPO assays, and the iodinated product has been identified by thin-layer chromatography and autoradiography.[30,31]

Solubilization and Purification Procedures

Preparation of Thyroid "Microsomes"

Frozen porcine thyroid glands are routinely used for starting material, although the procedures described below have also been applied to fresh tissue. Since it is unlikely that clean subcellular organelles are isolated from frozen tissue, we use quotation marks when referring to the "microsomal" pellet. Porcine thyroid glands (2–3 kg) are placed in 3 liters of 1.15% KCl, $10^{-4} M$ KI and allowed to thaw overnight at room temperature. Accessible fat and blood clots are removed from the tissue, and the tissue is soaked in ice-cold 1.15% KCl, $10^{-4} M$ KI for 30–60 min. The cleaned tissue is ground in an Oster meat grinder and washed with 1.15% KCl, $10^{-4} M$ KI (1 part ground tissue to 3 parts solution) by centrifugation at 13,000 rpm in a Sorvall GSA rotor for 20 min at 5°. The upper layer of white fatty material is removed, and the solution is decanted without disturbing the brown layer on top of the crude pellet. The sedimented material is washed until the supernatant has changed from red to nearly colorless (usually 4–6 washes). A crude homogenate is obtained by suspending 1 part of washed tissue in 4 parts of 0.25 M sucrose, $10^{-4} M$ KI and homogenizing in a Sorvall Omni-Mixer at 80% full speed for a total homogenization period of 4 min at 5°. An on/off cycle of 15 sec of homogenization followed by 15 sec of rest is used to minimize overheating; the

[29] M. L. Coval and A. Taurog, *J. Biol. Chem.* **242,** 5510 (1967).
[30] M. M. Krinsky and J. S. Fruton, *Biochem. Biophys. Res. Commun.* **43,** 935 (1971).
[31] B. Davidson, J. T. Neary, H. V. Strout, F. Maloof, and M. Soodak, *Biochim. Biophys. Acta* **522,** 318 (1978).

cycle is repeated 16 times. The mixture is filtered through three layers of cheesecloth into a beaker placed in ice, and the residue is rehomogenized with 1 part of 0.25 M sucrose, 10^{-4} M KI and filtered as above. Both filtrates are combined to form the crude homogenate. The filtrate is centrifuged at 7500 rpm in a Sorvall GSA rotor for 15 min (all centrifugation steps are performed at 5°). The supernatant is decanted and saved, and the pellet is resuspended in 1100 ml of 0.25 M sucrose, 10^{-4} M KI, homogenized in a Sorvall Omni-Mixer at 80% full speed (12 cycles of 15 sec on/15 sec off), and centrifuged as before. The final pellet is resuspended in 1100 ml of 0.25 M sucrose, 10^{-4} M KI and centrifuged again. The three supernatants are combined and centrifuged in a Beckman No. 19 rotor at 19,000 rpm for 3 hr. The pellets are transferred to plastic petri dishes, frozen on dry ice, and stored at $-70°$. No TPO activity is detected in the final supernatant.

The yield of "microsomes" from 2–3 kg of thyroid glands is 110–160 g (wet weight). Total TPO activity has ranged from 6100 to 17,700 guaiacol units with an average of 10,600 guaiacol units for 10 preparations. The frozen "microsomes" retain about 85% of the initial TPO activity after 3–4 months. The TPO activity is more stable in the frozen "microsomes" than in frozen thyroid glands. Use of porcine thyroid glands stored frozen for 3–4 months has resulted in low yields of TPO activity in the "microsomes" and low specific activities of the final purified preparation.

Preparation of Proteolytically Solubilized TPO

Step 1. Removal of Loosely Bound and Occluded Proteins. Unless otherwise noted, all steps are performed at 5°. "Microsomes" (25 g wet weight) are thawed and homogenized in a motor-driven Teflon–glass homogenizer in 350 ml of 1 M NaCl, 10 mM Tris, 10^{-4} M KI, pH 7.4, stirred gently for 1 hr, and centrifuged at 105,000 g for 1 hr. The supernatant contains about 25% of the protein and ≤1% TPO activity. The pellet is washed twice in 10^{-4} M KI by resuspending it in a similar volume of 10^{-4} M KI, stirring gently for 1 hr, and centrifuging at 105,000 g for 1 hr. About 20% of the protein is removed in the KI washes; at least 95% of the TPO activity remains in the pellet. Hence, the hypertonic and hypotonic washes remove 40–45% of the protein while retaining at least 95% of the TPO activity in the pellet. Most of the thyroglobulin is removed in this step as determined by double immunodiffusion using anti-porcine thyroglobulin.

Step 2. Solubilization with Trypsin. The pellet from step 1 is resuspended in 275 ml of 5 mM CaCl$_2$, 10^{-4} M KI, 20 mM Tris, pH 8.0. Trypsin (5 μg/mg "microsomal" protein; chymotrypsin-free, Calbiochem) is added, and the suspension is stirred gently at 37° for 1 hr. Soybean trypsin inhibitor (2 mg per milligram of added trypsin) is added, stirring is continued for an additional 10 min, and the suspension is centrifuged at 105,000 g for 1 hr. The extent of TPO solubilization ranges from 30 to 60%.

Step 3. DEAE-Cellulose Column. The supernatant from step 2 is applied to a 1 × 40 cm DEAE-cellulose column (Whatman DE-52) equilibrated in 25 mM KCl, 10^{-4} M KI, 15 mM Tris, pH 7.4. The column is washed with 200–300 ml of the equilibration buffer, and TPO activity is eluted by applying a linear KCl gradient (25 to 225 mM KCl in 10^{-4} M KI, 15 mM Tris, pH 7.4). The total volume of the gradient is 500 ml. TPO activity elutes between 150 and 275 ml, with the peak of activity in the range of 0.13 to 0.15 M KCl. The pooled fractions are concentrated by ultrafiltration on a UM-20E membrane (Amicon).

Step 4. Sephacryl S-200 Chromatography. The concentrated fraction from step 3 is applied to a 1.5 × 80 cm Sephacryl S-200 (Pharmacia) column equilibrated in 25 mM sodium citrate–citric acid buffer, pH 5.6. Potassium phosphate buffer (50 mM), pH 7.2, can also be used at this step. (In this column, KI has been omitted in order to prepare I^--free TPO for use in iodinating intermediate studies. No loss in activity has been observed when I^- is omitted at this step; similar recoveries have been obtained when 10^{-4} M KI is included in the potassium phosphate, pH 7.2, equilibrating buffer.) TPO activity elutes between 60 and 80 ml, and the pooled fractions are concentrated by ultrafiltration on a UM-20E membrane. Aliquots are stored frozen at $-70°$. Similar results have been obtained with a Sephadex G-200 column.

A summary of the purification procedure is shown in Table II. For 10 preparations, the average yield of TPO activity was 30% (range = 11 to 46%). About 5 mg of protein can be obtained from 25 g of "microsomes," which represents about 400 g of frozen porcine thyroids. The specific activity of the purified enzyme has ranged from 70 to 200 units of guaiacol activity per milligram of protein. The specific activity of preparations in the lower end of the range can be increased approximately 2-fold by employing a second DEAE-cellulose column. The R_z value of the purified preparations ranges from 0.11 to 0.23. Based on electrophoresis studies, TPO obtained by this procedure is at least 25% pure.[14] On nondenaturing gels, the major protein band coincides with TPO activity and a carbohydrate stain.[14] The advantage of the procedure described here is that it gives a high yield of a relatively purified, stable enzyme in a few steps. The purification procedures described by Taurog,[9] Alexander,[5] and Nunez[1,11,12] and their associates yield preparations with higher specific activity than that reported here, although in some cases[5,11] the highly

TABLE II
PURIFICATION OF PROTEOLYTICALLY SOLUBILIZED TPO FROM ca. 400 g
OF FROZEN PORCINE THYROIDS

Step	Total protein (mg)	Total activity	Specific activity[a]	Yield (%)
"Microsomes"	2210	2424	1.10	100
NaCl-washed "microsomes"	1581	2405	1.52	99
KI-washed "microsomes"	1024	2350	2.29	97
Trypsin extract[b]	362	1100	3.04	45
DEAE eluate	77.6	950	12.2	39
Sephacryl-200 eluate	5.5	761	138	31

[a] Micromoles of guaiacol oxidized per minute per milligram of protein.
[b] Five micrograms of trypsin per milligram of "microsomal" protein.

purified preparations are not stable. Tyrosine–agarose affinity chromatography has been utilized to obtain a partially purified TPO.[32]

Preparation of Triton X-100-Solubilized TPO

Step 1. Washed "Microsomes." Unless otherwise mentioned, all procedures are conducted at 5°. The "microsomal" pellet is washed in hypertonic (1 M NaCl) and hypotonic (10^{-4} M KI) solutions to release loosely bound and soluble proteins, in particular thyroglobulin and residual hemoglobin. The "microsomal" pellet (15 g wet weight) is suspended in 1 M NaCl, 10^{-4} M KI, 10 mM Tris, pH 7.4 (ca. 1 g of particles per 20 ml of solution), and the washing is conducted as described in step 1 of the preceding section.

Step 2. Solubilization by Triton X-100. The washed pellet from step 1 is resuspended in 225 ml of 0.1% Triton X-100, 10^{-4} M KI, 10 mM Tris, pH 7.4, and the suspension is mixed gently overnight at 4° on a wrist-action shaker and centrifuged at 105,000 g for 1 hr. The supernatant is concentrated to 12–14 ml by ultrafiltration on a UM-20E membrane. The TPO activity solubilized by this procedure has ranged from 60 to 95% in 35 experiments using frozen thyroids obtained from abattoirs in Iowa and Pennsylvania. Similar amounts of TPO have been solubilized from "microsomes" prepared either from frozen or fresh thyroid tissue or from "microsomes" in which the 1 M NaCl wash step was omitted. The extent

[32] K. Yamamoto and L. J. DeGroot, *J. Biochem.* (*Tokyo*) **91**, 775 (1982).

of TPO solubilization is dependent on protein and detergent concentrations, pH, and time of mixing; optimum solubilization conditions are 0.1–0.2% Triton X-100, 0.7–2.0 mg of protein per milliliter, pH 7.4–8.4, and overnight mixing.[16] Brij,[16] another nonionic detergent, and deoxycholate[13,33] have also been used to solubilize TPO.

Step 3. Sepharose 6B Chromatography. The concentrated supernatant from step 2 is applied to a 2.6 × 90 cm Sepharose 6B column (Pharmacia) equilibrated in 0.1% Triton X-100, 10^{-4} M KI, 10 mM Tris, pH 7.4. The column is washed in the equilibrating buffer, and TPO activity elutes in two peaks; one peak (I) is in the region of void volume (155–170 ml), and the second peak (II) elutes between 185 and 220 ml. Peak I contains 25–30% of the recovered TPO activity, and peak II has 70–75% of the recovered activity. Peak II contains phospholipids.[34] It is concentrated by ultrafiltration on a UM-20E membrane and stored at −20°. The preparation retains 75% of its initial activity after 1 month of storage at −20 or −70° and 40% of its activity when stored at 4° for 1 month.

A summary of the purification procedure is shown in Table III. For 9 preparations, the average yield of TPO activity was 33% (range = 11 to 64%). The specific activity of the preparation has ranged from 4.3 to 10.2 units of guaiacol activity per milligram of protein. To date, it has not been possible to purify the detergent-solubilized TPO to the extent that has been achieved for the trypsin-solubilized TPO. The specific activity of the Sepharose 6B peak II, Triton X-100-solubilized TPO can be increased an additional 2-fold by sucrose density centrifugation (5 to 20% sucrose gradient, Beckman SW-41 rotor, 41,000 rpm, 17 hr); 75% of the activity is recovered in a region one-third of the distance from the top of the gradient. Con A–agarose has been used to obtain a partially purified, detergent-solubilized TPO.[33] In our laboratory, other attempts at purification have included additional gel filtration chromatography (Sephacryl S-200; Ultrogel AcA 34), ion-exchange chromatography [DEAE-cellulose; (carboxymethyl)cellulose], hydrophobic chromatography (octyl- and dodecylagarose), and immunoaffinity chromatography (antibodies to trypsin-purified TPO linked to agarose); none of these procedures gave a purification greater than 2- to 3-fold.

The purification of intact TPO has been difficult because of its integral membrane nature. TPO appears to be an amphipathic protein that is partially submerged in the endoplasmic membrane of the thyroid. The native protein, as solubilized by detergents from thyroid microsomes, has an M_r

[33] S. Ohtaki, H. Nakagawa, and I. Yamazaki, *FEBS Lett.* **109,** 71 (1980).
[34] J. T. Neary, B. Davidson, A. Armstrong, M. Soodak, and F. Maloof, *Fed. Proc. Fed. Am. Soc. Exp. Biol.* **35,** 1629 (1976).

TABLE III

PURIFICATION OF TRITON X-100-SOLUBILIZED THYROID PEROXIDASE
FROM ca. 250 g OF FROZEN PORCINE THYROIDS

Step	Total protein (mg)	Total activity	Specific activity[a]	Yield (%)
"Microsomes"	852	1123	1.32	100
NaCl-washed "microsomes"	647	1109	1.71	99
KI-washed "microsomes"	530	1080	2.04	96
Triton X-100 supernatant	200	865	4.33	77
Sepharose 6B peak II	40	340	8.50	30

[a] Micromoles of oxidized guaiacol per minute per milligram of protein.

of about 135,000. A heme protein preparation of $\leq 103,000\ M_r$ is released from the endoplasmic membrane by mild proteolysis of thyroid microsomes with trypsin. This preparation, which can be regarded as a fragmented TPO, is water soluble and contains the catalytic site, heme, and carbohydrate, but not the hydrophobic portion responsible for membrane attachment. In the near future, one might anticipate that the solubilization and purification of native TPO will be reminiscent of the solubilization and purification of cytochrome b_5, an integral membrane protein that has been purified from liver microsomes but is also present in thyroid microsomes (33 pmol per milligram of microsomal protein vs 66 pmol of TPO per milligram of microsomal protein).[35] From 1956 to 1968, proteolysis was the main treatment used to solubilize cytochrome b_5 from liver microsomes. This procedure yielded a fragmented protein with an M_r of 11,000.[36,37] When detergent solubilization was employed, cytochrome b_5 was reported to have an M_r of 25,000[17] and subsequently was shown to have an M_r of 16,700.[18,38] The detergent-solubilized protein contains an extremely hydrophobic segment of 40 amino acids that is not found in the fragmented protein. The intact cytochrome b_5 is an amphipathic molecule in which the nonpolar region promotes polymerization in aqueous solution and apparently provides for the firm attachment to the phospholipid matrix of the endoplasmic reticulum, thus orienting the hydrophobic heme-containing portion at the membrane surface. It seems reasonable to predict that as progress is made in purifying and characterizing the deter-

[35] T. Hosoya and M. Morrison, *Biochemistry* **6**, 1021 (1967),
[36] P. Strittmatter and S. F. Velich, *J. Biol. Chem.* **221**, 253 (1956).
[37] P. Strittmatter, *J. Biol. Chem.* **235**, 2492 (1960).
[38] P. Strittmatter, M. J. Rogers, and L. Spatz, *J. Biol. Chem.* **247**, 7188 (1972).

gent-solubilized TPO, a pattern similar to that of cytochrome b_5 will emerge.

Properties

Molecular Weight and Homogeneity. The molecular weight of TPO in highly purified preparations ranges from 92,000 to 103,000,[1,5,9,11] and the purity of one preparation has been estimated to be 80–95%.[9] It is generally agreed that the native enzyme is probably altered by proteolytic solubilization.[5,9,12] The amounts of trypsin (1 to 10 μg/mg protein) used to solubilize TPO lead to fragmentation of thyroid membrane proteins,[15] and the wide range of molecular weights (see Table I) is probably due to variations in the conditions of proteolytic solubilization.[5] Estimates of molecular weight of detergent-solubilized, partially purified TPO have ranged up to 450,000. When correction was made for the amount of bound Triton X-100,[39] a value of 135,000 was obtained for the detergent-solubilized enzyme.[14] Rawitch *et al.*[9] have reported that trypsin-solubilized TPO is composed of 2 subunits with apparent M_r of 60,000 and 24,000, although Alexander reported[5] that TPO did not disaggregate when analyzed by SDS–gel electrophoresis. Both intact and fragmented TPO contain carbohydrate as determined by binding to Con A,[14] and carbohydrate analysis indicates that TPO is a glycoprotein containing about 10% by weight of carbohydrate.[9] The isoelectric point of trypsin-solubilized TPO is 5.75.[9]

TPO Stability. TPO prepared in the laboratory of Taurog showed no significant losses in enzymatic activity after storage at −20° for 4 years.[9] Other highly purified TPO preparations are more labile,[5,11] and one was reported to be inactive after 2–3 days.[5] The trypsin-solubilized, partially purified TPO reported here is stable for several months at −70°, and the Triton X-100-solubilized TPO is stable for a month at −70°.

Over the years, investigators have used the substrate KI, SCN, or thiourea with the aim of preserving TPO activity.[28] It has been shown that at 37° 10^{-5} *M* KI protects against losses of tyrosine iodinating activity.[40] However, KI, which appears to protect TPO in the partially purified form, does not prevent loss of activity of the highly purified TPO.[5] Systematic studies are needed to determine the optimal conditions of TPO stability during storage. It should be noted that while low concentrations of KI

[39] S. Clarke, *J. Biol. Chem.* **250,** 5459 (1975).
[40] L. J. DeGroot, J. E. Thompson, and A. D. Dunn, *Endocrinology* **76,** 632 (1965).

may protect TPO, high concentrations of KI (4 mM) inhibit protein iodination.[24,41]

Ammonium sulfate has been used in the purification of TPO in some cases, but there are reports that addition of ammonium sulfate to some TPO preparations leads to loss of TPO activity.[6] It has been shown that a glucosidase from thyroid microsomes is activated by ammonium sulfate.[42] It would be of interest to determine whether the loss of activity in some TPO preparations is due to activation of a glucosidase, which in turn removes glycoid units from TPO.

Effects of Protease Treatment. It has been reported that trypsin treatment abolishes stimulation of iodide oxidative activation by hematin.[43] There is also a report that trypsin treatment of the partially purified TPO reduces iodinating activity more than peroxidative activity.[7]

pH Optima. Maximum guaiacol and iodide oxidation occurs at pH 7.4 and 7.0, respectively.[5] The pH optimum for peptide and protein iodination is 7.0.[31,44]

Catalytic Activities. K_m values are 10^{-4} M for protein iodination, 3×10^{-6} M for Glu Tyr Glu iodination, and 6×10^{-3} M for iodide oxidation.[9,11,31] The turnover number for guaiacol oxidation is 50,000 μmol/min per micromole of enzyme, and that for triiodide formation is 32,000 μmol/min per micromole of enzyme, based on an M_r of 92,000.[5] The detergent-solubilized TPO appears to require lipids for full expression of activity.[34]

Regulation of TPO Activity by Diiodotyrosine. Diiodotyrosine (DIT) can both inhibit and stimulate TPO activity, depending on the concentration of DIT.[45] Nunez and associates found that $\geq 10^{-6}$ M DIT inhibits both iodination and coupling, whereas 10^{-8} M DIT stimulates coupling but has no effect on iodination. Free DIT could be an endogenous regulator of thyroxine biosynthesis, especially since the maximum stimulatory concentration (10^{-7} M) is in the range of free DIT normally found in the thyroid. It may be more than coincidence that the enzyme that synthesizes DIT (TPO) and the enzyme that deiodinates DIT (iodotyrosine deiodinase) are integral membrane proteins anchored in the endoplasmic reticulum membrane. In addition, TPO is inhibited by reducing agents

[41] J. Pommier, D. Deme, and J. Nunez, *Eur. J. Biochem.* **37,** 406 (1973).

[42] R. G. Spiro, M. J. Spiro, and V. D. Bhoyroo, *J. Biol. Chem.* **254,** 7659 (1979).

[43] N. M. Alexander, *in* "Further Advances in Thyroid Research" (K. Fellinger and R. P. Höfer, eds.), Vol. 1, p. 423. Verlangder-Viker Mediginischen Akademie, 1971.

[44] A. Taurog, *Recent Prog. Horm. Res.* **26,** 189 (1970).

[45] D. Deme, E. Fimiani, J. Pommier, and J. Nunez, *Eur. J. Biochem.* **51,** 329 (1975).

TABLE IV
HEME GROUP AND NATURE OF BINDING OF THYROID PEROXIDASE
AND OTHER PEROXIDASES

Peroxidase	Heme group	Binding	References
Thyroid peroxidase	Protoporphyrin IX	Noncovalent	4, 6, 46, 47
Thyroid peroxidase	Not protoporphyrin IX	Noncovalent	8, 9, 44
Myeloperoxidase	Deuteroporphyrin IX	Covalent	48
Chloroperoxidase	Protoporphyrin IX	Noncovalent	49
Lactoperoxidase	Mesoprotoporphyrin IX	Covalent	50
Horseradish peroxidase	Protoporphyrin IX	Noncovalent	51, 52

whereas the deiodinase is stabilized by reducing agents.[2] Further studies are needed to investigate the possibility that these two enzymes act in concert to regulate iodothyronine concentrations.

Absorption Spectrum. There is general agreement that TPO contains a heme prosthetic group, but the nature of the heme group has not been resolved (see Table IV).[4,6,8,9,44,46–52] TPO has an absorption band at 413 nm[10] that is similar to that of lactoperoxidase and intestinal peroxidase but different from horseradish peroxidase and chloroperoxidase, which have Soret peaks at 403 nm. TPO also has absorption bands at 502, 541, 592, and 630 nm.[10] It has been shown that the detergent-solubilized enzyme reacts with H_2O_2 to form compounds **I**, **II**, and **III**.[33,53] Several investigators[4,6,46,47] have suggested that the heme group is protoporphyrin IX. However, Taurog and colleagues report differences between the pyridine hemochromogens of TPO and horseradish peroxidase and suggest that the heme in TPO is not protoporphyrin IX.[8,9,44] Most hemoproteins have a shift in their spectrum upon reduction, but the spectral data reported by Rawitch *et al.*[9] did not show the usual shift. It may be that the prosthetic group in TPO is protoporphyrin IX, but that there are anomalies in its binding to the protein.[10] In order to clarify the nature of the heme group of TPO, it will be necessary to conduct chemical identification studies such as those conducted by Chang and Dolphin,[54] which involve ^{13}C nuclear magnetic resonance or high-pressure liquid chromatography.

[46] F. Maloof and M. Soodak, *Endocrinology* **76,** 555 (1965).
[47] M. M. Krinsky and N. M. Alexander, *J. Biol. Chem.* **246,** 4755 (1971).
[48] N. C. Wu and J. Schultz, *FEBS Lett.* **60,** 141 (1975).
[49] D. R. Morris and L. P. Hager, *J. Biol. Chem.* **241,** 1763 (1966).
[50] D. E. Hultquist and M. Morrison, *J. Biol. Chem.* **238,** 2843 (1963).
[51] D. Keilin and E. F. Hartree, *Biochem. J.* **49,** 88 (1951).
[52] L. M. Shannon, E. Kay, and J. Y. Lew, *J. Biol. Chem.* **241,** 2166 (1966).
[53] S. Ohtaki, H. Nakagawa, S. Kimura, and I. Yamazaki, *J. Biol. Chem.* **256,** 805 (1981).
[54] C. K. Chang and D. Dolphin, *Proc. Natl. Acad. Sci. U.S.A.* **73,** 3338 (1976).

TABLE V
EFFECT OF HEMOPROTEIN INHIBITORS ON THYROID PEROXIDASE

Inhibitor	Iodination		Thiocyanate oxidation		Thiourea desulfuration	
	Conc. (M)	Inhibition (%)	Conc. (M)	Inhibition (%)	Conc. (M)	Inhibition (%)
Azide	5×10^{-6}	95	10^{-4}	70	10^{-3}	84
Cyanide	5×10^{-4}	90	5×10^{-5}	90	10^{-4}	90

Inhibitors. The classical heme protein inhibitors[55] such as azide and cyanide are potent inhibitors for the thyroid peroxidase-catalyzed reactions, namely (1) the iodination of thyroid protein,[29] (2) the oxidation of thiocyanate,[56] and (3) the desulfuration of thiourea[57] (see Table V). While it is clear that azide and cyanide inhibit the TPO, there are differences of opinion among investigators when evaluating the effect of inhibitors of sulfhydryl groups or of disulfide groups in the thyroid peroxidase. Hence, the effects of these inhibitors will be detailed, especially since it has been suggested that an —SH group[58] or an —S—S— group[28] may be progenitors for the formation of a sulfenyl iodide species and that an —S—S— group is essential for TPO activity.

Table VI shows the results of experiments in which various compounds were tested for possible inhibition of the desulfuration of thiourea[57] or of the oxidation of thiocyanate.[56] The experiments shown in Table VI were conducted as follows. The enzyme system (10–15 mg of sheep thyroid microsomes) was preincubated with various inhibitors (0.01 M) and buffer in a final volume of 1.0 ml for 15 min at 4°, dialyzed overnight against water to remove excess inhibitors, and then evaluated for its ability to oxidize thiocyanate or desulfurate thiourea. The enzyme with buffer and no inhibitor was also dialyzed overnight and served as the control. In general, agents that cleave disulfide bonds such as cysteine, thioglycolic acid, bisulfite, and borohydride (BH_4) inhibit TPO, whereas sulfhydryl group inhibitors such as *p*-chloromercuribenzoate (PCMB) or methylmercuric nitrate are ineffective (10^{-2} M).

[55] R. Lemberg and J. W. Legge (eds.). "Hematin Compounds and Bile Pigments." Wiley (Interscience), New York, 1949.
[56] F. Maloof and M. Soodak, *J. Biol. Chem.* **239**, 1995 (1964).
[57] F. Maloof and L. Spector, *J. Biol. Chem.* **234**, 949 (1959).
[58] L. Jirousek and L. W. Cunningham, *Biochim. Biophys. Acta* **170**, 160 (1968).

TABLE VI
EFFECTS OF SULFHYDRYL GROUP AND DISULFIDE BOND INHIBITORS
ON THYROID PEROXIDASE

Compound	Conc. (M)	Inhibition (%)	
		Thiocyanate oxidation	Thiourea desulfuration
Cysteine	10^{-3}	—	76
Bisulfite	10^{-2}	92	80
Thioglycolic acid	10^{-2}	—	67
PCMB	10^{-2}	0	9
BH$_4$[a]	10^{-2}	50	—
BH$_4$ + PCMB[a]	10^{-2}	89	—
Iodoacetamide	5×10^{-2}	—	1
Methylmercuric nitrate	10^{-2}	—	2

[a] BH$_4$, borohydride; PCMB, p-chloromercuribenzoate.

The experiments with BH$_4$ require special comment. After treatment with BH$_4$ as described above, the peroxidase activity was 50% of normal, whereas pretreatment of TPO with BH$_4$ and PCMB led to 89% inhibition. It is clear that once the disulfide bonds were disrupted with BH$_4$, PCMB could alkylate the sulfhydryl groups and prevent the oxidation of —SH to —S—S— in contrast to the experiments with BH$_4$ alone. The inhibition of TPO activity following its reduction with BH$_4$ could be due to a reduction of its disulfide bonds or to alteration of its secondary and tertiary structure. The fact that approximately half of the activity was recovered by dialysis of excess BH$_4$ and aeration suggests that the inhibitory effect was not due to the destruction of the secondary and tertiary structure.[59] A full interpretation of the behavior of the —S—S— bonds of TPO will depend upon a knowledge of both the sequence and the crystallographic structure of a homogeneous TPO.

Somewhat different results have been obtained by others. The studies of DeGroot,[60] using a particulate preparation of calf thyroid tissue or a trypsin plus deoxycholate solubilized enzyme, would lead one to believe that —SH or —S—S— groups were involved in TPO activity. The data of Coval and Taurog[29] from a trypsin-solubilized porcine thyroid preparation suggested that an —SH group was essential for activity. Mahoney and

[59] C. B. Anfinsen and F. H. White, Jr., in "The Enzymes" (P. D. Boyer, H. Lardy, and K. Myrbäck, eds.), 2nd ed., Vol. 5, p. 95. Academic Press, New York, 1961.
[60] L. J. DeGroot, Biochim. Biophys. Acta 136, 364 (1967).

Igo,[13] using a deoxycholate-solubilized peroxidase from bovine thyroid tissue, concluded that an —S—S— bond was not the prosthetic group of TPO. However, more recent studies with a partially purified, trypsin-solubilized porcine TPO confirm the early results of Maloof and Soodak concerning the importance of an —S—S— bond in TPO (see following section).

Iodinating Intermediate. Although it is generally agreed that iodide, concentrated by thyroid tissue, must be oxidized to a higher oxidation state before displacing hydrogen from tyrosyl residues, the chemical nature of the oxidized iodide has been a subject of controversy. It has been difficult to differentiate between the possibilities that (1) TPO catalyzes peroxidation of iodide and also participates catalytically in the iodination of tyrosyl residues of thyroglobulin; (2) TPO catalyzes peroxidation of iodide; the product is molecular iodine, which may nonenzymatically iodinate thyroglobulin. The latter is a well-known chemical reaction of iodine and is, in fact, the basis of many currently used protein-labeling procedures. Thus, there is no need, on chemical grounds, to invoke direct participation of TPO in iodination. However, enzyme participation might well be necessary to confer specificity and control of iodination *in vivo*.

In the past several years, evidence obtained in several laboratories,[24,61] including our own,[31] supports the view that TPO has a catalytic function in both peroxidation and iodination. A site for an iodide acceptor substrate on TPO and also on horseradish peroxidase has been demonstrated by acceptor competition studies by the work of Nunez and his associates.[11,41,45,61a] Peroxidase-catalyzed iodination of protein has been shown to have different iodide[24,41,44,61a] and pH requirements than does nonenzymatic iodination. Enzymatic iodination is inhibited at those iodide concentrations necessary for enzymatic I_2 formation. In addition, enzymatic iodination proceeds at pH values that do not support chemical iodination by I_3^-.[31] Davidson *et al.*[31] carried out a series of experiments comparing the pH dependence of three reactions: the chemical iodination of Glu-Tyr-Glu (by triiodide), the enzymatic iodination of Glu-Tyr-Glu (by TPO), and the enzymatic generation of triiodide (also by TPO). They found that the pH dependence of TPO-catalyzed iodination does not follow either the pH profile of TPO catalyzed triiodide production or of chemical iodination of Glu-Tyr-Glu. Thus, these data favor the conclusion that free I_2 is not the primary iodinating species.

In studies using a model system, Jirousek and Soodak[62] showed that *n*-iodosuccinimide, an I^+ reagent, reacts with thionamide and other antithy-

[61] J. Nunez and J. Pommier, *Vitam. Horm.* (*N.Y.*) **39**, 175 (1982).
[61a] J. Pommier, L. Sokoloff, and J. Nunez, *Eur. J. Biochem.* **38**, 497 (1973).
[62] L. Jirousek and M. Soodak, *Proc. Natl. Acad. Sci. U.S.A.* **70**, 1401 (1973).

roid drugs to a degree that approximately parallels their rank as goitrogens. By contrast, molecular iodine, at a lower oxidation potential than the I^+ of n-iodosuccinimide, did not react with all the drugs tested. Jirousek and Soodak concluded that the I^+ of n-iodosuccinimide is a better model for thyroidal "active iodine" than is molecular iodine.

If molecular iodine is not the iodinating species, then what is the nature of the iodinating intermediate? Several investigators have suggested than an iodinimum species (I^+) may be the iodinating intermediate. The innovative work of Libby et al.[63] using a highly purified chloroperoxidase (CPO) demonstrated that Br_2 or Cl_2 are not formed as intermediates in the enzymatic halogenation of organic halogen acceptor substrates, and they proposed that halogenation involves the initial formation of compound **I** and its subsequent conversion into an iron(III) hypohalite halogenating intermediate, Cl^+ or Br^+. For TPO, Ohtaki and colleagues[33,53] have obtained spectrophotometric evidence for compounds **I** (417 nm), **II** (421 nm), and **III** (420 nm); from the kinetic data they concluded that the reaction between compound **I** and iodide occurred by way of a two-electron transfer, and the rate constant was estimated approximately at $2.1 \times 10^7 \, M^{-1} \, sec^{-1}$. The primary oxidation product of iodide was suggested to be the iodinium cation (I^+) in the case of either thyroid peroxidase or lactoperoxidase.[53]

The I^+ species may then iodinate tyrosine directly or may, by either oxidation of an —SH group or by oxidative cleavage of a disulfide bond, form a sulfenyliodide species $(S—I^+)$, which then iodinates tyrosine by covalent catalysis, according to Spector.[63a] In order to investigate the chemical nature of the TPO iodinating intermediate, Davidson and co-workers[31] have generated a TPO-associated intermediate, isolated it by column chromatography, and shown that the isolated intermediate can iodinate the tripeptide Glu-Tyr-Glu without additional H_2O_2 and $^{125}I^-$. The TPO-associated iodinating species (TPO-I_{oxid}) is generated by incubation of approximately equimolar amounts of the enzyme, ^{125}I, and H_2O_2 at pH 5.6. This labeled peroxidase intermediate was then separated from the low molecular weight reaction components by gel filtration and shown capable of iodinating Glu-Tyr-Glu, a tripeptide (apparent $K_m = 3 \times 10^{-6} \, M$) first described by Krinsky and Fruton.[30] Approximately 0.4 nmol ^{125}I was associated with 1 nmol TPO. From 25 to 50% of this label was transferred to Glu-Tyr-Glu. Monoiodo-Glu-Tyr-Glu was identified by means of thin-layer chromatography and paper electrophoresis. These

[63] R. D. Libby, J. A. Thomas, L. W. Kaiser, and L. P. Hager, *J. Biol. Chem.* **257**, 5030 (1982).
[63a] L. B. Spector, "Covalent Catalysis by Enzymes." Springer Publ., New York, 1982.

studies provide evidence for the existence of an enzyme-associated iodinating intermediate.

In order to characterize the iodinating site of the iodinating intermediate, site-specific reagents have been used to alter its activity. Early studies, performed in this laboratory using a particulate sheep or beef thyroid preparation, led to the suggestion that an active-site disulfide of TPO might participate in the formation of a sulfenyl iodide, which could be the iodinating species in thyroid tissue.[28] More recently, the sulfenyl iodide hypothesis was evaluated using a partially purified trypsin-solubilized TPO and reagents that react with sulfur-containing amino acids. Davidson et al. have confirmed some of the early results, which indicate the importance of a disulfide bond in the TPO.[28] Three parameters of TPO activity were measured: peroxidation, incorporation of ^{125}I into TPO, and transfer of TPO·I_{oxid} from the enzyme to Glu-Tyr-Glu (iodination). The enzyme was preincubated either with a 50-fold molar excess of p-chloromercuribenzoate, a 50-fold molar excess of sodium bisulfite, or a 5-fold molar excess of dithiothreitol for 1 hr. Excess reagent was removed by dialysis or gel filtration before proceeding with the experiment. Controls received the same treatment except that reagent was omitted. As shown in Table VII, preincubation of TPO with the sulfhydryl reagent, p-chloromercuribenzoate, did not alter the measured parameter, compared to a control. In contrast, treatment of TPO with two reagents that reduce disulfide bonds, such as sodium sulfite and dithiothreitol, resulted in substantial decreases in (1) peroxidative activity, (2) generation of labeled TPO, and (3) ability to transfer I_{oxid} to Glu-Tyr-Glu. These data suggest that an intact disulfide is essential for the formation of the iodinating intermediate.

Working on the premise that an intact disulfide bond is required for activity of the thyroid peroxidase and may possibly be the active site, we have postulated in the past that iodide (I^-) or an I^+ species may cleave a disulfide bond in thyroid tissue and produce a sulfenyl iodide intermediate, which may well be the active iodinating species in thyroid tissue.[28] This former suggestion was based in part on the studies by Klotz et al.,[64] who reported that I^- per se, under appropriate conditions, was able to cleave a disulfide bond in bovine serum albumin. However, it now appears unlikely that I^- can cleave a disulfide bond. Moreover, based on more recent studies with TPO, it now appears that I^+ is the primary oxidation product of I^-,[53,63] and it is possible that I^+ could cleave a disulfide bond to yield the sulfide–iodide species for iodination.

[64] I. M. Klotz, J. M. Urguhart, T. A. Klotz, and J. Ayers, J. Am. Chem. Soc. 77, 1919 (1955).

TABLE VII

EFFECTS OF —SH AND —S—S— GROUP REAGENTS ON THYROID PEROXIDASE (TPO)-ASSOCIATED IODINATING INTERMEDIATE

Reagent	Inhibitor conc. (M)	Peroxidation of guaiacol; (% of control)	Nanomoles of ^{125}I incorporated per mole of TPO	Iodination of Glu-Tyr-Glu by (TPO-I$_{oxid}$)	Method of detection of iodo-Glu-Tyr-Glu
None (control)	—	100	0.35	Yes	TLC Electrophoresis
Sodium p-chloromercuri-benzoate (PCMB) $\left(\text{TPO—SH} + \text{PCMB} \rightarrow \begin{array}{c} \text{TPO} \\ | \\ \text{S—PCMB} \end{array} \right)$	0.01	100	0.60	Yes	TLC Electrophoresis
Dithiothreitol (DTT) $\left(\text{TPO—}\begin{array}{c}\text{S}\\|\\\text{S}\end{array} + \text{DTT} \rightarrow \begin{array}{c}\text{TPO—SH}\\|\\\text{SH}\end{array} \right)$	0.01	50	0.09	No	TLC Electrophoresis
Sodium sulfite, CuSO$_4$ $\left(\text{TPO}\begin{array}{c}\text{S}\\ \diagdown \\ \text{S}\end{array} \xrightarrow[\text{Cu}^{2+}]{\text{SO}_3^-} \text{TPO}\begin{array}{c}\text{S—SO}_3\\ \diagdown \\ \text{S—SO}_3\end{array} \right)$	0.01	20	0.10	No	TLC Electrophoresis

That a sulfenyl iodide could serve as an iodinating species was first suggested by Kharasch in 1955.[65] This concept was strengthened by the work of Fraenkel-Conrat,[66] in which he clearly demonstrated that the iodination of tobacco mosaic virus led to the formation of a stable sulfenyl iodide group in this protein. Further evidence that such a species may exist and may serve as an iodinating intermediate has been obtained.[58,67,68] In 1964, Cunningham iodinated β-lactoglobulin and interpreted spectrophotometric changes at 355 nm as demonstrating the presence of a sulfenyl iodide group.[67] This sulfenyl iodide species was stable to reactions with cysteine or mercaptoethanol, yet reacted very rapidly with other reagents, such as thiourea, thiouracil, and related compounds, which are known to exert characteristic effects on thyroid metabolism. The reaction of thiouracil or thiourea with the sulfenyl iodide group led to the formation of the respective mixed disulfide with β-lactoglobulin. Similar results were obtained after the iodination of thyroid protein by Jirousek and Cunningham.[58,68] The similarity of this product to the intermediate proposed by Maloof and Soodak,[28] in the thyroid-catalyzed desulfuration of thiourea, led Cunningham[67] to support the thesis that a sulfenyl iodide might function as a reactive intermediate in thyroidal iodine metabolism.

Further evidence that a sulfenyl iodide species may serve as the iodinating intermediate in thyroid tissue has been presented by Fawcett.[69,70] When crude preparations of mammalian thyroid tissue were dialyzed against chlorinated water and subsequently treated with ^{131}I, the radioactive iodide was found to be bound covalently to the thyroid protein. Spectrophotometric changes suggested that a protein sulfenyl iodide species was formed, which then apparently served to iodinate some protein-bound tyrosyl residues to form monoiodotyrosine and diiodotyrosine. On the basis of pretreating the thyroid tissue with radioactively labeled, chlorinated (^{36}Cl) water, and identifying a protein-bound ^{36}Cl species by chromatography on Sephadex, Fawcett interpreted his data as suggesting that a protein-bound sulfenyl chloride intermediate had formed. The ^{131}I was then able to replace the ^{36}Cl bound to protein, thus forming a sulfenyl iodide species.

If sulfenyl iodide is the iodinating intermediate in thyroid tissue, what is the mechanism for its formation? The data of Cunningham[67] and Jirousek[68] suggest that this species may result from iodination of an SH

[65] N. Kharasch, *J. Chem. Educ.* **32,** 192 (1955).
[66] H. Fraenkel-Conrat, *J. Biol. Chem.* **217,** 373 (1955).
[67] L. W. Cunningham, *Biochemistry* **3,** 1629 (1964).
[68] L. Jirousek, *Biochim. Biophys. Acta* **170,** 152 (1968).
[69] D. M. Fawcett, *Can. J. Biochem.* **44,** 1669 (1966).
[70] D. M. Fawcett, *Can. J. Biochem.* **46,** 1433 (1968).

group in thyroid protein. The data of Maloof and Soodak[28,56] suggest that an SH group is probably not involved since —SH inhibitors such as PCMB, iodacetamide, N-ethylmaleimide, or methylmercuric nitrate do not inhibit the reaction involving the peroxidase-catalyzed oxidation of thiourea, or thiocyanate, or the iodination of tyrosyl groups in thyroid tissue. An active-site disulfide bond is another possibility based on our previous data using a partially purified sheep TPO.[56] However, assigning a specific role for a requisite —S—S— bond in the thyroid peroxidase, as essential for the formation of a sulfenyl iodide species, awaits further studies on the chemical composition of the active site of the iodinating intermediate.

In regard to the requirement for an —S—S— bond for TPO activity, it is known that there is a disulfide bond rearranging enzyme in thyroid microsomes and in the high speed supernatant that can reactivate a reduced form of ribonuclease.[71,72] There are 13 half-cystine residues per mole of TPO,[9] and the disulfide rearranging enzyme may serve to re-form the essential disulfide, which could be cleaved oxidatively to form the S—I^+ species.

Antithyroid Drugs. The extensive and wide-ranging studies of the chemical nature of compounds that inhibit the function of the thyroid gland, by Astwood[73] in 1943, set the groundwork for investigations of the mechanism of action of these compounds by scientists for the next 40 years.

In Vivo. Initial studies by Shulman and Keating[74,75] demonstrated that there was an uptake and oxidation of [^{35}S]thiourea by thyroid tissue. Similar studies were subsequently carried out by Maloof and Soodak utilizing noncyclic and cyclic thioureylene compounds, such as thiourea, thiouracil, and thioacetamide, labeled with radioactive sulfur.[76,77] The thioureylene grouping (—NHCSNH—) was previously suggested to be essential for antithyroid activity; however, thioacetamide (CH_3CSNH_2) and goitrin (5-vinyl-2-thiooxazolidine), which contain the NH—CS—O grouping,[78] have also been found to inhibit iodination in the thyroid. Hence, it appears that only the thioamide grouping is essential for the antithyroid activity of this type of compound. Initial studies with ^{35}S-

[71] F. DeLorenzo and G. Molea, *Biochim. Biophys. Acta* **146,** 593 (1967).

[72] R. Moras and I. H. Goldberg, *Endocrinology* **88,** 742 (1971).

[73] E. B. Astwood, *J. Pharmacol.* **78,** 79 (1943).

[74] J. Shulman, Jr. and R. P. Keating, *J. Biol. Chem.* **183,** 215 (1950).

[75] J. Shulman, Jr., *J. Biol. Chem.* **186,** 717 (1950).

[76] F. Maloof and M. Soodak, *Endocrinology* **61,** 555 (1957).

[77] F. Maloof and M. Soodak, *Endocrinology* **68,** 125 (1961).

[78] M. A. Greer and J. M. Deeny, *J. Clin. Invest.* **38,** 1465 (1959).

labeled thiourea, thiouracil, and thioacetamide[76] indicated that these compounds reach their site of action in the thyroid within minutes of their intraperitoneal injection into rats, but there is no concentration gradient for thiourea or thiouracil between the thyroid and serum. Since there appears to be a variability in the capacity of various antithyroid drugs to concentrate in the thyroid, we would like to review these data. Six hours after the administration of 39 μmol of thiouracil or 20 μmol of thiourea to normal rats, Maloof and Soodak found no significant concentration gradient in the thyroid over the serum.[76] However, several investigators have found that 6-propyl-2-thiouracil (PTU)[79] or 1-methyl-2-mercaptoimidazole (MMI)[80] are concentrated by the thyroid. After a single dose of 6 μmol of PTU to normal rats, Marchant et al.[79] found that it was concentrated by thyroid tissue (thyroid/serum ratio = T/S) with a peak at 8 hr of 65/1 and 33/1 at 20 hr. Marchant and Alexander[80] administered a single dose of [^{35}S]MMI to normal rats and found the peak concentration gradient for MMI in the thyroid at 8 hr to be 26/1. Eight hours after the administration of 1.2 μmol of thiouracil, PTU, or MMI to normal rats, Marchant et al.[81] found the following concentration gradients (T/S): 3.8/1 for thiouracil in comparison to 26/1 for MMI and 32/1 for PTU. Lees et al.[82] administered a large (39.0 μmol) and small (1.2 μmol) dose of thiouracil to normal rats. The concentration gradient (thyroid/serum ratio) at 15 hr was not impressive: 4/1 (high dose); 11/1 (low dose). That PTU is concentrated by thyroid tissue has been confirmed, using a radioimmunoassay for PTU, by Halpern et al.,[83] who found a T/S of 100 after the administration of PTU (0.1 μmol) per day for 1 week. It is evident from these data that there is a clear difference between thiouracil and the two drugs PTU and MMI. The latter two drugs are concentrated by the thyroid; a concentration gradient for thiouracil is less evident. The lower concentration gradient found for thiourea (20 μmol) or for two doses of thiouracil (39 or 1.2 μmol), as compared with PTU and MMI, needs further investigation. This difference may be explained in part by the interesting studies of Aungst et al.,[84] who showed a dose dependency in the uptake of PTU by the thyroid of normal rats.

[79] B. Marchant, W. D. Alexander, J. W. K. Robertson, and J. H. Lazarus, *Metabolism* **20,** 989 (1971).

[80] B. Marchant and W. D. Alexander, *Endocrinology* **91,** 747 (1972).

[81] B. Marchant, W. D. Alexander, J. H. Lazarus, J. Lees, and D. H. Clark, *J. Clin. Endocrinol.* **34,** 847 (1972).

[82] J. Lees, W. D. Alexander, and B. Marchant, *Endocrinology* **93,** 162 (1973).

[83] R. Halpern, D. S. Cooper, J. D. Kieffer, V. Saxe, H. Mover, F. Maloof, and E. C. Ridgway, *Endocrinology* **113,** 915 (1983).

[84] B. J. Aungst, E. S. Vesell, and J. R. Shapiro, *Biochem. Pharmacol.* **28,** 1479 (1979).

Thiourea, thiouracil, and thioacetamide are metabolized by thyroid tissue; the major sulfur products are sulfate, thiosulfate, and protein-bound sulfur.[28] Once the oxidation of the thiocarbamide is nearly complete and the concentration in thyroid tissue is less than 2.5×10^{-5} M, their inhibitory action is gone and iodination commences immediately. It appears that the rapidity with which the thiocarbamide compounds are metabolized by the thyroid is a major factor in determining the duration of their inhibitory action. The duration of action of equimolar doses of the thiocarbamides is thiouracil > thiourea > thioacetamide.

In Vitro. A system was found that simulated, *in vitro*, the metabolism of thiourea, *in vivo*.[57] Initially, the optimal requirements of this enzyme, *in vitro*, were found to be a cytoplasmic particulate fraction of thyroid tissue (recovered from the 8500 g supernatant by centrifugation at 105,000 g for 1 hr), thiocyanate (10^{-3} M), ascorbic acid (10^{-3} M), thiourea (5×10^{-5} M), Tris buffer (0.15 M), pH 7.4. Similar cytoplasmic particulate preparations of liver and kidney are inactive. The thyroid particulate fraction was subsequently found to contain the thyroid peroxidase. 2-D-Methyl-L-ascorbic acid is inactive in either replacing ascorbic acid or blocking its effect, thereby indicating that ascorbic acid is not functioning as a coenzyme in the reaction involving the enzyme glucosinolase, as studied by Ettlinger *et al.*[85] Ascorbic acid is probably serving as a source of hydrogen peroxide, since it can be replaced by hydrogen peroxide added directly (10^{-4} M) or generated enzymatically.

Subsequent studies revealed that catalytic concentrations of iodide are able to replace thiocyanate, providing glucose (10^{-2} M) and glucose oxidase (5 μg/ml) serve as the source of hydrogen peroxide.[86] Other halides, such as chloride, bromide, or fluoride, are ineffective at concentrations from 2×10^{-6} M to 1×10^{-3} M.

A catalytic role for iodide in the *in vitro* metabolism of thiourea by the thyroid was reported in 1965 by Maloof and Soodak.[86] In 1976, Taurog[87] found that the metabolism of two cyclic thioureylene-type antithyroid drugs (PTU and MMI) by thyroid tissue also required the halide iodide.

The fact that the oxidation of thiourea by the thyroid peroxidase *in vitro* requires iodide has been confirmed *in vivo*, since a low iodine diet to rats leads to a decreased uptake and metabolism of thiourea.[86] Marchant *et al.*[88] reported that a low-iodide diet in rats leads to a decrease in the

[85] M. G. Ettlinger, G. P. Dateo, Jr., B. W. Harrison, T. J. Mabry, and C. P. Thompson, *Proc. Natl. Acad. Sci. U.S.A.* **47**, 1875 (1961).
[86] F. Maloof and M. Soodak, *Curr. Top. Thyroid Res. Proc. Int. Thyroid. Conf. 5th, 1965*, p. 227 (1965).
[87] A. Taurog, *Endocrinology* **98**, 1031 (1976).
[88] B. Marchant, P. D. Papapetron, and W. D. Alexander, *Endocrinology* **97**, 154 (1975).

accumulation and oxidation of the MMI by the thyroid. The effect was reversible by the administration of iodide. These data confirm the early studies using thiourea.[86]

Reaction Products. Sulfur metabolites resulting from the desulfuration of thiourea *in vitro* are similar to those *in vivo,* namely sulfate, thiosulfate, and protein-bound sulfur. The latter predominates *in vitro,* whereas sulfate predominates *in vivo.* The nature of the protein to which ^{35}S was bound was determined by paper electrophoresis, and most of it did not behave like thyroglobulin. The nature of the sulfur of thiourea that is bound to thyroid protein was investigated.[57,89] From a series of experiments involving the release of sulfur from the thyroid protein and preincubation studies with nucleophilic anions, it appears that an intact disulfide bond in thyroid tissue is essential for the desulfuration of thiourea. Furthermore, the data lead one to conclude that the transsulfuration of the sulfur of thiourea to thyroid protein probably involves a thiol–disulfide interchange between thiourea and the thyroid protein in the presence of iodide or thiocyanate and a source of hydrogen peroxide.[28,75]

Mechanism of Action. Over the past several years, there have been innovative findings and real progress in regard to the interaction of antithyroid drugs with TPO and their mechanism of action. Davidson *et al.*[25] reported that MMI, PTU, and 2-thiouracil, only in the presence of H_2O_2 (1.6–3.5 nmol/ml), irreversibly inactivated the thyroid peroxidase (300–600 μg/ml) as measured by iodination and guaiacol peroxidation *in vitro.* Iodide (10^{-4} M) or thiocyanate (10^{-4} M), added simultaneously with TPO, the antithyroid drug, and H_2O_2, prevented the inactivation. Concentrations of MMI, PTU, and 2-thiouracil required for 50% inhibition were 0.38, 30, and 27 μM, respectively.

In 1963, Hosoya[22] reported that thiourea was a comparatively poor inhibitor of guaiacol peroxidation by the thyroid peroxidase; 50% inhibition required a concentration of 10^{-1} M thiourea compared to 10^{-5} M for 3-amino-1-2-4 triazole, 10^{-4} M for 2-thiouracil, and 10^{-4} M for PTU. Davidson *et al.*[90] pursued the observation of Hosoya[22] and found as he had that thiourea, at 5 mM, had no effect on guaiacol peroxidation by TPO in the presence of H_2O_2, but when they added iodide (0.1 mM) to thiourea (1.0 mM) in the incubation media, enzyme activity was inhibited. This latter inhibition was due to the reaction of thiourea and iodide forming cyanamide in the presence of H_2O_2 and TPO. Hence, thiourea exhibits different inhibitory properties than MMI, PTU, or thiouracil; thiourea is

[89] F. Maloof and M. Soodak, *J. Biol. Chem.* **236,** 1689 (1961).

[90] B. Davidson, M. Soodak, H. V. Strout, J. T. Neary, C. Nakamura, and F. Maloof, *Endocrinology* **104,** 919 (1979).

known to be a potent inhibitor of iodination and a potent goitrogen, yet it has no effect on TPO-catalyzed guaiacol peroxidation. Moreover, in contrast to the finding that iodide protects TPO from inactivation by thiouracil, PTU, and MMI,[25] adding iodide to guaiacol assays containing thiourea converts the thiourea into an inhibitor of TPO. This is an excellent example in which the metabolic product, cyanamide,[91] rather than the intact antithyroid drug, thiourea, inhibits TPO activity.

The administration of MMI, PTU, 2-thiouracil, thiourea, and cyanamide to rats reveals that iodination was completely inhibited, yet TPO activity as measured by guaiacol peroxidation was essentially normal for MMI and PTU, and partially inhibited by thiouracil (55%) and cyanamide (30%), demonstrating that MMI and PTU inhibit iodination *in vivo* without inhibiting TPO activity. As a result of these *in vitro* and *in vivo* data, Davidson *et al.*[25,90] reconfirmed the concept that these drugs, *in vivo,* inhibit iodination by competing with the nucleophile tyrosine for the iodinating intermediate.[28]

Engler *et al.*[92,93] confirmed the data of Davidson *et al.*[25] by showing that the addition of MMI (2 μM) or PTU (4 μM) with TPO (20 μg/ml) in the presence of H_2O_2 (2 μM) inactivated TPO. However, dilute solutions of their enzyme (<5 μg/ml) could be inactivated without H_2O_2; this was not true at concentrations of TPO of 20 μg/ml.

Engler *et al.*[92] have presented interesting spectral data on TPO and the changes that result from the addition of H_2O_2, MMI, or PTU and iodide. Native reduced TPO (20 μg/ml) exhibits a Soret band at 411 nm. Upon the addition of H_2O_2 (2 μM), the Soret band shifts to the red, where there is an increase in the absorption at 420 nm at 1 min and then a decrease in 4 min; upon the addition of MMI (2 μM), there is a shift to the blue with a decrease in the absorption at 414 nm. In the presence of H_2O_2, MMI inactivates the TPO. The addition of I^- (2 μM) to TPO, prior to the addition of H_2O_2 and MMI, prevents all the spectral changes, and the enzyme is not inactivated. The addition of MMI to TPO without H_2O_2 produced no spectral changes beyond 411 nm. These studies are consistent with those of Michot *et al.*,[94] who found no spectral changes with lactoperoxidase (LPO) (2.6 × 10^{-6} M) and methyl thiouracil (1.5 × 10^{-4} M) until they added H_2O_2 (2.9 × 10^{-5} M) to the incubation mixture.

These studies of Engler *et al.*[92,93] support those of Davidson *et al.*,[25] who reported that TPO–I_{oxid} is the form of TPO that oxidizes the antithy-

[91] G Toennies, *J. Biol. Chem.* **120,** 297 (1937).
[92] H. Engler, A. Taurog, and T. Nakashima, *Biochem. Pharmacol.* **31,** 3801 (1982).
[93] H. Engler, A. Taurog, C. Luthy, and M. L. Dorris, *Endocrinology* **112,** 86 (1983).
[94] J. L. Michot, J. Nunez, M. L. Johnson, G. Irace, and H. Edelhoch, *J. Biol. Chem.* **254,** 2205 (1979).

roid drugs and that competition between thioureylene drugs and tyrosyl residues in thyroglobulin for TPO–I_{oxid} is the basis for the acute inhibition of iodination by thioureylene drugs *in vivo.*

Using [^{35}S]MMI or [^{35}S]PTU, Engler *et al.*[92] found that 1 mol of MMI or PTU bound to 1 mol of TPO enzyme and that binding correlated with inactivation of the enzyme. Iodide (2 μM) inhibited the binding and also the inactivation. Binding was assumed to be covalent to the heme ligand based on dialysis experiments. Engler *et al.* have also reported that the interaction between MMI or PTU and the heme group of the thyroid peroxidase is responsible for the enzyme inactivation.[92] However, from studies with LPO, Michot *et al.*[94] suggested that the changes in the absorption band in the Soret region could result either from a modification in protein structure that altered the apoprotein–heme interaction or by a direct reaction with the heme group. These two possibilities will require further investigation for both TPO and LPO.

Using TPO solubilized by deoxycholate and trypsin from microsomes (7 μM TPO), Ohtaki *et al.*[10] reported that MMI (10 μM) in the presence of H_2O_2 (10 μM) accelerated the conversion of compound **I** to compound **II** and reacted with compound **II** to form an inactive enzyme, discernible spectrophotometrically. They postulated that this latter interaction might explain in part the inactivation of TPO by antithyroid drugs *in vitro.* This appears to be a reasonable concept; however, one would like to define whether an intact antithyroid drug or one of its metabolic products is bound to TPO. Then one will need to determine the nature and site of binding of the intact drug or its metabolic product. Transient and steady-state kinetic studies on the formation and nature of heme ligand in the thyroid peroxidase and the manner in which H_2O_2 and the halide ion (or iodinium cation) and the antithyroid drugs react sequentially with the enzyme should allow one to ascribe a specific role for I^- (or SCN^-) and H_2O_2 in the iodination reaction and its inhibition by the antithyroid drugs.

Comparison with Other Peroxidases. It is known that other peroxidases catalyze reactions similar to those previously described, although the efficiencies of peroxidation and iodination differ among the various enzymes. Sorbo and Ljunggren[95] found that thiourea, thiouracil, and thiocyanate, but not true thiol compounds such as cysteine, thioglycolic acid, or mercaptoethanol, are substrates for myeloperoxidase. The innovative and illuminating studies of Hager and associates using chloroperoxidase (CPO) has provided the thyroidologist with valuable baseline data concerning the mechanism of action of the antithyroid drugs[49,96] and

[95] B. Sorbo and J. G. Ljunggren, *Acta Chem. Scand.* **12**, 470 (1958).
[96] D. R. Morris and L. P. Hager, *J. Biol. Chem.* **241**, 3582 (1966).

a possible halogenating species for iodination.[63] Morris and Hager[49] have demonstrated that thiourea, thiouracil, and various antithyroid compounds are substrates for chloroperoxidase. These investigators have isolated, purified, and crystallized chloroperoxidase from the mold *Caldariomyces fumago,* which catalyzes the hydrogen peroxide-dependent oxidation of thiocarbamides only in the presence of a suitable halogen anion, such as iodide, bromide, or chloride.[96] The prosthetic group is ferri-protoporphyrin IX. On the basis of the heme content, the minimum molecular weight is 40,200. CPO is a glycoprotein; approximately 25–30% of the molecule is carbohydrate. Other peroxidases are also glycoproteins. Horseradish peroxidase,[97] Japanese radish peroxidase,[98] and lactoperoxidase[99] contain covalently linked carbohydrates, although cytochrome *c* peroxidase does not contain sugars.[100] Our studies[14] and those of Rawitch *et al.*[9] indicate that TPO is a glycoprotein.

It is appropriate to review the remarkable similarities between the thyroid peroxidase and chloroperoxidase. Chloroperoxidase can utilize Cl^-, Br^-, and I^- ions in the presence of H_2O_2 and catalyze the formation of a carbon–halogen bond in the presence of suitable acceptor molecules such as β-keto acids, cyclic β-diketones, and substituted phenols.[101] It can also serve to iodinate thyroglobulin.[102] According to the original hypothesis of Hager and associates,[103] the halogen anion is oxidized by the enzyme by the removal of a pair of electrons to form an electrophilic species as illustrated in the following equation.

$$X^- + H_2O_2 + \text{enzyme} \rightarrow \text{enzyme-}X^+ + 2\ OH^-$$

The X^- in this equation represents an oxidizable halogen anion that forms an enzyme-bound halogenium ion after removal of a pair of electrons. Further reaction of the enzyme-bound halogenium ion would depend upon the availability and reactivity of X^- ion and nucleophilic acceptor molecules. Halogen acceptors in the chloroperoxidase reaction, such as β-keto acids, cyclic β-diketones, and substituted phenols, share in common a nucleophilic character and hence would be expected to serve as acceptors for the halogenating species. Compounds containing the thioureylene group, such as thiourea and thiouracil, also share the common character-

[97] H. Theorell and A. Åkeson, *Ark. Kemi Mineral. Geol.* **16A,** 1 (1943).
[98] Y. Morita and K. Kameda, *Mem. Res. Inst. Food Sci. Kyoto Univ.* **14,** 49 (1958).
[99] W. A. Rombauts, W. A. Schroeder, and M. Morrison, *Biochemistry* **6,** 2965 (1967).
[100] N. Ellfolk, *Acta Chem. Scand.* **21,** 175 (1967).
[101] L. P. Hager, D. R. Morris, F. S. Brown, and H. Eberwein, *J. Biol. Chem.* **241,** 1769 (1966).
[102] A. Taurog and E. M. Howell, *J. Biol. Chem.* **241,** 1329 (1966).
[103] F. S. Brown and L. P. Hager, *J. Am. Chem. Soc.* **89,** 719 (1967).

istic of being extremely active nucleophiles. Hager and co-workers proposed that these thiocarbamide compounds inhibit the halogenation reaction by virtue of being extremely active nucleophiles and thus capable of reacting with an enzyme-bound halogenium ion and being oxidized.[96] They demonstrated that chloroperoxidase, when supplemented with a halogen anion (Cl^-, Br^-, or I^-) and H_2O_2 catalyzed the following reaction.

$$\text{Thiouracil} + H_2O_2 \xrightarrow[\text{halogen anion}]{\text{enzyme}} \text{uracil disulfide} \rightarrow \text{uracil} + S^0 + H_2O$$

On the basis of their data, they postulated that a sulfenyl halide of thiouracil might be considered as a likely intermediate in the formation of uracil disulfide and uracil. They proposed that a sulfenyl halide derived from thiouracil is relatively unstable, and they were not able to "trap" it. But in a subsequent study, Silverstein and Hager[104] demonstrated the formation of a sulfenyl-halide intermediate with 2-nitro-5-thiobenzoic acid, thereby indicating that a sulfenyl-halide intermediate may be formed in the halide-dependent oxidation of thiouracil.

Clearly, the fundamental mechanism proposed initially by Hager *et al.*,[105] Maloof and Soodak,[28] and Davidson *et al.*[25] for the inhibition of halogenation by the thionamides is similar, namely, that they act by competing with a nuclcophilic acceptor molecule for the iodinating intermediate. However, there are differences in the properties of the enzymes that should be noted.

1. The pH optimum of both enzymes varies; pH 2.75 for the chloroperoxidase halogenation reaction and pH 7.0 for the reaction by the thyroid peroxidase.
2. The halogen anion requirement for the chloroperoxidase to oxidize thiocarbamides is not specific, utilizing Cl^-, Br^-, or I , whereas of all the halides, only I^- is effective in thyroid tissue.
3. To date, the disulfide of thiourea has not been detected in the thyroid system.
4. The sulfur of thiourea becomes incorporated into the protein of the thyroid microsomes.
5. At present, the proposals for the iodinating intermediate may be somewhat similar in the two systems, namely, an enzyme-bound —I^+ species for the chloroperoxidase, and an enzyme-bound sulfenyl iodide in thyroid peroxidase.

[104] R. M. Silverstein and L. P. Hager, *Biochemistry* **13,** 5069 (1974).
[105] L. P. Hager, J. A. Thomas, and D. R. Morris, *in* "Biochemistry of the Phagocytic Process" (J. Schultz, ed.), p. 67. North-Holland Publ., Amsterdam, 1970.

Studies by Edelhoch *et al.*[94,106] using lactoperoxidase have provided useful information concerning (1) spectral data of native, reduced LPO and its reaction with H_2O_2 to form compounds **I** and **II**; (2) the effect of various antithyroid drugs [MMI, 6-methyl-2-thiouracil (MTU), 2-mercaptopyrimidine, thiobarbituric acid, and aminotriazole] in inactivating LPO in the presence of H_2O_2 and the respective spectral changes; (3) the effect of I^- in preventing the inactivation of LPO by some of the antithyroid drugs and H_2O_2; and (4) the role of I^- in stimulating the metabolism of antithyroid drugs. These effects were characterized by specific spectral changes. Lactoperoxidase ($2.63 \times 10^{-6}\,M$) has a Soret band at 412 nm; the addition of H_2O_2 ($2.94 \times 10^{-5}\,M$) causes a shift to the red (422 nm) and a decrease in the absorption; upon the addition of MTU ($1.50 \times 10^{-5}\,M$) there is a shift in the absorption band to the blue from 422 to 414 nm (and a further decrease in the absorption) with a rapid loss of enzymatic activity.[94] MMI and 2-mercaptopyrimidine also inactivate LPO. The inactivation by MTU, MMI, and 2-mercaptopyrimidine is prevented by the addition of iodide to LPO prior to the addition of H_2O_2 or the drug. Aminotriazole[94] also inactivates LPO, but the inhibition is not prevented by iodide. By contrast, thiobarbituric acid is a relatively poor inhibitor of LPO (19% inhibition), even in the presence of H_2O_2. I^- has a slight protective effect on the inhibition of LPO by thiourea. This is in contrast to the situation with TPO, where the addition of I^- causes inactivation of TPO in the presence of thiourea.

Iodide is required for the oxidation of MTU, mercaptopyrimidine, and thiobarbituric acid. However, it should be noted that 12% of MMI can be oxidized without iodide.

Nunez[61] stated, on the basis of his studies with Edelhoch, that the "iodide stimulatory effect on drug oxidation by LPO is not observed with all the thioureylene drugs tested. In addition, the relative importance of drug oxidation and LPO inactivation greatly depends on the chemical structure of the drug, i.e., on the ternary complex formed between the enzyme, H_2O_2, iodide and the drug." We agree with this statement.

Addendum

A review of the literature reveals several interesting observations:

Nakagawa and Ohtaki[107] evaluated the orientation of thyroid peroxidase in hog thyroid microsomes by trypsin treatment, gel filtration, bind-

[106] H. Edelhoch, G. Irace, M. L. Johnson, J. L. Michot, and J. Nunez, *J. Biol. Chem.* **254,** 11822 (1979).
[107] H. Nakagawa and S. Ohtaki, *J. Biochem.* **94,** 155 (1983).

ing to Concanavalin A-Sepharose, and iodination of thyroglobulin. Their results suggested that thyroid peroxidase is oriented toward the luminal side of the microsomal vesicles. This differs from cytochrome b_5, which is oriented toward the cytoplasmic side of the RER.

Novikoff et al.[108] have shown the presence of peroxisomes in thyroid tissue which appear to be attached to specific areas of the endoplasmic reticulum. Peroxisomes contain D-amino oxidase, which may serve as a source of H_2O_2 generated by D-amino oxidase and uricase; the peroxisomes also contain catalase, which must serve as a controlling influence over the concentration of H_2O_2.[109] This is an example of compartmentalization for control and preservation, since free-circulating TPO + $H_2O_2{\to}$Cpd I is a powerful oxidant, which obviously needs to be cloistered.

Meier and Meyer[110] reported that the synthesis of cytochrome P_{450} requires the functional and structural interaction of mitochondria and membranes of the RER. Mitochondrial RER complexes may operate as functional units in cytochrome P_{450} synthesis, regulating the extramitochondrial transport of heme. This system may well apply to thyroid tissue and will be evaluated.

Acknowledgment

We wish to acknowledge Morris Soodak (Brandeis University) for his inspiration and direction not only from the initiation but throughout the progress of these studies over many years; Leonard Spector (Rockefeller University) for his support, encouragement, and advice in regard to enzyme mechanisms; to Thomas Hollocher (Brandeis University) for his unselfish counsel concerning his expertise in the subject of heme proteins; and to James P. Dennehy (Notre Dame University) for his helpful discussions concerning sulfur chemistry. We also gratefully acknowledge Betty Davidson, Ann Armstrong, and Vincent Strout, who made valuable contributions to this work. We thank Ruth DiBlasi, Sharon Melanson, and Jane Leighton for expert secretarial service. This work was supported in part by U.S. Public Health Service Grants AM-13916 and AM-21945.

[108] A. B. Novikoff, P. M. Novikoff, M. Ma, W.-Y. Shin, and N. Quintant, *Adv. in Cytopharmacol.* **2**, 349 (1974).

[109] C. DeDuve, H. Beaufay, P. Jacques, Y. Rahman-Li, O. Z. Sellinger, R. Wattiaux, and S. DeConich, *Biochim. Biophys. Acta* **40**, 186 (1960).

[110] P. J. Meier and U. A. Meyer, *Hoppe-Seyler's Zeitschrift Physiol. Chemie* **357**, 1041 (1971).

[30] Thyroid Hormones: Mechanism of Phenoxy Ether Formation

By J. Nunez

Thyroid hormone synthesis is the result of two successive oxidative reactions that are catalyzed by the same enzyme, thyroid peroxidase. In the first reaction, i.e., iodination, a limited number of the tyrosine residues (~30 out of ~130) that are present within the polypeptide backbone of thyroglobulin are either monoiodinated or diiodinated by the enzyme.

$$\text{Tyr—R} + \text{I}^- \rightarrow \text{monoiodo-Tyr—R} \tag{1}$$
$$\text{Monoiodo-Tyr—R} + \text{I}^- \rightarrow \text{diiodo-Tyr—R} \tag{2}$$

In a second reaction catalyzed by the same enzyme some of these iodinated residues (~8 out of ~30) are coupled to thyroid hormones, i.e., to thyroxine (or tetraiodothyronine) and 3,5,3'-triiodothyronine

$$2 \text{ Diiodo-Tyr—R} \rightarrow \text{thyroxine} \tag{3}$$
$$1 \text{ Monoiodo-Tyr—R} + 1 \text{ diiodo-Tyr—R} \rightarrow 3,5,3'\text{-triiodothyronine} \tag{4}$$

Both the iodination and the coupling reaction can be catalyzed by other peroxidases (lactoperoxidase or horseradish peroxidase) and in nonenzymatic conditions (by ICI, I_2, chloramine-T + iodide, etc.). However, the experimental conditions as well as the amino acid distribution and the hormone yield change greatly depending on the iodinating agent.

Enzymatic Iodination and Coupling

Thyroid peroxidase and lactoperoxidase catalyze both the iodination and coupling reactions in the presence of hydrogen peroxide. Like all peroxidases,[1] these two enzymes form with H_2O_2 an addition compound that contains two oxidizing equivalents above the native state. It is still not clear[2] whether enzymatic iodination proceeds via an electrophilic substitution or a free-radical mechanism. In the electrophilic substitution, the two oxidating equivalents of the enzyme–H_2O_2 derivative are utilized to remove *two* electrons from iodide with the formation of the iodinium ion I^+. This iodine-oxidized species then easily substitutes the hydrogen

[1] P. George, *J. Biol. Chem.* **201,** 427 (1953).
[2] J. Nunez and J. Pommier, *Vitam. Horm.* (*N.Y.*) **39,** 175 (1982).

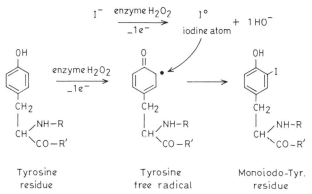

present in position 3 (or 5) of a Tyr residue of thyroglobulin in a nonenzymatic electrophilic reaction (Fig. 1).

In the radical mechanism *one* electron is removed from iodide with the formation of the iodine atom I°, whereas a second electron is removed from a Tyr residue of thyroglobulin with the production of the free radical Tyr° (Fig. 2). Monoiodo-Tyr is formed in this case by radical addition between I° and Tyr°. Thus, whatever the mechanism, 1 mol of enzyme–H$_2$O$_2$ addition compound, i.e., two oxidizing equivalents, is utilized for each mole of monoiodo-Tyr formed. The same stoichiometry applies to convert monoiodo-Tyr to diiodo-Tyr.

FIG. 2. Free-radical addition.

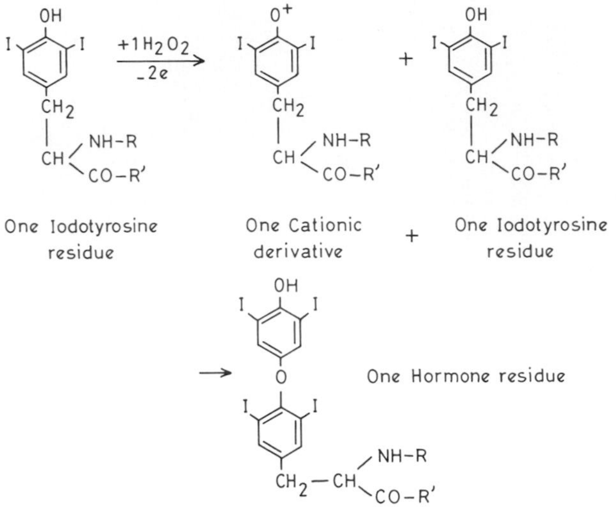

FIG. 3. Ionic coupling.

The coupling reaction might also proceed through either an ionic or a radical mechanism.[3] In nonenzymatic conditions the coupling mechanism is probably ionic, but it is not clear whether enzymatic coupling implies or not the formation of free radicals. However, two electrons must be removed whatever the mechanism; either one from each hormonogenic iodo-Tyr residue in the radical mechanism or two from one of them in the ionic process (Fig. 3). The formation of 1 mol of hormone therefore also requires the consumption of 1 mol of H_2O_2 (Fig. 4).

Thyroglobulin may contain varying iodine and hormone contents (*in vivo* 0–50 iodine atoms per mole of protein, and 0–70 iodine atoms after *in vitro* iodination).[4] When thyroglobulin is maximally iodinated, ~20 Tyr residues are monoiodinated, ~10 are diiodinated, and 8 are converted to thyroid hormones (3 mol of thyroxine; 0.5 mol of triiodothyronine). This means that 60 mol of H_2O_2 are utilized to iodinate thyroglobulin maximally and to couple all the hormonogenic iodotyrosine residues; 4 mol of H_2O_2, i.e., only 6% of this amount of H_2O_2, is required for the coupling reaction.

In most cases H_2O_2 is provided by a H_2O_2-generating system; the concentration of this system must be adjusted taking into account[5] that

[3] J. M. Gavaret, H. J. Cahnmann, and J. Nunez, *J. Biol. Chem.* **256**, 9167 (1981).
[4] J. Nunez, *in* "Thyroid Hormones" (M. de Visscher, ed.), p. 39. Raven, New York, 1980.
[5] J. Pommier, J. Tourniaire, B. Rahmoun, D. Deme, D. Pallo, H. Bornet, and J. Nunez, *J. Clin. Endocrinol. Metab.* **42**, 319 (1976).

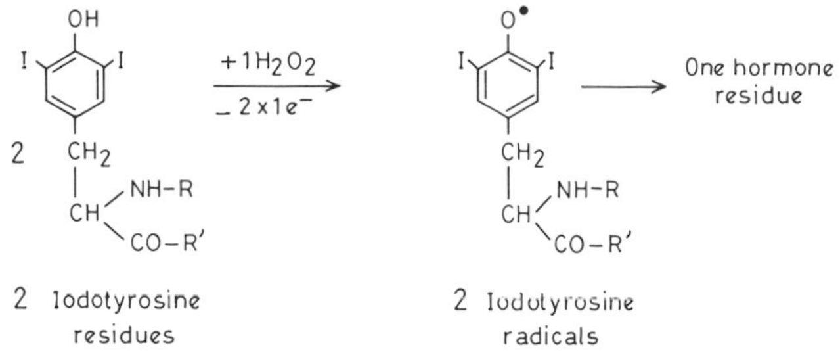

FIG. 4. Radical coupling.

(1) a nonlimiting concentration of H_2O_2 must be maintained during the time course of the reaction; (2) depending on the substrate concentration the rate of the reaction varies and so varies the rate of H_2O_2 utilization; (3) the presence of an excess of H_2O_2 must also be avoided, since peroxidases under these conditions form inactive derivatives. The simplest method to obtain a proper concentration of H_2O_2 during the reaction is to maintain constant the concentration of the substrates and of the H_2O_2 generating system while varying the amount of peroxidase: if the amount of H_2O_2 generating system is neither limiting nor in excess, the initial rate of iodination is constant.

Thyroid Peroxidase

Since thyroid peroxidase is a particulate enzyme, two types of preparations can be made. Treating the particulate material with detergents, such as deoxycholate, digitonin, or triton, yields a preparation that contains the native enzyme but is difficult to purify. A completely solubilized peroxidase[6] can be obtained also by treating twice the particulate material obtained from the homogenate (30,000 $g \times 2$ hr) with trypsin (45 mg per gram of tissue for 2 hr at 20°). The insoluble material, which contains 90–100% of the initial enzyme activity, is washed twice and then treated with digitonin (1% final, 24 hr at 20°). Most of the enzymatic activity that becomes soluble upon this treatment is collected by centrifugation and precipitated by ammonium sulfate (35% final). The precipitate is dissolved in Tris-HCl, 0.05 M, pH 7.2, and dialyzed. Purification of the enzyme is achieved by DEAE-cellulose column chromatography. A pH gradient 7.2 → 8.0 is first established, and the activity is eluted with an NaCl (0 to 0.5 M) linear gradient. Further purification is achieved by a second DEAE-cellulose chromatography, gel filtration on Sephadex G-100, and

[6] J. Pommier, S. de Prailauné, and J. Nunez, *Biochimie* **54,** 483 (1972).

sucrose gradient ultracentrifugation (5–20%, 35,000 g, 15 hr). The specific activity of the purified enzyme varies between 600 and 1100 U/mg (50–80% purity as determined by polyacrylamide gel electrophoresis).

Enzyme activity can be estimated by measuring the oxidation of guaiacol to tetraguaiacol at 470 nM[7] or the oxidation of iodide to iodine[8] at 353 nm. One unit of thyroid peroxidase activity is defined as that required to result in an increased of 1 unit of absorption per minute at 353 nm when the oxidation of iodide to iodine is used as an assay.[6]

Thyroglobulin

Noniodinated or poorly iodinated thyroglobulin can be prepared[9] from human goiters as follows. The tissue is homogenized in 1 volume (v/w) of 50 mM Tris-HCl, pH 7.2; the 30,000 g supernatant (2 hr, Sorvall centrifuge) of this homogenate is fractionated by gel filtration on an Ultrogel ACA 34 column equilibrated with 50 mM phosphate buffer, pH 7.2. The large excluded peak of proteins is then rechromatographed on an ACA 22 Ultrogel column. The thyroglobulin peak is pooled and dialyzed against 2 mM ammonium bicarbonate, pH 8.0. This preparation is 95% pure, as shown by SDS (0.1%)–polyacrylamide gel (4% acrylamide and 4% bis acrylamide relative to acrylamide) electrophoresis. Human goiters from patients with "iodide organification defect," "Pendred syndrome," some thyroid cancers, or "cold nodules" can be used as sources of very poorly iodinated thyroglobulin (less than 2 iodine atoms per mole of protein); simple "colloid" goiters can be used to prepare poorly iodinated thyroglobulin (2–8 iodine atoms/mole of protein).

[^{14}C]- or [^{3}H]Tyrosine-Labeled (Noniodinated) Thyroglobulin[10]

Fresh hog thyroid slices are incubated in an atmosphere of O_2/CO_2 (95 : 5) with L-[U-^{14}C]tyrosine (450–470 mCi/mmol) in tyrosine-free Earle's No. 2 medium (containing glutamic acid and supplemented with penicillin/streptomycin) for 20 hr at 37°. An antithyroid drug, 2-mercapto-1-imidazole, (10^{-4} M), is added to the medium to inhibit completely the iodination reaction. The slices are then homogenized with 50 mM sodium phosphate buffer, pH 7.4, and the homogenate is centrifuged for 1 hr at 150,000 g. The [^{14}C]thyroglobulin is purified by Sephadex G-200 or ACA 34 column chromatography of the 105,000 g supernatant. [^{3}H]Thyroglobu-

[7] B. Chance and A. C. Maehly, this series, Vol. 2, p. 770.
[8] N. M. Alexander, Anal. Biochem. 4, 341 (1962).
[9] A. Virion, D. Deme, and J. Nunez, Eur. J. Biochem. 112, 1 (1980).
[10] J. M. Gavaret, H. J. Cahnmann, and J. Nunez, J. Biol. Chem. 254, 11218 (1979).

lin can be prepared also by using [³H]tyrosine labeled in the alanine side chain.[11]

Enzymatic Iodination

With thyroid peroxidase the pH optimums for iodination and coupling[12] are identical (~7.5). With lactoperoxidase the pH optimum for iodination (~6.5) differs from that for coupling (~7.5); however, at pH 7.4 the iodoamino acid distribution is very similar whatever the level of thyroglobulin iodination for thyroid peroxidase and lactoperoxidase. This is not true for horseradish peroxidase, an enzyme that catalyzes the iodination reaction with an acidic pH optimum (~5.0), whereas two alkaline pH optima are seen for the coupling reaction (~7.5 and ~8.5).

With thyroid peroxidase the K_m for iodide for the iodination reaction is 0.1 mM. Above 0.2 mM iodide, the enzyme catalyzes another reaction, the oxidation of iodide to iodine, whereas the iodination reaction is inhibited.[13] The maximal concentration of iodine that can be used with thyroid peroxidase is, therefore, 0.2 mM. As little as 1 iodine atom per mole of thyroglobulin and a maximum of 70 iodine atoms can be incorporated in this protein: varying the concentration of iodide from 0.1 μM to 20 μM therefore allows the preparation of thyroglobulin samples containing 1–70 iodine atoms per mole. In the range of iodide concentration 10 to 100 μM the same amounts of peroxidase (0.8–1 U/ml) and of H_2O_2 generating system (1 μg of glucose oxidase, 6 mM glucose) can be used. The K_m for thyroglobulin is 1 μM; this concentration is usually used in the iodination assay. Iodide is generally labeled with ¹²⁵I or ¹³¹I; iodide labeling greatly facilitates the measurement of the iodine and iodoamino acid content. The reaction is carried out at 20° in a phosphate buffer, 0.05 M, pH 7.4, and is arrested, when required, by the addition of an NaHSO₃ solution. Under optimal conditions the iodination reaction levels off after 10 min of incubation and is completed after 30 min. In the presence of a slight excess of H_2O_2 the rate of iodination is faster during the first minutes of incubation, but the reaction levels off earlier and the final iodine content is decreased.

In the presence of a concentration of the H_2O_2-generating system that is optimal for the iodination reaction, the coupling reaction always begins with a lag period, i.e., when the iodination reaction levels off.[14] This lag is constant whatever the concentration of iodide, thyroglobulin, or enzyme. The presence of such a lag is of both theoretical (see below) and practical

[11] J. M. Gavaret and J. Nunez, unpublished results, 1980.
[12] A. Virion, J. Pommier, and J. Nunez, *Eur. J. Biol.* **102,** 549 (1979).
[13] J. Pommier, D. Deme, and J. Nunez, *Eur. J. Biochem.* **37,** 406 (1973).
[14] D. Deme, J. Pommier, and J. Nunez, *Eur. J. Biochem.* **70,** 435 (1976).

interest: it allows one to prepare thyroglobulin samples that are iodinated and contain iodotyrosine residues potentially able to couple, but no hormones. Such preparations have been referred to as "preiodinated" thyroglobulin.

Measurement of the Iodine and Hormone Content [14]

The different thyroglobulin samples obtained in the presence of varying iodide concentrations are analyzed for their iodine, iodotyrosines, and hormone content. The total iodine content is easily estimated when ^{125}I- or ^{131}I-labeled iodide of known specific activity was used during the iodination reaction. The percentage of iodide incorporated in each assay is measured after separation from excess free iodide by paper chromatographic techniques. Knowing the specific activity of the labeled iodide used in the assay and the concentration of thyroglobulin, the number of iodine atoms per mole of protein can be easily calculated.

Spectrofluorometric methods have been used also to follow the time course of iodination. These methods are based on the observation that the emission of fluorescence of tyrosine is quenched upon iodination. However, the product of iodination, diiodotyrosine, also quenches the fluorescence of tyrosine, thus precluding the use of this technique for quantitative estimation of the level of iodination.[15]

The iodotyrosine content (monoiodotyrosine and diiodotyrosine) and the hormone content is measured after enzymatic hydrolysis of the iodinated thyroglobulin samples.[10] Excess iodide is removed by dialysis and the samples are hydrolyzed by Pronase (3.3 mg, 15 hr phosphate buffer, 0.05 M, pH 7.6). Hydrolysis may be completed by further treatment with leucine aminopeptidase (4 units, 9 hr), and with carboxypeptidase A (2 units, 15 hr) in 0.2 M Tris-HCl, pH 8.0. The free iodoamino acids can be analyzed and quantitated by paper chromatographic techniques [solvents: ethanol–ammonium carbonate, 0.2 M (1/0.5, v/v) for iodotyrosines, and pentane-3-ol saturated with ammonia, 2 N, for the hormones]. Thyroid hormones can also be estimated by other procedures, including radioimmunoassay.

With horseradish peroxidase (5–10 μg/ml) the iodoamino acid distribution for the same total iodine content markedly differs from that obtained either with thyroid peroxidase or lactoperoxidase.[16] In addition, with horseradish peroxidase the iodination reaction is inhibited in the presence of low iodide concentrations, for instance, with tracer-labeled iodide.[17]

[15] B. Wilson and J. Nunez, unpublished results, 1980.
[16] D. Deme, J. Pommier, and J. Nunez, *Biochim. Biophys. Acta* **540,** 73 (1978).
[17] J. Pommier, L. Sokoloff, and J. Nunez, *Eur. J. Biochem.* **38,** 497 (1973).

This enzyme cannot therefore be used to prepare radioiodinated proteins with very high specific activities.

Nonenzymatic Iodination and Coupling

These reactions can be also achieved in a variety of nonenzymatic conditions, for instance with iodine + iodide, with chloramine-T + iodide, or with iodine monochloride.[3,18]

With iodine + iodide the pH of the reaction must be slightly alkaline. With chloramine-T + iodide or ICl the iodination reaction can proceed easily at acidic or neutral pH, but little or no hormone is produced under these conditions.

Iodine Plus Iodide System

A solution of iodine + iodide is prepared in 50 mM phosphate buffer, pH 7.4, by mixing ^{125}I-labeled potassium iodide, potassium iodate, and sulfuric acid in molar ratios $5 : 1 : 3$. The iodination and coupling reaction are started by addition of varying amounts of iodide labeled with ^{125}I to thyroglobulin (1 μM) in 50 μM phosphate buffer, pH 7.4, at 0°. Slow addition (1 hr) and efficient stirring is maintained throughout the incubation.

Chloramine-T Plus Iodide System

The incubation medium contains thyroglobulin (1 μM), ^{125}I-labeled iodide (10–100 μM) in 50 mM phosphate buffer, pH 7.2. The reaction is started by the addition of chloramine-T and performed at room temperature with efficient stirring.

Iodine Monochloride

A solution of freshly redistilled iodine monochloride (1 μmol) in 3 volumes of 50 mM sodium phosphate at pH 5.5, 7.4, or 8.1 is added slowly (1 hr) at 2° to 1 ml of a solution of thyroglobulin (6.6 mg) in the same buffer. After another hour the thyroglobulin is precipitated with ammonium sulfate and dialyzed.

Dissociation between the Iodination and the Coupling Reactions

Several methods are now available that allow a clear dissociation between the iodination and the coupling reactions.

[18] A. Virion, D. Deme, J. Pommier, and J. Nunez, *Eur. J. Biochem.* **118,** 239 (1981).

As indicated above, the coupling reaction occurs only after a lag period when the concentration of H_2O_2 generating system is optimal for the iodination step. This allows one to prepare "preiodinated thyroglobulin" samples containing varying iodine contents but no hormones. However, iodotyrosine residues potentially able to couple are already present in these samples. This has been shown[14] by labeling thyroglobulin with a first iodine (^{125}I) isotope and removing excess iodide from the medium after addition of $NaHSO_3$ to stop the reaction before the end of the lag period. ^{125}I-labeled thyroglobulin (containing no ^{125}I-labeled hormones) was then incubated again with fresh enzyme and a second iodine isotope, ^{131}I. Under such conditions, ^{125}I-labeled hormones were formed during the second incubation. Such observations made it possible to study the coupling reaction separately. For instance, it has been possible to show that (1) the iodination reaction and the coupling reaction are catalyzed by two different enzyme–H_2O_2 species[19]; (2) iodide is not only a substrate for the iodination reaction, but is also required for the coupling reaction to occur.[14] SCN^-, an anion with the same size as iodide, mimics this effect of iodide[9] at very low concentrations (μM); (3) free diiodotyrosine increases the rate of the coupling reaction at very low concentration ($0.1–1$ μM).

Data[19] obtained by using lactoperoxidase have shown that the iodination reaction is catalyzed by the enzyme–H_2O_2 species, which is formed readily upon addition of hydrogen peroxide (i.e., compound **I**). This form of peroxidase contains two oxidizing equivalents above the native state and is rapidly transformed in the absence of substrate to another species, compound **II**, which is spectrally similar to horseradish peroxidase compound **II**,[1] but contains two oxidizing equivalents above the native state. Lactoperoxidase compound **I** and compound **II** probably differ from one another in the localization of one of the two oxidizing equivalents. Lactoperoxidase compound **II** seems therefore similar in these respects to cytochrome c peroxidase compound ES.[20] Lactoperoxidase compound **II** very efficiently catalyzes the coupling of preformed iodotyrosine residues but is unable to catalyze iodination. Moreover, free diiodotyrosine stimulates the transfer of electrons from the hormonogenic iodotyrosine residues of thyroglobulin to compound **II**, but is without effect on the iodination reaction. Finally iodide (and SCN^-) stimulates the coupling reaction by preventing the accumulation of lactoperoxidase compound **III**, an inactive enzyme–H_2O_2 species that is formed in the presence of an excess of H_2O_2.

[19] F. Courtin, D. Deme, A. Virion, J. L. Michot, J. Pommier, and J. Nunez, *Eur. J. Biochem.* **124**, 603 (1982).
[20] A. F. Coulson, J. E. Erman, and T. Yonetani, *J. Biol. Chem.* **246**, 917 (1971).

Other methods have allowed a clear dissociation between the iodination and the coupling reaction. As indicated above[12] horseradish peroxidase, when used at pH 5.0, catalyzes only very poorly the coupling reaction; increasing the pH to 8.2 results in the coupling of those hormonogenic iodotyrosine residues that were preformed at pH 5.0. Similar results were obtained by performing the iodination reaction at pH 5.0 with ICl and then treating the preiodinated thyroglobulin obtained in such conditions either at pH 7.4 with thyroid peroxidase or again with ICl at slightly alkaline pH.[10]

Products of the Coupling Reaction

The coupling reaction consists of (1) an oxidation step where two iodotyrosine residues properly positioned in the tertiary structure of thyroglobulin are oxidized by the enzyme; (2) the formation of a charge transfer complex between these two residues that might have the structure of an unstable quinol ether intermediate; (3) the decomposition of this intermediate to two products, a hormone residue and an alanine side chain ("lost side chain"):

$$2 \text{ Iodo-Tyr—R} \rightarrow \text{charge transfer complex} \rightarrow 1 \text{ hormone} + \text{—CH}_2\text{CH} \Big\langle^{\text{NH}_2}_{\text{COOH}}$$

The "fate" of the lost chain has been a controversial issue for the last 40 years. It has been solved only recently[3,10,21] by using [^{14}C]Tyr— or [^{3}H]Tyr-labeled thyroglobulin iodinated *in vitro* enzymatically and nonenzymatically. It has been found that the "lost side chain" decomposes spontaneously, yielding 1 mol of dehydroalanine and 1 mol of protons

$$\text{—CH}_2\text{—CH} \Big\langle^{\text{NH—R}}_{\text{CO—R}'} \rightarrow \text{CH}_2\text{=C} \Big\langle^{\text{NH—R}}_{\text{CO—R}'} + \text{H}^+ \tag{5}$$

1 alanine radical → 1 dehydroalanine + 1 proton

Identification of Dehydroalanine

Dehydroalanine can be identified by several techniques. Upon proteolytic hydrolysis of [^{14}C]Tyr-labeled thyroglobulin (for conditions, see above under Measurement of the Iodine and Hormone Content) dehy-

[21] J. M. Gavaret, J. Nunez, and H. Cahnmann, *J. Biol. Chem.* **255**, 5281 (1980).

droalanine decomposes to [^{14}C]pyruvic acid. Upon acid hydrolysis (with constant-boiling HCl for 24 hr at 105° in evacuated sealed glass tubes) it decomposes to [^{14}C]acetic acid

$$[^{14}\text{C}]\text{Dehydroalanine}\text{—R} \overset{\nearrow}{\underset{\searrow}{}} \begin{array}{l} [^{14}\text{C}]\text{pyruvic acid (enzymatic proteolysis)} \\ \\ [^{14}\text{C}]\text{acetic acid (HCl hydrolysis)} \end{array} \tag{6}$$

[^{14}C]Pyruvic acid and [^{14}C]acetic acid are easily purified by ion-exchange chromatography (AG 50 W-X4).

Dehydroalanine can be also converted before hydrolysis to a [^{14}C]alanine residue (after reduction by sodium borohydride of [^{14}C]Tyr-labeled thyroglobulin) or to a benzylcysteine residue (after treatment of nonlabeled thyroglobulin with benzylmercaptan). Both [^{14}C]alanine and benzylcysteine can be identified by ion-exchange chromatography (AG 50W-X4) after enzymatic proteolysis.

$$[^{14}\text{C}]\text{Dehydroalanine—R} \xrightarrow{\text{BH}_4} [^{14}\text{C}]\text{alanine—R} \tag{7}$$

$$\text{Dehydroalanine—R} \xrightarrow{\text{C}_6\text{H}_5\text{SH}} \text{C}_6\text{H}_5\text{—S—CH}_2 \overset{\text{NH—R}}{\underset{\text{CO—R}'}{<}} \tag{8}$$

Quantitative measurements showed that these dehydroalanine derivatives were formed in a molar ratio of 1 mol of "side chain" per mole of hormone synthesized.

Reaction of Iodinated Thyroglobulin with Benzylmercaptan. Iodinated [^{14}C]thyroglobulin (13 mg) was dissolved in 200 μl of N-ethylmorpholine–acetic acid buffer, pH 8.5 (10% aqueous N-ethylmorpholine was adjusted at pH 8.5 with 2 M acetic acid and filtered through a 0.22 μM Millipore filter) and 200 μl of water. Absolute ethanol (700 μl) and benzylmercaptan (50 μl) were then added. The turbid reaction mixture was freed from oxygen and then sealed in a nitrogen-filled glass tube. After 10 days the mixture was dialyzed several times against 30% ethanol and then against water, until most droplets of benzyl-mercaptan had disappeared. The solution was then lyophilized, and the residue, after addition of 2 mg of phenol, was hydrolyzed with 2 ml of constant-boiling HCl.

Reaction of Iodinated [U-^{14}C]Tyr-Labeled Thyroglobulin with Sodium Borohydride. Iodinated [U-^{14}C]thyroglobulin (3.3 mg) was dissolved in 1 ml of freshly prepared sodium borate–guanidine buffer, pH 8.5 (guanidine hydrochloride was added to 0.1 M sodium borate buffer, pH 8.5, to a final guanidine concentration of 6 M; the pH was then readjusted to 8.5 with NaOH). Sodium borohydride (50 mg) was then added, and the solution was kept at room temperature for 3 days with occasional low stirring.

During the reaction a precipitate formed. The pH dropped first to ~7.5 and then slowly rose again to 9.5–10. It was readjusted to 8.5 with hydrochloric acid. After another 5 hr, unreacted sodium borohydride was decomposed by careful addition of 1 ml of constant-boiling hydrochloric acid, and the reaction mixture was concentrated in a rotating evaporator to about half its volume and finally dried by lyophilization. The residue was hydrolyzed with 3 ml of constant-boiling hydrochloric acid.

Measurement of the Proton[11]

One mole of proton was also released during the coupling reaction per mole of hormone formed. This was measured by preparing thyroglobulin labeled with [³H]Tyr. During the coupling reaction the ³H-labeled hydrogen present on the asymmetric carbon of the "lost side chain" is released in the medium and exchanges with water. At the end of the time course of iodination, the solution was distilled and the radioactivity was counted in an aliquot of the distillate. The data showed that one proton is released for each mole of hormone formed.

Comments

In conclusion, the mechanism of phenoxy ether formation seems to depend both on the properties of the enzyme and on the spatial properties of the protein, thyroglobulin, which is the "matrix" of thyroid hormone formation.

The identification of the "lost side chain" as dehydroalanine has made it possible to present a model for the reaction. According to this model the withdrawal of two electrons from a pair of hormonogenic iodotyrosine residues formed during the iodination of thyroglobulin will lead to the formation of a charge-transfer complex. This complex is a zwitterion resonance hybrid no matter whether the oxidation step leading to it is an ionic or a free-radical process (Fig. 5).

The enzyme–H_2O_2 species that catalyzes this reaction has the spectrum of horseradish peroxidase compound **II**[1] but contains two oxidizing equivalents above the native state as cytochrome c peroxidase compound ES.[20] A different enzyme–H_2O_2 species, compound **I**, seems to catalyze the iodination reaction. These two enzyme–H_2O_2 species do not seem to differ insofar as the number of oxidizing equivalents is concerned; rather, they contain a different distribution between the heme and the apoprotein of these two oxidizing equivalents, as proposed by Coulson et al.[20] for cytochrome c peroxidase compound ES. According to these assumptions, each enzyme–H_2O_2 species would interact specifically with only one of the

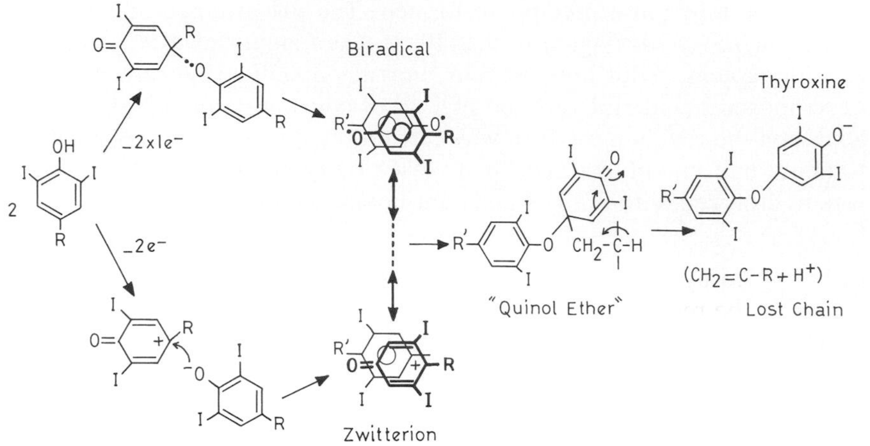

FIG. 5. Mechanism of the coupling reaction.

two different substrates, either a tyrosine residue that undergoes iodination or two iodotyrosine residues that undergo coupling.

Another parameter that is of utmost importance in describing the mechanism of phenoxy ether formation is the structure of thyroglobulin. The conformation of this protein in solution (which should be different for different degrees of iodination[22]) should be such that certain iodotyrosine residues must not only be close to each other in space, but also be oriented in such a way[23] that upon oxidation a charge-transfer complex can be formed.

[22] C. Marriq, C. Arnaud, M. Rolland, and S. Lissitzky, *Eur. J. Biochem.* **111,** 33 (1980).
[23] H. Cahnmann, J. Pommier, and J. Nunez, *Proc. Natl. Acad. Sci. U.S.A.* **74,** 5333 (1977).

[31] Iodotyrosine Deiodinase from Bovine Thyroid

By ISADORE N. ROSENBERG and AJIT GOSWAMI

$$\text{3,5-L-DIT} + \text{NADPH} \rightarrow \text{3-L-MIT} + \text{I}^- + \text{NADP}^+ \qquad (1)$$
$$\text{3-L-MIT} + \text{NADPH} \rightarrow \text{tyrosine} + \text{I}^- + \text{NADP}^+ \qquad (2)$$

Mono- and diiodotyrosine[1] are formed by the iodination of tyrosyl residues in thyroglobulin, the major iodinated protein of the thyroid gland. The iodotyrosines account for two-thirds of the iodine in thyro-

[1] DIT, Diiodo-L-tyrosine; MIT, monoiodo-L-tyrosine.

globulin and serve as precursors for the formation of the thyroid hormones thyroxine (T_4) and triiodothyronine (T_3). The molar ratio of iodotyrosine : iodothyronine in thyroglobulin is approximately 4 : 1.[2] Secretion of T_4 and T_3 by the thyroid requires proteolysis of thyroglobulin, in the course of which the free amino acids MIT and DIT are released from peptide linkage. Free MIT and DIT cannot be reutilized as such for thyroglobulin synthesis, but are enzymatically deiodinated, via the reactions shown above; the enzymatic deiodination is a reductive process that leads to the formation of I^- and tyrosine, both of which can be reutilized for thyroglobulin synthesis. Thus, the iodotyrosine deiodinase represents a mechanism for glandular iodine conservation, the importance of which is evident from the occurrence of goiter and hypothyroidism if the enzyme is genetically deficient or impaired.[3]

The enzyme occurs principally in the particulate (light mitochondrial and heavy microsomal) fraction of the thyroid; it is found also in liver and kidney. The naturally occurring cofactor is NADPH. Early studies[4] showed that dithionite could replace NADPH in this reaction. Treatment of the particulate fraction with steapsin and subsequent purification has led to the isolation of a soluble flavoprotein enzyme, essentially homogeneous on gel electrophoresis, that catalyzes the deiodination of MIT and DIT. However, unlike the native particulate enzyme, which can utilize both NADPH and dithionite for the deiodination of the iodotyrosines, the soluble flavoprotein enzyme does not respond to NADPH but is activated by dithionite and by other strong reducing agents, such as reduced flavins, viologens, and ferredoxins. Although a crude detergent-solubilized enzyme preparation that is active with NADPH has been prepared from the bovine thyroid particulate fractions,[5] further purification of this NADPH-responsive enzyme has not been successful. The procedures described below are, therefore, for the preparation of the purified dithionite-responsive enzyme.

Assay Methods

Principle. The assay measures the amount of inorganic iodide generated by action of the enzyme in the presence of [^{125}I]iodotyrosines of known specific activity; iodide is separated from the unconverted substrate by using a cation-exchange resin column at low pH. The method is

[2] M. R. Rolland, R. Aquaron, and S. Lissitzky, *Anal. Biochem.* **33,** 307 (1970).
[3] J. B. Stanbury, A. A. H. Kassenaar, and J. Meijer, *J. Clin. Endocrinol. Metab.* **16,** 848 (1956).
[4] I. N. Rosenberg and C. S. Ahn, *Endocrinology* **84,** 727 (1969).
[5] I. N. Rosenberg and A. Goswami, *J. Biol. Chem.* **254,** 12318 (1979).

applicable to assays of crude tissue preparations as well as purified enzyme.

Reagents

L-[^{125}I]Iodotyrosines: Labeled DIT and MIT are prepared as described below

Stock solutions of substrate containing 10 μM DIT and 0.025–0.05 μCi ^{125}I/ml: The solutions are made by adding stable L-DIT to an appropriate quantity of the high-specific-activity DIT in 10 mM Tris buffer (pH 7.4) containing 5 μM methimazole; the methimazole is added to minimize photocatalyzed oxidative deiodination during storage. The solution can be stored frozen for several weeks and should be monitored for the presence of inorganic iodide-^{125}I, which should not exceed 2–3%, by passage through ion-exchange columns (see below)

NADPH, 1 mM in 1% NaHCO$_3$

Sodium dithionite: 200-mg portions are dissolved in 3 ml of 5% NaHCO$_3$ just before use

Potassium phosphate buffer, 0.5 M, pH 7.4

KCl, 2 M

2-Mercaptoethanol, 0.5 M

Flavin adenine dinucleotide, 250 μg/ml

0.1% L-DIT in 0.1 N NaOH

Labeled iodotyrosines are conveniently prepared by the exchange labeling method described by Cahnman.[6] A typical labeling procedure employed in our laboratory is as follows: to 100 μl of 0.1 M L-DIT in 0.1 M ammonia (adjusted to pH 4 with dilute acetic acid) in a 2-ml glass-stoppered tube with a conical bottom, are added 1 mCi of carrier-free Na-^{125}I (in a volume of 1–2 μl) and 20 μl of a solution of 0.05 M I$_2$ in methanol. The reaction mixture is allowed to stand for 10 min with occasional shaking and then decolorized by the addition of a drop of a saturated aqueous solution of sodium metabisulfite. The solution is then applied to a column (4 × 1 cm) of cation-exchange resin [Bio-Rad AG 50W-X8 (mesh size 100–200)] equilibrated with 10% acetic acid (see below). After washing the column with 10% acetic acid until no further radioactivity appears in the eluate (25–30 ml), the DIT is eluted with 5 successive 2.5-ml portions of 1 N NH$_4$OH. Maximum counts usually appear in the third tube, which is used for preparing the standard solution. The concentration of DIT in the peak tube is determined spectrophotometrically from the absorbance at

---

[6] H. J. Cahnman, *in* "Methods in Investigative and Diagnostic Endocrinology" (S. A. Berson, ed.), Vol. 1, p. 27. American Elsevier, New York, 1972.

224 nm using the molar extinction coefficient 27,300. The specific activity of the DIT ranges between 1 and 5 $\mu Ci/\mu mol$. The preparations are stored in 50% propylene glycol at $-25°$. A suitable aliquot is purified by ascending paper chromatography (butanol–acetic acid–water, 78 : 5 : 17) before making the standard solution.

The procedure for exchange-labeling MIT is essentially the same as that outlined above for DIT, except for the use of a lower I_2 : iodoamino acid ratio (usually 1 : 20) so as to minimize concomitant formation of DIT.

Cation-Exchange Resin Columns. In our earlier work,[4] we used Rexyn 101(H), 40–100 mesh, purchased from Fisher Scientific Company. This resin is no longer marketed in this mesh size; a satisfactory substitute is Bio-Rad AG 50W X8 (mesh size 100–200). The resin columns are made in the inverted barrels of plastic 5-ml syringes with glass wool plugs in the outlet; the slurried resin, previously equilibrated with 10% acetic acid, is poured so that the top of the column is at the 4-ml mark. We usually use a column 4–5 times, washing with two 5-ml portions of 10% acetic acid after each use. If incomplete retention of iodotyrosines by the resin is suspected, the column is regenerated by successive washes with 2 N NaOH, water, and 10% acetic acid.

Procedure. The standard assay mixture contains 200 μl of potassium phosphate buffer, variable amounts (10–100 μl) of the enzyme suspension, and 100 μl each of 2-mercaptoethanol, FAD, KCl, and stock ^{125}I-DIT solution. When tissue (thyroid, liver) homogenates or particulate preparations are assayed, 100 μl of 15 mM methimazole is also included in the mixture to prevent organic binding of liberated iodide. A blank tube is prepared in which an appropriate volume of potassium phosphate buffer is substituted for the enzyme preparation. The volume is made to 900 μl with deionized glass-distilled water. The reaction is started by the addition of 0.1 ml of NADPH or dithionite, and the mixture is incubated for 10 min at 37°. The reaction is terminated by the addition of 100 μl of 0.1% DIT solution. Then 250 μl of the mixture is transferred to a counting tube (S), and the volume is made 5.0 ml by adding 10% acetic acid; this tube is used to determine the total ^{125}I in each incubation vessel. Another 500-μl aliquot of the mixture is applied to a cation-exchange column, and the effluent is collected in another counting tube (A). The column is then eluted with 4.5 ml of 10% acetic acid, the eluate being collected in tube A. A second elution is then made with 5 ml of 10% acetic acid, and this effluent is collected in another tube (B). Under these conditions, virtually all the [^{125}I]iodide is recovered in tubes A and B, and practically all the ^{125}I present as iodotyrosines is retained in the column. Radioactivity in effluents and aliquots of the incubation medium is measured in a gamma counter equipped with a well crystal.

Calculation and Definition of Units. Using the aliquots of the incubation mixtures described above [0.25 ml for the standard tube (S) and 0.5 ml applied to the column], the term

$$\frac{\text{cpm in A} + \text{cpm in B}}{2 \times \text{cpm in S}}$$

is the fraction (F_1) of labeled iodide in the incubation mixture. F_1 is corrected by subtracting the corresponding fraction (F_0) obtained for the blank (nonenzyme-containing) incubation tube. Under the above conditions—i.e., initial amount of substrate = 1 nmol, initial volume = 1.0 ml, and final volume = 1.1 ml, and 10 min of incubation—the rate of iodide production (nanomoles of I$^-$ per hour) is obtained from

$$(F_1 - F_0) \times \frac{1.1}{0.5} \times \frac{60}{10} \times 2$$

The factor 2 is necessary in the calculation when DIT is the substrate for the following reason. The atoms of ^{125}I in the labeled DIT molecules may be considered statistically to be in either the 3 or the 5 position, but not in both; hence, the molar-specific activity of the iodide released on deiodination of DIT is half that of the DIT itself, and the product of the fraction of labeled iodide and the initial concentration of DIT must be multiplied by 2 to yield the moles of iodide produced. When [^{125}I]MIT is used as the substrate, the factor 2 is not used (the specific activity of the iodide produced being equal to the specific activity of the substrate MIT).

One unit of enzyme activity is defined as the amount of enzyme that catalyzes the release of 1 nmol of iodide per hour at an initial concentration of 1 μM DIT. Specific activity is expressed as units per milligram of protein.

Purification of the Dithionite-Responsive Iodotyrosine Deiodinase

Fresh thyroid glands obtained from the local abattoir are dissected free of connective tissue and stored at $-25°$ until a sufficient quantity (usually 1 kg) has been accumulated.

All purification steps are carried out at 0–4° unless otherwise indicated. Protein is determined by the method of Lowry *et al.*[7]

Step 1. Homogenate. The frozen glands are thawed, minced with fine scissors, and soaked overnight in an equal volume of 0.25 M sucrose (containing 50 mM Tris, pH 7.4, 1 mM EDTA, and 0.1 mM mercapto-

[7] O. H. Lowry, N. J. Rosebrough, A. L. Farr, and R. J. Randall, *J. Biol. Chem.* **193,** 265 (1951).

ethanol). This step releases most of the thyroglobulin and facilitates subsequent homogenization. After discarding the supernatant, the tissue is resuspended in 3 volumes of the sucrose solution and disrupted by two 10-sec bursts in a Waring blender. The disrupted tissue suspension is then transferred, in 50-ml batches, to a motor-driven Potter–Elvehjem Teflon–glass homogenizer and homogenized by four to six passes of the piston. Use of a Polytron homogenizer (Brinkmann) has also been found to be satisfactory: 3–4 10-sec bursts at full speed usually suffice to yield a preparation that is comparable to the Potter–Elvehjem Teflon–glass homogenate.

Step 2. Particulate Fractions. The homogenate is transferred to 250-ml polycarbonate centrifuge bottles and centrifuged at 1800 g for 30 min. The supernatant is then apportioned into 30-ml Oak Ridge-type polycarbonate centrifuge tubes and centrifuged for 90 min in a Spinco L2-65 centrifuge at 30,000 rpm (RCF$_{av}$ 78,480 g) using the No. 30 Rotor. The supernatants, which contain no activity, are discarded, and the pellets are resuspended in the sucrose medium by vigorous agitation on a vortex mixer. After recentrifugation at 30,000 rpm for 90 min, the pellets are combined, suspended in 0.1 M potassium phosphate buffer (pH 7.4), and adjusted to a protein concentration of 10–15 mg/ml (fraction A).

Step 3. Solubilization. To an appropriate volume of fraction A is added, with stirring, 0.1 volume of a solution of 0.7% steapsin (pancreatic lipase, obtained from Nutritional Biochemicals) and 0.01 volume of 0.1 M EDTA (pH 7.0). The mixture is stirred gently overnight at 4° and then centrifuged at 100,000 g for 2 hr. The clear yellow supernatant (fraction B) contains about 20% of the protein and nearly all the deiodinating activity as assayed with dithionite. This fraction is unresponsive to NADPH, but is active with dithionite or with NADPH plus methyl viologen, presumably through NADPH-mediated reduction of methyl viologen by NADPH–cytochrome c reductase, which is present in this fraction. The pellet has very little or no activity, either with NADPH or dithionite.

Step 4. Ammonium Sulfate Treatment. Fraction B is brought to 50% saturation by the slow addition of solid ammonium sulfate with continuous stirring. After standing overnight at 4°, the suspension is centrifuged at 100,000 g for 60 min. The pellet is dissolved in 10 mM potassium phosphate buffer (pH 7.4)–0.3 mM EDTA and dialyzed[8] for 24 hr against

---

[8] Spectrapor 1 dialysis tubing (purchased from Fisher Scientific Co.) may be used without further treatment. Some commercially available brands of cellophane tubing must be specially treated to remove heavy metals and other contaminants (P. McPhie, this series, Vol. 22 [4]). Dialysis in untreated cellophane bags may cause rapid and profound loss of enzymatic activity.

100 volumes of the same buffer at 4° with 3 or 4 changes until tests of bag contents reveal no sulfate (as judged by absence of acid-insoluble turbidity with $BaCl_2$). The dialyzed preparation (fraction C) usually contains about 40% of the proteins in fraction B and shows approximately a 2-fold increase in specific activity.

Step 5. DEAE-Cellulose Chromatography. Fraction C is applied to a column of DEAE-cellulose (DE-52; 1.5 × 25 cm) previously equilibrated with 10 mM potassium phosphate (pH 7.4) and washed with 125 ml of the same buffer. The column is then eluted with a linear gradient generated from 250 ml of 10 mM potassium phosphate buffer (pH 7.4) containing 50 mM KCl and 250 ml of the same buffer containing 0.5 M KCl; 2.5-ml fractions are collected. Dithionite-dependent deiodinase activities are monitored using 50-μl aliquots of each fraction. The peak activity is eluted at approximately 0.15 M KCl (Fig. 1). The active fractions are combined, concentrated to about 40 ml on an Amicon PM-10 membrane filter, and then dialyzed overnight against 10 mM potassium phosphate (pH 7.4) with two changes. The resultant preparation (fraction D) has about one-third of the proteins in fraction C and shows a 3-fold increase in specific activity. The DEAE-cellulose chromatography separates the deiodinase enzyme from NADPH–cytochrome c reductase (Fig. 1) and renders it unresponsive to methyl viologen in the presence of NADPH. Recombina-

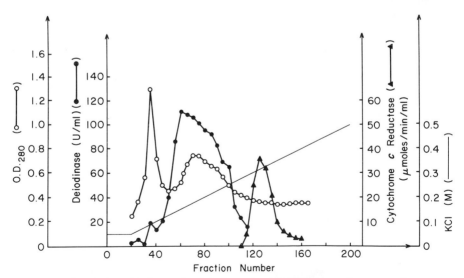

FIG. 1. Separation of iodotyrosine deiodinase from NADPH–cytochrome c reductase on DEAE-cellulose. The column was developed with a linear gradient of 0.05 to 0.5 M KCl in 10 mM potassium phosphate (pH 7.4), and 2.5-ml fractions were collected. Aliquots (50 μl) from each fraction were used for monitoring enzyme activities.

tion of fraction D with the reductase peak restores the response to viologen dyes in the presence of NADPH.

Step 6. First Hydroxyapatite Chromatography. Dialyzed fraction D is
adsorbed on a hydroxyapatite (BioGel HT, obtained from Bio-Rad Laboratories) column (2.5 × 20 cm), equilibrated with 10 mM potassium phosphate buffer (pH 7.4), and washed with 100 ml of the same buffer. The
effluent contains about one-third of the total proteins, but no activity.
This is followed by eluting with 100 ml of 0.1 M potassium phosphate
buffer, which removes another one-third of proteins and 1% of total activity. The column is then washed with 100 ml of 0.4 M potassium phosphate
buffer. This results in elution of more than 90% of the enzymatic activity
with a 10-fold increase in specific activity. The active fractions (E-1A) are
combined and dialyzed against 0.05 M potassium phosphate buffer (pH
7.4).

Step 7. Second Hydroxyapatite Chromatography. This is a scaled
down version of step 6. Pooled and dialyzed active fractions from step 6
are applied to a hydroxyapatite column (1.5 × 10 cm) equilibrated with
0.05 M potassium phosphate (pH 7.4). The column is first washed with 25
ml of 0.1 M potassium phosphate (pH 7.4), and then the enzymatic activities are eluted with 25 ml of 0.4 M potassium phosphate (pH 7.4). The
active fractions are combined and diluted to 0.05 M potassium phosphate
(fraction E-1B).

Step 8. Third Hydroxyapatite Chromatography (Gradient Elution).
Fraction E-1B is then applied to a third hydroxyapatite column (1 × 10
cm) equilibrated with 0.05 M potassium phosphate (pH 7.4). The column
is washed with 25 ml of the same buffer and then developed with a linear
gradient generated by 50 ml of 0.075 M potassium phosphate and 50 ml of
0.3 M potassium phosphate; 1-ml fractions are collected. The peak activity is eluted at about 0.25 M potassium phosphate. The active fractions
are pooled, concentrated on an Amicon PM-10 membrane filter,[9] and
equilibrated with 0.25 M potassium phosphate by overnight dialysis (fraction E-2).

Step 9. Chromatography on Sephadex G-75. Fraction E-2 (volume
approximately 10 ml) is passed through a column of Sephadex G-75 (2.5 ×
40 cm) equilibrated with 0.25 M potassium phosphate. Protein is eluted
with the same buffer at a flow rate of 40 ml/hr. Fractions of 1 ml are
collected, and 10-μl aliquots are assayed for deiodinase activity. At this

---

[9] Another frequently used method for concentrating large volumes of the sample at this or
the subsequent stages is to dilute it to <0.1 M and pass it through a small hydroxyapatite
column (0.5–1 ml). The adsorbed activity can then be eluted with a few milliliters of 0.4 M
potassium phosphate buffer.

PURIFICATION OF IODOTYROSINE DEIODINASE FROM BOVINE THYROID MICROSOMES

Fraction	Total protein (mg)	Total activity (units)	Specific activity (U/mg protein)
A. Mitochondrial-microsomal pellet	19,930	129,560	6.5[a]
B. Steapsin supernatant	4750	123,310	26[b]
C. Ammonium sulfate precipitate	1910	81,600	43
D. DEAE-cellulose eluate	618	73,100	118
E-1. Hydroxyapatite eluate			
A	60	63,060	1051
B	18	33,800	1867
E-2. Hydroxyapatite eluate	6.1	28,173	4600
F. Sephadex G-75 eluate	0.75	5863	7817

[a] As assayed with NADPH, freshly prepared particulate fraction A often has a specific activity approximately 25% higher than that obtained with dithionite. The NADPH response, however, is more labile, and, in stored preparations, the dithionite-supported activity tends to be greater than in presence of NADPH.

[b] The preparation, at this stage of purification as well as all subsequent ones, is unresponsive to NADPH but responds to dithionite and other strong reducing agents, such as viologen dyes and ferredoxins. The activities given here were obtained with dithionite.

step, the protein is resolved into two broad bands. The fractions in the first are devoid of deiodinase activity. The second band, the peak tube of which is eluted with a V_e/V_0 ratio of approximately 1.34, contains all the enzymatic activity. The peak enzyme fractions are pooled and concentrated on an Amicon PM-10 membrane.

A summary of a representative purification of enzyme from 1 kg of thyroid tissue is presented in the table.

Properties

Stability. The dithionite-responsive deiodinase activity is stable stored at $-25°$ for several months. The microsomal NADPH-responsive deiodinase activity is labile and falls rapidly on storage, showing 60–70% of the dithionite-responsive activity after about 3–4 weeks.

Purity. The purified enzyme preparation is homogeneous by polyacrylamide gel electrophoresis giving a single band at a loading of 100 μg of protein per gel.[5] The purest preparations have a specific activity of approximately 7500 U/mg.

Flavoprotein Nature. The purified enzyme shows the characteristic absorption spectrum of a flavoprotein (Fig. 2) with maxima at 274,

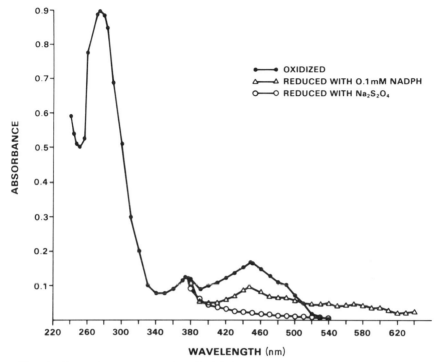

FIG. 2. Absorption spectrum of purified iodotyrosine deiodinase in 10 mM potassium phosphate (pH 7.4).

373, and 447 nm with a shoulder at 490 nm.[10] The flavin peak at 447 nm is completely reduced by dithionite. Anaerobic photoreduction by 0.1 mM NADPH generates a half-reduced species, with broad absorption bands between 510 and 610 nm characteristic of neutral flavin semiquinones (Fig. 2). The fluorescence emission spectrum (emission peak at 530 nm upon excitation at 450 nm, eliminated by dithionite) is characteristic of a flavoprotein. The flavin liberated on heating the enzyme has been identified as FMN by paper electrophoresis.[5] The fluorescence of the flavoenzyme is only slightly less than that of free FMN or FAD, indicating a relative paucity of neighboring aromatic residues.

An apoenzyme, prepared by treatment of preparation F with 1.5 M guanidine-HCl, is devoid of flavin fluorescence and catalytic activity.[10] The activity and the spectral properties of the apoenzyme can be fully restored by recombination with FMN ($K_a = 1.47 \times 10^8 \ M^{-1}$) and, to a lesser extent, with FAD ($K_a = 0.577 \times 10^8 \ M^{-1}$) but not with riboflavin.

[10] A. Goswami and I. N. Rosenberg, *J. Biol. Chem.* **254,** 12326 (1979).

Molecular Weight and Sedimentation Characteristics. The minimum molecular weight (M_r) based on flavin analysis is 41,700, which agrees well with an M_r of 42,000 obtained by gel filtration. Sodium dodecyl sulfate–polyacrylamide gel electrophoresis after reduction of disulfide bonds gives a single band with a molecular weight of 22,700, indicating the presence of two identical subunits within the enzyme molecule. The analytical data of 1 mol of flavin per mole of enzyme would thus indicate that the flavin is bound to only one of the subunits or that a single flavin molecule is shared cooperatively by two identical subunits.

The purified enzyme appears to form a dimer in concentrated sucrose solutions and has a sedimentation coefficient of 4.5 S in 5 to 20% sucrose gradients.

Substrate Specificity. The enzyme is specific for mono- and diiodotyrosines and does not deiodinate iodothyronines. The K_m values for DIT and MIT are approximately 2.5 and 50 μM, respectively. High concentrations (>5 μM) of L-DIT, but not L-MIT, are inhibitory. The enzyme has an exceptionally low turnover number, approximately 10 for L-DIT and 20 for L-MIT.

pH Optimum. The enzymatic activity shows a maximum at pH 6.3 with a sharp decline at more acid pH and above pH 8. The apparent pK for the enzyme-bound flavin has been determined to be 6.24.

Activators. The enzyme requires high ionic strength (equivalent to 0.2 M potassium phosphate) for optimal activity. Thiol compounds (mercaptoethanol, dithiothreitol) strongly activate the enzyme. Supplementation with micromolar concentrations of FMN or FAD, but not riboflavin, results in a 3- to 4-fold stimulation of basal activity. The enzyme can be partially activated by anaerobic photoreduction in the presence of FMN and, to a lesser extent, of FAD, but cannot be photoactivated in absence of added flavins. The enzyme-catalyzed reaction is a reductive process, and the reaction rates are the same in O_2 as in a N_2 atmosphere. The enzyme has an absolute requirement for strong reductants (dithionite, reduced viologens, reduced ferredoxins) for its catalytic activity.

Inhibitors. Substrate inhibition is observed at high (>5 μM) concentrations of L-DIT, but not with L-MIT. The nitro analogs 3-nitro-L-tyrosine and 3,5-dinitro-L-tyrosine are potent competitive inhibitors of the NADPH-supported microsomal deiodinase activity,[11] but have no effects on the dithionite-responsive activities, probably owing to the immediate reduction of the nitro groups by dithionite. 3,5-Dibromotyrosine is also inhibitory, whereas the 3-methyl, propyl, isopropyl, or amino derivatives have no effects.[11] The iron chelators dipyridyl and 1,10-phenanthroline,

[11] W. L. Green, *Endocrinology* **83**, 336 (1968).

which inhibit the NADPH-responsive microsomal activity, do not affect the dithionite-responsive purified enzyme, suggesting involvement of an iron-containing intermediate in the NADPH-mediated microsomal deiodination.[12] Preincubation with N-ethylmaleimide (NEM) almost completely inhibits the deiodinase activity with reduced methyl viologen or reduced ferredoxin, but does not affect the activity with dithionite, presumably owing to a displacement of bound NEM by dithionite with a restoration of enzyme SH groups.[12]

Redox Properties. The enzyme has a highly negative oxidation–reduction potential (−412 mV) and requires full reduction of the flavin for catalytic activity. NADPH or photoreduction of the enzyme generates a half-reduced species that is catalytically inactive. Reduction with dithionite ($E'_0 = 660$ mV[13]), reduced methyl viologen ($E'_0 = -446$ mV[14]), reduced clostridial ferredoxin ($E'_0 = -420$ mV[14]), and, to a lesser extent, reduced adrenodoxin ($E'_0 = -360$ mV[15]), results in the formation of a fully reduced species and in considerable activation.[16]

Concluding Remarks

The enzyme-catalyzed deiodination is a reductive process and is distinct from the heat-stable, photoactivated, oxygen-requiring microsomal deiodinating systems that have been described in various tissues, including the thyroid,[17,18] which seem to be peroxidative reactions of a probably artifactual, nonphysiological nature.

The highly negative oxidation–reduction potential of the enzyme and the inhibition of the NADPH-supported microsomal deiodination by iron chelators suggest the involvement of an intermediate, probably a non-heme iron protein, in the NADPH-mediated deiodination *in situ*. This putative intermediate is probably not extracted by steapsin digestion, or is destroyed by proteases contained in the steapsin preparations used for solubilization. Ferredoxin and NADPH-ferredoxin reductase have been identified in bovine thyroids and have been shown to catalyze NADPH-dependent deiodination of DIT.[16] It is conceivable that the solubilized and

[12] A. Goswami and I. N. Rosenberg, *Endocrinology* **101,** 331 (1977).

[13] S. G. Mayhew, *Eur. J. Biochem.* **85,** 535 (1978).

[14] H. R. Mahler and E. H. Cordes, "Biological Chemistry." Harper & Row, New York, 1971.

[15] R. W. Estabrook, K. Suzuki, J. I. Mason, J. Baron, W. E. Taylor, E. R. Simpson, J. Purvis, and J. McCarthy, *in* "Iron-Sulfur Proteins" (W. Lovenberg, ed.), Vol. 1, p. 193. Academic Press, New York, 1973.

[16] A. Goswami and I. N. Rosenberg, *J. Biol. Chem.* **256,** 893 (1981).

[17] J. B. Stanbury, *Ann. N. Y. Acad. Sci.* **86,** 417 (1960).

[18] S. Lissitzky and M. R. Roques, *C. R. Seances Soc. Biol. Ses Filiales* **152,** 1333 (1958).

purified deiodinase could be a derivative of NADPH–ferredoxin reductase, perhaps formed by the action of proteolytic enzymes present in steapsin, which has lost its ability to interact with thyroid ferredoxin for the transfer of the reducing equivalents of NADPH to the enzyme flavin. Dithionite or other effective strong reductants may activate the enzyme by fully reducing the enzyme flavin directly.

Section III

Miscellaneous Derivatives

[32] γ-Carboxyglutamic Acid: Assay Methods

By Gary L. Nelsestuen

Historically, the identification of γ-carboxyglutamic acid in proteins was greatly complicated by its decarboxylation under acid conditions to produce a quantitative yield of glutamic acid. Base hydrolysis releases γ-carboxyglutamic acid from peptides, but often produces ninhydrin-positive materials that coelute with γ-carboxyglutamic acid from the amino acid analyzer using standard programs.[1,2] One unpublished study concluded that EDTA used in sample preparation may be the source of the interfering material.[3] Nevertheless, base hydrolysis of the protein followed by use of cation-exchange resins and the amino acid analyzer has remained the most common approach for identification of γ-carboxyglutamic acid. Isolation of the γ-carboxyglutamic acid peak and decarboxylation to glutamic acid is considered necessary for positive identification. Since this approach has been presented in this series,[4] it will not be covered here.

Owing to the difficulties cited above and the desire to develop more specific single-step methods for assay of γ-carboxyglutamic acid, a number of techniques have been reported in the literature. Aside from cation or anion exchange with the amino acid analyzer, no single method has been widely adopted.

Anion Exchange on the Amino Acid Analyzer

Tabor and Tabor[5] reported the use of the automated amino acid analyzer and an anion exchange resin for quantitation of γ-carboxyglutamate. The column (0.9 × 12.5 cm) was packed with Aminex A-27 or A-28 (Bio-Rad) and eluted isocratically with 0.3 M sodium acetate buffer, pH 4.6. Buffer flow was 70 ml/hr at 30°. The protein samples were hydrolyzed with 100 μl of 15% NaOH, and the pH was adjusted with 0.78 mmol of acetic acid before application to the column. γ-Carboxyglutamic acid eluted at 55 min and gave about a 40% color yield compared to glutamic acid.

[1] P. V. Hauschka, *Anal. Biochem.* **80,** 212 (1977).
[2] S. E. Hamilton, G. King, D. Tesch, P. W. Riddles, D. T. Keough, J. Jell, and B. Zerner, *Biochem. Biophys. Res. Commun.* **108,** 610 (1982).
[3] J. B. Howard, personal communication.
[4] P. A. Price, this series, Vol. 91, p. 13.
[5] H. Tabor and C. W. Tabor, *Anal. Biochem.* **78,** 554 (1977).

METHODS IN ENZYMOLOGY, VOL. 107

This approach appears to be very convenient and can be used with standard equipment. However, elution position does not provide an unambiguous identification of γ-carboxyglutamic acid.[6] Others[7] have used this technique with separate runs of prothrombin hydrolysates as the standards to determine recovery. Prothrombin contains 10 γ-carboxyglutamic acid residues per 72,000 daltons. Fernlund[8] has also used anion exchange with the automated amino acid analyzer for quantitation of γ-carboxyglutamic acid in urine using β-carboxyglutamic acid as an internal standard.[9]

Other Chromatographic Approaches

One of the first methods used in identification of γ-carboxyglutamic acid was diborane reduction of the carboxyl groups to primary alcohols and analysis of the product, 5,5'-dihydroxyleucine, by standard amino acid analysis.[10] This product elutes just prior to serine under these conditions. This technique has the drawbacks of being essentially qualitative (since carboxyl reduction is not quantitative) and of requiring the use of [^{3}H]diborane if high sensitivity is desired. When tested on a wide range of crude substrates, this approach gave reliable identifications of γ-carboxyglutamate.[11]

In the absence of an amino acid analyzer, γ-carboxyglutamic acid can be isolated by anion exchange on a Dowex 1-X8 column (6 × 0.7 ml) equilibrated in 0.02 M HEPES buffer (pH 10).[12] Successive washes with 40 ml of 0.02 M HEPES buffer (pH 5) followed by 25 ml of 0.02 M HEPES-20 mM MgCl$_2$ (pH 4.5) at a flow rate of 25 ml/hr were discarded. The γ-carboxyglutamate was eluted during the next 25 ml of the latter buffer. The sample must be adjusted to pH 11.5 before application to the column. The γ-carboxyglutamic acid was quantitated fluorometrically by mixing 0.5 ml of column effluent with 0.5 ml 6.0 mM o-phthalaldehyde in 0.40 M borate buffer (pH 9.7). A standard curve was used for quantitation. Fluorescence excitation was at 340 nm and emission at 455 nm. Radiolabeled γ-carboxyglutamic acid can be used as an internal standard.

[6] L. B. James, *J. Chromatogr.* **175,** 211 (1979).
[7] R. A. DiScipio and E. W. Davie, *Biochemistry* **18,** 899 (1979).
[8] P. Fernlund, *Clin. Chim. Acta* **72,** 147 (1976).
[9] P. Fernlund, *in* "Vitamin K Metabolism and Vitamin K-Dependent Proteins" (J. W. Suttie, ed.), p. 166. Univ. Park Press, Baltimore, Maryland, 1980.
[10] T. H. Zytkovicz and G. L. Nelsestuen, *J. Biol. Chem.* **250,** 2968 (1975).
[11] T. H. Zytkovicz and G. L. Nelsestuen, *Biochim. Biophys. Acta* **444,** 344 (1976).
[12] C. M. Gundberg, J. B. Lian, and P. M. Gallop, *Anal. Biochem.* **98,** 219 (1979).

A method for identification of γ-carboxyglutamic acid by high-performance liquid chromatography has been reported.[13] This also involved anion-exchange chromatography with product detection by fluorescence intensity after automated postcolumn derivatization with o-phthalaldehyde. This method is therefore closely analogous to the previous one, which was a manual method. While both methods appeared to be specific for γ-carboxyglutamic acid from protein hydrolysates or urine samples, the total range of samples and sources of interference was not tested.

A gas chromatographic method for identification of γ-carboxyglutamic acid has been developed.[14] The N-isobutyloxycarbonyl trimethyl ester derivative was used to obtain a volatile compound. This method was tested with urine samples only and required prior preparation of γ-carboxyglutamic acid by ion-exchange methods. The potential for interference from other compounds in other types of samples has not been tested so that the specificity of this method for all types of samples is unknown. The use of combined gas chromatography–mass spectrometry would solve selectivity problems.

Analysis during Automated Sequenator Analysis

Initial sequence analysis of peptides containing γ-carboxyglutamic acid reported a low yield of glutamate at those positions containing γ-carboxyglutamic acid.[15] This is due to decarboxylation under conditions of repetitive cycles. Therefore, less than quantitative yields of γ-carboxyglutamic acid were obtained even when precautions were taken.[14] Qualitative detection of γ-carboxyglutamic acid was reported by mass spectral analysis of the methyl esters of the phenylisothiocyanate derivatives.[15] Fernlund and Stenflo[16] have reported a modified method using n-butanol instead of chlorobutane to extract the thiazolinone derivatives. This produced a much higher yield of the γ-carboxyglutamic acid derivative, which was then identified by standard high-performance liquid chromatography methods.

Analysis by Peptide Mass Spectrometry

This approach is useful only when defined peptides have been obtained. Mass spectrometry was used in the initial identifications of γ-

[13] M. Kuwada and K. Katayama, *Anal. Biochem.* **117**, 259 (1981).

[14] S. Matsu-ura, S. Yamamoto, and M. Makita, *Anal. Biochem.* **114**, 371 (1981).

[15] P. Fernlund, J. Stenflo, P. Roepstorff, and J. Thomsen, *J. Biol. Chem.* **250**, 6125 (1975).

[16] P. Fernlund and J. Stenflo, *in* "Vitamin K Metabolism and Vitamin K-Dependent Proteins" (J. W. Suttie, ed.), p. 161. Univ. Park Press, Baltimore, Maryland, 1980.

carboxyglutamic acid.[17–19] The methods for N-acetylation and carboxyl esterification are mild enough so that no decarboxylation was detected.[18] This technique suffers from the low volatility of large peptides. Mild methylation with methyl iodide, which produced no N-methyl derivatives, also allowed retention of both γ-carboxyl groups[17] but also suffered from the low volatility of larger peptides.

Permethylation needed to volatilize larger peptides resulted in extensive decarboxylation of γ-carboxyglutamic acid.[19] Consequently, identification of γ-carboxyglutamic acid by analysis of permethylated peptides is not entirely straightforward. γ-Carboxyglutamic acid is found in four different molecular forms after permethylation. This approach has nevertheless been extensively used to identify the γ-carboxyglutamic acids in three proteins.[20–22]

Carr et al.[23] also used mass spectrometry to sequence osteocalcin. These authors decarboxylated the γ-carboxyglutamate residues first under conditions that resulted in deuterium incorporation at the γ-position. They then reduced the peptide to the polyamino alcohol. Like the other mass spectrometry studies, this approach requires peptides of quality adequate for sequence analysis. It will be useful for well-defined peptides.

Other Methods

Low et al.[24] used the dansylated derivatives and two-dimensional thin-layer chromatography to identify γ-carboxyglutamic acid. This method did not prove to be quantitative.

Pecci and Cavallini[25] reported a direct colorimetric procedure that depends on formaldehyde or acetaldehyde cyclization of γ-carboxyglutamic acid to form a proline derivative, which is then quantitated as a

[17] J. Stenflo, P. Fernlund, W. Egan, and P. Roepstorff, Proc. Natl. Acad. Sci. U.S.A. 71, 2730 (1974).
[18] G. L. Nelsestuen, T. H. Zytkovicz, and J. B. Howard, J. Biol. Chem. 249, 6347 (1974).
[19] S. Magnusson, L. Sottrup-Jensen, T. E. Peterson, H. R. Morris, and A. Dell, FEBS Lett. 44, 189 (1974).
[20] H. R. Morris, A. Dell, T. E. Petersen, L. Sottrup-Jensen, and S. Magnusson, Biochem. J. 153, 663 (1976).
[21] H. C. Thogersen, T. E. Petersen, L. Sottrup-Jensen, S. Magnusson, and H. R. Morris, Biochem. J. 175, 613 (1978).
[22] P. Hojrup, P. Roepstorff, and T. E. Petersen, Eur. J. Biochem. 126, 343 (1982).
[23] S. A. Carr, P. V. Hauschka, and K. Biemann, J. Biol. Chem. 256, 9944 (1981).
[24] M. Low, J. J. Van Buskirk, and W. M. Kirsch, in "Vitamin K Metabolism and Vitamin K-Dependent Proteins" (J. W. Suttie, ed.), p. 150. Univ. Park Press, Baltimore, Maryland, 1980.
[25] L. Pecci and D. Cavallini, Anal. Biochem. 118, 70 (1981).

secondary amine. Samples containing γ-carboxyglutamic acid (50–500 nmol) plus 50 μl of acetaldehyde in 1 ml of water were heated at 100° for 30 min. Buffer (1 ml of 0.1 M Na$_2$B$_4$O$_7$–0.2 M Na$_2$CO$_3$, pH 9.8) was added along with 0.5 ml of secondary amine reagent (1% sodium nitroprusside in 10% acetaldehyde). After 10 min the absorbance at 590 nm was read and compared to a standard curve. Proline and cysteine will interfere with this assay. Unfractionated urine was also not a satisfactory sample for analysis. Overall specificity of this assay with various sample sources appears to be untested.

Overall it is apparent that unambiguous identification of γ-carboxyglutamic acid is still not a simple process and probably requires more than one approach.

[33] Vitamin K-Dependent Plasma Proteins

By Gary L. Nelsestuen

Prothrombin,[1,2] blood clotting factors X,[3] IX,[4] VII,[5] and protein C[6] are vitamin K-dependent plasma proteins that have been reviewed in this series. Two other proteins are protein S[7,8] and protein Z.[9,10] The comments in this section are intended to provide generalizations about these proteins with regard to their biological roles, the role of γ-carboxyglutamic acid, and to provide new methods for study of structure–function relationships.

Isolation

The most unusual feature in the isolation of this family of proteins is the use of barium citrate or barium sulfate adsorption. This simple pro-

---

[1] K. G. Mann, this series, Vol. 45, p. 123.

[2] K. G. Mann, J. Elion, R. J. Butkowski, M. Downing, and M. E. Nesheim, this series, Vol. 80, p. 286.

[3] K. Fujikawa and E. W. Davie, this series, Vol. 45, p. 89.

[4] K. Fujikawa and E. W. Davie, this series, Vol. 45, p. 74.

[5] R. Radcliffe and Y. Nemerson, this series, Vol. 45, p. 49.

[6] W. Kisiel and E. W. Davie, this series, Vol. 80, p. 321.

[7] R. G. DiScipio, M. A. Hermodson, S. G. Yates, and E. W. Davie, *Biochemistry* **16**, 698 (1977).

[8] B. Dahlbeck and J. Stenflo, *Proc. Natl. Acad. Sci. U.S.A.* **78**, 2512.

[9] C. V. Prowse and M. P. Esnouf, *Biochem. Soc. Trans.* **5**, 255 (1977).

[10] P. Hojrup, P. Roepstorff, and T. E. Petersen, *Eur. J. Biochem.* **126**, 343 (1982).

cess produces preparations in nearly quantitative yields that are virtually free of other plasma proteins. The γ-carboxyglutamic acid side chains apparently interact strongly with a structure in the barium crystals. Barium citrate is formed by adding excess $BaCl_2$ to citrated plasma.[1,5] $BaSO_4$ adsorption[4] requires oxalate as the anticoagulant, since citrate inhibits protein binding to $BaSO_4$. The latter phenomenon is useful for elution of proteins from the $BaSO_4$.[4] The barium salt must be washed to remove nonadsorbed proteins. Ammonium sulfate is often used to elute these proteins from barium citrate[2,5]; this forms $BaSO_4$ precipitate, and the released citrate prevents protein binding. This latter procedure also precipitates some proteins and removes the major form of protein S, a complex with complement C_4-binding protein, from solution.[8] Isolation of the protein S—C_4-binding protein complex therefore requires dissolving the barium citrate precipitate with EDTA[8] so that all proteins remain in solution.

After elution from the barium precipitate, the proteins are dialyzed against the starting buffer and chromatographed on a DEAE ion-exchange column. An example of the elution profile obtained from bovine plasma is shown in Fig. 1. This profile also helps to illustrate the relative abundance of several proteins. The protein activities and their elution positions shown include factors IX, protein C, prothrombin, and two ionic forms of factor X designated factors X_1 and X_2. Factor VII is found in the latter part of the prothrombin peak[5] and is present in the lowest amounts. Protein Z is purified from the factor X_2[9] peak, and free protein S[7] as well as its complex with C_4-binding protein[8] elute before the prothrombin peak but are not visible in this elution profile. This elution profile is very reproducible, and we generally pool the protein fractions without specific assay of each activity. It seems probable that other, as yet unidentified, vitamin K-dependent proteins are to be found in this elution profile. For example, without a specific enzyme activity assay for factor VII, it is unlikely that this trace protein would have been found.

Several protein fragments have been isolated, but none has been more heavily used than prothrombin fragment 1, the amino-terminal 156 residues of bovine prothrombin. This peptide contains the γ-carboxyglutamic acid residues and binds to phospholipids in a manner similar to prothrombin. Prothrombin (5 mg/ml) is mixed with thrombin (100 : 1, w/w) at pH 8 and incubated at 37° for 3 or 4 hr. The product is dialyzed against 0.02 M Tris buffer (pH 7.5)–0.15 M NaCl and applied to a DEAE-Sephadex column equilibrated with the same buffer (protein : gel, w/v, ratio similar to that shown in Fig. 1). The column is eluted with a linear gradient (6–10 times the column volume) of the starting buffer to a final buffer of 0.066 M Tris (pH 7.5)–0.5 M NaCl.[1] The second of two peaks eluting from the

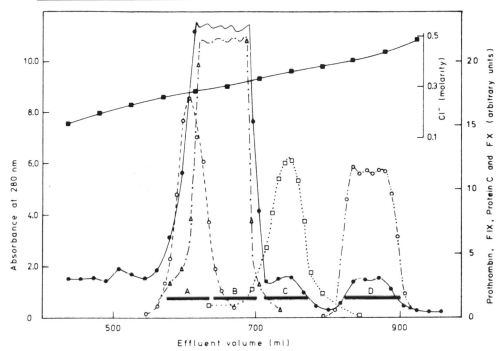

FIG. 1. Elution profile of bovine plasma barium citrate elute from DEAE-Sephadex. The protein (2.3 g) was applied to a column of DEAE-Sephadex (2.5 × 40 cm) in 0.1 M phosphate buffer, 1 mM benzamidine · HCl (pH 6.0). A linear gradient of NaCl (0.15 to 0.55 M) was applied as shown. The symbols used include: ●, $A_{280\,nm}$; ■, Cl⁻ concentration; ○ (peak A), factor IX; △, prothrombin; □, protein C; ○ (peak D), factor X. The broadened factor X peak is due to two ionic forms designated factors X_1 (first half) and X_2 (second half of the peak). This elution profile is from J. Stenflo [*J. Biol. Chem.* **251,** 355 (1976)].

column contains about 30% as much UV absorbance as the first peak and contains fragment 1.[1] This material is concentrated to an appropriate volume and rechromatographed on a column of Sephadex G-100 (80 cm in length). A peak of high molecular weight material separates from fragment 1 on this column. The elution position of fragment 1 corresponds to that of a globular protein of about $M_r = 50,000$.

Several other protein fragments have been isolated including prothrombin peptide 12–44,[11,12] prothrombin peptide 1–39,[13] and factor X pep-

[11] G. L. Nelsestuen and J. W. Suttie, *Proc. Natl. Acad. Sci. U.S.A.* **70,** 3366 (1973).

[12] B. C. Furie, M. Blumenstein, and B. Furie, *J. Biol. Chem.* **254,** 12521 (1979).

[13] H. C. Marsh, M. M. Sarasua, D. A. Madar, R. G. Hiskey, and K. A. Koehler, *J. Biol. Chem.* **256,** 7863 (1981).

tide 3–43.[14,15] These peptides also display quantitative adsorption to barium salts but do not appear to bind to phospholipid membranes. Factor X that has had the γ-carboxyglutamic acid region removed has also been isolated.[16] These structures may all be useful in the study of γ-carboxyglutamic acid function.

Structure

The vitamin K-dependent plasma proteins display a close sequence homology in portions of their structures. The sequence data for the amino-terminal 42 residues are shown in Fig. 2.[10,17–24] These sequences are highly homologous, which suggests that the genetic material for these segments is descended from a common ancestral gene. The amino-terminal 42 residues contain all the γ-carboxyglutamic acid residues found in these proteins. The total sequences of prothrombin,[17] factor X,[18,19] factor IX,[20] and protein C[21,22] are known. The carboxy-terminal sequence of these four proteins are all similar to that of the family of serine proteases. Intermediate between these protein regions, prothrombin shows considerable differences from the other three proteins. Prothrombin contains two segments of a sequence of about 80 amino acids with three disulfides, which Magnusson et al.[17] have called a "kringle." Factors X, IX, and protein C do not have a closely analogous structure. Thus, as in many other systems, it appears that the vitamin K-dependent proteins were formed by evolutionary recombination of short gene sequences to produce complete proteins with different function.

[14] J. B. Howard and G. L. Nelsestuen, *Proc. Natl. Acad. Sci. U.S.A.* **72**, 1281 (1975).

[15] G. L. Nelsestuen, M. Broderius, T. H. Zytkovicz, and J. B. Howard, *Biochem. Biophys. Res. Commun.* **65**, 233 (1975).

[16] T. Morita and C. M. Jackson, *in* "Vitamin K Metabolism and Vitamin K-Dependent Proteins" (J. W. Suttie, ed.), p. 124. Univ. Park Press, Baltimore, Maryland, 1980.

[17] S. Magnusson, T. E. Petersen, L. Sottrup-Jensen, and H. Claeys, *in* "Proteases and Biological Control" (E. Reich, D. E. Rifkin, and E. Shaw, eds.), Vol. 2, p. 123. Cold Spring Harbor Laboratory, Cold Spring Harbor, New York, 1975.

[18] D. L. Enfield, L. H. Ericsson, K. Fujikawa, K. A. Walsh, H. Neurath, and K. Titani, *Biochemistry* **19**, 659 (1980).

[19] H. C. Thogersen, T. E. Petersen, L. Sottrup-Jensen, S. Magnusson, and H. R. Morris, *Biochem. J.* **175**, 613 (1978).

[20] E. Katayama, L. H. Ericsson, D. L. Enfield, K. A. Walsh, H. Neurath, E. W. Davie, and K. Titani, *Proc. Natl. Acad. Sci. U.S.A.* **76**, 4990 (1979).

[21] P. Fernlund and J. Stenflo, *J. Biol. Chem.* **257**, 12170 (1982).

[22] J. Stenflo and P. Fernlund, *J. Biol. Chem.* **257**, 12180 (1982).

[23] W. Kisiel and E. W. Davie, *Biochemistry* **14**, 4928 (1975).

[24] J. Stenflo and M. Jonsson, *FEBS Lett.* **101**, 377 (1979).

	1	5	10	15	20
Prothrombin	Ala-Asn-Lys-Gly-Phe-Leu-Gla-Gla-Val-Arg-Lys-Gly-Asn-Leu-Gla-Arg-Gla-Cys-Leu-Gla-Gla-Pro-Cys-Ser-				
Factor IX	Tyr-Asn-Ser-Gly-Lys-Leu-Gla-Gla-Phe-Val-Arg-Gly-Asn-Leu-Gla-Arg-Gla-Cys-Lys-Gla-Gla-Lys-Cys-Ser-				
Factor X	Ala-Asn-Ser- - -Phe-Leu-Gla-Gla-Val-Lys-Gln-Gly-Asn-Leu-Gla-Arg-Gla-Cys-Leu-Gla-Gla-Ala-Cys-Ser-				
Protein C	Ala-Asn-Ser- - -Phe-Leu-Gla-Gla-Leu-Arg-Pro-Gly-Asn-Val-Gla-Arg-Gla-Cys-Ser-Gla-Gla-Val-Cys-Gla-				
Protein Z	Ala-Gly-Ser-Tyr-Leu-Leu-Gla-Gla-Leu-Phe-Gla-Gly-His-Leu-Gla-Lys-Gla-Cys-Trp-Gla-Gla-Ile-Cys-Val-				
Factor VII	Ala-Asn-Gly- - -Phe-Leu-Gla-Gla-Leu-Leu-Pro-Gly-Ser-Leu-				
Protein S	Ala-Asn-Thr- - -Leu-Leu-Gla-Gla Thr-Lys-Lys-Gly-Asn-Leu-				

	25	30	35	40
Prothrombin	Arg-Gla-Gla-Ala-Phe-Gla-Ala-Leu-Gla-Ser-Leu-Ser-Ala-Thr-Asp-Ala-Phe-Trp-			
Factor IX	Phe-Gla-Gla-Ala-Arg-Gla-Val-Phe-Gla-Asn-Thr-Gla-Lys-Thr Thr Gla-Phe-Trp-			
Factor X	Leu-Gla-Gla-Ala-Arg-Gla-Val-Phe-Gla-Asp-Ala-Gla-Gln-Thr-Asp-Gla-Phe-Trp-			
Protein C	Phe-Gla-Gla-Ala-Arg-Gla-Ile-Phe-Gln-Asn-Thr-Gla-Asp-Thr-Met-Ala-Phe-Trp-			
Protein Z	Tyr-Gla-Gla-Ala-Arg-Gla-Val-Phe-Gla-Asp-Asp-Gla-Thr-Thr-Asp-Gla-Phe-Trp-			

FIG. 2. Amino-terminal sequence of vitamin K-dependent plasma proteins. Gla designates γ-carboxyglutamic acid. The sources of these sequences are prothrombin,[17] factor X,[18,19] factor IX,[20] protein C,[21,22] protein Z,[10] factor VII,[23] and protein S.[24]

It appears likely that factor VII will have a sequence analogous to factors X and IX. Proteins S and Z have not been shown to be protease zymogens, and it is possible that their remaining sequences are quite different from any of the other vitamin K-dependent proteins.

Function

Calcium Binding. The only known function of γ-carboxyglutamic acid is to serve in calcium binding. Calcium binding by the plasma proteins is necessary for subsequent binding to appropriately constituted phospholipid membranes where activation of the coagulation cascade occurs. Factors IX, X, and prothrombin serve as the substrates and factors IXa, Xa, and VIIa serve as the enzymes[25] for these membrane-bound zymogen activations. In each step a second membrane-binding cofactor protein is required. These latter proteins are not vitamin K-dependent (reviewed by Jackson and Nemerson[25]).

The calcium-binding properties of the vitamin K-dependent plasma proteins are summarized in the table.[25-33] While considerable controversy

[25] C. M. Jackson and Y. Nemerson, *Annu. Rev. Biochem.* **49,** 727 (1980).
[26] J. Stenflo and P.-O. Ganrot, *Biochem. Biophys. Res. Commun.* **50,** 98 (1973).
[27] G. L. Nelsestuen, *J. Biol. Chem.* **251,** 5648 (1976).

FUNCTIONAL PROPERTIES OF VITAMIN K-DEPENDENT PROTEINS

Protein	Calcium binding				K_D^c (μM)	n^d	Binding to phospholipids		
	Sites	$Ca_{0.5}^a$ (mM)	Hill coefficient	Source[b]			$Ca_{0.5}$ ($-Mn^{2+}$) (mM)	$Ca_{0.5}$ ($+Mn^{2+}$) (mM)	Source[b]
Prothrombin	10	0.6	1.5	25–28	0.6	70	0.3	0.08	33
Fragment 1	6	0.4	2.6	29	—	100	—	—	—
Factor X	20	0.63	1.27	30	0.2	—	0.5	0.18	33
Factor IX	16	0.73	1.0	31	<1	—	1.0	0.2	33
Protein C	16	0.87	1.0	32	17	—	0.75	0.22	33
Protein S	—	—	—	—	<1	—	0.5	0.17	33
Factor VII	—	—	—	—	17	—	0.47	0.18	33

[a] $Ca_{0.5}$ is the free calcium concentration at half saturation. Protein–membrane binding was titrated in the presence and the absence of manganese. This ion catalyzes a necessary protein conformational change but does not cause protein–membrane binding.

[b] Numbers refer to text footnotes.

[c] K_D refers to the binding of protein to a specific phospholipid composition (20% phosphatidylserine–80% phosphatidylcholine) at 2 mM calcium.

[d] n is the number of protein molecules binding to a vesicle of $M_r = 3.6 \times 10^6$.

has surrounded some of these numbers, most of the deviations have not been great. The affinity for calcium is intermediate so that these proteins would not bind calcium at normal intracellular calcium concentrations. Several studies have shown that an *intact* protein secondary/tertiary structure is needed for proper calcium binding to the proteins.[11,15,30] Denatured proteins bind calcium with lower affinity,[11,30] and short protein segments, e.g., prothrombin residue 12–44[11,15] and factor X residues 3–43,[11,14] which contain the γ-carboxyglutamic acid residues, do not appear to bind to membrane surfaces.[15]

Analysis of the function of γ-carboxyglutamic acid is a physical measurement rather than an enzymatic measurement. In studying the properties of γ-carboxyglutamic acid there appears to be a need for a standard of comparison. This is difficult in the absence of an enzyme assay. Prothrombin fragment 1 has been extensively used as a model of γ-carboxyglutamic acid-containing proteins. A simple standard of comparison that, in our experience, appeared to provide a reliable measure of protein function is calcium-dependent protein fluorescence quenching. With the protein in a fluorescence cuvette, the fluorescence intensity (excitation at 280 nm, emission at 340 nm) is measured and 5 mM calcium is added. For bovine fragment 1, calcium causes a slow conformational change accompanied by a 55% quenching of fluorescence.[29] To reach equilibrium, 20–30 min at room temperature is required. The best practice is to compare the fluorescence intensity at equilibrium with that immediately after addition of excess EDTA. In this way the readings are taken over a short time interval. Our experience to date has been that this simple test designates a protein that still retains its native conformation[34] and still binds fully to phospholipid vesicles. The single exception to this rule was certain lysine modifications where membrane binding was more severely affected than was the fluorescence quenching.[33a] We have occasionally observed 57–60% quenching by calcium, but these results have not been consistent. In any case, it seems apparent that proteins that have lower extents of fluo-

[28] S. P. Bajaj, R. J. Butkowski, and K. G. Mann, *J. Biol. Chem.* **250,** 2150 (1975).

[29] G. L. Nelsestuen, R. M. Resnick, G. J. Wei, C. H. Pletcher, and V. A. Bloomfield, *Biochemistry* **20,** 351 (1981).

[30] R. A. Henriksen and C. M. Jackson, *Arch. Biochem. Biophys.* **170,** 149 (1975).

[31] G. W. Amphlett, W. Kisiel, and F. J. Castellino, *Arch. Biochem. Biophys.* **208,** 576 (1981).

[32] G. W. Amphlett, W. Kisiel, and F. J. Castellino, *Biochemistry* **20,** 2156 (1981).

[33] G. L. Nelsestuen, W. Kisiel, and R. G. DiScipio, *Biochemistry* **17,** 2134 (1978).

[33a] C. H. Pletcher, *Fed. Proc. Fed. Am. Soc. Exp. Biol.* **40,** 1585 (1981).

[34] C. H. Pletcher, R. M. Resnick, G. J. Wei, V. A. Bloomfield, and G. L. Nelsestuen, *J. Biol. Chem.* **255,** 7433 (1980).

rescence quenching contain nonfunctional protein molecules or contaminating proteins.

Membrane Binding. Analysis of the membrane binding of vitamin K-dependent proteins can be accomplished by a simple 90° light-scattering procedure.[36] Phospholipid vesicles can be prepared by any method that produces homogeneous single-bilayer vesicles of small diameter. We have used sonication followed by gel filtration[35,36] or ethanol injection.[37] For consistency in the latter technique it was necessary to add 5-μl aliquots of ethanol containing 1.0 mg of phospholipid per milliliter to 10 ml of buffer until the phospholipid concentration was 100 μg/ml. It is important that homogeneity be optimized to avoid large vesicles, which have considerable angular dependence in their molecular weights determined by light-scattering intensity.

The vesicles are placed in the light-scattering photometer at an appropriate concentration, which is determined by the protein–membrane binding constant. We have often used a fluorescence spectrophotometer with excitation and emission wavelengths set at the same value, usually 320–400 nm. It is desirable to test a range of phospholipid concentrations that give a range of free protein concentrations. The light-scattering intensity measurements can be converted to bound protein by the following relationship.[36]

$$\frac{I_{s2}}{I_{s1}} = \left(\frac{\partial n_2/\partial c_2}{\partial n_1/\partial c_1}\right)^2 \left(\frac{M_{r2}}{M_{r1}}\right)^2 \tag{1}$$

where I_{s2}/I_{s1} is the ratio of light scattering intensity of the protein–vesicle complex to that of the vesicles alone; $\partial n/\partial c$ are the refractive index increments and M_{r2}/M_{r1} is the molecular weight ratio of the protein–membrane complex to that of the initial vesicles. The $\partial n/\partial c$ term is an approximation but contributes relatively little to the total signal change. For an M_{r2}/M_{r1} of 2.0, the ratio of the refractive index increments was estimated to be about 1.06.[36] Other M_{r2}/M_{r1} ratios are assumed to have a proportional change in the refractive index increment term. Small corrections for light scattering from free protein and from buffer must be subtracted before application of this equation. An example of the results is shown in Fig. 3. The light scattering intensity gives the M_{r2}/M_{r1} ratio and can be used directly to show the existence of membrane-binding protein. The amount of bound protein is compared to the theoretical value anticipated if all added protein were bound to the phospholipid (dashed line, Fig. 3). The

[35] C. Huang, *Biochemistry* **8**, 344 (1969).

[36] G. L. Nelsestuen and T. K. Lim, *Biochemistry* **16**, 4164 (1977).

[37] J. M. H. Kremer, M. W. J. Esker, C. Pathmamanoharan, and P. H. Weirsema, *Biochemistry* **16**, 3932 (1977).

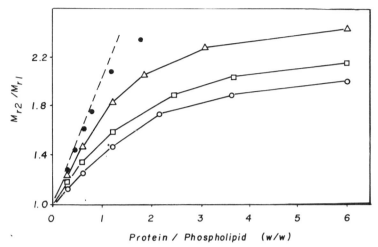

FIG. 3. Measurement of prothrombin–membrane binding. The weight average M_{r2}/M_{r1} ratio (protein–vesicle complex to vesicle weight ratio) was estimated from light-scattering intensity measurements as outlined in the text. The phospholipid concentrations used were ○, 23 μg/ml; □, 46 μg/ml; ∧, 93 μg/ml; ●, 1300 μg/ml. The calcium concentration was 5 mM in 0.05 M Tris (pH 7.5)–0.1 M NaCl. Prothrombin was added in the amounts indicated.

free protein concentration is obtained by the difference between experimental and theoretical values. The bound and free protein concentrations are then applied to a Scatchard plot or double-reciprocal plot to obtain an equilibrium constant and maximum protein binding capacity of the vesicles. Application of these equilibrium plots to the data requires demonstration that a reversible equilibrium exists. Two criteria that can be applied to these proteins are complete reversal of binding by excess EDTA[36] and variation of the phospholipid concentration to obtain varying ratios of free-to-bound protein.[36,38] Using high levels of phospholipid (1.3 mg/ml, Fig. 3), the estimated amount of bound protein reached 95% of the added prothrombin. This latter result requires that the protein and phospholipid concentration measurements were correct, that all the protein was capable of membrane binding, and that Eq. (1) was adequate to describe the interaction. The ratio of free-to-bound protein at the lower levels of phospholipid must vary in a manner predicted by the equilibrium constant.[36,38] This approach also appears to be applicable to virtually any protein–membrane interaction that occurs on preformed vesicles.

[38] G. J. Wei, V. A. Bloomfield, R. M. Resnick, and G. L. Nelsestuen, *Biochemistry* **21,** 1949 (1982).

Factor Xa, but not factor X, caused aggregation of the protein–vesicle complexes when its density on the membrane became sufficient. This difficulty precluded full titrations such as those shown in Fig. 3. The table shows some membrane-binding constants estimated for bovine vitamin K-dependent plasma proteins to membranes of 20% phosphatidylserine/80% phosphatidylcholine. The most striking feature is the low binding affinity of protein C and factor VII. These proteins are also unusual in the presence of proline at position 11 (Fig. 2).

[34] Osteocalcin: Isolation, Characterization, and Detection

By C. M. GUNDBERG, P. V. HAUSCHKA,
J. B. LIAN, and P. M. GALLOP

In 1974, several laboratories[1-3] demonstrated the existence in prothrombin of a previously unknown amino acid, 4-carboxyglutamic acid (Gla). Gla residues specifically bind calcium ions and are required for normal blood coagulation. Vitamin K is involved as a cofactor in the synthesis of Gla by posttranslational enzymatic carboxylation of certain glutamic acid residues in polypeptide chains.

Gla-containing proteins have also been found in a variety of tissues that are involved in calcium transport, deposition, and homeostasis (for a recent review, see Gallop et al.)[4] Much recent work has been directed toward establishing their biological functions, which are presumed to be related to their calcium binding properties. One of these, osteocalcin, is a small protein (5800 daltons) comprising 10–20% of the noncollagenous protein in bone.[5,6] The amino acid sequences of osteocalcin from human,[7]

[1] J. Stenflo, P. Fernlund, W. Egan, and P. P. Roepstorff, Proc. Natl. Acad. Sci. U.S.A. 71, 2730 (1974).
[2] G. L. Nelsestuen, T. H. Zytkovicz, and J. B. Howard, J. Biol. Chem. 249, 6347 (1974).
[3] S. Magnusson, L. Sottrup-Jensen, T. E. Peterson, H. R. Morris, and A. Dell, FEBS Lett. 44, 189 (1974).
[4] P. M. Gallop, J. B. Lian, and P. V. Hauschka, N. Engl. J. Med. 202, 1460 (1980).
[5] P. V. Hauschka, J. B. Lian, and P. M. Gallop, Proc. Natl. Acad. Sci. U.S.A. 72, 3925 (1975).
[6] P. A. Price, J. W. Poser, and N. Raman, Proc. Natl. Acad. Sci. U.S.A. 10, 3374 (1976).
[7] J. W. Poser, F. S. Esch, N. C. Lenz, and P. A. Price, J. Biol. Chem. 255, 8685 (1980).

METHODS IN ENZYMOLOGY, VOL. 107

monkey,[8] chicken,[9] cow,[6] rat,[10,11] and swordfish[12] show that many structural features are identical: Gla residues are located at positions 17, 21, and 24, and a disulfide bond joins Cys-23 and Cys-29. Osteocalcin, which possibly originates from a larger Gla-containing molecule,[13–15] is synthesized *de novo* in bone tissue[16,17] and appears to be coincident with the onset of mineralization.[13,18] Recent evidence also suggests that biosynthesis[19] and distribution[20] of osteocalcin may be affected by vitamin D.

Osteocalcin is distinguished by its specific calcium-binding properties. It has a moderate affinity for ionic calcium, binds strongly to hydroxyapatite, and inhibits the formation of hydroxyapatite from brushite *in vitro*.[21] In the presence of calcium, osteocalcin undergoes a transition to an α-helical conformation in which all Gla side chains are located on the same face of one α-helix. Gla residues are spaced at intervals of about 5.4 Å, closely paralleling the interatomic separation of Ca^{2+} in the hydroxyapatite lattice[22] and are apparently involved in the formation of a high-affinity mineral–protein complex.

While osteocalcin is predominantly associated with bone mineral, a small amount is found circulating in the blood. Measurement of osteocalcin in serum and Gla in urine has been shown to have clinical application in the evaluation of certain bone diseases.[23–26]

Although investigation into the synthesis and structure of osteocalcin

[8] P. V. Hauschka, S. A. Carr, and K. Biemann, *Biochemistry* **21,** 63 (1982).

[9] S. A. Carr, P. V. Hauschka, and K. Biemann, *J. Biol. Chem.* **256,** 9944 (1981).

[10] A. Linde, M. Bhown, and W. T. Butler, *J. Biol. Chem.* **255,** 5931 (1980).

[11] Y. Otawara, N. Hosoya, S. Moriuchi, H. Kass, and T. Okuyama, *Biomed. Res.* **2,** 442 (1980).

[12] P. A. Price, A. S. Otsuka, and J. W. Poser, *in* "Calcium Binding Proteins and Calcium Function" (R. H. Wasserman *et al.,* eds.), p. 333. Elsevier/North-Holland, Amsterdam, 1977.

[13] P. V. Hauschka, J. Frenkel, R. DeMuth, and C. M. Gundberg, *J. Biol. Chem.* **258,** 176 (1983).

[14] S. K. Nishimoto and P. A. Price, *J. Biol. Chem.* **255,** 6579 (1980).

[15] J. B. Lian and K. M. Heroux, *in* "Vitamin K Metabolism and Vitamin K-Dependent Proteins" (J. W. Suttie, ed.), p. 245. Univ. Park Press, Baltimore, Maryland, 1979.

[16] J. B. Lian and P. A. Friedman, *J. Biol. Chem.* **253,** 6623 (1978).

[17] S. K. Nishimoto and P. A. Price, *J. Biol. Chem.* **254,** 437 (1979).

[18] P. V. Hauschka and M. L. Reid, *J. Biol. Chem.* **253,** 9063 (1978).

[19] P. A. Price and S. A. Baukol, *J. Biol. Chem.* **255,** 11660 (1980).

[20] J. B. Lian, M. J. Glimcher, A. H. Roufosse, P. V. Hauschka, P. M. Gallop, L. Cohen-Solal, and B. Reit, *J. Biol. Chem.* **257,** 4999 (1982).

[21] P. V. Hauschka and P. M. Gallop, *in* "Calcium Binding Proteins and Calcium Function" (R. H. Wasserman *et al.,* eds.), p. 338. Elsevier/North-Holland, Amsterdam, 1977.

[22] P. V. Hauschka and S. A. Carr, *Biochemistry* **21,** 2538 (1982).

[23] C. M. Gundberg, J. B. Lian, and P. M. Gallop, *Clin. Chim. Acta* **128,** 1 (1983).

[24] P. A. Price, J. G. Parthemore, and L. J. Deftos, *J. Clin. Invest.* **66,** 878 (1980).

has progressed significantly in the past several years, there is still much that is not known concerning the function and metabolism of osteocalcin in bone. Several hypotheses currently exist as to the function of osteocalcin. First, the tight binding of osteocalcin to hydroxyapatite surfaces and its inhibition of the brushite–hydroxyapatite conversion suggest a regulatory role in mineral deposition and crystal growth. Second, osteocalcin may also play a role in the local control of calcium deposition or removal in bone. Fluctuating concentrations of extracellular calcium may alter the adsorption equilibrium of the protein with bone mineral and permit calcium movement in or out of bone. Third, since osteocalcin is found both temporally and topologically coincident with the first appearance of calcium in developing bone, it may have a role in mineralization. Fourth, the protein may play an informational role in mediating the action of vitamin D on bone. Finally, osteocalcin has chemotactic activity toward human peripheral monocytes *in vitro*.[27] Therefore it may be involved in recruiting cells to sites of bone resorption.

In view of the structure of osteocalcin, its striking hydroxyapatite binding properties, its synthesis and abundance in bone, and the high degree of sequence conservation throughout evolution, it seems likely that osteocalcin plays an important role in bone metabolism.

Isolation and Purification of Osteocalcin

Because osteocalcin is tightly adsorbed to the hydroxyapatite mineral phase of bone, thorough extraction is best achieved by dissolving the mineral of finely pulverized bone. Hydroxyapatite is soluble in neutral 0.5 M EDTA as well as in a variety of mineral and organic acids. Uniformly high yields of 1–2 mg of osteocalcin per gram of dry bone are obtained with EDTA, whereas acid procedures often leave 10–25% of the protein behind, presumably because of precipitation during excursions through the osteocalcin isoelectric point ($pI = 4.0 \pm 0.5$ depending on species and the number of Gla residues).

Bone Powder Preparation

Many techniques have evolved for grinding bone to a fine powder. The ultimate goal for rapid extraction is a dry powder of uniform particle size

[25] D. M. Slovick, C. M. Gundberg, J. B. Lian, and R. M. Neer, *Calcif. Tissue Int.* **34s,** 515 (1982).

[26] C. M. Gundberg, D. E. C. Cole, J. B. Lian, T. M. Reade, and P. M. Gallop, *J. Clin. Endocrinol. Metab.* **56,** 1063 (1983).

[27] J. D. Malone, S. Teetelbaum, G. L. Griffin, R. M. Senior, and A. J. Kahn, *J. Cell Biol.* **92,** 227 (1982).

less than 100 mesh (149 μm, A.S.T.M.). The tremendous variety of osse-ous tissues ranging from soft, rice grain sized embryonic chick bones to large, intractable cow femurs necessitates a pragmatic approach to grind-ing. Regardless of method, adherence to two important principles will maximize yield and minimize proteolytic degradation: (1) all wet proce-dures must be carried out below 4° in the presence of protease inhibitors, and for a minimum of time; (2) pulverization or milling of bone fragments must be done at liquid N_2 temperatures ($-196°$) with rapid processing to avoid frictional heating.

Freshly dissected bones are immersed in ice-cold Tris-buffered prote-ase inhibitor cocktail (TPIC: benzamidine 5 mM; 6-aminocaproic acid, 10 mM; p-hydroxymercuribenzoic acid, 100 μM; phenylmethylsulfonyl fluo-ride, 30 μM; and Tris-HCl, 20 mM, pH 7.8) and cleaned of adhering fibrous soft tissue. Marrow can be scraped out after splitting, but for small bones it is more effectively removed by direct wet grinding. Preliminary size reduction to less than 2 cm is achieved by sawing (Sawzall with No. 1064 blade, Milwaukee Tool Co.) or cutting with bone or pruning shears. Wet grinding in 5–10 volumes of ice-cold TPIC is carried out first with a stainless steel blender (Waring) and then with a Polytron PTA-20S genera-tor (Brinkmann Instruments, Westbury, NY), using 20-sec bursts of power. Homogenates are diluted with TPIC, and bone particles are al-lowed to sediment out in a beaker for a few minutes. Decanting of the supernatant and further homogenization of the granular bone sediment (>2 mm particle size) are repeated as often as 10 times until the supernatant is clear and colorless. A Polytron PTA 35/4 generator gives a very fine grind during these final stages. The white bone fragments, which are now substantially free of blood and marrow elements, are settled once through cold deionized H_2O and lyophilized. The principal advantage of wet grinding is that it instantaneously exposes virtually all of the bone matrix to proteolytic inhibitors, while washing away many contaminating proteins.

Liquid N_2 pulverization is carried out on lyophilized bone with a mag-netically driven mill (Spex Industries, Metuchen, NJ) using 3–5 cycles of 30 sec on, 60 sec off. About 60% of the resulting powder passes a 100-mesh sieve, and 80% passes a 60-mesh sieve (250 μm). Ideally, the bone has been first treated by wet grinding as described above. Some types of adult bone that are too tough for wet grinding can be lyophilized and then reduced in size prior to Spex milling by chilling in liquid N_2 and crushing while still brittle. A homemade mortar and pestle device that facilitates this consists of a solid steel base cylinder (3 in. o.d. × 3 in. high) that can be precooled in liquid N_2 and a 12-kg solid steel crushing cylinder ($2\frac{1}{4}$ in. o.d. × 24 in. high), both with flat, polished end faces. A piece of plastic pipe (3 in. i.d. × 18 in. high) fits over the base cylinder and prevents bone chips from flying across the laboratory. If quantities greater than 500 g of dry bone powder are required, industrial grinding equipment can be utilized. For crushing of large chunks (4 × 4 × 4 cm) down to 2 mm, a hammer mill (Model 25Z, American Alpine, Natick, MA) will accept 1 kg/min. Further reduction to a particle-size distribution equivalent to the Spex mill product is achieved with a fan beater mill at 17,000 rpm (Model 100LU, American Alpine), which processes 70 g/min. Both mills are compatible with liquid N_2-chilled material.

Extraction of Osteocalcin

Bone powder (100 g dry weight) is first washed by stirring in 2 liters of TPIC for 1 hr. This and all subsequent procedures are carried out at 0–4°. This wash removes extraneous serum proteins that were not accessible during the wet grinding process. After centrifugation (2000 g, 20 min) and discarding of the supernatant, the bone is suspended in 500 ml of EPIC (0.5 M ammonium EDTA, pH 6.2, EDTA acid neutralized with NH_4OH containing the same protease inhibitors as TPIC) and stirred for 16 hr. Centrifugation and reextraction with an additional 500 ml of EPIC for 8 hr yields virtually all the available osteocalcin (100–200 mg) in 1 liter of pooled EPIC supernatants. Dialysis against 20 liters of TPIC (2 days) and then four changes of 20 liters of H_2O (4 days) using half-filled bags of Spectrapor 1 membrane tubing (Spectrum Medical Industries, Los Angeles, CA; M_r cutoff = 6000–8000) is attended by frequent hand mixing of bag contents. Although osteocalcin is a small protein (M_r = 5200–5900 depending on species), its large negative charge causes it to behave on sizing gels and columns as though it were 10,000–17,000. Hence, it is not readily lost by dialysis. Unfortunately, the long dialysis time is necessitated by the slow diffusion of calcium–EDTA complexes having a mean Stokes radius equivalent to a polypeptide of M_r = 3000.[28]

Acid extraction of osteocalcin is also possible. Because of the buffering capacity of hydroxyapatite, some 20 mEq of acid per 1 mEq of hydroxyapatite is required to completely acidify a bone powder suspension. Extraction of 100 g of bone powder is carried out with 1 liter of 10% HCl[28] or 10% HCOOH.[29] After acid extraction, calcium and phosphate must be removed by dialysis against H_2O or dilute acid. Dialysis conditions that allow calcium-containing precipitates to form inside the bag must be avoided, as these precipitates trap osteocalcin. Column desalting of acid[29] and EDTA[13] extracts is also useful for small samples.

[28] P. V. Hauschka, unpublished data.
[29] P. A. Price, J. W. Lothringer, and S. K. Nishimoto, *J. Biol. Chem.* **255**, 2938 (1980).

Obviously the extraction rate and final yield depend on particle size. Kinetic experiments in which soluble osteocalcin was quantitated by direct high-pressure liquid chromatography of the centrifuged extract have demonstrated that EDTA and HCl give their maximum yields within 20 min for sub-200 mesh powder (<74 μm) whereas 80% yields are obtained in 3 hr with 250–500 μm powder.[28] Extracting for weeks or months, as is customary for phosphoproteins and some other bone matrix constituents,[30] does not significantly increase the yield of osteocalcin, except from very coarse bone particles, and is deleterious to the integrity of the protein. Other variables affecting yield include age, species, and anatomical region of bone. Embryonic bones are known to have lower levels of extractable osteocalcin than predicted from their content of Gla.[13] The osteocalcin and total Gla levels in adult bone are linearly related to the mineral content,[31] thus diaphyseal bone is often a richer source of osteocalcin than metaphyseal bone. In some species, such as monkey and cow, osteocalcin is relatively monodisperse, and a single polypeptide can be isolated in high yield. In other species, multiple components representing major proteolytic fragments of the intact osteocalcin may account for 30% (chicken) to 70% (rat) of the total Gla-containing proteins extracted.[9,10,21,28]

Column Purification

Sephadex G-75 Gel Filtration. The dialyzed, lyophilized EPIC extract from 100 g of bone powder (1–3 g, of which 100–200 mg is osteocalcin and a substantial amount may be EDTA) is dissolved in 100 ml of 0.05 M NH_4HCO_3 for 1 hr, centrifuged (35,000 g for 10 min), and applied to a column of Sephadex G-75 (Pharmacia; 10 × 100 cm) equilibrated with the same solvent. Elution at 240 ml/hr is monitored by A_{276}, and 24-ml fractions are collected, beginning at 1.8 liters after sample application.[21] Typically, osteocalcin elutes as the major absorbance peak between 4.3 and 5.5 liters ($v_0 = 2.1$ liters, $v_i = 7.5$ liters). The pooled fractions are lyophilized, weighed, and then further purified by ion-exchange chromatography. Other gel filtration media, such as Sephacryl S-200 and S-300, have been employed with 4–6 M guanidine-HCl for this initial purification step.[7,17]

DEAE—Cellulose Ion Exchange. Crude osteocalcin (100 mg) is dissolved in 5 ml of 0.07 M NH_4HCO_3 and dialyzed against the same buffer.

[30] L. Cohen-Solal, J. B. Lian, D. Kossiva, and M. J. Glimcher, *Biochem. J.* **177,** 81 (1979).
[31] J. B. Lian, A. H. Roufosse, B. Reit, and M. J. Glimcher, *Calcif. Tissue Int.* **34,** S82 (1982).

The sample is applied to a column (DE-53, Whatman; 1.5 × 50 cm) previously equilibrated with 0.07 M NH$_4$HCO$_3$ and eluted at 35 ml/hr with a 700-ml linear gradient of 0.07 M to 0.7 M NH$_4$HCO$_3$.[9] Fractions of 5 ml are collected, and protein is monitored by A_{276}. The salt gradient is established by conductivity measurements. Osteocalcin elutes as a single peak (or cluster of peaks depending on species) at 0.25–0.45 M NH$_4$HCO$_3$ with a typical band width of 0.015 M salt. As column dimensions are increased, the molarity required for elution increases. Protein peaks are pooled and lyophilized. Rechromatography at least twice on DE-53 using shallower gradients (e.g., 0.2 M to 0.5 M NH$_4$HCO$_3$) is advisable to eliminate minor contaminants and to obtain a symmetrical peak of material. Total recovery averages 90–95% per column run. DEAE-Sephadex and DEAE-Sepharose have also been used successfully for ion-exchange purification of osteocalcin.[10,21]

Hydroxyapatite. Hydroxyapatite chromatography is a useful final purification procedure, especially for chicken and rat osteocalcin and other species where heterogeneity is noted during the DE-53 step. Fifty milligrams of osteocalcin is dissolved in 5 ml of 0.1 M NH$_4$HCO$_3$ and dialyzed against 1 mM KH$_2$PO$_4$–KOH, pH 7.2. This buffer and other concentrations are most conveniently prepared by dilution from stock 500 mM KH$_2$PO$_4$–KOH, pH 6.8, without further adjustment of pH; the pH rises slightly with dilution. A column (1.6 × 24 cm) of encapsulated hydroxyapatite (Ultrogel HA, LKB) is poured in 500 mM KH$_2$PO$_4$–KOH and cycled at least three times between 500 mM and 1 mM until the A_{276} of the effluent reaches baseline values. Equilibration at 1 mM is checked by conductivity and pH. The sample is applied and washed on with 50 ml of 1 mM buffer and then eluted with a 700-ml linear gradient of 1 mM to 250 mM buffer at 45 ml/hr. The two buffer reservoirs must be in hydrostatic equilibrium to ensure a smooth initiation of the gradient; hence in a typical chamber gradient maker the meniscus of the denser 250 mM solution will be a few millimeters lower than the 1 mM level. Fractions of 7 ml are collected and pooled according to the A_{276} profile. Dialysis against 0.05 M NH$_4$HCO$_3$ and lyophilization yield fluffy white protein with a recovery of 85%. Chicken osteocalcin elutes at about 110 mM KH$_2$PO$_4$–KOH in this system with minor Gla-containing fragments eluting between 50 and 150 mM.[28,32] Some osteocalcins, such as cow, elute very early and are not effectively resolved by hydroxyapatite. The affinity of osteocalcin for hydroxyapatite is reduced substantially by 2 mM Mg^{2+} ion. Because its phosphate salts are relatively soluble, inclusion of 2 mM Mg^{2+} ion in the

---

[32] F. W. Wians, K. Krech, and P. V. Hauschka, *Magnesium* **2**, 83 (1983).

buffers is possible, and this specifically shifts the elution of chicken osteocalcin to 40 mM KH$_2$PO$_4$–KOH, providing a useful trick for further purification.[32] Hydroxyapatite chromatography also resolves chicken osteocalcin according to its number of Gla residues. After limited thermal decarboxylation, osteocalcin with no Gla residues elutes at 5 mM KH$_2$PO$_4$–KOH, followed by species with 1 Gla, 2 Gla, and 3 Gla.[28]

Techniques for Monitoring Osteocalcin Purification

Because osteocalcin has no known function or bioassay as of this writing, quantitation of the protein has depended on standard chemical and physical methods. The hallmarks of osteocalcin are its abundance in mineralized tissues, small size, highly anionic character, and content of Gla.[5,21] Historically, purification was continuously monitored by Gla amino acid analysis[21] of extracts, residues, and chromatography column fractions, and one criterion for purity was high and invariant Gla content. Now that the behavior of osteocalcin is well established, purification is straightforward. Two rapid methods for assessing purity of osteocalcin are gel electrophoresis and high-pressure liquid chromatography (HPLC).

Gel Electrophoresis. Osteocalcin is the most abundant, small, anionic protein species in bone, and it migrates rapidly in most polyacrylamide gel systems, including 20% at pH 8.9[17]; 10% in 8 M urea–Tris–glycine, pH 8.2, with or without 0.1% SDS; and 15% in 6 M urea–0.1 M sodium phosphate, pH 7.2–0.1% SDS.[28,33] The high anodal mobility is relatively insensitive to gel concentration over the range 4–20%. Apparent M_r values calculated from standardized SDS gels range from 7000 to as high as 17,000 even if Ferguson plots are employed.[28] This is considerably greater than the true M_r established by sequencing (see the table) and results from polyelectrolyte effects rather than aggregation of osteocalcin.

Agarose gel electrophoresis resolves charge variants of osteocalcin within 30 min and requires only 5–20 μg of protein.[21] One hazard of this and other gel methods is adequate fixation and staining of osteocalcin. The best fixative appears to be 12.5% trichloroacetic acid (TCA) at 0°,[21] which causes osteocalcin to precipitate rapidly in sharp white bands in 0.8% agarose gels. Conventional methanol–acetic acid fixation actually solubilizes osteocalcin from agarose and some types of polyacrylamide gels, even after TCA treatment. Staining of agarose gels is done in fresh 0.1% amino black in 2.5% TCA containing 50% methanol, while polyacrylamide gels may be stained with Coomassie Blue in 0.1% cupric ace-

[33] A. L. Shapiro, E. Viñuela, and J. B. Maizel, *Biochem. Biophys. Res. Commun.* **28,** 815 (1967).

AMINO ACID COMPOSITION OF OSTEOCALCIN

Amino Acid	Species[a]					
	Human	Monkey	Cow	Chicken	Swordfish	Rat
Gla	2	3	3	3	3	3
Glu	5	4	3	7	3	2
Asp	5	5	6	5	4	8–9
Hyp	0	1	1	0	0	1
Pro	7	6	6	5	2	5
Thr	0	0	0	0	3	2
Ser	0	0	0	2	1	0–1
Gly	3	3	3	4	3	4
Ala	3	4	4	6	8	4
Cys 1/2	2	2	2	2	2	2
Val	3	2	2	3	3	2
Met	0	0	0	0	1	0
Ile	1	1	1	1	3	2
Leu	5	5	5	3	5	5
Tyr	5	5	4	3	3	4
Phe	2	2	2	2	1	1
His	1	1	2	1	0	2
Lys	0	1	1	0	0	1
Trp	1	1	1	0	0	0
Arg	4	3	3	3	2	2
NH_3	(3)	(3)	(2)	(6)	(3)	(ND)
Residues	49	49	49	50	47	50–52
M_r	5879	5889	5850	5669	5194	—
$E^{1\%}$	—	—	12.8^b	8.17^b	—	10.2^b

[a] References (numbers refer to text footnotes): human,[7] monkey,[8] cow,[6] chicken,[9] swordfish,[12] rat.[10,11]

[b] Absorbance at 276 nm minus absorbance at 320 nm.

tate, 10% acetic acid, 25% isopropanol or with Stains-All. Caution must be exercised in using gel methods as proof of purity of osteocalcin and other bone matrix proteins because of fixation and staining losses.

HPLC. Reverse-phase HPLC is rapid and provides quantitative data regarding osteocalcin and its purity.[8] An analytical column (μBondapak C_{18}, 4.4 × 300 mm; Waters Associates, Milford, MA) is operated at 2 ml/min, using an aqueous mobile phase (solvent A) of either 0.1% (v/v) trifluoroacetic acid (TFA) (Pierce, Rockford, IL) or 0.1% (v/v) H_3PO_4 in HPLC grade H_2O (Waters Associates). TFA is preferred because of its volatility. Solvent B is prepared by mixing 700 ml of glass-distilled acetonitrile (Burdick and Jackson, Muskegon, MI) with 300 ml of solvent A.

Solvents A and B are warmed to 45° and degassed by evacuation for 3 min at 10 torr with vigorous magnetic stirring. Linear or convex gradients (curves 5 and 6, Solvent Programmer Model 660, Waters Associates) between 0% solvent B and 70% solvent B are run for 10 min or longer, with at least 3 min of equilibration at 0% solvent B (i.e., 100% solvent A) after each run. An automatic sample injector (WISP, Waters Associates) is ideally suited to these analyses because of precisely reproducible onset of injection, gradient, and equilibration times, allowing overlay comparison of separate absorbance tracings. Samples containing 1 μg or more of osteocalcin in 1–150 μl of virtually any solvent (H_2O, 0.1% TFA, 0.05 M HCL, 0.05 NH_4HCO_3, 0.5 M EDTA, 4 M guanidine-HCl) can be injected. Appropriate solvent blanks must also be run.

A standard curve (1–100 μg) can be constructed after injecting aliquots of pure osteocalcin stock solution. Elution of osteocalcin is monitored by absorbance at 210 nm or 276 nm (Variable Absorbance Detector, Model 440, Waters Associates). Despite the extinction coefficient increase of 15- to 25-fold at 210 nm (depending on Tyr content), the steeply sloping solvent baseline (decreasing, if solvent A is 0.1% TFA; increasing if 0.1% H_3PO_3) precludes operation at full-scale sensitivities below 0.4 Å. However, at 276 nm, solvent baselines are relatively flat, allowing a full-scale sensitivity of 0.02 Å to be used. Both wavelengths have their advantages: peptide fragments devoid of aromatic residues are best detected at 210 nm[34] while 276 nm accentuates protein peaks and usually minimizes peaks of other contaminating substances. At 276 nm, 1 μg (0.176 nmol) of chicken osteocalcin gives a peak height of 0.0015 A, which varies with flow rate and gradient slope. Recovery of labeled osteocalcin is better than 90% by this method.[35] A stringent criterion for homogeneity of osteocalcin is symmetrical peak shape on a 60-min linear gradient (0% solvent B to 70% solvent B). The versatility of this HPLC method is shown in Fig. 1, where osteocalcin is quantitatively resolved in a centrifuged EDTA extract corresponding to the total soluble protein from 2 mg of calf mandible powder. Obviously, column fractions from preparative procedures can also be monitored for osteocalcin in this fashion with the advantage that the elution position of osteocalcin can be established from the HPLC profile of the original crude extract. Protease digestion kinetics,[9,36] peptide isolation,[9,34] and isolation of small amounts of osteocalcin from tissue extracts are included in the HPLC repertoire.

[34] C. Fullmer and R. Wasserman, *J. Biol. Chem.* **254,** 7208 (1979).

[35] P. V. Hauschka, *Biochemistry* **18,** 4994 (1979).

[36] P. V. Hauschka, *in* "Heritable Disorders of Connective Tissue" (W. H. Akeson, ed.), p. 195. Mosby, St. Louis, Missouri, 1982.

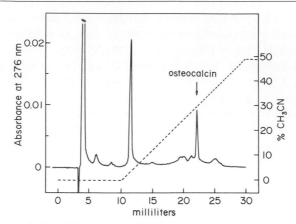

FIG. 1. Reversed-phase high-pressure liquid chromatography of osteocalcin in a crude extract of bone. Calf mandible powder was extracted with 0.5 M ammonium EDTA, pH 6.2 (0.40 ml/40 mg of bone) at 0° for 30 min. After centrifugation, 0.020 ml of the supernatant was injected directly into the chromatograph (see the text). A 2 ml/min linear gradient of CH_3CN in 0.1% trifluoroacetic acid caused the osteocalcin (~5 μg) to elute at ~30% (v/v) CH_3CN. The baseline of a blank injection has been subtracted. Pure cow osteocalcin elutes at the identical position as a single sharp peak.

Gel permeation chromatography using aqueous buffers and TSK-type columns (Varian) on a high-pressure liquid chromatograph is useful for characterizing osteocalcin. As above, crude 0.5 M EDTA, EPIC, or GE-PIC extracts of bone or osteocalcin at various stages of purity may be directly injected on the column (TSK 2000 SW, 7.5 × 500 mm, 0.1 M, NH_4HCO_3, isocratic, 1 ml/min) and monitored at 210 nm. A 0.5 M EDTA extract shows four well-resolved principal peaks at 9.3 ml (large matrix constituents), 12.0 ml (osteocalcin), 14.7 ml (Ca^{2+}–EDTA complex), and 21.0 ml (EDTA). Molecular weights can be established accurately by gel permeation on a tandem TSK 3000 SW and TSK 2000 SW column (7.5 × 2200 mm) eluted at 1 ml/min with 6 M guanidine HCl, 60 mM Tris HCl, pH 8.0.[13] Calibration with proteins of known molecular weight[37] shows the expected linear relationship between $K^{1/3}$ and $M^{0.555}$.[38] Chicken osteocalcin (M_r = 5670) and its 44-residue tryptic fragment (M_r = 4950) are readily resolved by this method,[9] and an M_r of 5600 ± 400 was determined for monkey osteocalcin, in close agreement with the sequencing value of 5889.[8]

[37] N. Ui, *Anal. Biochem.* **97,** 65 (1979).
[38] W. W. Fish, K. G. Mann, and C. Tanford, *J. Biol. Chem.* **244,** 4989 (1969).

Determination of Osteocalcin Concentration. Osteocalcin purified to homogeneity is used as a standard for establishing the extinction coefficient. Because absorbance shifts occur in the presence of Ca ions, the metal-free protein (5 mg/ml) is first prepared by dialysis against 0.05 M HCl at 0° for 3 hr, followed by dialysis overnight into 20 mM Tris-HCl, pH 7.6. The centrifuged stock solution is diluted (by weight for accuracy) to about 0.5 mg/ml. Absorbance against 20 mM Tris-HCl, pH 7.6, is recorded, and aliquots of this solution are subjected to acid hydrolysis and quantitative amino acid analysis for computation of protein concentration. Extinction values are listed in the table. Concentrated stock solutions (1–3 mg/ml) are stored at −20° for future use as standards for HPLC quantitation and radioimmunoassay.

Structural Properties of Osteocalcin

The amino acid composition of osteocalcin from six species has been established (see the table). Bound carbohydrate and phosphate account for an insignificant fraction of the weight of pure osteocalcin. Amino acid sequences have been determined for human,[7] monkey,[8] cow,[6] chicken,[9] and swordfish[12] osteocalcins, and partial sequences of the rat protein have been published.[10,11] A summary of the primary structural features has appeared.[22] Sequencing of Gla by gas chromatography–mass spectrometry of osteocalcin fragments has been facilitated by specific conversion of Gla to dideuterioglutamyl residues by decarboxylation in the presence of DCl under vacuum.[9] Osteocalcin may also be labeled by tritium exchange-decarboxylation creating 4-[³H]glutamyl residues of high specific activity at the three former Gla positions.[35] This derivative functions as a substrate in the bone microsome carboxylation system, releasing tritium from C4 during recarboxylation in a vitamin K-dependent fashion.[39] The three Gla residues of osteocalcin occur invariably at positions 17, 21, and 24 (except in human, where Glu occurs at 17) and are apparently involved in cation-binding and hydroxyapatite adsorption.[21,32,40] Osteocalcin binds Ca^{2+} at specific sites under physiological conditions: chicken, two sites, $K_d = 0.8$ mM[21], cow, 3 sites, $K_d = 3$ mM.[12] Calcium binding induces α-helix in osteocalcin, and a structural model has been developed.[22]

[39] J. B. Lian, P. V. Hauschka, and P. M. Gallop, *Fed. Proc. Fed. Am. Soc. Exp. Biol.* **37**, 2615 (1978).
[40] J. W. Poser and P. A. Price, *J. Biol. Chem.* **254**, 431 (1979).

Quantitation of Osteocalcin by Radioimmunoassay

The use of radioimmunoassay for the measurement of osteocalcin offers the advantages of specificity, sensitivity, and technical simplicity. The method can easily detect nanogram quantities of the protein in bone extracts,[13] cell cultures,[14] and serum.[24] The assay is based on the competition of the radioactively labeled antigen and an identical nonlabeled antigen for binding to a specific antibody. The amount of labeled antigen bound to the antibody is inversely proportional to the amount of unlabeled antigen present in the system (see footnotes 41–43 for reviews).

Materials

Radioiodide is purchased from New England Nuclear as carrier-free ^{125}I in NaOH at pH 10; Freund's complete and incomplete adjuvants are obtained from Difco, goat anti-rabbit IgG from Gibco, and rabbit anti-goat IgG from Pelfreeze. Bovine serum albumin (RIA grade) and chemicals (reagent grade or better) are purchased from Sigma.

Clarified serum is prepared by adding 35 g of activated charcoal to 1 liter of normal serum. The mixture is agitated for 16 hr at ambient temperature, centrifuged at 2000 g for 20 min, and filtered through consecutive 20-, 10-, and 1-μm pore filters. The clarified serum is adjusted to pH 7.4 and preserved with 0.5 g of sodium azide per liter.

Preparation of Antigens

Since small peptides are less antigenic than large proteins,[44] many small peptide hormones have been covalently linked to larger protein molecules[45] to produce an immune response. However, antibodies produced against such conjugated peptides often do not recognize the free peptide in solution.[46] Adsorption of peptides to large polymers often increases their antigenicity.[47] The procedure we have used is that of adsorption to polyvinylpyrrolidone (PVP) and emulsification with Freund's adjuvant.

For injection of rabbits, osteocalcin is dissolved in 0.15 M NaCl to a concentration of 2–10 mg/ml. This is diluted 1:5 (v/v) with 50% (w/v)

[41] J. J. Langone and H. Van Vunakis, eds., this series, Vol. 73.
[42] S. A. Berson and R. S. Yalow, *Clin. Chim. Acta* **22,** 51 (1968).
[43] W. M. Hunter, *in* "Handbook of Experimental Immunology" (D. M. Weir, ed.), Vol. 1, p. 14. Blackwell, Oxford, 1978.
[44] F. Haurowitz, *Physiol. Rev.* **45,** 1 (1965).
[45] L. A. Frohman, M. Reichlin, and J. E. Sokal, *Endocrinology* **87,** 1055 (1970).
[46] E. Haber, L. B. Page, and G. A. Jacoby, *Biochemistry* **4,** 693 (1968).
[47] R. B. Worobec, J. H. Wallace, and C. G. Huggins, *Immunochemistry* **9,** 229 (1972).

PVP, thoroughly mixed and, after 2 hr at 23°, emulsified with an equal volume of Freund's complete adjuvant (for primary injections) or incomplete adjuvant (for subsequent injections). This antigen preparation contains 0.2–1 mg of osteocalcin per milliliter.

Preparation of Antisera

Antibodies are produced in male New Zealand white rabbits or in pygmy goats. Rabbits are injected intradermally at multiple sites with 1 ml of the appropriate antigen preparation. One milligram of osteocalcin is used for the primary injection; 6 and 10 weeks later, the rabbits are boosted with 0.5 mg of osteocalcin. Rabbits are maintained on a booster schedule every 3 months with 0.2 mg of protein. Rabbits are bled from the marginal ear vein 10–12 days after each injection. Goats are injected with 2.5 mg and then 1.0 mg of osteocalcin and bled from the jugular vein. Serum is harvested after clotting, and the titer is determined. Positive antisera are preserved with 10 μl of 10% NaN$_3$ per milliliter of serum and stored in 1-ml aliquots at −85°.

Preparation of Tracers

[^{125}I]Osteocalcin is prepared by either chloramine-T[48] or lactoperoxidase coupled to Sepharose beads[49] using a New England Nuclear kit. A protein stock solution is prepared and used exclusively for iodinations. Pure lyophilized osteocalcin is weighed, dissolved in 0.3 M phosphate, pH 7.5, and dialyzed exhaustively against the same buffer. Protein concentration is determined by absorbance (see the table). Stock solutions (0.70 mg/ml) are kept frozen in 50-μl aliquots at −85°. For the lactoperoxidase method, 10 μl of stock solution is mixed with 130 μl of phosphate buffer; 100 μl (5 μg) of this protein solution is used in the iodination. For chloramine-T iodination, 10 μl of protein is mixed with 25 μl of 0.3 M phosphate, pH 7.5; 25 μl (5 μg) of the prepared osteocalcin is transferred to a polystyrene vial containing 2.0 mCi of Na^{125}I in 20 μl of alkaline solution. Ten microliters of fresh chloramine-T (1 mg/ml in 0.3 M phosphate, pH 7.5) is added, and the reaction mixture is shaken for 30 sec. Iodination is terminated with 50 μl of sodium metabisulfite (2.5 mg/ml in 0.3 M phosphate, pH 7.5). Labeled protein is separated from the other reactants by gel filtration on Sephadex G-25 (fine mesh; 1 × 30 cm). The column is equilibrated and eluted with a buffer consisting of 0.01 M NaH$_2$PO$_4$, 0.125 M NaCl, 0.025 M Na$_4$EDTA, 0.1% Tween 20, at pH 7.4.

[48] W. M. Hunter and F. C. Greenwood, *Nature* (*London*) **194**, 495 (1962).
[49] G. S. David and R. A. Reisfeld, *Biochemistry* **13**, 1014 (1974).

The [^{125}I]osteocalcin is diluted in this buffer to 2×10^6 cpm per milliliter and stored at $-20°$ and may be used for 8–10 weeks.

Radioimmunoassay Procedure

All dilutions are made in assay buffer that contains 0.01 M NaH$_2$PO$_4$, 0.125 M NaCl, 0.1% Tween, 0.025 M Na$_4$EDTA, and 0.1% bovine serum albumin, pH 7.4. Polystyrene or glass tubes may be used. Assays consist of 100 μl of appropriately diluted antiserum, carrier nonimmune serum[43] if using second antibody as a separation technique, a known concentration of standard osteocalcin or appropriate dilutions of the unknown sample, 2×10^4 cpm of [^{125}I]osteocalcin in 100 μl of assay buffer, and assay buffer to a final volume of 600 μl. Antiserum is diluted such that 60–70% of fresh [^{125}I]osteocalcin is bound to the antibody in the absence of unlabeled osteocalcin. For assay of human or bovine osteocalcin, each lot of BSA must be checked for any contaminating osteocalcin. For equilibrium assay, all components are combined and incubated at 25° for 20 hr with constant shaking. For nonequilibrium assay, all components except the [^{125}I]osteocalcin are incubated as above. After 20 hr [^{125}I]osteocalcin is added, followed by a 24-hr incubation. Nonspecific binding (NSB) of ^{125}I is determined for each assay by omitting antiserum from the reaction mixture. Assays are terminated by the addition of charcoal–dextran in 0.05 M phosphate, pH 7.4–0.1% human serum albumin, or by the addition of a second antibody.[13]

Standard curves are constructed using B/B_0 versus nanograms of osteocalcin standard per milliliter. B is defined as [^{125}I]osteocalcin cpm bound to antibody minus NSB, and B_0 as that bound to antibody in the absence of unlabeled osteocalcin minus NSB. Six separate determinations of NSB and B_0 are made for each assay. Total binding is determined for each assay by B_0/total counts added. All unknowns are assayed in triplicate or by multiple dilution.

Separation Procedures

Each preparation of antiserum may require a different concentration of charcoal–dextran for optimal separation of bound and free fractions.[50] Our various antisera require 300 μl of a 1.5% solution. After the addition of charcoal, each tube is rapidly mixed for 2 min and centrifuged at 600 g for 15 min, and the supernatant fluid is removed. Radioactivity can be determined on either the supernatant (bound fraction) or charcoal pellet (free fraction). If charcoal pellets are counted, they are washed once

---

[50] K. Lau, C. W. Gottlieb, and V. Herbert, *Proc. Soc. Exp. Biol. Med.* **123,** 126 (1966).

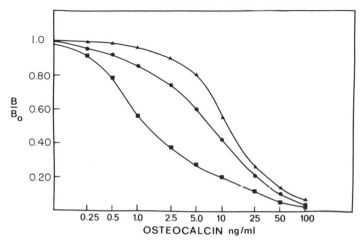

FIG. 2. Radioimmunoassay of cow osteocalcin. ●, Typical binding curves for the equilibrium assay; ■, nonequilibrium assay; ▲, equilibrium assay in the absence of EDTA.

before counting with 0.5 ml of 0.05 M phosphate, pH 7.4, 0.1% BSA, centrifuged, decanted, and counted.

For separation by the second antibody technique, either goat anti-rabbit IgG for antisera made in a rabbit or rabbit anti-goat IgG for antisera produced in a goat is added to each tube. The amount of free antibody added and the time of incubation varies with each preparation.[46] After incubation, the resulting precipitate is centrifuged for 20 min at 2000 g and decanted. The precipitate is washed with 0.5 ml of distilled water. Radioactivity is determined on the washed precipitate.

Assay Characteristics

Typical binding curves for the bovine assay are shown in Fig. 2. Human osteocalcin identically cross-reacts with antibody raised against purified cow osteocalcin.[23,51,52] Purification of cow osteocalcin is more convenient than isolation of human osteocalcin; thus, all human assays are performed using rabbit antiserum to cow osteocalcin. Rabbit antibodies to cow osteocalcin cross-react not only to human, but also to monkey, horse, goat, and elk.[13] Antisera raised to rat osteocalcin in rabbit,[29] guinea pig,[52] and goat[13] have been shown to cross-react with rat and possibly the

[51] P. A. Price and S. K. Nishimoto, *Proc. Natl. Acad. Sci. U.S.A.* **77**, 2234 (1980).
[52] P. Patterson-Allen, C. E. Brautigan, R. E. Gundeland, C. W. Asling, and P. X. Callahan, *Anal. Biochem.* **120**, 1 (1982).

mouse and dog.[52] Rabbit antisera to chicken osteocalcin will bind osteocalcin from several other avian species.[13]

Antibodies may differ in the antigenic site recognized. The COOH-terminal region of cow osteocalcin is required for antibody binding.[53] The antibody to chicken osteocalcin apparently requires both the carboxyl- and the amino-terminal ends for binding.[13]

Although the equilibrium assay is adequate for most purposes, we routinely use the nonequilibrium technique because the amount of sample required is 5-fold less. In the case of serum assay, this also means that nonspecific binding of the serum proteins is reduced. If more than 25 μl of undiluted serum is used in either assay, nonspecific binding is determined on each serum sample. The nonspecific counts are then subtracted from the sample-bound counts, and concentration is determined. For serum samples of less than 25 μl, nonspecific binding becomes negligible and separate NSB tubes can be omitted.

For all serum assays, standard curves are run in the presence of species-specific clarified serum in the same concentration as unknown samples. If more than one dilution of unknown sample is used, all samples are diluted in clarified serum to a uniform volume.

For the assay of osteocalcin in bone, 1–10 mg of ground, sieved bone powder is placed into 1.5-ml microcentrifuge tubes. Then 100 μl of EPIC is added for each 10 mg of bone powder and vortexed for 30 sec. Osteocalcin is extracted by end-over-end rotation in the cold for 18 hr; the solutions are then centrifuged and the supernatants are removed. Supernatants are diluted in RIA buffer and assayed as above. The charcoal method of separation is suitable for bone osteocalcin determinations.

The presence of EDTA is required in all assay buffers. Removal of the chelator results in an alteration of the antigen–antibody binding curve (Fig. 2). Since osteocalcin conformation is dependent upon calcium concentration,[22] antibody subpopulations are presumably present that are specific for the α-helical form of osteocalcin and/or the random-coil configuration. In order to eliminate this heterogeneity of binding, EDTA is added to all buffers.

Several other radioimmunoassays have been described for osteocalcin,[51,52,54] and several kits for the bovine assay are now commercially available (Seragen Corp., Boston, MA; Immunonuclear Inc., Stillwater, MN).

[53] C. M. Gundberg, unpublished observations.
[54] P. D. Delmas, D. Stenner, H. W. Wahner, K. G. Mann, and B. L. Riggs, *J. Clin. Invest.* **71**, 1316 (1983).

Osteocalcin Concentration in the Blood

In humans, circulating levels of osteocalcin are 6.8 ± 0.5 and 5.8 ± 0.5 ng/ml in adult males and females, respectively. These values are higher in children, averaging 25–30 ng/ml in a 1-year-old child and declining to the adult level at the age of puberty.[23] A similar pattern has been observed in rats. Circulating osteocalcin also exhibits circadian variations. In humans, circulating levels are lowest in the late afternoon and reach a peak nocturnally.[55] The values in adult males fluctuate from 5 ng/ml to 10 ng/ml in a 24-hr period. In rats, the pattern is reversed.[56] As a consequence, it is most important to regulate the time of blood sampling for osteocalcin determinations.

Elevations in circulating osteocalcin have been observed in a variety of metabolic bone diseases characterized by increased bone turnover.[24,25] Administration of 1,25-$(OH)_2D_3$ results in an increase in serum osteocalcin in normal individuals[55] and those with hereditary rickets.[26] The anticonvulsant drug phenytoin (diphenylhydantoin) also causes circulating osteocalcin to be increased.[57] In contrast, individuals on glucocorticoid therapy have a significantly lowered osteocalcin level.[25] Calcitonin therapy of Paget's disease also lowers the high levels normally observed in this disease.[58] Studies such as these indicate that the measurement of serum osteocalcin is a valuable tool in following the course of bone disease and therapeutic intervention.

Quantitation of 4-Carboxyglutamate

4-Carboxyglutamate (Gla) is found not only in the bone protein osteocalcin, but also in the other vitamin K-dependent calcium-binding proteins formed in liver,[59] kidney,[60,61] placenta,[62] and spleen[63] and in a variety

---

[55] C. M. Gundberg, M. Markowitz, and J. Rosen, *Calcif. Tissue Int.* (abstract) (1983).

[56] E. Freymiller and P. V. Hauschka, unpublished observations.

[57] D. A. Keith, C. M. Gundberg, A. Japour, J. Aronoff, N. Alvarez, and P. M. Gallop, *Clin. Pharmacol. Ther.* **34**, 529 (1983).

[58] L. J. Deftos, J. G. Parthemore, and P. A. Price, *Calcif. Tissue Int.* **34**, 121 (1982).

[59] C. T. Esmon, J. A. Sadowski, and J. W. Suttie, *J. Biol. Chem.* **250**, 4744 (1975).

[60] P. V. Hauschka, P. A. Friedman, H. P. Traverso, and P. M. Gallop, *Biochem. Biophys. Res. Commun.* **71**, 1207 (1976).

[61] P. A. Friedman, W. E. Mitch, and P. Silves, *J. Biol. Chem.* **257**, 110374 (1982).

[62] P. A. Friedman, P. V. Hauschka, M. A. Shia, and J. K. Wallace, *Biochim. Biophys. Acta* **583**, 261 (1979).

[63] S. D. Buchthal and R. G. Bell, *in* "Vitamin K Metabolism and Vitamin K-Dependent Proteins" (J. W. Suttie, ed.), p. 299. Univ. Park Press, Baltimore, Maryland, 1979.

of pathological calcifications.[64–66] The free amino acid is also found in the urine and presumably represents the final degradation product of all the Gla-containing proteins.

Several methods have been published for the analysis of Gla in protein hydrolysates and in urine. These methods employ either automated amino acid analysis,[67] HPLC,[68] or manual ion-exchange chromatography followed by fluorescence[69] or absorption spectroscopy.[70] The methods described here, which are the ones used in our laboratory, include automated cation-exchange amino acid chromatography and an isotope dilution method employing a single anion-exchange column and fluorescence detection.

Materials

Alkali-resistant tubes (13 × 100 mm) are either glass (Corning No. 7280) or Teflon (Tefzel 3113-0065). Dowex 1 is obtained from Bio-Rad Laboratories as AG 1-X8, <400-mesh, chloride form. Gla is obtained from Sigma, and 4-[*carboxy*-^{14}C]Gla was prepared by New England Nuclear as a custom synthesis. *o*-Phthalaldehyde is purchased from Eastman Kodak Company. All other chemicals are reagent grade or better.

Base Hydrolysis

4-Carboxyglutamate (Gla) is a malonic acid derivative that readily decarboxylates to glutamic acid under the acidic conditions used for protein hydrolysis. Therefore, for quantitative recovery of Gla from protein hydrolysates alkaline conditions are employed. Either pure protein, lyophilized tissues, or liquid samples in resistant tubes are made 2.0 M in KOH (0.05–50 mg dry weight in 0.3–0.8 ml of 2 M KOH). These in turn are placed inside large Pyrex tubes containing a small amount of 2 M KOH to prevent evaporation. After evacuation, the tubes are sealed with a torch under an atmosphere of nitrogen. Alternatively, a desiccator method that is more convenient for a large number of samples has been described.[67] The samples are heated for 22 hr at 110° and then placed on ice. Fresh, saturated $KHCO_3$ (110 μl per milliliter of 2 M KOH) is added

[64] R. J. Levy, J. B. Lian, and P. M. Gallop, *Biochem. Biophys. Res. Commun.* **91,** 41 (1979).
[65] J. B. Lian, M. S. Skinner, M. J. Glimcher, and P. M. Gallop, *Biochim. Biophys. Res. Commun.* **73,** 349 (1976).
[66] J. B. Lian, E. L. Prien, Jr., M. J. Glimcher, and P. M. Gallop, *J. Clin. Invest.* **59,** 1151 (1977).
[67] P. V. Hauschka, *Anal. Biochem.* **80,** 212 (1977).
[68] M. Kuwada and I. Katayama, *Anal. Biochem.* **117,** 259 (1981).
[69] C. M. Gundberg, J. B. Lian, and P. M. Gallop, *Anal. Biochem.* **98,** 219 (1979).
[70] L. Pecci and P. Cavallini, *Anal. Biochem.* **118,** 70 (1981).

followed by 3.3 M citric acid (22 μl per milliliter of 2 M KOH). Ice-cold 70% HClO$_4$ is added dropwise with rapid vortexing until the solution has reached pH 7 (spot checked on dual-range pH indicator paper). The tubes are allowed to stand on ice for 15 min and then centrifuged for 2 min at 1000 g to remove the precipitate. The supernatant fluid is stored at $-20°$. Because the KClO$_4$ precipitate can trap some of the free Gla after hydrolysis, recovery can be monitored by adding an internal standard of [^{14}C]Gla or [^{14}C]alanine. To determine total urinary Gla (free, conjugated, and protein bound), 0.8 ml of urine is mixed with 0.2 ml of 10 M KOH and hydrolyzed as above. Protein-bound Gla is determined by exhaustively dialyzing 50 ml of urine in Spectrapor 3 (M_r cutoff-3500) against distilled water. The dialysate is lyophilized and hydrolyzed as above.

Automated Amino Acid Analysis

Several procedures have been described to quantitate Gla in hydrolyzed samples. The method currently used in our laboratory is an adaptation of Method 2 of Hauschka.[67] This method may be combined with standard runs to obtain total amino acid composition. It is the method of choice when analyzing for Gla in heterogeneous samples because it separates Gla from all known acidic, ninhydrin-positive interfering compounds. The procedure has been developed for a Beckman 121M amino acid analyzer but is adaptable to other systems.

Alkaline hydrolysates are diluted in 0.2 M sodium citrate buffer, pH 2.2. The pH is checked on indicator paper (short range alkacid, pH 0.0–3.0) and adjusted to pH 2.2 with 6 M HCl if necessary. Gla is stable for several days at 20° under these conditions.[71]

All urines for free Gla determination are screened for protein by Chemstrip (Boehringer Mannheim). Protein is removed from any positive urines by the addition of 100 μl of 10% sulfosalicylic acid to 1.0 ml of urine. Urines are centrifuged at 600 g for 20 min to remove any precipitate and then diluted 1 : 3 or 1 : 4 in 0.2 M sodium citrate buffer, pH 2.2. Any precipitate after dilution is removed by centrifugation. The pH of the diluted urine is then adjusted to 2.2 with 6 N HCl if necessary.

A microbore column (2.8 × 330 mm) containing Beckman AA-20 resin at 51° is used in conjunction with nanobore tubing (Beckman). The buffer flow rate is 8.0 ml/hr, and the effluent stream is monitored by mixing with ninhydrin–DMSO reagent (Beckman) flowing at 4.0 ml/hr. All buffers contain 0.2 M citrate, 0.01% (v/v) caprylic acid, and 0.25% (v/v) thiodiglycol. The column is equilibrated with pH 2.68 citrate buffer. A 50-μl sam-

[71] P. V. Hauschka, E. B. Henson, and P. M. Gallop, *Anal. Biochem.* **108,** 57 (1980).

ple is injected, and the following buffer program cycle is used: pH 2.68 equilibration (16.0 min), sample injection (0.1 min), pH 2.68 (4.0 min), pH 3.10 (35.0 min), 0.2 M NaOH, 0.1% EDTA column wash (5.0 min), reequilibration at pH 2.68 (23.1 min), and restart of the cycle. Elution time for Gla is 33.0 min after injection. Glutamic acid can also be determined by lengthening the pH 3.10 buffer cycle.

Total and protein-bound urinary Gla can be determined after base hydrolysis using this program. For determination of free urinary Gla the same program is used, but at a lower pH to resolve Gla adequately from interfering substances. Gla elution time increases about 5–7 min for each decrease of 0.1 pH of the starting buffer.

Several unidentified compounds found in alkaline hydrolysates of crude tissues elute in the Gla region. These may erroneously be assumed to be Gla. To prove that the putative peak is Gla, the peak must be present in alkaline hydrolysates but not in acid hydrolysates, and furthermore the kinetics of decarboxylation from Gla to glutamic acid must match those of the authentic standard.[71] Acid conversion of Gla to glutamic acid results in a 2.38-fold increase in relative ninhydrin peak area.[71]

Isotope Dilution and Anion Exchange

We have also developed a rapid, sensitive procedure for Gla analysis that employs isotope dilution coupled with a single anion-exchange column and fluorescent detection.[69] This method is particularly useful for analysis of free urinary Gla, but it can be used also for the detection of Gla in alkaline hydrolysates.

For determination of free Gla in urine, 5000 cpm of [^{14}C]Gla (40 pmol) is added to 2.0 ml of urine. The sample is adjusted to pH 11.5 with 6 N KOH and rapidly centrifuged to remove any precipitate. A small aliquot is removed and counted. Then 1.0 ml of the urine is applied to a 6×0.7 cm column filled with Dowex 1-X8 (<400 mesh) previously equilibrated with 0.01 M HEPES, pH 11.5; 50 ml of 0.02 M HEPES, pH 5.0, is passed through the column and discarded. This eluate contains the basic and neutral amino acids. Then 50 ml of 0.020 M HEPES–0.020 M MgCl$_2$, pH 4.5, is passed through the column and collected in 5-ml fractions. These contain most of the acidic amino acids. Gla, free of any other primary amines, elutes at approximately 40 ml of this buffer. To determine precise location and recovery, 0.5 ml of each fraction is removed for counting. Gla is quantitated by mixing a second 0.5-ml aliquot of each fraction with 0.5 ml of 6.0 mM o-phthalaldehyde in 0.40 M boric acid, pH 9.7. Fluorescence is determined in a fluorometer with excitation and emission wavelengths of 340 and 455 nm, respectively. Concentration is determined from a 0.1 to 10 nmol/ml standard curve of Gla. Final Gla concentration is determined by correcting for recovery, which is generally >80%. Column fractions immediately adjacent to the Gla region are free of interfering components.

This procedure can be used also for determining Gla in KOH hydrolysates. The neutralized protein hydrolysates are brought to pH 11.5, and [^{14}C]Gla is added as described above. The sample is treated as above except that a larger column is used (10 × 0.7 cm) with 80 ml of the first buffer (0.02 M HEPES, pH 5.0). Results from this procedure and direct automated amino acid analysis yield values generally consistent to within 10%.

4-Carboxyglutamate Excretion

Bone contains the most abundant quantities of Gla, most of which is in osteocalcin. The amount of Gla in the bone varies with species, anatomical location, age, and mineral density (see below). The final step in the catabolism of all Gla-containing proteins is the excretion of free Gla and Gla-peptides. The Gla is not further metabolized, but is excreted quantitatively.[72] Our investigations demonstrated that 80–100% of the Gla in the urine is free, and only 0–20% is conjugated or peptide bound.[23]

High levels of urinary Gla are observed in growing children similar to the pattern observed with circulating osteocalcin. Free Gla decreases from a high of 150 ± 20 μmol per gram of creatinine at 1 year until puberty, beyond which time the excretion stabilizes at 44 ± 4 μmol per gram of creatinine.[23] Excretion of urinary Gla is related to the catabolism of all the vitamin K-dependent proteins. The turnover of the liver-derived vitamin K-dependent clotting factors is rapid. Consequently, the contribution to urinary Gla excretion from these factors is acutely affected by Warfarin therapy. Adult patients undergoing Warfarin therapy have a significantly decreased mean Gla excretion, and the percentage of active plasma prothrombin as measured by one-stage prothrombin assay correlated directly with the urinary Gla excretion.[73] Therefore, it is striking that urinary Gla excretion parallels the levels of circulating osteocalcin in growing children. The greater amounts of circulating osteocalcin and urinary Gla most likely reflect synthesis and rapid remodeling of bone during growth.

We have investigated the excretion of Gla in a variety of clinical situations. Urinary Gla is elevated in individuals with subcutaneous soft

[72] P. Fernlund, *Clin. Chim. Acta* **72,** 147 (1976).
[73] R. J. Levy and J. B. Lian, *Clin. Pharmacol. Ther.* **25,** 562 (1978).

tissue calcifications and in active bone resorption.[74,75] Therefore, the measurement of urinary Gla excretion in conjunction with serum osteocalcin can provide useful information regarding abnormalities of calcium homeostasis and bone metabolism.

Normalization Parameters for Bone and Osteocalcin

Bone is a heterogeneous tissue containing many cell types. In addition, dispersion occurs as a function of age in the fractional content of various matrix proteins and the composition of the mineral phase. Since a unique property of osteocalcin is its specific binding to hydroxyapatite, the most convenient normalization parameter is mineral content. In normal bone, Gla and osteocalcin parallel the mineral content of the tissue.[30]

Of use in evaluating the degree of bone mineralization and maturation is the technique of density centrifugation.[76–78] We have applied this method to the examination of Gla and osteocalcin in bone.

Preparation of Bone Samples

Samples of bone and calcific deposits to be analyzed for mineral and osteocalcin content in density fractions cannot be exposed to aqueous solutions during preparation. This not only would alter the distribution of the mineral phase, but in very young bone could remove a significant fraction of the osteocalcin. There are gradations of mineral and osteocalcin content throughout bone and calcified cartilage.[21,31] In order to increase homogeneity, analysis should be limited to an anatomically uniform sample, e.g., the diaphysis, avoiding any contamination by calcified cartilage.

After excision, the bone is split, rolled on Whatman 3 MM filter paper, and rinsed briefly in ethanol to remove contaminants. The bones are immediately frozen, lyophilized, and further dried in a vacuum desiccator at 4° (young bone) or 20° (mature bone). With large samples, the dry bone is ground to a fine powder in a liquid N_2 mill as described above. Small

[74] C. M. Gundberg, J. B. Lian, P. M. Gallop, and J. J. Steinberg, *J. Clin. Endocrinol. Metab.* **57,** 1221 (1983).

[75] J. B. Lian, L. M. Packman, C. M. Gundberg, R. E. H. Partridge, and M. C. Maryjowski, *Arthritis Rheum.* **25,** 1094 (1982).

[76] L. J. Richelle and C. Onkelinx, *in* "Mineral Metabolism" (C. L. Comar and F. Bronner, eds.), Vol. 3, p. 123. Academic Press, New York, 1969.

[77] A. H. Roufosse, W. J. Landis, W. K. Sabine, and M. J. Glimcher, *J. Ultrastruct. Res.* **68,** 235 (1979).

[78] L. C. Bonar, A. H. Roufosse, W. K. Sabine, M. D. Grympas, and M. J. Glimcher, *Calcif. Tissue Int.* **35,** 202 (1983).

samples are ground by hand with a mortar and pestle. It is critical that the bone powder be ground and sieved to a uniform size (75–150 μm) for routine analysis. However, for density fractionation a much finer powder (5–10 μm) is produced by extensive liquid N_2 milling; particle size and distribution are measured by light and electron microscopy.[77] Sonic sieving is also useful for obtaining particles in the micrometer range.

Mineral Parameters

Mineral composition analysis requires a constant initial dry weight that is achieved by vacuum desiccation over P_2O_5 at 23° for 18 hr. Lyophilization alone approaches 96–98% of dry weight, but water may be regained during prolonged storage. Ash content is calculated from the residual weight after heating 100 mg of dry bone at 600° for 24 hr. Bone ash is dissolved in 3 N HCl–5.86% La_2O_3 and is diluted appropriately for determination of calcium and phosphorus. Alternatively, samples can be digested (wet ashing) in a porcelain crucible with boiling concentrated nitric acid–70% perchloric acid (4 : 1 v/v) and then diluted for analysis. Calcium is quantitated by atomic absorption[18] and phosphorus by the method of Chen *et al.*[79] Calcium ranges from 15 to 19% of bone dry weight in embryonic bone, 20 to 25% in young bone, and 25 to 28% in mature bone. Phosphorus is 9–12% of dry weight. The calcium : phosphorus molar ratio is generally lower than the value of 1.67, which characterizes pure hydroxyapatite.

Density Fractionation

Density of whole bone can be determined by the Archimedes principle of volume displacement in water. A more precise measurement of bone density, which fractionates the bone into densities reflecting the maturation of mineralizated bone tissue, is the differential centrifugation technique originally described by Richelle.[76] Modification of the original technique[77,78] has allowed for density centrifugation of bone in a simple stepwise fashion for increased yields and ease of separation of the fractions. After grinding and sieving, the 5-μm bone particles are suspended in ethanol by sonication and then centrifuged. The procedure is repeated several times to eliminate colloidal particles. Fractions are separated using centrifugation in bromoform–toluene mixtures spanning the range of densities from 1.4 to 2.3 g/ml. Mixtures of bromoform (density 2.89 g/ml; Fisher Scientific) and toluene (density 0.867 g/ml) are prepared at intervals of 0.1 g/ml. The fluid densities are then measured to 0.02 g/ml accu-

[79] P. S. Chen, Jr., T. V. Toribaya, and H. Warner, *Anal. Chem.* **28**, 1756 (1956).

racy with calibrated hydrometers and density standards (R. P. Cargyle Laboratories, Inc.). Each mixture is stable for several weeks, and the density does not change during centrifugation.

To begin centrifugation, 200 mg of bone powder is suspended by sonication in a 35-mm polyallomer centrifuge tube partially filled with the fluid of a given density. The density chosen for the first run should allow about 50% of the particles to sediment. Thus, for young bone or pathologic calcifications, a density of 1.7 or 1.8 g/ml is chosen; for mature or old bone, a density of 2.0 is initially used. Centrifugation is for 30 min at 10,000 rpm in a Beckman SW27 rotor. Equilibrium conditions that allow all particles denser than the liquid to reach the bottom of the tube must be set for each sample and rotor size. After centrifugation, the supernatant is decanted, modified to the next lower density by the addition of toluene, and recentrifuged. Each precipitate obtained from the progressively decreasing density solutions is washed several times in ethanol to remove toluene and bromoform. The precipitate obtained with the initial centrifugation is resuspended in a solution of density 0.1 g/ml higher. Successive centrifugation at progressively increasing densities (0.1 g/ml steps) provides higher-density fractions. All fractions are dried to constant weight under vacuum at 23°. From weight determinations, the fractional distribution of the various bone densities is calculated. Osteocalcin, total Gla, and mineral determinations are performed on the individual density fractions.

Correlations of Osteocalcin, Gla, and Mineral After Density Fractionation

Fractionation of bone powder has provided a powerful tool for assessing the extent of maturation of bone in a fashion that X-ray diffraction cannot detail. The technique unmasks defects in bone tissue turnover that cannot be revealed by analysis of whole bone, particularly when disease or abnormal bone metabolism affects only a small percentage of the total bone mass. Specific examples include the precise amount of osteoid in vitamin D deficiency[80,81] and the relative amount of unresorbed bone in osteopetrosis.[82] Regarding bone matrix proteins, different distributions of total Gla and osteocalcin were revealed in the fractions obtained from vitamin D-deficient undermineralized bone. Gla was enriched in low-density fractions, representing newly synthesized matrix, and immunoreactive osteocalcin was decreased in the mature bone, high-density fractions.

[80] J. B. Lian, B. H. Roufosse, P. V. Hauschka, B. Reit, P. M. Gallop, and M. J. Glimcher, *J. Biol. Chem.* **257**, 4999 (1981).
[81] I. R. Dickson and E. Kodicek, *Biochem. J.* **182**, 429 (1979).
[82] A. L. Boskey, B. Dick, J. Lane, and M. Marks, *Int. Assoc. Dent. Res.* (abstract) (1983).

These results were interpreted as a defect in processing of osteocalcin from precursor protein.[80] Measurement of Gla and osteocalcin in other diseases affecting bone mineralization should be informative for protein functional relationships.

In most examples of pathologically calcified tissues, total Gla and calcium concentration correlate, although immunoreactive osteocalcin levels vary depending on the calcific deposit.[83,84] In tumoral calcinosis deposits,[85] the average density ranges from 2.4 to 2.7 g/ml whereas cortical bone in the same patient is 2.0 to 2.1 g/ml. In this pathological calcification, Gla concentration was low and osteocalcin was absent.

In Vitro Synthesis of Osteocalcin

Labeling of Gla and osteocalcin in vitro[14–17] can provide a means of examining the biosynthetic pathway, from induction of precursor protein[15,17] to the processing of extracellular osteocalcin. Gla synthesis can be monitored by vitamin K-dependent $^{14}CO_2$ incorporation into endogenous substrate or by synthesis from previously incorporated [^{14}C]glutamic acid residues. Additionally, osteocalcin can be labeled with amino acids such as leucine or proline followed by appropriate separation for quantitation. Osteocalcin synthesis has been studied extensively in chick embryonic bone in an organ culture system and in microsomal preparations for several reasons. Both Gla and osteocalcin are detectable in developing chick embryo bones as early as day 8,[13] and Gla synthesis can be inhibited by warfarin injection of chick embryos[18] with accumulation of endogenous precursor that can be carboxylated in a vitamin K-dependent bone microsome system.[16]

Bone Preparation

Studies can be carried out in bone from normal and Warfarin-treated embryos. For inhibition of Gla synthesis, embryos are injected at days 11 and 12 with 50 μl of sterile saline containing 100 μg of sodium Warfarin (Endo Laboratories, Garden City, NY). On days 13 and 14, 200 μg (100 μl volume) is injected. This treatment results in an 80% reduction of Gla content, a 3- to 5-fold increase in microsomal protein and a 70% survival

[83] J. B. Lian, G. Boivin, A. Patterson, M. D. Grynpas, and C. Walzer, Calcif. Tissue Int. 35, 555 (1983).
[84] A. L. Boskey, M. D. Vigorita, and J. B. Lian, Fed. Proc. Fed. Am. Soc. Exp. Biol. (in press) (1983).
[85] A. L. Boskey, M. D. Vigorita, O. Senser, M. D. Stuchin, and J. M. Lane, Clin. Orthop. Relat. Res. 178, 258 (1983).

rate. Calvariae, long bones, and mandibles are excised from the 15-day chick embryos on ice in Hanks balanced salt solution.

Organ Culture Method

The medium is Dulbecco's modified Eagle's medium, supplemented with 25 mM HEPES, 2 mM glutamine, 1 mg/ml powdered bovine serum albumin (Sigma Chemical Co.), 0.5 mg/ml ascorbic acid, 0.15 mg/ml aspartic and glutamic acids, 10 μg/ml Aqua MEPHYTON (Merck, Sharp and Dohme, West Point, PA), and 10 μCi/ml NaH^{14}CO$_3$ (New England Nuclear, Boston, MA). Bones are preincubated for 15 min in medium without vitamin K and bicarbonate, then pulsed for periods up to 24 hr in roller tubes with 1 bone per milliliter, such that the bone adheres to the wall of the glass tube and is bathed once every 20–30 sec. These conditions minimize resorption, and [^{14}C]Gla is found only in bone, not in the medium.[15] To label only endogenous accumulated substrate in Warfarin-treated embryonic bone, 200 μg of cycloheximide per milliliter is included in the medium and incubated for 2 hr. Incubations are stopped by rinsing the bone briefly in protease inhibitors (TPIC). In cultures of normal bones, amino acids become labeled, and thus [^{14}C]Gla must be separated after alkaline hydrolysis by ion-exchange chromatography. For measurement of osteocalcin, labeled bones are homogenized on ice in 12.5% TCA, which both demineralizes and removes excess bicarbonate and labeled free amino acids. The precipitated labeled proteins are analyzed by slab gel electrophoresis[15] on a 5 to 15% acrylamide gradient.

Bone Microsome Method

The carboxylase enzyme system is localized in the microsomal fraction of bone cells.[16] Synthesis of carboxylated osteocalcin from endogenous substrate appears to have similar requirements to the liver microsomal synthesis of prothrombin[86,87] and other vitamin K-dependent clotting proteins. One exception, however, is that bone microsomes prepared from Warfarin-treated chicks are found to have a higher vitamin K requirement (50 μg/ml) for maximal carboxylation than liver microsomes from the same chicks.[15] In microsomal incubations in the absence of vitamin K, there is no ^{14}C incorporation into protein from NaH^{14}CO$_3$, and in the presence of vitamin K, [^{14}C]Gla is the only labeled amino acid.

[86] J. W. Suttie, L. M. Canfield, and D. V. Shah, this series, Vol. 67, p. 180.
[87] B. C. Johnson, this series, Vol. 67, p. 165.

Osteocalcin and an M_r 70,000 protein have been identified as the products of vitamin K- and CO_2-dependent synthesis in bone microsomes. The M_r 70,000 protein occurs also in whole bone cultures from Warfarin-treated chicks and has been termed putative proosteocalcin.[13,15]

Preparation of Microsomes. An average preparation consists of 144 Warfarin-treated embryos sacrificed at day 16, yielding 300 mg of microsomal pellet (dry weight). At sacrifice, long bones without cartilaginous ends and calvariae are excised from embryos and kept on ice in buffer I (250 mM sucrose, 5 mM KCl, 1 mM $MgCl_2$, 50 mM Tris-HCl, pH 7.2). The bones are minced in 3–5 volumes (based on wet weight) of buffer I and homogenized for 10 sec with a Polytron (Brinkmann, Westbury, NY, or Tekmar Model STD, Cincinnati, OH) tissue homogenizer. This homogenate is then subjected to 10 strokes in a motor-driven Teflon glass homogenizer or a Ten Broeck glass homogenizer. The homogenate is centrifuged at 500 g for 5 min to remove debris, followed by continued centrifugation of the resultant supernatant for 10 min at 2000 g, 10 min at 15,000 g, and for 60 min at 100,000 g to bring down microsomes. At each step the pellet is washed in 5 ml of buffer I adding the wash to the supernatant. All centrifugations are carried out at 4°. The final microsomal pellet is reconstituted to 20 mg of protein per milliliter of buffer II (250 mM sucrose, 8 mM KCl, 5 mM Mg acetate, 25 mM imidazole, pH 7.2, with 0.3 mg of dithiothreitol per milliliter). Microsomal pellets from liver preparations are reconstituted in 1 ml of buffer per gram wet weight of the original tissue. However, in our experience the bone tissue microsomes require a suspension 3–4 times more concentrated based on tissue wet weight for equivalent protein concentration.

Intact microsomes are required for endogenous substrate carboxylation. To allow carboxylation of synthetic peptides, microsomes can be modified by disruption with Triton X-100[86] or by preparing an acetone powder.[15] In solubilized bone microsomes, synthetic pentapeptides homologous to residues 5–9 of prothrombin are active.[39] The details for soluble microsome preparations have been presented for the liver system.[87]

Incubation Conditions. The assay is carried out in 13 × 75 mm capped tubes in a final volume of 0.4 ml containing 0.2 ml of intact microsomes (4 mg of protein), 1 mM NADH, 2 mM dithiothreitol, vitamin K_1 (Sigma Chemical Co., St. Louis, MO) added in 10 μl of ethanol. 5 μCi of $NaH^{14}CO_3$ ($NaCO_3$ can also be used) is diluted in buffer I. The reaction is initiated by the addition of vitamin K_1 and allowed to proceed for 60 min with constant shaking. Maximum carboxylation occurs at 25°. The reaction is terminated by adding 4 N acetic acid to a final concentration of 0.4 N and placing the samples under reduced pressure to remove liberated

$^{14}CO_2$ with a KOH trap. Further removal of non-protein-bound ^{14}C is accomplished by the addition of 0.2 ml of 1 M sodium carbonate.

Endogenous microsomal protein can be recovered by different procedures depending on the ultimate goal of the study. For direct measurement of incorporated radioactivity, 15 ml of 10% trichloroacetic acid is added after 300 μg of bovine serum albumin carrier for efficient precipitation. After 30 min at 4°, the precipitate is collected, the supernatant is discarded, and the pellet is dissolved in either Protosol (New England Nuclear) or NCS solubilizer in a scintillation vial for counting. Values of carboxylation of endogenous substrate are expressed as the ratio of ^{14}C in the plus vitamin K samples compared to the minus vitamin K controls. If the protein products are to be examined, the acid-precipitated proteins can be reconstituted in the buffer suitable for the separation procedure. Absolute proof that the incorporated radioactivity is due to the formation of Gla can be obtained by alkaline hydrolysis and amino acid analysis for [^{14}C]Gla.[60] Acid hydrolysis of an aliquot of the alkaline hydrolysate destroys Gla and causes 50% recovery of the ^{14}C in glutamic acid.[60]

[35] Specific Tritium Labeling of γ-Carboxyglutamic Acid Residues in Proteins

By PAUL A. PRICE

The method for specific tritium labeling of proteins containing γ-carboxyglutamic acid (Gla) is based on the discovery that the γ-proton of Gla exchanges rapidly with water in acid but not in base.[1] Since all other protons in a protein that exchange rapidly in acid also exchange readily in base, equilibration of a Gla-containing protein in acidic ^3H_2O followed by desalting into unlabeled base incorporates ^3H specifically into the Gla residues of the protein (Fig. 1). The incorporated label is quite stable above pH 7 and may be used as a marker for Gla-containing proteins in any separation procedure carried out under alkaline conditions.[1] Under more acidic conditions, the out exchange of ^3H from Gla is dependent upon the degree to which the γ-carboxyl groups of Gla are protonated. Out exchange is consequently blocked by interactions that stabilize the ionized form of the γ-carboxylate groups, such as binding to Ca^{2+} or to

---

[1] P. A. Price, M. K. Williamson, and D. J. Epstein, *J. Biol. Chem.* **256**, 1172 (1981).

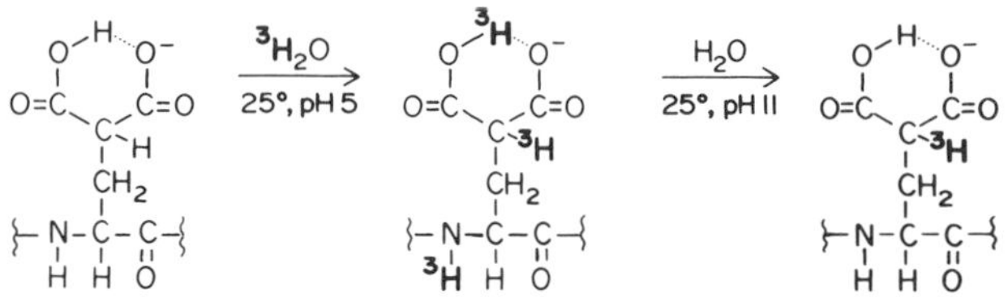

FIG. 1. Specific tritium labeling of γ-carboxyglutamic acid in proteins.

mineral surface. The effect of such binding interactions on ³H exchange from γ-proton labeled, Gla-containing proteins can therefore provide a direct measure of the role of Gla residues in the binding complex.

Labeling Procedure

In a typical experiment 2 mg of purified protein or tissue extract is first placed in 200 μl of 50 mM sodium acetate buffer at pH 5.0 with 7.5 M guanidine-HCl and incubated in a water bath at 25°. Tritium labeling is then initiated by the addition of 50 μl of ³H₂O (5 Ci/g). After the desired labeling interval at 25°, the exchange is quenched by the addition of 50 μl of 8 M NH₄OH in 6 M guanidine HCl, and the reaction mixture is then desalted over a 12-ml Sephadex G-25 column equilibrated with 0.1 M NH₄OH in 6 M guanidine HCl at 25°. During desalting it is useful to halt column flow for 1 hr when the protein has traveled one-quarter of the way through the column in order to facilitate more complete exchange of ³H label from nonspecific sites.

The labeling interval used in a given experiment depends on the desired objective. Since the half-time for exchange of ³H into the γ-position of Gla is 90 min at pH 5 and 25°, equilibrium labeling takes about 20 hr. At this point, the number of tritium atoms incorporated equals the number of Gla residues in the protein (Table I[1-3]). The long equilibration time required for stoichiometric labeling of Gla residues results in the incorporation of ³H into non-Gla-containing proteins at a level about 1% of that in prothrombin (Table I). This is a disadvantage in applications where relative specificity of labeling is important but stoichiometric labeling is not. For such applications, a reduction in the labeling interval from 20 hr to 0.5 hr will reduce the ³H incorporation into non-Gla-containing proteins to 0.13% of that in prothrombin.[1]

The stability of ³H label in Gla-containing proteins is critically dependent on the solution pH (Table II). Since the first ionization of the malonyl

[2] P. A. Price, this series, Vol. 91, p. 13.
[3] P. A. Price, this volume [36].

TABLE I
TRITIUM LABELING OF Gla-CONTAINING AND CONTROL PROTEINS[a,b]

Protein	Cpm/mg × 10⁻⁶	³H/molecule	Gla/molecule[c]
Bone Gla protein	5.01	2.6	2.6
Decarboxylated bone Gla protein[d]	0.58	0.3	0.3
Prothrombin	1.51	9.0	9.5
α-Chymotrypsin	0.05	0.09	0.0
RNase	0.02	0.02	0.0
Lysozyme	0.07	0.09	0.0
Serum albumin	0.04	0.15	0.0
Ovalalbumin	0.04	0.11	0.0

[a] From Price et al.,[1] with permission.
[b] Proteins were labeled with ^3H_2O (10^{13} cpm/mol proton) for 20 hr at 25° in 40 mM acetate buffer at pH 5 with 6 M guanidine-HCl and then desalted into 0.1 M NH$_4$OH.
[c] Determined by amino acid analysis.[2]
[d] After thermal decarboxylation of Gla to Glu.[3]

side chain of Gla is at pK_1 = 2.03, the essentially identical exchange rates at pH 1.5 and 3.1 indicate that both the diprotonated and the monoprotonated side chains allow rapid γ-proton exchange. The rate declines 10-fold as the pH is increased to 5.4, demonstrating that the second ionization, at pK_2 = 4.37, is accompanied by a decrease in γ-proton exchange. Finally, the rate of γ-proton exchange at pH 7.5 is over 200-fold slower than at pH 1.5, a result that is consistent with the hypothesis that the rate of ³H exchange from the fully ionized side chain is extremely slow.

Since the tritium labeling procedure requires handling of a highly radioactive and volatile reagent,³H₂O, it is imperative that all steps in the labeling procedure be carried out in a fume hood and that the experimenter wear disposable gloves that are impervious to water. All transfer steps must be carried out with disposable pipettes or pipette tips, and, immediately after use, all contaminated labware should be immersed in water in order to dilute the specific radioactivity of the label. Desalting eluent should be delivered to the column with a disposable pipette in order to avoid back contamination of buffer tubing and reservoir. Trial desalting experiments should be carried out separately with protein and with weakly radiolabeled water to establish elution volumes in a test column identical to the one to be used in the labeling experiment. In the labeling experiment the filtration column, typically a 10-ml plastic disposable pipette, should be sealed securely after sufficient eluent has flowed through the column to ensure the emergence of the protein and before any labeled water has eluted. All radioactive materials should then be sealed

TABLE II
STABILITY OF TRITIUM IN BONE Gla PROTEIN[a,b]

Buffer pH	Half-time for loss of ^{3}H from bone Gla protein (min)
1.5	33 ± 5
3.1	40 ± 10
5.4	400 ± 40
7.5	7800 ± 200
8.8	12,900 ± 200

[a] From Price et al.,[1] with permission.
[b] Samples of tritiated bone Gla protein (5×10^6 cpm/mg, see Table I) were placed in 2 ml of 6 M guanidine-HCl buffered to the indicated pH at 25°. Half-times were determined from a first-order plot of the counts remaining in protein.

from air and, together with the sealed column, be promptly removed by the appropriate health safety official. The labeled protein can be removed from the hood at this point and used as a marker in purification experiments carried out under alkaline conditions outside the fume hood.

Applications of ^{3}H Labeling Procedures

The exchange labeling of Gla with tritium provides the only known specific, nondegradative method of radiolabeling Gla-containing proteins. This method is particularly useful for radiolabeling Gla-containing proteins in a tissue extract in order subsequently to monitor their position during purification procedures carried out above pH 7. Since the exchange labeling of nonproteinaceous tissue components by this procedure has not yet been evaluated, however, the validity of ^{3}H labeling of Gla proteins should always be verified by comparing the incorporation before and after the specific thermal decarboxylation of Gla to Glu.[3]

Since incorporation of tritium into Gla is stoichiometric, this procedure yields a direct estimate of the Gla content of purified proteins. The stoichiometric ^{3}H-labeling of Gla in purified Gla-containing proteins can be used also to probe the interaction between the malonyl side chain of Gla and Ca^{2+} or Ca^{2+}-containing surfaces. Since such interactions retard ^{3}H exchange by stabilizing the fully ionized malonyl side chain of Gla, the extent of binding can be inferred from the degree to which γ-proton exchange has been retarded. We have observed that the tight binding of bone Gla protein to hydroxyapatite retards the out exchange of ^{3}H label from the bone Gla protein, and Bajaj and Epstein[4] have shown that the exchange of ^{3}H from prothrombin is dramatically reduced by the presence of Ca^{2+}.[4] For such applications, guanidine should be eliminated from buffers in order to avoid protein denaturation. Since tritium exchange from native proteins is shown, it may be necessary to repeat gel filtration in base after a suitable exchange interval in order fully to remove the base-labile, nonspecific ^{3}H in the protein.

[4] S. P. Bajaj and D. J. Epstein, personal communication.

[36] Decarboxylation of γ-Carboxyglutamic Acid Residues in Proteins

By PAUL A. PRICE

The method for specifically decarboxylating γ-carboxyglutamic acid (Gla) to glutamic acid is based on the well-known sensitivity of malonic acid and its derivatives to thermal decarboxylation. Because such decarboxylation reactions occur readily in the absence of water, it is possible to dry Gla-containing proteins from an acidic solution and then to decarboxylate thermally the protonated γ-carboxyl groups of Gla to Glu without hydrolytic side reactions.[1] Since dry proteins are far more stable to thermal denaturation, the decarboxylation of Gla to Glu can be carried out under conditions that affect neither the tertiary structure nor the enzymatic activity of the freeze-dried protein samples.[1,2]

The primary utility of this specific protein modification is to reverse the vitamin K-dependent reaction that forms Gla from Glu and so create a model for the abnormal, non-γ-carboxylated protein synthesized in vitamin K-deficient or Warfarin-treated animals. Such synthetic decarboxylated derivatives of Gla-containing proteins have been used to investigate the role of Gla residues in the binding of proteins to Ca^{2+}, hydroxyapatite, and phospholipid and to assess the importance of Gla residues to the immunoreactivity and structure of proteins.

[1] J. W. Poser and P. A. Price, J. Biol. Chem. **254**, 431 (1979).
[2] S. P. Bajaj, P. A. Price, and W. A. Russell, J. Biol. Chem. **257**, 3726 (1982).

Decarboxylation Procedures

There are two principal methods for the preparation of freeze-dried proteins in which the γ-carboxyl groups of Gla residues are protonated (Fig. 1). The direct procedure entails freeze-drying the Gla-containing protein from 50 mM HCl, a solution sufficiently acidic to protonate both γ-carboxyl groups of Gla. The indirect procedure involves freeze-drying the Gla-containing protein from 0.1 M ammonium bicarbonate or 0.1 M ammonium acetate to give the ammonium salt of the γ-carboxylate ion. Since subsequent heating of the ammonium salt of the Gla-containing protein yields a decarboxylation rate nearly identical to that of the protonated side chain,[1] it seems probable that the ammonium salt decomposes to ammonia and the protonated side chain prior to decarboxylation (Fig. 1). Both decarboxylation procedures are highly specific, and no other changes can be detected in the composition of treated proteins.[1] In addition, representative non-Gla-containing enzymes retain over 90% of their activity when their ammonium salts are subjected to the 5 hr at 110° needed to effect 98% decarboxylation in a Gla-containing protein.[1]

In a representative decarboxylation experiment, 5 mg of a Gla-containing protein is dissolved in 1 ml of 50 mM HCl, 0.1 M NH$_4$HCO$_3$, or 0.1 M ammonium acetate and then freeze-dried in a Pyrex test tube. The test tube is sealed under vacuum and heated at 110° to decarboxylate Gla to Glu. In our experience, decarboxylation is typically first order in Gla with a half-time of about 1 hr (Fig. 2). The degree of decarboxylation should be

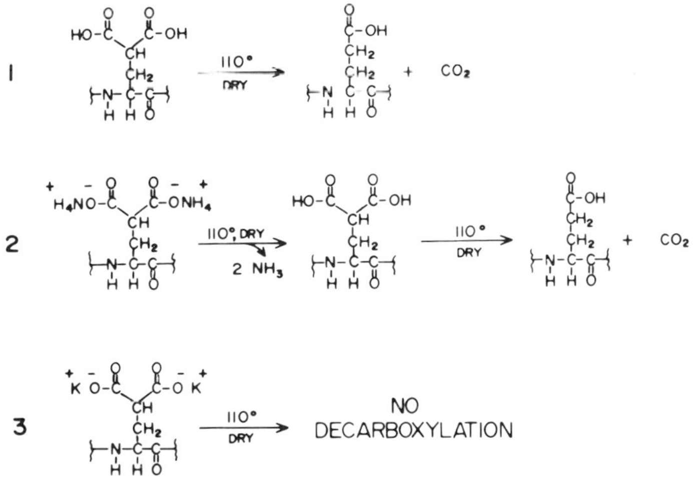

FIG. 1. Models for the thermal decarboxylation of γ-carboxyglutamic acid to glutamic acid in dry salts of Gla-containing proteins.

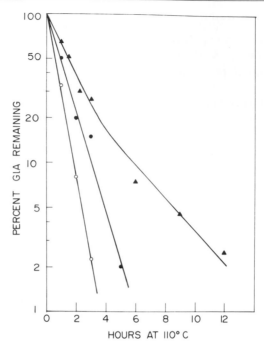

FIG. 2. Decarboxylation of Gla residues in Gla-containing proteins during heating under vacuum at 110°. ○——○, Calf bone Gla protein dried from 0.05 M HCl; ●——●, calf bone Gla protein dried from 0.1 M NH$_4$HCO$_3$; ▲——▲, human prothrombin dried from 0.1 M NH$_4$HCO$_3$. From Price and Poser[1] and Bajaj et al.,[2] with permission.

verified by amino acid analysis[3] of an alkaline hydrolysate of the treated sample. Failure to decarboxylate fully is most commonly caused by the presence of contaminating mono- or divalent metal ions that form a salt with the γ-carboxylate groups and so reduce the fraction of γ-carboxyl groups that are protonated. Although this problem is best treated by removal of the contaminating metal ion prior to heating, we have found that complete decarboxylation can usually be achieved by redissolving the treated sample in the appropriate buffer followed by freeze-drying and heating at 110° for an additional 5 hr under vacuum.

Protection against Decarboxylation

When a Gla-containing protein is dried from buffers such as ammonium acetate or ammonium bicarbonate, the γ-carboxylate groups must

---

[3] P. A. Price, this series, Vol. 91, p. 13.

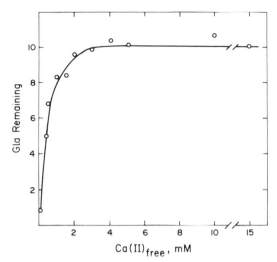

FIG. 3. Effect of Ca^{2+} binding on the thermal decarboxylation of Gla residues in human prothrombin. Protein samples were freeze-dried from 0.1 M ammonium acetate, pH 7.5, containing the indicated free concentrations of Ca^{2+} and heated under vacuum for 6 hr at 110°. The number of Gla residues remaining in each sample after heating was determined by alkaline hydrolysis and amino acid analysis. From Bajaj et al.,[2] with permission.

become ammonium salts for decarboxylation of Gla to Glu to occur. Any interaction that displaces the ammonium ion from the γ-carboxylate group, such as binding to Ca^{2+} or Ca^{2+}-containing salts (e.g., hydroxyapatite), will therefore protect Gla from decarboxylation.[1,2] This provides a useful method to assess the number of Gla residues involved in a given binding interaction. In a typical experiment, the Gla-containing protein is equilibrated with the desired concentration of free metal ion or metal ion salt in 0.1 M ammonium acetate and then freeze-dried and heated at 110° under vacuum. As can be seen in Fig. 3, the degree of protection is a function of the number of metal ions bound and, therefore, a function of the free metal ion concentration.

[37] Glutamate Carboxylase: Assays, Occurrence, and Specificity

By TOM BRODY and J. W. SUTTIE

The formation of γ-carboxyglutamate in proteins is catalyzed by a microsomal enzyme system that requires vitamin K as a cofactor. Early studies of this enzymatic activity have been previously reviewed in this series.[1,2]

Typical Assay Conditions

Enzyme assays are typically performed as follows: 200 μl of a Triton-solubilized microsomal preparation (see below) is added to 13 × 100 mm glass tubes in an ice bath followed by 300 μl of buffer A [0.5 M KCl, 10% glycerol, 1 mM dithiothreitol (DTT), and 25 mM imidazole-Cl, pH 7.0]. Buffer A also contains 16 mM MnCl$_2$, 1.6 mM peptide substrate, and 10 μCi [^{14}C]bicarbonate (60 mCi/mmol) and any other variable components desired. The mixture is brought to 17°, and the reaction is started by addition of 10 μl of vitamin KH$_2$ (5 mg/ml in ethanol). If the reduced vitamin is mixed with the other buffer components before introduction into the enzyme solution, it must be done just prior to this introduction to prevent oxidation of the vitamin, and 0.1% Triton X-100 must be added to keep the vitamin in solution.

The complete assay mixture is agitated briefly, incubated in a shaking water bath 30 min at 17°, and protected from the light. The reaction is terminated by addition of 2 ml of 10% trichloroacetic acid and the tube is centrifuged to form a firm pellet. The supernatant is removed and placed in a clean 13 × 100 mm glass tube. Unreacted [^{14}C]bicarbonate is discharged from the supernatant by bubbling with a slow and steady current of air for 15 min at room temperature. An aliquot of the supernatant is then mixed with Aquasol (New England Nuclear, Boston, MA), and the ^{14}CO$_2$ fixed into the acid-soluble peptide substrate is measured in a liquid scintillation spectrometer. For convenience, a large number of assay mixtures can be assembled rapidly on ice, and all reactions can be started simultaneously by placing in a water bath at 17°. It should be noted, however, that peptide carboxylase activity is significant at 4°.

[1] J. W. Suttie, L. M. Canfield, and D. V. Shah, this series, Vol. 67, p. 180.
[2] B. C. Johnson, this series, Vol. 67, p. 165.

METHODS IN ENZYMOLOGY, VOL. 107

To measure $^{14}CO_2$ incorporation into endogenous microsomal proteins, the precipitated pellet is dispersed with 1 ml of 1 M sodium bicarbonate and re-formed by gradual addition of 4 ml of 10% trichloroacetic acid. This process is performed three times to remove unreacted [^{14}C]bicarbonate. The washed moist pellet is then solubilized with 0.5–1.0 ml of NCS tissue solubilizer (Amersham Corp., Arlington Heights, IL) and mixed with a nonaqueous scintillation fluid for measurement of protein carboxylase activity. The protein pellet must be processed without delay as it loses dispersability within 2 days or so. Enzyme assay blanks are conducted without vitamin K addition. The exact conditions used in different laboratories have varied, and a summary of typical assay conditions appears in the table.

Preparation of Microsomal Enzyme

Male, 200–300 g vitamin K-deficient rats that have been fasted overnight are decapitated; the livers are quickly removed, minced, and homogenized with a Teflon pestle in two parts (w/v) of ice-cold 0.25 M sucrose, 25 mM imidazole Cl, pH 7.0, 1 mm of DTT. A postmitochondrial

ASSAY CONDITIONS

Component or variable	Quantity
Microsomal protein	0.6[a]–2 mg
Peptide substrate, tBoc-Glu-Glu-Leu-OMe	1 mM
Vitamin K hydroquinone	50 μg in 10 μl of ethanol
[^{14}C]Bicarbonate[b]	10–20 μCi (60 mCi/mmol)
Oxygen	Air
Imidazole-Cl or sodium MOPS, pH 7.0	25 mM
KCl	0.5 M
MnCl$_2$	10 mM
Triton X-100[c]	0.2–0.6%
Glycerol or sucrose	10–15%
Incubation time, protein carboxylase[d]	60–90 sec
Incubation time, peptide carboxylase[d]	30 min
Temperature	17°
Total volume	0.5 ml

[a] Amount of unpurified protein required for detection of enzyme activity.

[b] The concentration of bicarbonate present in distilled water is about 1 mM.

[c] The concentration will be dependent on the amount of protein used, where the protein is solubilized in Triton.

[d] Approximate maximal time at which activity is linear with time.

supernatant is obtained by centrifugation of the homogenate at 10,000 g for 10 min. Microsomes are prepared from the postmitochondrial supernatant by centrifugation at 105,000 g for 60 min, and the microsomal pellet is resuspended with 8 strokes of a loose-fitting Dounce homogenizer (Kontes type A pestle) in 1 M KCl, 25 mM imidazole-Cl, pH 7.0, 1 mm DTT (buffer B) followed by repelleting of the microsomes. The washed microsomes are solubilized at 4° by suspension with the glass homogenizer in buffer A containing 1.0% Triton X-100. The volume of buffer used is equal to that of the original postmitochondrial supernatant. The suspension is left at 4° for 30 min, and the solubilized protein is clarified by ultracentrifugation. This solubilized preparation of the carboxylase is 1.5-fold purified and contains about 6 mg of protein per milliliter.

Factors Influencing Activity

Type and Concentration of Detergent

After solubilization, the concentration of detergent may be lowered by dilution or attempts to purify carboxylase, and a detergent requirement for expression of maximal enzyme activity may become apparent. Carlisle and Suttie[3] observed maximal activity of Triton-permeabilized microsomes in assay mixtures containing 0.2% Triton X-100 with lesser activities expressed at greater or lower Triton concentrations. Similar effects can be seen when microsomes were initially solubilized in a buffer containing 1% Triton (Fig. 1). Observations (unpublished) in our laboratory show that all bile salts tested solubilized essentially the same amount of microsomal carboxylase as did Triton X-100. When microsomal pellets were suspended in 1% CHAPS, sodium cholate, sodium glycocholate, or sodium deoxycholate containing 0.5 M KCl, 10% glycerol, 25 mM imidazole Cl, pH 7.0, and 1 mM DTT, similar enzyme activities were detected. All solubilized pellets were clarified by ultracentrifugation, and assays were supplemented with 100 μl of heat-denatured microsomes (see below). The final incubations contained 0.15 mg of protein, and the final bile salt concentration was limited to 0.05% in an effort to impose uniform conditions in all assays.

Although the heat-denatured microsomal preparation contains some Triton, carboxylase activity can be supported by bile salts alone. We have found that CHAPS and sodium glycocholate were capable of supporting similar enzyme activity as Triton X-100. Maximal activity occurred at specific and unique concentrations of these detergents, and

[3] T. L. Carlisle and J. W. Suttie, *Biochemistry* **19**, 1161 (1980).

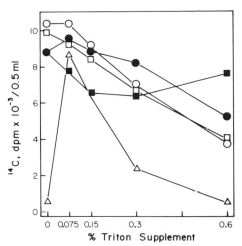

Fig. 1. Effect of detergent concentration on carboxylase activity. Incubation mixtures (0.5 ml) contained 1.5 mM tripeptide, 50 μg of vitamin K hydroquinone, 20 μCi of [^{14}C]bicarbonate (60 mCi/ml), 0.09 mg of Triton-solubilized protein, 10 mM MnCl$_2$, 0.5 M KCl, 15% glycerol, 25 mM imidazole Cl, pH 7.0, and 1 mM DTT. Heat-denatured microsomal extract, egg yolk phosphatidylcholine, and additional Triton X-100 were included where indicated. Reactions were started by adding 10 μl of 5 mg/ml vitamin K hydroquinone with 50 mM DTT in ethanol, and enzyme activity is expressed as ^{14}C dpm of acid-soluble product formed during the 30-min enzyme incubation. The enzyme preparation provided a basal concentration of 0.03% Triton, which was supplemented with 0–0.6% Triton as indicated. The incubations contained no additional supplement (—△—), 50 μl (—○—) or 100 μl (—●—) of heat-denatured microsomal extract, or 0.4 mg (—□—) or 2.0 mg (—■—) of egg yolk phosphatidylcholine in 100 μl of 1% Triton. The heat-denatured extract was prepared by thorough suspension of microsomal pellets with a Dounce homogenizer in 1% Triton X-100, 0.5 M KCl, 10% glycerol, and 1 mM DTT followed by warming for 10 min at 80°. The material was then cooled, and insoluble material was removed by ultracentrifugation. The volume of the Triton solution used was that of the original postmitochondrial supernatant and yielded a protein concentration of about 10 mg/ml prior to the heat denaturation step. The extract was stored at 4°.

activity curves resembled those obtained in a series of assays containing various concentrations of Triton (Fig. 1). The claim that Triton solubilizes less activity from microsomes than does CHAPS[4] has not been observed in our laboratory. Sodium cholate was less effective, and sodium deoxycholate proved to be incapable of supporting carboxylase activity. In general, we have found it difficult to get predictable and maximal carboxylase activity in the presence of bile salt detergents. Because of this we limit the final concentration of bile salt in all assays to about 0.05% and supplement these assays with Triton and phospholipid.

[4] J.-M. Girardot and B. C. Johnson, Anal. Biochem. 121, 315 (1982).

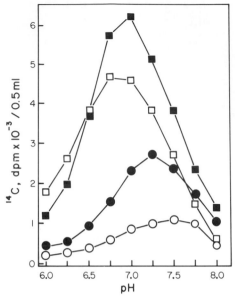

FIG. 2. Effect of pH on carboxylase activity. Incubation mixtures were as in Fig. 1 except that assay mixtures contained 1 mM tripeptide, 0.3 mg of Triton-solubilized protein, and 75 μl of heat-denatured microsomal extract. The reaction mixtures were completed by addition of 95 μl of a mixture of vitamin, heat-denatured extract, and $^{14}CO_2$. MnCl$_2$ and pyridoxal phosphate (PLP) were included only where indicated, and the pH was that of a 1 M stock solution of imidazole Cl. Assays contained no additive (—○—), 1 mM PLP (—●—), 10 mM MnCl$_2$ (—□—), or 1 mM PLP and 10 mM MnCl$_2$ (—■—).

pH and Buffer Salts

Maximal carboxylase activity occurs at or near pH 7.0.[5] Assays are often supplemented with 10 mM MnCl$_2$ or 1 mM pyridoxal phosphate (PLP), and these additions substantially increase the activity of the enzyme and shift the pH optimum to a slightly lower pH when determined in 25 mM imidazole Cl (Fig. 2). Maximal carboxylase activity can be expressed in imidazole Cl, sodium MOPS, sodium phosphate, and HEPES Cl, at pH 7.0, in the presence of 10 mM MnCl$_2$. All other buffers tested were unacceptably inhibitory. Buffer effects are minimal in MOPS (Fig. 3A). Imidazole inhibits severely at higher concentrations (Fig. 3A). In the absence of MnCl$_2$ activity is stimulated 2- to 3-fold by increasing concentrations of sodium phosphate or HEPES Cl. Imidazole and MOPS (25–50 mM) at pH 7.0 with 10 mM MnCl$_2$ present are suitable for routine use.

[5] L. M. Canfield, T. A. Sinsky, and J. W. Suttie, *Arch. Biochem. Biophys.* **202,** 515 (1980).

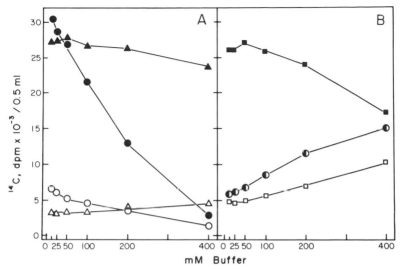

FIG. 3. Effect of different buffers in carboxylase activity. Incubations were as in Fig. 1 except that assay mixtures contained 0.5 mM tripeptide, 0.3 mg of Triton-solubilized protein, and reactions were started by adding 110 μl of a mixture of vitamin and egg yolk phosphatidylcholine (0.25 mg/assay) suspended in 1% Triton, 15% glycerol, and 1 mM DTT at pH 7.0. Buffers and MnCl₂ were as indicated below. Imidazole Cl (—○—); imidazole Cl, 10 mM MnCl₂ (—●—); Na MOPS (—△—); Na MOPS, 10 mM MnCl₂ (—▲—); HEPES Cl (—□—); HEPES Cl, 10 mM MnCl₂ (—■—); Na phosphate (—◐—) buffers, pH 7.0, at 12.5–400 mM were present where indicated. Livers were homogenized and enzyme was prepared in the absence of buffers.

Sodium phosphate forms a precipitate with Mn(II) resulting in variable decreases in the pH of the incubation medium (Fig. 3B). Therefore sodium phosphate and MnCl₂ should not be present together in enzyme assays.

Salt Concentration

Maximal activity of rat liver carboxylase occurs at 0.5 M KCl. Although the presence of 0.5 M KCl is required for solubilization of microsomes with 1% Triton X-100 or 1% sodium cholate, the carboxylase reaction rate is not very sensitive to changes in the KCl concentration. However, dicoumarol-treated calf and adult ox carboxylases are more sensitive to the KCl concentration, where the latter enzyme is completely inhibited by 0.5 M KCl.[6]

[6] L. Uotila and J. W. Suttie, *Biochem. J.* **201**, 249 (1982).

Apparent K_m of Substrates

The K_m for vitamin K hydroquinone is 10 μg/ml and 45 μg/ml for partially purified rat liver[7] and adult ox enzyme,[6] respectively. Standard assay mixtures usually contain a 0.5–1.0 mM low-molecular-weight peptide substrate such as tBoc-Glu-Glu-Leu-OMe or Phe-Leu-Glu-Glu-Leu. Most studies with rat, bovine, or equine enzyme have revealed the K_m of various peptide substrates to be 2–5 mM.[5–9]

The peptide corresponding to residues 13–29 of decarboxy bovine prothrombin has been purified[10] and claimed to have a K_m of 0.001 mM. This is much lower than that reported for other low-molecular-weight substrates, although a K_m of 0.01 mM has been observed for a peptide substrate covalently attached to the 3 position of vitamin K.[11]

The carboxylase is nearly saturated by O_2 at a concentration of 20% in the atmosphere, and incubations are routinely performed under air. Half-maximal carboxylase activity occurs at 5–10% O_2 in the atmosphere.[12]

Carbon dioxide rather than bicarbonate is the preferred substrate as determined by initial rates of carboxylation using intact and solubilized microsomes as the sources of protein.[13] These studies did not eliminate the possibility that bicarbonate functions as a substrate, but only established that CO_2 was preferred.

Inhibition and Stimulation of Activity

Chloro-K (2-chloro-3-phytyl-1,4-naphthoquinone)[14] and tetrachloro-4-pyridinol[15] are potent inhibitors of carboxylase with half-maximal inhibition occurring at under 0.5 μg/ml[5,16] and at 10 μM,[17] respectively. Although Warfarin toxicity in $vivo$ does not result from inhibition of

[7] R. Wallin and J. W. Suttie, $Arch.$ $Biochem.$ $Biophys.$ **214,** 155 (1982).

[8] D. H. Rich, S. R. Lehrman, M. Kawai, H. L. Goodman, and J. W. Suttie, in "Vitamin K Metabolism and Vitamin K-Dependent Proteins" (J. W. Suttie, ed.), p. 471. Univ. Park Press, Baltimore, Maryland, 1980.

[9] C. Vermeer and M. Ulrich, $Thromb.$ $Res.$ **28,** 171 (1982).

[10] B. A. M. Soute, M. de Metz, and C. Vermeer, $FEBS$ $Lett.$ **146,** 365 (1982).

[11] J. W. Suttie and P. Preusch, $Hemostasis$ (in press).

[12] J. W. Suttie, J. McTigue, A. E. Larson, and R. Wallin, $Ann.$ $N.Y.$ $Acad.$ $Sci.$ **370,** 271 (1981).

[13] J. P. Jones, E. J. Gardner, T. G. Cooper, and R. E. Olson, $J.$ $Biol.$ $Chem.$ **252,** 7738 (1977).

[14] J. Lowenthal and M. N. R. Chowdhury, $Can.$ $J.$ $Chem.$ **48,** 3957 (1970).

[15] F. N. Marshall, $Proc.$ $Soc.$ $Exp.$ $Biol.$ $Med.$ **139,** 223 (1972).

[16] J. W. Suttie, S. R. Lehrman, L. O. Geweke, J. M. Hageman, and D. H. Rich, $Biochem.$ $Biophys.$ $Res.$ $Commun.$ **86,** 500 (1979).

[17] P. A. Friedman and A. E. Griep, $Biochemistry$ **19,** 3381 (1980).

carboxylase,[18] 50% inhibition of bovine[6] and rat liver[18] carboxylase were observed at 1–10 mM Warfarin.

Manganese(II) stimulates rat liver peptide carboxylase 1.4-fold and other divalent cations stimulate to a lesser extent.[19] This effect is more pronounced where the enzyme had been treated with EDTA. The calf liver enzyme is more responsive and can be stimulated 22-fold by $MnCl_2$.[6] This effect is pH dependent, and stimulation of the rat liver enzyme by 10 mM $MnCl_2$ is 9- to 10-fold at pH 6.0–6.5 and only 25–50% at more alkaline pH values (Fig. 2). Pyridoxal phosphate (PLP) at a concentration of 3 mM stimulates peptide carboxylase activity about 3-fold[16] but has no effect on the rate of protein carboxylase activity.[20] A similar effect is observed with the dicoumarol-treated calf enzyme.[6] Stimulation by 1 mM PLP is fairly constant (2- to 3-fold) between pH 6 and 8 (Fig. 2). Where reaction mixtures contain 10 mM $MnCl_2$, the addition of PLP further stimulates activity at some pH values but not at others. The mechanisms of stimulation by Mn(II) and PLP are unknown and are probably independent of each other.

Carboxylase activity is slightly increased in the presence of 1 mM DTT,[16] and DTT appears to be necessary for maximal stimulation by manganese, especially in assays of stored enzyme.[6] The effect of DTT in preventing oxidation of the substrate vitamin K hydroquinone during the course of enzyme incubations is not clear. Microsomal vitamin K quinone and vitamin K epoxide reductases utilize DTT as a substrate, but these functions of DTT are not significant in carboxylase assays where vitamin K hydroquinone is used as the substrate. Stimulation of carboxylase activity by 30–50% occurs with supplementation of incubation mixtures with 15–20% glycerol or sucrose. Ethylene glycol and dimethyl sulfoxide (DMSO) stimulate to lesser extents, with maximal stimulation at about 10%. The bovine enzyme is also stimulated by organic solvents.[21]

Structural Specificity

trans-Phylloquinol (vitamin K_1 hydroquinone) is the most commonly used substrate for carboxylase. The enzyme requires the reduced form of the vitamin,[22] and either chemically reduced vitamin is used or DTT and

[18] E. F. Hildebrandt and J. W. Suttie, *Biochemistry* **21**, 2406 (1982).

[19] A. E. Larson and J. W. Suttie, *FEBS Lett.* **118**, 95 (1980).

[20] J. W. Suttie, L. O. Geweke, J. L. Finnan, S. R. Lehrman, and D. H. Rich, *in* "Vitamin K Metabolism and Vitamin K-Dependent Proteins" (J. W. Suttie, ed.), p. 450. Univ. Park Press, Baltimore, Maryland, 1980.

[21] M. de Metz, B. A. M. Soute, H. C. Hemker, and C. Vermeer, *Biochem. J.* **209**, 719 (1983).

[22] J. A. Sadowski, H. K. Schnoes, and J. W. Suttie, *Biochemistry* **16**, 3856 (1977).

NADH-requiring reductases present in some preparations of carboxylase allow the quinone to be used as a substrate. There is no strict requirement for the substituent at the 3-position although the unnatural cis-isomer appears not to be a substrate.[23] Menadione (2-methyl-1,4-naphthoquinone) is not a substrate, but 3-alkyl, 3-ethoxy, and 3-thiolmethane menadione, or the 3-adduct of menadione and DTT all function as substrates, once reduced to the hydroquinone form.[24,25] Menaquinones such as MK-2 or MK-3 were found to be better[23,26,27] or poorer[28] substrates than vitamin K at subsaturating concentrations of vitamin. Little difference was found at saturating concentrations, probably because the protein carboxylase reaction had reached completion earlier here and obscured differences in substrate specificity. In addition, the use of the quinone rather than the hydroquinone form of the vitamin makes these results difficult to interpret. The low-molecular-weight peptides commonly used as substrates contain Glu-Glu sequences, and available evidence suggests that only the first of these is carboxylated.[29,30] Rich et al.[31] have carried out an extensive investigation of the substrate activity of low-molecular-weight peptides. Compounds with a single Glu residue are not particularly good substrates, and there do not appear to be any unique sequences that result in very active substrates. The available data would suggest that macromolecular recognition sites are more important than any primary sequence consideration in determining the specificity of the enzyme for particular Glu residues in a protein.

Species, Tissue, and Subcellular Localization

Rat and bovine liver microsomes have been used as the source of enzyme in most studies of the vitamin K-dependent carboxylase. The

[23] P. A. Friedman and M. Shia, *Biochem. Biophys. Res. Commun.* **70,** 647 (1976).
[24] D. O. Mack, M. Wolfensberger, J.-M. Girardot, J. A. Miller, and B. C. Johnson, *J. Biol. Chem.* **254,** 2656 (1979).
[25] R. E. Olson, R. M. Houser, M. T. Searcy, E. J. Gardner, J. Scheinbuks, G. N. SubbaRao, J. P. Jones, and A. L. Hall, *Fed. Proc. Fed. Am. Soc. Exp. Biol.* **37,** 2610 (1978).
[26] P. A. Friedman and M. A. Shia, *Biochim. Biophys. Acta* **616,** 362 (1980).
[27] C. S. Yen and D. O. Mack, *Proc. Soc. Exp. Biol. Med.* **165,** 306 (1980).
[28] J. P. Jones, A. Fausto, R. M. Houser, E. J. Gardner, and R. E. Olson, *Biochem. Biophys. Res. Commun.* **72,** 589 (1976).
[29] P. Decottignies-Le Marechal, H. Rikong-Adie, and R. Azerad, *Biochem. Biophys. Res. Commun.* **90,** 700 (1979).
[30] J. L. Finnan and J. W. Suttie, *in* "Vitamin K Metabolism and Vitamin K-Dependent Proteins" (J. W. Suttie, ed.), p. 509. Univ. Park Press, Baltimore, Maryland, 1980.
[31] D. H. Rich, S. R. Lehrman, M. Kawai, H. L. Goodman, and J. W. Suttie, *J. Med. Chem.* **24,** 706 (1981).

hamster and Warfarin-resistant rat appear to have the highest levels of enzyme, and the chick the lowest.[32] The enzyme is located in the microsomes[33] where the specific activity is higher in the rough microsomes than in the smooth ones.[3,34–36] Enzymatic activity has also been detected in tissues other than liver, such as testes and spleen,[37] placenta,[10] and lung.[38,39]

The specific activity of peptide carboxylase increases 2- to 3-fold in the hypoprothrombinemic vitamin K-deficient state,[32,40] and thus the Warfarin-treated, dicoumarol-treated, or vitamin-deficient animal has been the usual source of enzyme. The specific activity of protein carboxylase concomitantly increases 10-fold or more in the hypoprothrombinemic state. This change reflects the apparent increases in carboxylase protein itself, as determined by peptide carboxylase assays, as well as in increases in the clotting factor precursors that accumulate in microsomes of most hypoprothrombinemic animals.

Problems in Assay of Dilute or Purified Protein

Recovery of enzyme activity in assays containing dilute or purified enzyme may be quite low and may lead to the erroneous conclusion that the enzyme has been denatured. A variety of studies have shown that maximal activity can be expressed by supplementing assays with phospholipid in the presence of various detergents.[7,41–43] It has not yet been distinguished, however, whether the phospholipid fulfills a true requirement for enzyme activity or merely prevents inhibition by excess concentrations of detergent.

Maximal activity may also be expressed in the absence of added phospholipid by use of specific concentrations of detergent. The data in Fig. 1 illustrate a peak of carboxylase activity at 0.075% Triton in the absence of added phospholipid. This value is not a constant, as it varies considerably

[32] D. V. Shah and J. W. Suttie, *Proc. Soc. Exp. Biol. Med.* **161**, 498 (1979).

[33] C. T. Esmon, J. A. Sadowski, and J. W. Suttie, *J. Biol. Chem.* **250**, 4744 (1975).

[34] L. Helgeland, *Biochim. Biophys. Acta* **499**, 181 (1977).

[35] K. Egeberg and L. Helgeland, *Biochim. Biophys. Acta* **627**, 255 (1980).

[36] R. Wallin and H. Prydz, *Thromb. Haemostasis* **41**, 529 (1979).

[37] C. Vermeer, H. Hendrix, and M. Daemen, *FEBS Lett.* **148**, 317 (1982).

[38] R. G. Bell, *Arch. Biochem. Biophys.* **203**, 58 (1980).

[39] S. D. Buchthal and R. G. Bell, *Biochemistry* **22**, 1077 (1983).

[40] D. V. Shah and J. W. Suttie, *Arch. Biochem. Biophys.* **191**, 571 (1978).

[41] J. C. Swanson and J. W. Suttie, *Biochemistry* **21**, 6011 (1982).

[42] M. de Metz, C. Vermeer, B. A. M. Soute, and H. C. Hemker, *J. Biol. Chem.* **256**, 10843 (1981).

[43] J.-M. Girardot, *J. Biol. Chem.* **257**, 15008 (1982).

with protein concentration. The severe inhibition of carboxylase activity observed at higher Triton concentrations can be prevented by various supplements of lipid suspended or dissolved in Triton. Activity remains most constant with variations in Triton concentration where the supplement is a heat-denatured extract of microsomes (Fig. 1), but supplements of egg yolk L-α-phosphatidylcholine in 1% Triton also support carboxylase at all Triton concentrations. Because of the practical difficulty in establishing the optimal concentration of Triton in routine assays of purified preparations of the carboxylase, the use of some type of lipid supplement appears to be desirable.

Detergent–lipid interactions are most important at low protein concentrations, and dependence on Triton and phospholipid supplements is relieved when assays contain per milliliter a few milligrams of microsomal protein such as those described in the section on typical assay conditions. When incubation mixtures contain 0.25–0.4 ml of microsomal protein solution (6 mg of protein per milliliter) prepared in buffer containing 1% Triton, 0.5 M KCl, and 10% glycerol, supplements of Triton and phospholipid are not necessary and need not be used.

Conditions that have been found useful for assaying small amounts of protein are as follows: assays of 10–100 μl of crude or purified microsomal protein (under 1 mg of protein) are supplemented with 100 μl of heat-denatured microsomal extract as described in the legend to Fig. 1. Where samples contain bile salt, the bile salt concentration should be limited to 0.05% or less, as noted in the section on type and concentration of detergent. Where well-defined incubation conditions are desired, egg yolk phosphatidylcholine suspended in Triton can be used instead of the heat-denatured microsomal extract.

Other Carboxylase Preparations

Microsomal acetone powders have also been used as a source of carboxylase, and both protein and peptide carboxylase activities are expressed in these preparations, which do not contain detergent.[26] The degree of phospholipid removal by acetone extraction is dependent on a number of factors[44] and was not assessed for this preparation. Association of Triton-solubilized bovine carboxylase with immobilized antifactor X yielded a 115-fold purified enzyme with a recovery of 20%.[45] The gel-bound enzyme is active, and a portion of the activity could be resolubilized by permitting

[44] R. L. Lester and S. Fleischer, *Biochem. Biophys. Acta* **47,** 358 (1961).
[45] M. de Metz, C. Vermeer, B. A. M. Soute, G. J. M. Van Scharrenburg, A. J. Slotboom, and H. C. Hemker, *FEBS Lett.* **123,** 215 (1981).

the enzyme to turn over,[45] perhaps in the presence of exogenously added Factor X.[42] Gel-bound rat liver carboxylase has also been prepared by association with immobilized rat antiprothrombin antibody.[41] Association of carboxylase from Triton-solubilized hypoprothrombinemic rat liver microsomes with concanavalin A–Sepharose (Pharmacia) results in a preparation of immobilized enzyme[46] that is 10- to 20-fold purified with a recovery of 50–120%.

The purification of carboxylase has proved to be a difficult task, not because of instability of the enzyme, but because of the difficulty in separating inactive protein from enzyme, and perhaps because assays of partially purified enzyme were not supplemented with phospholipid. A 400-fold purification of rat liver carboxylase has been achieved using the conventional methods of molecular sieve and ionic gel chromatography in CHAPS detergent.[43] The methods of selective extraction of microsomal proteins[5] followed by column chromatography[7] resulted in a 7- to 8-fold removal of inactive protein.

[46] T. Brody and J. W. Suttie, Fed. Proc. Fed. Am. Soc. Exp. Biol. 41, 467 (1982).

[38] β-Carboxyaspartic Acid

By Tad H. Koch, M. Robert Christy, Robert M. Barkley, Richard Sluski, Denise Bohemier, John J. Van Buskirk, and Wolff M. Kirsch

β-Carboxyaspartic acid (Asa) is a congener of γ-carboxyglutamic acid (Gla), formed by the vitamin-K mediated posttranslational γ-carboxylation of glutamyl residues.[1] Asa has been identified as an amino acid residue of the ribosomal proteins of *Escherichia coli*.[2] Although the biosynthesis of Asa is presently unknown, work of Hamilton, Tesch, and Zerner suggests that it might also be a vitamin K-dependent process.[3]

[1] J. Stenflo, P. Fernlund, W. Egan, and P. Roepstorff, *Proc. Natl. Acad. Sci. U.S.A.* 71, 2730 (1974); G. L. Nelsestuen, T. H. Zythovicz, and J. B. Howard, *J. Biol. Chem.* 249, 6347 (1974); S. Magnusson, L. Sottrup-Jensen, T. E. Peterson, H. R. Morris, and A. Dell, *FEBS Lett.* 44, 189 (1974).
[2] M. R. Christy, R. M. Barkley, T. H. Koch, J. J. Van Buskirk, and W. M. Kirsch, *J. Am. Chem. Soc.* 103, 3935 (1981).
[3] S. E. Hamilton, D. Tesch, and B. Zerner, *Biochem. Biophys. Res. Commun.* 107, 246 (1982).

The identification of Asa in bacterial proteins was facilitated by the organic synthesis of racemic Asa.[2] Our experimental methods for the synthesis of Asa and the identification of natural Asa as well as some chemical properties of Asa are described here. Other synthetic methods for the preparation of racemic Asa include the synthesis and hydrolysis of diethyl 5-hydantoin malonate by Henson, Gallop, and Hauschka[4] and the synthesis, reduction, and hydrolysis of diethyl(3-carboxyaspartato) tetraaminecobalt(III) perchlorate by Dixon and Sageson.[5] These latter syntheses, however, do not describe the isolation and characterization of the pure racemic Asa zwitterion.

Laboratory Synthesis of Racemic Asa (7)

Asa is synthesized in 200-mg quantities and 46% overall yield by transesterification of 1,1,2-tris(carbomethoxy)ethane (1) with benzyl alcohol followed by bromination at the 1-position, dehydrobromination with triethylamine, Michael addition of hydrazoic acid, and simultaneous catalytic hydrogenolysis of the benzyl esters and azide functional groups.

1,1,2-Tris(carbomethoxy)ethane is prepared in 47% yield by reaction of sodium dimethyl malonate with methyl chloroacetate using a slight modification of the procedure of Hall and Ykman.[6] These reactions have been performed by three different researchers in our laboratories. A summary of the transformations appears in Scheme 1.

1,1,2-Tris(carbomethoxy)ethane (**1**). 1,1,2-Tris(carbomethoxy)ethane is synthesized by a substitution reaction of methyl chloroacetate with sodium dimethyl malonate, prepared *in situ* by reaction of sodium methoxide with dimethyl malonate. Both methyl chloroacetate and dimethyl malonate are commercially available from the Aldrich Chemical Co. Dimethyl malonate, 188.4 g (1.545 mol, distilled prior to use) in 1500 ml of reagent grade, dry, absolute methanol, is placed in a 3-liter three-neck flask fitted with a mechanical overhead stirrer, reflux condenser, and refrigerated bath maintained at $0 \pm 2°$. Sodium (34.5 g, 1.50 g-atoms) is added slowly in small pieces to the cooled solution. Note that hydrogen is a by-product of this step. When all the sodium has reacted, 162.8 g of methyl chloroacetate (1.50 mol) is added dropwise with an addition funnel over a 30-min period, maintaining the reaction temperature below 2°. After 10 hr of reaction at 0°, the solvent is rotary evaporated, and 350 ml of water is added. The organic layer is separated, and the water layer is

[4] E. B. Henson, P. M. Gallop, and P. V. Hauschka, *Tetrahedron* **37**, 2561 (1981).
[5] N. E. Dixon and A. M. Sargeson, *J. Am. Chem. Soc.* **104**, 6716 (1982).
[6] H. K. Hall, Jr. and P. Ykman, *J. Am. Chem. Soc.* **97**, 800 (1975).

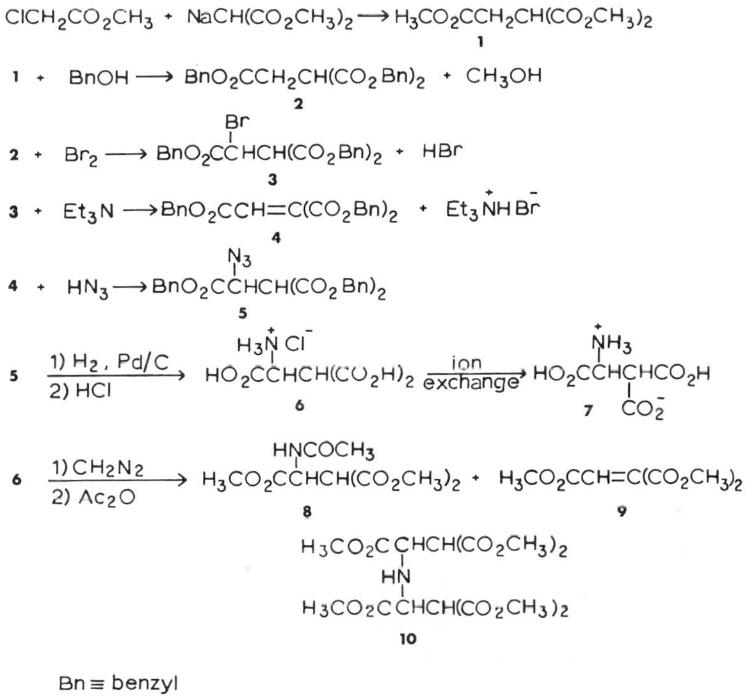

$$ClCH_2CO_2CH_3 + NaCH(CO_2CH_3)_2 \longrightarrow H_3CO_2CCH_2CH(CO_2CH_3)_2$$
$$\mathbf{1}$$

$$\mathbf{1} + BnOH \longrightarrow BnO_2CCH_2CH(CO_2Bn)_2 + CH_3OH$$
$$\mathbf{2}$$

$$\mathbf{2} + Br_2 \longrightarrow BnO_2C\overset{\overset{Br}{|}}{C}HCH(CO_2Bn)_2 + HBr$$
$$\mathbf{3}$$

$$\mathbf{3} + Et_3N \longrightarrow BnO_2CCH{=}C(CO_2Bn)_2 + Et_3\overset{+}{N}H\,\overset{-}{Br}$$
$$\mathbf{4}$$

$$\mathbf{4} + HN_3 \longrightarrow BnO_2C\overset{\overset{N_3}{|}}{C}HCH(CO_2Bn)_2$$
$$\mathbf{5}$$

$$\mathbf{5} \xrightarrow[\text{2) HCl}]{\text{1) } H_2 , \text{ Pd/C}} HO_2C\overset{\overset{\overset{+}{H_3N}\;\overset{-}{Cl}}{|}}{C}HCH(CO_2H)_2 \xrightarrow[\text{exchange}]{\text{ion}} HO_2C\overset{\overset{\overset{+}{NH_3}}{|}}{C}HCHCO_2H$$
$$\mathbf{6} \qquad\qquad\qquad\qquad \mathbf{7}\;\; \overset{-}{C}O_2^{-}$$

$$\mathbf{6} \xrightarrow[\text{2) Ac}_2\text{O}]{\text{1) } CH_2N_2} H_3CO_2C\overset{\overset{HNCOCH_3}{|}}{C}HCH(CO_2CH_3)_2 + H_3CO_2CCH{=}C(CO_2CH_3)_2$$
$$\mathbf{8} \qquad\qquad\qquad\qquad\qquad \mathbf{9}$$

$$H_3CO_2C\,\underset{\underset{H_3CO_2C\,CHCH(CO_2CH_3)_2}{\overset{|}{HN}}}{\overset{|}{C}}HCH(CO_2CH_3)_2$$
$$\mathbf{10}$$

Bn ≡ benzyl

SCHEME 1

extracted twice with 100-ml portions of ether. The combined organic layers are dried over magnesium sulfate and filtered. The ether is removed by rotary evaporation, and the product is purified by fractional distillation with a spinning band column (bp 83°, 0.3 torr) to give 123.8 g of tris(carbomethoxy)ethane (41% yield). Alternatively, fractional distillation with a 20 × 1.8 cm, heated, glass bead packed column gives 145 g of product (bp 88°, 0.5 torr, 47% yield) when the column heater is maintained at 100°.

1,1,2-Tris(carbomethoxy)ethane is characterized by the following spectral properties: IR (thin film), 5.81 and 5.92 μm; ^{1}H NMR (deuteriochloroform) δ 2.80 (doublet, $J = 7$ Hz, 2H), 3.53 (singlet, 3H), 3.60 (singlet, 6H), and 3.67 ppm (triplet, $J = 7$ Hz, 1H). A comparable yield of material suitable for the next step can be obtained also by precipitating the product from the combined ether extracts by cooling in a freezer. After collecting and drying by suction filtration, material obtained in this way has mp 32–33°. This synthetic procedure for preparation of (1) was adapted from Hall and Ykman[6] and reported here with permission from The American Chemical Society and the author, Professor H. K. Hall.

Transesterification of 1,1,2-Tris(carbomethoxy)ethane (**1**) *with Benzyl Alcohol.* 1,1,2-Tris(carbobenzyloxy)ethane (**2**) is synthesized by base-catalyzed reaction of 1,1,2-tris(carbomethoxy)ethane with excess benzyl alcohol, removing the methanol by-product as it is formed. To 0.444 g of 1,1,2-tris(carbomethoxy)ethane (2.2 mmol) dissolved in 5 ml of reagent grade benzyl alcohol is added 0.002 g of potassium hydroxide. The mixture is stirred magnetically for 45 min at ambient conditions and then for 3.75 hr at 60° and 0.05 torr, collecting the methanol by-product in a cold trap. The excess benzyl alcohol is removed by distillation with a pot temperature of 80° and similar vacuum. The residue is diluted with 20 ml of ether and extracted with 20 ml of distilled water. The aqueous layer is back-extracted with 20 ml of ether, and the combined ether layers are extracted with 10 ml of aqueous saturated sodium chloride solution. The resulting ethereal solution is dried over magnesium sulfate and filtered. Rotary evaporation of the ether yields 0.924 g (98%) of analytically pure 1,1,2-tris(carbobenzyloxy)ethane.

The product is a clear oil that crystallizes on standing to give a white solid with mp 40.5–42° and the following spectral and analytical properties: IR (chloroform) 5.78 μm; ^{1}H NMR (deuteriochloroform) δ 3.04 (doublet, $J = 7.5$ Hz, 2H), 3.98 (triplet, $J = 7.5$ Hz, 1H), 5.10 (singlet, 2H), 5.15 (s, 4H), 7.30 and 7.33 ppm (two singlets, 15H); mass spectrum (70 eV) m/z (relative intensity) 197(45), 107(74), 91(100), 78(10), 65(16); analysis calculated for $C_{26}H_{24}O_6$: C, 72.21; H, 5.59; found: C, 72.17; H, 5.62. This reaction can be run on a larger scale (15 g); however, the extraction procedure as described leads to an emulsion. The emulsion problem with this scale is minimized by diluting the crude product with 50 ml of ether, extracting with 50 ml of 0.5 M hydrochloric acid, and back-extracting the aqueous layer with 30 ml of ether. After drying the combined ether extracts with magnesium sulfate and rotary evaporating the ether, traces of residual benzyl alcohol are removed from the solid product by slurrying in 220 ml of water, filtering, and drying at high vacuum (0.01 torr) for 2 hr.

Bromination of 1,1,2-Tris(carbobenzyloxy)ethane (**2**). 1,1,2-Tris(carbobenzyloxy)ethane is regiospecifically brominated at the 2-position by reaction with bromine in the absence of light. In a 25-ml three-neck flask fitted with a reflux condenser and magnetic stirring apparatus in a fume hood is placed 0.471 g (1.09 mmol) of 1,1,2-tris(carbobenzyloxy)ethane in 5 ml of reagent grade carbon tetrachloride. Light is rigorously excluded by wrapping the apparatus completely with aluminum foil. Failure to eliminate light leads to free-radical bromination at the benzyl positions as indicated by the formation of benzaldehyde upon aqueous workup. With stirring, 4.36 ml of a 0.253 M solution of bromine in carbon tetrachloride (1.10 mmol) is added all at once, and the reaction is heated to 76° for 2 hr.

The solvent is then removed by rotary evaporation to yield 0.677 g of a yellow oil that is purified by thin-layer chromatography on three 20-cm by 20-cm by 2-mm Merck silica gel F254 preparative layer plates eluting with 3 : 1 (v/v) dichloromethane–ether. The band (R_f 0.58) is extracted to yield 0.002 g of the starting material (2), and a second band (R_f 0.78) is extracted to yield 0.47 g (88% yield, corrected for recovered compound 2) of pure 2-bromo-1,1,2-tris(carbobenzyloxy)ethane (3). For this scale or a larger scale reaction (30 g), an alternative method of purification is silica gel flash chromatography[7] eluting with 2.5 : 1 dichloromethane-n-hexane. The product is characterized by the following spectral and analytical properties: IR (chloroform) 5.73 and 12.70 μm; ^{1}H NMR (deuteriochloroform) δ 3.46 (singlet, 2H), 5.02 (singlet, 2H), 5.09 (singlet, 4II), 7.23 and 7.27 ppm (two singlets, 15H); mass spectrum (70 eV) m/z (relative intensity) 107(24), 91(11), 90(100), 64(13); analysis calculated for $C_{26}H_{25}BrO_6$: C, 61.01; H, 4.53; found: C, 61.15; H, 4.57.

Dehydrobromination of 2-Bromo-1,1,2-tris(carbobenzyloxy)ethane (3). Reaction of 3 with triethylamine gives 1,1,2-tris(carbobenzyloxy) ethylene (4) and triethylammonium bromide via elimination of hydrogen bromide. To 0.409 g of the bromo compound (3) (0.800 mmol) dissolved in 5 ml of reagent grade, anhydrous ether at 0° is added dropwise 0.112 ml of triethylamine (0.808 mmol) with magnetic stirring in a fume hood. The mixture is stirred for 3 hr at 0°, diluted with 20 ml of ether, and extracted with three 10-ml portions of distilled water. The aqueous layers are combined and back-extracted with three 10-ml portions of ether. The combined ether layers are dried over anhydrous magnesium sulfate and filtered. Rotary evaporation of the ether yields 0.344 g (100%) of pure 1,1,2-tris(carbobenzyloxy)ethylene as a colorless oil that crystallizes on standing to a white solid. The material is characterized by the following physical properties: mp 44–45°; IR (chloroform) 5.8 μm; ^{1}H NMR (deuteriochloroform) δ 5.11 (singlet, 4H), 5.19 (singlet, 2H), 6.89 (s, 1H), 7.19 ppm (multiplet, 15H); mass spectrum (70 eV) m/z (relative intensity) 430 (0.1), 107(11), 91(20), 90(100). A sample for elemental analysis is prepared by recrystallization from ethanol: analysis calculated for $C_{26}H_{22}O_6$: C, 72.55; H, 5.15; found: C, 72.47; H, 5.16. For larger scale reactions, workup is facilitated by removal of the bulk of the triethylammonium bromide by filtration prior to aqueous extraction.

Michael Addition of Hydrazoic Acid to 1,1,2-Tris(carbobenzyloxy) ethylene (4). 2-Azido-1,1,2-tris(carbobenzyloxy)ethane (5) is prepared by reaction of 4 with an excess of hydrazoic acid generated *in situ* with sodium azide and sulfuric acid. Note that hydrazoic acid is a highly toxic

---

[7] W. C. Still, M. Kahn, and A. Mitra, *J. Org. Chem.* **43**, 2923 (1978).

and volatile material and that this reaction and its workup must be conducted in an adequate fume hood. To 0.939 g (2.18 mmol) of 1,1,2-tris(carbobenzyloxy)ethylene and 1.15 g (17.5 mmol) of sodium azide in 10 ml of reagent grade tetrahydrofuran is added 50.0 ml of a 0.366 M solution of sulfuric acid (18.3 mmol). The reaction mixture is stirred for 2 hr at ambient temperature, cooled by addition of approximately 40 g of crushed ice, and extracted with three 75-ml portions of dichloromethane. The combined organic layers are dried over anhydrous sodium sulfate and filtered. Rotary evaporation of the solvent gives 0.945 g of **5** as a light yellow oil. Attempts to purify the product with silica gel or alumina chromatography or activated charcoal lead to quantitative elimination of hydrazoic acid giving the starting material (**4**); therefore, the azide (**5**) is used without purification. The unpurified azide is characterized by the following spectral and analytical properties: IR (chloroform) 4.74 and 5.73 μm; ^{1}H NMR (deuteriochloroform) δ 3.92 (doublet, $J = 9Hz$, 1H), 4.60 (doublet, $J = 9Hz$, 1H), (multiplet, 15H); mass spectrum (70 eV) m/z (relative intensity) 197(14), 107(34), 91(35), 90(100), 65(12); analysis calculated for $C_{26}H_{23}N_3O_6$: C, 65.95; H, 4.89; N, 8.87; found: C, 66.39; H, 5.01; N, 8.30.

DL-β-*Carboxyaspartic Acid* (**7**). Racemic Asa is synthesized by palladium on carbon-catalyzed hydrogenolysis of 2-azido-1,1,2-tris(carbobenzyloxy)ethane (**5**). By-products of the reaction are toluene and nitrogen, and side products are tricarboxyethane and aspartic acid. Tricarboxyethane results from elimination of hydrazoic acid from **5** followed by hydrogenation of the resulting carbon–carbon double bond as well as hydrogenolysis of the benzyl–oxygen bonds. Aspartic acid results from decarboxylation of Asa. A 0.200-g sample of 10% palladium on charcoal is suspended in 50 ml of absolute methanol with a nitrogen atmosphere. Note that mixing palladium on charcoal with methanol in the presence of oxygen can lead to ignition of the methanol. The suspended catalyst is cooled to 0° and prehydrogenated with magnetic stirring at 1 atm of hydrogen. A 0.945-g sample of **5** is dissolved in 40 ml of 10% acetic acid in absolute methanol. This solution is added to the suspended catalyst, and the hydrogenolysis is allowed to proceed at 0° and 1 atm of hydrogen with vigorous stirring until the uptake of hydrogen ceases (7.5 mmol, 3.8 equivalents, 1.2 hr). Maximum stirring rate is critical for optimum yield; slower stirring leads to incomplete hydrogenolysis products along with a higher proportion of tricarboxyethane. The catalyst is removed by filtration through Celite, and the solvent by rotary evaporation. Residual acetic acid and toluene by-products are removed azeotropically by addition of 75 ml of reagent grade benzene followed by rotary evaporation. This procedure is repeated two additional times, yielding 0.363 g of a white solid. A 0.355-g sample of the crude product in three equal portions is purified by cation-exchange, medium-pressure chromatography on a 6-mm by 10.5-cm column of Hamilton HC-X4.00 resin. The product is dissolved in 0.10 M HCl, injected onto the column, and eluted with 0.10 M hydrochloric acid at 12 ml/hr and a pressure of 35 psi. Fractions are collected every 10 min, and Asa hydrochloride appears in fractions 8–10. The combined yield from the three chromatographies is 0.232 g (46% from compound 1) of white crystalline material after high-vacuum (0.5 torr) rotary evaporation of the solvent. The material is characterized as the monohydrate hydrochloric salt of DL-β-carboxyaspartic acid (6) from the following spectral and analytical data: IR (potassium bromide) 2.8–4.8 (broad) and 4.8–6.6 μm (broad); ^{1}H NMR (hexadeuteriodimethyl sulfoxide) δ 4.11 (doublet, $J - 4.5$Hz, 1H), 4.29 (doublet, $J = 4.5$Hz, 1H), and 8–10 ppm (broad, 8H); ^{1}H NMR (deuterium oxide) δ 4.68 ppm (singlet); ^{13}C NMR (hexadeuteriodimethyl sulfoxide) δ 51.4, 52.2, 167.8, and 168.8 ppm; analysis calculated for $C_5H_{10}ClNO_7$: C, 25.92; H, 4.52; N, 6.05; found: C, 25.85; H, 4.52; N, 6.43.

The monohydrate of DL-β-carboxyaspartic acid is prepared by similar ion-exchange chromatography of the product mixture from hydrogenolysis eluting with distilled water. Pure Asa monohydrate is obtained by high-vacuum (0.5 torr) rotary evaporation of the solvent. Material of sufficient purity and crystal structure for X-ray analysis is obtained by recrystallization in distilled water at 2^v.[8] DL-β-Carboxyaspartic acid monohydrate is not hydroscopic and is much less prone to decarboxylation to aspartic acid than is the hydrochloride salt of Asa.

Asa and Asp as their zwitterions are easily distinguished by thin-layer chromatography with Merck cellulose F-254 TLC sheets, eluting with freshly prepared 50% butanol, 35% water, 15% acetic acid, and developing by spraying with 0.3% ninhydrin in 3% acetic acid in butanol. The R_f of Asa is 0.11, and the R_f of Asp is 0.25.

The mass spectrum of the zwitterion is observed directly without derivatization by fast atom bombardment mass spectrometry, which gives an intense M + 1 peak at m/e 178.

Identification of Asa in E. coli Ribosomal Proteins. Escherichia coli (Coli Genetics Stock Center, No. 5550, strain PA340)[9] are grown aerobically in rich L-broth (1% Bactotryptone, 0.5% yeast extract, 0.25% sodium chloride, 0.5% glucose) at 37°, harvested in late logarithmic growth, washed twice in cold 0.14 M potassium chloride, and stored in pellet form at −20°. The cells are broken by grinding at 0–4° with twice their weight of

[8] B. Richey, M. R. Christy, R. C. Haltiwanger, T. H. Koch, and S. J. Gill, *Biochemistry* 21, 4819 (1982).
[9] There is no reason to believe that Asa is confined to this strain of *E. coli.*

alumina (Sigma, A-8503) and enough TM [50 mM tris(hydroxymethyl) aminomethane hydrochloride, 10 mM magnesium chloride, pH 7.5] to form a thick paste. As grinding continues, TM is added in increments to a final volume of 2.5 ml per gram of cells. Alumina and debris are removed by centrifugation (10,000 rpm, 15 min, 4°, Sorvall SS-34 rotor). The supernatants are clarified (18,000 rpm, 25 min, 3°, Spinco type 30 rotor) and decanted to new tubes. These are centrifuged to pelletize crude ribosomes (40,000 rpm, 6.5 hr, 3°, Spinco type 42.1 rotor). The pellets are suspended thoroughly in TMK-NH$_4$Cl [10 mM tris(hydroxymethyl)aminomethane hydrochloride, 10 mM magnesium chloride, 50 mM potassium chloride, 1.0 M ammonium chloride, pH 7.5], 1.6 ml per gram of cells, and left overnight at 4°. The suspensions are clarified (18,000 rpm, 20 min, 3°, Spinco type 30 rotor). The clear portions of the supernatants are removed by pipette and centrifuged (47,000 rpm, 7 hr, 3°, Spinco type 50 rotor) to sediment ribosomes. The final supernatants are aspirated off, and the tubes and the pellets are rinsed and swirled several times with cold water. During this step a yellowish layer is removed from the otherwise clean ribosomal pellets.

Ribosomal proteins are separated from RNA by the lithium chloride–urea method of Leboy and co-workers[10] and dialyzed against several changes of water for 36 hr at 4°. The proteins, most of which precipitate during dialysis, are lyophilized and stored dry at $-20°$. Samples of protein (5–15 mg) are hydrolyzed in 1.0–2.5 ml of 2 M potassium hydroxide free of oxygen and carbon dioxide for 24 hr at 110° under nitrogen. Hydrolysis takes place in Teflon-lined aluminum tubes with Teflon disks in screw caps. Addition of potassium hydroxide and sealing of the tubes is done in a glove box kept filled with nitrogen. The hydrolysates are diluted to about 4 ml with water in 20-ml beakers, cooled in ice, and brought to pH 6.0 with perchloric acid. Initially, 60% acid is added dropwise, with frequent agitation, until the pH reaches 8. Thereafter, 6% acid is used. After cooling in ice for 30 min, the entire mixture is centrifuged to remove potassium perchlorate (10,000 rpm, 20 min, 4°, 15-ml Corex tubes, Sorvall SS-34 rotor). The clear supernatants are degassed under vacuum and stored at $-20°$.

Since the relative abundance of Asa in these supernatants is approximately 1 mol of Asa per 9000 mol of other amino acids, it is desirable to remove the common amino acids before analysis of the mixture. This is accomplished by chromatography on a 1.5 × 11 cm column of basic anion-exchange resin (Bio-Rad, AG2-X8, 200–400 mesh, chloride, converted to the acetate form before use). Successive additions at ambient

[10] P. S. Leboy, E. C. Cox, and J. G. Flaks, *Proc. Natl. Acad. Sci. U.S.A.* **52,** 1367 (1964).

temperature are: supernatant, 4 ml; water, 60 ml; 1.0 M hydrochloric acid, 30–45 ml. The flow rate (29 ml/hr) is controlled by a tubing pump at the bottom of the column, which also directs the effluent to a fraction collector (5-min fractions). The common amino acids are washed through the column by the water. Asa and several other highly acidic compounds appear early in the hydrochloric acid eluate, which begins several minutes before an abrupt drop in pH. Typically all of the Asa is recovered in four fractions, of which the second exhibits the pH change. The fractions are stored at $-20°$. Groups of four fractions from five columns are combined, flash evaporated twice at 40°, and dissolved in 800 μl of 0.2 M sodium citrate, pH 2.2, for amino acid analysis. Alternatively, the dry residue is processed for gas chromatographic–mass spectrometric analysis as described below.

Analyses of 250-μl samples are performed with a Beckman Model 120C Analyzer at 52° using a 0.9 × 54 cm column of Beckman W-1 resin and maximum recorder sensitivity. Buffer and ninhydrin flow rates are 53.3 ml/hr and 34.3 ml/hr, respectively. Using pH 2.10 elution buffer in this system, the elution time of Asa is 44–46 min.

The methods described are not necessarily appropriate for analysis of proteins from sources other than $E.$ $coli.$ Valuable information on the analysis and decarboxylation of γ-carboxyglutamic, β-carboxyaspartic, and aminomalonic acids appears in papers by Hauschka and co-workers.[11,12]

Derivatization of Asa

For high-sensitivity detection of Asa by gas chromatography–mass spectrometry, transformation to a volatile derivative is necessary. This is accomplished by esterification of the three carboxyl groups of Asa hydrochloride with ethereal diazomethane followed by acetylation of the amino function with acetic anhydride and pyridine. Note that although ethereal diazomethane is a routine organic reagent, it is a highly toxic, volatile, and explosive material and must be generated and used in an adequate fume hood with proper safety shields. These reactions are summarized in Scheme 1 above.

1-(Acetylamino)-1,2,2,-tris(carbomethoxy)ethane (**8**). The monohydrated hydrochloride salt of Asa (0.131 g, 0.57 mmol) is dissolved in 4 ml of anhydrous methanol at 0° in a small flask (50 ml) without ground glass joints or surface imperfections and equipped with magnetic stirring appa-

[11] P. V. Hauschka, *Anal. Biochem.* **80,** 212 (1977).
[12] P. V. Hauschka, E. G. Henson, and P. M. Gallop, *Anal. Biochem.* **108,** 57 (1980).

ratus. Ethereal diazomethane is generated in a fume hood from Diazald, *N*-methyl-*N*-nitroso-*p*-toluenesulfonamide, available from the Aldrich Chemical Co. according to instructions from Aldrich using special apparatus (or its equivalent) also available from Aldrich. In particular, note that the apparatus must contain no standard ground glass joints or scratches because rough or sharp surfaces initiate rapid decomposition of diazomethane. The ethereal diazomethane is distilled directly into the flask containing the solution of Asa hydrochloride at 0° with stirring until the reaction mixture remains yellow, the color of unreacted diazomethane. Acetic acid is added dropwise to the reaction mixture until a colorless end point is achieved to scavenge excess diazomethane. All solvents are removed by rotary evaporation, and 1 ml of dry pyridine is added to the residue to give a yellow solution. Addition of 2 ml of freshly distilled acetic anhydride followed by stirring for 12 hr at ambient temperature, excluding moisture with a drying tube, results in a peach-colored solution. Crushed ice (1–2 g) is added, and the mixture is stirred for an additional 30 min. The mixture is then extracted with three successive 25-ml portions of dichloromethane. The combined organic layers are extracted successively with two 20-ml portions of 0.1 N hydrochloric acid and one 25-ml portion of aqueous, saturated sodium chloride solution. After drying over anhydrous magnesium sulfate and filtering, the solvent is removed by rotary evaporation to yield 0.137 g of a brown oil. Gas chromatographic analysis at 177° on a 0.64 cm × 1.8 m column of 5% SE-30 on high-performance Chromosorb W with a helium flow rate of 60 ml/min and thermal conductivity detection indicates that the volatile portion of the brown oil is a mixture of 1-(acetylamino)-1,2,2,-tris(carbomethoxy)ethane (**8**, 60%), 1,2,2-tris(carbomethoxy)ethylene (**9**, 25%), di[1,2,2-tris(carbomethoxy)ethyl]amine (**10**, 4%), and partially esterified material (11%), uncorrected for differences in detector response. This mixture solidifies after pumping to high vacuum for several days to give a yellow solid. The solid is suspended in anhydrous ether, and the crystals are collected by suction filtration. This process is repeated two additional times to afford 0.066 g (45%) of **8**. The product is a white solid with the following physical properties: mp 77–78°; IR (chloroform) 2.90, 5.73, 5.94, 6.64, 6.93 μm; ^{1}H NMR (deuteriochloroform) δ 1.98 (singlet, 3H), 3.67 (singlet, 6H), 3.69 (singlet, 3H), 4.12 (doublet, $J = 4$Hz, 1H), 5.24 (doublet of doublets, $J = 4$ and 9Hz, 1H), 6.53 ppm (broad doublet, $J = 9$Hz, 1H); mass spectrum (chemical ionization) m/z 262; mass spectrum (70 eV) m/z (relative intensity) 261(0.2), 202(61), 171(13), 160(100), 128(18), 101(13), 88(11), 70(23). An analytical sample of **8** is prepared by one recrystallization from dichloromethane-anhydrous ether: analysis calculated for $C_{10}H_{15}NO_7$: C, 45.98; H, 5.79; N, 5.36; found: C, 46.02; H, 5.79; N, 5.35.

Derivatization of Asa in Partially Purified Hydrolysates of E. coli Ribosomal Proteins. A sample of partially purified *E. coli* ribosomal protein hydrolysates prepared as described above is diluted with 30 μl of anhydrous methanol in a 0.6-ml Pierce Reacti-Vial. This results in a brown solution containing some insoluble material. To this are added three drops of ethereal diazomethane (prepared as described above), which produces a vigorous evolution of gas and the formation of a white precipitate. The mixture is agitated for 5 min with a vortex stirrer; then three more drops of ethereal diazomethane are added followed by 5 min of agitation. The solvents are evaporated with a stream of nitrogen, and the resulting brown oil is dissolved in 50 μl of anhydrous methanol and allowed to react with an additional 0.1 ml of ethereal diazomethane to ensure complete esterification. This mixture is agitated for 5 min and then sonicated for 5 min. The solvents are removed with a stream of nitrogen; then residual solvents are removed under reduced pressure. This gives a tan solid that is partially dissolved in 0.1 ml of anhydrous pyridine and 0.1 ml of freshly distilled acetic anhydride. The mixture is agitated for 1 hr, then sonicated for 10 min. The pyridine and acetic anhydride are removed with a stream of nitrogen. The resulting brown oil is diluted with 50 μl of spectral grade dichloromethane and agitated for 5 min. Dichloromethane is removed under a stream of nitrogen, then under reduced pressure for 15 min. This gives a brown oil that is dissolved in 100 μl of spectral grade dichloromethane for gas chromatographic–mass spectrometric analysis.

Gas Chromatographic–Mass Spectrometric Analysis of Derivatized Natural and Synthethic Asa. The presence of Asa in derivatized base hydrolysates of *E. coli* ribosomal proteins is unambiguously established by gas chromatography–mass spectrometry (GC–MS). The conditions for the gas chromatographic separation are critical for the successful determination of nanogram quantities of Asa in a complex biological matrix. A capillary chromatographic system is required. The instrument used in these experiments is a Hewlett–Packard model 5982A GC–MS–data system modified for capillary chromatography with direct connection of the column to the ion source of the mass spectrometer. A fused silica open tubular column (0.3 mm i.d. × 30 m) coated with a 0.25 μm film of SE-52 (J and W Scientific) is employed. Samples are injected onto the column via a Grob-type splitless injector using a cold-trap technique.[13] Injections are made at a column temperature of 100° with a subsequent temperature program from 160 to 200° at a rate of 2°/min. At optimum carrier gas flow, the retention times of the derivative of synthetic and natural Asa prepared

[13] K. Grob and K. Grob, Jr., *HRC CC, J. High Resolut. Chromatogr. Chromatogr. Commun.* **1,** 57 (1978).

REACTIVITY OF β-CARBOXYASPARTIC ACID AS A
FUNCTION OF pH[a,b]

pH	Reaction composition (%)		
	Unreacted Asa	Asp	Tricarboxyethylene
1.8	0	100	0
4.6	35	50	15
10.1	76	0	24
12.2	99	0	1

[a] Adapted from Christy and Koch,[15] with permission of The American Chemical Society.
[b] Samples were heated to 60 ± 0.2° for 64 hr in sealed, Teflon-lined vessels.

as described above, as well as a mixture of the two, are identical at 5.7 min.

The 70 eV electron impact mass spectrum of the derivative of Asa from *E. coli* ribosomal proteins appears as m/z (relative intensity): 232(45), 171(23), 160(100), 128(17), 101(10), 88(11), and 70(30). The spectrum of synthetic derivative is described above. Even though very good chromatographic resolution is obtained under the conditions described, the background and overlapping peaks presented in the sample make interpretation of data difficult. Various data analysis techniques are employed to make a positive identification of Asa in the natural sample. These include visual inspection of mass chromatograms, background subtraction, and computerized data reduction using the mass-resolution technique developed by Biller and Biemann.[14] In all cases the resulting spectrum for the derivative of natural Asa is identical to the spectrum for synthetic material within the experimental error of the method.

Chemical Properties of Asa and a Pseudopeptide Derivative

β-Carboxyaspartic acid is the most acidic natural amino acid presently known. Its four pK_a values corrected to zero ionic strength at 25° are 0.8 ± 0.2, 2.5 ± 0.1, 4.7 ± 0.1, and 10.9 ± 0.1. The pK_a data, analyzed with empirical additivity relationships, are consistent with initial ionization of a β-carboxyl group to yield the zwitterion. The β-carboxyl zwitterion is approximately four times as prevalent as the α-carboxyl zwitterion. X-ray crystallographic data suggest that a β-carboxyl group is also

[14] J. E. Biller and K. Biemann, *Anal. Lett.* **7,** 515 (1974).

partially ionized in the solid state. 5-Hydantoinmalonic acid (**11**), a pseudopeptide derivative of Asa, has pK_a values at 25° of 1.85 ± 0.05, and 10.20 ± 0.05, also corrected to zero ionic strength.[8]

Asa efficiently decarboxylates to aspartic acid upon heating in acidic medium[2,12,15] and eliminates ammonia to give tricarboxyethylene upon heating in moderately basic medium.[15] In strong base Asa is reasonably stable even at 100° for 24 hr. These reactions as a function of pH are summarized in the table. Acid-catalyzed decarboxylation of Asa to Asp is a qualitative test for the presence of Asa.[2,12]

5-Hydantoinmalonic acid (**11**), a model for peptide-bound Asa, decarboxylates only 0.04 times as fast as Asa in 1 M hydrochloric acid at 70°; however, it is much less stable than Asa with respect to elimination to tricarboxyethylene in strong base.[15] The reactivity of **11** upon heating in 2 M sodium hydroxide suggests that some Asa (at least 50%) is lost during the *E. coli* ribosomal protein hydrolysis. Once Asa is released from the protein it is probably stable to the basic hydrolysis conditions. The reactivity of free and peptide bound Asa as a function of pH in basic medium has been discussed in terms of the leaving groups.[15]

11

Acknowledgment

We thank the National Cancer Institute (Grant CA-24665) and the General Medical Institute (Grant GM-24965) of the NIH and the VA Research Organization (4964) for financial assistance. T. H. K. also thanks the University of Colorado Council on Research and Creative Work for a faculty fellowship. This work was made possible in part by Professor Robert Sievers through the generous donation of time and assistance with his HP 5982A gas chromatography–mass spectrometry–data system. We also acknowledge the University of Nebraska Regional Mass Spectrometry Center for the FAB spectrum of Asa.

[15] M. R. Christy and T. H. Koch, *J. Am. Chem. Soc.* **104,** 1771 (1982).

[39] Occurrence and Characterization of Selenocysteine in Proteins

By THRESSA C. STADTMAN

Identification of selenocysteine as the organoselenium compound present in the selenoprotein A component of the clostridial glycine reductase complex[1] provided the first example of the normal occurrence of this selenoamino acid in a protein. Our initial attempts to isolate this labile selenium compound from acid hydrolysates and enzymatic digests of the native protein were unsuccessful. The reported acid lability of selenocysteine[2] accounted for the failure to detect the selenoamino acid residue in acid hydrolysates. Susceptibility of unprotected selenocysteine to spontaneous oxidation and elimination of selenium also was responsible for large losses during preparation and analysis of enzymatic digests of the native protein. Only after the selenocysteine residue in the protein was reduced and converted to an *Se*-alkyl derivative prior to hydrolysis or digestion was it possible to determine the amount and identity of the selenium moiety present. As will be pointed out later, the alkyl selenoether derivatives also are prone to spontaneous oxidation to the corresponding selenones and selenoxides, which then tend to decompose and eliminate selenium. In veiw of our experience in analysis of selenocysteine in bacterial selenoproteins and that of other investigators in glutathione peroxidase, occasional claims in the literature prior to 1975 of the detection of selenocysteine in acid hydrolysates of native proteins are to be viewed with scepticism.

Occurrence of Selenocysteine

Four selenium-dependent enzymes that contain essential selenium in the form of selenocysteine residues have been identified to date. Three of these are of bacterial origin (e.g., glycine reductase, certain formate dehydrogenases, and a hydrogenase), and the fourth, glutathione peroxidase, is found in mammals and birds. These four enzymes catalyze coupled oxidation–reduction reactions and can be considered to be typical redox enzymes.

---

[1] J. E. Cone, R. Martín del Río, J. N. Davis, and T. C. Stadtman, *Proc. Natl. Acad. Sci. U.S.A.* **73,** 2659 (1976).
[2] R. E. Huber and R. S. Criddle, *Arch. Biochem. Biophys.* **122,** 164 (1967).

METHODS IN ENZYMOLOGY, VOL. 107

The glycine reductase complex[3–5] is composed of a small acidic seleno-protein subunit (M_r ~12,000) and two larger protein components (M_r ~200,000 and M_r ~250,000). The selenoprotein subunit contains one se-lenocysteine and two cysteine residues within the polypeptide chain.[6,7] The amino terminus is blocked and the protein is glycosylated.

Selenium-dependent formate dehydrogenases have been isolated from *Escherichia coli*,[8,9] *Methanococcus vannielii*,[10] and *Clostridium ther-moaceticum*.[11] One of the *E. coli* enzymes is a ~600,000 M_r complex with an α_4, β_4, γ_2, or γ_4 structure.[8] Per mole of protein there are 4 g-atoms of Se, 4 of Mo, 56 of nonheme iron, and 52 of acid-labile sulfide. There are also 4 β-type heme equivalents per mole of enzyme. The selenium is located exclusively in the 110,000 α-subunits, presumably one per sub-unit. This species of formate dehydrogenase appears to be associated in the cell with nitrate reductase. Another formate dehydrogenase, also present in *E. coli*, contains an 80,000 M_r selenoprotein subunit rather than the 110,000 M_r species.[9] This formate dehydrogenase may be associated with a hydrogenase as part of the so-called formate hydrogen lyase com-plex. A selenium dependent formate dehydrogenase from *M. vannielii* (M_r ~500,000) is made up of selenoprotein subunits and molybdenum–iron–sulfur protein subunits (M_r 105,000). The selenoprotein subunit (M_r in the range of 100,000) contains selenium in the form of selenocysteine.[12] By analogy it is likely that the *E. coli* selenoprotein subunits also contain selenocysteine, although this has not been demonstrated. A selenium-containing NADP-dependent formate dehydrogenase (M_r ~340,000) iso-lated from *C. thermoaceticum* has an $\alpha_2\beta_2$ subunit composition. The 2 g-atoms of selenium per mole of enzyme are located in the 96,000 M_r α-subunits. The enzyme also contains tungsten and high amounts of non-heme iron and acid-labile sulfide.

A novel hydrogenase isolated from *M. vannielii* also proved to be a selenoenzyme.[13] Homogeneous preparations of this hydrogenase (M_r 340,000) contain 4 g-atoms of Se, 2 g-atoms of Ni, and 18–20 g-atoms of

[3] D. C. Turner and T. C. Stadtman, *Arch. Biochem. Biophys.* **154**, 366 (1973).

[4] H. Tanaka and T. C. Stadtman, *J. Biol. Chem.* **254**, 447 (1979).

[5] T. C. Stadtman, this series, Vol. 53, p. 373.

[6] J. E. Cone, R. Martín del Río, and T. C. Stadtman, *J. Biol. Chem.* **252**, 5337 (1977).

[7] G. L. Dilworth and T. C. Stadtman, unpublished observations.

[8] H. G. Enoch and R. L. Lester, *J. Biol. Chem.* **250**, 6693 (1975).

[9] J. C. Cox, E. S. Edwards, and J. A. DeMoss, *J. Bacteriol.* **145**, 1317 (1981).

[10] J. B. Jones and T. C. Stadtman, *J. Biol. Chem.* **256**, 656 (1981).

[11] I. Yamamoto, T. Saiki, S.-M. Liu, and L. G. Ljungdahl, *J. Biol. Chem.* **258**, 1826 (1983).

[12] J. B. Jones, G. L. Dilworth, and T. C. Stadtman, *Arch. Biochem. Biophys.* **195**, 255 (1979).

[13] S. Yamazaki, *J. Biol. Chem.* **257**, 7926 (1982).

Fe per mole of enzyme.[14] The enzyme is made up of three types of subunits, M_r 42,000, 35,000, and 27,000. The selenium, located exclusively in the 42,000 M_r subunits, is present as selenocysteine residues.

The only known mammalian selenoenzyme is glutathione peroxidase. This enzyme has been isolated from bovine,[15] ovine,[16] and human erythrocytes,[17] from rat liver,[18,19] from lens of bovine eye,[20] and it has been detected in numerous tissues of mammals and birds. The native protein (M_r 80,000–90,000) consists of four apparently identical subunits and contains 4 g-atoms of Se present as selenocysteine residues within the subunit polypeptide chains.[19,21]

Selenium-containing polypeptides of unknown catalytic function also have been found in various animal tissues. A 75,000 M_r species present in liver and kidney[22,23] that turns over rapidly has been reported to contain selenocysteine.[23]

Function of Selenocysteine in Proteins

The four enzymes in which selenium occurs in the form of selenocysteine residues are redox catalysts, and it is reasonable to suppose that in each of these the selenium serves as a redox center. Treatment of the glycine reductase selenoprotein with dithiothreitol, the electron donor for the *in vitro* reaction, converts both the selenocysteine and the cysteine residues to forms reactive with alkylating agents. During catalysis with molecular hydrogen as substrate, the selenium-containing hydrogenase is markedly sensitive to inactivation with alkylating agents, and under these conditions the corresponding alkyl seleno ethers of the selenocysteine residues are formed.[13] Similarly, glutathione peroxidase, after reaction with reduced glutathione, becomes susceptible to inactivation by iodoacetate.[19] The target groups in this case are the selenocysteine residues in the reduced native protein. Electron density changes observed at the selenocysteine sites by X-ray analysis of the oxidized and reduced forms of the

---

[14] S. Yamazaki, unpublished observations.

[15] L. Flohé, W. A. Günzler, and H. H. Schock, *FEBS Lett.* **32,** 132 (1973).

[16] J. T. Rotruck, A. L. Pope, H. E. Ganther, A. B. Swanson, D. G. Hafeman, and W. G. Hoekstra, *Science* **179,** 588 (1973).

[17] Y. C. Awasthi, E. Beutlar, and S. K. Srivastava, *J. Biol. Chem.* **250,** 5144 (1975).

[18] W. Nakamura, S. Hosoda, and K. Hayashi, *Biochim. Biophys. Acta* **358,** 251 (1974).

[19] J. W. Forstrom, J. J. Zakowski, and A. L. Tappel, *Biochemistry* **17,** 2639 (1978).

[20] N. J. Holmberg, *Exp. Eye Res.* **7,** 570 (1968).

[21] R. Ladenstein, O. Epp, and A. Wendel, *in* "Structural and Functional Aspects of Enzyme Catalysis" (H. Eggerer and R. Huber, eds.), p. 104. Springer-Verlag, Berlin, 1981.

[22] R. F. Burk and P. E. Gregory, *Arch. Biochem. Biophys.* **213,** 73 (1982).

[23] M. A. Motsenbocker and A. L. Tappel, *Biochim. Biophys. Acta* **704,** 253 (1982).

crystalline enzyme indicate that the selenolate anion is present in the GSH-reduced enzyme.[21] The marked sensitivity of the *M. vannielii* formate dehydrogenase to low levels of alkylating agents also is suggestive of a reactive selenolate anion as target,[10] but since borohydride reduction followed by alkylation was employed for identification of selenocysteine in this protein, the actual redox state of the selenoamino acid during catalysis is uncertain.[12]

Analysis of Selenocysteine Residues in Proteins

Selenocysteine residues in proteins are labile to the procedures commonly employed to cleave proteins to their constituent amino acids. However, if the selenocysteine residues are reduced and alkylated prior to acid hydrolysis or enzymatic digestion, they can be detected and estimated quantitatively. Identification of the derivatives is greatly facilitated if the proteins can be prelabeled with [75]Se. Derivatization of selenocysteine residues in proteins is accomplished by slight modification of the standard procedures used for alkylation of sulfhydryl groups. For example, to the protein in a solution buffered at about pH 8 containing 1 mM dithiothreitol is added a few grains of solid KBH_4 (amount sufficient to give about 5 mM) under argon. After 5–10 min solid iodoacetamide or deoxygenated neutralized iodoacetic acid is added to 20 mM final concentration (under argon) and the solution is stored in the dark at room temperature under argon for several hours or overnight. The reaction is terminated by the addition of 2-mercaptoethanol (20–30 mM final concentration). The protein is freed of reagents by dialysis under anaerobic conditions or (for low molecular weight proteins) by gel filtration on a column equilibrated with 1 mM dithiothreitol. A similar procedure has been used for preparation of the *Se*-carboxyethyl derivative using 3-bromopropionate and of the *Se*-aminoethyl derivative using ethylenimine.[1]

The choice of reagent may be dictated by its relative accessibility to the selenocysteine residue when native proteins are modified. Although standard acid hydrolysis procedures (6 M HCl) can be employed, mercaptoethanesulfonic acid (3 M; available from Pierce Chemical Co.) has the advantage of affording protection from oxidation. When acid hydrolysates or enzymatic digests of alkylated selenoprotein samples are chromatographed on an amino acid analyzer column using standard citrate buffer systems, *Se*-carboxymethylselenocysteine emerges just after aspartic acid in the position normally identified as methionine sulfone, *Se*-carboxyethylselenocysteine is eluted with pH 3.25 citrate at the back of the proline peak, and *Se*-aminoethylselenocysteine is eluted from the "short" column with pH 5.25 citrate near histidine and between *S*-aminoethyl-cysteine and ammonia.[1] Each of these selenocysteine alkyl ethers emerges from the amino acid analyzer column a few minutes after the corresponding sulfur analog. Comparable elution sequence relationships are observed when *Se*-methylselenocysteine (elution position near alanine) is compared to *S*-methylcysteine, or selenomethionine to methionine. Seleno ethers readily undergo spontaneous oxidation upon exposure to oxygen or traces of peroxides. In the case of the *Se*-carboxymethyl derivative of selenocysteine, the oxidation products were eluted from the amino acid analyzer column as discrete peaks just after the breakthrough volume. Treatment of the compounds from these peaks with KBH$_4$ reduced them to a product indistinguishable from the original seleno ether. By analogy with alkyl sulfur ether oxidation products, it is likely that the selenoxide and selenone derivatives of *Se*-carboxymethylselenocysteine had been formed.[1]

Some cellulose thin-layer chromatographic systems that we have used for analysis of *Se*-alkylselenocysteine derivatives together with typical R_f values are listed in the table. In general a selenium compound and its corresponding sulfur analog exhibit virtually identical R_f values in these systems and thus are not separated.

The lower pK_a of selenocysteine (5–6) as compared to cysteine (8–9) is such that at physiological pH the R-SeH will be anionic whereas ionization of the R-SH group should be slight. In the case of the selenoprotein A component of the glycine reductase complex, this property could be exploited for a spectrophotometric assay of the selenol content of the pro-

THIN-LAYER CHROMATOGRAPHY

Compound	R_f value in solvent[a]			
	A	B	C	D
Se-Carboxymethylselenocysteine	0.35	0.47	—	0.15
Se-Carboxamidomethylselenocysteine	0.18	—	—	0.28
Se-Carboxyethylselenocysteine	—	0.5	—	—
Se-Aminoethylselenocysteine	—	—	0.79	0.46
Se-Methylselenocysteine	—	—	—	0.66
Se-Methionine	—	0.86	—	0.7
Selenocystine	0.05–0.1	0.1	0.72	—
Methionine sulfone	—	0.36	—	0.4

[a] Ascending chromatography; ninhydrin spray detection; solvent compositions are as follows: (A) 2-propanol:88% HCOOH:H$_2$O (60:3:15); (B) *tert*-butyl alcohol:methyl ethyl ketone:88% HCOOH:H$_2$O (40:30:15:15); (C) 2-propanol:28% NH$_4$OH:H$_2$O (60:1.5:30); and (D) CHCl$_3$:CH$_3$OH:15% NH$_4$OH (40:40:10).

tein.[6] The increase in absorbancy at 243 nm observed upon reduction of the protein with KBH_4 at pH 7–8 was equivalent[24] to the known content of selenocysteine (1 mol per mole of protein).

The selenocysteine residues in glutathione peroxidase have been determined by X-ray crystallographic analysis.[21]

Synthesis of Radiolabeled Selenocysteine

Radioactive L-selenocysteine labeled with ^{75}Se, ^3H, or ^{14}C can be prepared from appropriately labeled O-acetyl-L-serine or $H_2{}^{75}Se$ using cysteine synthetase (O-acetylserine sulfhydrylase) from *Salmonella typhimurium*.[25] In this procedure labeled or unlabeled serine is converted to O-acetylserine by reaction with HCl-saturated glacial acetic acid. The O-acetyl derivative then is converted to selenocysteine by incubation under anaerobic conditions with H_2Se and O-acetylserine sulfhydrylase in a phosphate buffer at pH 7.2. The product is separated from the reaction mixture by absorption to Dowex 50-H^+ and elution with NH_4OH. The oxidized form of the product (selenocystine; R_f 0.08) was separated from contaminating serine (R_f 0.3) by thin-layer chromatography in $CHCl_3 : CH_3CH_2OH$: glacial acetic acid : H_2O (50 : 32 : 10 : 8).

Enzymatic Degradation of Selenocysteine

An enzyme, selenocysteine lyase, that specifically decomposes L-selenocysteine to stoichiometric amounts of alanine and H_2Se has been purified to homogeneity from pig liver.[26] The enzyme (M_r ~85,000) is composed of 48,000 M_r subunits and contains pyridoxal phosphate as coenzyme. L-Cysteine was not decomposed by the enzyme but was found to behave as a competitive inhibitor (K_i 1.0 mM) with DL-selenocysteine as substrate. The K_m value reported for L-selenocysteine (0.83 mM) is orders of magnitude higher than the normal amount of the selenoamino acid in tissues. This raises the interesting possibility that the natural substrate for the enzyme is a compound other than free selenocysteine.

[24] Based on a molar extinction coefficient of 5000 at 243 nm (pH 8) for the ionized selenol of selenocysteine.[2]

[25] G. L. Dilworth, *J. Labelled Compd. Radiopharm.* **19,** 1197 (1982).

[26] N. Esaki, T. Nakamura, H. Tanaka, and K. Soda, *J. Biol. Chem.* **257,** 4386 (1982).

[40] Identification of Selenocysteine by High-Performance Liquid Chromatography and Mass Spectrometry

By H. E. Ganther, R. J. Kraus, and S. J. Foster

Selenium is an essential component of glutathione peroxidase in animal tissues[1] and of a number of microbial enzymes.[2] The form of selenium in most of these enzymes is selenocysteine.[3] A large proportion of the selenium in animal tissues is apparently associated with unidentified proteins that may also contain selenocysteine.[4] In most cases the primary means of identification of selenocysteine has been cochromatography of *Se*-carboxylmethyl or *Se*-aminoethyl derivatives of the enzyme selenium and selenocysteine following hydrolysis of the alkylated selenoprotein.[3,4]

Because of the need for a more definitive method to characterize selenocysteine in selenoproteins, we developed a convenient procedure based on derivatization of selenocysteine by Sanger's reagent (1-fluoro-2,4-dinitrobenzene) that can be used for identification of selenocysteine by mass spectrometry. The method is also well suited for thin-layer chromatography and high-performance liquid chromatography (HPLC).

Materials

1-Fluoro-2,4-dinitrobenzene (FDNB), selenocystine, *O*-DNP-tyrosine, ε-DNP-lysine, α-DNP-arginine, *N*,*S*-di-DNP-L-cysteine, and methyl iodide are obtained from Sigma Chemical Co. *im*-DNP-histidine was a gift of Dr. Vincent Massey. Di-DNP-monoselenide and di-DNP-diselenide, synthesized by the procedure of Twiss,[5] were kindly provided by Dr. T. Hori, and Dr. K. B. Sharpless. [^{75}Se]Selenocystine is available from Amersham; *Se*-methylselenocysteine was from Cyclo Chemical. All solvents used for HPLC or mass spectrometry are redistilled.

General Procedures

Thin-layer chromatography (TLC) and electrophoresis (TLE) are done on precoated silica gel and cellulose plates (Brinkmann), respec-

[1] J. T. Rotruck, A. L. Pope, H. E. Ganther, A. B. Swanson, D. G. Hafeman, and W. G. Hoekstra, *Science* **179,** 588 (1973).

[2] T. C. Stadtman, *Annu. Rev. Biochem.* **49,** 93 (1980).

[3] T. C. Stadtman, this volume [39].

[4] A. L. Tappel, *et al.*, this volume [42].

[5] D. F. Twiss, *J. Chem. Soc.* **105,** 1672 (1914).

R_f VALUES OF DNP DERIVATIVES

Compound	R_f values in solvent systems[a]		
	A	B	C
Se-DNP-selenocysteine	0.47	0.0	0.0
S-DNP-cysteine	0.47	0.0	0.0
N,S-Di-DNP-cysteine	0.62	0.17	0.0
Se-Methyl-N-DNP-selenocysteine	0.68	0.39	0.11
Se-Methyl-N-DNP-O-methyl selenocysteine	0.74	0.78	0.55
2,4-Dinitrophenol	0.73	0.78	0.08
2,4-Dinitrofluorobenzene	0.73	0.83	0.47
Di-DNP-monoselenide	0.71	0.78	0.76
Di-DNP-diselenide	0.71	0.84	0.83
O-DNP-tyrosine	0.52	0.02	0.0
ε-DNP-lysine	0.44	0.017	0.0
α-DNP-arginine	0.44	0.0	0.0
im-DNP-histidine	0.35	0.0	0.0

[a] Solvents: A, n-butanol–acetic acid–water (4:1:1); B, chloroform–methanol–acetic acid (90:8:2); C, benzene–pyridine (80:20). Silica gel plates were used.

tively. The TLC solvent systems and R_f values for the various compounds are described in the table. For TLE, buffers of pH 1.6 (formic acid–acetic acid–water, 150:100:750, v/v/v), and pH 5.3 (pyridine–acetic acid–water, 20:5:2000), are used, and separations are conducted at approximately 10° using a water-cooled support, 20 V/cm. For visualization of compounds, the plates are illuminated with an ultraviolet light or sprayed with ninhydrin.[6] To increase the sensitivity for detecting DNP compounds and to make a permanent photocopy, the plate is covered with print paper[7] and placed over a source of 366 nm light.[8] The photocopy technique can easily detect DNP compounds in the amount of several nanomoles. If a region of the plate is to be scraped and eluted to recover selenium components, this should be done prior to making the photocopy. For TLE, DNP-ethanolamine, DNP-alanine, and ε-DNP-lysine are used as markers to establish relative mobilities of zero, −1, or +1.

Almost any HPLC apparatus can be used for separating the DNP derivatives on commercially available columns using isocratic elution.

[6] I. Smith (Ed.). "Chromatographic and Electrophoretic Techniques," Vol. I, p. 143. Wiley (Interscience), New York, 1960.
[7] Graphic Arts Proof Paper 503-1, DuPont.
[8] TLC Photocopier, Brinkmann.

The separations described here were mostly done on a 4×250 mm C_{18} Bondapak (10 μm) reversed-phase column made by Waters Associates, monitored with an ISCO V_4 variable wavelength detector. For larger amounts of samples, we used a Partisil M9 (10 $\times$ 250 mm) ODS reversed-phase column (Whatman), monitored at 254 nm with an ISCO UA-2 UV analyzer. The methanol–water systems used for HPLC are degassed by evacuation with a water aspirator, then filtered through a Metricel GN-6 filter, pore size 0.45 μm (Gelman). Fractions are collected, and aliquots are analyzed directly for Se by graphite furnace atomic absorption spectrometry or by fluorometric analysis after wet ashing.[9]

Mass spectra are obtained by direct probe analysis at 70 eV ionization energy using 8 KeV acceleration on a Kratos MS-9 instrument equipped with a DS-50 data system. The probe temperature is 165°.

To purify compounds for mass spectrometry, apply the reaction mixture to a silica gel plate (about 10 μg of the desired compound) and chromatograph in the benzene–pyridine system (see the table). Allow the plate to dry, then scrape the plate at the desired R_f and elute the scrapings with a 1 : 1 mixture of benzene–pyridine. The sample is concentrated by evaporation at room temperature and transferred to a glass capillary. About 1–2 μg of compound is needed to obtain a mass spectrum.

The usual precautions[6] should be taken when handling DNP-amino acids. They are susceptible to degradation by light, and therefore should be handled in the dark or under dim light whenever possible. DNP-amino acids are known to undergo decomposition in alkali and are almost completely destroyed in hot alkali. Even under mildly basic conditions, *im*-DNP-histidine, *O*-DNP-tyrosine, *S*-DNP-cysteine, and *Se*-DNP-selenocysteine decompose. DNP-amino acids should be stored in neutral or slightly acidic solutions, at 4°, in the dark, and never in chloroform.[6] It should be noted that storing in alcoholic solutions under acidic conditions may lead to the formation of esters.

Preparation of *Se*-DNP-selenocysteine

Dissolve 20 mg of selenocystine in 0.2 ml of 1 N HCl, then add 2 ml of water. For synthesis of the ^{75}Se-labeled compound add [^{75}Se]selenocystine to achieve the desired specific activity. Adjust the pH to approximately 4 with 1 N KOH, start purging the solution with N_2, and add 3 mg of NaBH$_4$. Reduction of the diselenide to the selenol is indicated by loss of the yellow color. Carefully adjust the pH to 6.0 (adjust to pH 7.0 when preparing *S*-DNP-cysteine) and add 3 ml of 0.5 M FDNB (freshly pre-

[9] S.-H. Oh, H. E. Ganther, and W. G. Hoekstra, *Biochemistry* **13**, 1825 (1974).

pared in ethanol). Add several extra milliliters of ethanol completely to dissolve the FDNB. The reaction is carried out for 2 hr at 25° in the dark, then stopped with several drops of concentrated HCl. Add 4 ml of benzene and flood the reaction mixture with water until the organic and aqueous phases separate. Extract three more times with benzene. About 80% of the ^{75}Se remains with the aqueous phase, while 20% is benzene extractable. TLC of the aqueous phase on silica gel in solvent system A shows that 95% or more of the ^{75}Se is Se-DNP-selenocysteine (UV-absorbing and ninhydrin-positive) and the remaining ^{75}Se is split equally between Se,N-di-DNP-selenocysteine and selenocystine. TLC of the benzene phase in the same systems shows 100% of the ^{75}Se moving as Se,N-di-DNP-selenocysteine. To purify Se-DNP-selenocysteine, concentrate the aqueous phase by rotoevaporation at 37°, then apply it to a Sephadex G-10 column (2.5 × 12 cm), equilibrated and eluted with 20% aqueous ethanol at a flow rate of 20 ml/hr. Se-DNP-selenocysteine (about 95% of the radioactivity) is eluted at $V_e/V_0 = 2.3$.

For further purification, about 5 mg of the Se-DNP-selenocysteine eluted from the Sephadex G-10 column is subjected to HPLC on the Whatman M9 Partisil column using 10% methanol–90% H_2O. A single UV-absorbing peak is eluted that contains 98% of the ^{75}Se. TLE at pH 1.3 shows a single compound, UV- and ninhydrin-positive, moving toward the cathode with a net charge of +1, and containing all of the ^{75}Se.

Identification of Se-DNP-selenocysteine

Absorption Spectrum. The absorption spectrum of Se-DNP-selenocysteine is shown in Fig. 1. The λ_{max} is at 335 nm and the $\varepsilon_{335\ nm}$ is 1.42 × $10^4\ M^{-1}\ cm^{-1}$. These values are comparable to those reported[10] for S-DNP-cysteine (λ_{max} 330 nm and $\varepsilon_{330} = 1.10 × 10^4\ M^{-1}\ cm^{-1}$).

For most N-DNP-amino acids the λ_{max} is approximately 360 nm and ε_{max} is 1.7 to 1.8 × $10^4\ M^{-1}\ cm^{-1}$.[6]

Identification by Nuclear Magnetic Resonance (NMR). Approximately 5 mg of compound is freeze-dried for 24 hr, then dissolved in 2 ml of solvent, transferred to an NMR tube, and tightly sealed. Proton NMR measurements (obtained on a Bruker-270 instrument) for Se-DNP-selenocysteine

$$\text{DNP---Se---}\overset{\overset{\displaystyle H_\beta}{|}}{\underset{\underset{\displaystyle H_{\beta'}}{|}}{C}}\text{---}\overset{\overset{\displaystyle H_\alpha}{|}}{\underset{\underset{\displaystyle NH_2}{|}}{C}}\text{---}\overset{\overset{\displaystyle O}{\|}}{C}\text{---O}^- \quad (CD_3COCD_3 : D_2O, 1:1):$$

[10] H. Kondo, F. Moriuchi, and J. Sunamoto, *J. Org. Chem.* **46,** 1333 (1981).

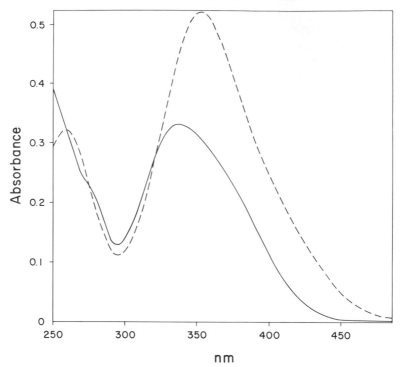

FIG. 1. Absorption spectrum of *Se*-DNP-selenocysteine before and after the Smiles rearrangement. ——, 2.32×10^{-5} *M* *Se*-DNP-selenocysteine in methanol; ------, 10 min after addition of 1.67 m*M* sodium barbital. $T = 25°$, path length = 1 cm.

δ 8.92 (d, $J = 2.2$ Hz, 1 H, H$_3$), 8.34 (dd, $J = 2.6$, $J = 8.8$ Hz, 1 H, H$_5$), 7.99 (d, $J = 8.8$ Hz, 1 H, H$_6$), 3.96 (dd, $J = 5.1$, $J = 8.1$ Hz, 1 H, H$_\alpha$), 3.96 (dd, $J = 5.1$, $J = 13.2$ Hz, 1 H, H$_{\beta'}$), 3.37 (dd, $J = 13.2$, $J = 8.1$ Hz, 1 H, H$_\beta$).

Derivatization for Mass Spectrometry

Initial attempts to identify *Se*-DNP-selenocysteine by mass spectrometry were unsuccessful. Direct probe analysis of the underivatized compound, the trimethylsilyl (TMS) derivative, or various alkylated derivatives, did not produce any selenium-containing fragments in the mass spectra. The use of gas chromatography–mass spectrometry (GC–MS) for the TMS and alkyl derivatives was also unsuccessful. One intense fragment ($m/e = 209$) appeared in direct probe analysis of both derivatized and underivatized *Se*-DNP-selenocysteine, corresponding to CH$_2$-CH-NH-DNP. The origin of this fragment appears to be a Se → N intermolecular rearrangement (Smiles rearrangement) that takes place on the probe, forming a selenol intermediate; selenium is eliminated from the selenol, thus accounting for the absence of selenium fragments in the spectra.

The Smiles rearrangement occurs easily with S-DNP-cysteine[10] and with Se-DNP-selenocysteine upon the addition of a weak base. Figure 1 shows spectroscopic evidence for the rearrangement. Compared to the spectrum of Se-DNP-selenocysteine, the spectrum recorded 10 min after the addition of sodium barbital shows that the absorption is intensified and λ_{max} is shifted from 335 nm to the 360 nm region, as expected for the formation of an N-DNP amino acid.

We modified the Smiles rearrangement by carrying out the reaction under anaerobic conditions in the presence of methyl iodide, thus trapping the liberated selenol as the stable product, Se-methyl-N-DNP-selenocysteine, in high yield.

$$DNP-Se-CH_2-\underset{\underset{NH_2}{|}}{CH}-COO^- \xrightarrow[N_2]{B^-} \left[^-Se-CH_2-\underset{\underset{NH-DNP}{|}}{CH}-COO^- \right] \xrightarrow[N_2]{CH_3I}$$

$$CH_3-Se-CH_2-\underset{\underset{NH-DNP}{|}}{CH}-COO^-$$

To enhance the volatility of Se-methyl-N-DNP-selenocysteine for mass spectrometry, the free carboxyl group is esterified.

$$CH_3-Se-CH_2-\underset{\underset{NH-DNP}{|}}{CH}-COO^- \xrightarrow{MeOH/HCl} CH_3-Se-CH_2-\underset{\underset{NH-DNP}{|}}{CH}-COOCH_3$$

The procedure is as follows: Begin by degassing methanol, using a water aspirator. Care must be taken to work under anaerobic conditions. Dissolve 1.75 mg of Se-DNP-selenocysteine in 1 ml of the degassed methanol and transfer to a rubber-stoppered reaction vessel. Pass N_2 (purified by means of an Oxy-Clear trap) (Pierce Chemical) over the top of the stirred solution for 30 min. Next add 50 μl of CH_3I through the stopper, followed by 200 μl of 0.1 N sodium barbital prepared in degassed methanol. Allow to react for 1 hr at room temperature in the dark. Take the solution to dryness by evacuating with a water aspirator. The progress of the reaction can be monitored by TLC (Fig. 2) and by TLE.

To esterify the product, dissolve the reaction mixture in 2 ml of methanol plus 30 μl of concentrated HCl, and allow to react overnight at room temperature in the dark, then reflux for 2 hr at 75°. As shown in Fig. 3, about 85% of the selenium is present as Se-methyl-N-DNP-O-methylselenocysteine after the esterification. Chromatography in solvent system C

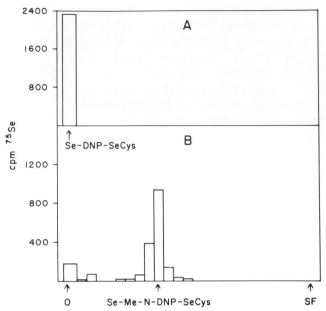

FIG. 2. Conversion of *Se*-DNP-selenocysteine to *Se*-methyl-*N*-DNP-selenocysteine. The reaction mixture (see text) was subjected to TLC on silica gel in solvent system B. (A) [^{75}Se]*Se*-DNP-selenocysteine before reaction (ninhydrin-positive and UV-positive). (B) After addition of sodium barbital and methyl iodide (see text); 81% of the ^{75}Se is associated with a ninhydrin-negative, UV-positive compound having the same R_f (0.39) as *Se*-methyl-*N*-DNP-selenocysteine.

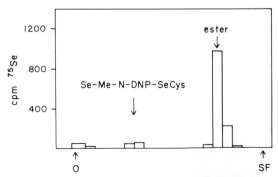

FIG. 3. Esterification of *Se*-methyl-*N*-DNP-selenocysteine. Thin-layer chromatography (TLC) on silica gel in solvent B. [^{75}Se]*Se*-methyl-*N*-DNP-selenocysteine was treated with methanol–HCl (see text), and the reaction mixture was subjected to TLC on silica gel in solvent system B; 87% of the ^{75}Se is associated with a UV-positive compound having the same R_f (0.78) as the *O*-methyl ester of *Se*-methyl-*N*-DNP-selenocysteine.

(data not shown) confirms that the product has the same R_f (0.55) as the authentic ester. Purify the ester by applying the reaction mixture as a band on a silica gel plate and chromatograph in solvent system C, then dry the plate, scrape the band at $R_f = 0.55$, and elute the compound with benzene–pyridine (50 : 50).

Standard Se-methyl-N-DNP-selenocysteine can also be prepared by allowing FDNB to react with Se-methylselenocysteine.

$$CH_3—Se—CH_2—\underset{\underset{NH_2}{|}}{CH}—COO^- + FDNB \rightarrow CH_3—Se—CH_2—\underset{\underset{NH—DNP}{|}}{CH}—COO^-$$

Dissolve 20 mg of Se-methylselenocysteine in 2 ml of water, add 40 mg of $NaHCO_3$ followed by 2.5 ml of 0.5 N FDNB (freshly prepared in ethanol). Allow to react for 1 hr at 40°, then to stand at room temperature overnight, all in the dark. Stop the reaction with several drops of concentrated HCl, then extract with benzene until the extracts are clear. Dry the benzene extract by rotoevaporation, then redissolve in chloroform. To purify Se-methyl-N-DNP-selenocysteine apply the extract to a silica gel H column (1.5 × 8.0 cm) equilibrated in chloroform. Wash first with chloroform until all the 2,4-dinitrophenol and excess FDNB are eluted, then wash with chloroform–methanol (50 : 50) to elute Se-methyl-N-DNP-selenocysteine. HPLC and TLC of the chloroform–methanol eluate show a single compound. For identification by NMR, freeze-dry the sample in the dark, then dissolve about 10 mg in a 1 : 1 mixture of trifluoroacetic acid–acetone-d_6.

$$\text{NMR:}\quad CH_3—Se—\underset{\underset{H_{\beta'}}{\overset{H_\beta}{|}}}{C}—\underset{\underset{NH—DNP}{\overset{H_\alpha}{|}}}{C}—\overset{\overset{O}{\|}}{C}—O^- \quad (CF_3CO_2D : CD_3COCD_3, 1 : 1)$$

δ 8.96 (d, $J = 2.6$ Hz, 1 H, H_3), 8.21 (dd, $J = 1.8$, $J = 2.6$ Hz, 1 H, H_5), 7.01 (d, $J = 9.6$ Hz, 1 H, H_6), 4.81 (t, $J = 5.1$ Hz, 1 H, H_α), 3.15 (ABX, $J = 4.8$, $J = 5.1$, $J = 13.2$ Hz, 2 H, $H_{\beta,\beta'}$), 1.91 (s, 3 H, $SeCH_3$).

Mass Spectrometry. About 200 μg of the compound is taken and identified by high- or low-resolution mass spectrometry. Figure 4 shows the low-resolution mass spectrum of Se-methyl-N-DNP-O-methylselenocysteine.

Application to Selenoproteins

The procedure has been successfully applied to the selenoenzyme, GSH peroxidase.[11] Se-DNP-selenocysteine was isolated after derivatiza-

[11] R. J. Kraus, S. J. Foster, and H. E. Ganther, *Biochemistry* **22**, 5853 (1983).

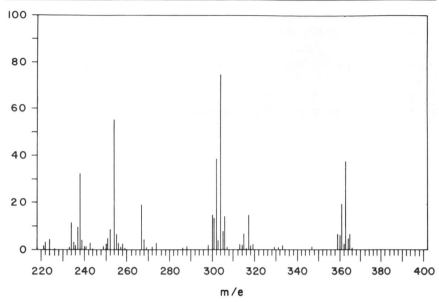

FIG. 4. Mass spectrum of *Se*-methyl-*N*-DNP-selenocysteine, methyl ester. The spectrum was obtained by electron impact ionization (70 eV), using direct probe technique (probe heated to 165°). The molecular ion ($m/e = 363$) is quite intense and shows the typical isotope distribution for selenium. Other Se-containing fragments are seen at $m/e = 317$ (loss of NO_2) and 304 (loss of CO_2CH_3). The fragment at $m/e = 254$ corresponds to loss of CH_3—Se—CH_2. A complete mass spectrum is presented elsewhere.[11]

tion of the borohydride-reduced enzyme with FDNB in 4 M guanidine, followed by dialysis and acid hydrolysis. *Se*-DNP-selenocysteine was separated from other water-soluble DNP amino acids by reversed-phase HPLC and identified by mass spectrometry following its conversion to *Se*-methyl-*N*-DNP-selenocysteine methyl ester. Some information relevant to the identification of selenocysteine in proteins is described below, and additional details are available in the original report.[11]

Stability to Acid Hydrolysis. This question was explored using [^{75}Se]*Se*-DNP-selenocysteine. Under stringent anaerobic conditions (purging with N_2, then evacuating with a mechanical pump prior to acid hydrolysis), *Se*-DNP-selenocysteine shows good stability to 6 N HCl at 110° for 20 hr, with 85–90% recovery of the starting compound in the aqueous phase after benzene extraction of the acid hydrolysate. Acid hydrolysis in an atmosphere of air caused considerable decomposition. Only 65% of the ^{75}Se was in the aqueous phase, and much of this was not *Se*-DNP-selenocysteine; the other 35% of the ^{75}Se was extractable into benzene and was identified by mass spectrometry as di-DNP-mono-

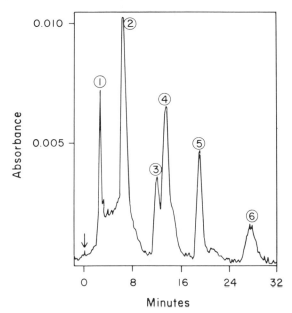

Fig. 5. Separation of water-soluble DNP-amino acids by reversed-phase HPLC. A mixture of approximately 1.9 nmol of each compound was chromatographed on a Waters Associates 10 μm C_{18} Bondapak column (4 × 250 mm) at 25° using isocratic elution with methanol–H_2O (30:70, v/v) at 1.25 ml/min. Absorbance at 254 nm was monitored with an ISCO Model V_4 detector, 0.01 A (full scale). 1, S-DNP-cysteine; 2, im-DNP-histidine; 3, Se-DNP-selenocysteine; 4, α-DNP-arginine; 5, ε-DNP-lysine; 6, O-DNP-tyrosine.

selenide and di-DNP-diselenide. Part of the selenium in the aqueous phase may be elemental selenium, since it could not be eluted when chromatographed on a Sephadex G-10 column. Good recovery of the selenium in GSH-peroxidase was obtained following acid hydrolysis.[11]

Separation of Water-Soluble DNP Amino Acids. Besides selenocysteine, the amino acids in proteins having side chains that can react with FDNB include cysteine, histidine, lysine, and tyrosine. In addition, N-terminal arginine forms a water-soluble DNP amino acid, since the guanidinium moiety does not react with FDNB. As shown in Fig. 5, reversed-phase HPLC, using isocratic elution in methanol–water, easily separates Se-DNP-selenocysteine from S-DNP-cysteine and the other DNP derivatives of amino acids having side chains that react with FDNB. DNP-arginine is not completely separated from Se-DNP-selenocysteine in this isocratic system, but could probably be resolved by use of gradient elution.

For isolation of *Se*-DNP-selenocysteine from the acid hydrolysate of ovine erythrocyte GSH peroxidase, gel filtration was used as the initial step, to separate the DNP amino acids from the bulk of the amino acids in the hydrolysate. Pure *Se*-DNP-selenocysteine was obtained following separation of the DNP amino acids in two steps on semipreparative and analytical reversed phase HPLC columns.[11] The high extinction coefficient of the DNP derivative of selenocysteine, and the ease with which it can be identified by mass spectrometry, are properties that make this method very sensitive for the identification of selenocysteine in selenoproteins, and could probably be used in the range of 1 mg of selenoprotein or less. The use of ^{14}C-labeled FDNB[12] would further enhance the sensitivity.

Comments

Although the reactions described here were done on a rather large scale, in order to obtain sufficient amounts of compounds for NMR spectrometry or other purposes, it should be emphasized that the methodology is fully suitable for handling amounts of selenium compounds in the microgram range. The derivatization of *Se*-DNP-selenocysteine for mass spectrometry is easy, and the yield of purified product after workup is very good. No special techniques are needed for introduction of sample or ionization of the derivative in the mass spectrometer. It is important to have the *Se*-DNP-selenocysteine quite pure before subjecting it to the Smiles rearrangement, otherwise intermolecular reactions may compete with the intramolecular arrangement. If the starting material is a purified selenoprotein, the purification from other amino acids is accomplished quite easily by HPLC. The method has not yet been applied to determining selenocysteine in crude animal or plant samples, but such applications may be possible.

Acknowledgments

Contribution from the College of Agricultural and Life Sciences, University of Wisconsin, Madison. Research support was also provided by the National Institutes of Health (Am 14184). We thank Dr. Walter Hargraves of the University of Wisconsin Food Research Institute and Dr. Heinrich Schnoes of the Department of Biochemistry for running the mass spectra, and Dr. H. Reich of the Department of Chemistry for obtaining the NMR spectra. S. J. Foster is a postdoctoral trainee in Environmental Toxicology and Pathology.

---

[12] Available from Amersham (Arlington Heights, IL) as the uniformly labeled compound.

[41] Oxidation States of Glutathione Peroxidase

By H. E. GANTHER and R. J. KRAUS

Selenium is a component of several animal and microbial enzymes, generally catalyzing some type of oxidation–reduction process. Selenium can exist in numerous oxidation states. Selenols (RSeH) undergo progressive two-electron oxidations to selenenic (RSeOH) and seleninic ($RSeO_2H$) acids. These can undergo reduction by thiols to form various derivatives containing selenium bonded to sulfur, such as RSe(O)SR and RSeSR.[1]

Physical chemical measurements such as X-ray photoelectron spectroscopy and ultraviolet spectroscopy should be useful in studying redox changes in the selenium moiety of selenoenzymes. However, the results of such investigations have often been conflicting, as discussed elsewhere,[1] particularly in regard to forms produced upon oxidation of glutathione peroxidase (EC 1.11.1.9). In this chapter a description is given of methods we have developed for the reproducible isolation of different oxidized forms of glutathione peroxidase which can be differentiated by their relative stability and by their reactivity with cyanide. In a companion chapter a method for identifying selenocysteine by HPLC and mass spectrometry is described.[2]

Enzyme Purification Procedure

Materials

Ammonium sulfate, 3.71 *M*. Dissolve 1961 g of ultra-pure (Schwarz/Mann) $(NH_4)_2SO_4$ in distilled, deionized water and bring the volume to 4 liters

Continuous-flow rotor. Model TZ-28 rotor, with adapters for continuous flow operation, used in a Model RC-5B refrigerated centrifuge (DuPont/Sorvall Instruments)

CM-32 (carboxymethyl)cellulose and DEAE-32-cellulose (from Whatman). Recycle before use by the manufacturer's recommended procedure (0.5 *N* NaOH, water, 0.5 *N* HCl, water, for CM-32; 0.5 *N* HCl, water, 0.5 *N* NaOH, water, for DEAE-32)

[1] H. E. Ganther and R. J. Kraus, *in* "Selenium in Biology and Medicine" (J. E. Spallholz, J. L. Martin, and H. E. Ganther, eds.), p. 54. Avi, Westport, Connecticut, 1981.

[2] H. E. Ganther, R. J. Kraus, and S. Foster, this volume [40].

BioGel HT hydroxyapatite (Bio-Rad Laboratories)
Sephadex G-150, particle size 40–120 μm (Pharmacia)

Assay

Glutathione peroxidase is assayed by a coupled method using glutathione reductase at 25°, pH 7.5, in a 2-ml volume using 1 mM GSH as described previously.[3] A more complete discussion of the principles of the method is given by Wendel.[4] After preincubation in the presence of glutathione for 10 min, the reaction is started by the addition of cumene hydroperoxide (final concentration 0.1 mM). One unit of enzyme activity equals 1 μmol of GSH oxidized per minute.

Enzyme Source

Ovine blood is the preferred enzyme source because it is richer in GSH-peroxidase than blood of other species. It is best to use sheep that have been maintained on a selenium-adequate diet for several months, since the enzyme activity in blood of animals fed only forages of low selenium content can be lower by a factor of two. Blood (about 1700 ml) is collected from the jugular vein of 6 sheep using sodium citrate (5 mg per milliliter of blood) as the anticoagulant. The blood is cooled on ice, and all subsequent operations are conducted at 4°. The blood is centrifuged at 5000 g for 20 min, and the plasma is discarded. The red cells are suspended in buffered saline (1 part of 0.1 M potassium phosphate, pH 7, plus 9 parts of 0.145 M NaCl) and centrifuged at 13,000 g for 20 min; the wash is discarded. The packed cells are hemolyzed by suspending in 5.67 volumes of water. The homoglobin concentration is determined by the Drabkin procedure[4] and adjusted to 30 mg/ml by addition of water.

Procedure

Step 1. Ammonium Sulfate Fractionation. In this step, ammonium sulfate is added to 20% saturation and the pellet is discarded; then ammonium sulfate is added to 55% saturation and the pellet is collected. Rapid processing through this step by use of a continuous-flow rotor results in better recovery of enzyme activity. Starting with hemolysate (30 mg of hemoglobin per milliliter), the volume of hemolysate is divided by 4 to determine the volume of 3.71 M ammonium sulfate needed for 20% saturation. This volume is added slowly with stirring and left stirring over-

[3] R. J. Kraus, J. R. Prohaska, and H. E. Ganther, *Biochim. Biophys. Acta* **615,** 19 (1980).
[4] A. Wendel, this series, Vol. 77, p. 325.

night. The mixture is pumped through a continuous-flow rotor at 80 ml/ min at 30,000 g, and the supernatant is collected at 4°. The volume of supernatant is multiplied by 7/9 to determine the volume of 3.71 M ammonium sulfate to add to reach 55% saturation. This is added as before; after overnight stirring, the pellet is collected using the continuous-flow rotor. About 95% of the hemoglobin remains in the supernatant fluid, as well as the endogenous glutathione. From this point on, glutathione (GSH) is added to stabilize the enzyme. Using an Oxford multiple dialyzer, the pellet is dialyzed extensively in two stages: (1) In 16 liters of pH 7 phosphate buffer (10 mM potassium phosphate, 0.2 mM GSH, and 1 μM EDTA) for 24 hr; (2) in 16 liters of pH 6.3 phosphate buffer (10 mM potassium phosphate, 0.5 mM GSH, and 1 μM EDTA) for 24 hr. Extensive dialysis is needed to get rid of ammonium sulfate prior to ion-exchange chromatography, but excessively long standing at pH 6.3 may cause loss of activity.

Step 2. (Carboxymethyl)cellulose. A column (5 × 10 cm) is prepared using Whatman CM-32 cellulose, and the column is equilibrated with pH 6.3 phosphate buffer (10 mM potassium phosphate, 0.5 mM GSH, 1 μM EDTA). The dialyzed enzyme is applied to the column and eluted at 60 ml/hr with the pH 6.3 phosphate buffer; glutathione peroxidase comes straight through, leaving hemoglobin bound to the column. The active fractions are pooled and dialyzed against 4 liters of pH 7.5 phosphate buffer (10 mM potassium phosphate, 0.5 mM GSH, 1 μM EDTA), then against 16 liters of 5 mM potassium phosphate, pH 7.5, containing 0.5 mM GSH and 1 μM EDTA.

Step 3. DEAE-Cellulose. A 5 × 25 cm column is prepared using Whatman DEAE-32, and the column is equilibrated with starting buffer (5 mM potassium phosphate, pH 7.5, containing 0.5 mM GSH and 1 μM EDTA). The enzyme is applied, and the column is washed with 200 ml of starting buffer. A linear gradient is begun using 600 ml of starting buffer and 600 ml of starting buffer containing 0.1 M KCl. Fractions of 10 ml are collected at 60 ml/hr. GSH-peroxidase is eluted ahead of a large peak of contaminating protein.

Step 4. Sephadex G-150. A 5 × 90 cm column is prepared and equilibrated in pH 6.9 phosphate buffer (10 mM, containing 0.5 mM GSH, and 1 μM EDTA). The enzyme eluted from the DEAE column (volume about 300 ml) is applied directly and eluted with starting buffer at 42 ml/hr. Fractions of 8.5 ml are collected.

Step 5. Hydroxyapatite. A 2.5 × 6 cm column is prepared and the column is washed with 0.3 M phosphate buffer (pH 6.9), then equilibrated with starting buffer (10 mM phosphate buffer, pH 6.9, containing 0.5 mM GSH and 1 μM EDTA). The enzyme eluted from the Sephadex G-150 column (about 400 ml) is applied directly, then eluted with a linear gradient of 100 ml starting buffer and 100 ml of 200 mM potassium phosphate (pH 6.9, containing 0.5 mM GSH and 1 μM EDTA) at a flow rate of 50 ml/hr, collecting 2.7-ml fractions. GSH-peroxidase is eluted after a small peak of contaminating protein. The active fractions are pooled and dialyzed against 2 liters of 50 mM potassium phosphate (pH 7.2) containing 5% ethanol. Final dialysis is in 2 liters of 50 mM potassium phosphate (pH 7.2) containing 10% ethanol. The enzyme is stored in this buffer at 4°.

Comments

The above procedure is convenient for isolating substantial amounts of GSH-peroxidase from relatively small volumes of blood (1–2 liters), using conventional isolation procedures. Removal of hemoglobin is accomplished without the use of organic solvents. The sequence of operations is designed to eliminate the need for concentration of the enzyme by ultrafiltration, a procedure that may cause extensive loss of activity. In the final stages of chromatography, the inclusion of 10% ethanol stabilizes the enzyme when it is in an oxidized form. The use of a continuous-flow rotor in the ammonium sulfate fractionation step has made it possible to obtain highly active enzyme with yields in the range of 60% or more (Table I), well above the yields obtained previously.[3] A second Sephadex G-150 chromatography following the hydroxyapatite chromatography (step 5) gives a slightly purer form of enzyme.[3] The ratio of enzyme activity to Se content attains a maximum early in the purification and remains constant (Table I). This ratio provides a means of detecting con-

TABLE I

PURIFICATION OF GLUTATHIONE PEROXIDASE FROM OVINE ERYTHROCYTES

			Specific activity			
Step	Total protein (mg)	Total enzyme (units)	Units per mg protein	Units per μg Se	Purification (fold)	Enzyme yield (%)
Hemolysate	264,500	50,600	0.19	340	1	100
1. Ammonium sulfate	6000	40,193	6.7	475	35	79.4
2. CM-cellulose	2438	37,620	15.4	482	81	74.3
3. DEAE-cellulose	57.8	35,369	612	470	3221	69.9
4. Sephadex G-150	44.8	35,004	781	480	4110	69.1
5. Hydroxyapatite	22.4	31,511	1407	470	7405	62.3

version of the enzyme to a form containing selenium in nonfunctional forms (such as elemental selenium) in the course of purification.

Preparation of Enzyme in Various Oxidation States[3]

Reduced GSH-peroxidase. GSH-peroxidase in buffer A (potassium phosphate, 50 mM, pH 7.2, containing 10% ethanol) is reduced with 0.20 mM GSH (molar ratio of GSH/Se = 4) for 20 min at 25°. Three different oxidized forms can be prepared from the reduced enzyme.

Oxidized Form A. Reduced GSH-peroxidase is immediately chromatographed at 4° on a 2.5 × 45 cm column of Sephadex G-150 in buffer A, 30 ml/hr, gas phase air. Chromatography and collection of fractions must be done at 4° because the enzyme is eluted in a thermal-labile form.

Oxidized Form B. Reduced GSH-peroxidase in buffer A is treated with a 50 : 1 molar ratio of H_2O_2 to Se for 20 min at 25°. If it is desired to remove the reagents, the enzyme can be chromatographed on a 2.5 × 45 cm column of Sephadex G-150 in buffer A at 4°.

Oxidized Form C. Reduced GSH-peroxidase in buffer A is dialyzed for 3 days at 4° against buffer A (three changes), then chromatographed on a 2.5 × 45 cm column of Sephadex G-150 in buffer A at 4°.

Differentiation of Various Oxidation States

Principle

Inhibition by iodoacetate of cyanide under mild conditions, along with temperature sensitivity, can be used to differentiate reduced GSH-peroxidase and several oxidized forms.[3] Cyanide sensitivity is highest for an oxidized enzyme containing glutathione (see Table II).

Reagents

Iodoacetate, 10 mM. Dissolve 0.0208 g of the sodium salt of iodoacetic acid (Sigma) and make to 10 ml with water.

Potassium cyanide, 100 mM. *Prepare this reagent in a hood.* Dissolve 0.0651 g of KCN in approximately 5 ml of water. Adjust the pH to 7.5 by the careful addition of 1 M HCl and make to 10 ml with water. Store at 4° in a closed container.

Procedure

Reduced GSH-peroxidase. Samples of stock enzyme solution are diluted to a final concentration of approximately 4 × 10^{-7} M Se in 20 mM HEPES buffer, pH 7.5, containing bovine serum albumin (50 μg/ml), and

TABLE II
SUMMARY OF VARIOUS FORMS OF GSH-PEROXIDASE

Designated form	Preparation	Stability at 25°	Effect with		Glutathione in enzyme
			Iodoacetate	Potassium cyanide	
Reduced	Reduce with GSH	Good	Inactivated	No effect	—
Oxidized A	Reduce with GSH, immediately gel filter	Poor	No effect[a]	Slight effect[b]	No
Oxidized B	Reduce with GSH, oxidize with H_2O_2	Good	No effect	Moderate effect	No
Oxidized C	Reduce with GSH, dialyze, then gel filter	Good	No effect	Rapid inactivation	Yes

[a] No difference between rate of loss with sodium iodoacetate as compared to sodium acetate control.
[b] Slightly greater rate of inactivation with KCN compared to KCl control.

incubated at 25° with 1 mM sodium iodoacetate or 1 mM sodium acetate (control) for 20 min. Small aliquots (10 μl) are taken at zero time and at 20 min and assayed for GSH-peroxidase activity (final volume = 2 ml). The reduced oxidation state of the enzyme is defined as the loss of 90% or more of the activity within 20 min, with little or no loss of activity in the control.

Oxidized Forms of GSH-peroxidase. Samples of enzyme (final concentration approximately 4×10^{-7} M Se) are added to 20 mM HEPES buffer, pH 7.5, containing 10 mM KCN (pH 7.5) or 10 mM KCl (control) and incubated in sealed tubes at 25° for up to 4 hr. Small aliquots (10 μl) are taken at zero time and at intervals of 20 min and assayed for GSH-peroxidase activity (final volume 2 ml).

Form A: Oxidized Form A is characterized by its thermal instability. This form of enzyme is stable at 15° or below, but it loses 50% or more of its activity when incubated for 1–2 hr at 25°. The rate of loss of activity for the sample incubated with KCN is only slightly greater than for the control incubated with KCl, in contrast to oxidized forms B and C (see below).

Form B: When GSH-peroxidase is reduced with glutathione and then oxidized with an excess of H_2O_2, iodoacetate sensitivity is lost and a fairly stable form of enzyme is produced; this form undergoes progressive inactivation when incubated with 10 mM KCN but not with KCl. However, the rate of loss with KCN is considerably slower than for oxidized Form C.

Form C: A fairly rapid loss of activity (80% or more within 60–120 min) in the presence of KCN with minimal loss of activity (about 10%) in the presence of KCl (control) is characteristic of Form C (cyanide sensitive). This sensitivity is correlated with the presence of a tightly bound glutathione moiety in a 1:1 molar ratio to enzyme Se.[3] Glutathione is probably present as the selenenyl sulfide (E-Se-SG); treatment with cyanide releases the glutathione moiety in the form of S-cyanoglutathione and converts the enzyme to a reduced (iodoacetate-sensitive) form.[5] Oxidized Forms A and B do not contain glutathione.[3]

Other Substances Tested. Under the same conditions in which GSH-peroxidase (Form C) is inhibited by 10 mM KCN at pH 7.5, little or no inhibition occurred with 10 mM KOCN, KSCN, KSeCN, KBr, KI, NaN$_3$, and β-aminopropionitrile.[6] Hydroxylamine (10 mM) does cause loss of activity, apparently through a mechanism of action similar to cyanide (see below), involving reduction of the enzyme selenium (E. Gartzke and H. Ganther, 1977, unpublished observation).

Mechanism For Inactivation by Cyanide.[5] Under the mild conditions used in the present procedure, cyanide inactivation of oxidized Form C does not cause loss of selenium from the enzyme. Inactivation was shown to involve the formation of reduced enzyme and GSCN, consistent with the cleavage of a selenenyl sulfide moiety.

$$E\text{—}Se\text{—}SG + CN^- \rightarrow E\text{—}Se^- + GSCN \tag{1}$$

The reduced enzyme undergoes spontaneous oxidation to Form A, which rapidly loses activity upon further incubation at 25°. The nature of the selenium in Form A is not established, but it may be a selenenic acid formed by air oxidation of a selenol in the reduced enzyme

$$E\text{—}Se^- \xrightarrow{O_2} E\text{—}SeOH \tag{2}$$

As an alternative to Eq. (1), it is possible for a selenenyl sulfide to undergo a reaction with cyanide in which selenium is converted to the selenocyanate

$$E\text{—}SeSG + CN^- \rightarrow E\text{—}SeCN + GS^- \tag{3}$$

This pathway would not account for the irreversible inactivation of the enzyme by cyanide, however, because the preincubation with glutathione

[5] R. J. Kraus and H. E. Ganther, *Biochem. Biophys. Res. Commun.* **96**, 1116 (1980).

[6] J. R. Prohaska, S.-H. Oh, W. G. Hoekstra, and H. E. Ganther, *Biochem. Biophys. Res. Commun.* **74**, 64 (1977).

prior to enzyme assay would be expected to convert the enzyme back to the active, reduced form:

$$\text{E—SeCN} + \text{GS}^- \rightarrow \text{E—Se}^- + \text{GSCN} \tag{4}$$

Under alkaline conditions it is possible for cyanide to inactivate GSH-peroxidase by a mechanism involving loss of Se from the enzyme. In one study, treatment of [^{75}Se]GSH-peroxidase with unneutralized KCN released 92% of the ^{75}Se in a low molecular weight form.[6] Doubly labeled selenocyanate was shown[7] to be one of the products released from [^{75}Se]GSH-peroxidase treated with K^{14}CN. Soaking crystalline GSH-peroxidase with 50 mM KCN at pH 8.2 was successfully employed to prepare the apoenzyme form of GSH-peroxidase for crystallographic studies.[8] It is likely that the loss of Se under basic conditions involves β-elimination of SeCN$^-$ from the selenocysteine moiety of the enzyme:

$$\text{E—SeCn} + \text{OH}^- \rightarrow \text{apoenzyme} + \text{SeCN}^- \tag{5}$$

Dehydroalanine would be expected to be formed in the apoenzyme after the elimination of SeCN$^-$, but this has not been verified. It should be noted that E—SeCN could be formed by the reaction of cyanide with either a selenenyl sulfide [Eq. (3)] or with a selenenic acid form of the enzyme:

$$\text{E—SeOH} + \text{CN}^- \rightarrow \text{E—SeCN} + \text{OH}^- \tag{6}$$

An analogous reaction of sulfenic acid derivatives of sulfhydryl enzymes such as papain has been reported to occur.[9] Still another possible mechanism of inactivation would involve cyclization of E—SeCN with cleavage of the polypeptide chain, analogous to cleavage of proteins containing the S-cyanocysteine moiety.[10]

Comments

When isolated by our procedure, GSH-peroxidase is in oxidized Form C. After storage at 4° for 4–8 weeks in buffer containing 10% ethanol, it loses about half of its original activity. Stored enzyme becomes relatively less sensitive to cyanide and more sensitive to thermal inactivation,[1] suggesting that the enzyme has released bound glutathione and been converted to Forms A and B. Whether the stabilization by storage in 10%

[7] H. E. Ganther, J. R. Prohaska, S.-H. Oh, and W. G. Hoekstra, *Fed. Proc. Fed. Am. Soc. Exp. Biol.* **36**, 1094 (1977).

[8] R. Ladenstein, O. Epp, K. Bartels, A. Jones, R. Huber, and A. Wendel, *J. Mol. Biol.* **134**, 199 (1979).

[9] W. S. Allison, *Acc. Chem. Res.* **9**, 293 (1976).

[10] Y. Degani and A. Patchornik, *Biochemistry* **13**, 1 (1974).

ethanol has anything to do with possible esterification of the selenium is not known.

Our working hypothesis for the redox reactions of selenium in GSH-peroxidase is consistent with the current knowledge of the enzyme; E—SeH represents the selenocysteine residue of the enzyme.

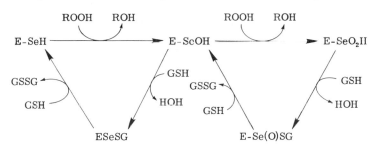

It is known that GSH-peroxidase reduced with GSH substrate is alkylated by iodoacetate to form Se-carboxymethylselenocysteine.[11] Similarly, enzyme reduced with borohydride reacts with FDNB to form Se-DNP-selenocysteine, and the formation of Se-DNP-selenocysteine has been confirmed using mass spectrometry.[2] These results indicate that the Se in reduced GSH-peroxidase is at the oxidation state of a selenol when an excess of thiol or other reducing agent is present. Spectral studies regarding the presence of a selenol in GSH-peroxidase indicate that a more complex situation exists compared to the case for some other selenoproteins.[1,12,13]

Upon oxidation with substrate peroxides, Se may be oxidized to a selenenic acid (E—SeOH) or a seleninic acid (E—SeO₂H), depending on the concentration of peroxide. These may correspond to Form A and Form B, respectively. Selenenic acids are known to be quite unstable at room temperature compared to seleninic acids, and Form A is less stable than Form B. It is possible that spontaneous oxidation of reduced enzyme by oxygen gives Form A (when GSH is absent) or Form C (when GSH is present), whereas Form B is produced by oxidation with peroxides. The presence or the absence of glutathione when spontaneous oxidation occurs determines whether Form C or Form A is produced. Form C contains glutathione, and its ease of reaction with cyanide is consistent with the enzyme being in a form such as ESeSG or ESe(O)SG.

[11] J. W. Forstrom, J. J. Zakowski, and A. L. Tappel, *Biochemistry* **17**, 2639 (1978).

[12] J. E. Cone, R. Martin del Río, J. N. Davis, and T. C. Stadtman, *Proc. Natl. Acad. Sci. U.S.A.* **73**, 2659 (1976).

[13] R. Ladenstein, O. Epp, R. Huber, and A. Wendel, *in* "Selenium in Biology and Medicine" (J. E. Spallholz, J. L. Martin, and H. E. Ganther, eds.), p. 33. Avi, Westport, Connecticut, 1981.

Instrumental studies of oxidation states of Se in GSH-peroxidase have not been particularly successful and are often in disagreement, as reviewed elsewhere.[1] Photoelectron spectroscopy studies by Wendel's group led to the conclusion that, when reduced enzyme was oxidized with peroxide, a form of the enzyme containing oxygen bonded to Se was produced, whereas Tappel's group concluded that selenium was not bonded to oxygen.[1] In view of the multiplicity of oxidized forms of GSH-peroxidase that can be formed, such problems can be expected. The chemical tests described here can help in differentiating various oxidized forms and should be used in conjunction with any instrumental study to aid in interpretation of the results. Also, it is believed that discrepancies in regard to the content of cysteine[13] and other amino acids could certainly have their origin in the analysis of different forms of the enzyme, containing bound glutathione in some oxidized states but not in others.

Acknowledgment

Contribution from the College of Agricultural and Life Sciences, University of Wisconsin-Madison, Wisconsin. Research support was also provided to the authors by the National Institutes of Health (AM14184).

[42] Selenocysteine-Containing Proteins and Glutathione Peroxidase

By AL L. TAPPEL, WAYNE C. HAWKES, ERIC C. WILHELMSEN, and MARVIN A. MOTSENBOCKER

Although selenium in biological systems has long been known to be associated with protein, it has only recently been shown[1,2] that almost all the selenium in the rat is present as selenocysteine in a limited number of discrete proteins. It seems clear that the biological effects of selenium must be mediated through these selenoproteins. Since only some of these selenoproteins have glutathione peroxidase activity, the others presumably have different biological or enzymatic activities. The following sections present the methods currently used in this laboratory to study glu-

---

[1] E. C. Wilhelmsen, W. C. Hawkes, M. A. Motsenbocker, and A. L. Tappel, in "Selenium in Biology and Medicine" (J. E. Spallholz, J. L. Martin, and H. E. Ganther, eds.), p. 535. Avi, Westport, Connecticut, 1981.

[2] W. C. Hawkes, E. C. Wilhelmsen, and A. L. Tappel, Fed Proc. Fed. Am. Soc. Exp. Biol. **42**, 928 (1983).

tathione peroxidase and other selenoproteins of rat tissues. The principal techniques described are the *in vivo* labeling of selenoproteins with ^{75}Se and the chromatographic fractionation of either native or denatured selenoproteins from various tissues for identification by elution pattern and apparent molecular weight. Also described is the use of tryptic digests of selenoproteins to obtain selenopeptides that allow further definition of the various selenoproteins.

Isotopic Equilibration

Short-term labeling of proteins with ^{75}Se has been a very useful technique for identification of mammalian selenoproteins.[3-7] However, short-term labeling has the important drawback that the tissue ^{75}Se radioactivity is not necessarily proportional to the actual selenium concentration. By equilibrating rat tissues *in vivo* to a constant specific activity of ^{75}Se, this shortcoming is overcome. Isotopic equilibration allows ^{75}Se radioactivity to be used as a specific elemental assay that is subject to virtually no interference from other elements. The sensitivity is limited only by the specific activity of the ^{75}Se added to the rat's drinking water as the sole source of selenium.

Method for Isotopic Equilibration

[^{75}Se]Selenious acid is mixed with unlabeled sodium selenite to a specific activity of 50 mCi of ^{75}Se per millimole and is stored in 0.5 M HCl. The radiolabeled selenite is diluted fresh each week in distilled water to a concentration of 0.2 ppm Se and is supplied as drinking water for the duration of the equilibration period (10–20 weeks). Weanling rats (21 days old, 40–60 g) are maintained on a Se-deficient, *Torula* yeast-based diet.[8] The diet and ^{75}Se-containing drinking water are supplied *ad libitum*. The selenium concentration in tissue and protein samples obtained from isotopically equilibrated rats is measured by counting the ^{75}Se in a gamma counter and calculating the ratio of measured counts to counts of a ^{75}Se standard prepared from the same ^{75}Se solution used to label the drinking water. Since all of the ^{75}Se decays at the same rate, the ratio cpm sample : cpm standard is always proportional to the amount of selenium

[3] K. P. McConnell and D. M. Roth, *Biochim. Biophys. Acta* **62,** 503 (1962).

[4] K. R. Millar, *N. Z. J. Agric. Res.* **15,** 547 (1972).

[5] J. T. Rotruck, A. L. Pope, H. E. Ganther, A. B. Swanson, D. G. Hafeman, and W. G. Hoekstra, *Science* **179,** 588 (1973).

[6] R. F. Burk and P. E. Gregory, *Arch. Biochem. Biophys.* **213,** 73 (1982).

[7] M. A. Motsenbocker and A. L. Tappel, *Biochim. Biophys. Acta* **704,** 253 (1982).

[8] C. J. Dillard, R. E. Litov, and A. L. Tappel, *Lipids* **13,** 396 (1978).

present. That the tissues of the rats become essentially equilibrated with ^{75}Se can be seen from the weight gain over the equilibration period (50 g to >250 g) and from estimates of the biological half-life of selenium in the rat of 20 days[9,10] and 56 days.[11]

Comments

With currently available [^{75}Se]selenite (35 Ci/mmol), the maximum sensitivity of this method is 40 ± 4 fmol of selenium. A specific activity of 50 mCi/mmol gives an acceptable compromise between sensitivity and safety; however, some analytical procedures require use of higher specific activity ^{75}Se; e.g., gel electrophoresis and HPLC.

Selenocysteine Analysis

While the standard techniques for carboxymethylation of cysteine in protein can be applied to selenocysteine, the yields of carboxymethylselenocysteine after acid hydrolysis can be quite low owing either to oxidation of selenocysteine or insolubility of the selenoproteins. Since selenium is much more susceptible to dialkylation by iodoacetate than is sulfur, the extent of derivatization cannot be improved by raising the iodoacetate concentration. In fact, selenocysteine is dialkylated at concentrations of iodoacetate normally employed for the preparation of carboxymethylcysteine. This is not a problem with soluble forms of glutathione peroxidase, and complete recovery of carboxymethylselenocysteine can be routinely achieved by derivatizing the reduced form of the enzyme (prepared by incubating the enzyme for 5 min at 37° with 0.25 mM glutathione or 10 mM mercaptoethanol) with a 5–10 mM excess of sodium iodoacetate (over added thiol) at 37° until the enzyme activity is more than 98% inhibited (2–10 min). The procedure given below was developed as a general selenocysteine assay that is affected by neither the sample composition nor the solubilities of the selenoproteins, yet it gives acceptable yields of carboxymethylselenocysteine.

Quantitative Analysis of Selenocysteine

[75]Se-labeled crude homogenates (up to 5 ml of a 1 : 2 homogenate) or protein fractions are mixed with 0.6 g of guanidinium-HCl per milliliter

[9] M. Richold, M. F. Robinson, and R. D. H. Stewart, *Br. J. Nutr.* **38**, 19 (1977).
[10] R. F. Burk, D. G. Brown, R. J. Seely, and C. C. Scaief, *J. Nutr.* **102**, 1049 (1972).
[11] D. G. Brown and R. F. Burk, *J. Nutr.* **102**, 102 (1972).

and one-half volume of 6 M guanidinium-HCl, 0.15 M Tris-HCl, pH 8.0, 15 mM glutathione, and 3 mM dithioerythritol and are dissolved by sonication for 5 min at room temperature. Sodium iodoacetate is added to a concentration of 10 mM (3 mM excess iodoacetate over added thiol), and the samples are incubated for 10 min at 37° before addition of potassium borohydride to a concentration of 2 mM and an additional 10-min incubation at 37°. The samples are then dialyzed against three changes of 100 volumes of distilled water, mixed with 1 volume of 12 M HCl and 0.4 μCi of [³H]carboxymethylselenocysteine prepared by the following procedure. Solid selenocystine is reduced to selenocysteine by adding potassium borohydride (2 mg/ml, pH 8) dropwise until the solids are dissolved. A 1.1 M excess of sodium iodoacetate mixed with [³H]iodoacetic acid is then added, and the reaction is continued for 60 min at room temperature. The excess ³H and reagents are removed by ion-exchange chromatography on Dowex AG-50 and AG-1. The reaction mixture is adjusted to pH 1 with concentrated HCl and applied to a 5-ml column of AG-50X8 (H⁺ form in 0.1 M HCl). The column is washed with 10 ml of 0.1 M HCl and 20 ml of water and then eluted with 20 ml of 2 M NH₄OH directly onto a 5-ml column of AG-1X8 (OH⁻ form in water). This column is washed with 20 ml of water and eluted with 20 ml of 2 M acetic acid. The eluent is evaporated to dryness, dissolved in 0.01 M HCl, and stored at −15°. The [³H]carboxymethylselenocysteine obtained in this manner contains >99% of the ³H in a single peak on amino acid analysis.

The [³H]carboxymethylselenocysteine-containing samples in 6 M HCl are placed in hydrolysis tubes, then flushed with nitrogen, evacuated, sealed, and hydrolyzed for 24 hr at 110°. The hydrolysates are evaporated to dryness, dissolved in 0.15 M lithium citrate, adjusted to pH 2.2 with HCl, filtered through a 0.45-μm Millipore filter, and then applied to a Beckman Model 120B amino acid analyzer. ³H and ⁷⁵Se are measured by liquid scintillation and gamma counting of 2-min fractions collected at the column outlet. The ⁷⁵Se at the elution time of carboxymethylselenocysteine (78 min) is integrated and divided by the fractional recovery of [³H]carboxymethylselenocysteine to yield the total ⁷⁵Se as selenocysteine in the samples.

Comments

This method is applicable to all tissue and protein sample types and has been used successfully with crude tissue samples from muscle, blood, liver, testes, epididymis, kidney, lung, and heart.[2] Many amino acid analyzers will not accept sample sizes as large as those stated above; however, even larger sample sizes may be accommodated by prepurifying the

[³H]carboxymethyl[⁷⁵Se]selenocysteine, as described above, after hydrolysis of the protein. This selenocysteine analysis can also be applied to unlabeled selenium samples by first derivatizing them with [³H]iodoacetic acid and then hydrolyzing them in the presence of carboxymethyl[⁷⁵Se]selenocysteine recovery marker. Since work using this technique has shown that >80% of the total selenium in the rat is in selenocysteine in protein,[2] this analysis is essentially a total selenoprotein assay.

Chromatographic Identification of Selenoproteins

Although all the selenocysteine-containing proteins in the rat presumably have an important biological function, the only mammalian selenoprotein that has an assay for its activity is glutathione peroxidase. At the present state of knowledge, the only way to assay other selenoproteins is to measure the selenium content of separated protein fractions. Since it is very convenient to follow selenium with ⁷⁵Se, the methods described below were developed to allow the simultaneous assay of several ⁷⁵Se-labeled selenoproteins from most types of mammalian tissues.[2]

Sephacryl S-300 Chromatography in Sodium Dodecyl Sulfate and Urea

Methods. Gel filtration of ⁷⁵Se-labeled sodium dodecyl sulfate (SDS) protein subunits is performed on a 1.5×90 cm column of Sephacryl S-300 in 0.2 M Tris-phosphate, pH 6.9, 7 M urea, and 5 mM SDS at a flow rate of 0.15 ml/min. Each 0.5-ml sample of crude homogenate or protein fraction is mixed with 0.5 g of urea, 0.24 ml of SDS at 180 mg/ml, 0.25 ml of mercaptoethanol, 25 mg of dithioerythritol, and 80 μl of 25% triethanolamine (v/v) and heated in boiling water for 10 min. Blue Dextran and cresol red dyes are added to each sample as internal markers for the column void and total volumes, respectively. The column eluent is collected in 1-ml fractions. The ⁷⁵Se peaks are characterized by the apparent partition coefficients of the peak ⁷⁵Se fractions. The apparent partition coefficient is defined as $K_a = (V_e - V_0)/(V_T - V_0)$, where V_e is the peak ⁷⁵Se fraction, V_0 is the peak fraction of Blue Dextran, and V_T is the peak fraction of cresol red.

A column prepared as described above was calibrated with molecular weight standards of phosphorylase (92,500), bovine serum albumin (66,200), ovalbumin (45,000), carbonic anhydrase (31,000), and lysozyme (14,400). Regression of the logarithm of the molecular weights against the apparent partition coefficients gave an $r^2 = 0.998$. The contribution from each subunit size to the selenium in the chromatograms is estimated by

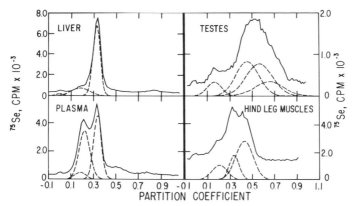

Fig. 1. Gel filtration of selenoprotein subunits on Sephacryl S-300 in sodium dodecyl sulfate (SDS) and urea. Samples of isotopically equilibrated rat tissues were solubilized in SDS, urea, mercaptoethanol, and dithioerythritol as described in the text. Chromatography was performed on a 1.5 × 90 cm column of Sephacryl S-300 in 0.2 M Tris-phosphate, pH 6.9, 7 M urea, 5 mM SDS. The apparent partition coefficients were defined as $K_a = (V_e - V_0)/(V_T - V_0)$, where V_e = the fraction number, V_0 = fraction number of the Blue Dextran peak, and V_T = fraction number of the cresol red peak. The dashed lines show the profiles of the individual selenoprotein subunits as estimated by the nonlinear least-squares procedure (see text).

summing the ^{75}Se in the fractions of those peaks that are completely separated and by nonlinear least-squares fitting of the complex peaks. The data are fit by a nonlinear least-squares method[12,13] by computer to a function that expresses a sum of gaussians centered at the appropriate partition coefficients. The gaussian equations used in the least-squares procedure have the means fixed and the peak shapes are set to be symmetrical while the height and width are varied to obtain the best fit to the data. The iterative least-squares procedure is continued until the solution converges with a standard deviation less than the background counts of ^{75}Se. The height and width of each component gaussian are then used to calculate the contribution from each subunit size to the ^{75}Se in the chromatogram (see Fig. 1).

Comments. If the protein concentration in the sample exceeds 60 mg/ml, then solid SDS should be added at a concentration of 1.4 mg of SDS per milligram of protein (over 60 mg/ml) before the sample is heated. The only practical limits to the amount of protein that can be analyzed by this method are the sample volume (<2% of the column volume) and the

[12] D. W. Marquardt, *J. Soc. Ind. Appl. Math.* **11**, 431 (1963).
[13] T. Tabata and R. Ito, *Comput. J.* **18**, 250 (1975).

viscosity of the sample. The sample viscosity can be significantly de-creased by passing the sample over a small column of Sephadex G-25 in the buffer used to equilibrate the Sephacryl S-300 column. All of the ^{75}Se peaks observed over the course of several experiments were separated into eight groups of partition coefficients.[2] A one-way analysis of variance and pairwise t tests showed that all the groups were significantly different at $p < 0.01$. The selenoprotein subunit molecular weights were >89,200 ($K_a = 0.01$), 46,400 ± 2300 ($K_a = 0.185$), 36,300 ± 1700 ($K_a = 0.245$), 26,500 ± 1100 ($K_a = 0.320$), 20,100 ± 800 ($K_a = 0.388$), 14,700 ± 900 ($K_a = 0.449$), 9780 ± 590 ($K_a = 0.565$), and 6790 ± 610 ($K_a = 0.650$) (see Fig. 1). Glutathione peroxidase was found to have a M_r 26,500 subunit.[2] These selenoprotein subunit sizes match well the previously reported M_r 45,000 selenoprotein subunit in plasma,[7] the M_r 17,000 subunit in sperm,[14] and the recently estimated rat liver glutathione peroxidase native molecu-lar weight of 105,000.[15]

DEAE-Sephacel Chromatography in Triton X-100 and Urea

Methods. Each 0.5-ml sample of ^{75}Se-labeled crude homogenate or protein fraction is mixed with 1 ml of column equilibration buffer (25 mM imidazole-HCl, pH 7.8, 7 M urea, and 0.1% Triton X-100), 0.2 ml of Triton X-100, 25 mg of dithioerythritol, and 0.75 g of urea and then heated with shaking for 15 min at 50°. The samples are centrifuged for 10 min in a clinical centrifuge, and the pellets obtained are suspended twice in 1 ml of equilibration buffer and recentrifuged. The supernatants are combined and adjusted to pH 7.8 with 1 M imidazole before they are applied to a 1.5 × 30 cm column of DEAE-Sephacel. The column is washed with 60 ml of equilibration buffer and eluted with a linear gradient of 0 to 0.5 M NaCl in 300 ml of the column equilibration buffer. Fractions (3 ml) are collected and counted for ^{75}Se, and the conductivities of the peak ^{75}Se fractions in the gradient are measured. The top 2 cm of the column is removed and counted for ^{75}Se. Each column is used only once, and the used gel is washed in 0.1 M HCl plus 1 M NaCl before reuse.

Comments. The selenium-containing proteins in the rat were sepa-rated into nine groups based on the elution positions and the conductivi-ties of the peak ^{75}Se fractions. These groups were designated as charge groups A through I. This designation as charge groups was used for con-venience only and is not meant to imply any specific knowledge of the actual net charge on these selenoproteins. The approximate elution vol-umes and NaCl concentrations of the selenoprotein charge groups were as

[14] H. A. Calvin, *J. Exp. Zool.* **204**, 445 (1978).
[15] D. E. Lyons, E. C. Wilhelmsen, and A. L. Tappel, *J. Liq. Chromatogr.* **4**, 2063 (1981).

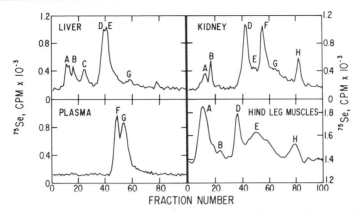

FIG. 2. Ion-exchange chromatography of selenoproteins. Samples of isotopically equilibrated rat tissues were prepared and analyzed on 1.5 × 30 cm columns of DEAE-Sephacel in 0.1% Triton X-100 and 7 M urea as described in the text. A gradient of 0 to 0.5 M NaCl in 300 ml of column buffer was started at fraction No. 22. Fractions (3 ml) were collected and counted for ^{75}Se. The selenoprotein charge groups A through H are described in the text.

follows: A, 42 ml; B, 54 ml; C, 65 ml; D (glutathione peroxidase), 102 ml (0.02 M); E, 120 ml (0.05 M); F, 156 ml (0.16 M); G, 195 ml (0.18 M); H, 240 ml (0.25 M); and I, which was characterized by irreversible adsorption to the column top (Fig. 2). The positions of the ^{75}Se peaks typically vary from these values by ca. ±10%. Since the ionic composition of the sample can significantly alter the elution of the earliest selenoprotein peaks, it is sometimes necessary to dialyze the samples against the column equilibration buffer before chromatography in order to obtain reproducible chromatograms.

Both of these chromatographic techniques can be applied to larger samples by increasing the sample size and reagents in the solubilization mixture proportionally and by dialyzing them against an appropriate dilution of the column equilibration buffer. The samples are then evaporated in the dialysis bags until the buffer is concentrated to approximately the original concentration.

Glutathione Peroxidase

Glutathione peroxidase is the best characterized rat selenoprotein. The purification and assay procedures were previously summarized in this series.[16] However, there have been several significant advances and improvements in methodology.

[16] A. L. Tappel, this series, Vol. 52, p. 506.

Purification

The purification of glutathione peroxidase previously reported in this series[16] has been improved in terms of yield and specific activity.[15] For a typical purification as outlined in Table I, three 350–400 g male Sprague-Dawley rats are each injected with 50 μCi of [^{75}Se]selenite. After 3 days, these rats are fasted overnight, anesthetized with ether, and exsanguinated, and the livers are removed. Twelve noninjected rats are similarly treated. All remaining steps are carried out at 4° unless otherwise stated. The 15 livers are washed and minced in 0.25 M sucrose. The minced liver is homogenized 1 : 4 (w/v) in 0.25 M sucrose, 11 mM glutathione for 30 sec in a Waring blender. The homogenate is centrifuged at 13,000 g for 30 min. The supernatant is adjusted to pH 7.6 with 100 mM Tris base, heated to 50° for 45 min, and cooled in ice. The cooled sample is made 52% acetone and held at −15° overnight. The acetone precipitate collected by centrifugation at 13,000 g for 20 min is resuspended in 10 mM Tris-HCl, 0.1 mM EDTA, 5 mM glutathione, pH 7.6, and the insoluble material is removed by centrifugation at 27,000 g for 30 min. The supernatant is chromatographed on a Sephadex G-100 column (5 × 100 cm) equilibrated with 10 mM Tris-HCl, 0.1 mM EDTA, pH 7.6 (buffer A), and the fractions that contain the ^{75}Se are pooled. The pooled fractions are made 7.1 mM in 2-mercaptoethanol and held for 20 min on ice. For the DEAE-

TABLE I
PURIFICATION OF GLUTATHIONE PEROXIDASE[a]

Fraction	Total protein (mg)	Total activity (units × 10^{-3})[b]	Specific activity[c]	Yield (%)	Purification (fold)
Homogenate	17,000	5.35	0.32	100	1
Supernatant, 13,000 g	6853	5.31	0.78	99	2.4
Heat-treated, 50°	5966	4.78	0.80	90	2.5
Acetone (52%) precipitate	3075	4.40	1.4	82	4.4
Sephadex G-100	752	3.51	4.7	66	15
DEAE-titration	147	3.44	23	64	72
DEAE-Sephacel	54	2.87	54	54	170
TSK-HPLC[d]	2.0	2.24	1100	42	3560

[a] Lyons et al.[15]
[b] Micromoles of NADPH oxidized per minute.
[c] Micromoles of NADPH oxidized per minute per milligram of protein.
[d] Following the third chromatography.

titration step shown in Table I, 10-ml aliquots of a 1 : 1 slurry of DEAE-Sephacel equilibrated in buffer A are added to the pooled fractions until there is no further change in the absorbance at 280 nm as monitored on 1–2-ml aliquots filtered through glass wool. If the glutathione peroxidase activity decreases, the additions of DEAE-Sephacel should be discontinued. The resultant mixture is filtered, and the DEAE-Sephacel is washed with 20 ml of buffer A. The filtrate and wash solutions are concentrated in an Amicon concentrator using a PM-10 membrane, and the concentrated sample is chromatographed isocratically on a DEAE-Sephacel column (2.5 × 100 cm) in buffer A. The fractions that contain the ^{75}Se are pooled and again concentrated in an Amicon concentrator with a PM-10 membrane to approximately 1 ml. The sample is injected as four separate 250-μl aliquots onto a Spherogel TSK-3000 SW HPLC column (7.5 × 600 mm) equilibrated with 10 mM sodium phosphate, 0.1 mM EDTA, pH 7. After the set of four injections and chromatographies, the fractions containing the enzyme from all four chromatographies are again pooled, concentrated, and rechromatographed twice on the same HPLC column. This purification has been modified by the addition of 0.5 mM glutathione to the buffers and the addition of 10% ethanol to the HPLC buffer.[17] These modifications improve the yield and stability of the enzyme.

Purification of Glutathione Peroxidase by Covalent Chromatography

The technique of covalent chromatography on Sepharose has been used to purify glutathione peroxidase.[18,19] This selenoenzyme has also been purified from red blood cells by the following chromatography procedure, which utilizes selenocysteine bound to a resin support. Selenocysteine serves as the ligand that binds glutathione peroxidase.

Procedure.[20] Selenocystine (125 mg) is dissolved in 25 ml of 25 mM sodium carbonate buffer, pH 9.2, at 50°. Saturated sodium carbonate solution is added to readjust the solution to pH 9.2. A 25-ml portion of Affi-Gel 10 resin is washed with water and then added to the selenocystine solution. After incubation for 1 hr at 37°, 0.1 ml of 1 M ethanolamine is added per milliliter of solution to neutralize any nonreacted column ligand. The resin is washed with 10 bed volumes of water to remove unreacted selenocystine. The selenocystinyl residues are activated to selenocysteine, and their concentration is determined as follows. The

[17] R. A. Condell and A. L. Tappel, *Biochim. Biophys. Acta* **709**, 304 (1983).
[18] P. L. Bergad, W. B. Rathbun, and W. Linder, *Exp. Eye Res.* **34**, 131 (1982).
[19] M. Yoshida, K. Iwami, and K. Yasumoto, *Agric. Biol. Chem.* **46**, 41 (1982).
[20] M. A. Motsenbocker, Doctoral Thesis, University of California, Davis, 1982.

resin is washed with 10 mM dithioerythritol, 10 mM Tris-HCl, pH 7.2. The absorbance of the eluted ionized selenol is measured at 248 nm and compared with the absorbance at 248 nm of a selenocysteine standard prepared in the same buffer. The amount of selenocysteine removed from the resin is equal to the amount of selenocysteine left on the resin.

Washed red blood cells are hypotonically lysed in a 40-fold excess of 10 mM Tris-HCl, pH 7.2. The red cell ghosts are removed by centrifugation for 20 min at 50,000 g. The protein solution is diluted to 2 mg/ml with 10 mM Tris-HCl, pH 7.2, degassed to remove dissolved oxygen and used directly for covalent chromatography.

The chromatography is carried out quickly at room temperature. Ten milliliters of Affi-Gel 10 resin is packed in a 2.5 × 2 cm column and washed with 3 bed volumes of nitrogen-saturated 10 mM dithioerythritol, 10 mM Tris-HCl, 0.1 mM EDTA buffer, pH 7.2. The dithioerythritol is immediately removed from the column with 2 bed volumes of nitrogen-saturated 10 mM Tris-HCl, 0.2 mM EDTA buffer, pH 7.2, at a flow rate of 2–10 ml/min. A 10-ml aliquot of the red cell protein is immediately applied to the column. The column is rinsed with 10 bed volumes of 100 mM Tris-HCl, 0.1 mM EDTA, pH 7.2. The bound protein is eluted with 3 bed volumes of 10 mM dithioerythritol, 10 mM Tris-HCl, 0.1 mM EDTA, pH 7.2.

Comments. The concentration of selenocysteine attached to the resin by this procedure is typically 1.5 mmol per liter of resin. Glutathione peroxidase has been purified an average of 340-fold, as measured by an increase in specific activity, from 20-mg samples of crude red blood cell protein in one step with this procedure. The average specific activity of the purified enzyme was 174 μmol of glutathione oxidized per minute per milligram of protein, and the average yield was 40%. The crude protein samples must be diluted to approximately 2 mg/ml because protein sulfhydryls can interfere with the binding of glutathione peroxidase to the resin. Although limited to small (20 mg) samples, this procedure is rapid. When the enzyme was analyzed by SDS–gel electrophoresis, only one band was observed.

Plasma Selenoprotein P

Plasma selenoprotein P was first studied by Herrman,[21] who found that a trace amount of ^{75}Se administered to rats was incorporated into an M_r 50,000 plasma selenopolypeptide. Rhesus monkey plasma and rat

---

[21] J. L. Herrman, *Biochim. Biophys. Acta* **500,** 61 (1977).

TABLE II
PURIFICATION OF PLASMA SELENOPROTEIN P

Fraction	^{75}Se (cpm/mg protein)	Purification (fold)	Recovery (%)
Plasma	55	1	100
NH$_4$SO$_4$ 35–50%	119	2.1	65
Affi-Gel blue chromatography	2500	45	45
DEAE anion-exchange chromatography	14,600	266	33

plasma selenoprotein P were both shown to contain selenocysteine.[7] This protein may transfer selenium from the liver to nonhepatic tissues during selenium nutritional deficiency in the rat.[22]

Purification

Table II summarizes a typical purification of plasma selenoprotein P.[23] The starting material is 100 ml of rat plasma. All manipulations are performed at 4°.

Plasma selenoprotein P is first labeled with ^{75}Se by intraperitoneal injection of a rat with 10 μCi of ^{75}Se as sodium [^{75}Se]selenite. Four hours after the injection, blood is removed with a citrate-rinsed syringe, and plasma is separated by centrifugation. More than 95% of the plasma ^{75}Se prepared in this way is attached to plasma selenoprotein P. The isotopically labeled plasma is pooled with plasma prepared from noninjected rats to a final volume of 100 ml.

Saturated NH$_4$SO$_4$ solution at 4° is added to the plasma to a final concentration of 36% saturation. After stirring for 2 hr, the precipitate is removed and the supernatant is brought to 50% saturation with saturated NH$_4$SO$_4$ solution and stirred for 3 hr. The 36–50% precipitate is redissolved in 50 ml of water.

The redissolved precipitate is applied to an Affi-Gel Blue column (2.5 × 35 cm) equilibrated with 50 mM Tris-HCl, pH 8.5. The column is washed with 125 ml of the equilibration buffer and then with 250 ml of the same buffer with 0.08 M sodium trichloroacetate. This second wash removes most of the adsorbed protein from the column. A 950-ml, 0.08 to

[22] M. A. Motsenbocker and A. L. Tappel, *Biochim. Biophys. Acta* **719,** 147 (1982).
[23] M. A. Motsenbocker and A. L. Tappel, *Fed. Proc. Fed. Am. Soc. Exp. Biol.* **42,** 533 (1983).

0.8 *M* gradient of sodium trichloroacetate in equilibration buffer is passed through the column. The ^{75}Se-labeled plasma selenoprotein P elutes at 0.4 *M* sodium trichloroacetate.

Fractions from Affi-Gel Blue chromatography that contain ^{75}Se are pooled (approximately 120 ml), diluted 10-fold with water, and then 35 g of loosely packed DEAE-Sephacel equilibrated in water is stirred in. The DEAE-Sephacel resin is recovered by filtration over sintered glass and packed on top of a 2.5 × 35 cm column of DEAE-Sephacel equilibrated in 50 m*M* Tris-HCl, pH 8.5. A 575-ml, 0.03 to 0.15 *M* gradient of sodium chloride in equilibration buffer is applied at a flow rate of 0.32 ml/min. The ^{75}Se-labeled plasma selenoprotein P elutes at a sodium chloride concentration of 0.08 *M*.

Comments. The purified protein is approximately 35% pure when analyzed by gel electrophoresis, and it consists of one 52,000 and one 35,000 M_r subunit. The selenocysteine in this protein exists in the 52,000 M_r subunit.[7] The purification procedure is carried out rapidly because the protein dissociates into its two subunits when it is removed from plasma. Because of its larger size, the 52,000 M_r selenopolypeptide of plasma selenoprotein P is easily separated from the 22,000 M_r subunit of glutathione peroxidase by SDS–polyacrylamide gel electrophoresis. Affi-Gel Blue can be used to separate these two selenoproteins chromatographically under nondenaturing conditions because glutathione peroxidase has a much lower affinity for this resin than does the plasma protein.

Kidney Selenoproteins

Four selenoproteins can be separated by gel filtration (Fig. 3) of a kidney supernatant fraction that is labeled with [75]Se.[24] One of the proteins is glutathione peroxidase. Another protein, selenoprotein P1, has a molecular weight of about 80,000. The selenium in selenoprotein P1 is in the amino acid selenocysteine in a polypeptide with a molecular weight of approximately 75,000. Selenoprotein P1 can also be obtained from liver tissue, and the amount of liver selenium in this protein is increased by feeding a high level (1 ppm) of selenium to rats.[25] The two smaller molecular weight selenoproteins that can be resolved by gel filtration of kidney do not exist in liver. These latter two selenoproteins we have named kidney selenoprotein 3 and kidney selenoprotein 4 because they are the third and fourth proteins eluted during gel filtration of rat kidney supernatant.

[24] M. A. Motsenbocker and A. L. Tappel, *Biochim. Biophys. Acta* **709,** 160 (1982).
[25] M. A. Motsenbocker and A. L. Tappel, *J. Nutr.* **114,** 279 (1984).

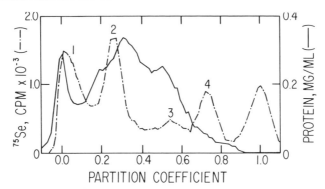

FIG. 3. Gel filtration of ^{75}Se-labeled rat kidney cytosol. A 2-ml sample was applied in phosphate-buffered saline to a 2.5 × 100 cm column of Sephadex G-150 and eluted at a flow rate of 0.16 ml/min.

Separation of Kidney Selenoproteins by Gel Filtration

Rats are maintained for 2 weeks on the dietary and ^{75}Se–water regimen described above under Isotopic Equilibration. Rats are lightly etherized. Kidney tissue is excised and homogenized 1:5 (w/v) in 0.25 M sucrose, 10 mM Tris-HCl, pH 7.6. The homogenate is centrifuged at 100,000 g for 60 min. A 2-ml aliquot of the supernatant is mixed with 100 μl of water that contains Blue Dextran and cresol red. This sample is then applied to a Sephadex G-150 column (2.5 × 85 cm) equilibrated with 0.8% sodium chloride, 50 mM sodium-potassium phosphate, pH 6.8. Equilibration buffer is applied at a flow rate of 0.16 ml/min.

Figure 3 depicts the chromatographic separation of kidney cytosolic protein by the above-described procedure. The four peaks from the chromatographic separation are pooled. The molecular weights of the seleno-polypeptides in each pooled peak are determined by gel filtration in the presence of SDS. Each pooled sample is concentrated by dialysis against 1% 2-mercaptoethanol, 10 mM Tris-HCl, pH 7.6, followed by lyophilization. The concentrated protein is denatured in 2% SDS, 10 mM dithioerythritol, 10% 2-mercaptoethanol, 10 mM Tris-HCl at 100° for 5 min. Then 100 μl of Blue Dextran with ^{3}H-labeled water solution is mixed with 0.4 ml of denatured sample as markers to determine the gel filtration void and total volumes, respectively. Each 0.5-ml sample is applied to a column (1.5 × 100 cm) of Sephacryl S-300 freshly equilibrated in 0.1% SDS, 5 mM dithioerythritol, 10 mM Tris-HCl, pH 7.6. The sample is eluted with equilibration buffer at a flow rate of 0.1 ml/min. As shown in Fig. 4, all of the ^{75}Se from peak 1 of the Sephadex G-150 gel filtration

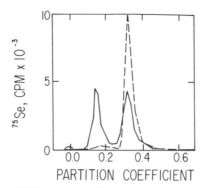

FIG. 4. Gel filtration of SDS-denatured [^{75}Se]polypeptides. ——, Pooled protein of the largest molecular weight fraction (peak 1) from Sephadex G-150 gel filtration of kidney; ——, pooled protein of the second largest molecular weight ^{75}Se-labeled fraction (peak 2) from Sephadex G-150 gel filtration of kidney.

elutes with an apparent molecular weight of 22,000 during the detergent Sephacryl S-300 chromatography. Peak 2 is separated as two ^{75}Se-labeled protein fractions of 22,000 and 75,000 M_r during the detergent Sephacryl S-300 chromatography. Peaks 3 and 4 from Sephadex G-150 filtration both elute as mixtures of 22,000 and approximately 15,000 M_r polypeptides from the detergent gel filtration analysis (not shown).

Comments. Peak 2 from gel filtration of kidney selenoproteins (Fig. 3) is a mixture of glutathione peroxidase and selenoprotein P1 since chromatography of this pooled sample over DEAE-Sephacel in 10 mM Tris-HCl, pH 7.3, at 4° results in the separation of ^{75}Se into the glutathione peroxidase polypeptide of 22,000 M_r, which does not bind to the DEAE resin, and the 75,000 M_r polypeptide of selenoprotein P1, which binds to the resin and is eluted with a 0.1 M sodium chloride solution.[24] A similar separation of liver selenoprotein ^{75}Se-P from glutathione peroxidase by DEAE ion-exchange chromatography has been reported.[6] Kidney selenoprotein P1 exhibits a native molecular weight similar to that of plasma selenoprotein P when chromatographed over Sephadex G-150 and also when chromatographed over DEAE ion-exchange resins. However, these two proteins behave differently when subjected to SDS gel filtration. When ^{75}Se-labeled selenoprotein in peak 3 of the chromatogram in Fig. 3 is concentrated and subjected to SDS–gel electrophoresis, the migration of the ^{75}Se is associated with a polypeptide that has a molecular weight the same as that of glutathione peroxidase selenopolypeptide. Kidney selenoproteins 3 and 4 have not been found in liver or testes.

Selenopeptide Analysis

Glutathione Peroxidase. The amino acid sequence around the active site of glutathione peroxidase has been determined.[17] This determination was made by sequencing the intact protein from the amino terminus and by sequencing the selenocysteine-containing tryptic peptide. In addition, a portion of the procedure used to obtain the tryptic peptide is a useful tool for studying subtle differences among selenoproteins. The selenium-containing tryptic peptide is obtained by first carboxymethylating the glutathione peroxidase with sodium iodoacetate until no enzymatic activity remains. The derivatized sample is made 10 mM $CaCl_2$. Trypsin treated with toluenesulfonylalaninechloromethyl ketone is added (trypsin–protein, 1:100, w/v), and the sample is incubated at 37°. Additional trypsin is added after 3 and 17 hr (trypsin–protein, 1:100, w/v). After 20 hr the sample is acidified to pH 2 with HCl. The tryptic digest is fractionated by reverse-phase HPLC on an Ultrasphere ODS C_{18} column (4.6 × 250 mm) equilibrated with 50 mM sodium phosphate, pH 6.5–acetonitrile, 90:10 (v/v). The peptides are eluted with a gradient of acetonitrile (0.75% per milliliter) at a flow rate of 0.7 ml/min. A portion of a chromatogram of a tryptic digest of glutathione peroxidase is shown in Fig. 5.

Plasma Selenoprotein and Kidney Selenoprotein

A low-molecular-weight selenoprotein from kidney (peak 3 in Fig. 3) exhibited a subunit size from SDS–gel electrophoresis analysis that was the same as the subunit size of glutathione peroxidase.[24] To establish

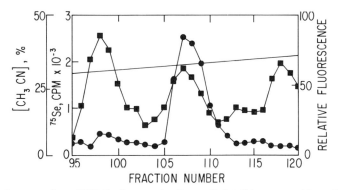

FIG. 5. Reverse-phase HPLC of a tryptic digest of glutathione peroxidase. The Ultrasphere ODS C_{18} column (4.6 × 250 mm) was used with an acetonitrile gradient. The 0.35-ml fractions were counted for ^{75}Se (●) and assayed for amino groups (■) with fluorescamine. A typical calibration curve gave approximately 1.3 μM amino groups per fluorescence unit.

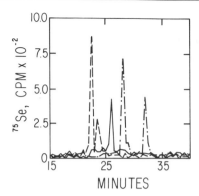

FIG. 6. Elution of ^{75}Se-labeled tryptic peptides from an ODS C_{18} column. ——, Peptide prepared from ^{75}Se-labeled glutathione peroxidase; ----, peptides prepared from ^{75}Se-labeled plasma selenoprotein P; ·—·—·, peptides prepared from a 22,000 M_r kidney ^{75}Se-labeled selenoprotein.

identity or nonidentity of Se-labeled selenopolypeptides, ^{75}Se-labeled tryptic fragments can be prepared and qualitatively separated by reverse-phase HPLC.

Procedure. Partially purified ^{75}Se-labeled selenoprotein is derivatized with iodoacetic acid and subjected to trypsin hydrolysis as indicated above for glutathione peroxidase.[17,26] Acetic acid and ethanol are each added to 10% (final volumes), and the mixture is centrifuged at low speed. A 0.5-ml portion of the supernatant is applied to a column of Sephadex G-25 (1.4 × 100 cm) followed by application of a 10% acetic acid, 10% ethanol solution. The major ^{75}Se-labeled peptide that elutes as a single peak from this gel filtration step is pooled and lyophilized, and the residue is taken up in 0.5 ml of 5% acetic acid, 10% acetonitrile. This redissolved sample is applied to an ODS C_{18} column (4.6 × 250 mm) equilibrated in 5% acetic acid, 10% acetonitrile. The ^{75}Se-labeled tryptic peptide is eluted in an increasing (10 to 60%) gradient of acetonitrile at 0.7 ml/min (Fig. 6). The chromatogram was confirmed by repetition of the procedure.

Comments. The ^{75}Se-labeled selenopeptides prepared from each ^{75}Se-labeled protein eluted at different times from the ODS C_{18} column owing to their different structures (Fig. 6). Therefore, the small-molecular-weight selenoprotein from kidney supernatant was not a monomer of glutathione peroxidase even though its selenopolypeptide size and native molecular weight indicated otherwise. This is further evidence that some

[26] J. J. Zakowski, J. W. Forstrom, R. A. Condell, and A. L. Tappel, *Biochem. Biophys. Res. Commun.* **84,** 248 (1978).

of the 22,000 M_r selenopolypeptide from rat tissue fractions may be in nonglutathione peroxidase selenoprotein.

Translational Synthesis of Selenoproteins

Although some research has suggested that selenium is incorporated into glutathione peroxidase via a posttranslational mechanism,[27] there has been no success in attempts to show the existence of an apoenzyme[19] or to identify the active precursor form of selenium. Furthermore, much evidence has become available to suggest that selenium may be incorporated into glutathione peroxidase by a translational mechanism.

1. The incorporation of ^{75}Se from [^{75}Se]selenite into glutathione peroxidase is inhibited 85–100% by cycloheximide.[22,28]

2. Rat liver contains a species of tRNA that is specific for selenocysteine relative to the 20 protein amino acids.[29]

3. The rat liver selenocysteine aminoacyl tRNA is more active in an *in vitro* protein synthesis system than is selenite, selenocysteine, or deacylated selenocysteyl-tRNA.[28]

4. The selenocysteine in glutathione peroxidase is located at residue 41 in the polypeptide backbone[17] and is not located in a side chain.[6]

These results are compatible with a translational mechanism. Since there is no known mechanism for the specific translational incorporation of selenocysteine, i.e., there is no codon assigned for selenocysteine, the mechanism of selenocysteine incorporation into protein has been grouped together with posttranslational mechanisms for the purposes of this presentation.

Acknowledgment

This research was supported by Research Grant AM-06424 from the National Institute of Arthritis, Diabetes, and Digestive and Kidney Diseases.

[27] R. A. Sunde and W. G. Hoekstra, *Biochem. Biophys. Res. Commun.* **93**, 1181 (1980).

[28] W. C. Hawkes and A. L. Tappel, *Biochim. Biophys. Acta* **739**, 225 (1983).

[29] W. C. Hawkes, D. E. Lyons, and A. L. Tappel, *Biochim. Biophys. Acta* **699**, 183 (1982).

[43] Characterization of Selenomethionine in Proteins

By MARK X. SLIWKOWSKI

Incorporation of selenium into protein occurs both specifically and nonspecifically. When selenium concentrations approach or exceed those of sulfur, random substitution of selenium for sulfur can occur.[1] In many instances selenium substitutes for sulfur at the amino acid level and is incorporated directly into newly synthesized polypeptide chains. For example, studies with β-galactosidase from *Escherichia coli* grown on 10 mM Na$_2$SeO$_4$ have shown 30–40% substitution of selenium for sulfur.[2] Notably, the selenium was found only as selenomethionine, and as many as 80 of the 150 methionine residues had been replaced with the selenium analog.[3] Similar nonspecific substitution of selenomethionine for methionine was reported with *E. coli* mutant 26 grown on 10 mM selenomethionine when supplemented with cysteine.[4] Additionally, [^{75}Se]selenomethionine can also be used to label proteins in cell culture, and a similar nonspecific incorporation mechanism is envisioned.[5]

In contrast, certain mammalian and bacterial enzymes exhibit a specific selenium requirement.[6–8] The chemical form of selenium in some of these proteins has been shown to be selenocysteine. Proteins that require selenium incorporate the element at high S : Se ratios. Thus, it is apparent that biochemical mechanisms exist that distinguish selenium from sulfur. This specific incorporation of selenium into proteins is assumed to be posttranslational, although direct evidence in support of this theory is lacking.

The fatty acid-producing anaerobe *Clostridium kluyveri* incorporates selenium into its acetoacetyl-CoA thiolase.[9] The chemical form of selenium has been identified as selenomethionine. The role of selenium in this protein is unknown; however, the occurrence as a selenoether moiety

---

[1] T. C. Stadtman, *Adv. Enzymol. Relat. Areas Mol. Biol.* **48**, 1 (1979).

[2] R. E. Huber, I. H. Segel, and R. S. Criddle, *Biochim. Biophys. Acta* **141**, 573 (1967).

[3] R. E. Huber and R. S. Criddle, *Biochim. Biophys. Acta* **141**, 587 (1967).

[4] E. H. Coch and R. C. Greene, *Biochim. Biophys. Acta* **230**, 223 (1971).

[5] C. G. Brooks, *J. Immunol. Methods* **22**, 23 (1978).

[6] J. T. Rostruck, A. L. Pope, H. E. Ganther, A. B. Swanson, D. G. Hafeman, and W. G. Hoekstra, *Science* **179**, 588 (1978).

[7] D. C. Turner and T. C. Stadtman, *Arch. Biochem. Biophys.* **154**, 366 (1973).

[8] J. E. Cone, R. Martín del Río, and T. C. Stadtman, *J. Biol. Chem.* **252**, 5337 (1977).

[9] M. G. N. Hartmanis and T. C. Stadtman, *Proc. Natl. Acad. Sci. U.S.A.* **79**, 4912 (1982).

allows for considerably less laborious experimental procedures, since the selenium is not easily oxidized or eliminated.

Incorporation of Selenium into Thiolase

Thiolase is isolated by a series of chromatographic and salt fractionation procedures.[9] The final step, chromatography on CM-Sephadex 50, results in a high yield of fully active enzyme.[10] Sheep antibodies raised against purified thiolase can be used to monitor the incorporation of selenium into the enzyme. L-Methionine (0.25–2 mM) was added to cultures of C. kluyveri in the presence of $Na_2{}^{75}SeO_3$ in order to determine whether the selenium levels of thiolase were affected. Coch and Greene[4] have estimated that 100 μM selenomethionine causes a 95% reduction in E. coli 26 methionine biosynthesis. Thus, the concentration of methionine added to the C. kluyveri cultures should be sufficient to suppress any de novo methionine biosynthesis and/or markedly to diminish any nonspecific incorporation of selenomethionine for methionine. In addition, amino acid analysis of the culture filtrates indicated that greater than 50% of the L-methionine was actively metabolized or incorporated into the cells.

The data in Table I indicate that the addition of L-methionine to cultures of C. kluyveri does not decrease the amount of ^{75}Se incorporated into thiolase.

Recently, however, it has been found by peptide mapping techniques that selenium in thiolase was distributed throughout the primary structure of the enzyme in a manner that paralleled methionine.[10] From these studies, it appears that a significant amount of selenium occurs in thiolase nonspecifically and is probably not required for biological activity.

Amino Acid Analysis of Selenomethionine-Containing Proteins

Several methods have been described for the resolution of selenium-containing amino acids using the amino acid analyzer.[11-13] In all these chromatographic procedures, selenomethionine elutes near or with leucine. Optimization of resolution of leucine from selenomethionine has been studied in great detail by Benson and Patterson.[12] These workers

[10] M. X. Sliwkowski and T. C. Stadtman, unpublished experiments, 1983.
[11] R. Walter, D. H. Schlesinger, and I. L. Schwartz, Anal. Biochem. 27, 231 (1969).
[12] J. W. Benson and J. A. Patterson, Anal. Biochem. 29, 130 (1969).
[13] J. L. Martin and M. L. Gerlach, Anal. Biochem. 29, 257 (1969).

TABLE I
IMMUNOPRECIPITATION OF THIOLASE FROM
Clostridium kluyveri[a]

L-Methionine added to culture (mM)	% Total ^{75}Se in extract	Cpm/1.6 thiolase units
None	35	16,446
0.25	22	17,045
0.50	27	24,082
1.00	13	19,658
2.00	20	32,023

[a] *Clostridium kluyveri* cultures (1 liter) were grown in a standard mixture supplemented with 0.5 mCi of H$_2$^{75}SeO$_3$ (194.9 mCi/mg). Cells were harvested by centrifugation and ruptured by sonication. Each incubation mixture (0.34 ml) contained 1.6 thiolase units and 32 μl of antithiolase IgG (20 mg/ml) in phosphate-buffered saline. IgG was prepared from immune serum by precipitation with 40% saturated ammonium sulfate. Reactions were allowed to proceed overnight at room temperature, and the immunoprecipitates were collected by centrifugation. The pellets were washed with 170 μl of H$_2$O, resuspended in 100 μl of H$_2$O, and assayed for radioactivity.

found that variation in temperature and pH affect the resolution of these two amino acids, whereas ionic strength is of little importance.

Thiolase was hydrolyzed using 3 M mercaptoethanesulfonic acid at 155° for 40 min.[14,15] The samples were then adjusted to pH 2.2 with NaOH and applied to a Dionex Model 300 amino acid component system equipped with *o*-phthalaldehyde detection. The jacketed column (0.4 × 15 cm) was packed with Dionex DC-5A resin (6 ± 0.5 μm particle diameter; Dionex Corp., Sunnyvale, CA 94086). Three conditions are presented (Table II) for the resolution of isoleucine, leucine, and selenomethionine, using authentic compounds. In agreement with the results of Benson and Patterson,[12] methods II and III have been found to give optimal resolution of selenomethionine derived from protein hydrolysates. Recovery of ^{75}Se associated with the selenomethionine peak appears to be about 60–70% of the total amount in the protein sample when hydrolysis is performed with 3 M mercaptoethanesulfonic acid.

[14] B. Penke, R. Ferenczi, and K. Kovacs, *Anal. Biochem.* **60**, 45 (1974).
[15] P. E. Hare, this series, Vol. 47, p. 3.

TABLE II
CHROMATOGRAPHY OF Ile, Leu, AND SELENOMETHIONINE
WITH AMINO ACID ANALYZER

	Elution times (min)		
Methods[a]	Ile	Leu	SeMet
I. 0.2 N Na$^+$, pH 4.26	15.4	18.3	21.1
II. 0.45 N Na$^+$, pH 3.5	26.4	32.9	35.8
III. 38 min: 0.2 N Na$^+$, pH 3.27;	50.6	53.7	55.5
45 min: 0.2 N Na$^+$, pH 4.26			

[a] All buffers were 67 mM in citrate.

Preparation and Identification of Se-Adenosyl[^{75}Se]selenomethionine

Previous workers have shown that selenomethione is a better substrate for S-adenosylmethionine synthetase than methionine.[16,17] This enzyme can be a useful tool in the determination of selenomethionine in proteins.

The ^{75}Se-labeled amino acid from thiolase, released by hydrolysis with 3 M mercaptoethanesulfonic acid for 40 min at 155° was isolated by chromatography on the amino acid analyzer column. This fraction was then incubated with ATP and S-adenosylmethionine synthetase for a total of 150 min.[18] The reaction mixture was analyzed by cellulose TLC in ethanol–2 M HCOONH$_4$, 70 : 30 (pH 3.5). The R_f values in this solvent are methionine–selenomethionine, 0.75; S-adenosylmethionine–Se-adenosylselenomethionine, 0.1. Alternatively, the reaction products were analyzed by ion-exchange chromatography on Bio-Rex 70.[18] By both methods, 17% (15,000 cpm) of the added ^{75}Se-labeled amino acid (92,000 cpm) was recovered in a product that was determined to be Se-adenosylselenomethionine. Control experiments with authentic [^{75}Se]selenomethionine gave a similar yield of a product that exhibited the same properties.

Acknowledgment

I would like to thank Dr. Thressa C. Stadtman for her advice and guidance throughout both the course of this work and the preparation of the manuscript. I am also grateful to Mr. J. Nathan Davis for details on the development of the amino acid analyzer program for the determination of selenomethionine in protein hydrolysates.

[16] S. H. Mudd and G. L. Cantoni, *Nature* (*London*) **180,** 1052 (1957).
[17] G. D. Markham, E. W. Hafner, C. W. Tabor, and H. Tabor, *J. Biol. Chem.* **255,** 9082 (1980).
[18] H. Tabor and C. W. Tabor, this series, Vol. 17B, p. 393.

[44] Citrulline in Proteins from the Enzymatic Deimination of Arginine Residues

By J. A. ROTHNAGEL and G. E. ROGERS

Before the advent of amino acid analysis and sequencing, claims were made for the presence of citrulline in proteins based on colorimetric reactions.[1,2] The demonstration of citrulline in a protein of hair follicle inner root sheath cells in 1958 was the first report[3] based on a reliable analytical procedure. Subsequent work conclusively showed that the citrulline was bound in normal peptide linkage in the proteins that are found in the differentiated cells of both the medulla of animal hair and the inner root sheath of the hair follicle.[4,5] From the evidence of amino acid composition and sequencing, it is clear that citrulline is not present in most proteins. Its occurrence is restricted to specialized proteins whose tissues of origin contain the requisite enzyme for the conversion of the guanido side chain of certain arginine residues to the ureido side chains of citrulline. The data of Table I give the occurrence of citrulline in proteins that have been described so far. The tissues of note are primarily vertebrate epidermis and the derived structures of it already mentioned, although it has been reported that citrulline is present in a protein of nerve myelin.[6]

Despite this apparent limitation to the occurrence of protein-bound citrulline, the enzymatic activity responsible for the conversion of arginine residues has already been described in hair follicles of guinea pig,[7] in the epidermis of cow[8] and newborn rat,[9] and also in vertebrate muscle[10] and brain.[11] It should be noted that citrulline residues per se have not been reported in the proteins of vertebrate muscle and brain. The enzyme is thought to act directly by the removal of the $>$NH from the guanido

[1] W. R. Fearon, *Biochem. J.* **33,** 902 (1939).

[2] V. N. Orekhovitch, Giornate Biochimiche, Italo-Franco-Elvitiche, Supplement to *Ric. Sci.* **25,** 457 (1955).

[3] G. E. Rogers and D. H. Simmonds, *Nature (London)* **182,** 186 (1958).

[4] G. E. Rogers, *Nature (London)* **194,** 1149 (1962).

[5] P. M. Steinert, H. W. J. Harding, and G. E. Rogers, *Biochim. Biophys. Acta* **175,** 1 (1969).

[6] P. R. Finch, D. D. Wood, and A. A. Moscarello, *FEBS Lett.* **15,** 145 (1971).

[7] G. E. Rogers, H. W. J. Harding, and I. J. Llewellyn Smith, *Biochim. Biophys. Acta* **495,** 159 (1977).

[8] J. Kubilus, R. W. Waitkus, and H. P. Baden, *Biochim. Biophys. Acta* **615,** 246 (1980).

[9] M. Fujisaki and K. Sugawara, *J. Biochem.* **89,** 257 (1981).

[10] K. Sugawara, Y. Oikawa, and T. Ouchi, *J. Biochem.* **91,** 1065 (1982).

[11] J. Kubilus and H. P. Baden, *Biochim. Biophys. Acta* **745,** 285 (1983).

METHODS IN ENZYMOLOGY, VOL. 107

TABLE I

EXAMPLES OF CITRULLINE RESIDUES IN PROTEINS

Tissue source of protein	Species	Citrulline content (moles/1000 mol)	Reference[a]
Stratum corneum	Guinea pig	1.9	(1)
	Cow	3.6	(1)
	Man	1.7	(1)
Hair medulla	Guinea pig	128	(2)
Hair follicle inner root sheath	Rat	38	(3)
Keratolinin from epidermal cell membrane	Cow	24	(4)
Myelin protein	Man	10	(5)

[a] Key to references: (1) J. Kubilus, R. W. Waitkus, and H. P. Baden, *Biochim. Biophys. Acta* **581,** 114 (1979); (2) H. W. J. Harding and G. E. Rogers, *Biochemistry* **10,** 624 (1971); (3) G. E. Rogers and D. H. Simmonds, *Nature (London)* **182,** 186 (1958); (4) C. J. Lobitz and M. M. Buxman, *J. Invest. Dermatol* **78,** 150 (1982); (5) P. R. Finch, D. D. Wood, and A. A. Moscarello, *FEBS Lett.* **15,** 145 (1971).

group as NH_3, and hence it is now referred to as a peptidylarginine deiminase. Enrichment of the enzyme activity has been achieved for epidermal tissue[8,9] and follicle tissue,[12] and, in two instances, muscle[10] and brain[11] purification to a single molecular species has been reported. Investigation of the enzyme and its mechanism of action has been retarded because of the use of protein substrates. Simple arginine-containing substrates have been introduced into more sensitive and reliable assays.[9–12]

The function of citrulline in its specialized locations, and hence the requirement for a converting enzyme, remains unclear. Presumably the hair follicle enzyme converts some arginine residues in strategic positions in a mature protein precursor, thereby possibly causing a conformational change. It is not known whether this is a general phenomenon in the maturation of other epidermal proteins found to contain citrulline or whether the arginine to citrulline conversion can take place during polypeptide growth on polysomes.

In studies *in vitro,* several unrelated proteins have been shown to be acted upon by the deiminase,[8,9,13] but no sequencing studies have yet been carried out to find the converted residues. A specificity of action based on

[12] G. E. Rogers and J. A. Rothnagel, *in* "Normal and Abnormal Epidermal Differentiation" (I. A. Bernstein and H. Ogawa, eds.) p. 171. Univ. of Tokyo Press, Tokyo, 1983.

[13] G. E. Rogers and L. D. Taylor, *Adv. Exp. Med. Biol.* **86A,** 283, 1977.

amino acid sequence environment, should it exist, would mean that the deiminase could become a useful protein reagent.

Identification and Estimation of Citrulline in Proteins

Histochemical Detection in Situ

Citrulline gives a yellow color when allowed to react with Erhlich reagent, p-dimethylaminobenzaldehyde (1% w/v) in 0.5–1.0 M HCl. In one instance[14] this reaction was applied to tissue sections of mammalian skin in the presence of 2 M NaCl to depress swelling, and protein-bound citrulline was detected in hair follicle cells. In a related study,[15] an improvement in the technique was obtained by using p-dimethylaminocinnamaldehyde under similar conditions. This compound yields a strong reddish color with substituted ureas.

Quantitative Estimation

Amino Acid Analysis. Citrulline can be estimated in acid hydrolysates of proteins and peptides using a standard amino acid analyzer. However, in order to elute the citrulline as a separate peak it is necessary to alter the conditions of elution.[16,17] The amino acid analyses by this procedure can be accurate provided corrections are applied to account for the partial loss of citrulline to ornithine during 16–24 hr hydrolysis in 6 M HCl.[5] This analytical approach is unattractive because it is slow and there is only a single result per analytical cycle.

Colorimetric Estimation

Principle. The colorimetric determination of citrulline in proteins and peptides utilizes the reaction of ureido groups with diacetyl monoxime described by Fearon[1] and developed by Archibald.[18] A widely used modification of the basic procedure is that of Guthöhrlein and Knappe,[19] but this method suffers from low sensitivity and nonlinearity of the calibration curve at low citrulline concentrations and from color instability. We find

[14] G. E. Rogers, *J. Histochem. Cytochem.* **11,** 700 (1963).
[15] G. E. Rogers and H. W. J. Harding, *in* "Biology and Disease of the Hair" (T. Kobori and W. Montagna, eds.), p. 411. Univ. of Tokyo Press, Tokyo, 1976.
[16] H. W. J. Harding and G. E. Rogers, *Biochemistry* **10,** 624 (1971).
[17] K. Sugawara, *Agric. Biol. Chem.* **43,** 2215 (1979).
[18] R. M. Archibald, *J. Biol. Chem.* **156,** 121 (1944).
[19] G. Guthöhrlein and J. Knappe, *Anal. Biochem.* **26,** 188 (1968).

the modification of Boyde and Rahmatullah[20] to be a distinct improvement because color development is achieved in 5 min, the sensitivity is at least twice that of the Guthörlein and Knappe method, the response is linear at low citrulline concentrations, and the color is stable in light.

Difficulties can be experienced with the quantitation of citrulline when colorimetric methods are applied directly to proteins. These difficulties include nonspecific color reactions and the appearance of trace amounts of suspended material in the reaction mixture. Consequently it is advantageous to precede the colorimetric reaction by acid hydrolysis of the proteins.[13]

Reagents

Reagent 1: To 550 ml of water add 250 ml of sulfuric acid (95–98%) and 200 ml of phosphoric acid (85%). Cool to room temperature and add 250 mg of $FeCl_3$

Reagent 2: Diacetyl monoxime (0.5% w/v), thiosemicarbazide, (0.01% w/v) in water

Reagent 3: Mix two parts of reagent 1 with one part of reagent 2 immediately before use

Reagent 4: Citrulline standard, 1 mM, 17.52 mg DL-citrulline in 100 ml of water

Procedure. This is essentially as described by Boyde and Rahmatullah.[20] However, to avoid interference in the color reaction, the protein is hydrolyzed (6 N HCl at 110° for 20 hr), and the hydrolysate is dried under vacuum. The residue is redissolved in 1 ml of 0.1 M HCl, and then the assay proceeds with the addition of 3 ml of reagent 3. After thorough mixing, the reaction is completed by immersion of the tubes (stoppered with a marble) in a water bath at 100° for 5 min. The tubes are cooled to room temperature, and the absorbance is measured at 530 nm. In our hands a linear response from 25–300 nmol of citrulline is consistently attainable.

Correction for hydrolytic loss of citrulline can be made by reference to a control of free citrulline, hydrolyzed under the standard conditions.

Peptidylarginine Deiminase and Its Assay

The first detection and attempted assay of the enzyme depended on the estimation of citrulline by amino acid analysis.[7] Hair follicle homogenates were incubated with a partially purified protein fraction, identified

---

[20] T. R. C. Boyde and M. Rahmatullah, *Anal. Biochem.* **107**, 424 (1980).

as the natural substrate, which had been prelabeled *in vivo* with [^{14}C]arginine. This procedure is tedious, requiring hydrolysis of the reaction products, and its sensitivity is limited by the attainable specific activity of the substrate.

A colorimetric assay was developed using the natural substrate and other arginine-rich proteins as substrates.[13] The reliability of the procedure is reduced by proteins but improved by acid hydrolysis. Nevertheless, the method still lacks sensitivity, and citrulline product formation is not reliably detected below 25 nmol.

It has been shown that peptidylarginine deiminase will act on simple substrates.[9–12] These molecules contain a high concentration of convertible arginine residues and do not require hydrolysis. The minimum requirement in the substrate is an α-N-blocked arginine. Enzymes from different tissues appear to have differing substrate specificities, in terms of functional groups neighboring the convertible arginine (see Table II).

Colorimetric Method

Assay for peptidylarginine deiminase using N-benzoyl-L-arginine.

Reagents

Substrate: N-Benzoyl-L-arginine is only slightly soluble in water, with an upper limit of approximately 11 mg/ml at 37°. For routine assays we usually use a stock solution of 5 mg/ml. It is possible to prepare a supersaturated solution by heating to 100°

Enzyme: Hair follicle enzyme from depilated tissue (e.g., guinea pig[7]) is extracted in a buffer containing 10 mM HEPES, pH 7.5, 2 mM EDTA, and 2 mM 2-mercaptoethanol[12]

Procedure. The incubation mix volume of 500 μl in Eppendorf tubes consisted of up to 200 μl of enzyme extract and final concentrations of 2.5 mM DTT, 10 mM CaCl$_2$, 20 mM HEPES, pH 7.5, and varying concentrations of N-benzoyl-L-arginine up to 40 mM. After incubation the reaction is stopped by the addition of 20 μl of 0.5 M EDTA. The mixture is deproteinized by adding 500 μl of methanol and held at $-20°$ for 1 hr. It is then centrifuged for 5 min at 12,000 g (Eppendorf), and the supernatant is lyophilized (Savant speedvac). Alternatively the proteins may be removed using an Amicon micropartition system.

The lyophilized material is dissolved in 1 ml of 0.1 M HCl and then assayed for citrulline as outlined above (colorimetric estimation).

Radioassay Method

This assay for peptidylarginine deiminase uses N-benzoyl-L-[*guanido*-^{14}C]arginine as the substrate.

TABLE II
COMPARISONS OF PEPTIDYLARGININE DEIMINASES

Enzyme (tissue source)	Molecular weight	Natural substrate	pH	Temperature (°C)	Other substrates[a]
Hair follicle					
Guinea pig[b]	≈50,000 (unpublished observation)	Trichohyalin[c]	7.0	37	BA, histone, polyarginine, keratin[d]
Epidermis					
Rat[e]	48,000	Epidermal cell	7.5	50	BA>BGA>BAA>BAEE[e]
Cow[f]	69,000	membrane[g–j]	7.1	37	Trichohyalin, keratins, polyarginine, histone[f]
Muscle					
Rabbit[k]	115,000	Not known	7.6	52	BAEE>BAA>BGA>BA[k]
Brain					
Cow[l]	125,000	Not known	7.5	50	BAEE>BAA>BA[l]

[a] BA: N-benzoylarginine; BGA: N-benzoylglycylarginine; BAA: N-benzoylarginine amide; BAEE: N-benzoylarginine ethyl ester
[b] Rogers and Rothnagel.[12]
[c] Rogers et al.[7]
[d] Rogers and Taylor.[13]
[e] Fujisaki and Sugawara.[9]
[f] Kubilus et al.[8]
[g] Sugawara.[17]
[h] J. Kubilus, R. W. Waitkus, and H. P. Baden, Biochim. Biophys. Acta 581, 114 (1979).
[i] P. M. Steinert and W. W. Idler, Biochemistry 18, 5664 (1979).
[j] C. J. Lobitz and M. M. Buxman, J. Invest. Dermatol. 78, 150 (1982).
[k] Sugawara et al.[10]
[l] Kubilus and Baden.[11]

Reagents. The radioactive substrate is prepared by benzoylation of [^{14}C]arginine to a specific activity of at least 50 μCi/mmol.[12]

Procedure. The enzyme and incubation mix is as outlined for the colorimetric method. The lyophilized supernatant after deproteinization is dissolved in 50 μl of acidified ethanol (0.2 M HCl in 70% ethanol). Some nonradioactive material remains undissolved in this buffer, and this can be removed by centrifugation (5 min at 12,000 g). The supernatant is applied as a band (0.5 × 2 cm) on Whatman 3 MM paper and dried. The amount loaded can be reduced for very high concentrations of N-benzoyl-L-[*guanido*-^{14}C]arginine and where high conversion rates are suspected. It is then subjected to high-voltage electrophoresis at approximately 15 V/cm for 2 hr in pyridine–acetic acid–water (20:20:960) pH 4.7. The electrophoretogram is dried in a forced-draft oven (100°) and a paper strip (3 × 6 cm) is cut by reference to a bromophenol blue marker band that has a mobility close to that of N-benzoyl-L-[*carbamoyl*-^{14}C]citrulline as shown

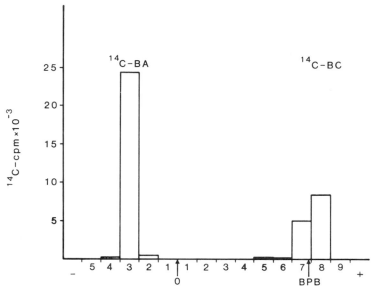

Distance from origin (cm)

FIG. 1. Histogram of counts per minute (cpm) versus distance from origin (0) constructed from 1 × 3 cm paper strips counted in toluene scintillation fluid. Separation of N-benzoyl-L-[*guanido*-14C]arginine (14C-BA) from N-benzoyl-L-[*carbamoyl*-14C]citrulline (14C-BC) by high-voltage paper electrophoresis in pyridine–acetate buffer, pH 4.7. The migration of bromophenol blue (BPB) is indicated; (+) anode; (−) cathode.

in Fig. 1. A second paper strip (3 × 6 cm) is cut from the cathode side of the origin. The appropriate area is counted after immersion of the paper roll in a toluene-based scintillation fluid (PPO, 0.35% w/v, POPOP, 0.035% w/v in toluene). The formation of citrulline is calculated from the percentage ratio (N-benzoyl-L-[*carbamoyl*-14C]citrulline cpm): (total cpm above background). The percentage conversion is recalculated to nanomoles using the value of initial substrate concentration.

Another radioassay has been developed that uses N-benzoylarginine [³H]ethyl ester.[11] The product is separated from the substrate by thin-layer chromatography.

Definition of Unit and Specific Activity. One unit of activity is equal to 1 nmol of N-benzoyl-L-citrulline formed per hour.

For a specific activity value the enzyme protein content can be estimated by the procedure of Bradford.[21] Care is necessary in the choice of

[21] M. M. Bradford, *Anal. Biochem.* **72,** 248 (1976).

the protein standard. Bovine serum albumin may not give the same dye-binding response as that of the protein of the tissue from which the enzyme is isolated.

Enzyme Properties Compared

All peptidylarginine deiminases described to date have an absolute requirement for Ca^{2+}, are active at neutral pH, and are activated by DTT. However, in contrast to the epidermal enzyme, the muscle enzyme is not inhibited by treatment with thiol blocking reagents.[10] Three reports show that the enzyme is active at unphysiological temperatures[9–11] (see Table II).

The muscle and brain enzymes can be differentiated from epidermal and hair follicle enzymes by their molecular weights and by their relative affinities for different substrates (see Table II). It should be noted that the choice of substrate would influence the amount of enzyme activity detected. This is an important consideration when making comparisons of enzyme activity between different tissues or species. Clearly more information is required on specific substrate requirements of the different peptidylarginine deiminases.

Author Index

Numbers in parentheses are footnote reference numbers and indicate that an author's work is referred to although the name is not cited in the text.

D

Dabrowska, R., 90
Dacha, M., 336, 338(38), 350(38)
Daemen, M., 561
Dahlbeck, B., 507, 508(8)
Dahlqvist-Edberg, U., 136, 145(7, 9)
Dahlqvist, U., 136
Dahmus, M. E., 97, 148(48)
Daile, P., 146(23)
D'Alessio, G., 346
Dancewicz, A. M., 379
Daniels, G. R., 158, 159, 160(30), 162(26), 163(26)
Danner, D. J., 435, 446, 447(7), 457(7)
Darigo, M. D., 262
Darrow, R. A., 190
Dasgupta, A., 9
Dateo, G. P., Jr., 468
Daugaard, P., 44
David, G. S., 529
Davidson, B., 446, 447(14), 448, 452(14), 454, 456(14, 15), 457(34), 462, 469, 470, 472(14), 473
Davidson, L. K., 305
Davie, E. W., 504, 507, 508(4), 510, 511(20, 23)
Davies, P. J. A., 145(8)
Davis, B. J., 319
Davis, J. N., 187, 190(24), 192, 576, 579(1), 601
Davis, R. H., 156
Dayhoff, M. O., 299
de Bree, P. K., 258
De Buysere, M. S., 94
De Conich, S., 475
Decottingnies-Le Marechal, P., 560
Dediukina, M. M., 148(54)
De Duve, C., 475
Deeny, J. M., 466
Defreyn, G., 106
Deftos, L. J., 517, 533
Degani, C., 29
Degani, Y., 600
De Gennaro, L. J., 221, 223(31)
Degens, E. T., 397
De Groot, L. J., 453, 456, 460
De Lange, R. J., 89, 102, 224
Del Boccio, G., 336(58) 340(58)
Dell, A., 506, 516, 563

Delmos, P. D., 532
De Lorenzo, F., 286, 288, 292, 301, 346, 466
De Lorenzo, R. J., 92, 151(78)
Demaille, J. G., 14, 91, 146(33)
Deme, D., 435, 457, 461(41, 45), 478, 480, 481, 482, 483, 484
Demetriou, J. A., 418
de Metz, M., 558, 559, 561, 562, 563(42, 45)
De Moss, J. C., 577
De Muth, R., 517, 526(13), 528(13), 530(13), 531(13), 532(13), 541(13), 543(13)
Dennily, D. T., 135
Dennis, S. C., 94
Denton, R. M., 150(72)
Deodhar, A. D., 263
De Paoli-Roach, A. A., 83, 90, 92, 94(17), 95, 150(66)
Depierre, J. W., 332
de Prailanne, S., 446, 447(11), 452(11), 456(11), 457(11), 461(11), 479, 480(6)
Desai, N. M., 369
Desjardins, P. R., 97
Deslauriers, R., 55
de Stefano, S., 438
Determan, H., 418
Deuel, T. F., 99
Deutsch, R., 258
de Verdier, C.-H., 3
De Vore, D. P., 377
Di Augustine, R. P., 246
Dick, B., 540
Dickson, I. R., 540
Di Cola, D., 336(58), 340(58)
Dill, K., 177, 181
Dillard, C. J., 603
Dillio, C., 336(58), 340(58)
Dillon, J., 377
Dills, W. L., Jr., 88
Dilworth, G. L., 577, 579(12), 581
Dipaolo, E. A., 228, 236(23), 237(23), 238(23), 240(23)
Dirden, B., 89
Di Sabato, G., 29
Di Scipio, R. G., 504, 507, 511(33)
Dixon, G. H., 138, 224, 226, 230(19)
Dixon, J. E., 289, 291, 292
Dixon, N. E., 564
Djerassi, C., 438
do Amaral, A. D., 415(9), 416
Dobson, C. M., 316

Subject Index

A

H

R

Rearrangease, 281
Reduced ribonuclease reactivating enzyme,
    281
Resilin, 377
Rheumatoid arthritis, 360
Rhizobium, glutamine synthetase, 190, 191
Rhodopsin, phosphorylation, 152
Rhodopsin kinase
    autophosphorylation, 152
    substrates, 152
Ribonuclease
    disulfide bond formation, analysis, 323,
        329
    dityrosine in, 379, 384
    reduction of disulfide bonds, by thiore-
        doxin, 299
    scrambled, renaturation, 281–285, 292
Ribonuclease A
    oxidized and reduced forms, covalent
        modifications, 307
    oxidative renaturation, 303
Ribonucleotide reductase, 295
RNA polymerase, phosphorylation,
    146, 148
Rous sarcoma virus, pp60^{src}, 98

S

Saccharomyces cerevisiae, pyruvate car-
    boxylase synthetase, 275
Salmonella typhimurium
    adenylyltransferase, 198
    P$_{IID}$ and uridylyltransferase, assay, 200
Sarcoma virus
    Esh, tyrosine protein kinase, 98
    Fujinami, tyrosine protein kinase, 98
    Gardner-Arnstein feline, tyrosine pro-
        tein kinase, 98
    PRC II, tyrosine protein kinase, 98
    protein kinase, 98
    Synder-Theilen feline, tyrosine protein
        kinase, 98
    UR-1, tyrosine protein kinase, 98
    UR-2, tyrosine protein kinase, 98
    Y73, tyrosine protein kinase, 98
Scleroprotein
    byssus, 437
    dopa in, 398
    halogenated amino acids in, 415, 416

halogenated tyrosine derivatives, 413–
    438
    release, 414
halogenated tyrosine residues
    effect of alkaline hydrolysis, 431
    separation, 423–429
halogenation, 436, 437
hydrolysates, containing halogenated
    amino acids
    column chromatography, 418–420
    preparation, 414–418
Sclerotization, 398
Selenenic acid, 593, 601
Seleninic acid, 593, 601
Selenium
    in biological systems, 582, 593, 602
    enzymes dependent on, 576–578
    incorporation into protein, 620
    incorporation into thiolase, 621
    proteins requiring, 620
Selenocysteine, 620
    assay, 604
    derivatization, by 1-fluoro-2,4-dinitro-
        benzene, 582
    DNP derivatives
        separation, 583, 584
        thin-layer chromatography, 582, 583
    enzymatic degradation, 581
    identification, by high-performance liq-
        uid chromatography and mass
        spectrometry, 582–592
    in proteins, 576–581, 602
        function, 578, 579
        identification, 576
        isolation, 576
        occurrence, 576–578
        thin-layer chromatography, 580, 581
    quantitative analysis, 604–606
    radiolabeled, synthesis, 581
    residues, in proteins
        analysis, 579–581
        derivitization, 579
Se-Selenocysteine, DNP derivative,
    601
    absorption spectrum, 585
    derivatization for mass spectrometry,
        586–589, 592
    identification, 585–589
    NMR, 585, 586
    preparation, 584, 585